28

STUDENT'S SOLUTIONS MANUAL

GEX PUBLISHING SERVICES

BEGINNING ALGEBRA

TWELFTH EDITION

Margaret L. Lial

American River College

John Hornsby

University of New Orleans

Terry McGinnis

PEARSON

Boston Columbus Indianapolis New York San Francisco
Amsterdam Cape Town Dubai London Madrid Milan Munich Paris Montreal Toronto
Delhi Mexico City São Paulo Sydney Hong Kong Seoul Singapore Taipei Tokyo

ISBN-13: 978-0-321-96981-1
ISBN-10: 0-321-96981-2

1 2 3 4 5 6 RRD-H 18 17 16 15

www.pearsonhighered.com

PEARSON

Table of Contents

R Prealgebra Review **1**

R.1 Fractions • 1

R.2 Decimals and Percents • 8

1 The Real Number System **14**

1.1 Exponents, Order of Operations, and Inequality • 14

1.2 Variables, Expressions, and Equations • 17

1.3 Real Numbers and the Number Line • 21

1.4 Adding and Subtracting Real Numbers • 23

1.5 Multiplying and Dividing Real Numbers • 28

Summary Exercises Performing Operations with Real Numbers • 33

1.6 Properties of Real Numbers • 35

1.7 Simplifying Expressions • 39

Chapter 1 Review Exercises • 43

Chapter 1 Mixed Review Exercises • 48

Chapter 1 Test • 49

2 Linear Equations and Inequalities in One Variable **52**

2.1 The Addition Property of Equality • 52

2.2 The Multiplication Property of Equality • 56

2.3 More on Solving Linear Equations • 61

Summary Exercises Applying Methods for Solving Linear Equations • 67

2.4 Applications of Linear Equations • 70

2.5 Formulas and Additional Applications from Geometry • 78

2.6 Ratio, Proportion, and Percent • 83

2.7 Further Applications of Linear Equations • 89

2.8 Solving Linear Inequalities • 95

Chapter 2 Review Exercises • 102

Chapter 2 Mixed Review Exercises • 108

Chapter 2 Test • 110

Chapters R–2 Cumulative Review Exercises • 113

3 Linear Equations and Inequalities in Two Variables; Functions **116**

3.1 Linear Equations and Rectangular Coordinates • 116

3.2 Graphing Linear Equations in Two Variables • 122

3.3 The Slope of a Line • 130

3.4 Slope-Intercept Form of a Linear Equation • 135

3.5 Point-Slope Form of a Linear Equation and Modeling • 141

Summary Exercises Applying Graphing and Equation-Writing Techniques for Lines • 146

3.6 Graphing Linear Inequalities in Two Variables • 150

3.7 Introduction to Functions • 154

Chapter 3 Review Exercises • 157

Chapter 3 Mixed Review Exercises • 164

Chapter 3 Test • 166

Chapters R–3 Cumulative Review Exercises • 168

4 Systems of Linear Equations and Inequalities **171**
4.1 Solving Systems of Linear Equations by Graphing • 171
4.2 Solving Systems of Linear Equations by Substitution • 176
4.3 Solving Systems of Linear Equations by Elimination • 182
Summary Exercises Applying Techniques for Solving Systems of Linear Equations • 188
4.4 Applications of Linear Systems • 193
4.5 Solving Systems of Linear Inequalities • 202
Chapter 4 Review Exercises • 205
Chapter 4 Mixed Review Exercises • 213
Chapter 4 Test • 215
Chapters R–4 Cumulative Review Exercises • 220
5 Exponents and Polynomials **224**
5.1 The Product Rule and Power Rules for Exponents • 224
5.2 Integer Exponents and the Quotient Rule • 227
Summary Exercises Applying the Rules for Exponents • 231
5.3 Scientific Notation • 235
5.4 Adding, Subtracting, and Graphing Polynomials • 240
5.5 Multiplying Polynomials • 246
5.6 Special Products • 252
5.7 Dividing Polynomials • 256
Chapter 5 Review Exercises • 263
Chapter 5 Mixed Review Exercises • 270
Chapter 5 Test • 272
Chapters R–5 Cumulative Review Exercises • 274
6 Factoring and Applications **279**
6.1 The Greatest Common Factor; Factoring by Grouping • 279
6.2 Factoring Trinomials • 283
6.3 More on Factoring Trinomials • 289
6.4 Special Factoring Techniques • 294
Summary Exercises Recognizing and Applying Factoring Strategies • 305
6.5 Solving Quadratic Equations Using the Zero-Factor Property • 307
6.6 Applications of Quadratic Equations • 313
Chapter 6 Review Exercises • 318
Chapter 6 Mixed Review Exercises • 323
Chapter 6 Test • 325
Chapters R–6 Cumulative Review Exercises • 327

7 Rational Expressions and Applications **331**
7.1 The Fundamental Property of Rational Expressions • 331
7.2 Multiplying and Dividing Rational Expressions • 338
7.3 Least Common Denominators • 343
7.4 Adding and Subtracting Rational Expressions • 348
7.5 Complex Fractions • 357
7.6 Solving Equations with Rational Expressions • 364
Summary Exercises Simplifying Rational Expressions vs. Solving Rational Equations • 375
7.7 Applications of Rational Expressions • 379
7.8 Variation • 385
Chapter 7 Review Exercises • 388
Chapter 7 Mixed Review Exercises • 395
Chapter 7 Test • 398
Chapters R–7 Cumulative Review Exercises • 402
8 Roots and Radicals **406**
8.1 Evaluating Roots • 406
8.2 Multiplying, Dividing, and Simplifying Radicals • 411
8.3 Adding and Subtracting Radicals • 415
8.4 Rationalizing the Denominator • 418
8.5 More Simplifying and Operations with Radicals • 422
Summary Exercises Applying Operations with Radicals • 428
8.6 Solving Equations with Radicals • 430
Chapter 8 Review Exercises • 438
Chapter 8 Mixed Review Exercises • 444
Chapter 8 Test • 445
Chapters R–8 Cumulative Review Exercises • 447
9 Quadratic Equations **452**
9.1 Solving Quadratic Equations by the Square Root Property • 452
9.2 Solving Quadratic Equations by Completing the Square • 456
9.3 Solving Quadratic Equations by the Quadratic Formula • 464
Summary Exercises Applying Methods for Solving Quadratic Equations • 472
9.4 Graphing Quadratic Equations • 477
Chapter 9 Review Exercises • 483
Chapter 9 Mixed Review Exercises • 489
Chapter 9 Test • 491
Chapters R–9 Cumulative Review Exercises • 494

Chapter R
Prealgebra Review

R.1 Fractions

Now Try Exercises

N1.

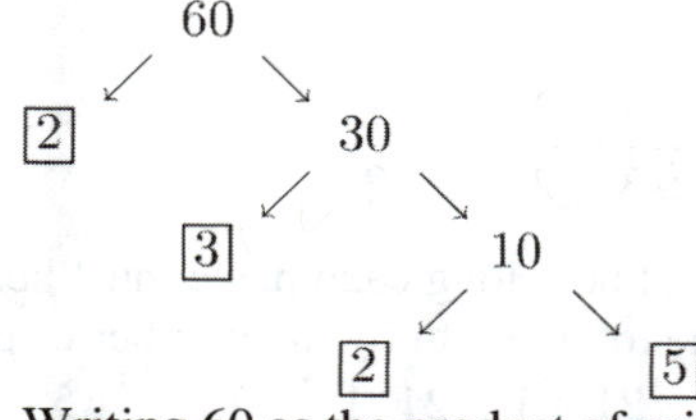

Writing 60 as the product of primes gives us $60 = 2 \cdot 2 \cdot 3 \cdot 5$.

N2. **(a)** $\dfrac{30}{42} = \dfrac{5 \cdot 6}{7 \cdot 6} = \dfrac{5 \cdot 1}{7 \cdot 1} = \dfrac{5}{7}$

(b) $\dfrac{10}{70} = \dfrac{1 \cdot 10}{7 \cdot 10} = \dfrac{1 \cdot 1}{7 \cdot 1} = \dfrac{1}{7}$

(c) $\dfrac{72}{120} = \dfrac{3 \cdot 24}{5 \cdot 24} = \dfrac{3 \cdot 1}{5 \cdot 1} = \dfrac{3}{5}$

N3. The fraction bar represents division. Divide the numerator of the improper fraction by the denominator.

$$\begin{array}{r} 18 \\ 5\overline{)92} \\ \underline{5} \\ 42 \\ \underline{40} \\ 2 \end{array}$$

Thus, $\dfrac{92}{5} = 18\dfrac{2}{5}$.

N4. Multiply the denominator of the fraction by the natural number and then add the numerator to obtain the numerator of the improper fraction.
$3 \cdot 11 = 33$ and $33 + 2 = 35$
The denominator of the improper fraction is the same as the denominator in the mixed number.

Thus, $11\dfrac{2}{3} = \dfrac{35}{3}$.

N5. **(a)** To multiply two fractions, multiply their numerators and then multiply their denominators. Then simplify and write the answer in lowest terms.

$$\frac{4}{7} \cdot \frac{5}{8} = \frac{4 \cdot 5}{7 \cdot 8} = \frac{4 \cdot 5}{7 \cdot 2 \cdot 4} = \frac{5}{14}$$

(b) To multiply two mixed numbers, first write them as improper fractions. Multiply their numerators and then multiply their denominators. Then simplify and write the answer as a mixed number in lowest terms.

$$3\frac{2}{5} \cdot 6\frac{2}{3} = \frac{17}{5} \cdot \frac{20}{3} = \frac{17 \cdot 20}{5 \cdot 3} = \frac{17 \cdot 5 \cdot 4}{5 \cdot 3} = \frac{68}{3}, \text{ or } 22\frac{2}{3}$$

N6. **(a)** To divide fractions, multiply by the reciprocal of the divisor.

$$\frac{2}{7} \div \frac{8}{9} = \frac{2}{7} \cdot \frac{9}{8} = \frac{2 \cdot 3 \cdot 3}{7 \cdot 2 \cdot 4} = \frac{9}{28}$$

(b) To divide fractions, multiply by the reciprocal of the divisor.

$$3\frac{3}{4} \div 4\frac{2}{7} = \frac{15}{4} \div \frac{30}{7} = \frac{15}{4} \cdot \frac{7}{30} = \frac{15 \cdot 7}{4 \cdot 2 \cdot 15} = \frac{7}{8}$$

N7. To find the sum of two fractions having the same denominator, add the numerators and keep the same denominator.

$$\frac{1}{8}+\frac{3}{8}=\frac{1+3}{8}=\frac{4}{8}=\frac{1\cdot 4}{2\cdot 4}=\frac{1}{2}$$

N8. **(a)** Since $12=2\cdot 2\cdot 3$ and $8=2\cdot 2\cdot 2$, the least common denominator must have three factors of 2 (from 8) and one factor of 3 (from 12), so it is $2\cdot 2\cdot 2\cdot 3=24$.
Write each fraction with a denominator of 24.

$$\frac{5}{12}=\frac{5\cdot 2}{12\cdot 2}=\frac{10}{24} \text{ and } \frac{3}{8}=\frac{3\cdot 3}{8\cdot 3}=\frac{9}{24}$$

Now add.

$$\frac{5}{12}+\frac{3}{8}=\frac{10}{24}+\frac{9}{24}=\frac{10+9}{24}=\frac{19}{24}$$

(b) Write each mixed number as an improper fraction.

$$3\frac{1}{4}+5\frac{5}{8}=\frac{13}{4}+\frac{45}{8}$$

The least common denominator is 8, so write each fraction with a denominator of 8.

$$\frac{45}{8} \text{ and } \frac{13}{4}=\frac{13\cdot 2}{4\cdot 2}=\frac{26}{8}$$

Now add.

$$\frac{13}{4}+\frac{45}{8}=\frac{26}{8}+\frac{45}{8}=\frac{26+45}{8}=\frac{71}{8}, \text{ or } 8\frac{7}{8}$$

N9. **(a)** Since $11=11$ and $9=3\cdot 3$, the least common denominator is $3\cdot 3\cdot 11=99$. Write each fraction with a denominator of 99.

$$\frac{5}{11}=\frac{5\cdot 9}{11\cdot 9}=\frac{45}{99} \text{ and } \frac{2}{9}=\frac{2\cdot 11}{9\cdot 11}=\frac{22}{99}$$

Now subtract.

$$\frac{5}{11}-\frac{2}{9}=\frac{45}{99}-\frac{22}{99}=\frac{23}{99}$$

(b) Write each mixed number as an improper fraction.

$$4\frac{1}{3}-2\frac{5}{6}=\frac{13}{3}-\frac{17}{6}$$

The least common denominator is 6. Write each fraction with a denominator of 6. $\frac{17}{6}$ remains unchanged, and $\frac{13}{3}=\frac{13\cdot 2}{3\cdot 2}=\frac{26}{6}$.

Now subtract.

$$\frac{13}{3}-\frac{17}{6}=\frac{26}{6}-\frac{17}{6}=\frac{26-17}{6}=\frac{9}{6}$$

Now reduce.

$$\frac{9}{6}=\frac{3\cdot 3}{2\cdot 3}=\frac{3}{2}, \text{ or } 1\frac{1}{2}$$

N10. To find out how long each piece must be, divide the total length by the number of pieces.

$$10\frac{1}{2}\div 4=\frac{21}{2}\div\frac{4}{1}=\frac{21}{2}\cdot\frac{1}{4}=\frac{21}{8}, \text{ or } 2\frac{5}{8}$$

Each piece should be $2\frac{5}{8}$ feet long.

N11. **(a)** In the circle graph, the sector for North America is the smallest, so North America had the least number of Internet users.

(b) A share of $\frac{11}{25}$ can be rounded to $\frac{1}{2}$, and the total number of Internet users, 2100 million, can be rounded to 2000 million (or 2 billion). Multiply $\frac{1}{2}$ by 2000.

$$\frac{1}{2}\cdot 2000=1000 \text{ million, or 1 billion}$$

(c) Multiply the actual fraction from the graph for Asia by the number of users.

$$\frac{11}{25}(2100)=924 \text{ million, or } 924{,}000{,}000$$

Exercises

1. True; the number above the fraction bar is called the numerator and the number below the fraction bar is called the denominator.

3. False; this is an improper fraction. Its value is 1.

5. False; the fraction $\frac{13}{39}$ can be written in lowest terms as $\frac{1}{3}$ since $\frac{13}{39}=\frac{13\cdot 1}{13\cdot 3}=\frac{1}{3}$.

7. False; *product* refers to multiplication, so the product of 10 and 2 is 20. The *sum* of 10 and 2 is 12.

9. $\frac{16}{24}=\frac{2\cdot 8}{3\cdot 8}=\frac{2}{3}$
Therefore, C is correct.

11. A common denominator for $\frac{p}{q}$ and $\frac{r}{s}$ must be a multiple of both denominators, q and s. Such a number is $q\cdot s$. Therefore, A is correct.

13. Since 19 has only itself and 1 as factors, it is a prime number.

15. $30=2\cdot 15$
$=2\cdot 3\cdot 5$
Since 30 has factors other than itself and 1, it is a composite number.

17. $64=2\cdot 32$
$=2\cdot 2\cdot 16$
$=2\cdot 2\cdot 2\cdot 8$
$=2\cdot 2\cdot 2\cdot 2\cdot 4$
$=2\cdot 2\cdot 2\cdot 2\cdot 2\cdot 2$
Since 64 has factors other than itself and 1, it is a composite number.

19. As stated in the text, the number 1 is neither prime nor composite, by agreement.

21. $57=3\cdot 19,$ so 57 is a composite number.

23. Since 79 has only itself and 1 as factors, it is a prime number.

25. $124=2\cdot 62$
$=2\cdot 2\cdot 31,$
so 124 is a composite number.

27. $500=2\cdot 250$
$=2\cdot 2\cdot 125$
$=2\cdot 2\cdot 5\cdot 25$
$=2\cdot 2\cdot 5\cdot 5\cdot 5,$
so 500 is a composite number.

29. $3458=2\cdot 1729$
$=2\cdot 7\cdot 247$
$=2\cdot 7\cdot 13\cdot 19$
Since 3458 has factors other than itself and 1, it is a composite number.

31. $\frac{8}{16}=\frac{1\cdot 8}{2\cdot 8}=\frac{1}{2}$

33. $\frac{15}{18}=\frac{3\cdot 5}{3\cdot 6}=\frac{5}{6}$

35. $\frac{64}{100}=\frac{4\cdot 16}{4\cdot 25}=\frac{16}{25}$

37. $\frac{18}{90}=\frac{1\cdot 18}{5\cdot 18}=\frac{1}{5}$

39. $\frac{144}{120}=\frac{6\cdot 24}{5\cdot 24}=\frac{6}{5}$

41. $7\overline{)12}$ quotient 1
$\underline{7}$
5

Therefore, $\frac{12}{7}=1\frac{5}{7}.$

43. $12\overline{)77}$ quotient 6
$\underline{72}$
5

Therefore, $\frac{77}{12}=6\frac{5}{12}.$

45. $11\overline{)83}$ quotient 7
$\underline{77}$
6

Therefore, $\frac{83}{11}=7\frac{6}{11}.$

47. Multiply the denominator of the fraction by the natural number and then add the numerator to obtain the numerator of the improper fraction.
$5\cdot 2=10$ and $10+3=13$
The denominator of the improper fraction is the same as the denominator in the mixed number.
Thus, $2\frac{3}{5}=\frac{13}{5}.$

49. Multiply the denominator of the fraction by the natural number and then add the numerator to obtain the numerator of the improper fraction.
$8\cdot 10=80$ and $80+3=83$
The denominator of the improper fraction is the same as the denominator in the mixed number.
Thus, $10\frac{3}{8}=\frac{83}{8}.$

51. Multiply the denominator of the fraction by the natural number and then add the numerator to obtain the numerator of the improper fraction.
$5 \cdot 10 = 50$ and $50 + 1 = 51$
The denominator of the improper fraction is the same as the denominator in the mixed number.
Thus, $10\frac{1}{5} = \frac{51}{5}$.

53. $\frac{4}{5} \cdot \frac{6}{7} = \frac{4 \cdot 6}{5 \cdot 7} = \frac{24}{35}$

55. $\frac{2}{3} \cdot \frac{15}{16} = \frac{2 \cdot 15}{3 \cdot 16} = \frac{2 \cdot 3 \cdot 5}{3 \cdot 2 \cdot 8} = \frac{5}{8}$

57. $\frac{1}{10} \cdot \frac{12}{5} = \frac{1 \cdot 12}{10 \cdot 5} = \frac{1 \cdot 2 \cdot 6}{2 \cdot 5 \cdot 5} = \frac{6}{25}$

59.
$$\frac{15}{4} \cdot \frac{8}{25} = \frac{15 \cdot 8}{4 \cdot 25} = \frac{3 \cdot 5 \cdot 4 \cdot 2}{4 \cdot 5 \cdot 5} = \frac{3 \cdot 2}{5} = \frac{6}{5}, \text{ or } 1\frac{1}{5}$$

61.
$$21 \cdot \frac{3}{7} = \frac{21 \cdot 3}{1 \cdot 7} = \frac{3 \cdot 7 \cdot 3}{1 \cdot 7} = \frac{3 \cdot 3}{1} = 9$$

63. Change both mixed numbers to improper fractions.
$$3\frac{1}{4} \cdot 1\frac{2}{3} = \frac{13}{4} \cdot \frac{5}{3} = \frac{13 \cdot 5}{4 \cdot 3} = \frac{65}{12}, \text{ or } 5\frac{5}{12}$$

65. Change both mixed numbers to improper fractions.
$$2\frac{3}{8} \cdot 3\frac{1}{5} = \frac{19}{8} \cdot \frac{16}{5} = \frac{19 \cdot 16}{8 \cdot 5} = \frac{19 \cdot 2 \cdot 8}{8 \cdot 5} = \frac{38}{5}, \text{ or } 7\frac{3}{5}$$

67. To divide fractions, multiply by the reciprocal of the divisor.
$$\frac{5}{4} \div \frac{3}{8} = \frac{5}{4} \cdot \frac{8}{3} = \frac{5 \cdot 8}{4 \cdot 3} = \frac{5 \cdot 4 \cdot 2}{4 \cdot 3} = \frac{5 \cdot 2}{3} = \frac{10}{3}, \text{ or } 3\frac{1}{3}$$

69. To divide fractions, multiply by the reciprocal of the divisor.
$$\frac{32}{5} \div \frac{8}{15} = \frac{32}{5} \cdot \frac{15}{8} = \frac{32 \cdot 15}{5 \cdot 8} = \frac{8 \cdot 4 \cdot 3 \cdot 5}{1 \cdot 5 \cdot 8} = \frac{4 \cdot 3}{1} = 12$$

71. To divide fractions, multiply by the reciprocal of the divisor.
$$\frac{3}{4} \div 12 = \frac{3}{4} \cdot \frac{1}{12} = \frac{3 \cdot 1}{4 \cdot 12} = \frac{3 \cdot 1}{4 \cdot 3 \cdot 4} = \frac{1}{4 \cdot 4} = \frac{1}{16}$$

73. To divide fractions, multiply by the reciprocal of the divisor.

$$6 \div \frac{3}{5} = \frac{6}{1} \cdot \frac{5}{3} = \frac{6 \cdot 5}{1 \cdot 3} = \frac{2 \cdot 3 \cdot 5}{1 \cdot 3} = \frac{2 \cdot 5}{1} = 10$$

75. Change the first number to an improper fraction, and then multiply by the reciprocal of the divisor.

$$6\frac{3}{4} \div \frac{3}{8} = \frac{27}{4} \div \frac{3}{8} = \frac{27}{4} \cdot \frac{8}{3} = \frac{27 \cdot 8}{4 \cdot 3} = \frac{3 \cdot 9 \cdot 2 \cdot 4}{4 \cdot 3} = \frac{9 \cdot 2}{1} = 18$$

77. Change both mixed numbers to improper fractions, and then multiply by the reciprocal of the divisor.

$$2\frac{1}{2} \div 1\frac{5}{7} = \frac{5}{2} \div \frac{12}{7} = \frac{5}{2} \cdot \frac{7}{12} = \frac{5 \cdot 7}{2 \cdot 12} = \frac{35}{24}, \text{ or } 1\frac{11}{24}$$

79. Change both mixed numbers to improper fractions, and then multiply by the reciprocal of the divisor.

$$2\frac{5}{8} \div 1\frac{15}{32} = \frac{21}{8} \div \frac{47}{32} = \frac{21}{8} \cdot \frac{32}{47} = \frac{21 \cdot 32}{8 \cdot 47} = \frac{21 \cdot 8 \cdot 4}{8 \cdot 47} = \frac{21 \cdot 4}{47} = \frac{84}{47}, \text{ or } 1\frac{37}{47}$$

81. $\frac{7}{15} + \frac{4}{15} = \frac{7+4}{15} = \frac{11}{15}$

83. $\frac{7}{12} + \frac{1}{12} = \frac{7+1}{12} = \frac{8}{12} = \frac{2 \cdot 4}{3 \cdot 4} = \frac{2}{3}$

85. Since $9 = 3 \cdot 3$, and 3 is prime, the LCD (least common denominator) is $3 \cdot 3 = 9$.

$$\frac{1}{3} = \frac{1}{3} \cdot \frac{3}{3} = \frac{3}{9}$$

Now add the two fractions with the same denominator.

$$\frac{5}{9} + \frac{1}{3} = \frac{5}{9} + \frac{3}{9} = \frac{8}{9}$$

87. Since $8 = 2 \cdot 2 \cdot 2$ and $6 = 2 \cdot 3$, the LCD is $2 \cdot 2 \cdot 2 \cdot 3 = 24$.

$$\frac{3}{8} = \frac{3}{8} \cdot \frac{3}{3} = \frac{9}{24} \text{ and } \frac{5}{6} \cdot \frac{4}{4} = \frac{20}{24}$$

Now add fractions with the same denominator.

$$\frac{3}{8} + \frac{5}{6} = \frac{9}{24} + \frac{20}{24} = \frac{29}{24}, \text{ or } 1\frac{5}{24}$$

89. $3\frac{1}{8}=3+\frac{1}{8}=\frac{24}{8}+\frac{1}{8}=\frac{25}{8}$

$2\frac{1}{4}=2+\frac{1}{4}=\frac{8}{4}+\frac{1}{4}=\frac{9}{4}$

$3\frac{1}{8}+2\frac{1}{4}=\frac{25}{8}+\frac{9}{4}$

Since $8=2\cdot 2\cdot 2$ and $4=2\cdot 2,$ the LCD is $2\cdot 2\cdot 2$ or 8.

$$\begin{aligned}3\frac{1}{8}+2\frac{1}{4}&=\frac{25}{8}+\frac{9\cdot 2}{4\cdot 2}\\&=\frac{25}{8}+\frac{18}{8}\\&=\frac{43}{8},\text{ or }5\frac{3}{8}\end{aligned}$$

91. $3\frac{1}{4}=3+\frac{1}{4}=\frac{12}{4}+\frac{1}{4}=\frac{13}{4}$

$1\frac{4}{5}=1+\frac{4}{5}=\frac{5}{5}+\frac{4}{5}=\frac{9}{5}$

Since $4=2\cdot 2,$ and 5 is prime, the LCD is $2\cdot 2\cdot 5=20.$

$$\begin{aligned}3\frac{1}{4}+1\frac{4}{5}&=\frac{13\cdot 5}{4\cdot 5}+\frac{9\cdot 4}{5\cdot 4}\\&=\frac{65}{20}+\frac{36}{20}\\&=\frac{101}{20},\text{ or }5\frac{1}{20}\end{aligned}$$

93. $\frac{7}{9}-\frac{2}{9}=\frac{7-2}{9}=\frac{5}{9}$

95.
$$\begin{aligned}\frac{13}{15}-\frac{3}{15}&=\frac{13-3}{15}\\&=\frac{10}{15}\\&=\frac{2\cdot 5}{3\cdot 5}=\frac{2}{3}\end{aligned}$$

97. Since $12=4\cdot 3$ (12 is a multiple of 3), the LCD is 12.

$\frac{1}{3}\cdot\frac{4}{4}=\frac{4}{12}$

Now subtract fractions with the same denominator.

$\frac{7}{12}-\frac{1}{3}=\frac{7}{12}-\frac{4}{12}=\frac{3}{12}=\frac{1\cdot 3}{4\cdot 3}=\frac{1}{4}$

99. Since $12=2\cdot 2\cdot 3$ and $9=3\cdot 3,$ the LCD is $2\cdot 2\cdot 3\cdot 3=36.$

$\frac{7}{12}=\frac{7}{12}\cdot\frac{3}{3}=\frac{21}{36}$ and $\frac{1}{9}\cdot\frac{4}{4}=\frac{4}{36}$

Now subtract fractions with the same denominator.

$\frac{7}{12}-\frac{1}{9}=\frac{21}{36}-\frac{4}{36}=\frac{17}{36}$

101. $4\frac{3}{4}=4+\frac{3}{4}=\frac{16}{4}+\frac{3}{4}=\frac{19}{4}$

$1\frac{2}{5}=1+\frac{2}{5}=\frac{5}{5}+\frac{2}{5}=\frac{7}{5}$

Since $4=2\cdot 2,$ and 5 is prime, the LCD is $2\cdot 2\cdot 5=20.$

$$\begin{aligned}4\frac{3}{4}-1\frac{2}{5}&=\frac{19\cdot 5}{4\cdot 5}-\frac{7\cdot 4}{5\cdot 4}\\&=\frac{95}{20}-\frac{28}{20}\\&=\frac{67}{20},\text{ or }3\frac{7}{20}\end{aligned}$$

103. $6\frac{1}{4}=6+\frac{1}{4}=\frac{24}{4}+\frac{1}{4}=\frac{25}{4}$

$5\frac{1}{3}=5+\frac{1}{3}=\frac{15}{3}+\frac{1}{3}=\frac{16}{3}$

Since $4=2\cdot 2,$ and 3 is prime, the LCD is $2\cdot 2\cdot 3=12.$

$$\begin{aligned}6\frac{1}{4}-5\frac{1}{3}&=\frac{25}{4}-\frac{16}{3}\\&=\frac{25\cdot 3}{4\cdot 3}-\frac{16\cdot 4}{3\cdot 4}\\&=\frac{75}{12}-\frac{64}{12}\\&=\frac{11}{12}\end{aligned}$$

105. Observe that there are 24 dots in the entire figure, 6 dots in the triangle, 12 dots in the rectangle, and 2 dots in the overlapping region.

(a) $\frac{12}{24}=\frac{1}{2}$ of all the dots are in the rectangle.

(b) $\frac{6}{24}=\frac{1}{4}$ of all the dots are in the triangle.

(c) $\frac{2}{6}=\frac{1}{3}$ of the dots in the triangle are in the overlapping region.

(d) $\frac{2}{12}=\frac{1}{6}$ of the dots in the rectangle are in the overlapping region.

107. Multiply the number of cups of water per serving by the number of servings.

$$\frac{3}{4}\cdot 8 = \frac{3}{4}\cdot\frac{8}{1} = \frac{3\cdot 8}{4\cdot 1} = \frac{3\cdot 2\cdot 4}{4\cdot 1} = \frac{3\cdot 2}{1} = 6 \text{ cups}$$

For 8 microwave servings, 6 cups of water will be needed.

109. The difference in length is found by subtracting.

$$3\frac{1}{4} - 2\frac{1}{8} = \frac{13}{4} - \frac{17}{8} = \frac{13\cdot 2}{4\cdot 2} - \frac{17}{8} \quad \text{LCD} = 8$$

$$= \frac{26}{8} - \frac{17}{8} = \frac{9}{8}, \text{ or } 1\frac{1}{8}$$

The difference is $1\frac{1}{8}$ inches.

111. The difference between the two measures is found by subtracting, using 16 as the LCD.

$$\frac{3}{4} - \frac{3}{16} = \frac{3\cdot 4}{4\cdot 4} - \frac{3}{16} = \frac{12}{16} - \frac{3}{16} = \frac{12-3}{16} = \frac{9}{16}$$

The difference is $\frac{9}{16}$ inch.

113. The perimeter is the sum of the measures of the 5 sides.

$$196 + 98\frac{3}{4} + 146\frac{1}{2} + 100\frac{7}{8} + 76\frac{5}{8}$$

$$= 196 + 98\frac{6}{8} + 146\frac{4}{8} + 100\frac{7}{8} + 76\frac{5}{8}$$

$$= 196 + 98 + 146 + 100 + 76 + \frac{6+4+7+5}{8}$$

$$= 616 + \frac{22}{8} \quad \left(\frac{22}{8} = 2\frac{6}{8} = 2\frac{3}{4}\right)$$

$$= 618\frac{3}{4} \text{ feet}$$

The perimeter is $618\frac{3}{4}$ feet.

115. Divide the total board length by 3.

$$15\frac{5}{8} \div 3 = \frac{125}{8} \div \frac{3}{1} = \frac{125}{8}\cdot\frac{1}{3} = \frac{125\cdot 1}{8\cdot 3} = \frac{125}{24}, \text{ or } 5\frac{5}{24}$$

The length of each of the three pieces must be $5\frac{5}{24}$ inches.

117. To find the number of cakes the caterer can make, divide $15\frac{1}{2}$ by $1\frac{3}{4}$.

$$15\frac{1}{2} \div 1\frac{3}{4} = \frac{31}{2} \div \frac{7}{4} = \frac{31}{2}\cdot\frac{4}{7} = \frac{31\cdot 2\cdot 2}{2\cdot 7} = \frac{62}{7}, \text{ or } 8\frac{6}{7}$$

There is not quite enough sugar for 9 cakes. The caterer can make 8 cakes with some sugar left over.

119. Multiply the amount of fabric it takes to make one costume by the number of costumes.

$$2\frac{3}{8}\cdot 7=\frac{19}{8}\cdot\frac{7}{1}$$
$$=\frac{19\cdot 7}{8\cdot 1}$$
$$=\frac{133}{8},\text{ or }16\frac{5}{8}\text{ yd}$$

For 7 costumes, $16\frac{5}{8}$ yards of fabric would be needed.

121. Subtract the heights to find the difference.

$$10\frac{1}{2}-7\frac{1}{8}=\frac{21}{2}-\frac{57}{8}$$
$$=\frac{21\cdot 4}{2\cdot 4}-\frac{57}{8}\qquad \text{LCD}=8$$
$$=\frac{84}{8}-\frac{57}{8}$$
$$=\frac{27}{8},\text{ or }3\frac{3}{8}$$

The difference in heights is $3\frac{3}{8}$ inches.

123. The sum of the fractions representing the U.S. foreign-born population from Latin America, Asia, or Europe is

$$\frac{27}{50}+\frac{7}{25}+\frac{3}{25}=\frac{27}{50}+\frac{7\cdot 2}{25\cdot 2}+\frac{3\cdot 2}{25\cdot 2}$$
$$=\frac{27+14+6}{50}$$
$$=\frac{47}{50}.$$

So the fraction representing the U.S. foreign-born population from other regions is

$$1-\frac{47}{50}=\frac{50}{50}-\frac{47}{50}$$
$$=\frac{3}{50}.$$

125. Multiply the fraction representing the U.S. foreign-born population from Europe, $\frac{3}{25}$, by the total number of foreign-born people in the U.S., approximately 40 million.

$$\frac{3}{25}\cdot 40=\frac{3}{25}\cdot\frac{40}{1}=\frac{3\cdot 5\cdot 8}{5\cdot 5\cdot 1}=\frac{24}{5},\text{ or }4\frac{4}{5}$$

There were approximately $4\frac{4}{5}$ million (or 4,800,000) foreign-born people in the U.S. who were born in Europe.

127. Estimate each fraction. $\frac{14}{26}$ is about $\frac{1}{2}$, $\frac{98}{99}$ is about 1, $\frac{100}{51}$ is about 2, $\frac{90}{31}$ is about 3, and $\frac{13}{27}$ is about $\frac{1}{2}$.

Therefore, the sum is approximately

$$\frac{1}{2}+1+2+3+\frac{1}{2}=7.$$

The correct choice is C.

R.2 Decimals and Percents

Now Try Exercises

N1. **(a)** $0.8=\frac{8}{10}$

(b) $0.431=\frac{431}{1000}$

(c) $2.58=\frac{258}{100}$

N2. **(a)**

$$\begin{array}{r} 68.900 \\ 42.720 \\ +\ 8.973 \\ \hline 120.593 \end{array}$$

(b)

$$\begin{array}{r} 351.800 \\ -\ \ 2.706 \\ \hline 349.094 \end{array}$$

N3. **(a)**

$$\begin{array}{rl} 9.32 & \text{2 decimal places} \\ \times\ 1.4 & \text{1 decimal place} \\ \hline 3728 & \qquad\downarrow \\ 932 & \quad 2+1=3 \\ \hline 13.048 & \text{3 decimal places} \end{array}$$

(b)

$$\begin{array}{rl} 0.06 & \text{2 decimal places} \\ \times 0.004 & \text{3 decimal places} \\ \hline 24 & \quad 2+3=5 \\ \hline 0.00024 & \text{5 decimal places} \end{array}$$

N4. **(a)** To change the divisor 14.9 into a whole number, move each decimal point one place to the right. Move the decimal point straight up and divide as with whole numbers.

$$149\overline{)4514.7}\quad \text{quotient } 30.3$$

447
447
447
0

Therefore, $451.47 \div 14.9 = 30.3$.

(b) To change the divisor 1.3 into a whole number, move each decimal point one place to the right. Move the decimal point straight up and divide as with whole numbers.

$$13\overline{)73.340}\quad \text{quotient } 5.641$$

65
83
78
54
52
20
13
7

We carried out the division to 3 decimal places so that we could round to 2 decimal places. Therefore, $7.334 \div 1.3 \approx 5.64$.

N5. **(a)** Move the decimal point one place to the right because 10 has one 0.
$294.72 \times 10 = 2947.2$

(b) Move the decimal point two places to the left because 100 has two 0s. Insert a 0 in front of the 4 to do this.
$4.793 \div 100 = 0.04793$

N6. **(a)** Divide 20 by 17. Add a decimal point and as many 0s as necessary.

$$20\overline{)17.00}\quad \text{quotient } 0.85$$

160
100
100
0

Therefore, $\dfrac{17}{20} = 0.85$.

(b) Divide 2 by 9. Add a decimal point and as many 0s as necessary.

$$9\overline{)2.000...}\quad \text{quotient } 0.222...$$

18
20
18
20
18
2

Note that the pattern repeats. Therefore, $\dfrac{2}{9} = 0.\bar{2}$, or 0.222.

N7. **(a)** $23\% = 23 \cdot 1\% = 23 \cdot 0.01 = 0.23$

(b) $350\% = 350 \cdot 1\% = 350 \cdot 0.01 = 3.50$, or 3.5

N8. **(a)** $0.71 = 71 \cdot 0.01 = 71 \cdot 1\% = 71\%$

(b) $1.32 = 132 \cdot 0.01 = 132 \cdot 1\% = 132\%$

N9. **(a)** $52\% = 0.52$

(b) $2\% = 02\% = 0.02$

(c) $0.45 = 45\%$

(d) $3.5 = 3.50 = 350\%$

N10. **(a)** $20\% = 20 \cdot 1\% = 20 \cdot \dfrac{1}{100} = \dfrac{20}{100}$

In lowest terms,
$\dfrac{20}{100} = \dfrac{20 \cdot 1}{20 \cdot 5} = \dfrac{1}{5}$

(b) $160\% = 160 \cdot 1\% = 160 \cdot \dfrac{1}{100} = \dfrac{160}{100}$

In lowest terms,
$\dfrac{160}{100} = \dfrac{20 \cdot 8}{20 \cdot 5} = \dfrac{8}{5}$

N11. **(a)** $\dfrac{6}{25} \cdot 100\% = \dfrac{6}{25} \cdot \dfrac{100}{1}\% = \dfrac{6 \cdot 25 \cdot 4}{25}\% = 24\%$

(b) $\dfrac{7}{2} \cdot 100\% = \dfrac{7}{2} \cdot \dfrac{100}{1}\% = \dfrac{7 \cdot 50 \cdot 2}{2}\% = 350\%$

N12. The discount is 60% of \$120. The word *of* here means multiply.

$$\begin{array}{ccc} 60\% & \text{of} & 120 \\ \downarrow & \downarrow & \downarrow \\ 0.60 & \cdot & 120 = 72 \end{array}$$

The discount is \$72. The sale price is found by subtracting.

$\$120.00 - \$72 = \$48$

Exercises

1. 367.9412

(a) Tens: 6

(b) Tenths: 9

(c) Thousandths: 1

(d) Ones: 7

(e) Hundredths: 4

3. 46.249

(a) 46.25

(b) 46.2

(c) 46

(d) 50

5. $0.4 = \frac{4}{10}$

7. $0.64 = \frac{64}{100}$

9. $0.138 = \frac{138}{1000}$

11. $0.043 = \frac{43}{1000}$

13. $3.805 = \frac{3805}{1000}$

15.
$$\begin{array}{r} 25.320 \\ 109.200 \\ +\ \ 8.574 \\ \hline 143.094 \end{array}$$

17.
$$\begin{array}{r} 28.73 \\ -\ \ 3.12 \\ \hline 25.61 \end{array}$$

19.
$$\begin{array}{r} 43.50 \\ -28.17 \\ \hline 15.33 \end{array}$$

21.
$$\begin{array}{r} 3.87 \\ 15.00 \\ +\ \ 2.90 \\ \hline 21.77 \end{array}$$

23.
$$\begin{array}{r} 32.560 \\ 47.356 \\ +\ \ 1.800 \\ \hline 81.716 \end{array}$$

25.
$$\begin{array}{r} 18.000 \\ -\ \ 2.789 \\ \hline 15.211 \end{array}$$

27.
$$\begin{array}{rr} 12.8 & \text{1 decimal place} \\ \times\ 9.1 & \text{1 decimal place} \\ \hline 128 & \downarrow \\ 1152 & 1+1=2 \\ \hline 116.48 & \text{2 decimal places} \end{array}$$

29.
$$\begin{array}{rr} 0.2 & \text{1 decimal place} \\ \times 0.03 & \text{2 decimal places} \\ \hline 6 & 1+2=3 \\ \hline 0.006 & \text{3 decimal places} \end{array}$$

31.
$$\begin{array}{r} 7.15 \\ 11\overline{)78.65} \\ \underline{77} \\ 16 \\ \underline{11} \\ 55 \\ \underline{55} \\ 0 \end{array}$$

33. To change the divisor 9.74 into a whole number, move each decimal point two places to the right. Move the decimal point straight up and divide as with whole numbers.

$$\begin{array}{r} 2.05 \\ 974\overline{)1996.70} \\ \underline{1948} \\ 4870 \\ \underline{4870} \\ 0 \end{array}$$

Therefore, $19.967 \div 9.74 = 2.05$.

35. Move the decimal point two places to the right because 100 has two 0s.
$57.116 \times 100 = 5711.6$

37. Move the decimal point three places to the right because 1000 has three 0s.
$0.094 \times 1000 = 94$

39. Move the decimal point one place to the left because 10 has one 0.
$1.62 \div 10 = 0.162$

41. Move the decimal point two places to the left because 100 has two 0s.
$124.03 \div 100 = 1.2403$

43. Convert from a decimal to a percent.
$0.01 = 1\%$

Fraction in Lowest Terms (or Whole Number)	Decimal	Percent
$\frac{1}{100}$	0.01	1%

45. Convert from a decimal to a fraction.

$$0.05 = 5 \cdot 0.01 = 5 \cdot \frac{1}{100} = \frac{5}{100}$$

In lowest terms,

$$\frac{5}{100} = \frac{5 \cdot 1}{20 \cdot 5} = \frac{1}{20}$$

Fraction in Lowest Terms (or Whole Number)	Decimal	Percent
$\frac{1}{20}$	0.05	5%

47. Convert the decimal to a percent.
$0.125 = 12.5 \cdot 0.01 = 12.5 \cdot 1\% = 12.5\%$

Fraction in Lowest Terms (or Whole Number)	Decimal	Percent
$\frac{1}{8}$	0.125	12.5%

49. Convert to a decimal first. Divide 1 by 4. Add a decimal point and as many 0s as necessary.

$$\begin{array}{r} 0.25 \\ 4\overline{)1.00} \\ \underline{8} \\ 20 \\ \underline{20} \\ 0 \end{array}$$

Convert the decimal to a percent.
$0.25 = 25 \cdot 0.01 = 25 \cdot 1\% = 25\%$

Fraction in Lowest Terms (or Whole Number)	Decimal	Percent
$\frac{1}{4}$	0.25	25%

51. Convert the percent to a decimal first.
$50\% = 0.50$, or 0.5
Convert from a decimal to a fraction.

$$0.50 = 50 \cdot 0.01 = 50 \cdot \frac{1}{100} = \frac{50}{100}$$

In lowest terms,

$$\frac{50}{100} = \frac{50 \cdot 1}{50 \cdot 2} = \frac{1}{2}$$

Fraction in Lowest Terms (or Whole Number)	Decimal	Percent
$\frac{1}{2}$	0.5	50%

53. Convert the decimal to a percent first.
$0.75 = 75 \cdot 0.01 = 75 \cdot 1\% = 75\%$
Convert from a decimal to a fraction.
$0.75 = 75 \cdot 0.01 = 75 \cdot \frac{1}{100} = \frac{75}{100}$
In lowest terms,
$\frac{75}{100} = \frac{25 \cdot 3}{25 \cdot 4} = \frac{3}{4}$

Fraction in Lowest Terms (or Whole Number)	Decimal	Percent
$\frac{3}{4}$	0.75	75%

55. Divide 3 by 8. Add a decimal point and as many 0s as necessary.

$$\begin{array}{r} 0.375 \\ 8\overline{)3.000} \\ \underline{24} \\ 60 \\ \underline{56} \\ 40 \\ \underline{40} \\ 0 \end{array}$$

57. Divide 5 by 4. Add a decimal point and as many 0s as necessary.

$$\begin{array}{r} 1.25 \\ 4\overline{)5.00} \\ \underline{4} \\ 10 \\ \underline{8} \\ 20 \\ \underline{20} \\ 0 \end{array}$$

59. Divide 5 by 9. Add a decimal point and as many 0s as necessary.

$$\begin{array}{r} 0.555... \\ 9\overline{)5.000...} \\ \underline{45} \\ 50 \\ \underline{45} \\ 50 \\ \underline{45} \\ 5 \end{array}$$

Note that the pattern repeats. Therefore,
$\frac{5}{9} = 0.\overline{5}$, or about 0.556.

61. Divide 1 by 6. Add a decimal point and as many 0s as necessary.

$$\begin{array}{r} 0.166... \\ 6\overline{)1.000...} \\ \underline{6} \\ 40 \\ \underline{36} \\ 40 \\ \underline{36} \\ 4 \end{array}$$

Note that the pattern repeats. Therefore,
$\frac{1}{6} = 0.1\overline{6}$, or about 0.167.

63. $54\% = 0.54$

65. $7\% = 07\% = 0.07$

67. $117\% = 1.17$

69. $2.4\% = 02.4\% = 0.024$

71. $6\frac{1}{4}\% = 6.25\% = 06.25\% = 0.0625$

73. $0.8\% = 00.8\% = 0.008$

75. $0.79 = 79\%$

77. $0.02 = 2\%$

79. $0.004 = 0.4\%$

81. $1.28 = 128\%$

83. $0.40 = 40\%$

85. $6 = 6.00 = 600\%$

87. $51\% = 51 \cdot 1\% = 51 \cdot \frac{1}{100} = \frac{51}{100}$

89. $15\% = 15 \cdot 1\% = 15 \cdot \frac{1}{100} = \frac{15}{100}$
In lowest terms,
$\frac{15}{100} = \frac{3 \cdot 5}{20 \cdot 5} = \frac{3}{20}$

91. $2\% = 2 \cdot 1\% = 2 \cdot \frac{1}{100} = \frac{2}{100}$
In lowest terms,
$\frac{2}{100} = \frac{1 \cdot 2}{50 \cdot 2} = \frac{1}{50}$

93. $140\% = 140 \cdot 1\% = 140 \cdot \frac{1}{100} = \frac{140}{100}$

In lowest terms,

$\frac{140}{100} = \frac{20 \cdot 7}{20 \cdot 5} = \frac{7}{5}$, or $1\frac{2}{5}$

95. $7.5\% = 7\frac{1}{2} \cdot 1\% = \frac{15}{2} \cdot \frac{1}{100} = \frac{15}{200}$

In lowest terms,

$\frac{15}{200} = \frac{5 \cdot 3}{5 \cdot 40} = \frac{3}{40}$

97. $\frac{4}{5} \cdot 100\% = \frac{4}{5} \cdot \frac{100}{1}\% = \frac{4 \cdot 5 \cdot 20}{5}\% = 80\%$

99. $\frac{7}{50} \cdot 100\% = \frac{7}{50} \cdot \frac{100}{1}\% = \frac{7 \cdot 2 \cdot 50}{50}\% = 14\%$

101. $\frac{2}{11} \cdot 100\% = \frac{2}{11} \cdot \frac{100}{1}\% = \frac{200}{11}\% = 18.\overline{18}\%$

103. $\frac{9}{4} \cdot 100\% = \frac{9}{4} \cdot \frac{100}{1}\% = \frac{9 \cdot 4 \cdot 25}{4}\% = 225\%$

105. $\frac{13}{6} \cdot 100\% = \frac{13}{6} \cdot \frac{100}{1}\% = \frac{13 \cdot 2 \cdot 50}{2 \cdot 3}\% = 216.\overline{6}\%$

107. The word *of* here means multiply.

$$\begin{array}{ccc} 50\% & \text{of} & 320 \\ \downarrow & \downarrow & \downarrow \\ 0.50 & \cdot & 320 \end{array} = 160$$

109. The word *of* here means multiply.

$$\begin{array}{ccc} 6\% & \text{of} & 80 \\ \downarrow & \downarrow & \downarrow \\ 0.06 & \cdot & 80 \end{array} = 4.8$$

111. The word *of* here means multiply.

$$\begin{array}{ccc} 14\% & \text{of} & 780 \\ \downarrow & \downarrow & \downarrow \\ 0.14 & \cdot & 780 \end{array} = 109.2$$

113. The tip is 20% of \$89. The word *of* here means multiply.

$$\begin{array}{ccc} 20\% & \text{of} & \$89 \\ \downarrow & \downarrow & \downarrow \\ 0.20 & \cdot & \$89 \end{array} = \$17.80$$

The tip is \$17.80. The total bill is found by adding.

$\$89 + \$17.80 = \$106.80$

115. The discount is 15% of \$795. The word *of* here means multiply.

$$\begin{array}{ccc} 15\% & \text{of} & \$795 \\ \downarrow & \downarrow & \downarrow \\ 0.15 & \cdot & \$795 \end{array} = \$119.25$$

The amount of the discount is \$119.25. The sale price is found by subtracting.

$\$795 - \$119.25 = \$675.75$

117. The portion of the circle graph showing the number of travelers from Canada is 33% of the circle. Find 33% of 60 million.

$$\begin{array}{ccc} 33\% & \text{of} & 60 \text{ million} \\ \downarrow & \downarrow & \downarrow \\ 0.33 & \cdot & 60 \text{ million} \end{array} = 19.8 \text{ million}$$

119. First, find the portion of the circle graph that represents "Other."

$100\% - (33\% + 22\% + 20\% + 12\%) = 13\%$

The portion of the circle graph showing the number of travelers from "Other" countries is 13% of the circle.

Chapter 1
The Real Number System

1.1 Exponents, Order of Operations, and Inequality

Now Try Exercises

N1. **(a)** $6^2 = 6 \cdot 6 = 36$

(b) $\left(\frac{4}{5}\right)^3 = \underbrace{\frac{4}{5} \cdot \frac{4}{5} \cdot \frac{4}{5}} = \frac{64}{125}$

$\frac{4}{5}$ is used as a factor 3 times.

N2. **(a)** $15 - 2 \cdot 6$

$= 15 - 12$ Multiply.

$= 3$ Subtract.

(b) $8 + 2(5 - 1)$

$= 8 + 2(4)$ Subtract inside parentheses.

$= 8 + 8$ Multiply.

$= 16$ Add.

(c) $6(2 + 4) - 7 \cdot 5$

$= 6(6) - 7 \cdot 5$ Add inside parentheses.

$= 36 - 35$ Multiply.

$= 1$ Subtract.

(d) $8 \cdot 10 \div 4 - 2^3 + 3 \cdot 4^2$

$= 8 \cdot 10 \div 4 - 8 + 3 \cdot 16$ Apply exponents.

$= 80 \div 4 - 8 + 48$ Multiply.

$= 20 - 8 + 48$ Divide.

$= 12 + 48$ Subtract.

$= 60$ Add.

N3. **(a)** $7\left[\left(3^2 - 1\right) + 4\right]$

$= 7[(9 - 1) + 4]$ Apply exponents.

$= 7[8 + 4]$ Subtract inside parentheses.

$= 7[12]$ Add inside brackets.

$= 84$ Multiply.

(b) $\frac{9(14 - 4) - 2}{4 + 3 \cdot 6}$

$= \frac{9(10) - 2}{4 + 3 \cdot 6}$ Subtract inside parentheses.

$= \frac{90 - 2}{4 + 18}$ Multiply.

$= \frac{88}{22}$ Subtract and add.

$= 4$ Divide.

N4. **(a)** The statement $12 \neq 10 - 2$ is *true* because 12 *is not equal to* 8.

(b) The statement $5 > 4 \cdot 2$ is *false* because 5 *is less than* 8.

(c) The statement $7 \leq 7$ is *true* since $7 = 7$.

(d) Write the fractions with a common denominator. The statement $\frac{5}{9} > \frac{7}{11}$ is equivalent to the statement $\frac{55}{99} > \frac{63}{99}$. Since 55 is *less* than 63, the original statement is *false*.

N5. **(a)** "Ten is not equal to eight minus two" is written as $10 \neq 8 - 2$.

(b) "Fifty is greater than fifteen" is written as $50 > 15$.

(c) "Eleven is less than or equal to twenty" is written as $11 \leq 20$.

N6. $8 < 9$ may be written as $9 > 8$.

Exercises

1. False; $3^2 = 3 \cdot 3 = 9$.

3. False; a number raised to the first power is that number, so $3^1 = 3$.

5. False; the common error leading to 42 is adding 4 to 3 and then multiplying by 6. One must follow the rules for order of operations.

$4 + 3(8 - 2)$

$= 4 + 3(6)$

$= 4 + 18$

$= 22$

7. Additions and subtractions are performed in order from left to right.

$18 \underset{1}{\underbrace{-}} 2 \underset{2}{\underbrace{+}} 3$

9. Multiplications and divisions are performed in order from left to right, and then additions and subtractions are performed in order from left to right.

$\underbrace{2\cdot 8}_{1}\underbrace{-}_{3}\underbrace{6\div 3}_{2}$

11. Multiplications and divisions are performed in order from left to right, and then additions and subtractions are performed in order from left to right. If grouping symbols are present, work within them first, starting with the innermost.

$\underbrace{3\cdot 5}_{2}\underbrace{-}_{4}\underbrace{2}_{3}(4\underbrace{+}_{1}2)$

13. $3^2 = 3\cdot 3 = 9$

15. $7^2 = 7\cdot 7 = 49$

17. $12^2 = 12\cdot 12 = 144$

19. $4^3 = 4\cdot 4\cdot 4 = 64$

21. $10^3 = 10\cdot 10\cdot 10 = 1000$

23. $3^4 = 3\cdot 3\cdot 3\cdot 3 = 81$

25. $4^5 = 4\cdot 4\cdot 4\cdot 4\cdot 4 = 1024$

27. $\left(\frac{1}{6}\right)^2 = \frac{1}{6}\cdot\frac{1}{6} = \frac{1}{36}$

29. $\left(\frac{2}{3}\right)^4 = \frac{2}{3}\cdot\frac{2}{3}\cdot\frac{2}{3}\cdot\frac{2}{3} = \frac{16}{81}$

31. $(0.4)^3 = (0.4)(0.4)(0.4) = 0.064$

33.
$$\begin{aligned} 64\div 4\cdot 2 &= 16\cdot 2 && \text{Divide.}\\ &= 32 && \text{Multiply.}\end{aligned}$$

35.
$$\begin{aligned} 13+9\cdot 5 &= 13+45 && \text{Multiply.}\\ &= 58 && \text{Add.}\end{aligned}$$

37.
$$\begin{aligned} 25.2-12.6\div 4.2 &= 25.2-3 && \text{Divide.}\\ &= 22.2 && \text{Subtract.}\end{aligned}$$

39.
$$\begin{aligned} \frac{1}{4}\cdot\frac{2}{3}+\frac{2}{5}\cdot\frac{11}{3} &= \frac{1}{6}+\frac{22}{15} && \text{Multiply.}\\ &= \frac{5}{30}+\frac{44}{30} && \text{LCD} = 30\\ &= \frac{49}{30}, \text{ or } 1\frac{19}{30} && \text{Add.}\end{aligned}$$

41.
$$\begin{aligned} 9\cdot 4-8\cdot 3 &= 36-24 && \text{Multiply.}\\ &= 12 && \text{Subtract.}\end{aligned}$$

43.
$$\begin{aligned} 20-4\cdot 3+5 &= 20-12+5 && \text{Multiply.}\\ &= 8+5 && \text{Subtract.}\\ &= 13 && \text{Add.}\end{aligned}$$

45.
$$\begin{aligned} 10+40\div 5\cdot 2 &= 10+8\cdot 2 && \text{Divide.}\\ &= 10+16 && \text{Multiply.}\\ &= 26 && \text{Add.}\end{aligned}$$

47.
$$\begin{aligned} &18-2(3+4)\\ &= 18-2(7) && \text{Add inside parentheses.}\\ &= 18-14 && \text{Multiply.}\\ &= 4 && \text{Subtract.}\end{aligned}$$

49.
$$\begin{aligned} 3(4+2)+8\cdot 3 &= 3\cdot 6+8\cdot 3 && \text{Add.}\\ &= 18+24 && \text{Multiply.}\\ &= 42 && \text{Add.}\end{aligned}$$

51.
$$\begin{aligned} 18-4^2+3 &= 18-16+3 && \text{Apply exponents.}\\ &= 2+3 && \text{Subtract.}\\ &= 5 && \text{Add.}\end{aligned}$$

53.
$$\begin{aligned} 2+3[5+4(2)] &= 2+3[5+8] && \text{Multiply.}\\ &= 2+3[13] && \text{Add.}\\ &= 2+39 && \text{Multiply.}\\ &= 41 && \text{Add.}\end{aligned}$$

55.
$$\begin{aligned} 5\left[3+4\left(2^2\right)\right] &= 5[3+4(4)] && \text{Apply exponents.}\\ &= 5(3+16) && \text{Multiply.}\\ &= 5(19) && \text{Add.}\\ &= 95 && \text{Multiply.}\end{aligned}$$

57.
$$\begin{aligned} &3^2[(11+3)-4]\\ &= 3^2[14-4] && \text{Add inside parentheses.}\\ &= 3^2[10] && \text{Subtract.}\\ &= 9[10] && \text{Apply exponents.}\\ &= 90 && \text{Multiply.}\end{aligned}$$

59. Simplify the numerator and denominator separately, and then divide.
$$\begin{aligned} \frac{6\left(3^2-1\right)+8}{8-2^2} &= \frac{6(9-1)+8}{8-4}\\ &= \frac{6(8)+8}{4}\\ &= \frac{48+8}{4}\\ &= \frac{56}{4} = 14\end{aligned}$$

61. Simplify the numerator and denominator separately, and then divide.

$$\frac{4(6+2)+8(8-3)}{6(4-2)-2^2} = \frac{4(8)+8(5)}{6(2)-2^2}$$
$$= \frac{4(8)+8(5)}{6(2)-4}$$
$$= \frac{32+40}{12-4}$$
$$= \frac{72}{8} = 9$$

63. $3 \cdot 6 + 4 \cdot 2 = 60$

Listed below are some possibilities. Use trial and error until you get the desired result.

$(3 \cdot 6) + 4 \cdot 2 = 18 + 8 = 26 \neq 60$

$(3 \cdot 6 + 4) \cdot 2 = 22 \cdot 2 = 44 \neq 60$

$3 \cdot (6 + 4 \cdot 2) = 3 \cdot 14 = 42 \neq 60$

$3 \cdot (6 + 4) \cdot 2 = 3 \cdot 10 \cdot 2 = 30 \cdot 2 = 60$

65. $10 - 7 - 3 = 6$

$10 - (7 - 3) = 10 - 4 = 6$

67. $8 + 2^2 = 100$

$(8+2)^2 = 10^2 = 10 \cdot 10 = 100$

69. $9 \cdot 3 - 11 \leq 16$

$27 - 11 \leq 16$

$16 \leq 16$

The statement is true since $16 = 16$.

71. $5 \cdot 11 + 2 \cdot 3 \leq 60$

$55 + 6 \leq 60$

$61 \leq 60$

The statement is false since 61 *is greater than* 60.

73. $0 \geq 12 \cdot 3 - 6 \cdot 6$

$0 \geq 36 - 36$

$0 \geq 0$

The statement is true since $0 = 0$.

75. $45 \geq 2[2 + 3(2+5)]$

$45 \geq 2[2 + 3(7)]$

$45 \geq 2[2 + 21]$

$45 \geq 2[23]$

$45 \geq 46$

The statement is false since 45 *is less than* 46.

77. $[3 \cdot 4 + 5(2)] \cdot 3 > 72$

$[12 + 10] \cdot 3 > 72$

$[22] \cdot 3 > 72$

$66 > 72$

The statement is false since 66 *is less than* 72.

79. $\frac{3+5(4-1)}{2 \cdot 4 + 1} \geq 3$

$\frac{3+5(3)}{8+1} \geq 3$

$\frac{3+15}{9} \geq 3$

$\frac{18}{9} \geq 3$

$2 \geq 3$

The statement is false since 2 *is less than* 3.

81. $3 \geq \frac{2(5+1) - 3(1+1)}{5(8-6) - 4 \cdot 2}$

$3 \geq \frac{2(6) - 3(2)}{5(2) - 8}$

$3 \geq \frac{12-6}{10-8}$

$3 \geq \frac{6}{2}$

$3 \geq 3$

The statement is true since $3 = 3$.

83. "$5 < 17$" means "five is less than seventeen." The statement is true.

85. "$5 \neq 8$" means "five is not equal to eight." The statement is true.

87. "$7 \geq 14$" means "seven is greater than or equal to fourteen." The statement is false.

89. "$15 \leq 15$" means "fifteen is less than or equal to fifteen." The statement is true.

91. "$\frac{1}{3} = \frac{3}{10}$" means "one-third is equal to three-tenths." The statement is false.

93. "$2.5 > 2.50$" means "two and five-tenths is greater than two and fifty-hundredths." The statement is false.

95. "Fifteen is equal to five plus ten" is written as $15 = 5 + 10$.

97. "Nine is greater than five minus four" is written as $9 > 5 - 4$.

99. "Sixteen is not equal to nineteen" is written as $16 \neq 19$.

101. "One-half is less than or equal to two-fourths" is written as $\frac{1}{2} \leq \frac{2}{4}$.

103. $5 < 20$ becomes $20 > 5$ when the inequality symbol is reversed.

105. $\frac{4}{5} > \frac{3}{4}$ becomes $\frac{3}{4} < \frac{4}{5}$ when the inequality symbol is reversed.

107. $2.5 \geq 1.3$ becomes $1.3 \leq 2.5$ when the inequality symbol is reversed.

109. **(a)** Substitute "40" for "age" in the expression for women.
$14.7 - 40 \cdot 0.13$

(b) $14.7 - 40 \cdot 0.13 = 14.7 - 5.2$ Multiply.
$= 9.5$ Subtract.

(c) 85% of 9.5 is $0.85(9.5) = 8.075$.
Walking at 5 mph is associated with 8.0 METs, which is the table value closest to 8.075.

(d) Substitute "55" for "age" in the expression for men.
$14.7 - 55 \cdot 0.11$

$14.7 - 55 \cdot 0.11 = 14.7 - 6.05$ Multiply.
$= 8.65$ Subtract.

85% of 8.65 is $0.85(8.65) = 7.3525$.
Swimming is associated with 7.0 METs, which is the table value closest to 7.3525.

111. The states that had a number greater than 13.8 are Alaska (16.2), Texas (14.7), California (24.1), and Idaho (17.6).

113. The states that had a number *not* less than 13.8, which is the same as greater than or equal to 13.8, are Alaska (16.2), Texas (14.7), California (24.1), Idaho (17.6), and Missouri (13.8).

1.2 Variables, Expressions, and Equations

Now Try Exercises

N1. **(a)** $9x = 9 \cdot x$
$= 9 \cdot 6$ Replace x with 6.
$= 54$ Multiply.

(b) $4x^2 = 4 \cdot x^2$
$= 4 \cdot 6^2$ Replace x with 6.
$= 4 \cdot 36$ Square 6.
$= 144$ Multiply.

N2. **(a)** $3x + 4y = 3(4) + 4(7)$
$= 12 + 28$ Multiply.
$= 40$ Add.

(b) $\frac{6x - 2y}{2y - 9} = \frac{6(4) - 2(7)}{2(7) - 9}$
$= \frac{24 - 14}{14 - 9}$ Multiply.
$= \frac{10}{5} = 2$ Subtract; reduce.

(c) $4x^2 - y^2 = 4 \cdot 4^2 - 7^2$
$= 4 \cdot 16 - 49$ Use exponents.
$= 64 - 49$ Multiply.
$= 15$ Subtract.

N3. **(a)** Using x as the variable to represent the number, "the sum of a number and 10" translates as $x + 10$, or $10 + x$.

(b) "A number divided by 7" translates as $x \div 7$, or $\frac{x}{7}$.

(c) "The difference between 9 and a number" translates as $9 - x$. Thus, "the product of 3 and the difference between 9 and a number" translates as $3(9 - x)$.

N4. $8k + 5 = 61$

$8 \cdot 7 + 5 \stackrel{?}{=} 61$ Replace k with 7.

$56 + 5 \stackrel{?}{=} 61$ Multiply.

$61 = 61$ True

The number 7 is a solution of the equation.

N5. Using x as the variable to represent the number, "the sum of a number and nine is equal to the difference between 25 and the number" translates as $x+9=25-x$. Now try each number from the set $\{0, 2, 4, 6, 8, 10\}$.

$x=4$: $4+9\stackrel{?}{=}25-4$
$13=21$ False

$x=6$: $6+9\stackrel{?}{=}25-6$
$15=19$ False

$x=8$: $8+9\stackrel{?}{=}25-8$
$17=17$ True

Similarly, $x=0$, 2, or 10 result in false statements. Thus, 8 is the only solution.

N6. **(a)** $2x+5=6$ has an equals symbol, so this is an equation.

(b) $2x+5-6$ has no equals symbol, so this is an expression.

Exercises

1. The expression $8x^2$ means $8\cdot x\cdot x$. The correct choice is B.

3. The sum of 15 and a number x is represented by the expression $15+x$. The correct choice is A.

5. $2x^3=2\cdot x\cdot x\cdot x$, while $2x\cdot 2x\cdot 2x=(2x)^3$. The last expression is equal to $8x^3$.

7. The exponent 2 applies only to its base, which is x. (The expression $(5x)^2$ would require multiplying 5 by $x=4$ first.)

9. **(a)** $x+7=4+7$
$=11$

(b) $x+7=6+7$
$=13$

11. **(a)** $4x=4(4)=16$

(b) $4x=4(6)=24$

13. **(a)** $4x^2=4\cdot 4^2$
$=4\cdot 16$
$=64$

(b) $4x^2=4\cdot 6^2$
$=4\cdot 36$
$=144$

15. **(a)** $\frac{x+1}{3}=\frac{4+1}{3}$
$=\frac{5}{3}$

(b) $\frac{x+1}{3}=\frac{6+1}{3}$
$=\frac{7}{3}$

17. **(a)** $\frac{3x-5}{2x}=\frac{3\cdot 4-5}{2\cdot 4}$
$=\frac{12-5}{8}$
$=\frac{7}{8}$

(b) $\frac{3x-5}{2x}=\frac{3\cdot 6-5}{2\cdot 6}$
$=\frac{18-5}{12}$
$=\frac{13}{12}$

19. **(a)** $3x^2+x=3\cdot 4^2+4$
$=3\cdot 16+4$
$=48+4=52$

(b) $3x^2+x=3\cdot 6^2+6$
$=3\cdot 36+6$
$=108+6=114$

21. **(a)** $6.459x=6.459\cdot 4$
$=25.836$

(b) $6.459x=6.459\cdot 6$
$=38.754$

23. **(a)** $8x+3y+5=8(2)+3(1)+5$
$=16+3+5$
$=19+5$
$=24$

(b) $8x+3y+5=8(1)+3(5)+5$
$=8+15+5$
$=23+5$
$=28$

25. (a) $3(x+2y) = 3(2+2\cdot 1)$
$= 3(2+2)$
$= 3(4)$
$= 12$

(b) $3(x+2y) = 3(1+2\cdot 5)$
$= 3(1+10)$
$= 3(11)$
$= 33$

27. (a) $x+\frac{4}{y} = 2+\frac{4}{1}$
$= 2+4$
$= 6$

(b) $x+\frac{4}{y} = 1+\frac{4}{5}$
$= \frac{5}{5}+\frac{4}{5}$
$= \frac{9}{5}$

29. (a) $\frac{x}{2}+\frac{y}{3} = \frac{2}{2}+\frac{1}{3}$
$= \frac{6}{6}+\frac{2}{6}$
$= \frac{8}{6} = \frac{4}{3}$

(b) $\frac{x}{2}+\frac{y}{3} = \frac{1}{2}+\frac{5}{3}$
$= \frac{3}{6}+\frac{10}{6}$
$= \frac{13}{6}$

31. (a) $\frac{2x+4y-6}{5y+2} = \frac{2(2)+4(1)-6}{5(1)+2}$
$= \frac{4+4-6}{5+2}$
$= \frac{8-6}{7}$
$= \frac{2}{7}$

(b) $\frac{2x+4y-6}{5y+2} = \frac{2(1)+4(5)-6}{5(5)+2}$
$= \frac{2+20-6}{25+2}$
$= \frac{22-6}{27}$
$= \frac{16}{27}$

33. (a) $2y^2+5x = 2\cdot 1^2+5\cdot 2$
$= 2\cdot 1+5\cdot 2$
$= 2+10$
$= 12$

(b) $2y^2+5x = 2\cdot 5^2+5\cdot 1$
$= 2\cdot 25+5\cdot 1$
$= 50+5$
$= 55$

35. (a) $\frac{3x+y^2}{2x+3y} = \frac{3(2)+1^2}{2(2)+3(1)}$
$= \frac{3(2)+1}{4+3}$
$= \frac{6+1}{7}$
$= \frac{7}{7}$
$= 1$

(b) $\frac{3x+y^2}{2x+3y} = \frac{3(1)+5^2}{2(1)+3(5)}$
$= \frac{3(1)+25}{2+15}$
$= \frac{3+25}{17}$
$= \frac{28}{17}$

37. (a) $0.841x^2+0.32y^2$
$= 0.841\cdot 2^2+0.32\cdot 1^2$
$= 0.841\cdot 4+0.32\cdot 1$
$= 3.364+0.32$
$= 3.684$

(b) $0.841x^2 + 0.32y^2$

$= 0.841 \cdot 1^2 + 0.32 \cdot 5^2$

$= 0.841 \cdot 1 + 0.32 \cdot 25$

$= 0.841 + 8$

$= 8.841$

39. "Twelve times a number" translates as $12 \cdot x$, or $12x$.

41. "Added to" indicates addition. "Nine added to a number" translates as $x + 9$.

43. "Four subtracted from a number" translates as $x - 4$.

45. "A number subtracted from seven" translates as $7 - x$.

47. "The difference between a number and 8" translates as $x - 8$.

49. "18 divided by a number" translates as $\frac{18}{x}$.

51. "The product of 6 and four less than a number" translates as $6(x-4)$.

53. $4m + 2 = 6; 1$

$4(1) + 2 \stackrel{?}{=} 6$ Let $m = 1$.

$4 + 2 \stackrel{?}{=} 6$

$6 = 6$ True

Because substituting 1 for *m* results in a true statement, 1 is a solution of the equation.

55. $2y + 3(y-2) = 14; 3$

$2 \cdot 3 + 3(3-2) \stackrel{?}{=} 14$ Let $y = 3$.

$2 \cdot 3 + 3 \cdot 1 \stackrel{?}{=} 14$

$6 + 3 \stackrel{?}{=} 14$

$9 = 14$ False

Because substituting 3 for *y* results in a false statement, 3 is not a solution of the equation.

57. $6p + 4p + 9 = 11; \frac{1}{5}$

$6\left(\frac{1}{5}\right) + 4\left(\frac{1}{5}\right) + 9 \stackrel{?}{=} 11$ Let $p = \frac{1}{5}$.

$\frac{6}{5} + \frac{4}{5} + 9 \stackrel{?}{=} 11$

$\frac{10}{5} + 9 \stackrel{?}{=} 11$

$2 + 9 \stackrel{?}{=} 11$

$11 = 11$ True

The true result shows that $\frac{1}{5}$ is a solution of the equation.

59. $3r^2 - 2 = 46; 4$

$3(4)^2 - 2 \stackrel{?}{=} 46$ Let $r = 4$.

$3 \cdot 16 - 2 \stackrel{?}{=} 46$

$48 - 2 \stackrel{?}{=} 46$

$46 = 46$ True

The true result shows that 4 is a solution of the equation.

61. $\frac{3}{8}x + \frac{1}{4} = 1; 2$

$\frac{3}{8}(2) + \frac{1}{4} \stackrel{?}{=} 1$ Let $x = 2$.

$\frac{3}{4} + \frac{1}{4} \stackrel{?}{=} 1$

$1 = 1$ True

The true result shows that 2 is a solution of the equation.

63. $0.5(x-4) = 80; 20$

$0.5(20-4) \stackrel{?}{=} 80$ Let $x = 20$.

$0.5(16) \stackrel{?}{=} 80$

$8 = 80$ False

The false result shows that 20 is not a solution of the equation.

65. "The sum of a number and 8 is 18" translates as $x+8=18$. Try each number from the given set, $\{2, 4, 6, 8, 10\}$, in turn.

$x+8=18$ Given equation
$2+8=18$ False
$4+8=18$ False
$6+8=18$ False
$8+8=18$ False
$10+8=18$ True

The only solution is 10.

67. "One more than twice a number is 5" translates as $2x+1=5$. Try each number from the given set. The only resulting true equation is $2\cdot 2+1=5$, so the only solution is 2.

69. "Sixteen minus three-fourths of a number is 13" translates as $16-\frac{3}{4}x=13$. Try each number from the given set, $\{2, 4, 6, 8, 10\}$, in turn.

$16-\frac{3}{4}x=13$ Given equation
$16-\frac{3}{4}(2)=13$ False
$16-\frac{3}{4}(4)=13$ True
$16-\frac{3}{4}(6)=13$ False
$16-\frac{3}{4}(8)=13$ False
$16-\frac{3}{4}(10)=13$ False

The only solution is 4.

71. "Three times a number is equal to 8 more than twice the number" translates as $3x=2x+8$. Try each number from the given set.

$3x=2x+8$ Given equation
$3(2)=2(2)+8$ False
$3(4)=2(4)+8$ False
$3(6)=2(6)+8$ False
$3(8)=2(8)+8$ True
$3(10)=2(10)+8$ False

The only solution is 8.

73. There is no equals symbol, so $3x+2(x-4)$ is an expression.

75. There is an equals symbol, so $7t+2(t+1)=4$ is an equation.

77. There is an equals symbol, so $x+y=9$ is an equation.

79. $y=0.180x-283$
$=0.180(1960)-283$
$=69.8$

The life expectancy of an American born in 1960 is about 70 years.

81. $y=0.180x-283$
$=0.180(1995)-283$
$=76.1$

The life expectancy of an American born in 1995 is about 76 years.

1.3 Real Numbers and the Number Line

Now Try Exercises

N1. Since the deepest point is below the water's surface, the depth is -136.

N2.

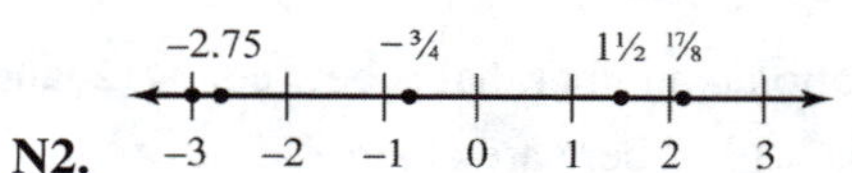

N3. **(a)** The whole numbers are 0 and 13.

(b) The integers are -7, 0, and 13.

(c) The rational numbers are $-7, -\frac{4}{5}, 0, 2.7,$ and 13.

(d) The irrational numbers are $\sqrt{3}$ and π.

N4. Since -8 lies to the right of -9 on the number line, -8 is greater than -9. Therefore, the statement $-8\le -9$ is *false*.

N5. **(a)** $|4|=4$

(b) $|-4|=-(-4)=4$

(c) $-|-4|=-[-(-4)]=-4$

N6. The category appliances is negative in both years.

Exercises

1. The number 0 is a whole number, but not a natural number.

3. The additive inverse of every negative number is a *positive* number.

5. A rational number is the quotient of two integers with the denominator not equal to 0.

7. **(a)** $|-9| = 9$ A

The distance between -9 and 0 on the number line is 9 units.

(b) $-(-9) = 9$ A

The opposite of -9 is 9.

(c) $-|-9| = -(9) = -9$ B

(d) $-|-(-9)| = -|9|$
$= -(9)$
$= -9$ (B)

9. The statement "Absolute value is always positive" is not true. The absolute value of 0 is 0, and 0 is not positive. We could say that absolute value is never negative, or absolute value is always nonnegative.

11. The only integer between 3.6 and 4.6 is 4.

13. There is only one whole number that is not positive and that is less than 1: the number 0.

15. An irrational number that is between $\sqrt{12}$ and $\sqrt{14}$ is $\sqrt{13}$. There are others.

17. True; every natural number is positive.

19. True; every integer is a rational number. For example, 5 can be written as $\frac{5}{1}$.

21. False; if a number is rational, it cannot be irrational, and vice versa.

23. Three examples of positive real numbers that are not integers are $\frac{1}{2}, \frac{5}{8},$ and $1\frac{3}{4}$. Other examples are $0.7, 4\frac{2}{3},$ and 5.1.

25. Three examples of real numbers that are not whole numbers are $-3\frac{1}{2}, -\frac{2}{3},$ and $\frac{3}{7}$. Other examples are $-4.3, -\sqrt{2},$ and $\sqrt{7}$.

27. Three examples of real numbers that are not rational numbers are $\sqrt{5}$, π, and $-\sqrt{3}$. All irrational numbers are real numbers that are not rational.

29. Use the integer 2,259,105 since "increased by 2,259,105" indicates a positive number.

31. Use the integer -3424 since "a decrease of 3424" indicates a negative number.

33. Use the rational number 46.77 since "closed up 46.77" indicates a positive number.

35. Graph 0, 3, -5, and -6.

Place a dot on the number line at the point that corresponds to each number. The order of the numbers from smallest to largest is $-6, -5, 0, 3$.

-6 -4 -2 0 2

37. Graph $-2, -6, -4, 3,$ and 4.

-6 -4 -2 0 2 4

39. Graph $\frac{1}{4}, 2\frac{1}{2}, -3.8, -4,$ and $-1\frac{5}{8}$.

-3.8 $-1\frac{5}{8}$ $\frac{1}{4}$ $2\frac{1}{2}$

-4 -2 0 2 4

41. **(a)** The natural numbers in the given set are 3 and 7, since they are in the natural number set $\{1, 2, 3, \ldots\}$.

(b) The set of whole numbers includes the natural numbers and 0. The whole numbers in the given set are 0, 3, and 7.

(c) The integers are the set of numbers $\{\ldots, -3, -2, -1, 0, 1, 2, 3, \ldots\}$. The integers in the given set are -9, 0, 3, and 7.

(d) Rational numbers are the numbers that can be expressed as the quotient of two integers, with denominators not equal to 0.
We can write numbers from the given set in this form as follows:

$-9 = \frac{-9}{1}, -1\frac{1}{4} = \frac{-5}{4}, -\frac{3}{5} = \frac{-3}{5}, 0 = \frac{0}{1},$
$0.\overline{1} = \frac{1}{9}, 3 = \frac{3}{1}, 5.9 = \frac{59}{10},$ and $7 = \frac{7}{1}$.

Thus, the rational numbers in the given set are $-9, -1\frac{1}{4}, -\frac{3}{5}, 0, 0.\overline{1}, 3, 5.9,$ and 7.

(e) Irrational numbers are real numbers that are not rational. $-\sqrt{7}$ and $\sqrt{5}$ can be represented by points on the number line but cannot be written as a quotient of integers. Thus, the irrational numbers in the given set are $-\sqrt{7}$ and $\sqrt{5}$.

(f) Real numbers are all numbers that can be represented on the number line. All the numbers in the given set are real.

43. **(a)** The additive inverse of -7 is found by changing the sign of -7. The additive inverse of -7 is 7.

(b) The absolute value of -7 is the distance between 0 and -7 on the number line, so $|-7|=7$.

45. **(a)** The additive inverse of 8 is -8.

(b) The distance between 0 and 8 on the number line is 8 units, so the absolute value of 8 is 8.

47. **(a)** The additive inverse of a number is found by changing the sign of a number, so the additive inverse of $-\frac{3}{4}$ is $\frac{3}{4}$.

(b) The distance between $-\frac{3}{4}$ and 0 on the number line is $\frac{3}{4}$ unit, so $\left|-\frac{3}{4}\right|=\frac{3}{4}$.

49. **(a)** The additive inverse of a number is found by changing the sign of a number, so the additive inverse of 5.6 is -5.6.

(b) The distance between 5.6 and 0 on the number line is 5.6 units, so $|5.6|=5.6$.

51. Since -6 is a negative number, its absolute value is the additive inverse of -6— that is, $|-6|=-(-6)=6$.

53. $-|12|=-(12)=-12$

55. $-\left|-\frac{2}{3}\right|=-\left[-\left(-\frac{2}{3}\right)\right]=-\left[\frac{2}{3}\right]=-\frac{2}{3}$

57. $|6-3|=|3|=3$

59. $-|6-3|=-|3|=-3$

61. Since -11 is located to the left of -3 on the number line, -11 is the lesser number.

63. Since -7 is located to the left of -6 on the number line, -7 is the lesser number.

65. Since $|-5|=5$, 4 is the lesser of the two numbers.

67. Since $|-3.5|=3.5$ and $|-4.5|=4.5$, $|-3.5|$ or 3.5 is the lesser number.

69. Since $-|-6|=-6$ and $-|-4|=-4$, $-|-6|$ is to the left of $-|-4|$ on the number line, so $-|-6|$ or -6 is the lesser number.

71. Since $|5-3|=|2|=2$ and $|6-2|=|4|=4$, $|5-3|$ or 2 is the lesser number.

73. Since -5 is to the *left* of -2 on the number line, -5 is *less than* -2, and the statement $-5<-2$ is true.

75. Since $-(-5)=5$ and $-4<5$, $-4\le-(-5)$ is true.

77. Since $|-6|=6$ and $|-9|=9$, and $6<9$, $|-6|<|-9|$ is true.

79. Since $-|8|=-8$ and $|-9|=-(-9)=9$, $-|8|<|-9|$, so $-|8|>|-9|$ is false.

81. Since $-|-5|=-5, -|-9|=-9$, and $-5>-9$, $-|-5|\ge-|-9|$ is true.

83. Since $|6-5|=|1|=1$ and $|6-2|=|4|=4$, $|6-5|<|6-2|$, so $|6-5|\ge|6-2|$ is false.

85. The number that represents the greatest percentage increase is 7.2, which corresponds to public transportation from 2010 to 2011.

87. The number with the smallest absolute value in the table is -0.3, so the least change corresponds to communication from 2009, to 2010.

1.4 Adding and Subtracting Real Numbers

Now Try Exercises

N1. **(a)** Start at 0 on a number line. Draw an arrow 3 units to the right to represent the positive number 3. From the right end of this arrow, draw a second arrow 5 units to the right to represent the addition of a positive number. The number below the end of this second arrow is 8, so $3+5=8$.

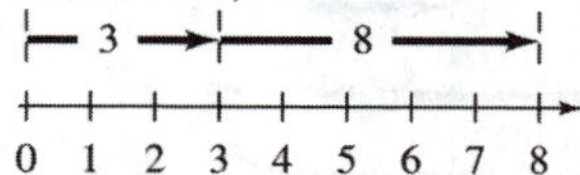

(b) Start at 0 on a number line. Draw an arrow 1 unit to the left to represent the negative number -1. From the left end of this arrow, draw a second arrow 3 units to the left. The number below the end of this second arrow is -4, so $-1+(-3)=-4$.

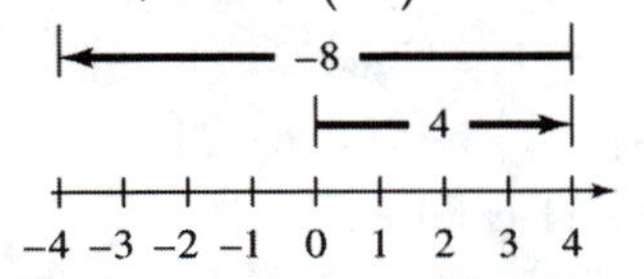

N2. **(a)** $-6+(-11)=-17$

The sum of two negative numbers is negative.

(b) $-\frac{2}{5}+\left(-\frac{1}{2}\right)=-\frac{9}{10}$

The sum of two negative numbers is negative.

N3. Start at 0 on a number line. Draw an arrow 4 units to the right. From the right end of this arrow, draw a second arrow 8 units to the left. The number below the end of this second arrow is -4, so $4+(-8)=-4$.

-8

4

-4 -3 -2 -1 0 1 2 3 4

N4. **(a)** Since the numbers have different signs, find the difference between their absolute values: $7-4=3$. Because 7 has the larger absolute value, the sum is negative: $7+(-4)=-3$.

(b) $\frac{2}{3}+\left(-2\frac{1}{9}\right)=-\frac{13}{9}$ or $-1\frac{4}{9}$

(c) $-5.7+3.7=-2$

(d) $-10+10=0$

N5. Use a number line to find the difference $6-2$.
Step 1 Start at 0 and draw an arrow 6 units to the *right*.
Step 2 From the right end of the first arrow, draw a second arrow 2 units to the *left* to represent the *subtraction*.

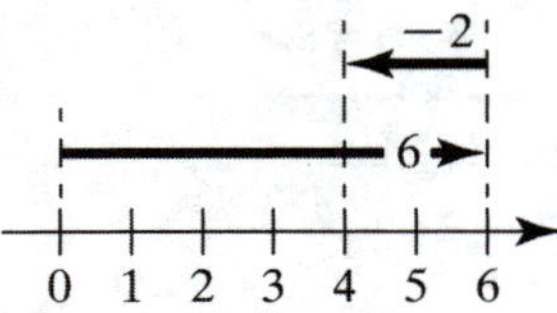

The number below the end of the second arrow is 4, so $6-2=4$.

N6. **(a)** $-5-(-11)=-5+(11)$ Add the opposite.
$=6$

(b) $4-15=4+(-15)$ Add the opposite.
$=-11$

(c) $-\frac{5}{7}-\frac{1}{3}=-\frac{5}{7}+\left(-\frac{1}{3}\right)$ Add the opposite.

$$=-\frac{15}{21}+\left(-\frac{7}{21}\right)$$
$$=-\frac{22}{21}, \text{ or } -1\frac{1}{21}$$

(d) $5.25-(-3.24)=8.49$

N7. **(a)** $8-[(-3+7)-(3-9)]$
$=8-[(4)-(3+(-9))]$
$=8-[4-(-6)]$
$=8-[4+6]$
$=8-10$
$=8+(-10)$
$=-2$

(b) $3|6-9|-|4-12|$
$=3|6+(-9)|-|4+(-12)|$
$=3|-3|-|-8|$
$=3(3)-8$
$=9-8$
$=1$

N8. "The sum of -3 and 7, increased by 10" is written $(-3+7)+10$.
$(-3+7)+10=4+10=14$

N9. **(a)** "The difference between 5 and -8, decreased by 4" is written $[5-(-8)]-4$.
$[5-(-8)]-4=[5+8]-4$
$=13-4$
$=9$

(b) "7 less than -2" is written $-2-7$.
$-2-7=-2+(-7)$
$=-9$

N10. The difference between a gain of 226 yards and a loss of 7 yards is given by
$226-(-7)=226+7$
$=233.$
The difference is 233 yards.

N11. Subtract the enrollment number for 1985 from the enrollment number for 1990.
$11.34 - 12.39 = -1.05$ million
A negative result indicates a decrease.

Exercises

1. The sum of two negative numbers will always be a *negative* number. In the illustration, we have $-2 + (-3) = -5$.

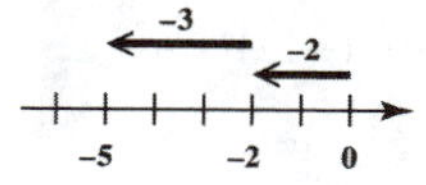

3. When adding a positive number and a negative number, where the negative number has the greater absolute value, the sum will be a *negative* number. In the illustration, the absolute value of -4 is larger than the absolute value of 2, so the sum is a negative number–that is, $-4 + 2 = -2$.

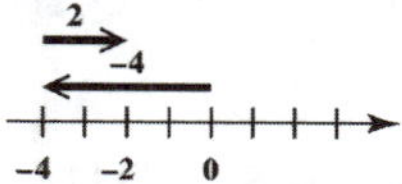

5. By the definition of subtraction, in order to perform the subtraction $-6 - (-8)$, we must add the opposite of $\underline{-8}$ to $\underline{-6}$ to obtain $\underline{2}$.

7. The expression $x - y$ would have to be positive since subtracting a negative number from a positive number is the same as adding a positive number to a positive number, which is a positive number.

9. $|x| = x$, since x is a positive number.
$y - |x| = y - x$, which is a negative number.

11. The sum of two negative numbers is negative.
$-6 + (-2) = -8$

13. Because the numbers have the same sign, add their absolute values: $5 + 7 = 12$. Because both numbers are negative, their sum is negative: $-5 + (-7) = -12$.

15. To add $6 + (-4)$, find the difference between the absolute values of the numbers.
$|6| = 6$ and $|-4| = 4$
$6 - 4 = 2$
Since $|6| > |-4|$, the sum will be positive: $6 + (-4) = 2$.

17. Since the numbers have different signs, find the difference between their absolute values: $6 - 4 = 2$. Because -6 has the larger absolute value, the sum is negative: $4 + (-6) = -2$.

19. Since the numbers have different signs, find the difference between their absolute values: $16 - 7 = 9$. Since -16 has the larger absolute value, the answer is negative: $-16 + 7 = -9$.

21. $6 + (-6) = 0$

23. $-\frac{1}{3} + \left(-\frac{4}{15}\right) = -\frac{5}{15} + \left(-\frac{4}{15}\right) = -\frac{9}{15} = -\frac{3}{5}$

25. $-\frac{1}{6} + \frac{2}{3} = -\frac{1}{6} + \frac{4}{6} = \frac{3}{6} = \frac{1}{2}$

27. Since $8 = 2 \cdot 2 \cdot 2$ and $12 = 2 \cdot 2 \cdot 3$, the LCD is $2 \cdot 2 \cdot 2 \cdot 3 = 24$.

$$\frac{5}{8} + \left(-\frac{17}{12}\right) = \frac{5 \cdot 3}{8 \cdot 3} + \left(-\frac{17 \cdot 2}{12 \cdot 2}\right) = \frac{15}{24} + \left(-\frac{34}{24}\right) = -\frac{19}{24}$$

29. $$2\frac{1}{2} + \left(-3\frac{1}{4}\right) = \frac{5}{2} + \left(-\frac{13}{4}\right) = \frac{10}{4} + \left(-\frac{13}{4}\right) = -\frac{3}{4}$$

31. $-3.5 + 12.4 = +(12.4 - 3.5) = 8.9$

33. $-2.34 + (-3.67) = -(2.34 + 3.67) = -6.01$

35. $4 + [13 + (-5)] = 4 + [8] = 12$

37. $8 + [-2 + (-1)] = 8 + [-3] = 5$

39. $-2 + [5 + (-1)] = -2 + [4] = 2$

41. $-6 + [6 + (-9)] = -6 + [-3] = -9$

43. $[(-9) + (-3)] + 12 = [-12] + 12 = 0$

45. $-6.1 + [3.2 + (-4.8)] = -6.1 + [-1.6] = -7.7$

47. $[-3 + (-4)] + [5 + (-6)] = [-7] + [-1] = -8$

49. $[-4+(-3)]+[8+(-1)]=[-7]+[7]$
$=0$

51. $[-4+(-6)]+[-3+(-8)]+[12+(-11)]$
$=([-10]+[-11])+[1]$
$=(-21)+1$
$=-20$

53. $4-7=4+(-7)=-3$

55. $5-9=5+(-9)=-4$

57. $-7-1=-7+(-1)=-8$

59. $-8-6=-8+(-6)=-14$

61. $7-(-2)=7+(2)=9$

63. $-6-(-2)=-6+(2)=-4$

65. $2-(3-5)=2-[3+(-5)]$
$=2-[-2]$
$=2+(2)$
$=4$

67. $\frac{1}{2}-\left(-\frac{1}{4}\right)=\frac{1}{2}+\frac{1}{4}$
$=\frac{2}{4}+\frac{1}{4}=\frac{3}{4}$

69. $-\frac{3}{4}-\frac{5}{8}=-\frac{3}{4}+\left(-\frac{5}{8}\right)$
$=-\frac{6}{8}+\left(-\frac{5}{8}\right)$
$=-\frac{11}{8}$, or $-1\frac{3}{8}$

71. $\frac{5}{8}-\left(-\frac{1}{2}-\frac{3}{4}\right)=\frac{5}{8}-\left[-\frac{1}{2}+\left(-\frac{3}{4}\right)\right]$
$=\frac{5}{8}-\left[-\frac{2}{4}+\left(-\frac{3}{4}\right)\right]$
$=\frac{5}{8}-\left(-\frac{5}{4}\right)$
$=\frac{5}{8}+\frac{5}{4}$
$=\frac{5}{8}+\frac{10}{8}$
$=\frac{15}{8}$, or $1\frac{7}{8}$

73. $3.4-(-8.2)=3.4+8.2$
$=11.6$

75. $-6.4-3.5=-6.4+(-3.5)$
$=-9.9$

77. $(4-6)+12=[4+(-6)]+12$
$=[-2]+12$
$=10$

79. $(8-1)-12=[8+(-1)]+(-12)$
$=[7]+(-12)$
$=-5$

81. $6-(-8+3)=6-(-5)$
$=6+5$
$=11$

83. $2+(-4-8)=2+[-4+(-8)]$
$=2+[-12]$
$=-10$

85. $|-5-6|+|9+2|=|-5+(-6)|+|11|$
$=|-11|+|11|$
$=-(-11)+11$
$=11+11$
$=22$

87. $|-8-2|-|-9-3|$
$=|-8+(-2)|-|-9+(-3)|$
$=|-10|-|-12|$
$=-(-10)-[-(-12)]$
$=10-[12]$
$=-2$

89. $\left(-\frac{3}{4}-\frac{5}{2}\right)-\left(-\frac{1}{8}-1\right)$
$=\left(-\frac{3}{4}-\frac{10}{4}\right)-\left(-\frac{1}{8}-\frac{8}{8}\right)$
$=-\frac{13}{4}-\left(-\frac{9}{8}\right)$
$=-\frac{26}{8}+\frac{9}{8}$
$=-\frac{17}{8}$, or $-2\frac{1}{8}$

91. $\left(-\frac{1}{2}+0.25\right)-\left(-\frac{3}{4}+0.75\right)$
$=\left(-\frac{1}{2}+\frac{1}{4}\right)-\left(-\frac{3}{4}+\frac{3}{4}\right)$
$=\left(-\frac{2}{4}+\frac{1}{4}\right)-0$
$=-\frac{1}{4},$ or -0.25

93. $-9+[(3-2)-(-4+2)]$
$=-9+[1-(-2)]$
$=-9+[1+2]$
$=-9+3$
$=-6$

95. $-3+[(-5-8)-(-6+2)]$
$=-3+[(-5+(-8))-(-4)]$
$=-3+[-13+4]$
$=-3+[-9]$
$=-12$

97. $-9.12+[(-4.8-3.25)+11.279]$
$=-9.12+[(-4.8+(-3.25))+11.279]$
$=-9.12+[-8.05+11.279]$
$=-9.12+3.229$
$=-5.891$

99. "The sum of -5 and 12 and 6" is written $-5+12+6$.
$-5+12+6=[-5+12]+6$
$=7+6=13$

101. "14 added to the sum of -19 and -4" is written $[-19+(-4)]+14$.
$[-19+(-4)]+14=(-23)+14$
$=-9$

103. "The sum of -4 and -10, increased by 12" is written $[-4+(-10)]+12$.
$[-4+(-10)]+12=-14+12$
$=-2$

105. "$\frac{2}{7}$ more than the sum of $\frac{5}{7}$ and $-\frac{9}{7}$" is written $\left[\frac{5}{7}+\left(-\frac{9}{7}\right)\right]+\frac{2}{7}$.
$\left[\frac{5}{7}+\left(-\frac{9}{7}\right)\right]+\frac{2}{7}=-\frac{4}{7}+\frac{2}{7}$
$=-\frac{2}{7}$

107. "The difference of 4 and -8" is written $4-(-8)$.
$4-(-8)=4+8=12$

109. "8 less than -2" is written $-2-8$.
$-2-8=-2+(-8)=-10$

111. "The sum of 9 and -4, decreased by 7" is written $[9+(-4)]-7$.
$[9+(-4)]-7=5+(-7)=-2$

113. "12 less than the difference of 8 and -5" is written $[8-(-5)]-12$.
$[8-(-5)]-12=[8+(5)]-12$
$=13-12$
$=13+(-12)$
$=1$

115. $[-4+1]+[-2+(-5)]=-3+(-7)$
$=-10$
The total score below par is 10, which can be represented by the number -10.

117. $[-2+9]+[7+(-4)]=7+3$
$=10$
The total score above par is 10, which can be represented by the number $+10$.

119. $[-5+(-4)]+(-3)=-9+(-3)$
$=-12$
The total number of seats that New York, Pennsylvania, and Ohio are projected to lose is 12, which can be represented by the signed number -12.

121. $0+(-130)+(-54)=-130+(-54)$
$=-184$
Their new altitude is 184 meters below the surface, which can be represented by the signed number -184 m.

123. The lowest temperature is represented by $-29°F$. The highest temperature is represented by $-29+149,$ or $120°F$.

125. $33°F$ lower than $-36°F$ can be represented as
$$-36-33=-36+(-33)$$
$$=-69.$$
The record low in Utah is $-69°F$.

127. Add the scores of the four turns to get the final score.
$$-19+28+(-5)+13=9+(-5)+13$$
$$=4+13$$
$$=17$$
Her final score for the four turns was 17.

129. **(a)** $4.4-(-0.5)=4.4+0.5$
$$=4.9$$
The difference is 4.9%.

(b) Americans spent more money than they earned, which means they had to dip into savings or increase borrowing.

131. $4906+788-154$
$=5694-154$ Add.
$=5540$ Subtract.
The average was \$5540.

133. Sum of withdrawals:
$$\$35.84+\$26.14+\$3.12=\$61.98+\$3.12$$
$$=\$65.10$$
Sum of deposits:
$$\$85.00+\$120.76=\$205.76$$
To obtain the final balance, add the deposits and subtract the withdrawals from the beginning balance.
$$\text{Final balance }=\$904.89-\$65.10+\$205.76$$
$$=\$839.79+\$205.76$$
$$=\$1045.55$$
Her account balance at the end of August was \$1045.55.

135. Linda starts with a debt of \$870.00, or $-\$870.00$. She returns two items, increasing the amount she has by $35.90+150.00=185.90$. She purchases three items, decreasing the amount she has by
$82.50+10.00+10.00=102.50$.
Finally, add the payment and subtract the finance charge to calculate how much money she has.

−870.00	Amount owed
+185.90	Two return credits
−684.10	
−102.50	Three purchases
−786.60	
+500.00	Payment
−286.60	
−37.23	Finance charge
−323.83	

She still owes \$323.83.

137. The percent return for 2009 is 26.00% and the percent return for 2010 is 14.97%. Thus, the change in percent returns is
$14.97-26.00=-11.03\%$ (a decrease).

139. The percent return for 2011 is 1.51% and the percent return for 2012 is 15.31%. Thus, the change in percent return is $15.31-1.51=13.8\%$ (an increase).

141. $17,400-(-32,995)=17,400+32,995$
$$=50,395$$
The difference between the height of Mt. Foraker and the depth of the Philippine Trench is 50,395 feet.

143. $-23,376-(-24,721)=-23,376+24,721$
$$=1345$$
The Cayman Trench is 1345 feet deeper than the Java Trench.

145. $14,246-14,110=14,246+(-14,110)$
$$=136$$
Mt. Wilson is 136 feet higher than Pikes Peak.

1.5 Multiplying and Dividing Real Numbers

Now Try Exercises

N1. **(a)** $-11(9)=-(11\cdot 9)=-99$

(b) $3.1(-2.5)=-(3.1\cdot 2.5)=-7.75$

N2. **(a)** $-8(-11)=8\cdot 11=88$

(b) $-\frac{1}{7}\left(-\frac{5}{2}\right)=\frac{1}{7}\cdot\frac{5}{2}=\frac{1\cdot 5}{7\cdot 2}=\frac{5}{14}$

N3. (a) $\frac{-10}{5} = -10 \cdot \left(\frac{1}{5}\right) = -2$

(b) $\frac{-1.44}{-0.12} = -1.44\left(-\frac{1}{0.12}\right) = 12$

(c) $-\frac{3}{8} \div \frac{7}{10} = -\frac{3}{\cancel{8}_4} \cdot \frac{\cancel{10}^5}{7} = -\frac{15}{28}$

N4. (a)
$$\begin{aligned} -4(6) - (-5)5 &= -24 - (-25) \\ &= -24 + 25 \\ &= 1 \end{aligned}$$

(b)
$$\begin{aligned} \frac{12(-4) - 6(-3)}{-4(7-16)} &= \frac{-48 - (-18)}{-4(-9)} \\ &= \frac{-48+18}{36} \\ &= \frac{-30}{36} = -\frac{5}{6} \end{aligned}$$

N5. Replace x with -4 and y with -3.

$$\begin{aligned} \frac{3x^2 - 12}{y} &= \frac{3(-4)^2 - 12}{-3} \\ &= \frac{3(16) - 12}{-3} \\ &= \frac{48 - 12}{-3} \\ &= \frac{36}{-3} = -12 \end{aligned}$$

N6. (a) "Twice the sum of -10 and 7" is written $2(-10+7)$.
$2(-10+7) = 2(-3) = -6$

(b) "40% of the difference of 45 and 15" is written $0.40(45-15)$.
$0.40(45-15) = 0.4(30) = 12.$

N7. "The quotient of 21 and the sum of 10 and -7" is written $\frac{21}{10+(-7)}$.

$$\frac{21}{10+(-7)} = \frac{21}{3} = 7$$

N8. (a) "The sum of a number and -4 is 7" is written $x + (-4) = 7$. Here, x must be 4 more than 7, so the solution is 11.
$11 + (-4) = 7, \ldots$

(b) "The difference of -8 and a number is -11" is written $-8 - x = -11$. If we start at -8 on a number line, we must move 3 units to the left to get to -11, so the solution is 3.

Exercises

1. The product or the quotient of two numbers with the same sign is greater than 0, since the product or quotient of two positive numbers is positive and the product or quotient of two negative numbers is positive.

3. If three negative numbers are multiplied, the product is less than 0, since a negative number times a negative number is a positive number, and that positive number times a negative number is a negative number.

5. If a negative number is squared and the result is added to a positive number, the result is greater than 0, since a negative number squared is a positive number, and a positive number added to another positive number is a positive number.

7. If three positive numbers, five negative numbers, and zero are multiplied, the product is equal to 0. Since one of the numbers is zero, the product is zero (regardless of what the other numbers are).

9. The quotient formed by any nonzero number divided by 0 is undefined, and the quotient formed by 0 divided by any nonzero number is 0. Examples include $\frac{1}{0}$, which is undefined, and $\frac{0}{1}$, which equals 0.

11. $5(-6) = -(5 \cdot 6) = -30$
Note that the product of a positive number and a negative number is negative.

13. $-5(-6) = 5 \cdot 6 = 30$
Note that the product of two negative numbers is positive.

15. $-10(-12) = 10 \cdot 12 = 120$

17. $3(-11) = -(3 \cdot 11) = -33$

19. $-0.5(0) = 0$

21. $-6.8(0.35) = -(6.8 \cdot 0.35) = -2.38$

23. $-\frac{3}{8}\cdot\left(-\frac{10}{9}\right)=\frac{3}{8}\left(\frac{10}{9}\right)$
$=\frac{3\cdot 10}{8\cdot 9}$
$=\frac{3\cdot(2\cdot 5)}{(4\cdot 2)\cdot(3\cdot 3)}$
$=\frac{3\cdot 2\cdot 5}{4\cdot 2\cdot 3\cdot 3}$
$=\frac{5}{4\cdot 3}=\frac{5}{12}$

25. $\frac{2}{15}\left(-1\frac{1}{4}\right)=\frac{2}{15}\left(-\frac{5}{4}\right)$
$=-\frac{2\cdot 5}{15\cdot 4}$
$=-\frac{2\cdot 5}{3\cdot 5\cdot 2\cdot 2}$
$=-\frac{1}{3\cdot 2}=-\frac{1}{6}$

27. $-8\left(-\frac{3}{4}\right)=8\left(\frac{3}{4}\right)=\frac{24}{4}=6$

29. Using only positive integer factors, 32 can be written as $1\cdot 32,\ 2\cdot 16,$ or $4\cdot 8$. Including the negative integer factors, we see that the integer factors of 32 are $-32, -16, -8, -4, -2, -1,$ 1, 2, 4, 8, 16, and 32.

31. The integer factors of 40 are $-40, -20, -10, -8, -5, -4, -2, -1,$ 1, 2, 4, 5, 8, 10, 20, and 40.

33. The integer factors of 31 are $-31, -1, 1,$ and 31.

35. $\frac{15}{5}=\frac{5\cdot 3}{5}=\frac{3}{1}=3$

37. $\frac{-42}{6}=-\frac{2\cdot 3\cdot 7}{2\cdot 3}=-7$
Note that the quotient of two numbers having different signs is negative.

39. $\frac{-32}{-4}=\frac{4\cdot 8}{4}=8$
Note that the quotient of two numbers having the same sign is positive.

41. $\frac{96}{-16}=-\frac{6\cdot 16}{16}=-6$

43. $\frac{-8.8}{2.2}=-\frac{4(2.2)}{2.2}=-4$

45. Dividing by a fraction (in this case, $-\frac{1}{8}$) is the same as multiplying by the reciprocal of the fraction (in this case, $-\frac{8}{1}$).
Note that dividing by a number with absolute value between 0 and 1 gives us a number larger than the original numerator.
$\left(-\frac{4}{3}\right)\div\left(-\frac{1}{8}\right)=\left(-\frac{4}{3}\right)\cdot\left(-\frac{8}{1}\right)$
$=\frac{4\cdot 8}{3\cdot 1}$
$=\frac{32}{3}$, or $10\frac{2}{3}$

47. $-\frac{5}{6}\div\frac{8}{9}=-\frac{5}{6}\cdot\frac{9}{8}$
$=-\frac{5\cdot 9}{6\cdot 8}$
$=-\frac{45}{48}$, or $-\frac{15}{16}$

49. $\frac{0}{-5}=0$, because 0 divided by any nonzero number is 0.

51. $\frac{11.5}{0}$ is undefined, because we cannot divide by 0.

53. $7-3\cdot 6=7-18$
$=-11$

55. $-10-(-4)(2)=-10-(-8)$
$=-10+8$
$=-2$

57. $-7(3-8)=-7[3+(-8)]$
$=-7(-5)=35$

59. $7+2(4-1)=7+2(3)$
$=7+6$
$=13$

61. $-4+3(2-8)=-4+3(-6)$
$=-4-18$
$=-22$

63. $(12-14)(1-4)=(-2)(-3)$
$=6$

65. $(7-10)(10-4)=(-3)(6)$
$=-18$

67. $(-2-8)(-6)+7=(-10)(-6)+7$
$=60+7$
$=67$

69. $3(-5)+|3-10|=-15+|-7|$
$=-15+7$
$=-8$

71. $\frac{-5(-6)}{9-(-1)}=\frac{30}{10}$
$=\frac{3\cdot 10}{10}=3$

73. $\frac{-21(3)}{-3-6}=\frac{-63}{-3+(-6)}$
$=\frac{-63}{-9}=7$

75. $\frac{-10(2)+6(2)}{-3-(-1)}=\frac{-20+12}{-3+1}$
$=\frac{-8}{-2}=4$

77. $\frac{3^2-4^2}{7(-8+9)}=\frac{9-16}{7(1)}=\frac{-7}{7}=-1$

79. $\frac{8(-1)-|(-4)(-3)|}{-6-(-1)}=\frac{-8-|12|}{-6+1}$
$=\frac{-8-12}{-5}$
$=\frac{-20}{-5}=4$

81. $\frac{-13(-4)-(-8)(-2)}{(-10)(2)-4(-2)}$
$=\frac{52-16}{-20-(-8)}$
$=\frac{36}{-20+8}$
$=\frac{36}{-12}=-3$

83. $3+2\times 4\div 2-3\times 7-4+47$
$=3+\left(\frac{2\times 4}{2}\right)-(3\times 7)-4+47$
$=3+\left(\frac{8}{2}\right)-(21)-4+47$
$=3+4-21-4+47$
$=29$

85. $5x-2y+3a=5(6)-2(-4)+3(3)$
$=30-(-8)+9$
$=30+8+9$
$=38+9$
$=47$

87. $(2x+y)(3a)=[2(6)+(-4)][3(3)]$
$=[12+(-4)](9)$
$=(8)(9)$
$=72$

89. $\left(\frac{1}{3}x-\frac{4}{5}y\right)\left(-\frac{1}{5}a\right)$
$=\left[\frac{1}{3}(6)-\frac{4}{5}(-4)\right]\left[-\frac{1}{5}(3)\right]$
$=\left[2-\left(-\frac{16}{5}\right)\right]\left(-\frac{3}{5}\right)$
$=\left(2+\frac{16}{5}\right)\left(-\frac{3}{5}\right)$
$=\left(\frac{10}{5}+\frac{16}{5}\right)\left(-\frac{3}{5}\right)$
$=\left(\frac{26}{5}\right)\left(-\frac{3}{5}\right)$
$=-\frac{78}{25}$

91. $(-5+x)(-3+y)(3-a)$
$=(-5+6)[-3+(-4)][3-3]$
$=(1)(-7)(0)$
$=0$

93. $-2y^2+3a=-2(-4)^2+3(3)$
$=-2(16)+9$
$=-32+9$
$=-23$

95. $\frac{2y-x}{a-3}=\frac{2(-4)-(6)}{3-3}$
$=\frac{(-8)-6}{0}$
$=\frac{-14}{0}$

The expression is undefined.

97. "The product of −9 and 2, added to 9" is written $9+(-9)(2)$.
$9+(-9)(2)=9+(-18)$
$=-9$

99. "Twice the product of −1 and 6, subtracted from −4" is written $-4-2[(-1)(6)]$.
$-4-2[(-1)(6)]=-4-2(-6)$
$=-4-(-12)$
$=-4+12=8$

101. "Nine subtracted from the product of 1.5 and −3.2" is written $(1.5)(-3.2)-9$.
$(1.5)(-3.2)-9=-4.8-9$
$=-4.8+(-9)$
$=-13.8$

103. "The product of 12 and the difference of 9 and −8" is written $12[9-(-8)]$.
$12[9-(-8)]=12[9+8]$
$=12(17)=204$

105. "The quotient of −12 and the sum of −5 and −1" is written $\frac{-12}{-5+(-1)}$, and

$\frac{-12}{-5+(-1)}=\frac{-12}{-6}=2$.

107. "The sum of 15 and −3, divided by the product of 4 and −3" is written $\frac{15+(-3)}{4(-3)}$, and

$\frac{15+(-3)}{4(-3)}=\frac{12}{-12}=-1$.

109. "Two-thirds of the difference of 8 and −1" is written $\frac{2}{3}[8-(-1)]$, and

$\frac{2}{3}[8-(-1)]=\frac{2}{3}[8+(1)]=\frac{2}{3}[9]=6$.

111. "20% of the product of −5 and 6" is written $0.20(-5\cdot 6)$, and
$0.20(-5\cdot 6)=0.20(-30)=-6$.

113. "The sum of $\frac{1}{2}$ and $\frac{5}{8}$, times the difference of $\frac{3}{5}$ and $\frac{1}{3}$" is written $\left(\frac{1}{2}+\frac{5}{8}\right)\left(\frac{3}{5}-\frac{1}{3}\right)$, and

$\left(\frac{1}{2}+\frac{5}{8}\right)\left(\frac{3}{5}-\frac{1}{3}\right)=\left(\frac{4}{8}+\frac{5}{8}\right)\left(\frac{9}{15}-\frac{5}{15}\right)$
$=\frac{9}{8}\left(\frac{4}{15}\right)$
$=\frac{3\cdot 3\cdot 4}{2\cdot 4\cdot 3\cdot 5}=\frac{3}{10}$.

115. "The product of $-\frac{1}{2}$ and $\frac{3}{4}$, divided by $-\frac{2}{3}$" is written $\frac{-\frac{1}{2}\left(\frac{3}{4}\right)}{-\frac{2}{3}}$. Simplifying gives us

$\frac{-\frac{1}{2}\left(\frac{3}{4}\right)}{-\frac{2}{3}}=\frac{-\frac{3}{8}}{-\frac{2}{3}}$
$=-\frac{3}{8}\cdot\left(-\frac{3}{2}\right)$
$=\frac{9}{16}$.

117. "The quotient of a number and 3 is −3" is written $\frac{x}{3}=-3$. The solution is −9, since $\frac{-9}{3}=-3$.

119. "6 less than a number is 4" is written $x-6=4$. The solution is 10, since $10-6=4$.

121. "When 5 is added to a number, the result is −5" is written $x+5=-5$. The solution is −10, since $-10+5=-5$.

123. Add the numbers and divide by 5.

$\frac{(23+18+13)+[(-4)+(-8)]}{5}$
$=\frac{54-12}{5}$
$=\frac{42}{5}$, or $8\frac{2}{5}$

125. Add the numbers and divide by 4.

$$\frac{(29+8)+[(-15)+(-6)]}{4}$$
$$=\frac{37-21}{4}$$
$$=\frac{16}{4}=4$$

127. Add the integers from -10 to 14.

$$(-10)+(-9)+\cdots+14=50$$

[The 3 dots indicate that the pattern continues.] There are 25 integers from -10 to 14 (10 negative, zero, and 14 positive). Thus, the average is $\frac{50}{25}=2$.

129. **(a)** 3,473,986 is divisible by 2 because its last digit, 6, is divisible by 2.

(b) 4,336,879 is not divisible by 2 because its last digit, 9, is not divisible by 2.

131. **(a)** 6,221,464 is divisible by 4 because the number formed by its last two digits, 64, is divisible by 4.

(b) 2,876,335 is not divisible by 4 because the number formed by its last two digits, 35, is not divisible by 4.

133. **(a)** 1,524,822 is divisible by 2 because its last digit, 2, is divisible by 2. It is also divisible by 3 because the sum of its digits, $1+5+2+4+8+2+2=24$, is divisible by 3. Because 1,524,822 is divisible by both 2 and 3, it is divisible by 6.

(b) 2,873,590 is divisible by 2 because its last digit, 0, is divisible by 2. However, it is not divisible by 3 because the sum of its digits, $2+8+7+3+5+9+0=34$, is not divisible by 3. Because 2,873,590 is not divisible by both 2 and 3, it is not divisible by 6.

135. **(a)** 4,114,107 is divisible by 9 because the sum of its digits, $4+1+1+4+1+0+7=18$, is divisible by 9.

(b) 2,287,321 is not divisible by 9 because the sum of its digits, $2+2+8+7+3+2+1=25$, is not divisible by 9.

Summary Exercises Performing Operations with Real Numbers

1. $14-3\cdot10=14-30$
$=14+(-30)$
$=-16$

2. $-3(8)-4(-7)=-24-(-28)$
$=-24+28$
$=4$

3. $(3-8)(-2)-10=(-5)(-2)-10$
$=10-10$
$=0$

4. $-6(7-3)=-6(4)$
$=-24$

5. $7+3(2-10)=7+3(-8)$
$=7-24$
$=-17$

6. $-4[(-2)(6)-7]=-4[-12-7]$
$=-4[-19]$
$=76$

7. $(-4)(7)-(-5)(2)=(-28)-(-10)$
$=-28+(10)$
$=-18$

8. $-5[-4-(-2)(-7)]=-5[-4-(14)]$
$=-5[-18]$
$=90$

9. $40-(-2)[8-9]=40-(-2)[-1]$
$=40-(2)$
$=38$

10. $\frac{5(-4)}{-7-(-2)}=\frac{-20}{-7+2}$
$=\frac{-20}{-5}=4$

11. $\frac{-3-(-9+1)}{-7-(-6)}=\frac{-3-(-8)}{-7+6}$
$=\frac{-3+8}{-1}$
$=\frac{5}{-1}=-5$

12. $\dfrac{5(-8+3)}{13(-2)+(-7)(-3)}=\dfrac{5(-5)}{-26+21}$
$=\dfrac{-25}{-5}=5$

13. $\dfrac{6^2-8}{-2(2)+4(-1)}=\dfrac{36-8}{-4+(-4)}$
$=\dfrac{28}{-8}$
$=-\dfrac{4\cdot 7}{2\cdot 4}=-\dfrac{7}{2},\ \text{or}\ -3\dfrac{1}{2}$

14. $\dfrac{16(-8+5)}{15(-3)+(-7-4)(-3)}=\dfrac{16(-3)}{-45+(-11)(-3)}$
$=\dfrac{-48}{-45+33}$
$=\dfrac{-48}{-12}=4$

15. $\dfrac{9(-6)-3(8)}{4(-7)+(-2)(-11)}=\dfrac{-54-24}{-28+22}$
$=\dfrac{-78}{-6}=13$

16. $\dfrac{2^2+4^2}{5^2-3^2}=\dfrac{4+16}{25-9}$
$=\dfrac{20}{16}=\dfrac{5}{4},\ \text{or}\ 1\dfrac{1}{4}$

17. $\dfrac{(2+4)^2}{(5-3)^2}=\dfrac{(6)^2}{(2)^2}$
$=\dfrac{36}{4}=9$

18. $\dfrac{4^3-3^3}{-5(-4+2)}=\dfrac{64-27}{-5(-2)}$
$=\dfrac{37}{10},\ \text{or}\ 3\dfrac{7}{10}$

19. $\dfrac{-9(-6)+(-2)(27)}{3(8-9)}=\dfrac{(54)+(-54)}{3(-1)}$
$=\dfrac{0}{-3}=0$

20. $|-4(9)|-|-11|=|-36|-11$
$=36-11$
$=25$

21. $\dfrac{6(-10+3)}{15(-2)-3(-9)}=\dfrac{6(-7)}{(-30)-(-27)}$
$=\dfrac{-42}{-30+27}$
$=\dfrac{-42}{-3}$
$=14$

22. $\dfrac{3^2-5^2}{(-9)^2-9^2}=\dfrac{9-25}{81-81}$
$=\dfrac{-16}{0}$, which is undefined.

23. $\dfrac{(-10)^2+10^2}{-10(5)}=\dfrac{100+100}{-50}$
$=\dfrac{200}{-50}=-4$

24. $-\dfrac{3}{4}\div\left(-\dfrac{5}{8}\right)=-\dfrac{3}{4}\cdot\left(-\dfrac{8}{5}\right)$
$=\dfrac{3\cdot 2\cdot 4}{4\cdot 5}$
$=\dfrac{3\cdot 2}{5}=\dfrac{6}{5},\ \text{or}\ 1\dfrac{1}{5}$

25. $\dfrac{1}{2}\div\left(-\dfrac{1}{2}\right)=\dfrac{1}{2}\cdot\left(-\dfrac{2}{1}\right)$
$=-\dfrac{2}{2}=-1$

26. $\dfrac{8^2-12}{(-5)^2+2(6)}=\dfrac{64-12}{25+12}$
$=\dfrac{52}{37},\ \text{or}\ 1\dfrac{15}{37}$

27. $\left[\dfrac{5}{8}-\left(-\dfrac{1}{16}\right)\right]+\dfrac{3}{8}=\left[\dfrac{10}{16}+\dfrac{1}{16}\right]+\dfrac{6}{16}$
$=\left[\dfrac{11}{16}\right]+\dfrac{6}{16}$
$=\dfrac{17}{16},\ \text{or}\ 1\dfrac{1}{16}$

28. $\left(\frac{1}{2}-\frac{1}{3}\right)-\frac{5}{6}=\left(\frac{3}{6}-\frac{2}{6}\right)-\frac{5}{6}$
$=\left(\frac{1}{6}\right)-\frac{5}{6}$
$=-\frac{4}{6}$
$=-\frac{2}{3}$

29. $-0.9(-3.7)=0.9(3.7)$
$=3.33$

30. $-5.1(-0.2)=5.1(0.2)$
$=1.02$

31. $|-2(3)+4|-|-2|=|-6+4|-2$
$=|-2|-2$
$=2-2=0$

32. $40+2[-5-3]=40+2[-8]$
$=40-16$
$=24$

33. $-x+y-3a=-(-2)+3-3(4)$
$=2+3-12$
$=5-12$
$=-7$

34. $(x-y)-(a-2y)=(-2-3)-(4-2\cdot3)$
$=(-5)-(4-6)$
$=-5-(-2)$
$=-5+2$
$=-3$

35. $\left(\frac{1}{2}x+\frac{2}{3}y\right)\left(-\frac{1}{4}a\right)$
$=\left(\frac{1}{2}(-2)+\frac{2}{3}(3)\right)\left(-\frac{1}{4}(4)\right)$
$=(-1+2)(-1)$
$=(1)(-1)$
$=-1$

36. $\frac{2x+3y}{a-xy}=\frac{2(-2)+3(3)}{4-(-2)(3)}$
$=\frac{-4+9}{4-(-6)}$
$=\frac{5}{4+6}$
$=\frac{5}{10}=\frac{1}{2}$

37. $\frac{x^2-y^2}{x^2+y^2}=\frac{(-2)^2-3^2}{(-2)^2+3^2}$
$=\frac{4-9}{4+9}$
$=\frac{-5}{13}=-\frac{5}{13}$

38. $-x^2+3y=-(-2)^2+3(3)$
$=-(4)+9$
$=5$

39. $\left(\frac{x}{y}\right)^3=\left(\frac{-2}{3}\right)^3=\left(-\frac{2}{3}\right)\left(-\frac{2}{3}\right)\left(-\frac{2}{3}\right)$
$=-\frac{8}{27}$

40. $\left(\frac{a}{x}\right)^2=\left(\frac{4}{-2}\right)^2=(-2)^2=(-2)(-2)=4$

1.6 Properties of Real Numbers

Now Try Exercises

N1. **(a)** $7+(-3)=-3+\underline{7}$

(b) $(-5)4=4\cdot\underline{(-5)}$

N2. **(a)** $-9+(3+7)=\underline{(-9+3)}+7$

(b) $5[(-4)\cdot9]=\underline{[5\cdot(-4)]\cdot9}$

N3. $5+(7+6)=5+(6+7)$
While the same numbers are grouped inside the two pairs of parentheses, the order of the numbers has been changed. This illustrates a commutative property.

N4. (a) $8+54+7+6+32$
$=(8+32)+(54+6)+7$
$=40+60+7$
$=100+7$
$=107$

(b) $5(37)(20)=5(20)(37)=100(37)=3700$

N5. (a) $\frac{2}{5}\cdot\underline{1}=\frac{2}{5}$ Multiplicative identity

(b) $8+\underline{0}=8$ Additive identity

N6. (a) $\frac{16}{20}=\frac{4\cdot 4}{5\cdot 4}$ Factor.
$=\frac{4}{5}\cdot\frac{4}{4}$ Write as a product.
$=\frac{4}{5}\cdot 1$ Divide.
$=\frac{4}{5}$ Identity property

(b) $\frac{2}{5}+\frac{3}{20}=\frac{2}{5}\cdot 1+\frac{3}{20}$ Identity property
$=\frac{2}{5}\cdot\frac{4}{4}+\frac{3}{20}$ Multiply by $\frac{4}{4}$.
$=\frac{8}{20}+\frac{3}{20}$ Multiply.
$=\frac{11}{20}$ Add.

N7. (a) $10+\underline{(-10)}=0$ Inverse property

(b) $-9\cdot\underline{\left(-\frac{1}{9}\right)}=1$ Inverse property

N8. $-\frac{1}{3}x+7+\frac{1}{3}x$
$=\left(-\frac{1}{3}x+7\right)+\frac{1}{3}x$ Order of operations
$=\left[7+\left(-\frac{1}{3}x\right)\right]+\frac{1}{3}x$ Commutative property
$=7+\left[\left(-\frac{1}{3}x\right)+\frac{1}{3}x\right]$ Associative property
$=7+0$ Inverse property
$=7$ Identity property

N9. (a) $2(p+5)=2(p)+2(5)$ Distributive prop.
$=2p+10$ Multiply.

(b) $-5(4x+1)=-5\cdot 4x+(-5\cdot 1)$
$=-20x-5$

(c) $6(2r+t-5z)=6(2r)+6t+6(-5z)$
$=12r+6t-30z$

N10. (a) $-(2-r)=-1(2-r)$
$=-2+r$

(b) $-(2x-5y-7)=-1(2x-5y-7)$
$=-2x+5y+7$

Exercises

1. (a) B, since 0 is the identity element for addition.

(b) F, since 1 is the identity element for multiplication.

(c) C, since $-a$ is the additive inverse of a.

(d) I, since $\frac{1}{a}$ is the multiplicative inverse, or reciprocal, of any nonzero number a.

(e) B, since 0 is the only number that is equal to its negative—that is, $0=-0$.

(f) D and F, since -1 has reciprocal $\frac{1}{(-1)}=-1$ and 1 has a reciprocal $\frac{1}{(1)}=1$—that is, -1 and 1 are their own multiplicative inverses.

(g) B, since the multiplicative inverse of a number a is $\frac{1}{a}$ and the only number that we cannot divide by is 0.

(h) A, because the equation $(5\cdot 4)\cdot 3=(3\cdot 4)\cdot 5$ is true by the associative property.

(i) G, since we can consider $(5\cdot 4)$ to be one number, $(5\cdot 4)\cdot 3$ is the same as $3\cdot(5\cdot 4)$ by the commutative property.

(j) H, because $5(4+3)$ is the same as $5\cdot 4+5\cdot 3$ by the distributive property.

3. "Washing your face" and "brushing your teeth" are commutative.

5. "Preparing a meal" and "eating a meal" are not commutative.

7. "Putting on your socks" and "putting on your shoes" are not commutative.

9. "(Foreign sales) clerk" is a clerk dealing with foreign sales, whereas "foreign (sales clerk)" is a sales clerk who is foreign.

11. $25-(6-2)=25-(4)$
$=21$
$(25-6)-2=19-2$
$=17$
Since $21 \neq 17$, this example shows that subtraction is not associative.

13. In general, a number and its additive inverse have *opposite* signs. A number and its multiplicative inverse have *the same* signs.

Number	Additive Inverse	Multiplicative Inverse
5	−5	$\frac{1}{5}$
−10	10	$-\frac{1}{10}$
$-\frac{1}{2}$	$\frac{1}{2}$	−2
$\frac{3}{8}$	$-\frac{3}{8}$	$\frac{8}{3}$
$x(x \neq 0)$	$-x$	$\frac{1}{x}$
$-y(y \neq 0)$	y	$-\frac{1}{y}$

15. $-15+9=9+\underline{(-15)}$ by the commutative property of addition.

17. $-8\cdot 3=\underline{3}\cdot(-8)$ by the commutative property of multiplication.

19. $(3+6)+7=3+(\underline{6}+7)$ by the associative property of addition.

21. $7\cdot(2\cdot 5)=(\underline{7}\cdot 2)\cdot 5$ by the associative property of multiplication.

23. $4+15=15+4$
The order of the two numbers has been changed, so this is an example of the commutative property of addition: $a+b=b+a$.

25. $5\cdot(13\cdot 7)=(5\cdot 13)\cdot 7$
The numbers are in the same order but grouped differently, so this is an example of the associative property of multiplication: $(ab)c=a(bc)$.

27. $-6+(12+7)=(-6+12)+7$
The numbers are in the same order but grouped differently, so this is an example of the associative property of addition: $(a+b)+c=a+(b+c)$.

29. $-9+9=0$
The sum of the two numbers is 0, so they are additive inverses (or opposites) of each other. This is an example of the additive inverse property: $a+(-a)=0$.

31. $\frac{2}{3}\left(\frac{3}{2}\right)=1$
The product of the two numbers is 1, so they are multiplicative inverses (or reciprocals) of each other. This is an example of the multiplicative inverse property: $a\cdot\frac{1}{a}=1(a \neq 0)$.

33. $1.75+0=1.75$
The sum of a number and 0 is the original number. This is an example of the identity property of addition: $a+0=a$.

35. $(4+17)+3=3+(4+17)$
The order of the numbers has been changed, but the grouping has not, so this is an example of the commutative property of addition: $a+b=b+a$.

37. $2(x+y)=2x+2y$
The number 2 outside the parentheses is "distributed" over the x and y. This is an example of the distributive property.

39. $-\frac{5}{9}=-\frac{5}{9}\cdot\frac{3}{3}=-\frac{15}{27}$
$\frac{3}{3}$ is a form of the number 1. We use it to rewrite $-\frac{5}{9}$ as $-\frac{15}{27}$. This is an example of the identity property of multiplication.

41. $4(2x)+4(3y)=4(2x+3y)$
This is an example of the distributive property. The number 4 is "distributed" over $2x$ and $3y$.

43. $$\begin{aligned}97+13+3+37&=(97+3)+(13+37)\\&=100+50\\&=150\end{aligned}$$

45. $$\begin{aligned}1999+2+1+8&=(1999+1)+(2+8)\\&=2000+10\\&=2010\end{aligned}$$

47. $$\begin{aligned}159+12+141+88&=(159+141)+(12+88)\\&=300+100\\&=400\end{aligned}$$

49. $$\begin{aligned}&843+627+(-43)+(-27)\\&=[843+(-43)]+[627+(-27)]\\&=800+600\\&=1400\end{aligned}$$

51. $5(47)(2)=5(2)(47)=10(47)=470$

53. $$\begin{aligned}-4\cdot5\cdot93\cdot5&=-4\cdot5\cdot5\cdot93\\&=-20\cdot5\cdot93\\&=-100\cdot93\\&=-9300\end{aligned}$$

55. $6t+8-6t+3$

$=6t+8+(-6t)+3$	Def. of subtraction
$=(6t+8)+(-6t)+3$	Order of operations
$=(8+6t)+(-6t)+3$	Commutative property
$=8+[6t+(-6t)]+3$	Associative property
$=8+0+3$	Inverse property
$=(8+0)+3$	Order of operations
$=8+3$	Identity property
$=11$	Add.

57. $\frac{2}{3}x-11+11-\frac{2}{3}x$

$=\frac{2}{3}x+(-11)+11+\left(-\frac{2}{3}x\right)$	
$=\left[\frac{2}{3}x+(-11)\right]+11+\left(-\frac{2}{3}x\right)$	
$=\frac{2}{3}x+(-11+11)+\left(-\frac{2}{3}x\right)$	Associative prop.
$=\frac{2}{3}x+0+\left(-\frac{2}{3}x\right)$	Inverse property
$=\left(\frac{2}{3}x+0\right)+\left(-\frac{2}{3}x\right)$	
$=\frac{2}{3}x+\left(-\frac{2}{3}x\right)$	Identity property
$=0$	Inverse property

59. $\left(\frac{9}{7}\right)(-0.38)\left(\frac{7}{9}\right)$

$=\left[\left(\frac{9}{7}\right)(-0.38)\right]\left(\frac{7}{9}\right)$	Order of operations
$=\left[(-0.38)\left(\frac{9}{7}\right)\right]\left(\frac{7}{9}\right)$	Commutative property
$=(-0.38)\left[\left(\frac{9}{7}\right)\left(\frac{7}{9}\right)\right]$	Associative property
$=(-0.38)(1)$	Inverse property
$=-0.38$	Identity property

61. $t+(-t)+\frac{1}{2}(2)$

$=t+(-t)+1$	Inverse property
$=[t+(-t)]+1$	Order of operations
$=0+1$	Inverse property
$=1$	Identity property

63. When distributing a negative number over a quantity, be careful not to "lose" a negative sign. The problem should be worked in the following way.
$$\begin{aligned}-3(4-6)&=-3(4)-3(-6)\\&=-12+18\\&=6\end{aligned}$$

65. $\frac{3}{4}=\frac{3}{4}\cdot 1=\frac{3}{4}\cdot\frac{3}{3}=\frac{9}{12}$

To rewrite $\frac{3}{4}$ as $\frac{9}{12}$, use the fact that $\frac{3}{3}$ is another name for the multiplicative identity element, 1.

67. $$\begin{aligned}5(9+8)&=5\cdot 9+5\cdot 8\\&=45+40\\&=85\end{aligned}$$

69. $$\begin{aligned}4(t+3)&=4\cdot t+4\cdot 3\\&=4t+12\end{aligned}$$

71. $$\begin{aligned}7(z-8)&=7[z+(-8)]\\&=7z+7(-8)\\&=7z-56\end{aligned}$$

73. $$\begin{aligned}-8(r+3)&=-8(r)+(-8)(3)\\&=-8r+(-24)\\&=-8r-24\end{aligned}$$

75. $$\begin{aligned}&-\frac{1}{4}(8x+3)\\&=-\frac{1}{4}(8x)+\left(-\frac{1}{4}\right)(3)\\&=\left[\left(-\frac{1}{4}\right)\cdot 8\right]x-\frac{3}{4}\\&=-2x-\frac{3}{4}\end{aligned}$$

77. $$\begin{aligned}-5(y-4)&=-5(y)+(-5)(-4)\\&=-5y+20\end{aligned}$$

79. $$\begin{aligned}2(6x+5)&=2(6x)+2(5)\\&=12x+10\end{aligned}$$

81. $$\begin{aligned}-3(2x-5)&=(-3)(2x)+(-3)(-5)\\&=-6x+15\end{aligned}$$

83. $$\begin{aligned}-6(8x+1)&=(-6)(8x)+(-6)(1)\\&=-48x-6\end{aligned}$$

85. $$\begin{aligned}&-\frac{4}{3}(12y+15z)\\&=-\frac{4}{3}(12y)+\left(-\frac{4}{3}\right)(15z)\\&=\left[\left(-\frac{4}{3}\right)\cdot 12\right]y+\left[\left(-\frac{4}{3}\right)\cdot 15\right]z\\&=-16y+(-20)z\\&=-16y-20z\end{aligned}$$

87. $$\begin{aligned}&8(3r+4s-5y)\\&=8(3r)+8(4s)+8(-5y) &&\text{Dist. prop.}\\&=(8\cdot 3)r+(8\cdot 4)s+[8(-5)]y &&\text{Assoc. prop.}\\&=24r+32s-40y &&\text{Multiply.}\end{aligned}$$

89. $$\begin{aligned}&-3(8x+3y+4z)\\&=-3(8x)+(-3)(3y)+(-3)(4z) &&\text{Dist. prop.}\\&=(-3\cdot 8)x+(-3\cdot 3)y+(-3\cdot 4)z &&\text{Assoc. prop.}\\&=-24x-9y-12z &&\text{Multiply.}\end{aligned}$$

91. $$\begin{aligned}&-(4t+3m)\\&=-1(4t+3m) &&\text{Identity property}\\&=-1(4t)+(-1)(3m) &&\text{Distributive property}\\&=(-1\cdot 4)t+(-1\cdot 3)m &&\text{Associative property}\\&=-4t-3m &&\text{Multiply.}\end{aligned}$$

93. $$\begin{aligned}&-(-5c-4d)\\&=-1(-5c-4d) &&\text{Identity property}\\&=-1(-5c)+(-1)(-4d) &&\text{Distributive property}\\&=(-1\cdot -5)c+(-1\cdot 4)d &&\text{Associative property}\\&=5c+4d &&\text{Multiply.}\end{aligned}$$

95. $$\begin{aligned}&-(-q+5r-8s)\\&=-1(-q+5r-8s)\\&=-1(-q)+(-1)(5r)+(-1)(-8s)\\&=(-1\cdot -1)q+(-1\cdot 5)r+(-1\cdot -8)s\\&=q-5r+8s\end{aligned}$$

1.7 Simplifying Expressions

Now Try Exercises

N1. **(a)** $$\begin{aligned}3(2x-4y)&=3(2x)-3(4y)\\&=(3\cdot 2)x-(3\cdot 4)y\\&=6x-12y\end{aligned}$$

(b) $-4-(-3y+5)=-4-1(-3y+5)$
$=-4-1(-3y)-1(5)$
$=-4+3y+(-5)$
$=-4+(-5)+3y$
$=-9+3y,$ or $3y-9$

N2. **(a)** $4x+6x-7x=(4+6-7)x=3x$

(b) $z+z=1z+1z=(1+1)z=2z$

(c) $4p^2-3p^2=(4-3)p^2=1p^2$, or p^2

N3. **(a)** $5k-6-(3-4k)$
$=5k-6-1(3-4k)$
$=5k-6-1(3)-1(-4k)$
$=5k-6-3+4k$
$=9k-9$

(b) $\frac{1}{4}x-\frac{2}{3}(x-9)=\frac{1}{4}x-\frac{2}{3}(x)-\frac{2}{3}(-9)$
$=\frac{3}{12}x-\frac{8}{12}x+6$
$=-\frac{5}{12}x+6$

N4. "Twice a number, subtracted from the sum of the number and 5" is written $(x+5)-2x$.
$(x+5)-2x=x+5-2x$
$=-x+5,$ or $5-x$

Exercises

1. $-(6x-3)=-1(6x-3)$
$=-1(6x)-1(-3)$
$=-6x+3$
The correct response is B.

3. Examples A, B, and D are pairs of *unlike* terms, since either the variables or their powers are different. Example C is a pair of *like* terms, since both terms have the same variables (r and y) and the same exponents (both variables are to the first power). Note that we can use the commutative property to rewrite $6yr$ as $6ry$.

5. The student made a sign error when applying the distributive property.
$7x-2(3-2x)=7x-2(3)-2(-2x)$
$=7x-6+4x$
$=11x-6$
The correct answer is $11x-6$.

7. $4r+19-8=4r+11$

9. $7(3x-4y)=7(3x)+7(-4y)$
$=21x-28y$

11. $5+2(x-3y)=5+2(x)+2(-3y)$
$=5+2x-6y$

13. $-2-(5-3p)=-2-1(5-3p)$
$=-2-1(5)-1(-3p)$
$=-2-5+3p$
$=-7+3p$

15. $6+(4-3x)-8=6+4-3x-8$
$=10-3x-8$
$=10-8-3x$
$=2-3x$

17. The numerical coefficient of the term $-12k$ is -12.

19. The numerical coefficient of the term $3m^2$ is 3.

21. Because xw can be written as $1\cdot xw$, the numerical coefficient of the term xw is 1.

23. Since $-x=-1x$, the numerical coefficient of the term $-x$ is -1.

25. Since $\frac{x}{2}=\frac{1}{2}x$, the numerical coefficient of the term $\frac{x}{2}$ is $\frac{1}{2}$.

27. Since $\frac{2x}{5}=\frac{2}{5}x$, the numerical coefficient of the term $\frac{2x}{5}$ is $\frac{2}{5}$.

29. Since $-0.5x^3=-0.5\cdot x^3$, the numerical coefficient of the term $-0.5x^3$ is -0.5.

31. The numerical coefficient of the term 10 is 10.

33. $8r$ and $-13r$ are like terms since they have the same variable with the same exponent (which is understood to be 1).

35. $5z^4$ and $9z^3$ are unlike terms. Although both have the variable z, the exponents are not the same.

37. All numerical terms (constants) are considered like terms, so 4, 9, and -24 are like terms.

39. x and y are unlike terms because they do not have the same variable.

41. $7y+6y=(7+6)y$
$=13y$

43. $-6x-3x=(-6-3)x$
$=-9x$

45. $12b+b=12b+1b$
$=(12+1)b$
$=13b$

47. $3k+8+4k+7=3k+4k+8+7$
$=(3+4)k+15$
$=7k+15$

49. $-5y+3-1+5+y-7$
$=(-5y+1y)+(3+5)+(-1-7)$
$=(-5+1)y+(8)+(-8)$
$=-4y+8-8$
$=-4y$

51. $-2x+3+4x-17+20$
$=(-2x+4x)+(3-17+20)$
$=(-2+4)x+6$
$=2x+6$

53. $16-5m-4m-2+2m$
$=(16-2)+(-5m-4m+2m)$
$=14+(-5-4+2)m$
$=14-7m$

55. $-10+x+4x-7-4x$
$=(-10-7)+(1x+4x-4x)$
$=-17+(1+4-4)x$
$=-17+1x$
$=-17+x$

57. $1+7x+11x-1+5x$
$=(1-1)+(7x+11x+5x)$
$=0+(7+11+5)x$
$=23x$

59. $-\frac{4}{3}+2t+\frac{1}{3}t-8-\frac{8}{3}t$
$=\left(2t+\frac{1}{3}t-\frac{8}{3}t\right)+\left(-\frac{4}{3}-8\right)$
$=\left(2+\frac{1}{3}-\frac{8}{3}\right)t+\left(-\frac{4}{3}-8\right)$
$=\left(\frac{6}{3}+\frac{1}{3}-\frac{8}{3}\right)t+\left(-\frac{4}{3}-\frac{24}{3}\right)$
$=-\frac{1}{3}t-\frac{28}{3}$

61. $6y^2+11y^2-8y^2=(6+11-8)y^2$
$=9y^2$

63. $2p^2+3p^2-8p^3-6p^3$
$=(2p^2+3p^2)+(-8p^3-6p^3)$
$=(2+3)p^2+(-8-6)p^3$
$=5p^2-14p^3$ or $-14p^3+5p^2$

65. $2(4x+6)+3=2(4x)+2(6)+3$
$=8x+12+3$
$=8x+15$

67. $-6-4(y-7)$
$=-6-4(y)+(-4)(-7)$ Distributive prop.
$=-6-4y+28$
$=-4y+22$

69. $13p+4(4-8p)=13p+4(4)+4(-8p)$
$=13p+16-32p$
$=-19p+16$

71. $3t-5-2(2t-4)=3t-5-2(2t)-2(-4)$
$=3t-5-4t+8$
$=-t+3$

73. $100[0.05(x+3)]$
$=[100(0.05)](x+3)$ Associative property
$=5(x+3)$
$=5(x)+5(3)$ Distributive prop.
$=5x+15$

75. $10[0.3(5-3x)]$
$=[10(0.3)](5-3x)$ Associative prop.
$=3(5-3x)$
$=3(5)+3(-3x)$ Distributive prop.
$=15-9x$

77. $-5(5y-9)+3(3y+6)$
$=-5(5y)+(-5)(-9)+3(3y)+3(6)$
$=-25y+45+9y+18$
$=(-25y+9y)+(45+18)$
$=(-25+9)y+63$
$=-16y+63$

79. $2(5r+3)-3(2r-3)$
$=2(5r)+2(3)+(-3)(2r)+(-3)(-3)$
$=10r+6-6r+9$
$=(10r-6r)+(6+9)$
$=(10-6)r+(15)$
$=4r+15$

81. $8(2k-1)-(4k-3)$
$=8(2k)+8(-1)+(-1)(4k)+(-1)(-3)$
$=16k-8-4k+3$
$=(16k-4k)+(-8+3)$
$=(16-4)k+(-5)$
$=12k-5$

83. $-\frac{4}{3}(y-12)-\frac{1}{6}y$
$=-\frac{4}{3}y-\frac{4}{3}(-12)-\frac{1}{6}y$
$=-\frac{4}{3}y+16-\frac{1}{6}y$
$=-\frac{4}{3}y-\frac{1}{6}y+16$
$=\left(-\frac{8}{6}-\frac{1}{6}\right)y+16$
$=-\frac{3}{2}y+16$

85. $\frac{1}{2}(2x+4)-\frac{1}{3}(9x-6)$
$=\frac{1}{2}(2x)+\frac{1}{2}(4)+\left(-\frac{1}{3}\right)(9x)+\left(-\frac{1}{3}\right)(-6)$
$=x+2-3x+2$
$=-2x+4$

87. $-\frac{2}{3}(5x+7)-\frac{1}{3}(4x+8)$
$=\left(-\frac{2}{3}\right)5x+\left(-\frac{2}{3}\right)7+\left(-\frac{1}{3}\right)4x+\left(-\frac{1}{3}\right)8$
$=-\frac{10x}{3}-\frac{14}{3}-\frac{4x}{3}-\frac{8}{3}$
$=-\frac{14x}{3}-\frac{22}{3}$

89. $-7.5(2y+4)-2.9(3y-6)$
$=-7.5(2y)-7.5(4)-2.9(3y)-2.9(-6)$
$=-15y-30-8.7y+17.4$
$=-23.7y-12.6$

91. $-2(-3k+2)-(5k-6)-3k-5$
$=-2(-3k)+(-2)(2)-1(5k-6)-3k-5$
$=6k-4+(-1)(5k)+(-1)(-6)-3k-5$
$=6k-4-5k+6-3k-5$
$=-2k-3$

93. $-4(-3x+3)-(6x-4)-2x+1$
$=-4(-3x+3)-1(6x-4)-2x+1$
$=12x-12-6x+4-2x+1$
$=(12x-6x-2x)+(-12+4+1)$
$=4x-7$

95. $(4x+8)+(3x-2)$
$=4x+8+3x-2$
$=7x+6$

97. $(5x+1)-(x-7)$
$=5x+1-x+7$
$=4x+8$

99. "Five times a number, added to the sum of the number and three" is written $(x+3)+5x$.
$(x+3)+5x=x+3+5x$
$=(x+5x)+3$
$=6x+3$

101. "A number multiplied by −7, subtracted from the sum of 13 and six times the number" is written $(13+6x)-(-7x)$.

$$(13+6x)-(-7x) = 13+6x+7x$$
$$= 13+13x$$

103. "Six times a number added to −4, subtracted from twice the sum of three times the number and 4" is written $2(3x+4)-(-4+6x)$.

$$2(3x+4)-(-4+6x)$$
$$= 2(3x+4)-1(-4+6x)$$
$$= 6x+8+4-6x$$
$$= 6x+(-6x)+8+4$$
$$= 0+12 = 12$$

105. For gizmos, the fixed cost is \$1000 and the variable cost is \$5 per gizmo, so the cost to produce x gizmos is $1000+5x$ (dollars).

107. The total cost to make x gizmos and y gadgets is $1000+5x+750+3y$ (dollars).

Chapter 1 Review Exercises

1. $5^4 = 5\cdot5\cdot5\cdot5 = 625$

2. $\left(\frac{3}{5}\right)^3 = \frac{3}{5}\cdot\frac{3}{5}\cdot\frac{3}{5} = \frac{27}{125}$

3. $\left(\frac{1}{8}\right)^2 = \left(\frac{1}{8}\right)\left(\frac{1}{8}\right)$

$$= \frac{1}{64}$$

4. $(0.1)^3 = (0.1)(0.1)(0.1)$

$$= 0.001$$

5. $8\cdot5-13 = 40-13 = 27$

6. $16+12\div4-2 = 16+(12\div4)-2$

$$= 16+3-2$$
$$= 19-2$$
$$= 17$$

7. $20-2(5+3) = 20-2(8)$

$$= 20-16$$
$$= 4$$

8. $7\left[3+6\left(3^2\right)\right] = 7\left[3+6(9)\right]$

$$= 7(3+54)$$
$$= 7(57)$$
$$= 399$$

9. $\frac{9\left(4^2-3\right)}{4\cdot5-17} = \frac{9(16-3)}{20-17}$

$$= \frac{9(13)}{3}$$
$$= \frac{3\cdot3\cdot13}{3} = 39$$

10. $\frac{6(5-4)+2(4-2)}{3^2-(4+3)} = \frac{6(1)+2(2)}{9-(4+3)}$

$$= \frac{6+4}{9-7}$$
$$= \frac{10}{2} = 5$$

11. $12\cdot3-6\cdot6 = 36-36 = 0$
Since $0 = 0$ is true, so is $0 \le 0$, and therefore, the statement "$12\cdot3-6\cdot6 \le 0$" is true.

12. $3\left[5(2)-3\right] = 3(10-3) = 3(7) = 21$

Therefore, the statement "$3\left[5(2)-3\right] > 20$" is true.

13. $4^2-8 = 16-8 = 8$
Since $9 \le 8$ is false, the statement "$9 \le 4^2-8$" is false.

14. "Thirteen is less than seventeen" is written $13 < 17$.

15. "Five plus two is not equal to ten" is written $5+2 \ne 10$.

16. "Two-thirds is greater than or equal to four-sixths" is written $\frac{2}{3} \ge \frac{4}{6}$.

17. $2x+6y = 2(6)+6(3)$

$$= 12+18 = 30$$

18. $4(3x-y) = 4\left[3(6)-3\right]$

$$= 4(18-3)$$
$$= 4(15) = 60$$

19. $\frac{x}{3}+4y=\frac{6}{3}+4(3)$
$=2+12=14$

20. $\frac{x^2+3}{3y-x}=\frac{6^2+3}{3(3)-6}$
$=\frac{36+3}{9-6}$
$=\frac{39}{3}=13$

21. "Six added to a number" translates as $x+6$.

22. "A number subtracted from eight" translates as $8-x$.

23. "Nine subtracted from six times a number" translates as $6x-9$.

24. "Three-fifths of a number added to 12" translates as $12+\frac{3}{5}x$.

25. $5x+3(x+2)=22;\ 2$
$5x+3(x+2)=5(2)+3(2+2)$ Let $x=2$.
$=5(2)+3(4)$
$=10+12=22$
Since the left side and the right side are equal, 2 is a solution of the given equation.

26. $\frac{t+5}{3t}=1;\ 6$
$\frac{t+5}{3t}=\frac{6+5}{3(6)}$ Let $t=6$.
$=\frac{11}{18}$
Since the left side, $\frac{11}{18}$, is not equal to the right side, 6 is not a solution of the equation.

27. "Six less than twice a number is 10" is written $2x-6=10$.
Letting x equal 0, 2, 4, 6, and 10 results in a false statement, so those values are not solutions. Since $2(8)-6=16-6=10$, the solution is 8.

28. "The product of a number and 4 is 8" is written $4x=8$. Since $4(2)=8$, the solution is 2.

29. $-4, -\frac{1}{2}, 0, 2.5, 5$
Graph these numbers on a number line. They are already arranged in order from smallest to largest.

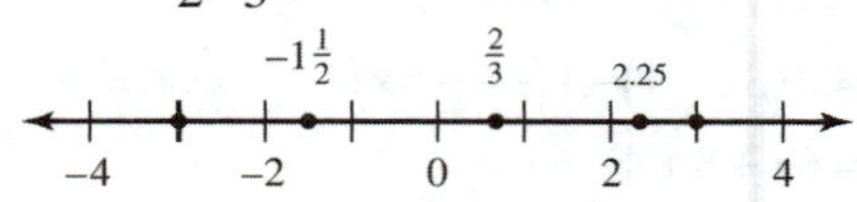

30. $-3, -1\frac{1}{2}, \frac{2}{3}, 2.25, 3$

$-1\frac{1}{2}$ $\frac{2}{3}$ 2.25
−4 −2 0 2 4

31. Since $\frac{4}{3}$ is the quotient of two integers, it is a rational number. Since all rational numbers are also real numbers, $\frac{4}{3}$ is a real number.

32. Since the decimal representation of $0.\overline{63}$ repeats, it is a rational number. Since all rational numbers are also real numbers, $0.\overline{63}$ is a real number.

33. Since 19 is a natural number, it is also a whole number and an integer. We can write it as $\frac{19}{1}$, so it is a rational number and, hence, a real number.

34. Since the decimal representation of $\sqrt{6}$ does not terminate or repeat, it is an irrational number. Since all irrational numbers are also real numbers, $\sqrt{6}$ is a real number.

35. Since any negative number is less than any positive number, -10 is the lesser number.

36. Since -9 is to the left of -8 on the number line, -9 is the lesser number.

37. To compare these fractions, use a common denominator.
$-\frac{2}{3}=-\frac{8}{12}, -\frac{3}{4}=-\frac{9}{12}$
Since $-\frac{9}{12}$ is to the left of $-\frac{8}{12}$ on the number line, $-\frac{3}{4}$ is the lesser number.

38. Since $-|23|=-23$ and $-23<0$, $-|23|$ is the lesser number.

39. The statement $12 > -13$ is true since 12 is to the right of -13 on the number line.

40. The statement $0 > -5$ is true since 0 is to the right of -5 on the number line.

41. The statement $-9 < -7$ is true since -9 is to the left of -7 on the number line.

42. The statement $-13 \geq -13$ is true since $-13 = -13$.

43. **(a)** The opposite of the number -9 is its negative—that is, $-9(-9) = 9$.

(b) Since $-9 < 0$, the absolute value of the number -9 is $|-9| = -(-9) = 9$.

44. **(a)** $-0 = 0$

(b) $|0| = 0$

45. **(a)** $-(6) = -6$

(b) $|6| = 6$

46. **(a)** $-\left(-\frac{5}{7}\right) = \frac{5}{7}$

(b) $\left|-\frac{5}{7}\right| = -\left(-\frac{5}{7}\right) = \frac{5}{7}$

47. $|-12| = -(-12) = 12$

48. $-|3| = -3$

49. $-|-19| = -[-(-19)] = -19$

50. $-|9-2| = -|7| = -7$

51. $-10 + 4 = -6$

52. $14 + (-18) = -4$

53. $-8 + (-9) = -17$

54. $\frac{4}{9} + \left(-\frac{5}{4}\right) = \frac{4 \cdot 4}{9 \cdot 4} + \left(-\frac{5 \cdot 9}{4 \cdot 9}\right)$ LCD = 36

$= \frac{16}{36} + \left(-\frac{45}{36}\right)$

$= -\frac{29}{36}$

55. $-13.5 + (-8.3) = -21.8$

56. $(-10+7) + (-11) = (-3) + (-11)$

$= -14$

57. $[-6 + (-8) + 8] + [9 + (-13)]$

$= \{[-6 + (-8)] + 8\} + (-4)$

$= [(-14) + 8] + (-4)$

$= (-6) + (-4) = -10$

58. $(-4+7) + (-11+3) + (-15+1)$

$= (3) + (-8) + (-14)$

$= [3 + (-8)] + (-14)$

$= (-5) + (-14) = -19$

59. $-7 - 4 = -7 + (-4) = -11$

60. $-12 - (-11) = -12 + (11) = -1$

61. $5 - (-2) = 5 + (2) = 7$

62. $-\frac{3}{7} - \frac{4}{5} = -\frac{3 \cdot 5}{7 \cdot 5} - \frac{4 \cdot 7}{5 \cdot 7}$

$= -\frac{15}{35} - \frac{28}{35}$ LCD = 35

$= -\frac{15}{35} + \left(-\frac{28}{35}\right)$

$= -\frac{43}{35}$, or $-1\frac{8}{35}$

63. $2.56 - (-7.75) = 2.56 + (7.75)$

$= 10.31$

64. $(-10-4) - (-2) = [-10 + (-4)] + 2$

$= (-14) + (2)$

$= -12$

65. $(-3+4) - (-1) = (-3+4) + 1$

$= 1 + 1$

$= 2$

66. $-(-5+6) - 2 = -(1) + (-2)$

$= -1 + (-2)$

$= -3$

67. "19 added to the sum of -31 and 12" is written

$(-31+12) + 19 = (-19) + 19$

$= 0.$

68. "13 more than the sum of -4 and -8" is written

$[-4 + (-8)] + 13 = -12 + 13$

$= 1.$

69. "The difference between −4 and −6" is written
$$-4-(-6)=-4+6 = 2.$$

70. "Five less than the sum of 4 and −8" is written
$$[4+(-8)]-5=(-4)+(-5) = -9.$$

71. $-23.75+50.00=26.25$
He now has a positive balance of \$26.25.

72. $-26+16=-10$
The high temperature was $-10°$F.

73.
$$\begin{aligned}-28+13-14&=(-28+13)-14\\&=(-28+13)+(-14)\\&=-15+(-14)\\&=-29\end{aligned}$$
His present financial status is −\$29.

74.
$$\begin{aligned}-3-7&=-3+(-7)\\&=-10\end{aligned}$$
The new temperature is −10°.

75.
$$\begin{aligned}8-12+42&=[8+(-12)]+42\\&=-4+42\\&=38\end{aligned}$$
The total net yardage is 38.

76. To get the closing value for the previous day, we can add the amount it was down to the amount at which it closed.
$14,810.31+30.64=14,840.95$

77. $(-12)(-3)=36$

78.
$$\begin{aligned}15(-7)&=-(15\cdot 7)\\&=-105\end{aligned}$$

79.
$$\begin{aligned}-\frac{4}{3}\left(-\frac{3}{8}\right)&=\frac{4}{3}\cdot\frac{3}{8}\\&=\frac{4\cdot 3}{3\cdot 4\cdot 2}\\&=\frac{1}{2}\end{aligned}$$

80. $(-4.8)(-2.1)=10.08$

81.
$$\begin{aligned}5(8-12)&=5[8+(-12)]\\&=5(-4)=-20\end{aligned}$$

82.
$$\begin{aligned}(5-7)(8-3)&=[5+(-7)][8+(-3)]\\&=(-2)(5)=-10\end{aligned}$$

83.
$$\begin{aligned}2(-6)-(-4)(-3)&=-12-(12)\\&=-12+(-12)\\&=-24\end{aligned}$$

84. $3(-10)-5=-30+(-5)=-35$

85. $\frac{-36}{-9}=\frac{4\cdot 9}{9}=4$

86. $\frac{220}{-11}=-\frac{20\cdot 11}{11}=-20$

87. $-\frac{1}{2}\div\frac{2}{3}=-\frac{1}{2}\cdot\frac{3}{2}=-\frac{3}{4}$

88. $\frac{-33.9}{-3}=\frac{11.3\cdot(-3)}{-3}=11.3$

89.
$$\begin{aligned}\frac{-5(3)-1}{8-4(-2)}&=\frac{-15+(-1)}{8-(-8)}\\&=\frac{-16}{8+8}\\&=\frac{-16}{16}=-1\end{aligned}$$

90.
$$\begin{aligned}\frac{5(-2)-3(4)}{-2[3-(-2)]-1}&=\frac{-10-12}{-2(3+2)-1}\\&=\frac{-10+(-12)}{-2(5)-1}\\&=\frac{-22}{-10+(-1)}\\&=\frac{-22}{-11}=2\end{aligned}$$

91.
$$\begin{aligned}\frac{10^2-5^2}{8^2+3^2-(-2)}&=\frac{100-25}{64+9+2}\\&=\frac{75}{75}=1\end{aligned}$$

92.
$$\begin{aligned}\frac{(0.6)^2+(0.8)^2}{(-1.2)^2-(-0.56)}&=\frac{0.36+0.64}{1.44+0.56}\\&=\frac{1.00}{2.00}=0.5\end{aligned}$$

93. $6x-4z=6(-5)-4(-3)$
$=-30-(-12)$
$=-30+12=-18$

94. $5x+y-z=5(-5)+(4)-(-3)$
$=(-25+4)+3$
$=-21+3=-18$

95. $5x^2=5(-5)^2$
$=5(25)$
$=125$

96. $z^2(3x-8y)=(-3)^2[3(-5)-8(4)]$
$=9(-15-32)$
$=9[-15+(-32)]$
$=9(-47)=-423$

97. "Nine less than the product of -4 and 5" is written
$-4(5)-9=-20+(-9)$
$=-29.$

98. "Five-sixths of the sum of 12 and -6" is written
$\frac{5}{6}[12+(-6)]=\frac{5}{6}(6)$
$=5.$

99. "The quotient of 12 and the sum of 8 and -4" is written
$\frac{12}{8+(-4)}=\frac{12}{4}=3.$

100. "The product of -20 and 12, divided by the difference of 15 and -15" is written
$\frac{-20(12)}{15-(-15)}=\frac{-240}{15+15}$
$=\frac{-240}{30}=-8.$

101. "8 times a number is -24" is written
$8x=-24.$
If $x=-3$, $8x=8(-3)=-24$. The solution is -3.

102. "The quotient of a number and 3 is -2" is written $\frac{x}{3}=-2$. If $x=-6$, $\frac{x}{3}=\frac{-6}{3}=-2$. The solution is -6.

103. The statement $6+0=6$ is an example of an identity property.

104. The statement $5\cdot 1=5$ is an example of an identity property.

105. The statement $-\frac{2}{3}\left(-\frac{3}{2}\right)=1$ is an example of an inverse property.

106. The statement $17+(-17)=0$ is an example of an inverse property.

107. The statement $5+(-9+2)=[5+(-9)]+2$ is an example of an associative property.

108. The statement $w(xy)=(wx)y$ is an example of an associative property.

109. The statement $3(x+y)=3x+3y$ is an example of the distributive property.

110. The statement $(1+2)+3=3+(1+2)$ is an example of a commutative property.

111. $7(y+2)=7y+7\cdot 2$
$=7y+14$

112. $-12(4-t)=-12(4)-(-12)(t)$
$=-48+12t$

113. $3(2s+5y)=3(2s)+3(5y)$
$=6s+15y$

114. $-(-4r+5s)=-1(-4r+5s)$
$=(-1)(-4r)+(-1)(5s)$
$=4r-5s$

115. $2m+9m=(2+9)m$
$=11m$

116. $15p^2-7p^2+8p^2$
$=(15-7+8)p^2$
$=16p^2$

117. $5p^2-4p+6p+11p^2$
$=(5+11)p^2+(-4+6)p$
$=16p^2+2p$

118. $-2(3k-5)+2(k+1)$
$=-6k+10+2k+2$
$=-4k+12$

119. $7(2m+3)-2(8m-4)$
$=14m+21-16m+8$
$=(14-16)m+29$
$=-2m+29$

120. $-(2k+8)-(3k-7)$
$=-1(2k+8)-1(3k-7)$
$=-2k-8-3k+7$
$=-5k-1$

121. "Seven times a number, subtracted from the product of -2 and three times the number" is written $-2(3x)-7x=-6x-7x=-13x$.

122. "A number multiplied by 8, added to the sum of 5 and four times the number" is written $(5+4x)+8x=5+(4x+8x)=5+12x$.

Chapter 1 Mixed Review Exercises

1. Complete the 1st row of the table.

Number	Absolute Value	Additive Inverse	Multiplicative Inverse
-3	3	3	$-\frac{1}{3}$

2. Complete the 2nd row of the table.

Number	Absolute Value	Additive Inverse	Multiplicative Inverse
12	12	-12	$\frac{1}{12}$

3. Complete the 3rd row of the table.

Number	Absolute Value	Additive Inverse	Multiplicative Inverse
$-\frac{2}{3}$	$\frac{2}{3}$	$\frac{2}{3}$	$-\frac{3}{2}$

4. Complete the 4th row of the table.

Number	Absolute Value	Additive Inverse	Multiplicative Inverse
0.2	0.2	-0.2	5

5. The repeating decimal $0.\overline{6}$ is a rational number. All rational numbers are real numbers, so it is also a real number.

6. $(x+6)^3-y^3$
$=(-2+6)^3-(3)^3$
$=(4)^3-(27)$
$=64-27$
$=37$

7. $\frac{6(-4)+2(-12)}{5(-3)+(-3)}=\frac{-24+(-24)}{-15+(-3)}$
$=\frac{-48}{-18}=\frac{8\cdot 6}{3\cdot 6}$
$=\frac{8}{3}$, or $2\frac{2}{3}$

8. $\frac{3}{8}-\frac{5}{12}=\frac{3\cdot 3}{8\cdot 3}-\frac{5\cdot 2}{12\cdot 2}$
$=\frac{9}{24}-\frac{10}{24}$
$=\frac{9}{24}+\left(-\frac{10}{24}\right)$
$=-\frac{1}{24}$

9. $\frac{8^2+6^2}{7^2+1^2}=\frac{64+36}{49+1}$
$=\frac{100}{50}=2$

10. $-\frac{12}{5}\div\frac{9}{7}=-\frac{12}{5}\cdot\frac{7}{9}$
$=-\frac{12\cdot 7}{5\cdot 9}$
$=-\frac{3\cdot 4\cdot 7}{5\cdot 3\cdot 3}$
$=-\frac{28}{15}$, or $-1\frac{13}{15}$

11. $2\frac{5}{6}-4\frac{1}{3}=\frac{17}{6}-\frac{13}{3}$
$=\frac{17}{6}-\frac{13\cdot 2}{3\cdot 2}$
$=\frac{17}{6}-\frac{26}{6}$
$=\frac{17}{6}+\left(-\frac{26}{6}\right)$
$=-\frac{9}{6}=-\frac{3}{2}$, or $-1\frac{1}{2}$

12. $\left(\frac{5}{6}\right)^2=\left(\frac{5}{6}\right)\left(\frac{5}{6}\right)$
$=\frac{25}{36}$

13. $[(-2)+7-(-5)]+[-4-(-10)]$
$=\{[(-2)+7]-(-5)\}+(-4+10)$
$=(5+5)+6$
$=10+6=16$

14. $-16(-3.5)-7.2(-3)$
$=56-[(7.2)(-3)]$
$=56-(-21.6)$
$=56+21.6$
$=77.6$

15. $-8+[(-4+17)-(-3-3)]$
$=-8+\{(13)-[-3+(-3)]\}$
$=-8+[13-(-6)]$
$=-8+(13+6)$
$=-8+19=11$

16. $-4(2t+1)-8(-3t+4)$
$=-4(2t)-4(1)-8(-3t)-8(4)$
$=-8t-4+24t-32$
$=16t-36$

17. $5x^2-12y^2+3x^2-9y^2$
$=(5x^2+3x^2)+(-12y^2-9y^2)$
$=(5+3)x^2+(-12-9)y^2$
$=8x^2-21y^2$

18. $(-8-3)-5(2-9)$
$=[-8+(-3)]-5[2+(-9)]$
$=-11-5(-7)$
$=-11-(-35)$
$=-11+35=24$

19. $118-165=118+(-165)$
$=-47$
The lowest temperature ever recorded in Iowa was $-47°$F.

20. $14,494-(-282)=14,494+282$
$=14,776$
The difference in elevation is 14,776 ft.

Chapter 1 Test

1. $4[-20+7(-2)]=4[-20+(-14)]$
$=4(-34)=-136$
Since $-136\le 135$, the statement "$4[-20+7(-2)]\le 135$" is true.

2. $\left(\frac{1}{2}\right)^2+\left(\frac{2}{3}\right)^2=\left(\frac{1}{2}+\frac{2}{3}\right)^2$
$\frac{1}{4}+\frac{4}{9}=\left(\frac{7}{6}\right)^2$
$\frac{25}{36}\neq\frac{49}{36}$
Since $\frac{25}{36}\neq\frac{49}{36}$ the statement
$\left(\frac{1}{2}\right)^2+\left(\frac{2}{3}\right)^2=\left(\frac{1}{2}+\frac{2}{3}\right)^2$ is false.

3. The number $-\frac{2}{3}$ can be written as a quotient of two integers with denominator not 0, so it is a rational number. Since all rational numbers are real numbers, it is also a real number.

4. Simplify $-|-8|$.
$-|-8|=-[-(-8)]$
$=-(8)$
$=-8$
The number -8 is less than 6, and thus $-|-8|$ is the lesser number.

5. "The quotient of -6 and the sum of 2 and -8" is written $\frac{-6}{2+(-8)}$, and $\frac{-6}{2+(-8)}=\frac{-6}{-6}=1$.

6. If a and b are both negative, $a+b$ is negative and $a\cdot b$ is positive. A positive number divided by a negative number is negative, and thus $\frac{a+b}{a\cdot b}$ is negative.

7. $-2-(5-17)+(-6)$
$=-2-[5+(-17)]+(-6)$
$=-2-(-12)+(-6)$
$=(-2+12)+(-6)$
$=10+(-6)=4$

8. $-5\frac{1}{2}+2\frac{2}{3}=-\frac{11}{2}+\frac{8}{3}$
$=-\frac{11\cdot 3}{2\cdot 3}+\frac{8\cdot 2}{3\cdot 2}$
$=-\frac{33}{6}+\frac{16}{6}$
$=-\frac{17}{6}$, or $-2\frac{5}{6}$

9. $-6.2-[-7.1+(2.0-3.1)]$
$=-6.2-[-7.1+(2.0-3.1)]$
$=-6.2-[-7.1-1.1]$
$=-6.2-[-8.2]$
$=-6.2+8.2$
$=2$

10. $4^2+(-8)-(2^3-6)$
$=16+(-8)-(8-6)$
$=[16+(-8)]-2$
$=8-2=6$

11. $(-5)(-12)+4(-4)+(-8)^2$
$=(-5)(-12)+4(-4)+64$
$=[60+(-16)]+64$
$=44+64=108$

12. $\frac{30(-1-2)}{-9[3-(-2)]-12(-2)}$
$=\frac{30(-3)}{-9(5)-(-24)}$
$=\frac{-90}{-45+24}$
$=\frac{-90}{-21}$
$=\frac{30\cdot 3}{7\cdot 3}=\frac{30}{7}$, or $4\frac{2}{7}$

13. $3x-4y^2$
$=3(-2)-4(4^2)$ Let $x=-2$, $y=4$.
$=3(-2)-4(16)$
$=-6-64=-70$

14. $\frac{5x+7y}{3(x+y)}$
$=\frac{5(-2)+7(4)}{3(-2+4)}$ Let $x=-2$, $y=4$.
$=\frac{-10+28}{3(2)}$
$=\frac{18}{6}=3$

15. The difference between the highest and lowest elevations is
$6960-(-40)=6960+40=7000$ meters.

16. 4 saves (3 points per save)
+3 wins (3 points per win)
+2 losses (-2 points per loss)
+1 blown save (-2 points per blown save)
$=4(3)+3(3)+2(-2)+1(-2)$
$=12+9-4-2$
$=15$ points
He has a total of 15 points.

17. $2.45-3.54=2.45+(-3.54)=-1.09$
As a signed number, the federal budget deficit is $-\$1.09$ trillion.

18. $3x+0=3x$ illustrates an identity property. The correct response is D.

19. $(5+2)+8=8+(5+2)$ illustrates a commutative property because the order of the numbers is changed, but the grouping is not. The correct response is A.

20. $-3(x+y)=-3x+(-3y)$ illustrates the distributive property. The correct response is E.

21. $-5+(3+2)=(-5+3)+2$ illustrates an associative property because the grouping of the numbers is changed, but the order is not. The correct response is B.

22. $-\frac{5}{3}\left(-\frac{3}{5}\right)=1$ illustrates an inverse property. The correct response is C.

23. $3(x+1)=3\cdot x+3\cdot 1$
$=3x+3$
The distributive property is used to rewrite $3(x+1)$ as $3x+3$.

24. **(a)** $-6[5+(-2)]=-6(3)=-18$

(b) $-6[5+(-2)]=-6(5)+(-6)(-2)$
$=-30+12=-18$

(c) The distributive property assures us that the answers must be the same, because $a(b+c)=ab+ac$ for all a, b, c.

25. $8x+4x-6x+x+14x$
$=(8+4-6+1+14)x$
$=21x$

26. $5(2x-1)-(x-12)+2(3x-5)$
$=5(2x-1)-1(x-12)+2(3x-5)$
$=10x-5-x+12+6x-10$
$=(10-1+6)x+(-5+12-10)$
$=15x-3$

Chapter 2
Linear Equations and Inequalities in One Variable

2.1 The Addition Property of Equality

Now Try Exercises

N1. $x-13=4$ Given

$x-13+13=4+13$ Add 13.

$x=17$ Combine like terms.

We check by substituting 17 for x in the *original* equation.

Check $x-13=4$ Original equation

$17-13\stackrel{?}{=}4$ Let $x=17$.

$4=4$ True

Since a true statement results, $\{17\}$ is the solution set.

N2. $t-5.7=-7.2$

$t-5.7+5.7=-7.2+5.7$ Add 5.7.

$t=-1.5$

Check $t=-1.5: -7.2=-7.2$ True

The solution set is $\{-1.5\}$.

N3. $-15=x+12$

$-15-12=x+12-12$ Subtract 12.

$-27=x$

Check $x=-27: -15=-15$ True

The solution set is $\{-27\}$.

N4. $x-5=2x$

$x-5-x=2x-x$ Subtract x.

$-5=x$ Combine terms.

Check $x=-5: -10=-10$ True

The solution set is $\{-5\}$.

N5. $\frac{2}{3}x+4=\frac{5}{3}x$

$\frac{2}{3}x+4-\frac{2}{3}x=\frac{5}{3}x-\frac{2}{3}x$ Subtract $\frac{2}{3}x$.

$4=x$ Combine terms.

Check $x=4: \frac{20}{3}=\frac{20}{3}$ True

The solution set is $\{4\}$.

N6. $6x-8=12+5x$

$6x-8-5x=12+5x-5x$ Subtract $5x$.

$x-8=12$ Combine terms.

$x-8+8=12+8$ Add 8.

$x=20$ Combine terms.

Check $x=20: 112=112$ True

The solution set is $\{20\}$.

N7. $5x-10-12x=4-8x-9$

$-7x-10=-8x-5$ Combine terms.

$-7x-10+8x=-8x-5+8x$ Add $8x$.

$x-10=-5$ Combine terms.

$x-10+10=-5+10$ Add 10.

$x=5$ Combine terms.

Check $x=5:\ -45=-45$ True

The solution set is $\{5\}$.

N8. $4(3x-2)-(11x-4)=3$

$4(3x-2)-1(11x-4)=3$ $-a=-1a$

$12x-8-11x+4=3$ Distributive prop.

$x-4=3$ Combine terms.

$x-4+4=3+4$ Add 4.

$x=7$

Check $x=7: 3=3$ True

The solution set is $\{7\}$.

Exercises

1. An equation includes an equality symbol, while an expression does not.

3. Equations that have exactly the same solution set are equivalent equations.

5. (a) $5x+8-4x+7$

This is an expression, not an equation, since there is no equals symbol. It can be simplified by rearranging terms and then combining like terms.

$5x+8-4x+7=5x-4x+8+7$

$=x+15$

(b) $-6y+12+7y-5$

This is an expression, not an equation, since there is no equals symbol. It can be simplified by rearranging terms and then combining like terms.

$-6y+12+7y-5=-6y+7y+12-5$

$=y+7$

(c) $5x+8-4x=7$
This is an equation because of the equals symbol.
$$5x+8-4x=7$$
$$x+8=7$$
$$x=-1$$
The solution set is $\{-1\}$.

(d) This is an equation because of the equals symbol.
$$-6y+12+7y=-5$$
$$y+12=-5$$
$$y=-17$$
The solution set is $\{-17\}$.

7. Equations A $\left(x^2-5x+6=0\right)$ and B $\left(x^3=x\right)$ are *not* linear equations in one variable because they cannot be written in the form $Ax+B=C$. Note that in a linear equation the exponent on the variable must be 1.

9.
$$x-3=9$$
$$x-3+3=9+3$$
$$x=12$$
Check this solution by replacing x with 12 in the original equation.
$$x-3=9$$
$$12-3\stackrel{?}{=}9 \quad \text{Let } x=12.$$
$$9=9 \quad \text{True}$$
Because the final statement is true, $\{12\}$ is the solution set.

11.
$$x-12=19$$
$$x-12+12=19+12$$
$$x=31$$
Check $x=31$
$$31-12\stackrel{?}{=}19 \quad \text{Let } x=31.$$
$$19=19 \quad \text{True}$$
Thus, $\{31\}$ is the solution set.

13.
$$x-6=-9$$
$$x-6+6=-9+6$$
$$x=-3$$
Checking yields a true statement, so $\{-3\}$ is the solution set.

15.
$$r+8=12$$
$$r+8-8=12-8$$
$$r=4$$
Checking yields a true statement, so $\{4\}$ is the solution set.

17.
$$x+28=19$$
$$x+28-28=19-28$$
$$x=-9$$
Checking yields a true statement, so $\{-9\}$ is the solution set.

19.
$$x+\frac{1}{4}=-\frac{1}{2}$$
$$x+\frac{1}{4}-\frac{1}{4}=-\frac{1}{2}-\frac{1}{4}$$
$$x=-\frac{2}{4}-\frac{1}{4}$$
$$x=-\frac{3}{4}$$
Check $x=-\frac{3}{4}$: $-\frac{1}{2}=-\frac{1}{2}$ True

The solution set is $\left\{-\frac{3}{4}\right\}$.

21.
$$7+r=-3$$
$$r+7=-3$$
$$r+7-7=-3-7$$
$$r=-10$$
The solution set is $\{-10\}$.

23.
$$2=p+15$$
$$2-15=p+15-15$$
$$-13=p$$
The solution set is $\{-13\}$.

25.
$$-4=x-14$$
$$-4+14=x-14+14$$
$$10=x$$
The solution set is $\{10\}$.

27.
$$-\frac{1}{3}=x-\frac{3}{5}$$
$$-\frac{1}{3}+\frac{3}{5}=x-\frac{3}{5}+\frac{3}{5}$$
$$-\frac{5}{15}+\frac{9}{15}=x$$
$$\frac{4}{15}=x$$
Check $x=\frac{4}{15}$: $-\frac{5}{15}=\frac{4}{15}-\frac{9}{15}$ True

The solution set is $\left\{\frac{4}{15}\right\}$.

29.
$$x-8.4=-2.1$$
$$x-8.4+8.4=-2.1+8.4$$
$$x=6.3$$
The solution set is $\{6.3\}$.

31.
$$t+12.3=-4.6$$
$$t+12.3-12.3=-4.6-12.3$$
$$t=-16.9$$
The solution set is $\{-16.9\}$.

33.
$$3x=2x+7$$
$$3x-2x=2x+7-2x \quad \text{Subtract } 2x.$$
$$1x=7 \quad \text{or} \quad x=7$$
Check $x=7$: $21=21$ True

The solution set is $\{7\}$.

35.
$$10x+4=9x$$
$$10x+4-9x=9x-9x \quad \text{Subtract } 9x.$$
$$1x+4=0$$
$$x+4-4=0-4 \quad \text{Subtract } 4.$$
$$x=-4$$
Check $x=-4$: $-36=-36$ True

The solution set is $\{-4\}$.

37.
$$8x-3=9x$$
$$8x-3-8x=9x-8x \quad \text{Subtract } 8x.$$
$$-3=x$$
Check $x=-3$: $8(-3)-3=9(-3)$ True

The solution set is $\{-3\}$.

39.
$$6t-2=5t$$
$$6t-2-5t=5t-5t \quad \text{Subtract } 5t.$$
$$t-2=0$$
$$t-2+2=2 \quad \text{Add } 2.$$
$$t=2$$
Check $t=2$: $6(2)-2=5(2)$ True

The solution set is $\{2\}$.

41.
$$\frac{2}{5}w-6=\frac{7}{5}w$$
$$\frac{2}{5}w-6-\frac{2}{5}w=\frac{7}{5}w-\frac{2}{5}w \quad \text{Subtract } \frac{2}{5}w.$$
$$-6=\frac{5}{5}w$$
$$-6=w$$
Check $w=-6$: $\frac{2}{5}(-6)-6=\frac{7}{5}(-6)$ True

The solution set is $\{-6\}$.

43.
$$\frac{1}{2}x+5=-\frac{1}{2}x$$
$$\frac{1}{2}x+\frac{1}{2}x+5=-\frac{1}{2}x+\frac{1}{2}x$$
$$x+5=0$$
$$x+5-5=0-5$$
$$x=-5$$
The solution set is $\{-5\}$.

45.
$$5.6x+2=4.6x$$
$$5.6x+2-4.6x=4.6x-4.6x$$
$$1.0x+2=0$$
$$x+2-2=0-2$$
$$x=-2$$
The solution set is $\{-2\}$.

47.
$$1.4x-3=0.4x$$
$$1.4x-3-0.4x=0.4x-0.4x$$
$$1.0x-3=0$$
$$1.0x-3+3=0+3$$
$$x=3$$
The solution set is $\{3\}$.

49.
$$5p=4p$$
$$5p-4p=4p-4p$$
$$p=0$$
The solution set is $\{0\}$.

51. $3x+7-2x=0$
$x+7=0$
$x+7-7=0-7$
$x=-7$
The solution set is $\{-7\}$.

53. $3x+7=2x+4$
$3x+7-2x=2x+4-2x$
$x+7=4$
$x+7-7=4-7$
$x=-3$
The solution set is $\{-3\}$.

55. $8t+6=7t+6$
$8t+6-7t=7t+6-7t$
$t+6=6$
$t+6-6=6-6$
$t=0$
The solution set is $\{0\}$.

57. $-4x+7=-5x+9$
$-4x+7+5x=-5x+9+5x$
$x+7=9$
$x+7-7=9-7$
$x=2$
The solution set is $\{2\}$.

59. $5-x=-2x-11$
$5-x+2x=-2x-11+2x$ Add $2x$.
$5+x-5=-11-5$ Subtract 5.
$x=-16$
The solution set is $\{-16\}$.

61. $1.2y-4=0.2y-4$
$1.2y-4-0.2y=0.2y-4-0.2y$
$1.0y-4=-4$
$y-4+4=-4+4$
$y=0$
The solution set is $\{0\}$.

63. $3x+6-10=2x-2$
$3x-4=2x-2$ Combine terms.
$3x-4-2x=2x-2-2x$ Subtract $2x$.
$x-4=-2$
$x-4+4=-2+4$
$x=2$
The solution set is $\{2\}$.

65. $5t+3+2t-6t=4+12$
$(5+2-6)t+3=16$
$t+3-3=16-3$
$t=13$
Check $t=13$: $16=16$ True
The solution set is $\{13\}$.

67. $6x+5+7x+3=12x+4$
$13x+8=12x+4$
$13x+8-12x=12x+4-12x$
$x+8=4$
$x+8-8=4-8$
$x=-4$
Check $x=-4$: $-44=-44$ True
The solution set is $\{-4\}$.

69. $5.2q-4.6-7.1q=-0.9q-4.6$
$-1.9q-4.6=-0.9q-4.6$
$-1.9q-4.6+0.9q=-0.9q-4.6+0.9q$
$-1.0q-4.6=-4.6$
$-1.0q-4.6+4.6=-4.6+4.6$
$-q=0$
$q=0$
Check $q=0$: $-4.6=-4.6$ True
The solution set is $\{0\}$.

71. $\frac{5}{7}x+\frac{1}{3}=\frac{2}{5}-\frac{2}{7}x+\frac{2}{5}$
$\frac{5}{7}x+\frac{1}{3}=\frac{4}{5}-\frac{2}{7}x$
$\frac{5}{7}x+\frac{2}{7}x+\frac{1}{3}=\frac{4}{5}-\frac{2}{7}x+\frac{2}{7}x$ Add $\frac{2}{7}x$.
$\frac{7}{7}x+\frac{1}{3}=\frac{4}{5}$ Combine terms.
$1x+\frac{1}{3}-\frac{1}{3}=\frac{4}{5}-\frac{1}{3}$ Subtract $\frac{1}{3}$.
$x=\frac{12}{15}-\frac{5}{15}$ LCD $=15$
$x=\frac{7}{15}$
Check $x=\frac{7}{15}$: $\frac{2}{3}=\frac{2}{3}$ True
The solution set is $\left\{\frac{7}{15}\right\}$.

73. $(5y+6)-(3+4y)=10$

$5y+6-3-4y=10$ Distributive prop.

$y+3=10$ Combine terms.

$y+3-3=10-3$ Subtract 3.

$y=7$

Check $y=7$: $10=10$ True

The solution set is $\{7\}$.

75. $2(p+5)-(9+p)=-3$

$2p+10-9-p=-3$

$p+1=-3$

$p+1-1=-3-1$

$p=-4$

Check $p=-4$: $-3=-3$ True

The solution set is $\{-4\}$.

77. $-6(2b+1)+(13b-7)=0$

$-12b-6+13b-7=0$

$b-13=0$

$b-13+13=0+13$

$b=13$

Check $b=13$: $0=0$ True

The solution set is $\{13\}$.

79. $10(-2x+1)=-19(x+1)$

$-20x+10=-19x-19$

$-20x+10+19x=-19x-19+19x$

$-x+10=-19$

$-x+10-10=-19-10$

$-x=-29$

$x=29$

Check $x=29$: $-570=-570$ True

The solution set is $\{29\}$.

81. $-2(8p+2)-3(2-7p)-2(4+2p)=0$

$-16p-4-6+21p-8-4p=0$

$p-18=0$

$p-18+18=0+18$

$p=18$

Check $p=18$: $0=0$ True

The solution set is $\{18\}$.

83. $4(7x-1)+3(2-5x)-4(3x+5)=-6$

$28x-4+6-15x-12x-20=-6$

$x-18=-6$

$x-18+18=-6+18$

$x=12$

Check $x=12$: $-6=-6$ True

The solution set is $\{12\}$.

85. Answers will vary. One example is $x-6=-8$.

87. "Three times a number is 17 more than twice the number."

$3x=2x+17$

$3x-2x=2x+17-2x$

$x=17$

The number is 17 and $\{17\}$ is the solution set.

89. "If six times a number is subtracted from seven times the number, the result is −9."

$7x-6x=-9$

$x=-9$

The number is −9 and $\{-9\}$ is the solution set.

2.2 The Multiplication Property of Equality

Now Try Exercises

N1. $8x=80$

$\frac{8x}{8}=\frac{80}{8}$ Divide by 8.

$x=10$

Check $x=10$: $80=80$ True

The solution set is $\{10\}$.

N2. $10x=-24$

$\frac{10x}{10}=-\frac{24}{10}$ Divide by 10.

$x=-\frac{24}{10}=-\frac{12}{5}$ Write in lowest terms.

Check $x=-\frac{12}{5}$: $-24=-24$ True

The solution set is $\left\{-\frac{12}{5}\right\}$.

N3. $7.02 = -1.3x$

$$\frac{7.02}{-1.3} = \frac{-1.3x}{-1.3} \quad \text{Divide by } -1.3.$$

$$x = -5.4$$

Check $x = -5.4$: $7.02 = 7.02$ True

The solution set is $\{-5.4\}$.

N4.

$$\frac{x}{5} = -7$$

$$\frac{1}{5}x = -7$$

$$5 \cdot \frac{1}{5}x = 5(-7) \quad \text{Multiply by 5.}$$

$$p = -35$$

Check $p = -35$: $-7 = -7$ True

The solution set is $\{-35\}$.

N5.

$$\frac{4}{7}z = -16$$

$$\frac{7}{4}\left(\frac{4}{7}z\right) = \frac{7}{4}(-16) \quad \text{Multiply by } \frac{7}{4}.$$

$$1 \cdot t = \frac{7}{4} \cdot \frac{-16}{1}$$

$$t = -28$$

Check $t = -28$: $-16 = -16$ True

The solution set is $\{-28\}$.

N6.

$$-x = -9$$

$$-1 \cdot x = -9 \qquad -x = -1 \cdot x$$

$$(-1)(-1) \cdot x = (-1)(-9) \quad \text{Multiply by } -1.$$

$$1 \cdot x = 9$$

$$x = 9$$

Check $x = 9$: $-9 = -9$ True

The solution set is $\{9\}$.

N7. $9n - 6n = 21$

$$3n = 21 \quad \text{Combine terms.}$$

$$\frac{3n}{3} = \frac{21}{3} \quad \text{Divide by 3.}$$

$$n = 7$$

Check $n = 7$: $21 = 21$ True

The solution set is $\{7\}$.

Exercises

1. **(a)** multiplication property of equality; to get x alone on the left side of the equation, multiply each side by $\frac{1}{3}$ (or divide each side by 3).

(b) addition property of equality; to get x alone on the left side of the equation, add -3 (or subtract 3) on each side.

(c) multiplication property of equality; to get x alone on the left side of the equation, multiply each side by -1 (or divide each side by -1).

(d) addition property of equality; to get x alone on the right side of the equation, add -6 (or subtract 6) on each side.

3. Choice B; to find the solution of $-x = -\frac{3}{4}$, multiply (or divide) each side by -1, or use the rule "If $-x = a$, then $x = -a$."

5. To get just x on the left side, multiply both sides of the equation by the reciprocal of $\frac{4}{5}$, which is $\frac{5}{4}$.

7. This equation is equivalent to $\frac{1}{10}x = 5$. To get just x on the left side, multiply both sides of the equation by the reciprocal of $\frac{1}{10}$, which is 10.

9. To get just x on the left side, multiply both sides of the equation by the reciprocal of $-\frac{9}{2}$, which is $-\frac{2}{9}$.

11. This equation is equivalent to $-1x = 0.75$. To get just x on the left side, multiply both sides of the equation by the reciprocal of -1, which is -1.

13. To get just x on the left side, divide both sides of the equation by the coefficient of x, which is 6.

15. To get just x on the left side, divide both sides of the equation by the coefficient of x, which is -4.

17. To get just x on the left side, divide both sides of the equation by the coefficient of x, which is 0.12.

19. This equation is equivalent to $-1x = 25$. To get just x on the left side, divide both sides of the equation by the coefficient of x, which is -1.

21. $6x = 36$

$\frac{6x}{6} = \frac{36}{6}$ Divide by 6.

$1x = 6$

$x = 6$

Check $x = 6: 36 = 36$ True

The solution set is $\{6\}$.

23. $2m = 15$

$\frac{2m}{2} = \frac{15}{2}$ Divide by 2.

$m = \frac{15}{2}$

Check $m = \frac{15}{2}: 15 = 15$ True

The solution set is $\left\{\frac{15}{2}\right\}$.

25. $4x = -20$

$\frac{4x}{4} = \frac{-20}{4}$ Divide by 4.

$x = -5$

Check $x = -5: -20 = -20$ True

The solution set is $\{-5\}$.

27. $-7x = 28$

$\frac{-7x}{-7} = \frac{28}{-7}$ Divide by -7.

$x = -4$

Check $x = -4: 28 = 28$ True

The solution set is $\{-4\}$.

29. $10t = -36$

$\frac{10t}{10} = \frac{-36}{10}$ Divide by 10.

$t = -\frac{36}{10} = -\frac{18}{5}$ Lowest terms

Check $t = -\frac{18}{5}: -36 = -36$ True

The solution set is $\left\{-\frac{18}{5}\right\}$, or $\{-3.6\}$.

31. $-6x = -72$

$\frac{-6x}{-6} = \frac{-72}{-6}$ Divide by -6.

$x = 12$

Check $x = 12: -72 = -72$ True

The solution set is $\{12\}$.

33. $4r = 0$

$\frac{4r}{4} = \frac{0}{4}$ Divide by 4.

$r = 0$

Check $r = 0$: $0 = 0$ True

The solution set is $\{0\}$.

35. $-x = 12$

$-1 \cdot (-x) = -1 \cdot 12$ Multiply by -1.

$x = -12$

Check $x = -12$: $12 = 12$ True

The solution set is $\{-12\}$.

37. $-x = -\frac{3}{4}$

$-1 \cdot (-x) = -1 \cdot \left(-\frac{3}{4}\right)$

$x = \frac{3}{4}$

Check $x = \frac{3}{4}: -\frac{3}{4} = -\frac{3}{4}$ True

The solution set is $\left\{\frac{3}{4}\right\}$.

39. $0.2t = 8$

$\frac{0.2t}{0.2} = \frac{8}{0.2}$

$t = 40$

Check $t = 40: 8 = 8$ True

The solution set is $\{40\}$.

41. $-0.3x = 9$

$\frac{-0.3x}{-0.3} = \frac{9}{-0.3}$

$x = -30$

Check $x = -30: 9 = 9$ True

The solution set is $\{-30\}$.

43. $0.6x = -1.44$

$\frac{0.6x}{0.6} = \frac{-1.44}{0.6}$ Divide by 0.6.

$x = -2.4$

Check $x = -2.4: -1.44 = -1.44$ True

The solution set is $\{-2.4\}$.

45. $-9.1 = -2.6x$

$\frac{-9.1}{-2.6} = \frac{-2.6x}{-2.6}$ Divide by -2.6.

$x = 3.5$

Check $x = 3.5: -9.1 = -9.1$ True

The solution set is $\{3.5\}$.

47. $-2.1m = 25.62$

$\frac{-2.1m}{-2.1} = \frac{25.62}{-2.1}$ Divide by -2.1.

$m = -12.2$

Check $m = -12.2: 25.62 = 25.62$ True

The solution set is $\{-12.2\}$.

49. $\frac{1}{4}x = -12$

$4 \cdot \frac{1}{4}x = 4(-12)$ Multiply by 4.

$1x = -48$

$x = -48$

Check $x = -48: -12 = -12$ True

The solution set is $\{-48\}$.

51. $\frac{z}{6} = 12$

$\frac{1}{6}z = 12$

$6 \cdot \frac{1}{6}z = 6 \cdot 12$

$z = 72$

Check $z = 72: 12 = 12$ True

The solution set is $\{72\}$.

53. $\frac{x}{7} = -5$

$\frac{1}{7}x = -5$

$7\left(\frac{1}{7}x\right) = 7(-5)$

$x = -35$

Check $x = -35: -5 = -5$ True

The solution set is $\{-35\}$.

55. $\frac{2}{7}p = 4$

$\frac{7}{2}\left(\frac{2}{7}p\right) = \frac{7}{2}(4)$ Multiply by $\frac{7}{2}$.

$p = 14$

Check $p = 14: 4 = 4$ True

The solution set is $\{14\}$.

57. $-\frac{5}{6}t = -15$

$-\frac{6}{5}\left(-\frac{5}{6}t\right) = -\frac{6}{5}(-15)$ Multiply by $-\frac{6}{5}$.

$t = 18$

Check $t = 18: -15 = -15$ True

The solution set is $\{18\}$.

59. $-\frac{7}{9}x = \frac{3}{5}$

$-\frac{9}{7}\left(-\frac{7}{9}x\right) = -\frac{9}{7} \cdot \frac{3}{5}$ Multiply by $-\frac{9}{7}$.

$x = -\frac{27}{35}$

Check $x = -\frac{27}{35}: \frac{3}{5} = \frac{3}{5}$ True

The solution set is $\left\{-\frac{27}{35}\right\}$.

61. $4x + 3x = 21$

$7x = 21$

$\frac{7x}{7} = \frac{21}{7}$

$x = 3$

Check $x = 3: 21 = 21$ True

The solution set is $\{3\}$.

63. $6r-8r=10$

$-2r=10$

$\frac{-2r}{-2}=\frac{10}{-2}$

$r=-5$

Check $r=-5: 10=10$ True

The solution set is $\{-5\}$.

65. $\frac{2}{5}x-\frac{3}{10}x=2$

$\frac{4}{10}x-\frac{3}{10}x=2$

$\frac{1}{10}x=2$

$10\cdot\frac{1}{10}x=10\cdot 2$

$x=20$

Check $x=20$: $8-6=2$ True

The solution set is $\{20\}$.

67. $7m+6m-4m=63$

$9m=63$

$\frac{9m}{9}=\frac{63}{9}$

$m=7$

Check $m=7: 63=63$ True

The solution set is $\{7\}$.

69. $-6x+4x-7x=0$

$-9x=0$

$\frac{-9x}{-9}=\frac{0}{-9}$

$x=0$

Check $x=0: 0=0$ True

The solution set is $\{0\}$.

71. $8w-4w+w=-3$

$5w=-3$

$\frac{5w}{5}=\frac{-3}{5}$

$w=-\frac{3}{5}$

Check $w=-\frac{3}{5}: -3=-3$ True

The solution set is $\left\{-\frac{3}{5}\right\}$.

73. $\frac{1}{3}x-\frac{1}{4}x+\frac{1}{12}x=3$

$\left(\frac{1}{3}-\frac{1}{4}+\frac{1}{12}\right)x=3$ Distributive property

$\left(\frac{4}{12}-\frac{3}{12}+\frac{1}{12}\right)x=3$ LCD = 12

$\frac{1}{6}x=3$ Lowest terms

$6\left(\frac{1}{6}x\right)=6(3)$ Multiply by 6.

$x=18$

Check $x=18: 6-4.5+1.5=3$ True

The solution set is $\{18\}$.

75. $0.9w-0.5w+0.1w=-3$

$0.5w=-3$ Combine terms.

$\frac{0.5w}{0.5}=\frac{-3}{0.5}$ Divide by 0.5.

$w=-6$

Check $w=-6: -3=-3$ True

The solution set is $\{-6\}$.

77. Answers will vary. One example is $\frac{3}{2}x=-6$.

79. "When a number is multiplied by 4, the result is 6."

$4x=6$

$\frac{4x}{4}=\frac{6}{4}$

$x=\frac{3}{2}$

The number is $\frac{3}{2}$ and $\left\{\frac{3}{2}\right\}$ is the solution set.

81. "When a number is divided by -5, the result is 2."

$\frac{x}{-5}=2$

$(-5)\left(-\frac{1}{5}x\right)=(-5)(2)$

$x=-10$

The number is -10 and $\{-10\}$ is the solution set.

2.3 More on Solving Linear Equations

Now Try Exercises

N1. *Step 1* (not necessary)
Step 2
$7+2m=-3$
$7+2m-7=-3-7$ Subtract 7.
$2m=-10$ Combine terms.
Step 3
$\frac{2m}{2}=\frac{-10}{2}$ Divide by 2.
$m=-5$
Step 4
Check $m=-5$: $7-10=-3$ True
The solution set is $\{-5\}$.

N2. *Step 1* (not necessary)
Step 2
$2q+3=4q-9$
$2q+3-2q=4q-9-2q$ Subtract $2q$.
$3=2q-9$ Combine terms.
$3+9=2q-9+9$ Add 9.
$12=2q$ Combine terms.
Step 3
$\frac{12}{2}=\frac{2q}{2}$ Divide by 2.
$6=q$
Step 4
Check $q=6$: $12+3=24-9$ True
The solution set is $\{6\}$.

N3. *Step 1*
$3(z-6)-5z=-7z+7$
$3z-18-5z=-7z+7$ Distributive property
$-2z-18=-7z+7$ Combine terms.
Step 2
$-2z-18+18=-7z+7+18$ Add 18.
$-2z=-7z+25$
$-2z+7z=-7z+25+7z$ Add $7z$.
$5z=25$
Step 3
$\frac{5z}{5}=\frac{25}{5}$ Divide by 5.
$z=5$
Step 4
Check $z=5$: $-28=-28$ True
The solution set is $\{5\}$.

N4. *Step 1*
$5x-(x+9)=x-4$
$5x-x-9=x-4$ Distributive property
$4x-9=x-4$
Step 2
$4x-9+9=x-4+9$ Add 9.
$4x=x+5$
$4x-x=x+5-x$ Subtract x.
$3x=5$
Step 3
$\frac{3x}{3}=\frac{5}{3}$ Divide by 3.
$x=\frac{5}{3}$
Step 4
Check $x=\frac{5}{3}$: $-\frac{7}{3}=-\frac{7}{3}$ True
The solution set is $\left\{\frac{5}{3}\right\}$.

N5. *Step 1*
$24-4(7-2t)=4(t-1)$
$24-28+8t=4t-4$ Distributive property
$-4+8t=4t-4$
Step 2
$-4+8t+4=4t-4+4$ Add 4.
$8t=4t$
$8t-4t=4t-4t$ Subtract $4t$.
$4t=0$
Step 3
$\frac{4t}{4}=\frac{0}{4}$ Divide by 4.
$t=0$
Step 4
Check $t=0$: $24-4(7)\stackrel{?}{=}4(-1)$
$-4=-4$ True
The solution set is $\{0\}$.

N6. $-3(x-7)=2x-5x+21$
$-3x+21=-3x+21$
$-3x+21-21=-3x+21-21$ Subtract 21.
$-3x=-3x$
$-3x+3x=-3x+3x$ Add $3x$.
$0=0$ True
The variable x has "disappeared," and a true statement has resulted. The original equation is an identity. This means that for every real

number value of x, the equation is true. Thus, the solution set is {all real numbers}.

N7.

$-4x+12=3-4(x-3)$	
$-4x+12=3-4x+12$	Distr. prop.
$-4x+12=-4x+15$	Combine.
$-4x+12+4x=-4x+15+4x$	Add $4x$.
$12=15$	False

The variable x has "disappeared," and a *false* statement has resulted. This means that for every real number value of x, the equation is false. Thus, the equation has no solution and its solution set is the empty set, or null set, symbolized $\varnothing$.

N8. *Step 1*

$$\frac{1}{2}x+\frac{5}{8}x=\frac{3}{4}x-6$$

The LCD of all the fractions in the equation is 8, so multiply each side by 8 to clear the fractions.

$$8\left(\frac{1}{2}x+\frac{5}{8}x\right)=8\left(\frac{3}{4}x-6\right)$$

$$8\left(\frac{1}{2}x\right)+8\left(\frac{5}{8}x\right)=8\left(\frac{3}{4}x\right)-8(6)$$

$$4x+5x=6x-48$$

$$9x=6x-48$$

Step 2

$9x-6x=6x-48-6x$ Subtract $6x$.

$$3x=-48$$

Step 3

$\frac{3x}{3}=\frac{-48}{3}$ Divide by 3.

$$x=-16$$

Step 4

Check $x=-16$: $-8-10\stackrel{?}{=}-12-6$

$-18=-18$ True

The solution set is $\{-16\}$.

N9. *Step 1*

$$\frac{2}{3}(x+2)-\frac{1}{2}(3x+4)=-4$$

$$6\left[\frac{2}{3}(x+2)-\frac{1}{2}(3x+4)\right]=6(-4)$$

$$6\left[\frac{2}{3}(x+2)\right]-6\left[\frac{1}{2}(3x+4)\right]=-24$$

$$4(x+2)-3(3x+4)=-24$$

$$4x+8-9x-12=-24$$

$$-5x-4=-24$$

Step 2

$-5x-4+4=-24+4$ Add 4.

$-5x=-20$ Combine like terms.

Step 3

$\frac{-5x}{-5}=\frac{-20}{-5}$ Divide by -5.

$$x=4$$

Step 4

Check $x=4$: $4-8=-4$ True

The solution set is $\{4\}$.

N10. *Step 1*

$$0.05(13-t)-0.2t=0.08(30)$$

To clear decimals, multiply both sides by 100.

$$100[0.05(13-t)-0.2t]=100[0.08(30)]$$

$$5(13-t)-20t=8(30)$$

$$65-5t-20t=240$$

$$65-25t=240$$

Step 2

$$65-25t-65=240-65$$

$$-25t=175$$

Step 3

$$\frac{-25t}{-25}=\frac{175}{-25}$$

$$t=-7$$

Step 4

Check $t=-7$: $1+1.4\stackrel{?}{=}-2.4$

$2.4=2.4$ True

The solution set is $\{-7\}$.

N11. First, suppose that the sum of two numbers is 18, and one of the numbers is 10. How would you find the other number? You would subtract 10 from 18. Instead of using 10 as one of the numbers, use m. This gives us the expression $18-m$ for the other number.

Exercises

1. Use the addition property of equality to subtract 8 from each side.

3. Clear the parentheses by using the distributive property.

5. Clear fractions by multiplying by the LCD, 6.

7. **(a)** $6=6$ (The original equation is a(n) identity.) This goes with choice B, {all real numbers}.

(b) $x = 0$ (The original equation is a(n) conditional.) This goes with choice A, $\{0\}$.

(c) $-5 = 0$ (The original equation is a(n) contradiction.) This goes with choice C, $\varnothing$.

9. $3x + 2 = 14$

$3x + 2 - 2 = 14 - 2$ Subtract 2.

$3x = 12$ Combine like terms.

$\frac{3x}{3} = \frac{12}{3}$ Divide by 3.

$x = 4$

Check $x = 4$: $12 + 2 = 14$ True

The solution set is $\{4\}$.

11. $-5z - 4 = 21$

$-5z - 4 + 4 = 21 + 4$ Add 4.

$-5z = 25$ Combine like terms.

$\frac{-5z}{-5} = \frac{25}{-5}$ Divide by -5.

$z = -5$

Check $z = -5$: $25 - 4 = 21$ True

The solution set is $\{-5\}$.

13. $4p - 5 = 2p$

$4p - 5 - 4p = 2p - 4p$ Subtract $4p$.

$-5 = -2p$ Combine like terms.

$\frac{-5}{-2} = \frac{-2p}{-2}$ Divide by -2.

$\frac{5}{2} = p$

Check $p = \frac{5}{2}$: $10 - 5 = 5$ True

The solution set is $\left\{\frac{5}{2}\right\}$.

15. $2x + 9 = 4x + 11$

$-2x + 9 = 11$ Subtract $4x$.

$-2x = 2$ Subtract 9.

$x = -1$ Divide by -2.

Check $x = -1$: $7 = 7$ True

The solution set is $\{-1\}$.

17. For this equation, Step 1 is not needed.

Step 2

$5m + 8 - 8 = 7 + 3m - 8$ Subtract 8.

$5m = 3m - 1$

$5m - 3m = 3m - 1 - 3m$ Subtract $3m$.

$2m = -1$

Step 3

$\frac{2m}{2} = \frac{-1}{2}$

$m = -\frac{1}{2}$

Step 4

Substitute $-\frac{1}{2}$ for m in the original equation.

$5m + 8 = 7 + 3m$

$5\left(-\frac{1}{2}\right) + 8 \stackrel{?}{=} 7 + 3\left(-\frac{1}{2}\right)$ Let $m = -\frac{1}{2}$.

$-\frac{5}{2} + 8 \stackrel{?}{=} 7 + \left(-\frac{3}{2}\right)$

$\frac{11}{2} = \frac{11}{2}$ True

The solution set is $\left\{-\frac{1}{2}\right\}$.

19. $-12x - 5 = 10 - 7x$

$-12x - 5 + 7x = 10 - 7x + 7x$ Add $7x$.

$-5x - 5 = 10$

$-5x - 5 + 5 = 10 + 5$ Add 5.

$-5x = 15$

$\frac{-5x}{-5} = \frac{15}{-5}$ Divide by -5.

$x = -3$

Check $x = -3$: $36 - 5 = 10 + 21$ True

The solution set is $\{-3\}$.

21. $12h - 5 = 11h + 5 - h$

$12h - 5 = 10h + 5$ Combine like terms.

$2h - 5 = 5$ Subtract $10h$.

$2h = 10$ Add 5.

$h = 5$ Divide by 2.

Check $h = 5$: $55 = 55$ True

The solution set is $\{5\}$.

23. $7r - 5r + 2 = 5r + 2 - r$

$2r + 2 = 4r + 2$ Combine like terms.

$2 = 2r + 2$ Subtract $2r$.

$0 = 2r$ Subtract 2.

$0 = r$ Divide by 2.

Check $r = 0$: $2 = 2$ True

The solution set is $\{0\}$.

25. $3(4x+2)+5x=30-x$

$12x+6+5x=30-x$ Distributive prop.

$17x+6=30-x$ Combine like terms.

$18x+6=30$ Add $1x$.

$18x=24$ Subtract 6.

$x=\frac{24}{18}=\frac{4}{3}$ Divide by 18.

Check $x=\frac{4}{3}$: $\frac{86}{3}=\frac{86}{3}$ True

The solution set is $\left\{\frac{4}{3}\right\}$.

27. $-2p+7=3-(5p+1)$

$-2p+7=3-5p-1$ Distributive property

$-2p+7=-5p+2$ Combine like terms.

$3p+7=2$ Add $5p$.

$3p=-5$ Subtract 7.

$p=-\frac{5}{3}$

Check $p=-\frac{5}{3}$: $\frac{31}{3}=\frac{31}{3}$ True

The solution set is $\left\{-\frac{5}{3}\right\}$.

29. $11x-5(x+2)=6x+5$

$11x-5x-10=6x+5$

$6x-10=6x+5$

$6x-10-6x=6x+5-6x$

$-10=5$

Since $-10=5$ is a false statement, the equation has no solution set, symbolized by $\varnothing$.

31. $6(3w+5)=2(10w+10)$

$18w+30=20w+20$

$18w=20w-10$ Subtract 30.

$-2w=-10$ Subtract $20w$.

$w=5$ Divide by -2.

Check $w=5$: $120=120$ True

The solution set is $\{5\}$.

33. $-(4x+2)-(-3x-5)=3$

$-1(4x+2)-1(-3x-5)=3$

$-4x-2+3x+5=3$

$-x+3=3$

$-x=0$

$x=0$

Check $x=0$: $3=3$ True

The solution set is $\{0\}$.

35. $3(2x-4)=6(x-2)$

$6x-12=6x-12$

$-12=-12$ Subtract $6x$.

$0=0$ Add 12.

The variable has “disappeared.” Since the resulting statement is a true one, any real number is a solution. We indicate the solution set as {all real numbers}.

37. $6(4x-1)=12(2x+3)$

$24x-6=24x+36$

$-6=36$ Subtract $24x$.

The variable has "disappeared," and the resulting equation is false. Therefore, the equation has no solution set, symbolized by $\varnothing$.

39. The least common denominator of all the fractions in the equation is 10.

$$10\left(\frac{3}{5}t-\frac{1}{10}t\right)=10\left(t-\frac{5}{2}\right)$$

$$10\left(\frac{3}{5}t\right)+10\left(-\frac{1}{10}t\right)=10t+10\left(-\frac{5}{2}\right)$$

$$6t-t=10t-25$$

$$5t=10t-25$$

$$-5t=-25$$

$$\frac{-5t}{-5}=\frac{-25}{-5}$$

$$t=5$$

Check $t=5$: $\frac{5}{2}=\frac{5}{2}$ True

The solution set is $\{5\}$.

41. The least common denominator of all the fractions in the equation is 12, so multiply both sides by 12 and solve for x.

$$12\left(\frac{3}{4}x-\frac{1}{3}x+5\right)=12\left(\frac{5}{6}x\right)$$
$$9x-4x+60=10x$$
$$5x+60=10x$$
$$60=5x$$
$$\frac{60}{5}=\frac{5x}{5}$$
$$12=x$$

Check $x=12: 9-4+5=10$ True

The solution set is $\{12\}$.

43. The least common denominator of all the fractions in the equation is 35, so multiply both sides by 35 and solve for x.

$$35\left[\frac{1}{7}(3x+2)-\frac{1}{5}(x+4)\right]=35(2)$$
$$5(3x+2)-7(x+4)=70$$
$$15x+10-7x-28=70$$
$$8x-18=70$$
$$8x=88$$
$$\frac{8x}{8}=\frac{88}{8}$$
$$x=11$$

Check $x=11: 5-3=2$ True

The solution set is $\{11\}$.

45. The LCD of all the fractions is 4.

$$4\left[-\frac{1}{4}(x-12)+\frac{1}{2}(x+2)\right]=4(x+4)$$
$$4\left(-\frac{1}{4}\right)(x-12)+4\left(\frac{1}{2}\right)(x+2)=4x+16$$
$$(-1)(x-12)+2(x+2)=4x+16$$
$$-x+12+2x+4=4x+16$$
$$x+16=4x+16$$
$$-3x+16=16$$
$$-3x=0$$
$$\frac{-3x}{-3}=\frac{0}{-3}$$
$$x=0$$

Check $x=0$: $4=4$ True

The solution set is $\{0\}$.

47. The least common denominator of all the fractions in the equation is 6, so multiply both sides by 6 and solve for k.

$$6\left[\frac{2}{3}k-\left(k-\frac{1}{2}\right)\right]=6\left[\frac{1}{6}(k-51)\right]$$
$$6\left(\frac{2}{3}k\right)-6\left(k-\frac{1}{2}\right)=6\left[\frac{1}{6}(k-51)\right]$$
$$4k-6k+3=1(k-51)$$
$$-2k+3=k-51$$
$$-3k+3=-51$$
$$-3k=-54$$
$$k=18$$

Check $k=18$: $-\frac{11}{2}=-\frac{11}{2}$ True

The solution set is $\{18\}$.

49. To clear the equation of decimals, we multiply by 100.

$$100(0.75x-3.2)=100(0.55-0.5x)$$
$$75x-320=55-50x$$
$$75x-320+50x=55-50x+50x$$
$$125x-320=55$$
$$125x-320+320=55+320$$
$$125x=375$$
$$\frac{125x}{125}=\frac{375}{125}$$
$$x=3$$

Check $x=3$: $-0.95=-0.95$ True

The solution set is $\{3\}$.

51. Solve the equation for t.

$$0.8t+0.15=2t-1.35$$
$$100(0.8t+0.15)=100(2t-1.35)$$
$$80t+15=200t-135$$
$$80t+15-80t=200t-135-80t$$
$$15=120t-135$$
$$15+135=120t-135+135$$
$$150=120t$$
$$\frac{150}{120}=\frac{120t}{120}$$
$$t=\frac{5}{4}$$

Check $t=\frac{5}{4}$: $1.15=1.15$ True

The solution set is $\left\{\frac{5}{4}\right\}$.

53. To eliminate the decimal in 0.2 and 0.1, we need to multiply both sides by 10. But to eliminate the decimal in 0.05, we need to multiply by 100, so we choose 100.

$$100\left[0.2(60)+0.05x\right]=100\left[0.1(60+x)\right]$$
$$20(60)+5x=10(60+x)$$
$$1200+5x=600+10x$$
$$1200-5x=600$$
$$-5x=-600$$
$$x=\frac{-600}{-5}=120$$

Check $x=120$: $18=18$ True

The solution set is $\{120\}$.

55. $1.00x+0.05(12-x)=0.10(63)$

To clear the equation of decimals, we multiply both sides by 100.

$$100[1.00x+0.05(12-x)]=100[0.10(63)]$$
$$100x+5(12-x)=10(63)$$
$$100x+60-5x=630$$
$$95x+60=630$$
$$95x=570$$
$$x=\frac{570}{95}=6$$

Check $x=6$: $6.3=6.3$ True

The solution set is $\{6\}$.

57. $0.6(10,000)+0.8x=0.72(10,000+x)$

$$60(10,000)+80x=72(10,000+x)$$
$$600,000+80x=720,000+72x$$
$$600,000+8x=720,000$$
$$8x=120,000$$
$$x=\frac{120,000}{8}$$
$$x=15,000$$

Check $x=15,000$: $18,000=18,000$ True

The solution set is $\{15,000\}$.

59. $10(2x-1)=8(2x+1)+14$

$$20x-10=16x+8+14$$
$$20x-10=16x+22$$
$$4x-10=22$$
$$4x=32$$
$$x=8$$

Check $x=8$: $150=150$ True

The solution set is $\{8\}$.

61. $24-4(7-2t)=4(t-1)$

$$24-28+8t=4t-4$$
$$-4+8t=4t-4$$
$$-4+8t-4t=4t-4-4t$$
$$-4+4t=-4$$
$$4t=0$$
$$t=0$$

Check $t=0$: $-4=-4$ True

The solution set is $\{0\}$.

63. $4(x+8)=2(2x+6)+20$

$$4x+32=4x+12+20$$
$$4x+32=4x+32$$
$$4x=4x$$
$$0=0$$

Since $0=0$ is a true statement, the solution set is {all real numbers}.

65. To clear fractions, multiply both sides by the LCD, which is 4.

$$4\left[\frac{1}{2}(x+2)+\frac{3}{4}(x+4)\right]=4(x+5)$$
$$4\left(\frac{1}{2}\right)(x+2)+4\left(\frac{3}{4}\right)(x+4)=4x+20$$
$$2(x+2)+3(x+4)=4x+20$$
$$2x+4+3x+12=4x+20$$
$$5x+16=4x+20$$
$$x+16=20$$
$$x=4$$

Check $x=4$: $9=9$ True

The solution set is $\{4\}$.

67. To eliminate the decimals, multiply both sides by 10.

$$10[0.1(x+80)+0.2x]=10(14)$$
$$1(x+80)+2x=140$$
$$x+80+2x=140$$
$$3x+80=140$$
$$3x=60$$
$$x=20$$

Check $x=20$: $14=14$ True

The solution set is $\{20\}$.

69. $9(v+1)-3v=2(3v+1)-8$
$$9v+9-3v=6v+2-8$$
$$6v+9=6v-6$$
$$9=-6$$
Because $9=-6$ is a false statement, the equation has no solution set, symbolized by $\varnothing$.

71. The sum of q and the other number is 11. To find the other number, you would subtract q from 11, so an expression for the other number is $11-q$.

73. The product of x and the other number is 9. To find the other number, you would divide 9 by x, so an expression for the other number is $\frac{9}{x}$.

75. If a baseball player gets 65 hits in one season, and h of the hits are in one game, then $65-h$ of the hits came in the rest of the games.

77. If Monica is x years old now, then 15 years from now she will be $x+15$ years old. Five years ago, she was $x-5$ years old.

79. Since the value of each quarter is 25 cents, the value of r quarters is $25r$ cents.

81. Since each bill is worth 5 dollars, the number of bills is $\frac{t}{5}$.

83. Since each adult ticket costs x dollars, the cost of 3 adult tickets is $3x$. Since each child's ticket costs y dollars, the cost of 2 children's tickets is $2y$. Therefore, the total cost is $3x+2y$ (dollars).

Summary Exercises Applying Methods for Solving Linear Equations

1. This is an equation since it has an equals sign.
$x+2=-3$
$x=-5$ Subtract 2.
Check $x=-5$: $-3=-3$ True
The solution set is $\{-5\}$.

2. This is an expression since it does not have an equals sign.
$4p-6+3p-8=7p-14$

3. This is an expression since it does not have an equals sign.
$$-(m-1)-(3+2m)=-m+1-3-2m$$
$$=-3m-2$$

4. This is an equation since it has an equals sign.
$$6q-9=12+3q$$
$$3q-9=12$$
$$3q=21$$
$$q=7$$
Check $q=7$: $33=33$ True
The solution set is $\{7\}$.

5. This is an equation since it has an equals sign.
$5x-9=3(x-3)$
$5x-9=3x-9$ Distributive property
$2x-9=-9$ Subtract $3x$.
$2x=0$ Add 9.
$x=0$ Divide by 2.
Check $x=0$: $-9=-9$ True
The solution set is $\{0\}$.

6. This is an equation since it has an equals sign. To clear fractions, multiply both sides by the LCD, which is 12.
$$12\left(\frac{2}{3}x+8\right)=12\left(\frac{1}{4}x\right)$$
$$8x+96=3x$$
$$5x+96=0$$
$$5x=-96$$
$$x=-\frac{96}{5}$$
Check $x=-\frac{96}{5}$: $-\frac{24}{5}=-\frac{24}{5}$ True
The solution set is $\left\{-\frac{96}{5}\right\}$.

7. This is an expression since it does not have an equals sign.
$$2-6(z+1)-4(z-2)-10$$
$$=2-6z-6-4z+8-10$$
$$=-10z-6$$

8. This is an equation since it has an equals sign.
$$7(p-2)+p=2(p+2)$$
$$7p-14+p=2p+4$$
$$8p-14=2p+4$$
$$6p-14=4$$
$$6p=18$$
$$p=3$$
Check $p=3$: $10=10$ True
The solution set is $\{3\}$.

9. This is an expression since it does not have an equals sign.

$$\frac{1}{2}(x+10)-\frac{2}{3}x=\frac{1}{2}x+5-\frac{2}{3}x$$
$$=\frac{3}{6}x+5-\frac{4}{6}x$$
$$=-\frac{1}{6}x+5$$

10. This is an expression since it does not have an equals sign.

$$-4(k+2)+3(2k-1)=-4k-8+6k-3$$
$$=2k-11$$

11. $-6z=-14$

$z=\frac{-14}{-6}$ Divide by -6.

$=\frac{7}{3}$

Check $z=\frac{7}{3}$: $-14=-14$ True

The solution set is $\left\{\frac{7}{3}\right\}$.

12. $2m+8=16$

$2m=8$ Subtract 8.

$m=4$ Divide by 2.

Check $m=4$: $16=16$ True

The solution set is $\{4\}$.

13. $12.5x=-63.75$

$x=\frac{-63.75}{12.5}$ Divide by 12.5.

$=-5.1$

Check $x=-5.1$: $-63.75=-63.75$ True

The solution set is $\{-5.1\}$.

14. $-x=-12$

$x=12$ Multiply by -1.

Check $x=12$: $-12=-12$ True

The solution set is $\{12\}$.

15. $\frac{4}{5}x=-20$

$x=\left(\frac{5}{4}\right)(-20)$ Multiply by $\frac{5}{4}$.

$=-25$

Check $x=-25$: $-20=-20$ True

The solution set is $\{-25\}$.

16. $7m-5m=-12$

$2m=-12$ Combine like terms.

$m=-6$ Divide by 2.

Check $m=-6$: $-12=-12$ True

The solution set is $\{-6\}$.

17. $-x=6$

$x=-6$ Multiply by -1.

Check $x=-6$: $6=6$ True

The solution set is $\{-6\}$.

18. $\frac{x}{-2}=8$

$-\frac{1}{2}x=8$

$x=-2(8)$ Multiply by -2.

$=-16$

Check $x=-16$: $8=8$ True

The solution set is $\{-16\}$.

19. $4x+2(3-2x)=6$

$4x+6-4x=6$

$6=6$

Since $6=6$ is a true statement, the solution set is {all real numbers}.

20. $x-16.2=7.5$

$x=23.7$ Add 16.2.

Check $x=23.7$: $7.5=7.5$ True

The solution set is $\{23.7\}$.

21. $7m-(2m-9)=39$

$7m-2m+9=39$

$5m+9=39$

$5m=30$

$m=6$

Check $m=6$: $39=39$ True

The solution set is $\{6\}$.

22. $2-(m+4)=3m-2$

$2-m-4=3m-2$

$-m-2=3m-2$

$-4m-2=-2$

$-4m=0$

$m=\frac{0}{-4}=0$

Check $m=0$: $-2=-2$ True

The solution set is $\{0\}$.

23. $-3(m-4)+2(5+2m)=29$

$-3m+12+10+4m=29$

$m+22=29$

$m=7$

Check $m=7$: $29=29$ True

The solution set is $\{7\}$.

24. To eliminate the decimals, multiply both sides by 10.

$10[-0.3x+2.1(x-4)]=10(-6.6)$

$-3x+21(x-4)=-66$

$-3x+21x-84=-66$

$18x-84=-66$

$18x=18$

$x=1$

Check $x=1$: $-0.3-6.3=-6.6$ True

The solution set is $\{1\}$.

25. To eliminate the decimals, multiply both sides by 100.

$100[0.08x+0.06(x+9)]=100(1.24)$

$8x+6(x+9)=124$

$8x+6x+54=124$

$14x+54=124$

$14x=70$

$x=5$

Check $x=5$: $0.4+0.84=1.24$ True

The solution set is $\{5\}$.

26. $3(m+5)-1+2m=5(m+2)$

$3m+15-1+2m=5m+10$

$5m+14=5m+10$

$14=10$

Because $14=10$ is a false statement, the equation has no solution set, symbolized by $\varnothing$.

27. $-2t+5t-9=3(t-4)-5$

$-2t+5t-9=3t-12-5$

$3t-9=3t-17$

$-9=-17$

Because $-9=-17$ is a false statement, the equation has no solution set, symbolized by $\varnothing$.

28. To eliminate the decimals, multiply both sides by 10.

$10[2.3x+13.7]=10[1.3x+2.9]$

$23x+137=13x+29$

$10x+137=29$

$10x=-108$

$x=-10.8$

Check $x=-10.8$: $-11.14=-11.14$ True

The solution set is $\{-10.8\}$.

29. To eliminate the decimals, multiply both sides by 10.

$10[0.2(50)+0.8r]=10[0.4(50+r)]$

$2(50)+8r=4(50+r)$

$100+8r=200+4r$

$100+4r=200$

$4r=100$

$r=25$

Check $r=25$: $10+20=30$ True

The solution set is $\{25\}$.

30. $r+9+7r=4(3+2r)-3$

$8r+9=12+8r-3$

$8r+9=8r+9$

$9=9$

Since $9=9$ is a true statement, the solution set is $\{\text{all real numbers}\}$.

31. $2(3+7x)-(1+15x)=2$

$6+14x-1-15x=2$

$-x+5=2$

$-x=-3$

$x=3$

Check $x=3$: $48-46=2$ True

The solution set is $\{3\}$.

32. To eliminate the decimals, multiply both sides by 10.

$10[0.6(100-x)+0.4x]=10[0.5(92)]$

$6(100-x)+4x=5(92)$

$600-6x+4x=460$

$600-2x=460$

$-2x=-140$

$x=70$

Check $x=70$: $18+28=46$ True

The solution set is $\{70\}$.

33. To clear fractions, multiply both sides by the LCD, which is 4.

$$4\left(\frac{1}{4}x-4\right)=4\left(\frac{3}{2}x+\frac{3}{4}x\right)$$
$$x-16=6x+3x$$
$$x-16=9x$$
$$-16=8x$$
$$x=-2$$

Check $x=-2$: $-4.5=-3-1.5$ True

The solution set is $\{-2\}$.

34. To clear fractions, multiply both sides by the LCD, which is 12.

$$12\left[\frac{3}{4}(z-2)-\frac{1}{3}(5-2z)\right]=12(-2)$$
$$9(z-2)-4(5-2z)=-24$$
$$9z-18-20+8z=-24$$
$$17z-38=-24$$
$$17z=14$$
$$z=\frac{14}{17}$$

Check $z=\frac{14}{17}$: $-\frac{15}{17}-\frac{19}{17}=-2$ True

The solution set is $\left\{\frac{14}{17}\right\}$.

2.4 Applications of Linear Equations

Now Try Exercises

N1. *Step 2*

Let $x=$ the number.

Step 3

The product of 7, and	a number increased by 3,	is	−63.
↓	↓	↓	↓
$7\cdot$	$(x+3)$	$=$	-63

Step 4

Solve the equation.

$7(x+3)=-63$

$7x+21=-63$ Distributive property

$7x=-84$ Subtract 21.

$x=-12$ Divide by 7.

Step 5

The number is -12.

Step 6

-12 plus 3 is -9, times 7 is -63, so -12 is the number.

N2. *Step 2*

Let $x=$ the number.

Step 3

If 5 is	added to	a number,	the result is	7 less than 3 times the number.
↓	↓	↓	↓	↓
5	+	x	=	$3x-7$

Step 4

Solve the equation.

$5+x=3x-7$

$5+x-5=3x-7-5$ Subtract 5.

$x=3x-12$

$x-3x=3x-12-3x$ Subtract $3x$.

$-2x=-12$

$\frac{-2x}{-2}=\frac{-12}{-2}$ Divide by -2.

$x=6$

Step 5

The number is 6.

Step 6

5 added to 6 is 11. 3 times 6 is 18, and 7 less than 18 is 11, so 6 is the number.

N3. *Step 2*

Let $x=$ the number of medals Great Britain won.

Let $x-21=$ the number of medals Germany won.

Step 3

The total	is	the number of medals Great Britain won	plus	the number of medals Germany won.
↓	↓	↓	↓	↓
109	=	x	+	$(x-21)$

Step 4

Solve the equation.

$x+(x-21)=109$

$2x-21=109$ Combine like terms.

$2x=130$ Add 21.

$x=65$ Divide by 2.

Step 5

Great Britain won 65 medals and Germany won $65-21=44$ medals.

Step 6
44 is 21 less than 65, and the sum of 65 and 44 is 109.

N4. *Step 2*
Let $x =$ the number of orders for chocolate scones.
Then $\frac{2}{3}x =$ the number of orders for bagels.

Step 3

The total	is	orders for chocolate scones	plus	orders for bagels.
↓	↓	↓	↓	↓
525	=	x	+	$\frac{2}{3}x$

Step 4
Solve the equation.

$$525 = 1x + \frac{2}{3}x \quad x = 1x$$
$$525 = \frac{3}{3}x + \frac{2}{3}x \quad \text{LCD} = 3$$
$$525 = \frac{5}{3}x \quad \text{Combine like terms.}$$
$$\frac{3}{5}(525) = \frac{3}{5}\left(\frac{5}{3}x\right) \quad \text{Multiply by } \frac{3}{5}.$$
$$315 = x$$

Step 5
The number of orders for chocolate scones was 315, so the number of orders for bagels was $\frac{2}{3}(315) = 210$.

Step 6
Two-thirds of 315 is 210, and the sum of 315 and 210 is 525.

N5. *Step 2*
Let $x =$ the number of residents.
Then $4x =$ the number of guests.
(If each resident brought four guests, there would be four times as many guests as residents.)

Step 3

Number of residents	plus	number of guests	is	the total in attendance.
↓	↓	↓	↓	↓
x	+	$4x$	=	175

Step 4
Solve the equation.

$$x + 4x = 175$$
$$5x = 175$$
$$\frac{5x}{5} = \frac{175}{5}$$
$$x = 35$$

Step 5
There were 35 residents and $4 \cdot 35 = 140$ guests.

Step 6
140 is four times as much as 35, and the sum of 35 and 140 is 175.

N6. *Step 2*
Let $x =$ the time spent practicing free throws.
Then $2x =$ the time spent lifting weights and $x + 2 =$ the time spent watching game films.

Step 3

Free throws	plus	lifting weights	plus	watching films	is	total time.
↓	↓	↓	↓	↓	↓	↓
x	+	$2x$	+	$x+2$	=	6

Step 4
Solve the equation.

$$x + 2x + (x + 2) = 6$$
$$4x + 2 = 6$$
$$4x = 4$$
$$\frac{4x}{4} = \frac{4}{4}$$
$$x = 1$$

Step 5
The time spent practicing free throws is 1 hour, the time spent lifting weights is $2(1) = 2$ hours, and the time spent watching game films is $1 + 2 = 3$ hours.

Step 6
Since 2 hours is twice as much time as 1 hour, 3 hours is 2 more hours than 1 hour, and the sum of the times is $1 + 2 + 3 = 6$ hours (the total time spent), the answers are correct.

N7. *Step 2*
Let $x =$ the lesser page number.
Then $x+1 =$ the greater page number.
Step 3
Because the sum of the page numbers is 593, an equation is $x+(x+1)=593$.
Step 4
Solve the equation.
$2x+1=593$ Combine like terms.
$2x=592$ Subtract 1.
$x=296$ Divide by 2.
Step 5
The lesser page number is 296, and the greater page number is $296+1=297$.
Step 6
297 is one more than 296, and the sum of 296 and 297 is 593.

N8. Let $x =$ the lesser odd integer.
Then $x+2 =$ the greater odd integer.
From the given information, we have
$2\cdot x+3(x+2)=191$.
Solve the equation.
$2x+3x+6=191$
$5x+6=191$
$5x=185$
$x=37$
The lesser odd integer is 37 and the greater consecutive odd integer is $37+2=39$. Two times 37 is 74, three times 39 is 117, and 74 plus 117 is 191.

N9. Let $x =$ the degree measure of the angle.
Then $90-x =$ the degree measure of its complement.

The complement	is	twice the angle.
↓	↓	↓
$90-x$	$=$	$2x$

Solve the equation.
$90-x=2x$
$90=3x$
$30=x$
The measure of the angle is 30°.

N10. *Step 2*
Let $x =$ the degree measure of the angle.
Then $90-x =$ the degree measure of its complement, and $180-x =$ the degree measure of its supplement.
Step 3

The supplement	equals	46° less than 3 times its complement.
↓	↓	↓
$180-x$	$=$	$3\cdot(90-x)-46$

Step 4
Solve the equation.
$180-x=3(90-x)-46$
$180-x=270-3x-46$ Distributive prop.
$180-x=224-3x$ Combine terms.
$180+2x=224$ Add $3x$.
$2x=44$ Subtract 180.
$x=22$ Divide by 2.
Step 5
The measure of the angle is 22°.
Step 6
The complement of 22° is $90°-22°=68°$, and 46° less than 3 times 68° is
$3(68°)-46°=204°-46°=158°$.
The supplement is $180°-22°=158°$.

Exercises

1. Choice D, $6\frac{1}{2}$, is *not* a reasonable answer in an applied problem that requires finding the number of cars on a dealer's lot, since you cannot have $\frac{1}{2}$ of a car. The number of cars must be a whole number.

3. Choice A, -10, is *not* a reasonable answer since distance cannot be negative.

5. *Step 2*
Let $x =$ the number.
Step 3

The sum of a number and	9	is	−26.
↓	↓	↓	↓
$x+$	9	$=$	-26

Step 4
Solve the equation.
$x+9=-26$
$x=-35$

Step 5
The number is -35.
Step 6
The sum of -35 and 9 is -26.

7. *Step 2*
Let $x =$ the number.
Step 3

The product of 8,	and a number increased by 6,	is	104.
↓	↓	↓	↓
$8\cdot$	$(x+6)$	$=$	104

Step 4
Solve the equation.

$$8(x+6)=104$$
$$8x+48=104$$
$$8x=56$$
$$x=7$$

Step 5
The number is 7.
Step 6
Check that 7 is the correct answer by substituting this result into the words of the original problem. 7 increased by 6 is 13. The product of 8 and 13 is 104, so 7 is the number.

9. *Step 2*
Let $x =$ the unknown number. Then $5x+2$ represents "2 is added to five times a number," and $4x+5$ represents "5 more than four times the number."
Step 3
$5x+2=4x+5$

Step 4
Solve the equation.

$$5x+2=4x+5$$
$$x+2=5$$
$$x=3$$

Step 5
The number is 3.
Step 6
Check that 3 is the correct answer by substituting this result into the words of the original problem. Two added to five times a number is $2+5(3)=17$ and 5 more than four times the number is $5+4(3)=17$. The values are equal, so the number 3 is the correct answer.

11. *Step 2*
Let $x =$ the unknown number. Then $x-2$ is two subtracted from the number, $3(x-2)$ is triple the difference, and $x+6$ is six more than the number.
Step 3
$3(x-2)=x+6$
Step 4
Solve the equation.

$$3x-6=x+6$$
$$2x-6=6$$
$$2x=12$$
$$x=6$$

Step 5
The number is 6.
Step 6
Check that 6 is the correct answer by substituting this result into the words of the original problem. Two subtracted from the number is $6-2=4$. Triple this difference is $3(4)=12$, which is equal to 6 more than the number, since $6+6=12$.

13. *Step 2*

Let $x =$ the unknown number. Then $\frac{3}{4}x$ is $\frac{3}{4}$ of the number, and $\frac{3}{4}x+6$ is 6 added to $\frac{3}{4}$ of the number. $x-4$ is 4 less than the number.
Step 3
$\frac{3}{4}x+6=x-4$
Step 4
Solve the equation.

$$\frac{3}{4}x+6=x-4$$
$$-\frac{1}{4}x+6=-4$$
$$-\frac{1}{4}x=-10$$
$$-4\left(-\frac{1}{4}x\right)=-4(-10)$$
$$x=40$$

Step 5
The number is 40.

Step 6
Check that 40 is the correct answer by substituting this result into the words of the original problem. $\frac{3}{4}$ of 40 is 30. When 6 is added to 30, the sum is 36, which is 4 less than 40 because 6 is added to 30.

15. *Step 2*
Let $x =$ the unknown number. Then $3x$ is three times the number, $x+7$ is 7 more than the number, $2x$ is twice the number, and $-11-2x$ is the difference between -11 and twice the number.
Step 3
$3x+(x+7)=-11-2x$
Step 4
Solve the equation.
$$4x+7=-11-2x$$
$$6x+7=-11$$
$$6x=-18$$
$$x=-3$$
Step 5
The number is -3.
Step 6
Check that -3 is the correct answer by substituting this result into the words of the original problem. The sum of three times a number and 7 more than the number is $3(-3)+(-3+7)=-5$ and the difference between -11 and twice the number is $-11-2(-3)=-5$. The values are equal, so the number -3 is the correct answer.

17. *Step 1*
We must find the number of Democrats and the number of Republicans.
Step 2
Let $x =$ the number of Republicans.
Then $x-4 =$ the number of Democrats.
Step 3

Number of Democrats	+	number of Republicans	equals	total members.
↓	↓	↓	↓	↓
$(x-4)$	+	x	=	150

Step 4
Solve the equation.
$$(x-4)+x=150$$
$$2x-4=150$$
$$2x=154$$
$$x=77$$
Step 5
There were 77 Republicans and $77-4=73$ Democrats.
Step 6
Check that the numbers found are the correct answers by substituting the result into the words of the original problem. 73 is 4 fewer than 77, and $77+73=150$.

19. Let $x =$ the number of drive-in movie screens in Ohio.
Then $x+2 =$ the number of drive-in movie screens in New York.
Since the total number of screens was 56, we can write the equation $x+(x+2)=56$.
Solve the equation.
$$2x+2=56$$
$$2x=54$$
$$x=27$$
Since $x=27$, $x+2=29$.
There were 27 drive-in movie screens in Ohio and 29 in New York. Since 29 is 2 more than 27 and $27+29=56$, this answer checks.

21. Let $x =$ the number of Republicans.
Then $x+6 =$ the number of Democrats.
Since the total number of Democrats and Republicans was 98, we can write the equation $x+(x+6)=98$.
Solve the equation.
$$2x+6=98$$
$$2x=92$$
$$x=46$$
Since $x=46$, $x+6=52$.
There were 46 Republicans and 52 Democrats. Since $46+52=98$, this answer checks.

23. Let $x =$ revenue from ticket sales for Madonna. Then $x - 29 =$ the revenue from the ticket sales for Bruce Springsteen.
Since the total revenue from ticket sales was \$427 (all numbers in millions), we can write the equation $x + (x - 29) = 427$.
Solve the equation.

$$2x - 29 = 427$$
$$2x = 456$$
$$x = 228$$

Since $x = 228$, $x - 29 = 199$.

Madonna took in \$228 million and Bruce Springsteen took in \$199 million. Since 199 is 29 less than 228 and $228 + 199 = 427$, this answer checks.

25. Let $x =$ the number of games the Heat lost. Then $4x + 2 =$ the number of games the Heat won.
Since the total number of games played was 82, we can write the equation $x + (4x + 2) = 82$.
Solve the equation.

$$5x + 2 = 82$$
$$5x = 80$$
$$x = 16$$

Since $x = 16$, $4x + 2 = 66$.

The Heat won 66 games and lost 16 games. Since $66 + 16 = 82$, this answer checks.

27. Let $x =$ the number of mg of vitamin C in a one-cup serving of pineapple juice.
Then $4x - 3 =$ the number of mg of vitamin C in a one-cup serving of orange juice.
Since the total amount of vitamin C in a serving of the two juices is 122 mg, we can write $x + (4x - 3) = 122$.
Solve the equation.

$$5x - 3 = 122$$
$$5x = 125$$
$$x = 25$$

Since $x = 25$, $4x - 3 = 97$.

A one-cup serving of pineapple juice has 25 mg of vitamin C and a one-cup serving of orange juice has 97 mg of vitamin C. Since 97 is 3 less than four times 25 and $25 + 97 = 122$, this answer checks.

29. Let $x =$ the number of Blu-Ray discs sold.

Then $\frac{2}{3}x =$ the number of DVDs sold.

The total number of Blu-Rays and DVDs sold was 280, so $x + \frac{2}{3}x = 280$.

Solve the equation.

$$1x + \frac{2}{3}x = 280$$
$$\frac{5}{3}x = 280$$
$$\frac{3}{5}\left(\frac{5}{3}x\right) = \frac{3}{5}(280)$$
$$x = \frac{3}{\cancel{5}_{1}} \cdot \frac{\cancel{280}^{56}}{1} = 168$$

Since $x = 168$, $\frac{2}{3}x = \frac{2}{3}(168) = 112$.

There were 112 DVDs sold.

31. Let $x =$ the number of kg of onions.
Then $6.6x =$ the number of kg of grilled steak.
The total weight of these two ingredients was 617.6 kg, so
$x + 6.6x = 617.6$.
Solve the equation.

$$1x + 6.6x = 617.6$$
$$7.6x = 617.6$$
$$x = \frac{617.6}{7.6} \approx 81.3$$

Since $x = \frac{617.6}{7.6}$, $6.6x = 6.6\left(\frac{617.6}{7.6}\right) \approx 536.3$.

To the nearest tenth of a kilogram, 81.3 kg of onions and 536.3 kg of grilled steak were used to make the taco.

33. Let $x =$ the value of the 1945 nickel.
Then $2x =$ the value of the 1950 nickel.
The total value of the two coins is \$24.00, so
$x + 2x = 24$.
Solve the equation.

$$3x = 24$$
$$x = 8 \quad \text{Divide by 8.}$$

Since $x = 8$, $2x = 2(8) = 16$.

The value of the 1945 Philadelphia nickel is \$8.00 and the value of the 1950 Denver nickel is \$16.00.

35. Let $x =$ the number of ounces of rye flour.
Then $4x =$ the number of ounces of whole-wheat flour.
The total number of ounces would be 32, so $x + 4x = 32.$
Solve the equation.
$5x = 32$
$x = \frac{32}{5} = 6.4$
Since $x = 6.4, 4x = 4(6.4) = 25.6.$ To make a loaf of bread weighing 32 oz, use 6.4 oz of rye flour and 25.6 oz of whole-wheat flour.

37. Let $x =$ the number of tickets booked on United Airlines.
Then $x + 7 =$ the number of tickets booked on American Airlines, and $2x + 4 =$ the number of tickets booked on Southwest Airlines.
The total number of tickets booked was 55, so $x + (x + 7) + (2x + 4) = 55.$
Solve the equation.
$4x + 11 = 55$
$4x = 44$
$x = \frac{44}{4} = 11$
Since $x = 11, x + 7 = 11 + 7 = 18$ and $2x + 4 = 2(11) + 4 = 26.$ He booked 18 tickets on American, 11 tickets on United, and 26 tickets on Southwest.

39. Let $x =$ the length of the shortest piece.
Then $x + 5 =$ the length of the middle piece, and $x + 9 =$ the length of the longest piece.
The total length is 59 inches, so $x + (x + 5) + (x + 9) = 59.$
Solve the equation.
$3x + 14 = 59$
$3x = 45$
$x = 15$
Since $x = 15, x + 5 = 20$ and $x + 9 = 24.$
The shortest piece should be 15 inches, the middle piece should be 20 inches, and the longest piece should be 24 inches. The answer checks since $15 + 20 + 24 = 59.$

41. Let $x =$ the number of silver medals.
Then $x =$ the number of bronze medals, and $x + 17 =$ the number of gold medals.
The total number of medals earned by the United States was 104, so $x + x + (x + 17) = 104.$
Solve the equation.
$3x + 17 = 104$
$3x = 87$
$x = 29$
Since $x = 29, \ x + 17 = 46.$
The United States earned 46 gold medals, 29 silver medals, and 29 bronze medals. The answer checks since $46 + 29 + 29 = 104.$

43. Let $x =$ the distance of Mercury from the sun.
Then $x + 31.2 =$ the distance of Venus from the sun, and $x + 57 =$ the distance of Earth from the sun.
Since the total of the distances from these three planets is 196.2 (all distances in millions of miles), we can write the equation $x + (x + 31.2) + (x + 57) = 196.2.$
Solve the equation.
$3x + 88.2 = 196.2$
$3x = 108$
$x = 36$
Mercury is 36 million miles from the sun, Venus is $36 + 31.2 = 67.2$ million miles from the sun, and Earth is $36 + 57 = 93$ million miles from the sun. The answer checks since $36 + 67.2 + 93 = 196.2.$

45. Let $x =$ the measure of angles A and B.
Then $x + 60 =$ the measure of angle C.
The sum of the measures of the angles of any triangle is 180°, so $x + x + (x + 60) = 180.$
Solve the equation.
$3x + 60 = 180$
$3x = 120$
$x = 40$
Angles A and B have measures of 40 degrees, and angle C has a measure of $40 + 60 = 100$ degrees. The answer checks since $40 + 40 + 100 = 180.$

47. Let $x =$ the number on the first locker.
Then $x + 1 =$ the number on the next locker.
Since the numbers have a sum of 137, we can write the equation $x + (x + 1) = 137.$
Solve the equation.
$2x + 1 = 137$
$2x = 136$
$x = \frac{136}{2} = 68$
Since $x = 68, x + 1 = 69.$
The lockers have numbers 68 and 69. Since $68 + 69 = 137,$ this answer checks.

49. Because the two pages are back-to-back, they must have page numbers that are consecutive integers.
Let $x =$ the lesser page number.
Then $x+1 =$ the greater page number.

$$x+(x+1)=203$$
$$2x+1=203$$
$$2x=202$$
$$x=101$$

Since $x=101$, $x+1=102$.

The page numbers are 101 and 102. This answer checks since the sum is 203.

51. Let $x =$ the lesser even integer.
Then $x+2 =$ the greater even integer.
"The lesser added to three times the greater gives a sum of 46" can be written as
$x+3(x+2)=46$.

$$x+3x+6=46$$
$$4x+6=46$$
$$4x=40$$
$$x=10$$

Since $x=10$, $x+2=12$.

The integers are 10 and 12. This answer checks since $10+3(12)=46$.

53. Let $x =$ the lesser odd integer.
Then $x+2 =$ the greater odd integer.
"59 more than the lesser is 4 times the greater" can be written as $x+59=4(x+2)$.

$$x+59=4x+8$$
$$59=3x+8$$
$$51=3x$$
$$17=x$$

Since $x=17$, $x+2=19$.

The integers are 17 and 19. This answer checks since $17+59=76$ and $4\cdot 19=76$.

55. Let $x =$ the lesser integer.
Then $x+1 =$ the greater integer.

$$x+3(x+1)=43$$
$$x+3x+3=43$$
$$4x+3=43$$
$$4x=40$$
$$x=10$$

Since $x=10$, $x+1=11$.

The integers are 10 and 11. This answer checks since $10+3(11)=43$.

57. Let $x =$ the first even integer.
Then $x+2 =$ the second even integer, and
$x+4 =$ the third even integer.

$$x+(x+2)+(x+4)=60$$
$$3x+6=60$$
$$3x=54$$
$$x=18$$

Since $x=18$, $x+2=20$ and $x+4=22$.

The first even integer is 18. This answer checks since $18+20+22=60$.

59. Let $x =$ the first odd integer.
Then $x+2 =$ the second odd integer, and
$x+4 =$ the third odd integer.

$$2[(x+4)-6]=[x+2(x+2)]-23$$
$$2(x-2)=x+2x+4-23$$
$$2x-4=3x-19$$
$$-4=x-19$$
$$15=x$$

Since $x=15$, $x+2=17$ and $x+4=19$.

The integers are 15, 17, and 19.

61. Let $x =$ the measure of the angle.
Then $90-x =$ the measure of its complement.
The phrase "complement is four times its measure" can be written as $90-x=4x$.
Solve the equation.

$$90=5x$$
$$x=\frac{90}{5}=18$$

The measure of the angle is 18°. The complement is $90°-18°=72°$, which is four times 18°.

63. Let $x =$ the measure of the angle.
Then $180-x =$ the measure of its supplement.
The phrase "supplement is eight times its measure" can be written as $180-x=8x$.
Solve the equation.

$$180=9x$$
$$x=\frac{180}{9}=20$$

The measure of the angle is 20°. The supplement is $180°-20°=160°$, which is eight times 20°.

65. Let $x =$ the measure of the angle. Then $90 - x =$ the measure of its complement, and $180 - x =$ the measure of its supplement.

Its supplement	measures	39°	more than	twice its complement.
↓	↓	↓	↓	↓
$180-x$	$=$	39	$+$	$2(90-x)$

Solve the equation.

$$180 - x = 39 + 2(90 - x)$$
$$180 - x = 39 + 180 - 2x$$
$$180 - x = 219 - 2x$$
$$x + 180 = 219$$
$$x = 39$$

The measure of the angle is 39°. The complement is $90° - 39° = 51°$. Now 39° more than twice its complement is $39° + 2(51°) = 141°$, which is the supplement of 39° since $180° - 39° = 141°$.

67. Let $x =$ the measure of the angle. Then $180 - x =$ the measure of its supplement, and $90 - x =$ the measure of its complement. Remember that difference means subtraction.

The supplement	minus	3 times its complement	is	10°.
↓	↓	↓	↓	↓
$(180-x)$	$-$	$3(90-x)$	$=$	10

Solve the equation.

$$(180 - x) - 3(90 - x) = 10$$
$$180 - x - 270 + 3x = 10$$
$$2x - 90 = 10$$
$$2x = 100$$
$$x = 50$$

The measure of the angle is 50°. The supplement is $180° - 50° = 130°$ and the complement is $90° - 50° = 40°$. The answer checks since $130° - 3(40°) = 10°$.

2.5 Formulas and Additional Applications from Geometry

Now Try Exercises

N1. $P = 2a + 2b$

$78 = 2(12) + 2b$ Let $P = 78$ and $a = 12$.

$78 = 24 + 2b$

$54 = 2b$ Subtract 24.

$27 = b$ Divide by 2.

N2. The fence will enclose the perimeter of the rectangular garden, so use the formula for the perimeter of a rectangle. Find the width of the garden by substituting $P = 160$ and $L = 2W - 10$ into the formula and solving for W.

$$P = 2L + 2W$$
$$160 = 2(2W - 10) + 2W$$
$$160 = 4W - 20 + 2W$$
$$180 = 6W$$
$$30 = W$$

Since $W = 30$, $L = 2(30) - 10 = 50$. The dimensions of the garden are 50 ft by 30 ft.

N3. Let $s =$ the length of the medium side, in feet; $s + 1 =$ the length of the longest side, and $s - 7 =$ the length of the shortest side. The perimeter is 30 feet, so $s + (s + 1) + (s - 7) = 30$.

$$3s - 6 = 30$$
$$3s = 36$$
$$s = 12$$

Since $s = 12$, $s + 1 = 13$ and $s - 7 = 5$. The lengths of the sides are 5, 12, and 13 feet. The perimeter is $5 + 12 + 13 = 30$, as required.

N4. Use the formula for the area of a triangle.

$A = \frac{1}{2}bh$

$77 = \frac{1}{2}(14)h$ Let $A = 77$, $h = 14$.

$77 = 7h$

$11 = h$

The height is 11 centimeters.

N5. Since the marked angles are vertical angles, they have equal measures.

$$6x+2=8x-8$$
$$2=2x-8$$
$$10=2x$$
$$5=x$$

If $x=5$, $6x+2=6(5)+2=32$ and $8x-8=8(5)-8=32$.

The measure of the angles is $32°$.

N6. Solve $W=Fd$ for F.

$$\frac{W}{d}=\frac{Fd}{d} \quad \text{Divide by } d.$$
$$\frac{W}{d}=F, \text{ or } F=\frac{W}{d}$$

N7. Solve $Ax+By=C$ for A.

$$Ax=C-By \quad \text{Subtract } By.$$
$$\frac{Ax}{x}=\frac{C-By}{x} \quad \text{Divide by } x.$$
$$A=\frac{C-By}{x}$$

N8. Solve $S=\frac{1}{2}(a+b+c)$ for a.

$$2S=a+b+c \quad \text{Multiply by 2.}$$
$$2S-b-c=a \quad \text{Subtract } b \text{ and } c.$$

N9. **(a)** Solve $5x+y=3$ for y.

$$5x+y-5x=3-5x \quad \text{Subtract } 5x.$$
$$y=3-5x$$

(b) Solve $x-2y=8$ for y.

$$x-2y-x=8-x \quad \text{Subtract } x.$$
$$y=\frac{8-x}{-2} \quad \text{Divide by } -2.$$
$$y=-4+\frac{1}{2}x, \text{ or } y=\frac{1}{2}x-4$$

Exercises

1. Carpeting for a bedroom covers the surface of the bedroom floor, so area would be used.

3. To measure fencing for a yard, use perimeter since you would need to measure the lengths of the sides of the yard.

5. Tile for a bathroom covers the surface of the bathroom floor, so area would be used.

7. To determine the cost of replacing a linoleum floor with a wood floor, use area since you need to know the measure of the surface covered by the wood.

9. $P=2L+2W$; $L=8, W=5$

$$P=2L+2W$$
$$=2(8)+2(5)$$
$$=16+10$$
$$P=26$$

11. $A=\frac{1}{2}bh$; $b=8$, $h=16$

$$A=\frac{1}{2}bh$$
$$=\frac{1}{2}(8)(16)$$
$$A=64$$

13. $P=a+b+c$; $P=12$, $a=3$, $c=5$

$$P=a+b+c$$
$$12=3+b+5$$
$$12=b+8$$
$$4=b$$

15. $d=rt$; $d=252$, $r=45$

$$d=rt$$
$$252=45t$$
$$\frac{252}{45}=\frac{45t}{45}$$
$$5.6=t$$

17. $A=\frac{1}{2}h(b+B)$; $A=91$, $h=7$, $b=12$

$$A=\frac{1}{2}h(b+B)$$
$$91=\frac{1}{2}(7)(12+B)$$
$$182=(7)(12+B)$$
$$12+B=\frac{1}{7}(182)$$
$$B=26-12=14$$

19. $C=2\pi r$; $C=16.328$, $\pi=3.14$

$$C=2\pi r$$
$$16.328=2(3.14)r$$
$$16.328=6.28r$$
$$2.6=r$$

21. $C = 2\pi r;\ C = 20\pi$

$C = 2\pi r$

$20\pi = 2\pi r$

$10 = r$ Divide by 2π.

23. $A = \pi r^2;\ r = 4,\ \pi = 3.14$

$A = \pi r^2$

$= 3.14(4)^2$

$= 3.14(16)$

$A = 50.24$

25. $S = 2\pi rh;\ S = 120\pi,\ h = 10$

$S = 2\pi rh$

$120\pi = 2\pi r(10)$

$120\pi = 20\pi r$

$6 = r$ Divide by 20π.

27. $V = LWH;\ L = 10,\ W = 5,\ H = 3$

$V = LWH$

$= (10)(5)(3)$

$V = 150$

29. $V = \frac{1}{3}Bh;\ B = 12,\ h = 13$

$V = \frac{1}{3}Bh$

$= \frac{1}{3}(12)(13)$

$V = 52$

31. $V = \frac{4}{3}\pi r^3;\ r = 12,\ \pi = 3.14$

$V = \frac{4}{3}\pi r^3$

$= \frac{4}{3}(3.14)(12)^3$

$= \frac{4}{3}(3.14)(1728)$

$V = 7234.56$

33. $I = prt;\ p = \$7500,\ r = 4\%,\ t = 2$ yr

$I = prt$

$= (\$7500)(0.04)(2)$

$= \$600$

35. $I = prt;\ I = \$33,\ r = 2\%,\ t = 3$ yr

$I = prt$

$\$33 = p(0.02)(3)$

$\$33 = (p)(0.06)$

$\$550 = p$

37. $I = prt;\ I = \$180,\ p = \$4800,\ r = 2.5\%$

$I = prt$

$\$180 = (\$4800)(0.025)(t)$

$\$180 = (\$120)(t)$

$1.5 \text{ yr} = t$

39. $P = 2L + 2W$

$54 = 2(W + 9) + 2W$ Let $L = W + 9$.

$54 = 2W + 18 + 2W$

$54 = 4W + 18$

$36 = 4W$

$9 = W$

The width is 9 inches and the length is $9 + 9 = 18$ inches.

41. $P = 2L + 2W$

$36 = 2(3W + 2) + 2W$ Let $L = 3W + 2$.

$36 = 6W + 4 + 2W$

$36 = 8W + 4$

$32 = 8W$

$4 = W$

The width is 4 meters and the length is $3(4) + 2 = 14$ meters.

43. Let $s =$ the length of the shortest side, in inches; $s + 2 =$ the length of the medium side, and $s + 3 =$ the length of the longest side. The perimeter is 20 inches, so $s + (s + 2) + (s + 3) = 20$.

$3s + 5 = 20$

$3s = 15$

$s = 5$

Since $s = 5$, $s + 2 = 7$ and $s + 3 = 8$. The lengths of the sides are 5, 7, and 8 inches. The perimeter is $5 + 7 + 8 = 20$, as required.

45. Let $s =$ the length of the two sides that have equal length, in meters, and $2s - 4 =$ the length of the third side.
The perimeter is 24 meters, so
$$s + s + (2s - 4) = 24.$$
$$4s - 4 = 24$$
$$4s = 28$$
$$s = 7$$
Since $s = 7$, $2s - 4 = 10$. The lengths of the sides are 7, 7, and 10 meters. The perimeter is $7 + 7 + 10 = 24$, as required.

47. The page is a rectangle with length 1.5 m and width 1.2 m, so use the formulas for the perimeter and area of a rectangle.
$$P = 2L + 2W$$
$$= 2(1.5)+2(1.2)$$
$$= 3+2.4$$
$$P = 5.4 \text{ meters}$$
$$A = LW$$
$$= (1.5)(1.2)$$
$$A = 1.8 \text{ square meters}$$

49. Use the formula for the area of a triangle with $A = 70$ and $b = 14$.
$$A = \frac{1}{2}bh$$
$$70 = \frac{1}{2}(14)h$$
$$70 = 7h$$
$$10 = h$$
The height of the sign is 10 feet.

51. The diameter of the circle is 443 feet, so its radius is $\frac{443}{2} = 221.5$ ft. Use the area of a circle formula to find the enclosed area.
$$A = \pi r^2$$
$$= \pi(221.5)^2$$
$$\approx 154{,}133.6 \text{ ft}^2,$$
or about 154,000 ft^2. (If 3.14 is used for π, the value is 154,055.465.)

53. To find the area of the drum face, use the formula for the area of a circle, $A = \pi r^2$.
$$A = \pi r^2$$
$$\approx (3.14)(7.87)^2$$
$$= (3.14)(61.9369)$$
$$A \approx 194.48$$
The area of the drum face is about 194.48 square feet.
Use the circumference of a circle formula.
$$C \approx 2(3.14)(7.87)$$
$$= 49.4236$$
The circumference is about 49.42 feet.

55. Use the formula for the area of a trapezoid with $B = 115.80$, $b = 171.00$, and $h = 165.97$.
$$A = \frac{1}{2}(B + b)h$$
$$= \frac{1}{2}(115.80 + 171.00)(165.97)$$
$$= \frac{1}{2}(286.80)(165.97)$$
$$= 23{,}800.098$$
To the nearest hundredth of a square foot, the combined area of the two lots is 23,800.10 square feet.

57. The girth is $4 \cdot 18 = 72$ inches. Since the length plus the girth is 108, we have
$$L + G = 108$$
$$L + 72 = 108$$
$$L = 36 \text{ in.}$$
The volume of the box is
$$V = LWH$$
$$= (36)(18)(18)$$
$$= 11{,}664 \text{ in.}^3$$

59. The two angles are supplementary, so the sum of their measures is 180°.
$$(x + 1) + (4x - 56) = 180$$
$$5x - 55 = 180$$
$$5x = 235$$
$$x = 47$$
Since $x = 47$, $x + 1 = 47 + 1 = 48$ and $4x - 56 = 4(47) - 56 = 132$.
The measures of the angles are 48° and 132°.

61. In the figure, the two angles are complementary, so their sum is 90°.
$$(8x - 1) + 5x = 90$$
$$13x - 1 = 90$$
$$13x = 91$$
$$x = 7$$
Since $x = 7$, $8x - 1 = 8(7) - 1 = 55$ and $5x = 5(7) = 35$.
The two angle measures are 55° and 35°.

63. The two angles are vertical angles, which have equal measures. Set their measures equal to each other and solve for x.

$$5x-129=2x-21$$
$$3x-129=-21$$
$$3x=108$$
$$x=36$$

Since $x=36$, $5x-129=5(36)-129=51$ and $2x-21=2(36)-21=51$.

The measure of each angle is $51°$.

65. The angles are vertical angles, so their measures are equal.

$$12x-3=10x+15$$
$$2x-3=15$$
$$2x=18$$
$$x=9$$

Since $x=9, 12x-3=12(9)-3=105$ and $10x+15=10(9)+15=105$.

The measure of each angle is $105°$.

67. Solve $d=rt$ for t.

$\frac{d}{r}=\frac{rt}{r}$ Divide by r.

$\frac{d}{r}=t$, or $t=\frac{d}{r}$

69. Solve $A=bh$ for b.

$\frac{A}{h}=\frac{bh}{h}$ Divide by h.

$\frac{A}{h}=b$, or $b=\frac{A}{h}$

71. Solve $C=\pi d$ for d.

$\frac{C}{\pi}=\frac{\pi d}{\pi}$ Divide by π.

$\frac{C}{\pi}=d$, or $d=\frac{C}{\pi}$

73. Solve $V=LWH$ for H.

$\frac{V}{LW}=\frac{LWH}{LW}$ Divide by LW.

$\frac{V}{LW}=H$, or $H=\frac{V}{LW}$

75. Solve $I=prt$ for r.

$\frac{I}{pt}=\frac{prt}{pt}$ Divide by pt.

$\frac{I}{pt}=r$, or $r=\frac{I}{pt}$

77. Solve $A=\frac{1}{2}bh$ for h.

$2A=2\left(\frac{1}{2}bh\right)$ Multiply by 2.

$2A=bh$

$\frac{2A}{b}=\frac{bh}{b}$ Divide by b.

$\frac{2A}{b}=h$, or $h=\frac{2A}{b}$

79. Solve $V=\frac{1}{3}\pi r^2h$ for h.

$3V=3\left(\frac{1}{3}\right)\pi r^2h$ Multiply by 3.

$3V=\pi r^2h$

$\frac{3V}{\pi r^2}=\frac{\pi r^2h}{\pi r^2}$ Divide by πr^2.

$\frac{3V}{\pi r^2}=h$, or $h=\frac{3V}{\pi r^2}$

81. Solve $P=a+b+c$ for b.

$P-a-c=a+b+c-a-c$ Subtract a and c.

$P-a-c=b$, or $b=P-a-c$

83. Solve $P=2L+2W$ for W.

$P-2L=2L+2W-2L$ Subtract $2L$.

$P-2L=2W$

$\frac{P-2L}{2}=\frac{2W}{2}$ Divide by 2.

$\frac{P-2L}{2}=W$, or $W=\frac{P-2L}{2}$

85. Solve $y=mx+b$ for m.

$y-b=mx+b-b$ Subtract b.

$y-b=mx$

$\frac{y-b}{x}=\frac{mx}{x}$ Divide by x.

$\frac{y-b}{x}=m$, or $m=\frac{y-b}{x}$

87. Solve $Ax+By=C$ for y.

$By=C-Ax$ Subtract Ax.

$\frac{By}{B}=\frac{C-Ax}{B}$ Divide by B.

$y=\frac{C-Ax}{B}$

89. Solve $M = C(1+r)$ for r.

$M = C + Cr$ Distributive property

$M - C = Cr$ Subtract C.

$\frac{M-C}{C} = \frac{Cr}{C}$ Divide by C.

$\frac{M-C}{C} = r$, or $r = \frac{M-C}{C}$

Alternative solution:

$M = C(1+r)$

$\frac{M}{C} = 1 + r$ Divide by C.

$\frac{M}{C} - 1 = r$ Subtract 1.

91. Solve $P = 2(a+b)$ for a.

$P = 2a + 2b$ Distributive property

$P - 2b = 2a$ Subtract $2b$.

$\frac{P-2b}{2} = \frac{2a}{2}$ Divide by 2.

$\frac{P-2b}{2} = a$

93. Solve $S = \frac{1}{2}(a+b+c)$ for b.

$2S = a + b + c$ Multiply by 2.

$2S - a - c = b$ Subtract a and c.

95. Solve $C = \frac{5}{9}(F-32)$ for F.

$C = \frac{5}{9}(F-32)$

$\frac{9}{5}C = \frac{9}{5}\cdot\frac{5}{9}(F-32)$

$\frac{9}{5}C = F - 32$

$\frac{9}{5}C + 32 = F - 32 + 32$

$\frac{9}{5}C + 32 = F$

97. Solve $6x + y = 4$ for y.

$6x + y - 6x = 4 - 6x$ Subtract $6x$.

$y = 4 - 6x$

99. Solve $5x - y = 2$ for y.

$5x - y - 5x = 2 - 5x$ Subtract $5x$.

$\frac{-y}{-1} = \frac{2-5x}{-1}$ Divide by -1.

$y = -2 + 5x$, or $y = 5x - 2$

101. Solve $-3x + 5y = -15$ for y.

$-3x + 5y + 3x = -15 + 3x$ Add $3x$.

$\frac{5y}{5} = \frac{-15+3x}{5}$ Divide by 5.

$y = -3 + \frac{3}{5}x$, or $y = \frac{3}{5}x - 3$

103. Solve $x - 3y = 12$ for y.

$x - 3y - x = 12 - x$ Subtract x.

$\frac{-3y}{-3} = \frac{12-x}{-3}$ Divide by -3.

$y = -4 + \frac{1}{3}x$, or $y = \frac{1}{3}x - 4$

2.6 Ratio, Proportion, and Percent

Now Try Exercises

N1. **(a)** The ratio of 7 inches to 4 inches is
$\frac{7 \text{ inches}}{4 \text{ inches}} = \frac{7}{4}.$

(b) 45 seconds $= \frac{45}{60} = \frac{3}{4}$ minute

The ratio of 45 seconds to 2 minutes is then
$\frac{45 \text{ seconds}}{2 \text{ minutes}} = \frac{3}{4} \div 2 = \frac{3}{4}\cdot\frac{1}{2} = \frac{3}{8}.$

N2. The results in the following table are rounded to the nearest thousandth.

Size	**Unit Cost** (dollars per oz)
75 oz	$\frac{\$8.94}{75} = \$0.119$ (*)
100 oz	$\frac{\$13.97}{100} = \$0.140$
150 oz	$\frac{\$19.97}{150} = \$0.133$

Because the 75-oz size produces the lowest unit cost, it is the best buy. The unit cost, to the nearest thousandth, is \$0.119 per oz.

N3. **(a)** $\frac{1}{3} = \frac{33}{100}$

Compare the cross products.

$1 \cdot 100 = 100$

$3 \cdot 33 = 99$

The cross products are *different*, so the proportion is *false*.

(b) $\frac{4}{13} = \frac{16}{52}$

Check to see whether the cross products are equal.

$4 \cdot 52 = 208$

$13 \cdot 16 = 208$

The cross products are *equal*, so the proportion is *true*.

N4. $\frac{9}{7} = \frac{x}{56}$

$7x = 9 \cdot 56$ Cross products

$x = \frac{9 \cdot 56}{7}$ Divide by 7.

$= \frac{9 \cdot 7 \cdot 8}{7}$ Factor.

$= 72$ Cancel.

The solution set is $\{72\}$.

Note: We could have multiplied $9 \cdot 56$ to get 504 and then divided 504 by 7 to get 72. This may be the best approach if you are doing these calculations on a calculator. The factor and cancel method is preferable if you're not using a calculator.

N5. $\frac{k-3}{6} = \frac{3k+2}{4}$

$4(k-3) = 6(3k+2)$ Cross products

$4k - 12 = 18k + 12$ Distributive property

$-14k - 12 = 12$ Subtract $18k$.

$-14k = 24$ Add 12.

$k = -\frac{24}{14}$ Divide by -14.

$= -\frac{12}{7}$

The solution set is $\left\{-\frac{12}{7}\right\}$.

N6. Let $x =$ the cost for 27 gallons.

$\frac{\$69.80}{20} = \frac{x}{27}$

$20x = 27(69.80)$ Cross products

$20x = 1884.60$ Multiply.

$x = 94.23$ Divide by 20.

It would cost $94.23.

N7. **(a)** What is 20% of 70?

$20\% \cdot 70 = 0.20 \cdot 70 = 14$

(b) 40% of what number is 130?

As in Example 7(b), let n denote the number.

$0.40 \cdot n = 130$

$n = \frac{130}{0.40}$ Divide by 0.40.

$= 325$ Simplify.

40% of 325 is 130.

(c) 121 is what percent of 484?

As in Example 7(c), let p denote the percent.

$121 = p \cdot 484$

$p = \frac{121}{484} = 0.25 = 25\%$

121 is 25% of 484.

N8. We can think of this problem as "48 is what percent of 120?"

Let p denote the percent.

$48 = p \cdot 120$

$p = \frac{48}{120} = 0.40 = 40\%$

The coat costs 40% of the regular price, so the savings is $100\% - 40\% = 60\%$ of the regular price.

Exercises

1. **(a)** 75 to 100 is $\frac{75}{100} = \frac{3}{4}$, or 3 to 4.

The answer is C.

(b) 5 to 4, or $\frac{5}{4} = \frac{5 \cdot 3}{4 \cdot 3} = \frac{15}{12}$, or 15 to 12.

The answer is D.

(c) $\frac{1}{2} = \frac{1 \cdot 50}{2 \cdot 50} = \frac{50}{100}$, or 50 to 100.

The answer is B.

(d) 4 to 5, or $\frac{4}{5} = \frac{4 \cdot 20}{5 \cdot 20} = \frac{80}{100}$, or 80 to 100.

The answer is A.

3. The ratio of 40 miles to 30 miles is

$\frac{40 \text{ miles}}{30 \text{ miles}} = \frac{40}{30} = \frac{4}{3}$.

5. The ratio of 120 people to 90 people is

$\frac{120 \text{ people}}{90 \text{ people}} = \frac{4 \cdot 30}{3 \cdot 30} = \frac{4}{3}$.

7. To find the ratio of 20 yards to 8 feet, first convert 20 yards to feet.

$$20 \text{ yards} = 20 \text{ yards} \cdot \frac{3 \text{ feet}}{1 \text{ yard}} = 60 \text{ feet}$$

The ratio of 20 yards to 8 feet is then

$$\frac{60 \text{ feet}}{8 \text{ feet}} = \frac{60}{8} = \frac{15 \cdot 4}{2 \cdot 4} = \frac{15}{2}.$$

9. Convert 2 hours to minutes.

$$2 \text{ hours} = 2 \text{ hours} \cdot \frac{60 \text{ minutes}}{1 \text{ hour}}$$
$$= 120 \text{ minutes}$$

The ratio of 24 minutes to 2 hours is then

$$\frac{24 \text{ minutes}}{120 \text{ minutes}} = \frac{24}{120} = \frac{1 \cdot 24}{5 \cdot 24} = \frac{1}{5}.$$

11. $2 \text{ yards} = 2 \cdot 3 = 6 \text{ feet}$

$6 \text{ feet} = 6 \cdot 12 = 72 \text{ inches}$

The ratio of 60 inches to 2 yards is then

$$\frac{60 \text{ inches}}{72 \text{ inches}} = \frac{5 \cdot 12}{6 \cdot 12} = \frac{5}{6}.$$

13. Find the unit price for each size.

Size	Price	Unit Cost (dollars per lb)
4 lb	$3.29	$\frac{\$3.29}{4} = \$0.823$
10 lb	$7.49	$\frac{\$7.49}{10} = \$0.749$ (*)

The 10-lb size is the best buy.

15. Find the unit price for each size.

Size	Price	Unit Cost (dollars per lb)
64 oz	$2.99	$\frac{\$2.99}{64} \approx \$0.047$ (*)
89 oz	$4.79	$\frac{\$4.79}{89} \approx \$0.054$
128 oz	$6.49	$\frac{\$6.49}{128} \approx \$0.051$

The 64-oz size is the best buy.

17. Find the unit price for each size.

Size	Price	Unit Cost (dollars per lb)
8.5 oz	$5.79	$\frac{\$5.79}{8.5} \approx \$0.681$
12.5 oz	$7.99	$\frac{\$7.99}{12.5} \approx \$0.639$
32 oz	$16.99	$\frac{\$16.99}{32} \approx \$0.531$ (*)

The 32-oz size is the best buy.

19. Find the unit price for each size.

Size	Price	Unit Cost (dollars per lb)
32 oz	$1.79	$\frac{\$1.79}{32} \approx \$0.056$ (*)
36 oz	$2.69	$\frac{\$2.69}{36} \approx \$0.075$
40 oz	$2.49	$\frac{\$2.49}{40} \approx \$0.062$
64 oz	$4.38	$\frac{\$4.38}{64} \approx \$0.068$

The 32-oz size is the best buy.

21. Find the unit price for each size.

Size	Price	Unit Cost (dollars per lb)
87 oz	$7.88	$\frac{\$7.88}{87} \approx \$0.091$
131 oz	$10.98	$\frac{\$10.98}{131} \approx \$0.084$
263 oz	$19.96	$\frac{\$19.96}{263} \approx \$0.076$ (*)

The 263-oz size is the best buy.

23. Check to see whether the cross products are equal.

$5 \cdot 56 = 280$

$35 \cdot 8 = 280$

The cross products are *equal*, so the proportion is *true*.

25. Check to see whether the cross products are equal.

$120 \cdot 10 = 1200$

$82 \cdot 7 = 574$

The cross products are *different*, so the proportion is *false*.

27. Check to see whether the cross products are equal.

$$\frac{1}{2}\cdot 10 = 5$$
$$5\cdot 1 = 5$$

The cross products are *equal*, so the proportion is *true*.

29.
$$20k = 4(175) \quad \text{Cross products}$$
$$20k = 700$$
$$\frac{20k}{20} = \frac{700}{20} \quad \text{Divide by 20.}$$
$$k = 35$$

The solution set is $\{35\}$.

31.
$$\frac{49}{56} = \frac{z}{8}$$
$$56z = 49(8) \quad \text{Cross products}$$
$$56z = 392$$
$$\frac{56z}{56} = \frac{392}{56} \quad \text{Divide by 56.}$$
$$z = 7$$

The solution set is $\{7\}$.

33.
$$\frac{x}{24} = \frac{15}{16}$$
$$16x = 24(15) \quad \text{Cross products}$$
$$16x = 360$$
$$\frac{16x}{16} = \frac{360}{16} \quad \text{Divide by 16.}$$
$$x = \frac{45\cdot 8}{2\cdot 8} = \frac{45}{2}$$

The solution set is $\left\{\frac{45}{2}\right\}$.

35.
$$\frac{z}{2} = \frac{z+1}{3}$$
$$3z = 2(z+1) \quad \text{Cross products}$$
$$3z = 2z + 2 \quad \text{Distributive property}$$
$$z = 2 \quad \text{Subtract } 2z.$$

The solution set is $\{2\}$.

37.
$$\frac{3y-2}{5} = \frac{6y-5}{11}$$
$$11(3y-2) = 5(6y-5) \quad \text{Cross products}$$
$$33y - 22 = 30y - 25 \quad \text{Distributive property}$$
$$3y - 22 = -25 \quad \text{Subtract } 30y.$$
$$3y = -3 \quad \text{Add 22.}$$
$$y = -1 \quad \text{Divide by 3.}$$

The solution set is $\{-1\}$.

39.
$$\frac{5k+1}{6} = \frac{3k-2}{3}$$
$$3(5k+1) = 6(3k-2) \quad \text{Cross products}$$
$$15k + 3 = 18k - 12 \quad \text{Distributive property}$$
$$-3k + 3 = -12 \quad \text{Subtract } 18k.$$
$$-3k = -15 \quad \text{Subtract 3.}$$
$$k = 5 \quad \text{Divide by } -3.$$

The solution set is $\{5\}$.

41.
$$\frac{2p+7}{3} = \frac{p-1}{4}$$
$$4(2p+7) = 3(p-1) \quad \text{Cross products}$$
$$8p + 28 = 3p - 3 \quad \text{Distributive property}$$
$$5p + 28 = -3 \quad \text{Subtract } 3p.$$
$$5p = -31 \quad \text{Subtract 28.}$$
$$p = -\frac{31}{5} \quad \text{Divide by 5.}$$

The solution set is $\left\{-\frac{31}{5}\right\}$.

43.
$$\frac{2(x-4)}{3} = \frac{4(x-3)}{5}$$
$$\frac{2x-8}{3} = \frac{4x-12}{5} \quad \text{Distributive property}$$
$$5(2x-8) = 3(4x-12) \quad \text{Cross products}$$
$$10x - 40 = 12x - 36 \quad \text{Dist. prop.}$$
$$-40 = 2x - 36 \quad \text{Subtract } 10x.$$
$$-4 = 2x \quad \text{Add 36.}$$
$$x = \frac{-4}{2} = -2 \quad \text{Divide by 2.}$$

The solution set is $\{-2\}$.

45. Let $x =$ the cost of 24 candy bars.
Set up a proportion.

$$\frac{x}{24} = \frac{\$20.00}{16}$$
$$16x = 24(20)$$
$$16x = 480$$
$$x = 30$$

The cost of 24 candy bars is \$30.00.

47. Let $x =$ the cost of 5 quarts of oil.
Set up a proportion.

$$\frac{x}{5} = \frac{\$14.00}{8}$$
$$8x = 5(14)$$
$$8x = 70$$
$$x = 8.75$$

The cost of 5 quarts of oil is \$8.75.

49. Let $x =$ the cost of 5 pairs of jeans.

$$\frac{9 \text{ pairs}}{\$121.50} = \frac{5 \text{ pairs}}{x}$$
$$9x = 5(121.50)$$
$$9x = 607.5$$
$$\frac{9x}{9} = \frac{607.5}{9}$$
$$x = 67.5$$

The cost of 5 pairs is \$67.50.

51. Let $x =$ the cost for filling a 15-gallon tank.
Set up a proportion.

$$\frac{x \text{ dollars}}{\$22.56} = \frac{15 \text{ gallons}}{6 \text{ gallons}}$$
$$6x = 15(22.56)$$
$$6x = 338.4$$
$$x = 56.4$$

It would cost \$56.40 to completely fill a 15-gallon tank.

53. Let $x =$ the number of fish in North Bay.
Set up a proportion with one ratio involving the sample and the other involving the total number of fish.

$$\frac{7 \text{ fish}}{700 \text{ fish}} = \frac{500 \text{ fish}}{x \text{ fish}}$$
$$7x = (700)(500)$$
$$7x = 350,000$$
$$x = 50,000$$

We estimate that there are 50,000 fish in North Bay.

55. Let $x =$ the distance between Memphis and Philadelphia on the map (in feet).
Set up a proportion with one ratio involving map distances and the other involving actual distances.

$$\frac{x \text{ feet}}{2.4 \text{ feet}} = \frac{1000 \text{ miles}}{600 \text{ miles}}$$
$$\frac{x}{2.4} = \frac{1000}{600}$$
$$600x = (2.4)(1000)$$
$$600x = 2400$$
$$x = 4$$

The distance on the map between Memphis and Philadelphia would be 4 feet.

57. Let $x =$ the number of inches between St. Louis and Des Moines on the map.
Set up a proportion.

$$\frac{8.5 \text{ inches}}{x \text{ inches}} = \frac{1040 \text{ miles}}{333 \text{ miles}}$$
$$1040x = 8.5(333)$$
$$1040x = 2830.5$$
$$x \approx 2.72$$

St. Louis and Des Moines are about 2.7 inches apart on the map.

59. Let $x =$ the number of inches between Moscow and Berlin on the globe.
Set up a proportion.

$$\frac{12.4 \text{ inches}}{x \text{ inches}} = \frac{10,080 \text{ km}}{1610 \text{ km}}$$
$$10,080x = 12.4(1610)$$
$$10,080x = 19,964$$
$$x \approx 1.98$$

Moscow and Berlin are about 2.0 inches apart on the globe.

61. Let $x =$ the number of cups of cleaner.
Set up a proportion with one ratio involving the number of cups of cleaner and the other involving the number of gallons of water.

$$\frac{x \text{ cups}}{\frac{1}{4} \text{ cup}} = \frac{10\frac{1}{2} \text{ gallons}}{1 \text{ gallons}}$$
$$x \cdot 1 = \frac{1}{4}\left(10\frac{1}{2}\right)$$
$$x = \frac{1}{4}\left(\frac{21}{2}\right) = \frac{21}{8}$$

The amount of cleaner needed is $2\frac{5}{8}$ cups.

63. Let $x =$ the number of U.S. dollars Ashley exchanged.
Set up a proportion.
$$\frac{\$1.3492}{x \text{ dollars}} = \frac{1 \text{ euro}}{300 \text{ euros}}$$
$$x \cdot 1 = 1.3492(300)$$
$$x = 404.76$$
She exchanged \$404.76.

65. $\frac{x}{12} = \frac{3}{9}$
$$9x = 12 \cdot 3 = 36$$
$$x = 4$$
Other possibilities for the proportion are
$\frac{12}{x} = \frac{9}{3}$, $\frac{x}{12} = \frac{5}{15}$, and $\frac{12}{x} = \frac{15}{5}$.

67. $\frac{x}{2} = \frac{12}{3}$
$$3x = 2 \cdot 12 = 24$$
$$x = 8$$

69. $\frac{x}{15} = \frac{12}{8}$ $\qquad$ $\frac{y}{17} = \frac{12}{8}$
$$8x = 15(12) = 180 \qquad 8y = 17(12) = 204$$
$$x = 22.5 \qquad y = 25.5$$

71. (a)

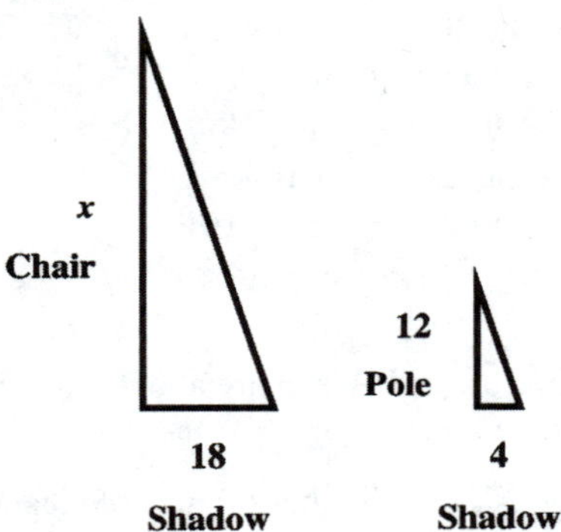

(b) These two triangles are similar, so their sides are proportional.
$$\frac{x}{12} = \frac{18}{4}$$
$$4x = 18(12)$$
$$4x = 216$$
$$x = 54$$
The chair is 54 feet tall.

73. Let $x =$ the 2001 price of electricity.
$$\frac{1999 \text{ price}}{1999 \text{ index}} = \frac{2001 \text{ price}}{2001 \text{ index}}$$
$$\frac{225}{166.6} = \frac{x}{177.1}$$
$$166.6x = 225(177.1)$$
$$x = \frac{225(177.1)}{166.6} \approx \$239$$
The 2001 price would be about \$239.

75. Let $x =$ the 2007 price of electricity.
$$\frac{1999 \text{ price}}{1999 \text{ index}} = \frac{2007 \text{ price}}{2007 \text{ index}}$$
$$\frac{225}{166.6} = \frac{x}{207.3}$$
$$166.6x = 225(207.3)$$
$$x = \frac{225(207.3)}{166.6} \approx \$280$$
The 2007 price would be about \$280.

77. (a) Find the total amount of medication by multiplying.
$$(375 \text{ mg/day})(7 \text{ days}) = 2625 \text{ mg}$$

(b) Let $x =$ the number of mL of suspension.
$$\frac{\text{mg of Amoxil}}{\text{mL of suspension}} = \frac{\text{total mg of Amoxil}}{\text{total mL of suspension}}$$
$$\frac{125 \text{ mg}}{5 \text{ mL}} = \frac{2625 \text{ mg}}{x}$$

(c) Solve the proportion.
$$\frac{125 \text{ mg}}{5 \text{ mL}} = \frac{2625 \text{ mg}}{x}$$
$$125x = 5(2625)$$
$$x = \frac{5(2625)}{125} = 105 \text{ mL}$$
Logan's pharmacist will make 105 mL of Amoxil suspension for the total course of treatment.

79. What is 18% of 780?
$$18\% \cdot 780 = 0.18 \cdot 780 = 140.4$$

81. 42% of what number is 294?
As in Example 7(b), let n denote the number.
$$0.42 \cdot n = 294$$
$$n = \frac{294}{0.42} \quad \text{Divide by 0.42.}$$
$$= 700 \quad \text{Simplify.}$$
42% of 700 is 294.

83. 120% of what number is 510?
As in Example 7(b), let n denote the number.
$1.20 \cdot n = 510$

$$n = \frac{510}{1.20} \quad \text{Divide by 1.20.}$$
$$= 425 \quad \text{Simplify.}$$

120% of 425 is 510.

85. 4 is what percent of 50?
As in Example 7(c), let p denote the percent.
$4 = p \cdot 50$

$$p = \frac{4}{50} = 0.08 = 8\%$$

4 is 8% of 50.

87. What percent of 30 is 36?
As in Example 7(c), let p denote the percent.
$36 = p \cdot 30$

$$p = \frac{36}{30} = 1.2 = 120\%$$

36 is 120% of 30.

89. 48 is what percent of 60?
As in Example 7(c), let p denote the percent.
$48 = p \cdot 60$

$$p = \frac{48}{60} = 0.08 = 80\%$$

He earned 80% of the total points.

91. Find the amount of the savings.
$\$700 - \$504 = \$196$
What percent of \$700 is \$196?
As in Example 7(b), let n denote the number.
$n \cdot 700 = 196$

$$n = \frac{196}{700} \quad \text{Divide by 700.}$$
$$= 0.28 \quad \text{Simplify.}$$

$0.28 = 28(0.01) = 28 \cdot 1\% = 28\%$

93. What percent of \$1500 is \$480?
As in Example 7(b), let n denote the number.
$n \cdot 1500 = 480$

$$n = \frac{480}{1500} \quad \text{Divide by 1500.}$$
$$= 0.32 \quad \text{Simplify.}$$

$0.32 = 32(0.01) = 32 \cdot 1\% = 32\%$
Tyler pays 32% of his income in rent.

95. 65% of what number is 1950?
As in Example 7(b), let n denote the number.
$0.65 \cdot n = 1950$

$$n = \frac{1950}{0.65} \quad \text{Divide by 0.65.}$$
$$= 3000 \quad \text{Simplify.}$$

She needs \$3000 for the car.

2.7 Further Applications of Linear Equations

Now Try Exercises

N1. **(a)** The amount of pure alcohol in 70 L of a 20% alcohol solution is

70 L	×	0.20	= 14 L.
↑		↑	↑
Amount of solution		Rate of concentration	Amount of pure alcohol

(b) If \$3200 is invested for one year at 2% simple interest, the amount of interest earned is

\$3200	×	0.02	= \$64.
↑		↑	↑
Principal		Interest rate	Interest earned

N2. Let $x =$ the number of ounces of seasoning that is 70% salt.
Then $x + 30 =$ the number of ounces of seasoning that is 50% salt.

Salt in 70% seasoning	plus	salt in 10% seasoning	is	salt in 50% seasoning.
↓	↓	↓	↓	↓
$0.70x$	$+$	$0.10(30)$	$=$	$0.50(x+30)$

Solve the equation.
$0.70x + 0.10(30) = 0.50x + 15$
Multiply by 10 to clear decimals.

$$7x + 1(30) = 5x + 150$$
$$7x + 30 = 5x + 150$$
$$2x + 30 = 150$$
$$2x = 120$$
$$x = 60$$

60 ounces of seasoning that is 70% salt is needed.

Check $x = 60$:

LS and RS refer to the left side and right side of the original equation.

LS: $0.70(60) + 0.10(30) = 42 + 3 = 45$

RS: $0.50(60 + 30) = 0.50(90) = 45$

N3. Let $x =$ the number of liters of 25% saline solution.

Then $15 - x =$ the number of liters of 10% saline solution.

25% solution	plus	10% solution	is	15% solution.
↓	↓	↓	↓	↓
$0.25x$	$+$	$0.1(15 - x)$	$=$	$0.15(15)$

Solve the equation.

$$0.25x + 0.1(15 - x) = 0.15(15)$$

Multiply by 100 to clear decimals.

$$25x + 10(15 - x) = 15(15)$$
$$25x + 150 - 10x = 225$$
$$15x + 150 = 225$$
$$15x = 75$$
$$x = 5$$

There needs to be 5 liters of the 25% saline solution.

Check $x = 5$:

LS and RS refer to the left side and right side of the original equation.

LS: $25(5) + 10(15 - (5)) = 125 + 100 = 225$

RS: $15(15) = 225$

N4. Let $x =$ the amount invested at 3%.

Then $x + 5000 =$ the amount invested at 4%.

Amount Invested (in dollars)	Rate of Interest	Interest for One Year
x	0.03	$0.03x$
$x + 5000$	0.04	$0.04(x + 5000)$

Since the total annual interest is \$410, the equation is $0.03x + 0.04(x + 5000) = 410$.

Multiply by 100 to eliminate the decimals.

$$3x + 4(x + 5000) = 100(410)$$
$$3x + 4x + 20{,}000 = 41{,}000$$
$$7x + 20{,}000 = 41{,}000$$
$$7x = 21{,}000$$
$$x = 3000$$

The financial advisor should invest \$3000 at 3% and \$8000 at 4%.

N5. Let $x =$ the number of dimes.

Then $x + 10 =$ the number of quarters.

The value of dimes	plus	the value of quarters	is	\$5.65.
↓	↓	↓	↓	↓
$0.10x$	$+$	$0.25(x + 10)$	$=$	\$5.65

Multiply by 100 to eliminate the decimals.

$$10x + 25(x + 10) = 565$$
$$10x + 25x + 250 = 565$$
$$35x + 250 = 565$$
$$35x = 315$$
$$x = 9$$

Since $x = 9$, $x + 10 = 19$.

Clayton has 19 quarters and 9 dimes.

N6. $r = \frac{d}{t} = \frac{400 \text{ miles}}{6 \text{ hours}} \approx 66.6667$

The rate was about 66.67 miles per hour.

N7. Let $t =$ the time it takes for the bicyclists to be 5 miles apart. Use the formula $d = rt$.

$$d_{\text{faster}} - d_{\text{slower}} = d_{\text{total}}$$
$$20t - 18t = 5$$
$$2t = 5$$
$$t = \frac{5}{2} = 2.5$$

It will take 2.5 hours for the bicyclists to be 5 miles apart.

N8. Let $x =$ the rate of the slower car.

Then $x + 6 =$ the rate of the faster car.

Use the formula $d = rt$ and the fact that each car travels for $\frac{1}{4}$ hour.

$$d_{\text{slower}} + d_{\text{faster}} = d_{\text{total}}$$
$$(x + 6)\left(\frac{1}{4}\right) + x\left(\frac{1}{4}\right) = 35$$
$$\frac{1}{4}x + \frac{3}{2} + \frac{1}{4}x = \frac{70}{2}$$
$$\frac{2}{4}x + \frac{3}{2} = \frac{70}{2}$$
$$\frac{1}{2}x = \frac{67}{2}$$
$$\frac{2}{1}\left(\frac{1}{2}x\right) = \frac{2}{1}\left(\frac{67}{2}\right)$$
$$x = 67$$

Since $x = 67$, $x + 6 = 73$. The slower car had a rate of 67 mph and the faster car had a rate of 73 mph.
Check: The slower car traveled
$67\left(\frac{1}{4}\right) = \frac{67}{4}$ miles and the faster car traveled
$73\left(\frac{1}{4}\right) = \frac{73}{4}$ miles. The total miles traveled is
$\frac{67}{4} + \frac{73}{4} = \frac{140}{4} = 35$, as required.

Exercises

1. The amount of pure alcohol in x liters of a 75% alcohol solution is 0.75 times the volume of solution, or $0.75x$ liters. So choice A is the correct answer.

3. Use $d = rt$, where $r = 55$ and t is the number of hours. $d = (55)t = 55t$ miles.
So choice C is the correct answer.

5. The concentration of the new solution could not be more than the strength of the stronger of the original solutions, so the correct answer is D, since 32% is stronger than both 20% and 30%.

7. To estimate the average rate of the trip, round 405 to 400 and 8.2 to 8.
Use $r = \frac{d}{t}$ with $d = 405$ and $t = 8.2$.
$$r = \frac{d}{t} = \frac{405}{8.2} \approx 49.4$$
The best estimate is choice A, 50 mph.

9. The amount of pure alcohol in 150 liters of a 30% alcohol solution is

150	×	0.30	= 45 liters.
↑		↑	↑
Amount of solution		Rate of concentration	Amount of pure alcohol

11. If $25,000 is invested at 3% simple interest for one year, the amount of interest earned is

$25,000	×	0.03	×	1	= $750.
↑		↑		↑	↑
Principal		Interest rate		Time	Interest earned

13. The monetary value of 35 half-dollars is

35	×	$0.50	= $17.50.
↑		↑	↑
Number of coins		Denomination	Monetary value

15. *Step 2*
Let $x =$ the number of liters of 25% acid solution to be used.
Step 3
Use the box diagram in the textbook to write the equation.

Pure acid in 25% solution	plus	pure acid in 40% solution	is	pure acid in 30% solution.
↓	↓	↓	↓	↓
$0.25x$	$+$	$0.40(80)$	$=$	$0.30(x+80)$

Step 4
Multiply by 100 to clear decimals.
$$25x + 40(80) = 30(x + 80)$$
$$25x + 3200 = 30x + 2400$$
$$25x + 800 = 30x$$
$$800 = 5x$$
$$160 = x$$
Step 5
160 liters of 25% acid solution must be added.
Step 6
25% of 160 liters plus 40% of 80 liters is 40 liters plus 32 liters, or 72 liters, of pure acid; which is equal to 30% of $(160 + 80)$ liters.
$[0.30(240) = 72]$

17. Let $x =$ the number of liters of 5% drug solution.

Pure drug in 10% solution	plus	pure drug in 5% solution	is	pure drug in 8% solution.
↓	↓	↓	↓	↓
$0.10(20)$	$+$	$0.05x$	$=$	$0.08(x+20)$

Solve the equation.
$$0.10(20) + 0.05x = 0.08(x + 20)$$
$$10(20) + 5x = 8(x + 20)$$
$$200 + 5x = 8x + 160$$
$$200 = 3x + 160$$
$$40 = 3x$$
$$x = \frac{40}{3} = 13\frac{1}{3}$$

The pharmacist needs $13\frac{1}{3}$ liters of 5% drug solution.

Check $x = 13\frac{1}{3}$:

LS and RS refer to the left side and right side of the original equation.

LS: $0.10(20) + 0.05\left(13\frac{1}{3}\right) = 2\frac{2}{3}$

RS: $0.08\left(13\frac{1}{3} + 20\right) = 2\frac{2}{3}$

19. Let $x =$ the number of liters of the 20% alcohol solution.

Complete the table.

Strength	Liters of Solution	Liters of Pure Alcohol
12%	12	$0.12(12) = 1.44$
20%	x	$0.20x$
14%	$x + 12$	$0.14(x + 12)$

From the last column, we can formulate an equation that compares the number of liters of pure alcohol. The equation is

$1.44 + 0.20x = 0.14(x + 12)$.

Solve the equation.

$$1.44 + 0.20x = 0.14(x + 12)$$
$$1.44 + 0.20x = 0.14x + 1.68$$
$$0.06x = 0.24$$
$$x = 4$$

4 L of the 20% alcohol solution is needed.

21. Let $x =$ the amount of water to be added.

Then $20 + x =$ the amount of 2% solution.

There is no minoxidil in water.

Pure minoxidil in x mL solution	plus	pure minoxidil in 4% solution	is	pure minoxidil in 2% solution.
↓	↓	↓	↓	↓
$0(x)$	$+$	$0.04(20)$	$=$	$0.02(20 + x)$

Solve the equation.

$$0x + 0.04(20) = 0.02(20 + x)$$
$$4(20) = 2(20 + x)$$
$$80 = 40 + 2x$$
$$40 = 2x$$
$$20 = x$$

20 milliliters of water should be used.

Check $x = 20$:

LS: $0(20) + 0.04(20) = 0.8$

RS: $0.02(20 + 20) = 0.8$

This answer should make common sense—that is, equal amounts of 0% and 4% solutions should produce a 2% solution.

23. Let $x =$ the number of liters of 60% acid solution.

Then $20 - x =$ the number of liters of 75% acid solution.

Pure acid in 60% solution	plus	pure acid in 75% solution	is	pure acid in 72% solution.
↓	↓	↓	↓	↓
$0.60x$	$+$	$0.75(20 - x)$	$=$	$0.72(20)$

Solve the equation.

$$0.60x + 0.75(20 - x) = 0.72(20)$$
$$60x + 75(20 - x) = 72(20)$$
$$60x + 1500 - 75x = 1440$$
$$1500 - 15x = 1440$$
$$-15x = -60$$
$$x = 4$$

4 liters of 60% acid solution must be used.

Check $x = 4$:

LS: $0.60(4) + 0.75(20 - 4) = 14.4$

RS: $0.72(20) = 14.4$

25. Let $x =$ the amount invested at 5% (in dollars).

Then $x - 1200 =$ the amount invested at 4% (in dollars).

Amount Invested (in dollars)	Rate of Interest	Interest for One Year
x	0.05	$0.05x$
$x - 1200$	0.04	$0.04(x - 1200)$

Since the total annual interest was \$141, the equation is

$0.05x + 0.04(x - 1200) = 141$.

$$5x + 4(x - 1200) = 100(141)$$
$$5x + 4x - 4800 = 14{,}100$$
$$9x - 4800 = 14{,}100$$
$$9x = 18{,}900$$
$$x = 2100$$

Since $x = 2100$, $x - 1200 = 900$.

Arlene invested \$2100 at 5% and \$900 at 4%.

27. Let $x =$ the amount invested at 6%.
Then $3x + 6000 =$ the amount invested at 5%.

$$0.06x + 0.05(3x + 6000) = 825$$
$$6x + 5(3x + 6000) = 100(825)$$
$$6x + 15x + 30,000 = 82,500$$
$$21x + 30,000 = 82,500$$
$$21x = 52,500$$
$$x = 2500$$

Since $x = 2500$, $3x + 6000 = 13,500$.
The artist invested \$2500 at 6% and \$13,500 at 5%.

29. Let $x =$ the amount Jamal invested at 8%.
Then $2500 - x =$ the amount invested at 2%.

$$0.08x + 0.02(2500 - x) = 152$$
$$8x + 2(2500 - x) = 15,200$$
$$8x + 5000 - 2x = 15,200$$
$$6x + 5000 = 15,200$$
$$6x = 10,200$$
$$x = 1700$$

Since $x = 1700$, $2500 - x = 800$.
Jamal invested \$1700 at 8% and \$800 at 2%.

31. Let $x =$ the number of nickels.
Then $x + 2 =$ the number of dimes.

The value of nickels	plus	the value of dimes	is	\$1.70.
↓	↓	↓	↓	↓
$0.05x$	$+$	$0.10(x+2)$	$=$	1.70

$$5x + 10(x + 2) = 100(1.70)$$
$$5x + 10x + 20 = 170$$
$$15x + 20 = 170$$
$$15x = 150$$
$$x = 10$$

The collector has 10 nickels.

33. Let $x =$ the number of 46-cent stamps.
Then $45 - x =$ the number of 20-cent stamps.
The value of the 46-cent stamps is $0.46x$ and the value of the 20-cent stamps is $0.20(45 - x)$.
The total value is \$15.50, so

$$0.46x + 0.20(45 - x) = 15.50$$
$$46x + 20(45 - x) = 1550$$
$$46x + 900 - 20x = 1550$$
$$26x + 900 = 1550$$
$$26x = 650$$
$$x = 25$$

Since $x = 25$, $45 - x = 20$.
She bought twenty-five 46-cent stamps valued at \$11.50 and twenty 20-cent stamps valued at \$4.00, for a total value of \$15.50.

35. Let $x =$ the number of pounds of Colombian Decaf beans.
Then $2x =$ the number of pounds of Arabian Mocha beans.

	Number of Pounds	Cost per Pound	Total Value (in \$)
Colombian Decaf	x	\$8.00	$8x$
Arabian Mocha	$2x$	\$8.50	$8.5(2x)$

The total value is \$87.50, so

$$8x + 8.5(2x) = 87.50.$$
$$8x + 17x = 87.50$$
$$25x = 87.50$$
$$x = 3.5$$

Since $x = 3.5$, $2x = 7$.
She can buy 3.5 pounds of Colombian Decaf and 7 pounds of Arabian Mocha.

37. Use the formula $d = rt$ with $r = 53$ and $t = 10$.

$$d = rt$$
$$= (53)(10)$$
$$= 530$$

The distance between Memphis and Chicago is 530 miles.

39. Use $d = rt$ with $d = 500$ and $r = 187.433$.

$$d = rt$$
$$500 = 187.433t$$
$$t = \frac{500}{187.433} \approx 2.668$$

His time was about 2.668 hours.

41. $r = \frac{d}{t} = \frac{200 \text{ meters}}{21.88 \text{ seconds}} \approx 9.14$

Her rate was about 9.14 meters per second.

43. $r = \frac{d}{t} = \frac{110 \text{ meters}}{12.92 \text{ seconds}} \approx 8.51$

His rate was about 8.51 meters per second.

45. Let $t =$ the number of hours Marco and Celeste traveled.
Make a chart using the formula $d = rt$.

	r	t	d
Marco	10	t	$10t$
Celeste	12	t	$12t$

Marco's distance minus Celeste's distance is 15.

$$12t - 10t = 15$$

Solve the equation.

$12t - 10t = 15$

$2t = 15$

$t = \frac{15}{2}$ or $7\frac{1}{2}$

They will be 15 miles apart in $7\frac{1}{2}$ hours.

47. Let $t =$ the number of hours until John and Pat meet.
The distance John travels and the distance Pat travels total 440 miles.

John's distance and Pat's distance is total distance.

$$60t + 28t = 440$$

Solve the equation.

$60t + 28t = 440$

$88t = 440$

$t = 5$

It will take 5 hours for them to meet.

49. Let $t =$ the number of hours until the trains are 315 kilometers apart.

Distance of northbound train plus distance of southbound train is total distance.

$$85t + 95t = 315$$

Solve the equation.

$85t + 95t = 315$

$180t = 315$

$t = \frac{315}{180} = \frac{7}{4}$

It will take $1\frac{3}{4}$ hours for the trains to be 315 kilometers apart.

51. Let $x =$ the rate of the westbound plane.
Then $x - 150 =$ the rate of the eastbound plane.
Using the formula $d = rt$ and the chart in the text, we see that

$d_{west} + d_{east} = d_{total}$

$x(3) + (x - 150)(3) = 2250$

$3x + 3x - 450 = 2250$

$6x = 2700$

$x = 450$

Since $x = 450$, $x - 150 = 300$.
The rate of the westbound plane is 450 mph and the rate of the eastbound plane is 300 mph.

53. Let $x =$ the rate of the slower car.
Then $x + 20 =$ the rate of the faster car.
Use the formula $d = rt$ and the fact that each car travels for 4 hours.

$d_{faster} + d_{slower} = d_{total}$

$(x + 20)(4) + (x)(4) = 400$

$4x + 80 + 4x = 400$

$8x = 320$

$x = 40$

The rate of the slower car is 40 mph and the rate of the faster car is 60 mph.

55. Let $x =$ Bob's current age.
Then $3x =$ Kevin's current age.
Three years ago, Bob's age was $x - 3$ and Kevin's age was $3x - 3$, and this sum was 22.

$(x - 3) + (3x - 3) = 22$

$4x - 6 = 22$

$4x = 28$

$x = 7$

Bob is 7 years old and Kevin is $3(7) = 21$ years old.

57. Let $w =$ the width of the table.
Then $3w =$ the length of the table.
If we subtract 3 feet from the length $(3w - 3)$ and add 3 feet to the width $(w + 3)$, then the length and the width would be equal.

$3w - 3 = w + 3$

$3w = w + 6$

$2w = 6$

$w = 3$

The width is 3 feet and the length is $3(3) = 9$ feet.

59. Let $x =$ her gross pay (pay before deductions).
gross pay − deduction = take-home pay

$$x - 0.10(x) = 585$$
$$0.90x = 585$$
$$x = \frac{585}{0.90} = 650$$

She was paid \$650 before deductions.

2.8 Solving Linear Inequalities

Now Try Exercises

N1. **(a)** The statement $x < -1$ says that x can represent any number less than -1. The interval is written as $(-\infty, -1)$. Graph this inequality by placing a parenthesis at -1 on a number line and drawing an arrow to the left.

−3 −2 −1 0 1 2 3

(b) The statement $-2 \le x$ is the same as $x \ge -2$. The interval is written as $[-2, \infty)$. Graph this inequality by placing a bracket at -2 on a number line and drawing an arrow to the right.

−3 −2 −1 0 1 2 3

N2.

$$5 + 5x \ge 4x + 3$$
$$5 + 5x - 5 \ge 4x + 3 - 5 \quad \text{Subtract 5.}$$
$$5x \ge 4x - 2$$
$$5x - 4x \ge 4x - 2 - 4x \quad \text{Subtract } 4x.$$
$$x \ge -2$$

Graph the solution set $[-2, \infty)$.
To graph this inequality, place a bracket at -2 on a number line and draw an arrow to the right.

−4 −3 −2 −1 0 1 2

N3.

$$-5k \ge 15$$
$$\frac{-5k}{-5} \le \frac{15}{-5} \quad \text{Divide by } -5.$$
$$k \le -3$$

Graph the solution set $(-\infty, -3]$.

−6 −5 −4 −3 −2 −1 0

N4.

$$6 - 2t + 5t \le 8t - 4$$
$$3t + 6 \le 8t - 4$$
$$3t + 6 - 8t \le 8t - 4 - 8t \quad \text{Subtract } 8t.$$
$$-5t + 6 \le -4$$
$$-5t + 6 - 6 \le -4 - 6 \quad \text{Subtract 6.}$$
$$-5t \le -10$$
$$\frac{-5t}{-5} \ge \frac{-10}{-5} \quad \text{Divide by } -5.$$
$$t \ge 2$$

Graph the solution set $[2, \infty)$.

−1 0 1 2 3 4 5

N5.

$$2x - 3(x - 6) < 4(x + 7)$$
$$2x - 3x + 18 < 4x + 28 \quad \text{Dist. prop.}$$
$$-x + 18 < 4x + 28$$
$$-x + 18 + x < 4x + 28 + x \quad \text{Add } x.$$
$$18 < 5x + 28$$
$$18 - 28 < 5x + 28 - 28 \quad \text{Subtract 28.}$$
$$-10 < 5x$$
$$\frac{-10}{5} < \frac{5x}{5} \quad \text{Divide by 5.}$$
$$-2 < x$$
$$x > -2$$

Graph the solution set $(-2, \infty)$.

−4 −3 −2 −1 0 1 2

N6.

$$\frac{1}{8}(x + 4) \ge \frac{1}{6}(2x + 8)$$
$$3(x + 4) \ge 4(2x + 8) \quad \text{Multiply by 24.}$$
$$3x + 12 \ge 8x + 32 \quad \text{Dist. prop.}$$
$$3x + 12 - 3x \ge 8x + 32 - 3x \quad \text{Subtract } 3x.$$
$$12 \ge 5x + 32$$
$$12 - 32 \ge 5x + 32 - 32 \quad \text{Subtract 32.}$$
$$-20 \ge 5x$$
$$\frac{-20}{5} \ge \frac{5x}{5} \quad \text{Divide by 5.}$$
$$-4 \ge x$$
$$x \le -4$$

Graph the solution set $(-\infty, -4]$.

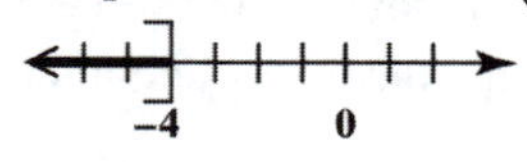

N7. Let $x =$ Will's score on the third test.

The average is at least 90.

$$\downarrow \quad \downarrow \quad \downarrow$$

$$\frac{98+85+x}{3} \geq 90$$

Solve the inequality. Combine like terms in the numerator, and multiply by 3 to eliminate the fraction.

$$3\left(\frac{183+x}{3}\right) \geq 3(90)$$

$$183+x \geq 270$$

$$183+x-183 \geq 270-183 \quad \text{Subtract 183.}$$

$$x \geq 87 \quad \text{Combine terms.}$$

He must score 87 or more on the third test to have an average of *at least* 90.

N8. The statement $0 \leq x < 2$ says that x can represent any number between 0 and 2, including 0 and excluding 2. To graph the inequality, place a bracket at 0 and a parenthesis at 2 and draw a line segment between them. The interval is written as $[0, 2)$.

N9.

$$-4 \leq \frac{3}{2}x-1 \leq 0$$

$$-4+1 \leq \frac{3}{2}x-1+1 \leq 0+1 \quad \text{Add 1.}$$

$$-3 \leq \frac{3}{2}x \leq 1$$

$$\frac{2}{3}(-3) \leq \frac{2}{3}\left(\frac{3}{2}x\right) \leq \frac{2}{3}(1) \quad \text{Multiply by } \frac{2}{3}.$$

$$-2 \leq x \leq \frac{2}{3}$$

Graph the solution set $\left[-2, \frac{2}{3}\right]$.

Exercises

1. When graphing an inequality, use a parenthesis if the inequality symbol is $\underline{>}$ or $\underline{<}$. Use a square bracket if the inequality symbol is $\underline{\geq}$ or $\underline{\leq}$.

Examples:
A parenthesis would be used for the inequalities $x < 2$ and $x > 3$. A square bracket would be used for the inequalities $x \leq 2$ and $x \geq 3$.

3. In interval notation, the set $\{x \mid x > 0\}$ is written $\underline{(0, \infty)}$.

5. The set of numbers graphed corresponds to the inequality $x > -4$.

7. The set of numbers graphed corresponds to the inequality $x \leq 4$.

9. The statement $z \leq 4$ says that z can represent any number less than or equal to 4. The interval is written as $(-\infty, 4]$. To graph the inequality, place a square bracket at 4 (to show that 4 is part of the graph) and draw an arrow extending to the left.

11. The statement $x < -3$ says that x can represent any number less than -3. The interval is written as $(-\infty, -3)$. To graph the inequality, place a parenthesis at -3 (to show that -3 is *not* part of the graph) and draw an arrow extending to the left.

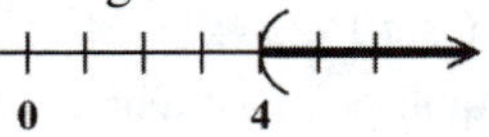

13. The statement $t > 4$ says that t can represent any number greater than 4. The interval is written $(4, \infty)$. To graph the inequality, place a parenthesis at 4 (to show that 4 is *not* part of the graph) and draw an arrow extending to the right.

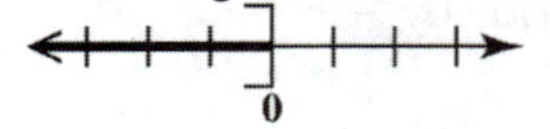

15. The statement $0 \geq x$ (or $x \leq 0$) says that x can represent any number less than or equal to 0. The interval is written as $(-\infty, 0]$. To graph the inequality, place a bracket at 0 (to show that 0 is part of the graph) and draw an arrow extending to the left.

17. The statement $-\frac{1}{2} \le x$ $\left(\text{or } x \ge -\frac{1}{2}\right)$ says that x can represent any number greater than or equal to $-\frac{1}{2}$. The interval is written $\left[-\frac{1}{2}, \infty\right)$. To graph the inequality, place a bracket at $-\frac{1}{2}$ (to show that $-\frac{1}{2}$ is part of the graph) and draw an arrow extending to the right.

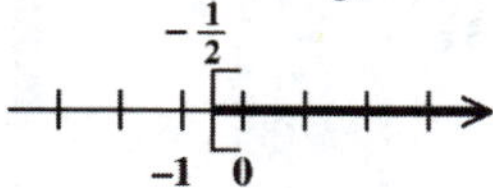

19.

$$z-8 \ge -7$$
$$z-8+8 \ge -7+8 \quad \text{Add 8.}$$
$$z \ge 1$$

Graph the solution set $[1, \infty)$.

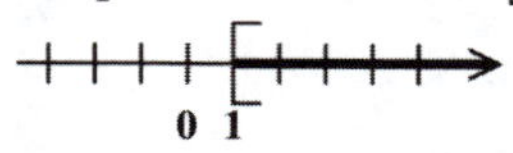

21.

$$2k+3 \ge k+8$$
$$2k+3-k \ge k+8-k \quad \text{Subtract } k.$$
$$k+3 \ge 8$$
$$k+3-3 \ge 8-3 \quad \text{Subtract 3.}$$
$$k \ge 5$$

Graph the solution set $[5, \infty)$.

23.

$$3n+5 < 2n-6$$
$$3n-2n+5 < 2n-2n-6 \quad \text{Subtract } 2n.$$
$$n+5 < -6$$
$$n+5-5 < -6-5 \quad \text{Subtract 5.}$$
$$n < -11$$

Graph the solution set $(-\infty, -11)$.

25. The inequality symbol must be reversed when one is multiplying or dividing by a negative number.

27. $3x < 18$

$$\frac{3x}{3} < \frac{18}{3} \quad \text{Divide by 3.}$$
$$x < 6$$

Graph the solution set $(-\infty, 6)$.

29. $2y \ge -20$

$$\frac{2y}{2} \ge \frac{-20}{2} \quad \text{Divide by 2.}$$
$$y \ge -10$$

Graph the solution set $[-10, \infty)$.

31. $-8t > 24$

$$\frac{-8t}{-8} < \frac{24}{-8} \quad \text{Divide by } -8.$$
$$t < -3$$

Graph the solution set $(-\infty, -3)$.

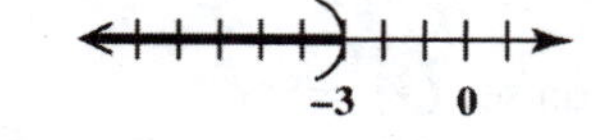

33. $-x \ge 0$

$$-1x \ge 0$$
$$\frac{-1x}{-1} \le \frac{0}{-1} \quad \text{Divide by } -1.$$
$$x \le 0$$

Graph the solution set $(-\infty, 0]$.

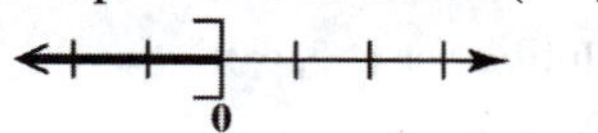

35. Multiply by $-\frac{4}{3}$, the reciprocal of $-\frac{3}{4}$; reverse the inequality sign.

$$-\frac{3}{4}r < -15$$
$$\left(-\frac{4}{3}\right)\left(-\frac{3}{4}r\right) > \left(-\frac{4}{3}\right)(-15)$$
$$r > 20$$

Graph the solution set $(20, \infty)$.

37. $-0.02x \le 0.06$

$$\frac{-0.02x}{-0.02} \ge \frac{0.06}{-0.02} \quad \text{Divide by } -0.02.$$
$$x \ge -3$$

Graph the solution set $[-3, \infty)$.

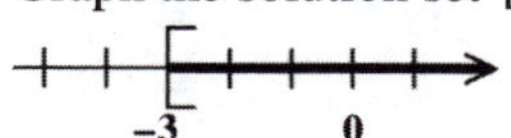

39. $8x+9\le -15$

$8x\le -24$ Subtract 9.

$\frac{8x}{8}\le \frac{-24}{8}$ Divide by 8.

$x\le -3$

Graph the solution set $(-\infty, -3]$.

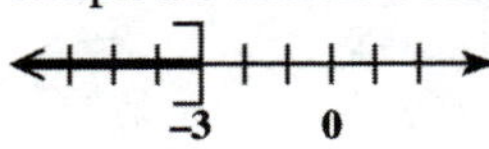

41. $-4x-3<1$

$-4x<4$ Add 3.

$\frac{-4x}{-4}>\frac{4}{-4}$ Divide by -4.

$x>-1$

Graph the solution set $(-1, \infty)$.

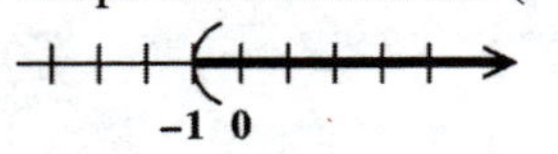

43. $5r+1\ge 3r-9$

$2r+1\ge -9$ Subtract $3r$.

$2r\ge -10$ Subtract 1.

$r\ge -5$ Divide by 2.

Graph the solution set $[-5, \infty)$.

–5 0

45. $5x-2\le -x+10$

$6x-2\le 10$ Add x.

$6x\le 12$ Add 2.

$x\le 2$ Divide by 6.

Graph the solution set $(-\infty, 2]$.

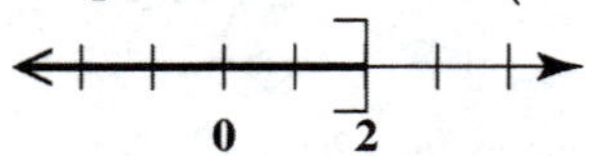

47. $-7x+4>-3x-2$

$4>4x-2$ Add $7x$.

$6>4x$ Add 2.

$\frac{6}{4}>x$ Divide by 4.

$\frac{3}{2}>x$

$x<\frac{3}{2}$

Graph the solution set $\left(-\infty, \frac{3}{2}\right)$.

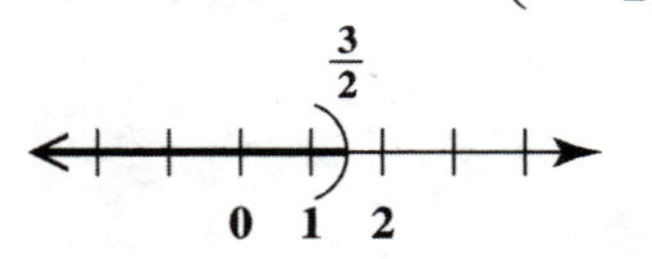

49. $6x+3+x<2+4x+4$

$7x+3<4x+6$ Combine like terms.

$3x+3<6$ Subtract $4x$.

$3x<3$ Subtract 3.

$x<1$ Divide by 3.

Graph the solution set $(-\infty, 1)$.

0 1

51. $-x+4+7x\le -2+3x+6$

$6x+4\le 4+3x$

$3x+4\le 4$

$3x\le 0$

$x\le 0$

Graph the solution set $(-\infty, 0]$.

0

53. $5(t-1)>3(t-2)$

$5t-5>3t-6$

$2t-5>-6$

$2t>-1$

$t>-\frac{1}{2}$

Graph the solution set $\left(-\frac{1}{2}, \infty\right)$.

$-\frac{1}{2}$

0 1

55. $5(x+3)-6x\le 3(2x+1)-4x$

$5x+15-6x\le 6x+3-4x$

$-x+15\le 2x+3$

$-3x+15\le 3$

$-3x\le -12$

$\frac{-3x}{-3}\ge \frac{-12}{-3}$ Divide by -3.

$x\ge 4$

Graph the solution set $[4, \infty)$.

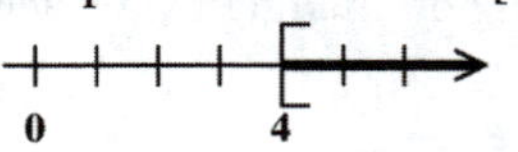

57.
$$\frac{1}{3}(5x-4)\geq\frac{2}{5}(x+3)$$
$$15\left(\frac{1}{3}\right)(5x-4)\geq 15\left(\frac{2}{5}\right)(x+3)$$
$$5(5x-4)\geq 6(x+3)$$
$$25x-20\geq 6x+18$$
$$19x\geq 38$$
$$x\geq 2$$

Graph the solution set $[2, \infty)$.

−1 0 1 2 3 4 5

59.
$$\frac{2}{3}(p+3)>\frac{5}{6}(p-4)$$
$$6\left(\frac{2}{3}\right)(p+3)>6\left(\frac{5}{6}\right)(p-4)$$
$$4(p+3)>5(p-4)$$
$$4p+12>5p-20$$
$$-p+12>-20$$
$$-p>-32$$
$$\frac{-p}{-1}<\frac{-32}{-1} \quad \text{Divide by } -1.$$
$$p<32$$

Graph the solution set $(-\infty, 32)$.

0 8 32

61.
$$\frac{4}{5}x-\frac{1}{2}(x+3)\leq\frac{3}{10}$$
$$10\left[\frac{4}{5}x-\frac{1}{2}(x+3)\right]\leq 10\left(\frac{3}{10}\right)$$
$$8x-5(x+3)\leq 3$$
$$8x-5x-15\leq 3$$
$$3x-15\leq 3$$
$$3x\leq 18$$
$$\frac{3x}{3}\leq\frac{18}{3}$$
$$x\leq 6$$

Graph the solution set $(-\infty, 6]$.

0 6

63.
$$4x-(6x+1)\leq 8x+2(x-3)$$
$$4x-6x-1\leq 8x+2x-6$$
$$-2x-1\leq 10x-6$$
$$-12x-1\leq -6$$
$$-12x\leq -5$$
$$\frac{-12x}{-12}\geq\frac{-5}{-12} \quad \text{Divide by } -12.$$
$$x\geq\frac{5}{12}$$

Graph the solution set $\left[\frac{5}{12}, \infty\right)$.

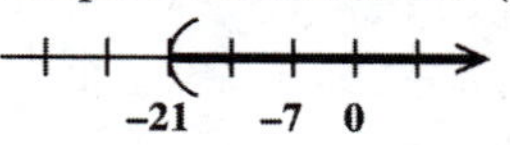

65.
$$5(2k+3)-2(k-8)>3(2k+4)+k-2$$
$$10k+15-2k+16>6k+12+k-2$$
$$8k+31>7k+10$$
$$k+31>10$$
$$k>-21$$

Graph the solution set $(-21, \infty)$.

−21 −7 0

67. The statement "You must be at least 18 yr old to vote" translates as $x\geq 18$.

69. The statement "Chicago received more than 5 in. of snow" translates as $x>5$.

71. The statement "Tracy could spend at most \$20 on a gift" translates as $x\leq 20$.

73. Let $x =$ the score on the third test.

The average of the three tests	is at least	80.
↓	↓	↓
$\frac{76+81+x}{3}$	$\geq$	80

Then solve the inequality.
$$\frac{157+x}{3}\geq 80$$
$$3\left(\frac{157+x}{3}\right)\geq 3(80)$$
$$157+x\geq 240$$
$$x\geq 83$$

In order to average at least 80, Christy's score on her third test must be 83 or greater.

75. Let $x =$ the amount of precipitation in December.

The average precipitation	exceeds	4.6 in.
↓	↓	↓
$\frac{5.7+4.3+x}{3}$	$>$	4.6

Then solve the inequality.

$$\frac{10+x}{3} > 4.6$$

$$3\left(\frac{10+x}{3}\right) > 3(4.6)$$

$$10+x > 13.8$$

$$x > 3.8$$

In order for the average monthly precipitation to exceed 4.6 in., more than 3.8 in. must fall in December.

77. Let $n =$ the number.

"When 2 is added to the difference between six times a number and 5, the result is greater than 13 added to five times the number" translates to $(6n-5)+2 > 5n+13$.

Solve the inequality.

$$6n-5+2 > 5n+13$$

$$6n-3 > 5n+13$$

$$n-3 > 13 \quad \text{Subtract } 5n.$$

$$n > 16 \quad \text{Add 3.}$$

All numbers greater than 16 satisfy the given condition.

79. The Fahrenheit temperature must correspond to a Celsius temperature that is greater than or equal to $-25°$.

$$C = \frac{5}{9}(F-32) \geq -25$$

$$\frac{9}{5}\left[\frac{5}{9}(F-32)\right] \geq \frac{9}{5}(-25)$$

$$F-32 \geq -45$$

$$F \geq -13$$

The temperature in Minneapolis on a certain winter day is never less than $-13°$ Fahrenheit.

81. $P = 2L+2W$; $P \geq 400$

From the figure, we have $L = 4x+3$ and $W = x+37$. Thus, we have the inequality $2(4x+3)+2(x+37) \geq 400$.

Solve this inequality.

$$8x+6+2x+74 \geq 400$$

$$10x+80 \geq 400$$

$$10x \geq 320$$

$$x \geq 32$$

The rectangle will have a perimeter of at least 400 if the value of x is 32 or greater.

83.

$$2+0.30x \leq 5.60$$

$$10(2+0.30x) \leq 10(5.60)$$

$$20+3x \leq 56$$

$$3x \leq 36$$

$$x \leq 12$$

Alan can use the phone for a maximum of 12 minutes.

85. "Revenue from the sales of the DVDs is \$5 per DVD less sales costs of \$100" translates to $R = 5x-100$, where x represents the number of DVDs to be produced.

87. $P = R-C$

$$= (5x-100)-(125+4x)$$

$$= 5x-100-125-4x$$

$$= x-225$$

We can use this expression for P to solve the inequality.

$$P > 0$$

$$x-225 > 0$$

$$x > 225$$

89. The graph corresponds to the inequality $-1 < x < 2$, excluding both -1 and 2.

91. The graph corresponds to the inequality $-1 < x \leq 2$, excluding -1 but including 2.

93. The statement $8 \leq x \leq 10$ says that x can represent any number between 8 and 10, including 8 and 10. To graph the inequality, place brackets at 8 and 10 (to show that 8 and 10 are part of the graph) and draw a line segment between the brackets. The interval is written as [8, 10].

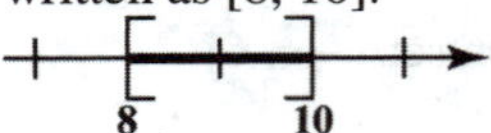

95. The statement $0 < y \leq 10$ says that y can represent any number between 0 and 10, excluding 0 and including 10. To graph the inequality, place a parenthesis at 0 and a bracket at 10 and draw a line segment between them. The interval is written as (0, 10].

0 5 10

97. The statement $4 > x > -3$ can be written as $-3 < x < 4$. Graph the solution set $(-3, 4)$.

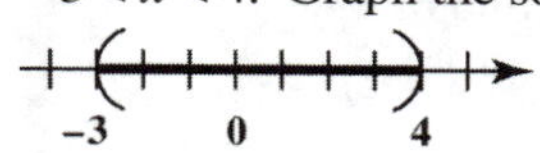

99.

$$\begin{array}{lll} -8 < & 4x & \leq 4 \\ \frac{-8}{4} < & \frac{4}{4}x & \leq \frac{4}{4} \quad \text{Divide by 4.} \\ -2 < & x & \leq 1 \end{array}$$

Graph the solution set $(-2, 1]$.

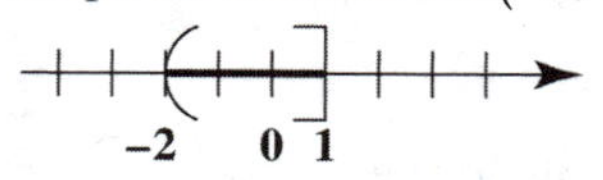

101.

$$\begin{array}{llll} -5 \leq & 2x-3 & \leq 9 & \\ -5+3 \leq & 2x-3+3 & \leq 9+3 & \text{Add 3.} \\ -2 \leq & 2x & \leq 12 & \\ \frac{-2}{2} \leq & 2x & \leq \frac{12}{2} & \text{Divide by 2.} \\ -1 \leq & x & \leq 6 & \end{array}$$

Graph the solution set $[-1, 6]$.

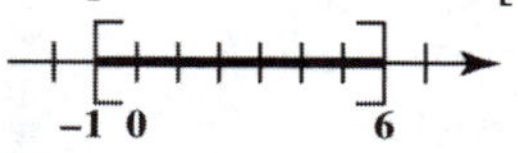

103.

$$\begin{array}{llll} 10 < & 7p+3 & < 24 & \\ 7 < & 7p & < 21 & \text{Subtract 3.} \\ 1 < & p & < 3 & \text{Divide by 7.} \end{array}$$

Graph the solution set $(1, 3)$.

0 1 3

105.

$$\begin{array}{llll} -4 < & -2x & < 12 & \\ \frac{-4}{-2} > & \frac{-2x}{-2} & > \frac{12}{-2} & \text{Divide by} -2. \\ 2 > & x & > -6 & \\ -6 < & x & < 2 & \end{array}$$

Graph the solution set $(-6, 2)$.

-6 0 2

107.

$$\begin{array}{llll} 5 < & 1-6m & < 12 & \\ 5-1 < & 1-6m-1 & < 12-1 & \text{Subtract 1.} \\ 4 < & -6m & < 11 & \\ \frac{4}{-6} > & \frac{-6m}{-6} & > \frac{11}{-6} & \text{Divide by} -6. \\ -\frac{2}{3} > & m & > -\frac{11}{6} & \\ -\frac{11}{6} < & m & < -\frac{2}{3} & \end{array}$$

Graph the solution set $\left(-\frac{11}{6}, -\frac{2}{3}\right)$.

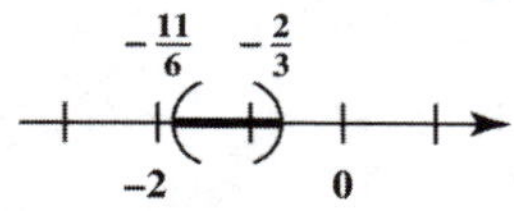

109.

$$\begin{array}{llll} 6 \leq & 3(x-1) & < 18 & \\ 6 \leq & 3x-3 & < 18 & \text{Distributive prop.} \\ 6+3 \leq & 3x-3+3 & < 18+3 & \text{Add 3.} \\ 9 \leq & 3x & < 21 & \\ \frac{9}{3} \leq & \frac{3x}{3} & < \frac{21}{3} & \text{Divide by 3.} \\ 3 \leq & x & < 7 & \end{array}$$

Graph the solution set $[3, 7)$.

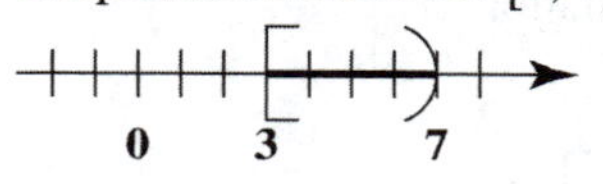

111.

$$\begin{array}{llll} -12 \leq & \frac{1}{2}z+1 & \leq 4 & \text{Multiply by 2.} \\ 2(-12) \leq & 2\left(\frac{1}{2}z+1\right) & \leq 2(4) & \\ -24 \leq & z+2 & \leq 8 & \text{Dist. prop.} \\ -24-2 \leq & z+2-2 & \leq 8-2 & \text{Subtract 2.} \\ -26 \leq & z & \leq 6 & \end{array}$$

Note: We could have started this solution by subtracting 1 from each part.

Graph the solution set $[-26, 6]$.

-26 6

-30 -20 -10 0 10

113.
$$
\begin{array}{llll}
1 \le & 3+\frac{2}{3}p & \le 7 & \\
3(1) \le & 3\left(3+\frac{2}{3}p\right) & \le 3(7) & \text{Multiply by 3.} \\
3 \le & 9+2p & \le 21 & \text{Dist. prop.} \\
3-9 \le & 9+2p-9 & \le 21-9 & \text{Subtract 9.} \\
-6 \le & 2p & \le 12 & \text{Divide by 2.} \\
-3 \le & p & \le 6 &
\end{array}
$$

Graph the solution set $[-3, 6]$.

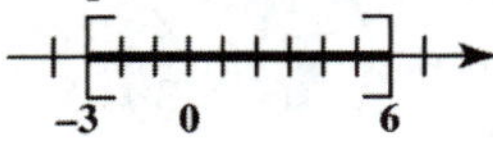

115.
$$
\begin{array}{llll}
-7 \le & \frac{5}{4}r-1 & \le -1 & \\
-7+1 \le & \frac{5}{4}r-1+1 & \le -1+1 & \text{Add 1.} \\
-6 \le & \frac{5}{4}r & \le 0 & \\
\frac{4}{5}(-6) \le & \frac{4}{5}\left(\frac{5}{4}r\right) & \le \frac{4}{5}(0) & \text{Multiply by } \frac{4}{5}. \\
-\frac{24}{5} \le & r & \le 0 &
\end{array}
$$

Graph the solution set $\left[-\frac{24}{5}, 0\right]$.

$-\frac{24}{5}$

–5 0

117. $3x+2=14$

$3x=12$

$x=4$

Solution set: $\{4\}$

1 2 3 4 5 6 7 8

119. $3x+2<14$

$3x<12$

$x<4$

Solution set: $(-\infty, 4)$

1 2 3 4 5 6 7

Chapter 2 Review Exercises

1. $x-5=1$

$x=6$ Add 5.

The solution set is $\{6\}$.

2. $x+8=-4$

$x=-12$ Subtract 8.

The solution set is $\{-12\}$.

3. $3t+1=2t+8$

$t+1=8$ Subtract $2t$.

$t=7$ Subtract 1.

The solution set is $\{7\}$.

4. $5z=4z+\frac{2}{3}$

$z=\frac{2}{3}$ Subtract $4z$.

The solution set is $\left\{\frac{2}{3}\right\}$.

5. $(4r-2)-(3r+1)=8$

$(4r-2)-1(3r+1)=8$

$4r-2-3r-1=8$ Distributive property

$r-3=8$

$r=11$ Add 3.

The solution set is $\{11\}$.

6. $3(2x-5)=2+5x$

$6x-15=2+5x$ Distributive property

$x-15=2$ Subtract $5x$.

$x=17$ Add 15.

The solution set is $\{17\}$.

7. $7x=35$

$x=5$ Divide by 7.

The solution set is $\{5\}$.

8. $12r=-48$

$r=-4$ Divide by 12.

The solution set is $\{-4\}$.

9. $2p-7p+8p=15$

$3p=15$

$p=5$ Divide by 3.

The solution set is $\{5\}$.

10. $\frac{x}{12}=-1$

$x=-12$ Multiply by 12.

The solution set is $\{-12\}$.

11. $\frac{5}{8}q = 8$

$\frac{8}{5}\left(\frac{5}{8}q\right) = \frac{8}{5}(8)$ Multiply by $\frac{8}{5}$.

$q = \frac{64}{5}$

The solution set is $\left\{\frac{64}{5}\right\}$.

12. $12m + 11 = 59$

$12m = 48$ Subtract 11.

$m = 4$ Divide by 12.

The solution set is $\{4\}$.

13. $3(2x+6) - 5(x+8) = x - 22$

$6x + 18 - 5x - 40 = x - 22$

$x - 22 = x - 22$

This is a true statement, so the solution set is {all real numbers}.

14. $5x + 9 - (2x - 3) = 2x - 7$

$5x + 9 - 2x + 3 = 2x - 7$

$3x + 12 = 2x - 7$

$x + 12 = -7$

$x = -19$

The solution set is $\{-19\}$.

15. $\frac{1}{2}r - \frac{r}{3} = \frac{r}{6}$

$6\left(\frac{1}{2}r\right) - 6\left(\frac{r}{3}\right) = 6\left(\frac{r}{6}\right)$ Multiply by 6.

$3r - 2r = r$

$r = r$

This is a true statement, so the solution set is {all real numbers}.

16. Multiply by 10 to clear decimals.

$10[0.1(x+80) + 0.2x] = 10(14)$

$(x+80) + 2x = 140$ Dist. prop.

$3x + 80 = 140$

$3x = 60$

$x = 20$

The solution set is $\{20\}$.

17. $3x - (-2x + 6) = 4(x - 4) + x$

$3x + 2x - 6 = 4x - 16 + x$

$5x - 6 = 5x - 16$

$-6 = -16$

This statement is false, so there is no solution set, symbolized by $\varnothing$.

18. Multiply both sides by 6 to eliminate fractions.

$6\left[\frac{1}{2}(x+3) - \frac{2}{3}(x-2)\right] = 6(3)$

$6\left[\frac{1}{2}(x+3)\right] - 6\left[\frac{2}{3}(x-2)\right] = 18$

$3(x+3) - 4(x-2) = 18$

$3x + 9 - 4x + 8 = 18$

$-x + 17 = 18$

$-x = 1$

$x = -1$

The solution set is $\{-1\}$.

19. Let x represent the number.

$5x + 7 = 3x$

$7 = -2x$ Subtract $5x$.

$-\frac{7}{2} = x$ Divide by -2.

The number is $-\frac{7}{2}$.

20. *Step 2*

Let x = the number of Republicans.

Then $x + 24$ = the number of Democrats.

Step 3

$x + (x + 24) = 118$

Step 4

$2x + 24 = 118$

$2x = 94$

$x = 47$

Step 5

Since $x = 47$, $x + 24 = 71$.

There were 71 Democrats and 47 Republicans.

Step 6

There are 24 more Democrats than Republicans and the total is 118.

21. *Step 2*
Let $x =$ the land area of Rhode Island.
Then $x + 5213 =$ the land area of Hawaii.
Step 3
The areas total 7637 square miles, so
$x + (x + 5213) = 7637.$
Step 4
$$\begin{aligned} 2x + 5213 &= 7637 \\ 2x &= 2424 \\ x &= 1212 \end{aligned}$$
Step5
Since $x = 1212$, $x + 5213 = 6425.$
The land area of Rhode Island is 1212 square miles and that of Hawaii is 6425 square miles.
Step 6
The land area of Hawaii is 5213 square miles greater than the land area of Rhode Island and the total is 7637 square miles.

22. *Step 2*
Let $x =$ the height of Twin Falls.
Then $\frac{5}{2}x =$ the height of Seven Falls.
Step 3
The sum of the heights is 420 feet, so
$x + \frac{5}{2}x = 420.$
Step 4
$$\begin{aligned} 2\left(x + \frac{5}{2}x\right) &= 2(420) \\ 2x + 5x &= 840 \\ 7x &= 840 \\ x &= 120 \end{aligned}$$
Step 5
Since $x = 120$, $\frac{5}{2}x = \frac{5}{2}(120) = 300.$
The height of Twin Falls is 120 feet and that of Seven Falls is 300 feet.
Step 6
The height of Seven Falls is $\frac{5}{2}$ the height of Twin Falls and the sum is 420.

23. *Step 2*
Let $x =$ the measure of the angle.
Then $90 - x =$ the measure of its complement and $180 - x =$ the measure of its supplement.
Step 3
$180 - x = 10(90 - x)$
Step 4
$$\begin{aligned} 180 - x &= 900 - 10x \\ 9x + 180 &= 900 \\ 9x &= 720 \\ x &= 80 \end{aligned}$$
Step5
The measure of the angle is $80°$.
Its complement measures $90° - 80° = 10°$, and its supplement measures $180° - 80° = 100°$.
Step 6
The measure of the supplement is 10 times the measure of the complement.

24. *Step 2*
Let $x =$ lesser odd integer.
Then $x + 2 =$ greater odd integer.
Step 3
$x + 2(x + 2) = (x + 2) + 24$
Step 4
$$\begin{aligned} x + 2x + 4 &= x + 26 \\ 3x + 4 &= x + 26 \\ 2x + 4 &= 26 \\ 2x &= 22 \\ x &= 11 \end{aligned}$$
Step5
Since $x = 11$, $x + 2 = 13.$
The consecutive odd numbers are 11 and 13.
Step 6
The lesser plus twice the greater is $11 + 2(13) = 37$, which is 24 more than the greater.

25. Solve for h.
$$\begin{aligned} A &= \frac{1}{2}bh \\ 44 &= \frac{1}{2}(8)h \\ 44 &= 4h \\ 11 &= h \end{aligned}$$

26. Solve for A.
$$\begin{aligned} A &= \frac{1}{2}h(b + B) \\ A &= \frac{1}{2}(8)(3 + 4) \\ &= \frac{1}{2}(8)(7) \\ &= (4)(7) \\ A &= 28 \end{aligned}$$

27. Solve for r.

$$C = 2\pi r$$
$$29.83 = 2(3.14)r$$
$$29.83 = 6.28r$$
$$\frac{29.83}{6.28} = \frac{6.28r}{6.28}$$
$$4.75 = r$$

28. Solve for V.

$$V = \frac{4}{3}\pi r^3$$
$$= \frac{4}{3}(3.14)(6)^3$$
$$= \frac{4}{3}(3.14)(216)$$
$$= \frac{4}{3}(678.24)$$
$$V = 904.32$$

29. Solve for h.

$$\frac{A}{b} = \frac{bh}{b} \quad \text{Divide by } b.$$
$$\frac{A}{b} = h, \quad \text{or} \quad h = \frac{A}{b}$$

30. Solve for h.

$$2A = 2\left[\frac{1}{2}h(b+B)\right] \quad \text{Multiply by 2.}$$
$$2A = h(b+B)$$
$$\frac{2A}{(b+B)} = \frac{h(b+B)}{(b+B)} \quad \text{Divide by } b+B.$$
$$\frac{2A}{b+B} = h, \quad \text{or} \quad h = \frac{2A}{b+B}$$

31. Solve for y.

$$x + y = 11$$
$$y = -x + 11$$

32. Solve for y.

$$3x - 2y = 12$$
$$-2y = -3x + 12$$
$$y = \frac{3}{2}x - 6$$

33. Because the two angles are supplementary, $(8x-1)+(3x-6)=180$.

$$11x - 7 = 180$$
$$11x = 187$$
$$x = 17$$

Since $x = 17, 8x-1 = 135,$ and $3x-6 = 45.$

The measures of the two angles are $135°$ and $45°$.

34. The angles are vertical angles, so their measures are equal.

$$3x + 10 = 4x - 20$$
$$10 = x - 20$$
$$30 = x$$

Since $x = 30, 3x+10 = 100$ and $4x - 20 = 100.$

Each angle has a measure of $100°$.

35. Let W = the width of the rectangle.
Then $W + 12$ = the length of the rectangle.
"The perimeter of the rectangle is 16 times the width" can be written as $2L + 2W = 16W$ since the perimeter is $2L + 2W$.
Because $L = W + 12,$ we have

$$2(W+12) + 2W = 16W.$$
$$2W + 24 + 2W = 16W$$
$$4W + 24 = 16W$$
$$-12W + 24 = 0$$
$$-12W = -24$$
$$W = 2$$

The width is 2 cm and the length is $2 + 12 = 14$ cm.

36. The sum of the three marked angles in the triangle is $180°$.

$$45° + (x+12.2)° + (3x+2.8)° = 180°$$
$$4x + 60 = 180$$
$$4x = 120$$
$$x = 30$$

Since $x = 30, (x+12.2)° = 42.2°$ and $(3x+2.8)° = 92.8°$.

37. The ratio of 60 centimeters to 40 centimeters is

$$\frac{60 \text{ cm}}{40 \text{ cm}} = \frac{3 \cdot 20}{2 \cdot 20} = \frac{3}{2}.$$

38. To find the ratio of 90 inches to 10 feet, first convert 10 feet to inches.

10 feet $= 10 \cdot 12 = 120$ inches

Thus, the ratio of 90 inches to 10 feet is

$$\frac{90}{120} = \frac{3 \cdot 30}{4 \cdot 30} = \frac{3}{4}.$$

39. $\frac{p}{21}=\frac{5}{30}$

$30p=105$ Cross products

$\frac{30p}{30}=\frac{105}{30}$ Divide by 30.

$p=\frac{105}{30}=\frac{7\cdot15}{2\cdot15}=\frac{7}{2}$

The solution set is $\left\{\frac{7}{2}\right\}$.

40. $\frac{5+x}{3}=\frac{2-x}{6}$

$6(5+x)=3(2-x)$ Cross products

$30+6x=6-3x$ Distributive property

$30+9x=6$ Add $3x$.

$9x=-24$ Subtract 30.

$x=\frac{-24}{9}=-\frac{8}{3}$

The solution set is $\left\{-\frac{8}{3}\right\}$.

41. Let $x=$ the tax on a $36.00 item.
Set up a proportion with one ratio involving sales tax and the other involving the costs of the items.

$\frac{x \text{ dollars}}{\$2.04}=\frac{\$36}{\$24}$

$24x=(2.04)(36)=73.44$

$x=\frac{73.44}{24}=3.06$

The sales tax on a $36.00 item is $3.06.

42. Let $x=$ the actual distance between the second pair of cities (in kilometers).
Set up a proportion with one ratio involving map distances and the other involving actual distances.

$\frac{x \text{ kilometers}}{150 \text{ kilomters}}=\frac{80 \text{ centimeters}}{32 \text{ centimeters}}$

$32x=(150)(80)=12{,}000$

$x=\frac{12{,}000}{32}=375$

The cities are 375 kilometers apart.

43. Let $x=$ the number of gold medals earned by Italy.

$\frac{x \text{ gold medals}}{28 \text{ medals}}=\frac{2 \text{ gold medals}}{7 \text{ medals}}$

$7x=2(28)=56$

$x=8$

At the 2012 Olympics, 8 gold medals were earned by Italy.

44. To find the best buy, divide the price by the number of units to get the unit cost. Each result was found by using a calculator and rounding the answer to three decimal places. The best buy (based on price per unit) is the smallest unit cost. The results in the following table are rounded to the nearest thousandth.

Size	Price	Unit Cost (dollar per oz)
9 oz	$3.49	$\frac{\$3.49}{9}\approx\$0.388$
14 oz	$3.99	$\frac{\$3.99}{14}=\$0.285$
18 oz	$4.49	$\frac{\$4.49}{18}\approx\$0.249$ (*)

Because the 18-oz size produces the lowest unit cost, it is the best buy. The unit cost, to the nearest thousandth, is $0.249 per oz.

45. What percent of 12 is 21?
Let p denote the percent.

$21=p\cdot12$

$p=\frac{21}{12}=1.75=175\%$

21 is 175% of 12.

46. 36% of what number is 900?
Let n denote the number.

$0.36\cdot n=900$

$n=\frac{900}{0.36}$ Divide by 0.36.

$=2500$ Simplify.

Thus, 36% of 2500 is 900.

47. Let $x =$ the number of liters of the 60% solution to be used.
Then $x+15 =$ the number of liters of the 20% solution.

10% solution	plus	60% solution	is	20% solution.
↓	↓	↓	↓	↓
$0.10(15)$	$+$	$0.60(x)$	$=$	$0.20(x+15)$

Multiply by 10 to clear decimals.

$$1(15)+6x = 2(x+15)$$
$$15+6x = 2x+30$$
$$15+4x = 30$$
$$4x = 15$$
$$x = \frac{15}{4} = 3.75$$

3.75 liters of 60% solution are needed.

48. Let $x =$ the amount invested at 5%.
Then $10,000 - x =$ the amount invested at 3%.

Interest at 5%	plus	interest at 3%	equals	\$400.
↓	↓	↓	↓	↓
$0.05x$	$+$	$0.03(10,000-x)$	$=$	400

Solve the equation.

$$5x+3(10,000-x) = 100(400)$$
$$5x+30,000-3x = 40,000$$
$$2x = 10,000$$
$$x = 5000$$

Robert invested \$5000 at 5% and $10,000-5000 = \$5000$ at 3%.

49. Use the formula $d = rt$ or $r = \dfrac{d}{t}.$

$$r = \frac{d}{t} = \frac{3150}{384} \approx 8.203$$

Rounded to the nearest tenth, the *Yorkshire's* average rate was 8.2 mph.

50. Let $t =$ the number of hours until the planes are 1925 miles apart.

The distance one plane flies north	plus	the distance the other plane flies south	equals	the distance between the planes.
↓	↓	↓	↓	↓
$350t$	$+$	$420t$	$=$	1925

Solve the equation.

$$350t+420t = 1925$$
$$770t = 1925$$
$$t = \frac{1925}{770} = \frac{5}{2} = 2\frac{1}{2}$$

The planes will be 1925 miles apart in $2\frac{1}{2}$ hours.

51. The statement $x \geq -4$ can be written as $[-4, \infty)$.

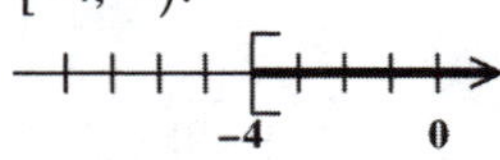

52. The statement $x < 7$ can be written as $(-\infty, 7)$.

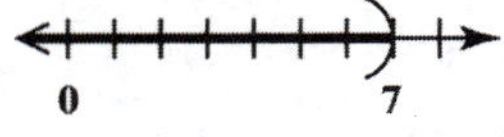

53. The statement $-5 \leq x < 6$ can be written as $[-5, 6)$.

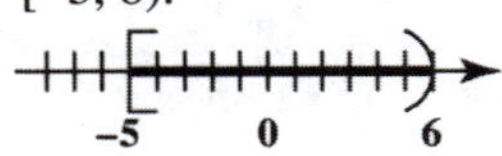

54. By examining the choices, we see that $-4x \leq 36$ is the only inequality that has a negative coefficient of x. Thus, B is the only inequality that requires a reversal of the inequality symbol when it is solved.

55. $x+6 \geq 3$

$x \geq -3$ Subtract 6.

Graph the solution set $[-3, \infty)$.

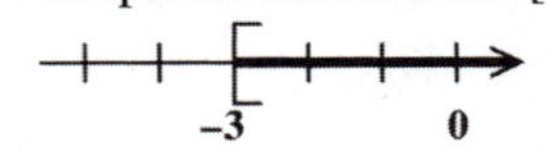

56. $5x < 4x+2$

$x < 2$ Subtract $4x$.

Graph the solution set $(-\infty, 2)$.

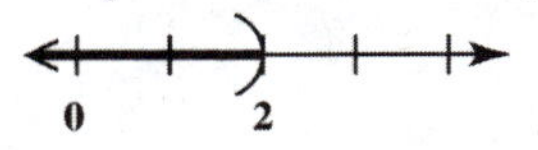

57. $-6x \leq -18$

$\dfrac{-6x}{-6} \geq \dfrac{-18}{-6}$ Divide by -6.

$x \geq 3$

Graph the solution set $[3, \infty)$.

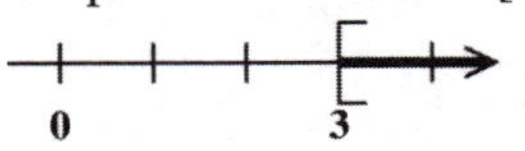

58. $8(x-5)-(2+7x)\geq 4$

$8x-40-2-7x\geq 4$

$x-42\geq 4$

$x\geq 46$

Graph the solution set $[46, \infty)$.

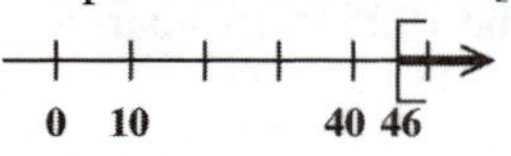

59. $4x-3x>10-4x+7x$

$x>10+3x$

$-2x>10$

$\frac{-2x}{-2}<\frac{10}{-2}$ Divide by -2.

$x<-5$

Graph the solution set $(-\infty, -5)$.

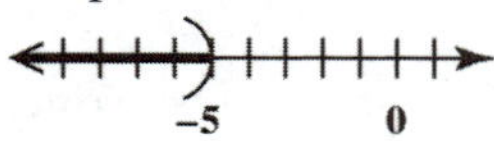

60. $3(2x+5)+4(8+3x)<5(3x+7)$

$6x+15+32+12x<15x+35$

$18x+47<15x+35$

$3x+47<35$

$3x<-12$

$x<-4$

Graph the solution set $(-\infty, -4)$.

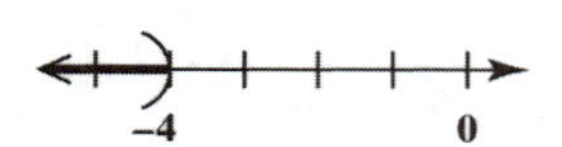

61. $-3\leq 2x+1\leq 4$

$-4\leq 2x\leq 3$ Subtract 1.

$-2\leq x\leq \frac{3}{2}$ Divide by 2.

Graph the solution set $\left[-2, \frac{3}{2}\right]$.

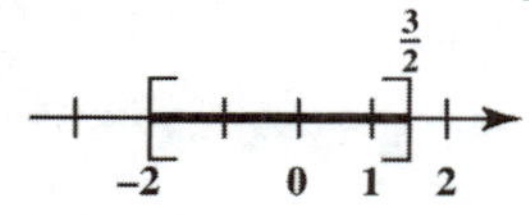

62. $9<3x+5\leq 20$

$4<3x\leq 15$ Subtract 5.

$\frac{4}{3}<x\leq 5$ Divide by 3.

Graph the solution set $\left(\frac{4}{3}, 5\right]$.

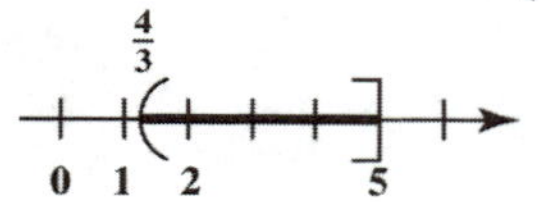

63. Let $x=$ the score on the third test.

The average of the three tests	is at least	90.
↓	↓	↓
$\frac{94+88+x}{3}$	$\geq$	90

$\frac{182+x}{3}\geq 90$

$3\left(\frac{182+x}{3}\right)\geq 3(90)$

$182+x\geq 270$

$x\geq 88$

In order to average at least 90, Awilda's score on her third test must be 88 or more.

64. Let $n=$ the number.
"If nine times a number is added to 6, the result is at most 3" can be written as $9n+6\leq 3$.
Solve the inequality.

$9n+6\leq 3$

$9n\leq -3$ Subtract 6.

$n\leq \frac{-3}{9}$ Divide by 9.

$n\leq -\frac{1}{3}$

All numbers less than or equal to $-\frac{1}{3}$ satisfy the given condition.

Chapter 2 Mixed Review Exercises

1. $\frac{x}{7}=\frac{x-5}{2}$

$2x=7(x-5)$ Cross products

$2x=7x-35$

$-5x=-35$

$x=7$

The solution set is $\{7\}$.

2. Solve $I=prt$ for r.

$\frac{I}{pt}=\frac{prt}{pt}$ Divide by pt.

$\frac{I}{pt}=r$, or $r=\frac{I}{pt}$

3. $-2x > -4$

$\frac{-2x}{-2} < \frac{-4}{-2}$ Divide by -2.

$x < 2$

The solution set is $(-\infty, 2)$.

4. $2k - 5 = 4k + 13$

$-2k - 5 = 13$ Subtract $4k$.

$-2k = 18$ Add 5.

$k = -9$ Divide by -2.

The solution set is $\{-9\}$.

5. $0.05x + 0.02x = 4.9$

To clear decimals, multiply both sides by 100.

$100(0.05x + 0.02x) = 100(4.9)$

$5x + 2x = 490$

$7x = 490$

$x = 70$

The solution set is $\{70\}$.

6. $2 - 3(x - 5) = 4 + x$

$2 - 3x + 15 = 4 + x$

$17 - 3x = 4 + x$

$17 - 4x = 4$

$-4x = -13$

$x = \frac{-13}{-4} = \frac{13}{4}$

The solution set is $\left\{\frac{13}{4}\right\}$.

7. $9x - (7x + 2) = 3x + (2 - x)$

$9x - 7x - 2 = 3x + 2 - x$

$2x - 2 = 2x + 2$

$-2 = 2$

Because $-2 = 2$ is a false statement, the given equation has no solution, symbolized by $\varnothing$.

8. $\frac{1}{3}s + \frac{1}{2}s + 7 = \frac{5}{6}s + 5 + 2$

$\frac{1}{3}s + \frac{1}{2}s = \frac{5}{6}s$ Subtract 7.

The least common denominator is 6.

$6\left(\frac{1}{3}s + \frac{1}{2}s\right) = 6\left(\frac{5}{6}s\right)$

$2s + 3s = 5s$

$5s = 5s$

Because $5s = 5s$ is a true statement, the solution set is $\{\text{all real numbers}\}$.

9. Let $x =$ the number of calories a 175-pound athlete can consume.
Set up a proportion with one ratio involving calories and the other involving pounds.

$\frac{x \text{ calories}}{50 \text{ calories}} = \frac{175 \text{ pounds}}{2.2 \text{ pounds}}$

$2.2x = 50(175)$

$x = \frac{8750}{2.2} \approx 3977.3$

To the nearest hundred calories, a 175-pound athlete in a vigorous training program can consume 4000 calories per day.

10. Let $x =$ the sales for DiGiorno, in millions of dollars.
Then $x - 399.9$ the sales for Red Baron, in millions of dollars.

$x + (x - 399.9) = 937.5$

$2x - 399.9 = 937.5$

$2x = 1337.4$

$x = 668.7$

Since $x = 668.7$, $x - 399.9 = 268.8$.

DiGiorno sold $668.7 million worth of frozen pizza and Red Baron sold $268.8 million worth of frozen pizza.

11. The results in the following table are rounded to the nearest thousandth.

Size	Price	Unit Cost (dollar per oz)
50 oz	$3.99	$\frac{\$3.99}{50} \approx \$0.080$
100 oz	$7.29	$\frac{\$7.29}{100} \approx \$0.073$
160 oz	$9.99	$\frac{\$9.99}{160} \approx \$0.062$ (*)

Because the 160-oz size produces the lowest unit cost, it is the best buy. The unit cost, to the nearest thousandth, is $0.062 per oz.

12. The angles make up a right angle, so the sum of their measures is $90°$.

$(3x)° + (8x + 2)° = 90°$

$11x + 2 = 90$

$11x = 88$

$x = 8$

Since $x = 8$, $(3x)° = 24°$ and $(8x + 2)° = 66°$.

13. Use the formula $d = rt$, or $t = \frac{d}{r}$, where $d = 819$ and $r = 63$.

$t = \frac{d}{r} = \frac{819}{63} = 13$

It took Janet 13 hours to drive from Louisville to Dallas.

14. Let $x =$ the rate of the slower train.
Then $x + 30 =$ the rate of the faster train.

	r	**t**	**d**
Slower Train	x	3	$3x$
Faster Train	$x+30$	3	$3(x+30)$

The sum of the distances traveled by the two trains is 390 miles, so $3x + 3(x+30) = 390$.
Solve the equation.

$$3x + 3(x+30) = 390$$
$$3x + 3x + 90 = 390$$
$$6x + 90 = 390$$
$$6x = 300$$
$$x = 50$$

Since $x = 50$, $x + 30 = 80$.

The rate of the slower train is 50 miles per hour and the rate of the faster train is 80 miles per hour.

15. Let $x =$ the length of the first side.
Then $2x =$ the length of the second side.
Use the formula for the perimeter of a triangle, $P = a + b + c$, with perimeter 96 and third side 30.

$$x + 2x + 30 = 96$$
$$3x + 30 = 96$$
$$3x = 66$$
$$x = 22$$

The sides have lengths 22 meters, 44 meters, and 30 meters. The length of the longest side is 44 meters.

16. Let $s =$ the length of a side of the square. The formula for the perimeter of a square is $P = 4s$.

The perimeter	cannot be greater than	200.
↓	↓	↓
$4s$	$\le$	200

$$4s \le 200$$
$$s \le 50$$

The length of a side is 50 meters or less.

Chapter 2 Test

1. $5x + 9 = 7x + 21$

$-2x + 9 = 21$ Subtract $7x$.

$-2x = 12$ Subtract 9.

$x = -6$ Divide by -2.

The solution set is $\{-6\}$.

2. $-\frac{4}{7}x = -12$

$$\left(-\frac{7}{4}\right)\left(-\frac{4}{7}x\right) = \left(-\frac{7}{4}\right)(-12)$$
$$x = 21$$

The solution set is $\{21\}$.

3. $7 - (x - 4) = -3x + 2(x+1)$

$$7 - x + 4 = -3x + 2x + 2$$
$$-x + 11 = -x + 2$$
$$11 = 2$$

Because the last statement is false, the equation has no solution set, symbolized by $\varnothing$.

4. To clear decimals, multiply both sides by 100.

$$100[0.06(x+20) + 0.08(x-10)] = 100(4.6)$$
$$6(x+20) + 8(x-10) = 460$$
$$6x + 120 + 8x - 80 = 460$$
$$14x + 40 = 460$$
$$14x = 420$$
$$x = 30$$

The solution set is $\{30\}$.

5. $-8(2x+4) = -4(4x+8)$

$$-16x - 32 = -16x - 32$$

Because the last statement is true, the solution set is {all real numbers}.

6. $2-3(x-5)=3+(x+1)$

$2-3x+15=3+x+1$ Distributive property

$-3x+17=4+x$

$17=4+4x$ Add $3x$.

$13=4x$ Subtract 4.

$\frac{13}{4}=x$

The solution set is $\left\{\frac{13}{4}\right\}$.

7. Let $x=$ the number of games the Cardinals lost.
Then $2x-33=$ the number of games the Cardinals won.
The total number of games played was 162.

$x+(2x-33)=162$

$3x-33=162$

$3x=195$

$x=65$

Since $x=65$, $2x-33=97$.

The Cardinals won 97 games and lost 65 games.

8. Let $x=$ the area of Kauai (in square miles).
Then $x+177=$ the area of Maui (in square miles), and $(x+177)+3293=x+3470=$ the area of Hawaii.

$x+(x+177)+(x+3470)=5300$

$3x+3647=5300$

$3x=1653$

$x=551$

Since $x=551$, $x+177=728$ and $x+3470=4021$.

The area of Hawaii is 4021 square miles, the area of Maui is 728 square miles, and the area of Kauai is 551 square miles.

9. Let $x=$ the measure of the angle.
Then $90-x=$ the measure of its complement, and $180-x=$ the measure of its supplement.

$180-x=3(90-x)+10$

$180-x=270-3x+10$

$180-x=280-3x$

$180+2x=280$

$2x=100$

$x=50$

The measure of the angle is $50°$. The measure of its supplement, $130°$, is $10°$ more than three times its complement, $40°$.

10. *Step 2*
Let $x=$ the lesser even integer.
Then $x+2=$ the greater even integer.
Step 3

$3x=20+2(x+2)$

Step 4

$3x=20+2(x+2)$

$3x=20+2x+4$

$3x=24+2x$

$x=24$

Step5
Since $x=24$, $x+2=26$.

The consecutive even numbers are 24 and 26.
Step 6
Three times the lesser is $3(24)=72$, and 20 more than twice the greater is $20+2(26)=72$.

11. **(a)** Solve $P=2L+2W$ for W.

$P-2L=2W$

$\frac{P-2L}{2}=W$, or $W=\frac{P-2L}{2}$

(b) Substitute 116 for P and 40 for L in the formula obtained in part (a).

$W=\frac{P-2L}{2}$

$=\frac{116-2(40)}{2}$

$=\frac{116-80}{2}=\frac{36}{2}=18$

12. $5x-4y=8$

$5x-4y-5x=8-5x$ Subtract $5x$.

$-4y=8-5x$

$\frac{-4y}{-4}=\frac{8-5x}{-4}$ Divide by -4.

$y=-\frac{8-5x}{4}$, or $y=\frac{5}{4}x-2$

(There are other correct forms.)

13. The angles are vertical angles, so their measures are equal.

$3x+15=4x-5$

$15=x-5$

$20=x$

Since $x=20$, $3x+15=75$ and $4x-5=75$.

Both angles have measure $75°$.

14. $\frac{z}{8}=\frac{12}{16}$

$16z=8(12)$ Cross products

$16z=96$

$\frac{16z}{16}=\frac{96}{16}$ Divide by 16.

$z=6$

The solution set is $\{6\}$.

15. $\frac{x+5}{3}=\frac{x-3}{4}$

$4(x+5)=3(x-3)$

$4x+20=3x-9$

$x+20=-9$

$x=-29$

The solution set is $\{-29\}$.

16. What percent of 65 is 26?
Let p denote the percent.
$26=p\cdot 65$

$p=\frac{26}{65}=0.4=40\%$

26 is 40% of 65.

17. The results in the following table are rounded to the nearest thousandth.

Size	Price	Unit Cost (dollar per oz)
8 oz	\$2.99	$\frac{\$2.99}{8}\approx \$0.374$
16 oz	\$3.99	$\frac{\$3.99}{16}\approx \$0.249$ (*)
48 oz	\$14.69	$\frac{\$14.69}{48}\approx \$0.306$

Because the 16-oz size produces the lowest unit cost, it is the best buy. The unit cost, to the nearest thousandth, is \$0.249 per oz.

18. Let $x=$ the actual distance between Seattle and Cincinnati.

$\frac{x\text{ miles}}{1050\text{ miles}}=\frac{92\text{ inches}}{42\text{ inches}}$

$42x=92(1050)=96,600$

$x=\frac{96,600}{42}=2300$

The actual distance between Seattle and Cincinnati is 2300 miles.

19. Let $x=$ the amount invested at 3%.
Then $x+6000=$ the amount invested at 4.5%.

Amount Invested (in dollars)	Rate of Interest	Interest for One Year
x	0.03	$0.03x$
$x+6000$	0.045	$0.045(x+6000)$

$0.03x+0.045(x+6000)=870$

To clear decimals, multiply both sides by 1000.

$30x+45x(x+6000)=870,000$

$30x+45x+270,000=870,000$

$75x+270,000=870,000$

$75x=600,000$

$x=8000$

Since $x=8000$, $x+6000=14,000$.

Carlos invested \$8000 at 3% and \$14,000 at 4.5%.

20. Use the formula $d=rt$ and let t be the number of hours they traveled.

	r	***t***	***d***
First Car	50	t	$50t$
Second Car	65	t	$65t$

First car's distance	and	second car's distance	is	total distance.
↓	↓	↓	↓	↓
$50t$	$+$	$65t$	$=$	460

Solve the equation.

$50t+65t=460$

$115t=460$

$t=4$

The two cars will be 460 miles apart in 4 hours.

21. **(a)** The set of numbers graphed corresponds to the inequality $x<0$.

(b) The set of numbers graphed corresponds to the inequality $-2<x\leq 3$.

22. $-3x>-33$

$x<11$ Divide by -3.

Graph the solution set $(-\infty, 11)$.

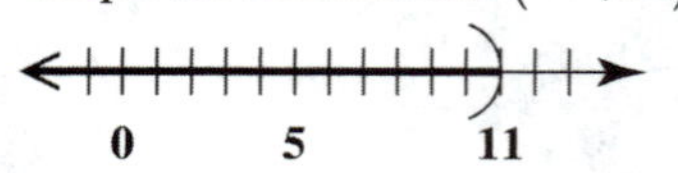

23. $-10 < 3x - 4 \le 14$

$-6 < 3x \le 18$ Add 4.

$-2 < x \le 6$ Divide by 3.

Graph the solution set $(-2, 6]$.

24. $-4x + 2(x-3) \ge 4x - (3+5x) - 7$

$-4x + 2x - 6 \ge 4x - 3 - 5x - 7$

$-2x - 6 \ge -x - 10$

$-x - 6 \ge -10$

$-x \ge -4$

$\frac{-1x}{-1} \le \frac{-4}{-1}$

$x \le 4$

Graph the solution set $(-\infty, 4]$.

25. Let x = the score on the third test.

The average of the three tests	is at least	80.
↓	↓	↓
$\frac{76+81+x}{3}$	$\ge$	80

$\frac{157+x}{3} \ge 80$

$3\left(\frac{157+x}{3}\right) \ge 3(80)$

$157 + x \ge 240$

$x \ge 83$

In order to average at least 80, Susan's score on her third test must be 83 or more.

Chapters R–2 Cumulative Review Exercises

1. $\frac{5}{6} + \frac{1}{4} - \frac{7}{15} = \frac{50}{60} + \frac{15}{60} - \frac{28}{60}$

$= \frac{65-28}{60}$

$= \frac{37}{60}$

2. $\frac{9}{8} \cdot \frac{16}{3} \div \frac{5}{8} = \frac{9}{8} \cdot \frac{16}{3} \cdot \frac{8}{5}$

$= \frac{3 \cdot 3 \cdot 16 \cdot 8}{8 \cdot 3 \cdot 5}$

$= \frac{48}{5}$

3. $4.8 + 12.5 + 16.73 = 34.03$

4. "The difference of half a number and 18" is written $\frac{1}{2}x - 18$.

5. "The quotient of 6 and 12 more than a number is 2" is written $\frac{6}{x+12} = 2$.

6. $\frac{8(7) - 5(6+2)}{3 \cdot 5 + 1} \ge 1$

$\frac{8(7) - 5(8)}{3 \cdot 5 + 1} \ge 1$

$\frac{56 - 40}{15 + 1} \ge 1$

$\frac{16}{16} \ge 1$

$1 \ge 1$

The statement is true.

7. $\frac{-4(9)(-2)}{-3^2} = \frac{-36(-2)}{-1 \cdot 3^2}$

$= \frac{72}{-9}$

$= -8$

8. $(-7-1)(-4) + (-4) = (-8)(-4) + (-4)$

$= 32 + (-4)$

$= 28$

9. $\frac{6(-4) - (-2)(12)}{3^2 + 7^2} = \frac{6(-4) - (-2)(12)}{9 + 19}$

$= \frac{-24 - (-24)}{9 + 19}$

$= \frac{0}{28} = 0$

10. Let $x=-2$, $y=-4$, and $z=3$.

$$\frac{3x^2-y^3}{-4z}=\frac{3(-2)^2-(-4)^3}{-4(3)}=\frac{3(4)-(-64)}{-12}=\frac{12+64}{-12}=\frac{76}{-12}=-\frac{19}{3}$$

11. $7(p+q)=7p+7q$

The multiplication of 7 is distributed over the sum, which illustrates the distributive property.

12. $7+(-7)=0$

A number added to its opposite is equal to 0. This illustrates the inverse property (of addition).

13. $3.5(1)=3.5$

A number multiplied by 1 is equal to itself. This illustrates the identity property (of multiplication).

14. $2r-6=8r$

$-6=6r$

$-1=r$

Check $r=-1$: $-8=-8$ True

The solution set is $\{-1\}$.

15. $4-5(s+2)=3(s+1)-1$

$4-5s-10=3s+3-1$

$-5s-6=3s+2$

$-8s-6=2$

$-8s=8$

$s=-1$

Check $s=-1$: $-1=-1$ True

The solution set is $\{-1\}$.

16. $\frac{2}{3}x+\frac{3}{4}x=-17$

$12\left(\frac{2}{3}x+\frac{3}{4}x\right)=12(-17)$ LCD = 12

$8x+9x=-204$

$17x=-204$

$x=-12$

Check $x=-12$: $-17=-17$ True

The solution set is $\{-12\}$.

17. $\frac{2x+3}{5}=\frac{x-4}{2}$

$(2x+3)(2)=(5)(x-4)$

$4x+6=5x-20$

$6=x-20$

$26=x$

Check $x=26$: $11=11$ True

The solution set is $\{26\}$.

18. Solve $3x+4y=24$ for y.

$4y=24-3x$

$y=\frac{24-3x}{4}$

19. Solve $P=a+b+c+B$ for c.
Subtract a, b, and B.

$P=a+b+c+B$

$P-a-b-B=c$

20. $6(r-1)+2(3r-5)\le -4$

$6r-6+6r-10\le -4$

$12r-16\le -4$

$12r\le 12$

$r\le 1$

Graph the solution set $(-\infty, 1]$.

0 1

21. $-18\le -9z<9$

$2\ge z>-1$ Divide; reverse the symbols.

or $-1<z\le 2$

Graph the solution set $(-1, 2]$.

-1 0 2

22. Let $x =$ the length of the middle-sized piece.
Then $3x =$ the length of the longest piece, and
$x - 5 =$ the length of the shortest piece.

$$x + 3x + (x - 5) = 40$$
$$5x - 5 = 40$$
$$5x = 45$$
$$x = 9$$

The length of the middle-sized piece is
9 centimeters, that of the longest piece is
27 centimeters, and that of the shortest piece is
4 centimeters.

23. Let $r =$ the radius and use 3.14 for π.
Using the formula for circumference, $C = 2\pi r$,
and $C = 78$, we have

$$2\pi r = 78$$
$$r = \frac{78}{2\pi} \approx 12.4204$$

To the nearest hundredth, the radius is
12.42 cm.

24. Let $x =$ the rate of the slower car.
Then $x + 20 =$ the rate of the faster car.
Use the formula $d = rt$.

$$d_{\text{slower}} + d_{\text{faster}} = d_{\text{total}}$$
$$(x)(4) + (x + 20)(4) = 400$$
$$4x + 4x + 80 = 400$$
$$8x + 80 = 400$$
$$8x = 320$$
$$x = 40$$

The rates are 40 mph and 60 mph.

25. **(a)** The segment of the circle representing white cars is 19% of the circle. What is 19% of 2.8 million?
$19\% \cdot 2.8 = 0.19 \cdot 2.8 = 0.532$
0.532 million cars, or 532,000 cars, are white.

(b) The segment of the circle representing silver cars is 18% of the circle. What is 18% of 2.8 million?
$18\% \cdot 2.8 = 0.18 \cdot 2.8 = 0.504$
0.504 million cars, or 504,000 cars, are silver.

(c) The segment of the circle representing red cars is 12% of the circle. What is 12% of 2.8 million?
$12\% \cdot 2.8 = 0.12 \cdot 2.8 = 0.336$
0.336 million cars, or 336,000 cars, are red.

Chapter 3
Linear Equations and Inequalities in Two Variables; Functions

3.1 Linear Equations and Rectangular Coordinates

Classroom Examples, Now Try Exercises

N1. **(a)** Move up from 2010 on the horizontal scale to the point plotted for 2010. Looking across at the vertical scale, this point is close to the line on the vertical scale for \$2.80. So, it cost about \$2.80 for a gallon of gasoline in 2010.

(b) The point for 2012 is about one-half of the way between the lines on the vertical scale for \$3.60 and \$3.80. So, it cost about \$3.70 for a gallon of gasoline in 2012. The increase is \$3.70 − \$2.80 = \$0.90.

N2. **(a)** $(3, 4)$

$$3x-7y=19$$
$$3(3)-7(4)\stackrel{?}{=}19 \quad \text{Let } x=3,\ y=4.$$
$$9-28\stackrel{?}{=}19$$
$$-19=19 \quad \text{False}$$

No, $(3, 4)$ is not a solution.

(b) $(-3, -4)$

$$3x-7y=19$$
$$3(-3)-7(-4)\stackrel{?}{=}19 \quad \text{Let } x=-3,\ y=-4.$$
$$-9+28\stackrel{?}{=}19$$
$$19=19 \quad \text{True}$$

Yes, $(-3, -4)$ is a solution.

N3. **(a)** In the ordered pair $(4, \_\_)$, $x=4$.

$$y=3x-12$$
$$y=3(4)-12 \quad \text{Let } x=4.$$
$$y=12-12$$
$$y=0$$

The ordered pair is $(4, 0)$.

(b) In the ordered pair $(\_\_, 3)$, $y=3$. Find the corresponding value of x by replacing y with 3 in the given equation.

$$y=3x-12$$
$$3=3x-12 \quad \text{Let } y=3.$$
$$15=3x \quad \text{Add 12.}$$
$$5=x \quad \text{Divide by 3.}$$

The ordered pair is $(5, 3)$.

N4. To complete the first ordered pair, let $x=0$.

$$5x-4y=20$$
$$5(0)-4y=20 \quad \text{Let } x=0.$$
$$-4y=20$$
$$y=-5$$

The first ordered pair is $(0, -5)$.

To complete the second ordered pair, let $y=0$.

$$5x-4y=20$$
$$5x-4(0)=20 \quad \text{Let } y=0.$$
$$5x=20$$
$$x=4$$

The second ordered pair is $(4, 0)$.

To complete the third ordered pair, let $x=2$.

$$5x-4y=20$$
$$5(2)-4y=20 \quad \text{Let } x=2.$$
$$10-4y=20$$
$$-4y=10$$
$$y=-\frac{5}{2}$$

The third ordered pair is $\left(2, -\frac{5}{2}\right)$.

The completed table of ordered pairs follows.

x	y
0	−5
4	0
2	$-\frac{5}{2}$

N5. To plot the ordered pairs, start at the origin in each case.

(a) $(-3, 1)$

Go 3 units to the left along the x-axis; then go up 1 unit, parallel to the y-axis.

(b) $(2, -4)$

Go 2 units to the right along the x-axis; then go down 4 units.

(c) $(0, -1)$

Go down 1 unit. This point is on the y-axis since the x-coordinate is 0.

(d) $\left(\frac{5}{2}, 3\right)$

Go 2.5 units to the right; then go up 3 units.

(e) $(-4, -3)$

Go 4 units to the left; then go down 3 units.

(f) $(-4, 0)$

Go 4 units to the left. This point is on the x-axis since the y-coordinate is 0.

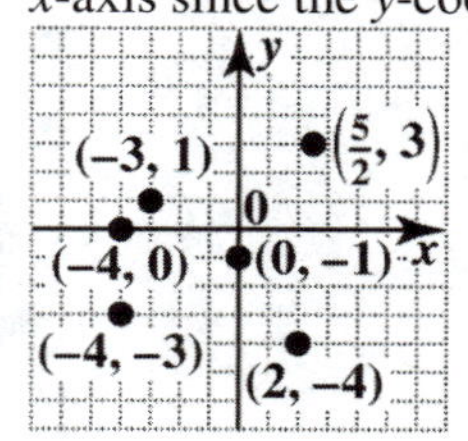

N6. To find y when $x = 2010$, substitute 2010 for x into the equation and use a calculator.

$y = -1.421x + 2989$

$y = -1.421(2010) + 2989$

$y \approx 133$

There were about 133,000 twin births in the United States in 2010.

Exercises

1. The symbol (x, y) *does* represent an ordered pair, while the symbols $[x, y]$ and $\{x, y\}$ *do not* represent ordered pairs. (Note that only parentheses are used to write ordered pairs.)

3. All points having coordinates in the form (negative, positive) are in quadrant II, so the point whose graph has coordinates $(-4, 2)$ is in quadrant $\underline{\text{II}}$.

5. All ordered pairs that are solutions of the equation $y = 3$ have y-coordinates equal to 3, so the ordered pair $(4, \underline{3})$ is a solution of the equation $y = 3$.

7. The point with coordinates (x, y) is in quadrant III if x is $\underline{\text{negative}}$ and y is $\underline{\text{negative}}$.

9. The point with coordinates (x, y) is in quadrant IV if x is $\underline{\text{positive}}$ and y is $\underline{\text{negative}}$.

11. If $xy < 0$, then either $x < 0$ and $y > 0$ or $x > 0$ and $y < 0$. If $x < 0$ and $y > 0$, then the point lies in quadrant II. If $x > 0$ and $y < 0$, then the point lies in quadrant IV.

13. The line segments between 2010 and 2011, and 2011 and 2012 fall, so these are the pairs of consecutive years in which the unemployment rate decreased.

15. For 2011, the unemployment rate was about 9%. For 2012, the unemployment rate was about 8%. The decline is $9\% - 8\% = 1\%$.

17. To determine whether $(0, 8)$ is a solution of the given equation, substitute 0 for x and 8 for y.

$x + y = 8$

$0 + 8 \stackrel{?}{=} 8$ Let $x = 0$, $y = 8$.

$8 = 8$ True

The result is true, so $(0, 8)$ is a solution of the given equation $x + y = 8$.

19. Substitute 3 for x and -1 for y.

$2x + y = 5$

$2(3) + (-1) \stackrel{?}{=} 5$ Let $x = 3$, $y = -1$.

$6 - 1 \stackrel{?}{=} 5$

$5 = 5$ True

The result is true, so $(3, -1)$ is a solution of $2x + y = 5$.

21. Substitute 5 for x and 2 for y.

$5x - 3y = 15$

$5(5) - 3(2) \stackrel{?}{=} 15$ Let $x = 5$, $y = 2$.

$25 - 6 \stackrel{?}{=} 15$

$19 = 15$ False

The result is false, so $(5, 2)$ is not a solution of $5x - 3y = 15$.

23. Substitute -8 for x and 2 for y.

$x = -4y$

$-8 \stackrel{?}{=} -4(2)$ Let $x = -8$, $y = 2$.

$-8 = -8$ True

The result is true, so $(-8, 2)$ is a solution of $x = -4y$.

25. Since x does not appear in the equation, we just substitute 2 for y.

$y = 2$

$2 = 2$ Let $y = 2$; true

The result is true, so $(4, 2)$ is a solution of $y = 2$.

27. Since y does not appear in the equation, we just substitute 4 for x.

$x-6=0$

$4-6\stackrel{?}{=}0$ Let $x=4$.

$-2=0$ False

The result is false, so $(4, 2)$ is not a solution of $x-6=0$.

29. In this ordered pair, $x=5$. Find the corresponding value of y by replacing x with 5 in the given equation.

$y=2x+7$

$y=2(5)+7$ Let $x=5$.

$y=10+7$

$y=17$

The ordered pair is $(5, \underline{17})$.

31. In this ordered pair, $y=-3$. Find the corresponding value of x by replacing y with -3 in the given equation.

$y=2x+7$

$-3=2x+7$ Let $y=-3$.

$-10=2x$

$-5=x$

The ordered pair is $(\underline{-5}, -3)$.

33. $y=-4x-4$

$0=-4x-4$ Let $y=0$.

$4=-4x$

$-1=x$

The ordered pair is $(\underline{-1}, 0)$.

35. $y=-4x-4$

$24=-4x-4$ Let $y=24$.

$28=-4x$

$-7=x$

The ordered pair is $(\underline{-7}, 24)$.

37. $4x+3y=24$

If $x=0$,

$4(0)+3y=24$

$0+3y=24$

$y=8$.

If $y=0$,

$4x+3(0)=24$

$4x+0=24$

$x=6$.

If $y=4$,

$4x+3(4)=24$

$4x+12=24$

$4x=12$

$x=3$.

The completed table of values is shown below.

x	y	ordered pair
0	8	$(0, 8)$
6	0	$(6, 0)$
3	4	$(3, 4)$

39. $4x-9y=-36$

If $y=0$,

$4x-9(0)=-36$

$4x-0=-36$

$4x=-36$

$x=-9$.

If $x=0$,

$4(0)-9y=-36$

$0-9y=-36$

$-9y=-36$

$y=4$.

If $y=8$,

$4x-9(8)=-36$

$4x-72=-36$

$4x=36$

$x=9$.

The completed table of values is shown below.

x	y	ordered pair
-9	0	$(-9, 0)$
0	4	$(0, 4)$
9	8	$(9, 8)$

41. $x=12$

No matter which value of y is chosen, the value of x will always be 12. Each ordered pair can be completed by placing 12 in the first position.

x	y	ordered pair
12	3	$(12, 3)$
12	8	$(12, 8)$
12	0	$(12, 0)$

43. $y = -10$

No matter which value of x is chosen, the value of y will always be -10. Each ordered pair can be completed by placing -10 in the second position.

x	y	ordered pair
4	−10	$(4, -10)$
0	−10	$(0, -10)$
−4	−10	$(-4, -10)$

45. The given equation, $y + 2 = 0$, may be written $y = -2$. For any value of x, the value of y will always be -2.

x	y	ordered pair
9	−2	$(9, -2)$
2	−2	$(2, -2)$
0	−2	$(0, -2)$

47. The given equation, $x - 4 = 0$, may be written $x = 4$. For any value of y, the value of x will always be 4.

x	y	ordered pair
4	4	$(4, 4)$
4	0	$(4, 0)$
4	−4	$(4, -4)$

49. No, the ordered pair $(3, 4)$ represents the point 3 units to the right of the origin and 4 units up from the x-axis. The ordered pair $(4, 3)$ represents the point 4 units to the right of the origin and 3 units up from the x-axis.

51. Point A was plotted by starting at the origin, going 2 units to the right along the x-axis, and then going up 4 units on a line parallel to the y-axis. The ordered pair for this point is $(2, 4)$, which is in quadrant I.

53. Point C was plotted by starting at the origin, going 5 units to the left along the x-axis, and then going up 4 units on a line parallel to the y-axis. The ordered pair for this point is $(-5, 4)$, which is in quadrant II.

55. Point E was plotted by starting at the origin, going 3 units to the right along the x-axis. The ordered pair for this point is $(3, 0)$, which is not in any quadrant.

57. Point G was plotted by starting at the origin, going 4 units to the right along the x-axis, and then going down 4 units on a line parallel to the y-axis. The ordered pair for this point is $(4, -4)$, which is in quadrant IV.

59. To plot $(6, 2)$, start at the origin, go 6 units to the right, and then go up 2 units.

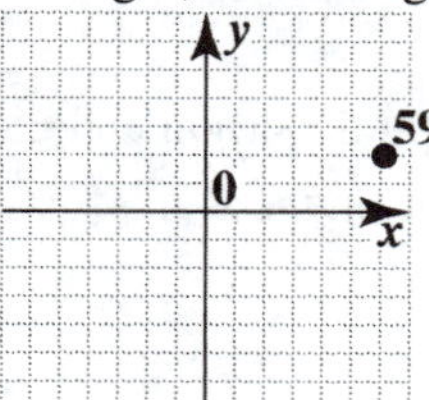

61. To plot $(-4, 2)$, start at the origin, go 4 units to the left, and then go up 2 units.

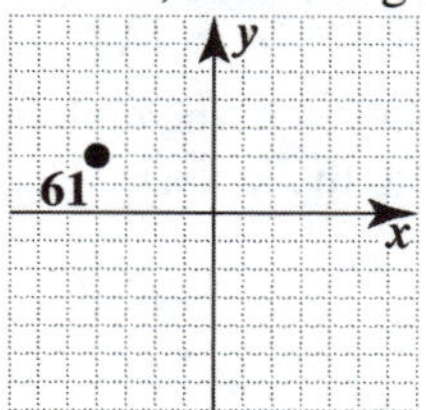

63. To plot $\left(-\frac{4}{5}, -1\right)$, start at the origin, go $\frac{4}{5}$ unit to the left, and then go down 1 unit.

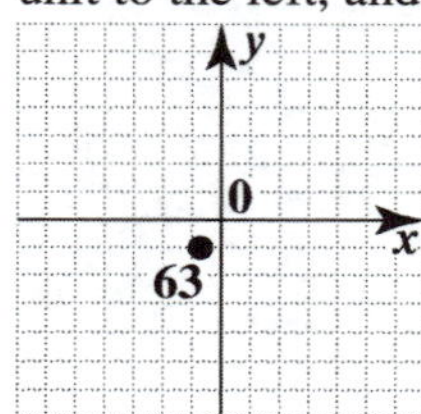

65. To plot $(3, -1.75)$, start at the origin, go 3 units to the right, and then go down $1.75\left(\text{or } 1\frac{3}{4}\right)$ units.

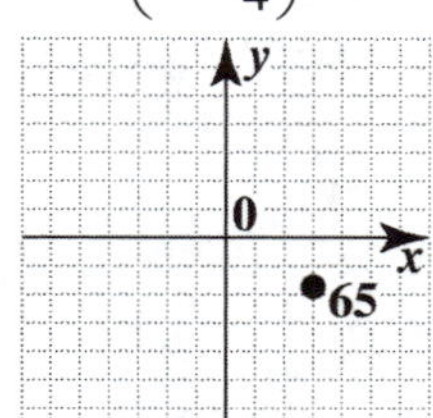

67. To plot $(0, 4)$, start at the origin and go up 4 units. The point lies on the y-axis.

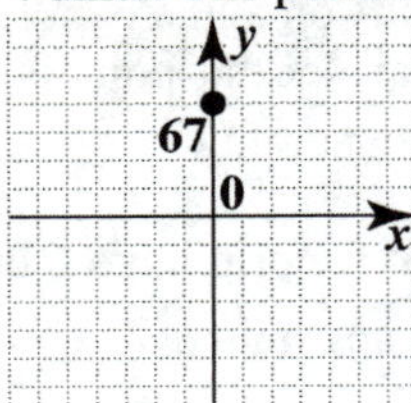

69. To plot $(4, 0)$, start at the origin and go right 4 units. The point lies on the x-axis.

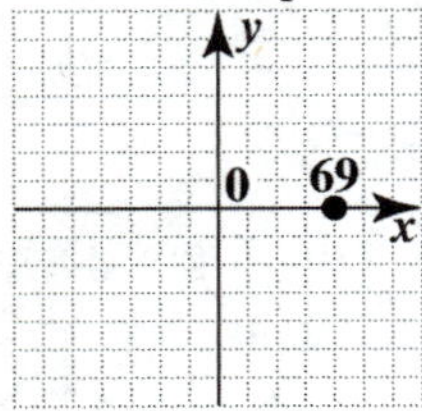

71. Substitute the given values in the table into $x - 2y = 6$ to complete the table.

Find y when $x = 0$.

$$\begin{aligned} x - 2y &= 6 \\ 0 - 2y &= 6 \quad \text{Let } x = 0. \\ -2y &= 6 \\ y &= -3 \end{aligned}$$

Find x when $y = 0$.

$$\begin{aligned} x - 2y &= 6 \\ x - 2(0) &= 6 \quad \text{Let } y = 0. \\ x - 0 &= 6 \\ x &= 6 \end{aligned}$$

Find y when $x = 2$.

$$\begin{aligned} x - 2y &= 6 \\ 2 - 2y &= 6 \quad \text{Let } x = 2. \\ -2y &= 4 \\ y &= -2 \end{aligned}$$

Find x when $y = -1$.

$$\begin{aligned} x - 2y &= 6 \\ x - 2(-1) &= 6 \quad \text{Let } y = -1. \\ x + 2 &= 6 \\ x &= 4 \end{aligned}$$

The completed table of values follows.

x	y
0	−3
6	0
2	−2
4	−1

Plot the points $(0, -3), (6, 0), (2, -2)$, and $(4, -1)$ on a coordinate system.

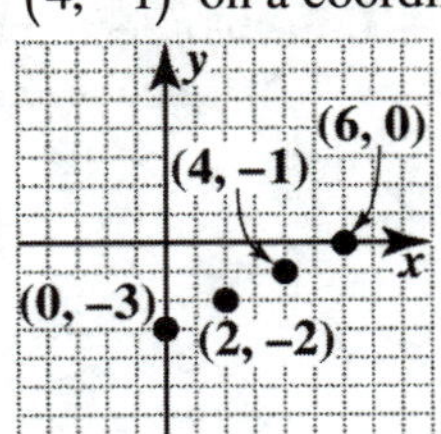

73. Substitute the given values in the table into $3x - 4y = 12$ to complete the table.

Find y when $x = 0$.

$$\begin{aligned} 3x - 4y &= 12 \\ 3(0) - 4y &= 12 \quad \text{Let } x = 0. \\ 0 - 4y &= 12 \\ -4y &= 12 \\ y &= -3 \end{aligned}$$

Find x when $y = 0$.

$$\begin{aligned} 3x - 4y &= 12 \\ 3x - 4(0) &= 12 \quad \text{Let } y = 0. \\ 3x - 0 &= 12 \\ 3x &= 12 \\ x &= 4 \end{aligned}$$

Find y when $x = -4$.

$$\begin{aligned} 3x - 4y &= 12 \\ 3(-4) - 4y &= 12 \quad \text{Let } x = -4. \\ -12 - 4y &= 12 \\ -4y &= 24 \\ y &= -6 \end{aligned}$$

Find x when $y = -4$.

$$\begin{aligned} 3x - 4y &= 12 \\ 3x - 4(-4) &= 12 \quad \text{Let } y = -4. \\ 3x + 16 &= 12 \\ 3x &= -4 \\ x &= -\frac{4}{3} \end{aligned}$$

The completed table is as follows.

x	y
0	−3
4	0
−4	−6
$-\frac{4}{3}$	−4

Plot the points $(0, -3), (4, 0), (-4, -6)$, and $\left(-\frac{4}{3}, -4\right)$ on a coordinate system.

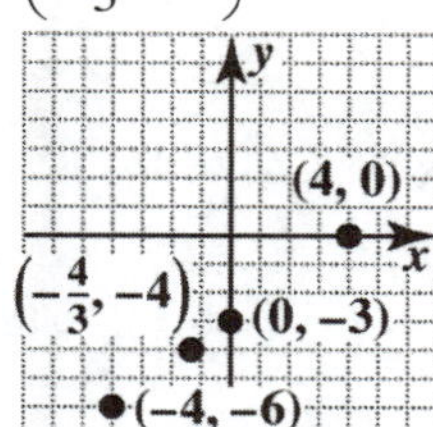

75. The given equation, $y + 4 = 0$, can be written as $y = -4$. So regardless of the value of x, the value of y is -4.

x	y
0	−4
5	−4
−2	−4
−3	−4

Plot the points $(0, -4), (5, -4), (-2, -4)$, and $(-3, -4)$ on a coordinate system.

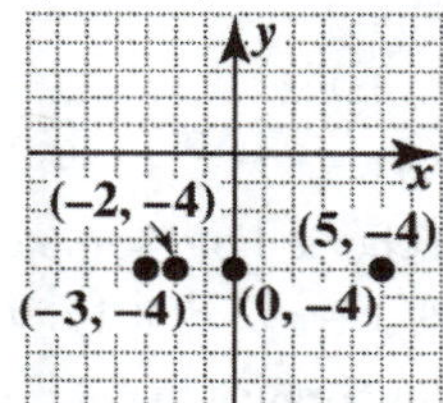

77. The points in each graph appear to lie on a straight line.

79. **(a)** When $x = 5$, $y = 45$. The ordered pair is $(5, 45)$.

(b) When $y = 50$, $x = 6$. The ordered pair is $(6, 50)$.

81. **(a)** We can write the results from the table as ordered pairs (x, y).

$(2008, 29.3), (2009, 28.3), (2010, 28.0), (2011, 26.9), (2012, 25.4), (2013, 22.5)$

(b) $(2013, 22.5)$ means that 22.5 percent of 2-year college students in 2013 received a degree within 3 years.

(c)

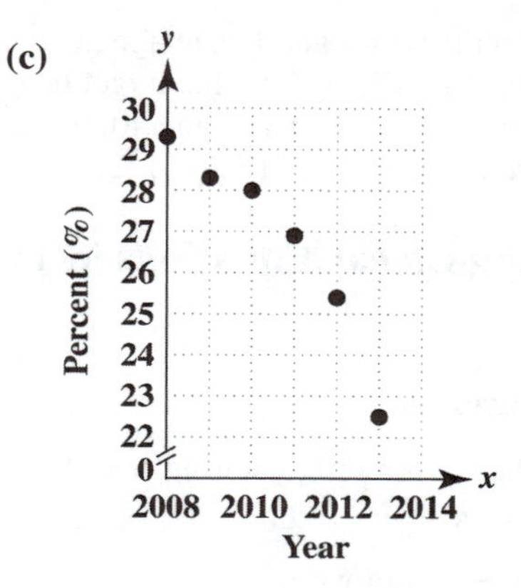

(d) The points lie approximately in a linear pattern. Rates at which 2-year college students complete a degree within 3 years were decreasing.

83. **(a)** Substitute $x = 20, 40, 60, 80$ in the equation $y = -0.65 + 143$.

$y = -0.65(20) + 143 = -13 + 143 = 130$

$y = -0.65(40) + 143 = -26 + 143 = 117$

$y = -0.65(60) + 143 = -39 + 143 = 104$

$y = -0.65(80) + 143 = -52 + 143 = 91$

The completed table follows.

Age	Heartbeats (per minute)
20	130
40	117
60	104
80	91

(b) In the ordered pairs (x, y), x represents age and y represents heartbeats.

$(20, 130), (40, 117), (60, 104), (80, 91)$

(c)

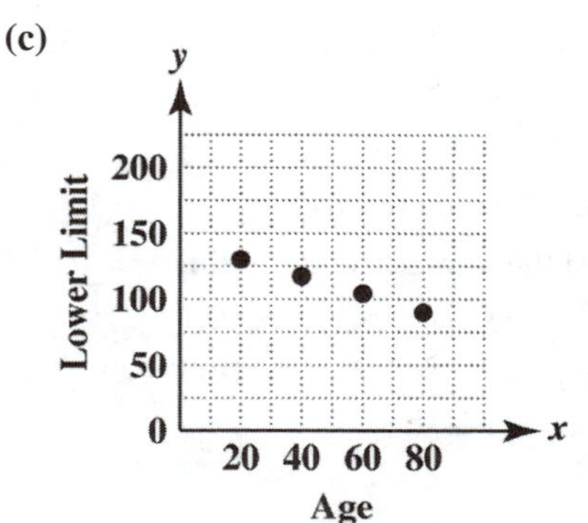

Yes, the points lie in a linear pattern.

85. Refer to the tables and locate age 20 and age 40 in each table. For age 20, the target heart rate is between 130 and 170. For age 40, the target heart rate is between 117 and 153.

3.2 Graphing Linear Equations in Two Variables

Now Try Exercises

N1. Find the x-intercept by letting $y = 0$.

$x + y = -5$

$x + 0 = -5$ Let $y = 0$.

$x = -5$

The x-intercept is $(-5, 0)$.

Find the y-intercept by letting $x = 0$.

$x + y = -5$

$0 + y = -5$ Let $x = 0$.

$y = -5$

The y-intercept is $(0, -5)$.

Get a third point as a check. For example, choosing $x = 1$ gives $y = -6$. Plot $(-5, 0)$, $(0, -5)$, and $(1, -6)$ and draw a line through them.

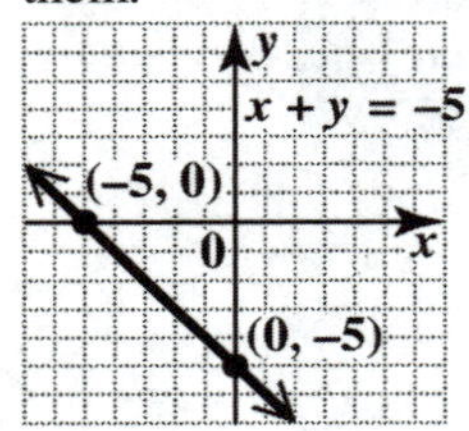

N2. Find the x-intercept by letting $y = 0$.

$2x - 4y = 8$

$2x - 4(0) = 8$ Let $y = 0$.

$2x - 0 = 8$

$2x = 8$

$x = 4$

The x-intercept is $(4, 0)$.

Find the y-intercept by letting $x = 0$.

$2x - 4y = 8$

$2(0) - 4y = 8$ Let $x = 0$.

$0 - 4y = 8$

$-4y = 8$

$y = -2$

The y-intercept is $(0, -2)$.

Get a third point as a check. For example, choosing $x = -2$ gives $y = -3$. Plot $(4, 0)$, $(0, -2)$, and $(-2, -3)$ and draw a line through them.

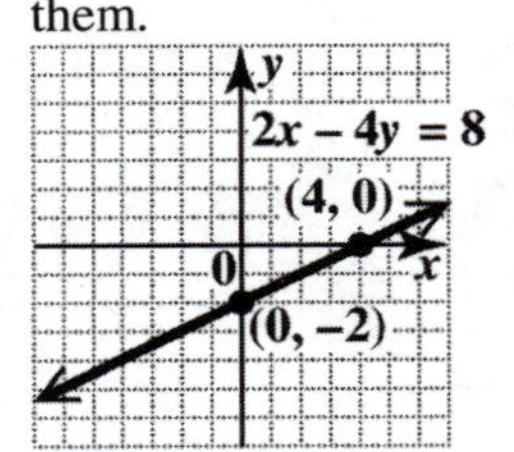

N3. Find the x-intercept by letting $y = 0$.

$x + 2y = 2$

$x + 2(0) = 2$ Let $y = 0$.

$x + 0 = 2$

$x = 2$

The x-intercept is $(2, 0)$.

Find the y-intercept by letting $x = 0$.

$x + 2y = 2$

$0 + 2y = 2$ Let $x = 0$.

$2y = 2$

$y = 1$

The y-intercept is $(0, 1)$.

Get a third point as a check. For example, choosing $x = 4$ gives $y = -1$. Plot $(2, 0)$, $(0, 1)$, and $(4, -1)$ and draw a line through them.

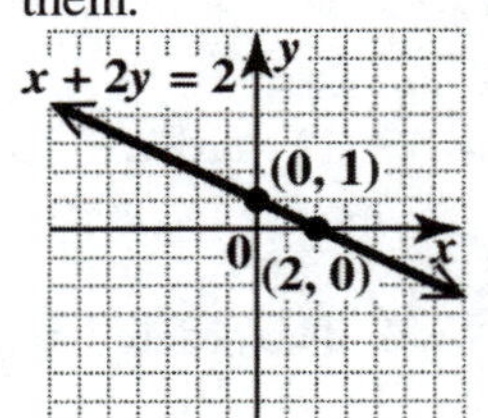

N4. Find the x intercept by letting $y = 0$.

$y = \frac{1}{3}x + 1$

$0 = \frac{1}{3}x + 1$ Let $y = 0$.

$-1 = \frac{1}{3}x$

$3(-1) = 3\left(\frac{1}{3}x\right)$ Multiply by 3.

$-3 = x$

The x-intercept is $(-3, 0)$.

Find the y-intercept by letting $x = 0$.

$$y = \frac{1}{3}x + 1$$

$$y = \frac{1}{3}(0) + 1 \quad \text{Let } x = 0.$$

$$y = 0 + 1$$

$$y = 1$$

The y-intercept is $(0, 1)$.

Get a third point as a check. For example, choosing $x = 3$ gives $y = 2$. Plot $(-3, 0)$, $(0, 1)$, and $(3, 2)$ and draw a line through them.

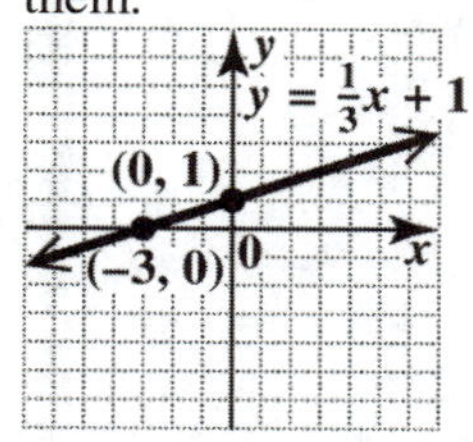

N5. The x-intercept and the y-intercept are the *same* point, $(0, 0)$. Select two other values for x or y to find two other points on the graph, as in the following table.

x	y
0	0
1	−2
−1	2

Graph the equation by plotting these points and drawing a line through them.

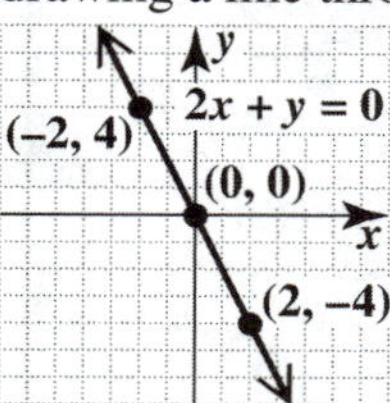

N6. No matter what value we choose for x, y is always 2, as shown in the table of ordered pairs.

x	y
0	2
−5	2
3	2

The graph is a horizontal line through $(0, 2)$.

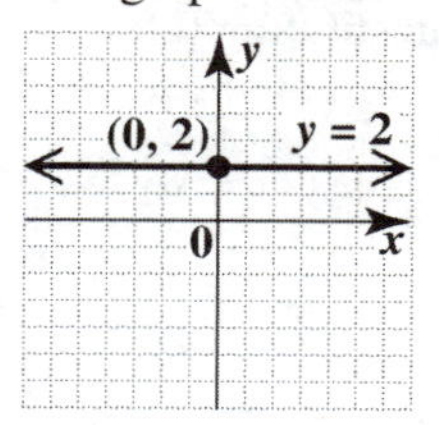

N7. x is always -4 regardless of the value of y, as shown in the table of ordered pairs.

x	y
−4	0
−4	−5
−4	2

The graph is a vertical line through $(-4, 0)$.

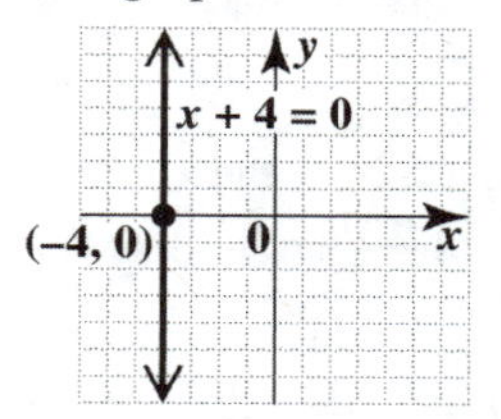

N8. **(a)** $x = 0$ represents 2000, so $x = 9$ represents 2009. On the graph, find 9 on the horizontal axis, move up to the graphed line, then across to the vertical axis. It appears that the weekly time spent online in 2009 was about 18 hours.

(b) To use the equation, substitute 6 for x.

$$y = 0.96x + 9.1$$

$$y = 0.96(9) + 9.1 \quad \text{Let } x = 9.$$

$$y = 17.7 \text{ hours}$$

Exercises

1. A linear equation in two variables x and y can be written in the form $Ax + \underline{By} = \underline{C}$, where A, B, and C are real numbers and A and B are not both $\underline{0}$.

3. **(a)** To determine which equation has y-intercept $(0, -4)$, set x equal to 0 and see which equation is equivalent to $y = -4$. Choice A is correct since

$$3x + y = -4$$

$$3(0) + y = -4$$

$$y = -4.$$

(b) If the graph of the equation goes through the origin, then substituting 0 for x and 0 for y will result in a true statement. Choice C is correct since

$y = 4x$

$0 = 4(0)$

$0 = 0$

is a true statement.

(c) Choice D is correct since the graph of $y = 4$ is a horizontal line.

(d) To determine which equation has x-intercept $(4, 0)$, set y equal to 0 and see which equation is equivalent to $x = 4$. Choice B is correct since

$x - 4 = 0$

$x = 4.$

5. The dot at 4 on the x-axis is the x-intercept, that is, $(4, 0)$. The dot at -4 on the y-axis is the y-intercept, that is, $(0, -4)$.

7. The dot at -2 on the x-axis is the x-intercept, that is, $(-2, 0)$. The dot at -3 on the y-axis is the y-intercept, that is, $(0, -3)$.

9. **(a)** The graph of $x = -2$ is a vertical line with x-intercept $(-2, 0)$, which is choice D.

(b) The graph of $y = -2$ is a horizontal line with y-intercept $(0, -2)$, which is choice C.

(c) The graph of $x = 2$ is a vertical line with x-intercept $(2, 0)$, which is choice B.

(d) The graph of $y = 2$ is a horizontal line with y-intercept $(0, 2)$, which is choice A.

11. If $x = 0$, | If $y = 0$,

$0 + y = 5$ | $x + 0 = 5$

$y = 5.$ | $x = 5.$

If $x = 2$,

$2 + y = 5$

$y = 3.$

The ordered pairs are $(0, \underline{5})$, $(\underline{5}, 0)$, and $(2, \underline{3})$.

Plot the corresponding points and draw a line through them.

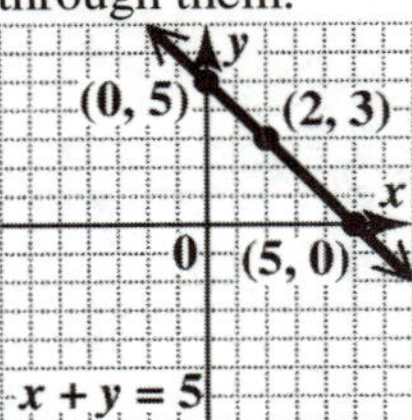

13. If $x = 0$, | If $x = 3$,

$y = \frac{2}{3}(0) + 1$ | $y = \frac{2}{3}(3) + 1$

$y = 0 + 1$ | $y = 2 + 1$

$y = 1.$ | $y = 3.$

If $x = -3$,

$y = \frac{2}{3}(-3) + 1$

$y = -2 + 1$

$y = -1.$

The ordered pairs are $(0, \underline{1})$, $(3, \underline{3})$, and $(-3, \underline{-1})$. Plot the corresponding points and draw a line through them.

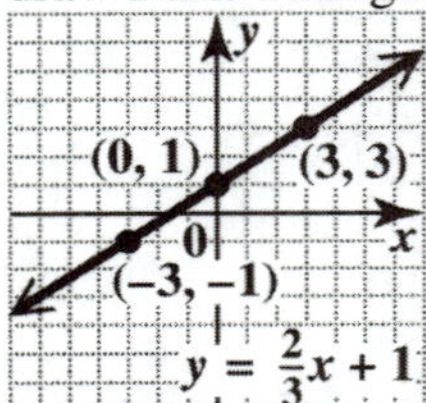

15. If $x = 0$, | If $y = 0$,

$3(0) = -y - 6$ | $3x = -0 - 6$

$0 = -y - 6$ | $3x = -6$

$y = -6.$ | $x = -2.$

If $x = -\frac{1}{3}$,

$3\left(-\frac{1}{3}\right) = -y - 6$

$-1 = -y - 6$

$y - 1 = -6$

$y = -5.$

The ordered pairs are $(0, \underline{-6})$, $(\underline{-2}, 0)$, and $\left(-\frac{1}{3}, \underline{-5}\right)$. Plot the corresponding points and draw a line through them.

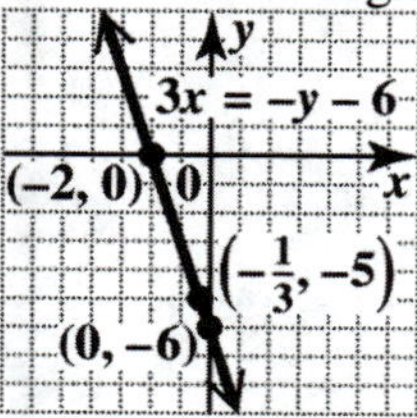

17. To find the x-intercept, let $y = 0$.

$x - y = 8$
$x - 0 = 8$
$x = 8$

The x-intercept is $(8, 0)$.

To find the y-intercept, let $x = 0$.

$x - y = 8$
$0 - y = 8$
$-y = 8$
$y = -8$

The y-intercept is $(0, -8)$.

19. To find the x-intercept, let $y = 0$.

$5x - 2y = 20$
$5x - 2(0) = 20$
$5x = 20$
$x = 4$

The x-intercept is $(4, 0)$.

To find the y-intercept, let $x = 0$.

$5x - 2y = 20$
$5(0) - 2y = 20$
$-2y = 20$
$y = -10$

The y-intercept is $(0, -10)$.

21. To find the x-intercept, let $y = 0$.

$x + 6y = 0$
$x + 6(0) = 0$
$x + 0 = 0$
$x = 0$

The x-intercept is $(0, 0)$. Since we have found the point with x equal to 0, this is also the y-intercept.

23. To find the x-intercept, let $y = 0$.

$y = -2x + 4$
$0 = -2x + 4$
$2x = 4$
$x = 2$

The x-intercept is $(2, 0)$.

To find the y-intercept, let $x = 0$.

$y = -2x + 4$
$y = -2(0) + 4$
$y = 0 + 4$
$y = 4$

The y-intercept is $(0, 4)$.

25. To find the x-intercept, let $y = 0$.

$y = \frac{1}{3}x - 2$
$0 = \frac{1}{3}x - 2$
$2 = \frac{1}{3}x$
$6 = x$

The x-intercept is $(6, 0)$.

To find the y-intercept, let $x = 0$.

$y = \frac{1}{3}x - 2$
$y = \frac{1}{3}(0) - 2$
$y = 0 - 2$
$y = -2$

The y-intercept is $(0, -2)$.

27. To find the x-intercept, let $y = 0$.

$2x - 3y = 0$
$2x - 3(0) = 0$
$2x - 0 = 0$
$2x = 0$
$x = 0$

The x-intercept is $(0, 0)$. Since we have found the point with x equal to 0, this is also the y-intercept.

29. $x - 4 = 0$ is equivalent to $x = 4$. This is an equation of a vertical line. Its x-intercept is $(4, 0)$ and there is no y-intercept.

31. $y = 2.5$ is an equation of a horizontal line. Its y-intercept is $(0, 2.5)$ and there is no x-intercept.

33. Find the intercepts.

$x - y = 4$ $\quad$ $x - y = 4$

$x - 0 = 4$ Let $y = 0$. $\quad$ $0 - y = 4$ Let $x = 0$.

$x = 4$ $\quad$ $y = -4$

The x-intercept is $(4, 0)$ and the y-intercept is $(0, -4)$. To find a third point, choose $y = 1$.

$x - y = 4$

$x - 1 = 4$ Let $y = 1$.

$x = 5$

This gives the ordered pair $(5, 1)$. Plot $(4, 0)$, $(0, -4)$, and $(5, 1)$ and draw a line through them.

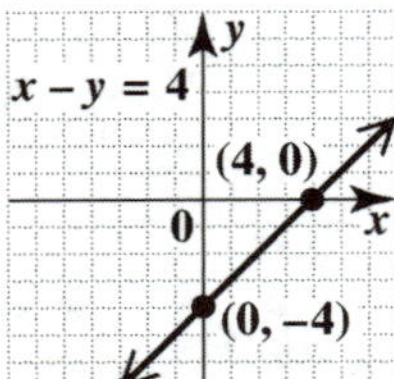

35. Find the intercepts.

$2x + y = 6$ $\quad$ $2x + y = 6$

$2x + 0 = 6$ Let $y = 0$. $\quad$ $2(0) + y = 6$ Let $x = 0$.

$2x = 6$ $\quad$ $0 + y = 6$

$x = 3$ $\quad$ $y = 6$

The x-intercept is $(3, 0)$ and the y-intercept is $(0, 6)$. To find a third point, choose $x = 1$.

$2x + y = 6$

$2(1) + y = 6$ Let $x = 1$.

$2 + y = 6$

$y = 4$

This gives the ordered pair $(1, 4)$. Plot $(3, 0)$, $(0, 6)$, and $(1, 4)$ and draw a line through them.

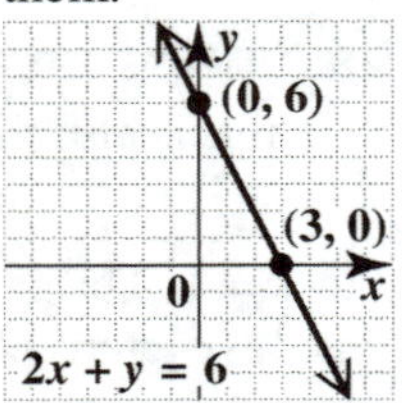

37. Find the intercepts.

$y = 2x - 5$ $\quad$ $y = 2x - 5$

$0 = 2x - 5$ Let $y = 0$. $\quad$ $y = 2(0) - 5$ Let $x = 0$.

$5 = 2x$ $\quad$ $y = 0 - 5$

$\frac{5}{2} = x$ $\quad$ $y = -5$

The x-intercept is $\left(\frac{5}{2}, 0\right)$ and the y-intercept is $(0, -5)$. To find a third point, choose $x = 1$.

$y = 2x - 5$

$y = 2(1) - 5$ Let $x = 1$.

$y = 2 - 5$

$y = -3$

This gives the ordered pair $(1, -3)$. Plot $\left(\frac{5}{2}, 0\right)$, $(0, -5)$, and $(1, -3)$ and draw a line through them.

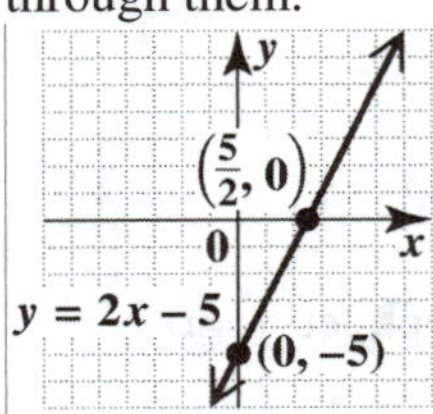

39. Begin by finding the intercepts.

$x = y + 2$ $\quad$ $x = y + 2$

$x = 0 + 2$ Let $y = 0$. $\quad$ $0 = y + 2$ Let $x = 0$.

$x = 2$ $\quad$ $-2 = y$

The x-intercept is $(2, 0)$ and the y-intercept is $(0, -2)$. To find a third point, choose $y = 1$.

$x = y + 2$

$x = 1 + 2$ Let $y = 1$.

$x = 3$

This gives the ordered pair $(3, 1)$. Plot $(2, 0)$, $(0, -2)$, and $(3, 1)$ and draw a line through them.

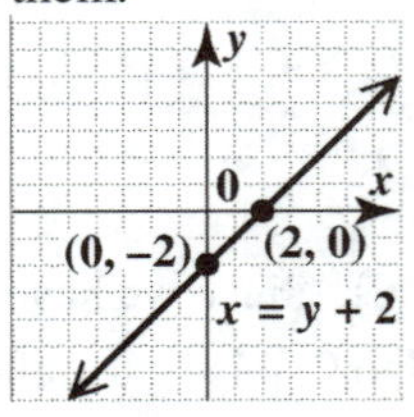

41. Find the intercepts. Find x when $y = 0$.

$$\begin{aligned} 2x-5y&=10 \\ 2x-5(0)&=10 \quad \text{Let } y=0. \\ 2x+0&=10 \\ 2x&=10 \\ x&=5 \end{aligned}$$

Find y when $x = 0$.

$$\begin{aligned} 2x-5y&=10 \\ 2(0)-5y&=10 \quad \text{Let } x=0. \\ 0-5y&=10 \\ -5y&=10 \\ y&=-2 \end{aligned}$$

The x-intercept is $(5, 0)$ and the y-intercept is $(0, -2)$. To find a third point, choose $x = 10$.

$$\begin{aligned} 2x-5y&=10 \\ 2(10)-5y&=10 \quad \text{Let } x=10. \\ 20-5y&=10 \\ -5y&=-10 \\ y&=2 \end{aligned}$$

This gives the ordered pair $(10, 2)$. Plot $(5, 0)$, $(0, -2)$, and $(10, 2)$. Draw a line through these three points.

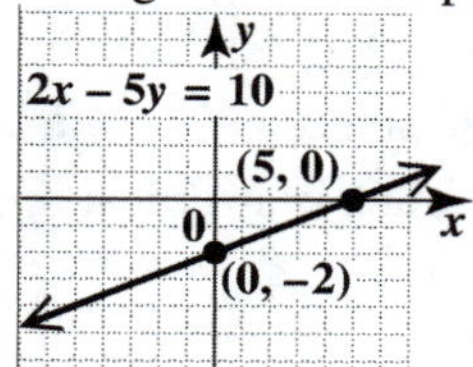

43. Find the intercepts. Find x when $y = 0$.

$$\begin{aligned} 3x+7y&=14 \\ 3x+7(0)&=14 \quad \text{Let } y=0. \\ 3x+0&=14 \\ 3x&=14 \\ x&=\frac{14}{3} \end{aligned}$$

Find y when $x = 0$.

$$\begin{aligned} 3x+7y&=14 \\ 3(0)+7y&=14 \quad \text{Let } x=0. \\ 0+7y&=14 \\ 7y&=14 \\ y&=2 \end{aligned}$$

The x-intercept is $\left(\frac{14}{3}, 0\right)$ and the y-intercept is $(0, 2)$. To find a third point, choose $x = 2$.

$$\begin{aligned} 3x+7y&=14 \\ 3(2)+7y&=14 \quad \text{Let } x=2. \\ 6+7y&=14 \\ 7y&=8 \\ y&=\frac{8}{7} \end{aligned}$$

This gives the ordered pair $\left(2, \frac{8}{7}\right)$. Plot $\left(\frac{14}{3}, 0\right)$, $(0, 2)$, and $\left(2, \frac{8}{7}\right)$. Writing $\frac{14}{3}$ as the mixed number $4\frac{2}{3}$ and $\frac{8}{7}$ as $1\frac{1}{7}$ will be helpful for plotting. Draw a line through these three points.

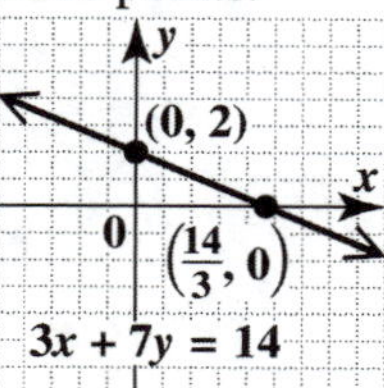

45. Find the x-intercept.

$$\begin{aligned} 0&=-\frac{3}{4}x+3 \\ \frac{3}{4}x&=3 \qquad \text{Let } y=0. \\ x&=3\left(\frac{4}{3}\right) \\ x&=4 \end{aligned}$$

The x-intercept is $(4, 0)$.

Find the y-intercept.

$$\begin{aligned} y&=-\frac{3}{4}x+3 \\ y&=-\frac{3}{4}(0)+3 \quad \text{Let } x=0. \\ y&=0+3 \\ y&=3 \end{aligned}$$

The y-intercept is $(0, 3)$.

To find a third point, choose $x = 2$.

$$\begin{aligned} y&=-\frac{3}{4}(2)+3 \\ y&=-\frac{3}{2}+3 \\ y&=\frac{3}{2} \end{aligned}$$

Plot $(4, 0), (0, 3)$, and $\left(2, \frac{3}{2}\right)$ and draw a line through these points.

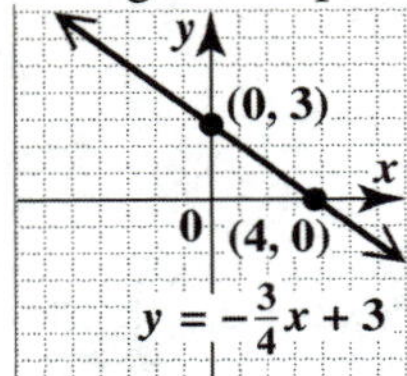

47. If $y = 0, x = 0$. Both intercepts are the origin, $(0, 0)$. Find two additional points.

$y - 2x = 0$
$y - 2(1) = 0$ Let $x = 1$.
$y - 2 = 0$
$y = 2$

Find y when $x = -3$.

$y - 2x = 0$
$y - 2(-3) = 0$ Let $x = -3$.
$y + 6 = 0$
$y = -6$

Plot $(0, 0), (1, 2)$, and $(-3, -6)$ and draw a line through them.

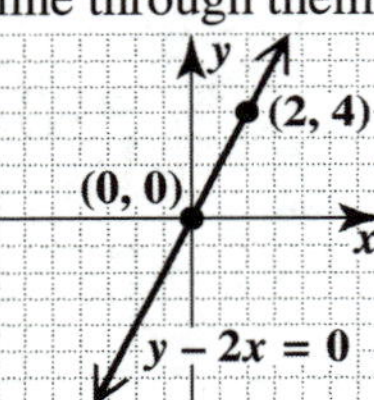

49. Find three points on the line.
If $x = 0$, $y = -6(0) = 0$.
If $x = 1$, $y = -6(1) = -6$.
If $x = -1$, $y = -6(-1) = 6$.
Plot $(0, 0), (1, -6)$, and $(-1, 6)$ and draw a line through these points.

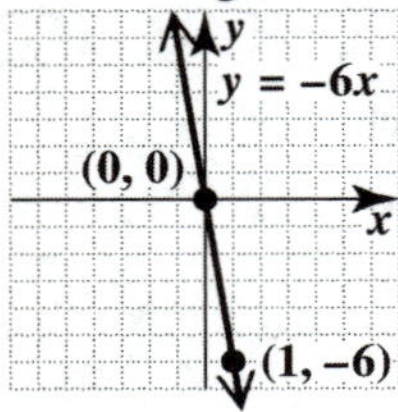

51. For any value of x, the value of y is -1. Three ordered pairs are $(-4, -1), (0, -1)$, and $(3, -1)$. Plot these points and draw a line through them. The graph is a horizontal line.

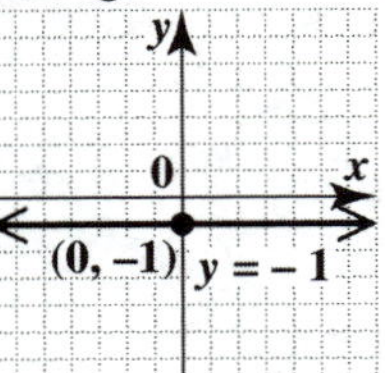

53. For any value of y, the value of x is 5. Three ordered pairs are $(5, 3), (5, 0)$, and $(5, -1)$. Plot these points and draw a line through them. The graph is a vertical line.

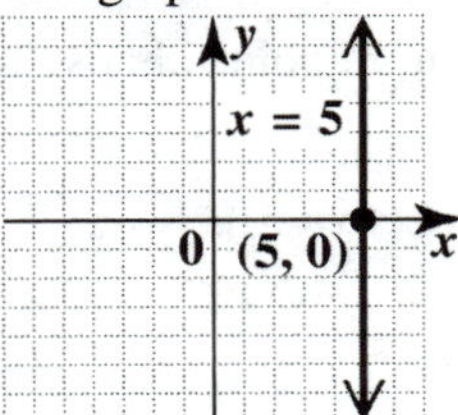

55. Solve for x.

$x + 2 = 0$
$x = -2$

For any value of y, the value of x is -2. Three ordered pairs are $(-2, 3), (-2, 0)$, and $(-2, -4)$. Plot these points and draw a line through them. The graph is a vertical line.

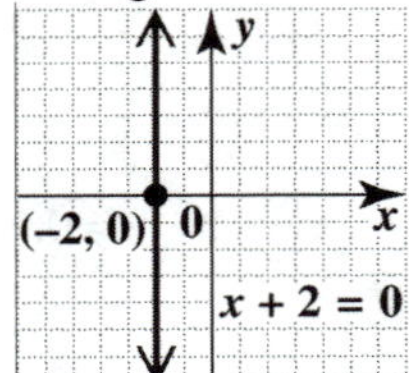

57. Solve for y.

$-3y = 15$
$y = -5$

For any value of x, the value of y is -5. Three ordered pairs are $(-2, -5), (0, -5)$, and $(1, -5)$. Plot these points and draw a line through them. The graph is a horizontal line.

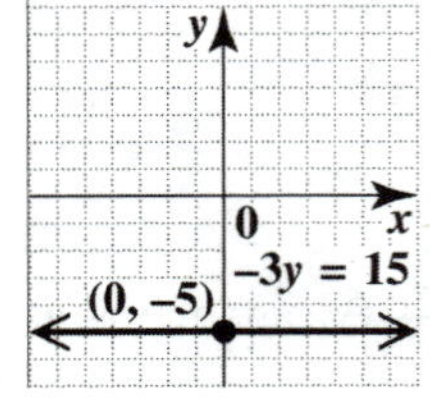

59. Solve for x.
$x+2=8$
$x=6$
For any value of y, the value of x is 6. Three ordered pairs are $(6,-2)$, $(6,0)$, and $(6,1)$. Plot these points and draw a line through them. The graph is a vertical line.

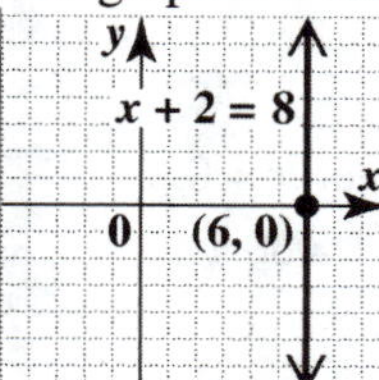

61. If $x=0$, $3(0)=y-9$, $0=y-9$, $9=y$.
If $y=0$, $3x=0-9$, $3x=-9$, $x=-3$.
The graph of this equation is a line with x-intercept $(-3,0)$ and y-intercept $(0,9)$.

62. If $x=0$, $2(0)=y-4$, $0=y-4$, $4=y$.
If $y=0$, $2x=0-4$, $2x=-4$, $x=-2$.
The graph of this equation is a line with x-intercept $(-2,0)$ and y-intercept $(0,4)$.

63. Solve for x.
$x-10=1$
$x=11$ Add 10.
The graph of this equation is a vertical line with x-intercept $(11,0)$.

65. Solve for y.
$3y=-6$
$y=-2$ Divide by 3.
The graph of this equation is a horizontal line with y-intercept $(0,-2)$.

67. $2x=4y$ can be written as $2x-4y=0$, so the graph of this equation passes through the origin, $(0,0)$, which is the x-intercept and the y-intercept. If we let $x=2$, then $4=4y$, so $y=1$; and if we let $x=4$, then $8=4y$, so $y=2$. Thus, the graph also passes through the points $(2,1)$ and $(4,2)$.

69. The points $(3,5)$, $(3,0)$, and $(3,-3)$ all have x-coordinate 3, so $x=3$ is the equation of the line.

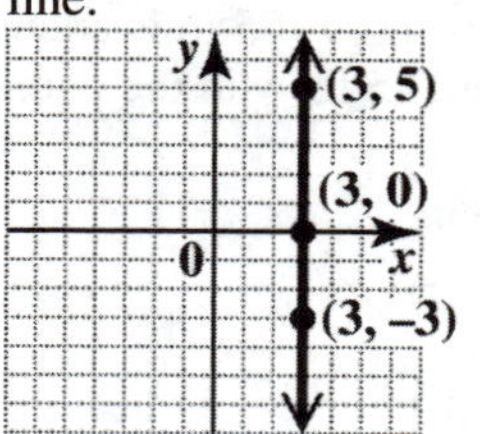

71. The points $(-3,-3)$, $(0,-3)$, and $(4,-3)$ all have y-coordinate -3, so $y=-3$ is the equation of the line.

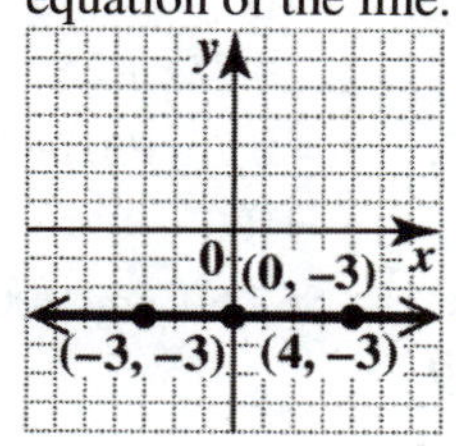

73. **(a)** $y=5.5x-220$
$y=5.5(62)-220=121$ Let $x=62$.
$y=5.5(66)-220=143$ Let $x=66$.
$y=5.5(72)-220=176$ Let $x=72$.
The approximate weights of men whose heights are 62 in., 66 in., and 72 in. are 121 lb, 143 lb, and 176 lb, respectively.

(b) The three ordered pairs from part (a) are $(62,121)$, $(66,143)$, and $(72,176)$.

(c) Plot the points $(62,121)$, $(66,143)$, and $(72,176)$ and connect them with a smooth line.

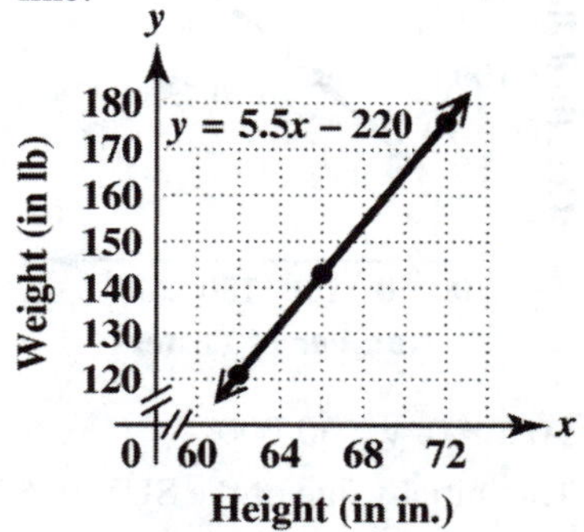

(d) Locate 155 on the vertical scale, then move across to the line, then down to the horizontal scale. From the graph, the height of a man who weighs 155 lb is about 68 in. Now substitute 155 for y in the equation.

$$y = 5.5x - 220$$
$$155 = 5.5x - 220$$
$$375 = 5.5x$$
$$x = \frac{375}{5.5} \approx 68.18$$

From the equation, the height is 68 in. to the nearest inch.

75. (a) Let $x = 50$. Let $x = 100$.

$$y = 0.75x + 25 \qquad y = 0.75x + 25$$
$$y = 0.75(50) + 25 \qquad y = 0.75(100) + 25$$
$$y = 62.50 \qquad y = 100$$

Thus, 50 posters cost $62.50 and 100 posters cost $100.

(b) Let $y = 175$. $175 = 0.75x + 25$

$$150 = 0.75x$$
$$x = \frac{150}{0.75} = 200$$

Thus, 200 posters cost $175.

(c) The three ordered pairs from parts (a) and (b) are $(50, 62.50)$, $(100, 100)$, and $(200, 175)$.

(d) Plot the points in part (c) and connect them with a smooth line.

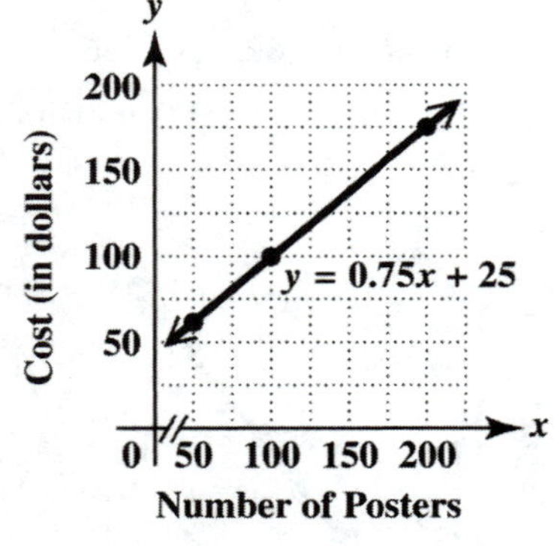

77. (a) At $x = 0$, $y = 30{,}000$.

The initial value of the SUV is $30,000.

(b) At $x = 3$, $y = 15{,}000$.

$$30{,}000 - 15{,}000 = 15{,}000$$

The depreciation after the first 3 years is $15,000.

(c) The line connecting consecutive years from 0 to 1, 1 to 2, 2 to 3, and so on, drops 5000 for each segment. Therefore, the annual depreciation in each of the first 5 years is $5000.

(d) $(5, 5000)$ means after 5 years the SUV has a value of $5000.

79. (a) $y = 0.307x + 30.1$

For the year 2000, let $x = 0$.

$$y = 0.307(0) + 30.1 = 30.1$$

For 2008, let $x = 8$.

$$y = 0.307(8) + 30.1 = 32.6$$

For 2012, let $x = 12$.

$$y = 0.307(12) + 30.1 = 33.8$$

The approximate per capita consumptions for 2000, 2008, and 2012 are 30.1lb, 32.6 lb, and 33.8 lb, respectively.

(b) Locate 0, 8, and 12 on the horizontal scale, then find the corresponding value on the vertical scale.
The approximate per capita consumptions for 2000, 2008, and 2012 are 29.8 lb, 32.8 lb, and 33.5 lb, respectively.

(c) The corresponding values are quite close.

(d) $y = 0.307x + 30.1$

For the year 2022, let $x = 22$.

$$y = 0.307(22) + 30.1 = 36.9$$

The approximation of 36.9 is very close to the USDA projection of 36.8.

3.3 The Slope of a Line

Now Try Exercises

N1. The given points are $(-2, -2)$ and $(-1, 1)$.

$$\text{slope} = \frac{\text{change in } y \text{ (rise)}}{\text{change in } x \text{ (run)}}$$
$$= \frac{1 - (-2)}{-1 - (-2)} = \frac{1 + 2}{-1 + 2} = \frac{3}{1} = 3$$

N2. Use the slope formula with $(4, -5)$ and $(-2, -4)$.

$$\text{slope } m = \frac{y_2 - y_1}{x_2 - x_1} = \frac{-4 - (-5)}{-2 - 4} = \frac{1}{-6} = -\frac{1}{6}$$

N3. Use the slope formula with $(1,-3)$ and $(4,-3)$.

$$m=\frac{y_2-y_1}{x_2-x_1}=\frac{-3-(-3)}{4-1}=\frac{0}{3}=0$$

N4. Use the slope formula with $(-2,1)$ and $(-2,-4)$.

$$m=\frac{y_2-y_1}{x_2-x_1}=\frac{-4-1}{-2-(-2)}=\frac{-5}{0},$$

which is *undefined.*

N5. Solve the equation for *y*.

$$3x+5y=-1$$

$$5y=-3x-1 \quad \text{Subtract } 3x.$$

$$y=-\frac{3}{5}x-\frac{1}{5} \quad \text{Divide by 5.}$$

The slope is given by the coefficient of *x*, so the slope is $-\frac{3}{5}$.

N6. Find the slope of each line by first solving each equation for *y*.

$$2x-3y=1 \qquad\qquad 4x+6y=5$$

$$-3y=-2x+1 \qquad\qquad 6y=-4x+5$$

$$y=\frac{2}{3}x-\frac{1}{3} \qquad\qquad y=-\frac{2}{3}x+\frac{5}{6}$$

Slope is $\frac{2}{3}$. Slope is $-\frac{2}{3}$.

The slopes are not equal, so the lines are not parallel. The product of the slopes is not -1, so the lines are not perpendicular. Thus, the lines are neither parallel nor perpendicular.

Exercises

1. Slope is a measure of the steepness of a line. Slope is the *vertical* change compared to the *horizontal* change while moving along the line from one point to another.

3. **(a)** Looking at the graph, the vertical change from -4 to 2 is 6 units up.

   (b) Looking at the graph, the horizontal change from -1 to 3 is 4 units to the right.

   (c) The quotient of the numbers found in parts (a) and (b) is the slope, the rise divided by the run.

$$\text{slope}=\frac{\text{change in } y\text{ (rise)}}{\text{change in } x\text{ (run)}}=\frac{6}{4}=\frac{3}{2}$$

5. **(a)** The indicated points have coordinates $(-1,2)$ and $(2,0)$.

$$\text{slope}=\frac{\text{change in } y\text{ (rise)}}{\text{change in } x\text{ (run)}}=\frac{0-2}{2-(-1)}=\frac{-2}{3}=-\frac{2}{3}\text{ (choice C)}$$

   (b) The indicated points have coordinates $(-1,2)$ and $(-4,0)$.

$$\text{slope}=\frac{\text{change in } y\text{ (rise)}}{\text{change in } x\text{ (run)}}=\frac{0-2}{-4-(-1)}=\frac{-2}{-4+1}=\frac{-2}{-3}=\frac{2}{3}\text{ (choice A)}$$

   (c) The indicated points have coordinates $(-1,2)$ and $(1,-1)$.

$$\text{slope}=\frac{\text{change in } y\text{ (rise)}}{\text{change in } x\text{ (run)}}=\frac{-1-2}{1-(-1)}=\frac{-3}{2}=-\frac{3}{2}\text{ (choice D)}$$

   (d) The indicated points have coordinates $(-1,2)$ and $(-3,-1)$.

$$\text{slope}=\frac{\text{change in } y\text{ (rise)}}{\text{change in } x\text{ (run)}}=\frac{-1-2}{-3-(-1)}=\frac{-3}{-2}=\frac{3}{2}\text{ (choice B)}$$

7. Negative slope
   Sketches will vary. The line must fall from left to right. One such line is shown in the following graph.

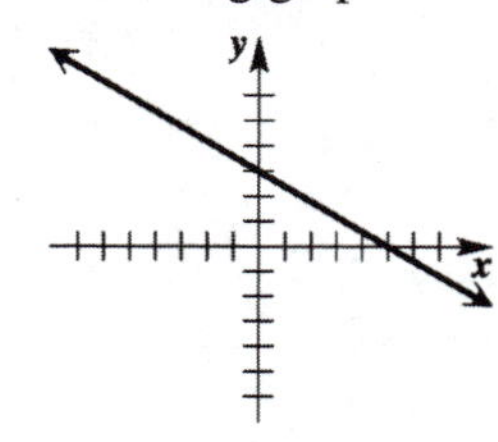

9. Undefined slope
Sketches will vary. The line must be vertical. One such line is shown in the following graph.

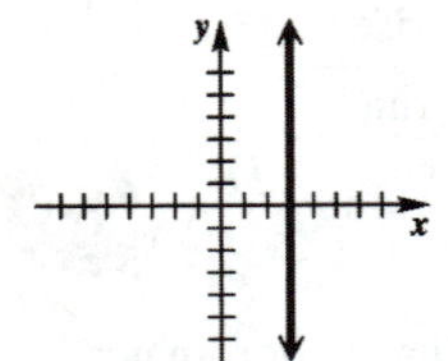

11. **(a)** Because the line falls from left to right, its slope is negative.

(b) Because the line intersects the y-axis at the origin, the y-value of its y-intercept is zero.

13. **(a)** Because the line rises from left to right, its slope is positive.

(b) Because the line intersects the y-axis below the origin, the y-value of its y-intercept is negative.

15. **(a)** The line is horizontal, so its slope is zero.

(b) The line intersects the y-axis below the origin, so the y-value of its y-intercept is negative.

17. Because he found the difference $3-5=-2$ in the numerator, he should have subtracted in the same order in the denominator to get $-1-2=-3$. The correct slope is $\frac{3-8}{-1-2}=\frac{-2}{-3}=\frac{2}{3}$.

19. The slope (or grade) of the hill is the ratio of the rise to the run, or the ratio of the vertical change to the horizontal change. Since the rise is 32 and the run is 108, the slope is $\frac{32}{108}=\frac{8\cdot 4}{27\cdot 4}=\frac{8}{27}$.

21.
$$\text{slope}=\frac{\text{vertical change (rise)}}{\text{horizontal change (run)}}$$
$$=\frac{-8}{12}\ \text{("drops" indicates negative)}$$
$$=-\frac{2}{3}$$

23. The indicated points have coordinates $(-1,-4)$ and $(1,4)$.
$$\text{slope}=\frac{\text{change in } y\text{ (rise)}}{\text{change in } x\text{ (run)}}$$
$$=\frac{4-(-4)}{1-(-1)}=\frac{8}{2}=4$$

25. The indicated points have coordinates $(-3,2)$ and $(5,-2)$.
$$\text{slope}=\frac{\text{change in } y\text{ (rise)}}{\text{change in } x\text{ (run)}}$$
$$=\frac{-2-2}{5-(-3)}=\frac{-4}{8}=-\frac{1}{2}$$

27. The indicated points have coordinates $(-2,-4)$ and $(4,-4)$.
$$\text{slope}=\frac{\text{change in } y\text{ (rise)}}{\text{change in } x\text{ (run)}}$$
$$=\frac{-4-(-4)}{4-(-2)}=\frac{-4+4}{4+2}=\frac{0}{6}=0$$

29. Use the slope formula with $(1,-2)=(x_1,y_1)$ and $(-3,-7)=(x_2,y_2)$.
$$\text{slope } m=\frac{y_2-y_1}{x_2-x_1}=\frac{-7-(-2)}{-3-1}=\frac{-5}{-4}=\frac{5}{4}$$

31. Use the slope formula with $(0,3)=(x_1,y_1)$ and $(-2,0)=(x_2,y_2)$.
$$\text{slope } m=\frac{y_2-y_1}{x_2-x_1}=\frac{0-3}{-2-0}=\frac{-3}{-2}=\frac{3}{2}$$

33. Use the slope formula with $(4,3)=(x_1,y_1)$ and $(-6,3)=(x_2,y_2)$.
$$\text{slope } m=\frac{y_2-y_1}{x_2-x_1}=\frac{3-3}{-6-4}=\frac{0}{-10}=0$$

35. Use the slope formula with $(-2,4)=(x_1,y_1)$ and $(-3,7)=(x_2,y_2)$.
$$\text{slope } m=\frac{y_2-y_1}{x_2-x_1}=\frac{7-4}{-3-(-2)}=\frac{3}{-1}=-3$$

37. Use the slope formula with $(-12, 3) = (x_1, y_1)$ and $(-12, -7) = (x_2, y_2)$.

$$\text{slope } m = \frac{y_2 - y_1}{x_2 - x_1} = \frac{-7-3}{-12-(-12)} = \frac{-10}{0}$$

The slope is undefined because of the 0 denominator.

39. Use the slope formula with $(4.8, 2.5) = (x_1, y_1)$ and $(3.6, 2.2) = (x_2, y_2)$.

$$\text{slope } m = \frac{y_2 - y_1}{x_2 - x_1} = \frac{2.2-2.5}{3.6-4.8} = \frac{-0.3}{-1.2} = \frac{1}{4}$$

41. Use the slope formula with $\left(-\frac{7}{5}, \frac{3}{10}\right) = (x_1, y_1)$ and $\left(\frac{1}{5}, -\frac{1}{2}\right) = (x_2, y_2)$.

$$\text{slope } m = \frac{y_2 - y_1}{x_2 - x_1}$$

$$= \frac{-\frac{1}{2} - \frac{3}{10}}{\frac{1}{5} - \left(-\frac{7}{5}\right)} = \frac{-\frac{5}{10} - \frac{3}{10}}{\frac{1}{5} + \frac{7}{5}} = \frac{-\frac{8}{10}}{\frac{8}{5}}$$

$$= \left(-\frac{8}{10}\right)\left(\frac{5}{8}\right) = -\frac{5}{10} = -\frac{1}{2}$$

43. Since the equation is already solved for y, the slope is given by the coefficient of x, which is 5. Thus, the slope of the line is 5.

45. Solve the equation for y.

$4y = x + 1$

$y = \frac{1}{4}x + \frac{1}{4}$ Divide by 4.

The slope of the line is given by the coefficient of x, so the slope is $\frac{1}{4}$.

47. Solve the equation for y.

$3x - 2y = 3$

$-2y = -3x + 3$ Subtract $3x$.

$y = \frac{3}{2}x - \frac{3}{2}$ Divide by -2.

The slope of the line is given by the coefficient of x, so the slope is $\frac{3}{2}$.

49. Solve the equation for y.

$-3x + 2y = 5$

$2y = 3x + 5$ Add $3x$.

$y = \frac{3}{2}x + \frac{5}{2}$ Divide by 2.

The slope of the line is given by the coefficient of x, so the slope is $\frac{3}{2}$.

51. Solve the equation for y.

$x + y = -4$

$y = -x - 4$ Subtract x.

The slope of the line is given by the coefficient of x, so the slope is -1.

53. This is an equation of a horizontal line. Its slope is 0. (This equation may be rewritten in the form $y = 0x - 5$, where the coefficient of x gives the slope.)

55. This is an equation of a vertical line. Its slope is undefined.

57. **(a)** Use the slope formula with the intercepts, $(5, 0)$ and $(0, 10)$. (Note that other points can be used.)

$$m = \frac{y_2 - y_1}{x_2 - x_1} = \frac{10-0}{0-5} = \frac{10}{-5} = -2$$

(b) Find the slope of the line by solving the equation for y.

$2x + y = 10$

$y = -2x + 10$ Subtract $2x$.

The slope is given by the coefficient of x, so the slope is -2.

59. **(a)** Use the slope formula with the intercepts, $(3, 0)$ and $(0, -5)$. (Note that other points can be used.)

$$m = \frac{y_2 - y_1}{x_2 - x_1} = \frac{-5-0}{0-3} = \frac{-5}{-3} = \frac{5}{3}$$

(b) Find the slope of the line by solving the equation for y.

$5x - 3y = 15$

$-3y = -5x + 15$ Subtract $5x$.

$y = \frac{-5}{-3}x + \frac{15}{-3}$ Divide by -3.

$y = \frac{5}{3}x - 5$ Simplify.

The slope is given by the coefficient of x, so the slope is $\frac{5}{3}$.

61. (a) Use the slope formula with the points $(-4, 0)$ and $(0, 4)$. Note that other points can be used.

$$m = \frac{y_2 - y_1}{x_2 - x_1} = \frac{4-0}{0-(-4)} = \frac{4}{4} = 1$$

(b) The x-intercept occurs when $y = 0$ in the given table. This occurs when $x = -4$. The x-intercept is the point $(-4, 0)$. The y-intercept occurs when $x = 0$ in the given table. This occurs when $y = 4$. The y-intercept is the point $(0, 4)$.

(c)

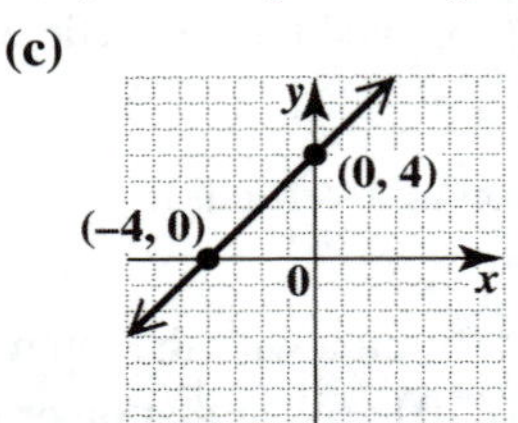

63. (a) Use the slope formula with the points $(0, -2)$ and $(-6, 0)$. Note that other points can be used.

$$m = \frac{y_2 - y_1}{x_2 - x_1} = \frac{0-(-2)}{-6-0} = \frac{2}{-6} = -\frac{1}{3}$$

(b) The x-intercept occurs when $y = 0$ in the given table. This occurs when $x = -6$. The x-intercept is the point $(-6, 0)$. The y-intercept occurs when $x = 0$ in the given table. This occurs when $y = -2$. The y-intercept is the point $(0, -2)$.

(c)

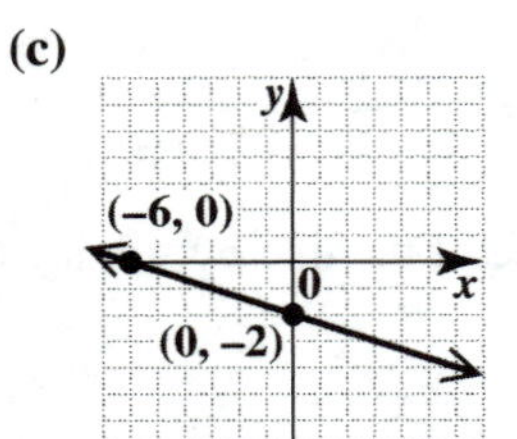

65. (a) Solve the equation for y.

$3x + y = 7$

$y = -3x + 7$ Subtract $3x$.

The slope of the given line is -3, so the slope of a line whose graph is parallel to the graph of the given line is also -3.

(b) Solve the equation for y.

$3x + y = 7$

$y = -3x + 7$ Subtract $3x$.

The slope of a line whose graph is perpendicular to the graph of the given line is the negative reciprocal of -3, that is, $\frac{1}{3}$.

67. If two lines are both vertical then they will both have an undefined slope. If two lines are both horizontal then they will both have a slope of 0. Either way, if two lines are both vertical or both horizontal, they will have the same slope. Therefore, they are parallel. Choice A is correct.

69. Find the slope of each line by solving the equations for y.

$2x + 5y = 4$

$5y = -2x + 4$ Subtract $2x$.

$y = -\frac{2}{5}x + \frac{4}{5}$ Divide by 5.

The slope of the first line is $-\frac{2}{5}$.

$4x + 10y = 1$

$10y = -4x + 1$ Subtract $4x$.

$y = -\frac{4}{10}x + \frac{1}{10}$ Divide by 10.

$y = -\frac{2}{5}x + \frac{1}{10}$ Lowest terms

The slope of the second line is $-\frac{2}{5}$.

The slopes are equal, so the lines are parallel.

71. Find the slope of each line by solving the equations for y.

$8x - 9y = 6$

$-9y = -8x + 6$ Subtract $8x$.

$y = \frac{8}{9}x - \frac{2}{3}$ Divide by -9.

The slope of the first line is $\frac{8}{9}$.

$8x + 6y = -5$

$6y = -8x - 5$ Subtract $8x$.

$y = -\frac{4}{3}x - \frac{5}{6}$ Divide by 6.

The slope of the second line is $-\frac{4}{3}$.

The slopes are not equal, so the lines are not parallel. The slopes are not negative reciprocals (the negative reciprocal of $\frac{8}{9}$ is $-\frac{9}{8}$), so the lines are not perpendicular. Thus, the lines are neither parallel nor perpendicular.

73. Find the slope of each line by solving the equations for y.

$3x-2y=6$

$-2y=-3x+6$ Subtract $3x$.

$y=\frac{3}{2}x-3$ Divide by -2.

The slope of the first line is $\frac{3}{2}$.

$2x+3y=3$

$3y=-2x+3$ Subtract $2x$.

$y=-\frac{2}{3}x+1$ Divide by 3.

The slope of the second line is $-\frac{2}{3}$.

The product of the slopes is

$\frac{3}{2}\left(-\frac{2}{3}\right)=-1,$

so the lines are perpendicular.

75. Find the slope of each line by solving the equations for y.

$5x-y=1$

$-y=-5x+1$ Subtract $5x$.

$y=5x-1$ Divide by -1.

The slope of the first line is 5.

$x-5y=-10$

$-5y=-x-10$ Subtract x.

$y=\frac{1}{5}x+2$ Divide by -5.

The slope of the second line is $\frac{1}{5}$.

The slopes are not equal, so the lines are not parallel. The slopes are not negative reciprocals (the negative reciprocal of 5 is $-\frac{1}{5}$), so the lines are not perpendicular. Thus, the lines are neither parallel nor perpendicular.

77. **(a)** The ordered pairs that represent music purchases are $(2004, 817)$ and $(2012, 1661)$.

(b) $m=\frac{1661-817}{2012-2004}=\frac{844}{8}=105.5$

(c) Music purchases increased by 844 million units in 8 years, or 105.5 million units per year.

79. We use the points with coordinates $(2000,\ 6.6)$ and $(2012,\ 12.6)$.

$m=\frac{6.6-12.6}{2012-2000}=\frac{6}{12}=0.5$

81. The increase is approximately 0.5% students per year.

83. The slope of the line in Figure B is *negative.* This means that during the period represented, the percent of freshmen planning to major in Business *decreased.*

3.4 Slope-Intercept Form of a Linear Equation

Now Try Exercises

N1. **(a)** $y=-\frac{3}{5}x-9$

The slope is the coefficient of x, that is, $-\frac{3}{5}$. The y-intercept has x-coordinate 0 and y-coordinate -9, that is, $(0,-9)$.

(b) $y=-\frac{x}{3}+\frac{7}{3}$ or $y=-\frac{1}{3}x+\frac{7}{3}$

The slope is $-\frac{1}{3}$ and the y-intercept is $\left(0,\frac{7}{3}\right)$.

N2. *Step 1*

Begin by solving for y.

$3x+2y=8$ Given equation

$2y=-3x+8$ Subtract $3x$.

$y=-\frac{3}{2}x+4$ Divide by 2.

Step 2

The y-intercept is $(0, 4)$. Plot this point.

Step 3

The slope is $-\frac{3}{4}$. By definition,

slope $m = \frac{\text{change in } y \text{ (rise)}}{\text{change in } x \text{ (run)}} = -\frac{3}{4}$.

Starting at the y-intercept, we count 3 units down and 2 units right to obtain another point on the graph, $(2, 1)$.

Step 4

Draw the line through the points $(0, 4)$ and $(2, 1)$ to obtain the graph of $3x + 2y = 8$.

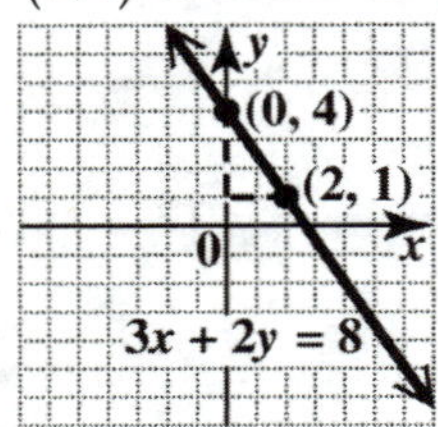

N3. To graph the line, remember the slope is $m = \frac{\text{change in } y \text{ (rise)}}{\text{change in } x \text{ (run)}} = \frac{5}{2}$.

Locate $(-3, -4)$. Count 5 units up and 2 units to the right to locate another point on the line. Draw a line through this point, $(-1, 1)$, and $(-3, -4)$.

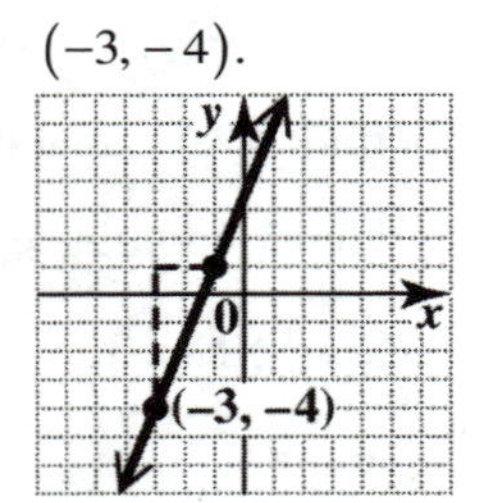

N4. **(a)** Since the point $(0, 2)$ is the y-intercept, $b = 2$. Substitute this value for b and the given slope directly into the slope-intercept form.

$y = mx + b$ Slope-intercept form

$y = -4x + 2$ Substitute.

The equation of the line is $y = -4x + 2$.

(b) Find the y-intercept by substituting $x = -2$ and $y = 1$ from the given point and the given slope $m = 3$ into the slope-intercept form.

$y = mx + b$ Slope-intercept form

$1 = 3(-2) + b$ Substitute.

$1 = -6 + b$ Multiply.

$7 = b$ Add 6.

Now substitute the values of m and b into slope-intercept form.

$y = mx + b$ Slope-intercept form

$y = 3x + 7$ Substitute.

The equation of the line is $y = 3x + 7$.

N5. **(a)** This line is vertical because it has undefined slope. To graph this line plot the point $(-3, 3)$ and draw a vertical line through it.

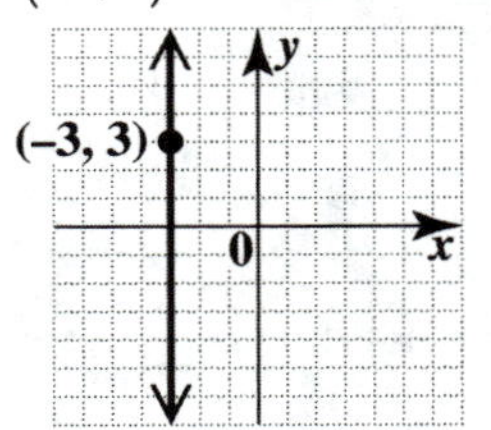

(b) This line is horizontal because it has slope 0. To graph this line plot the point $(3, -3)$ and draw a horizontal line through it.

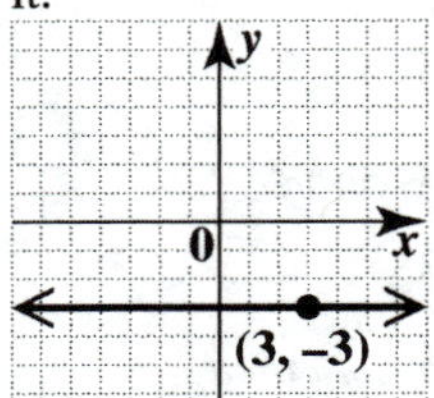

N6. **(a)** This line is vertical because it has undefined slope. A vertical line through the point (a, b) has the equation $x = a$. The x-coordinate of $(-1, 1)$ is -1, so the equation is $x = -1$.

(b) This line is horizontal because it has slope 0. A horizontal line through the point (a, b) has the equation $y = b$. The y-coordinate of $(-1, 1)$ is 1, so the equation is $y = 1$.

Exercises

1. In slope-intercept form $y = mx + b$ of the equation of a line, the slope is $\underline{m}$ and the y-intercept is $\underline{(0, b)}$.

3. **(a)** The graph of $y = x + 3$ is a line with slope 1 and y-intercept $(0, 3)$. The only graph which has a positive slope and intersects the y-axis above the origin is C.

(b) The graph of $y=-x+3$ is a line with slope -1 and y-intercept $(0, 3)$. The only graph which has a negative slope and intersects the y-axis above the origin is B.

(c) The graph of $y=x-3$ is a line with slope 1 and y-intercept $(0, -3)$. The only graph which has a positive slope and intersects the y-axis below the origin is A.

(d) The graph of $y=-x-3$ is a line with slope -1 and y-intercept $(0, -3)$. The only graph which has a negative slope and intersects the y-axis below the origin is D.

5. The slope of $y=\frac{5}{2}x-4$ is the coefficient of x, $\frac{5}{2}$, and the y-intercept is $(0, -4)$.

7. The slope of $y=-x+9$ is the coefficient of x, -1, and the y-intercept is $(0, 9)$.

9. The slope of $y=\frac{x}{5}-\frac{3}{10}$ is the coefficient of x, $\frac{1}{5}$, and the y-intercept is $\left(0, -\frac{3}{10}\right)$.

11. The slope of $y=3x+2$ is the coefficient of x, 3. By definition, $m=\frac{\text{change in } y}{\text{change in } x}=\frac{3}{1}$. The y-intercept is $(0, 2)$. Plot that point and count up 3 units and right 1 unit to get to the point $(1, 5)$. Draw a line through the points.

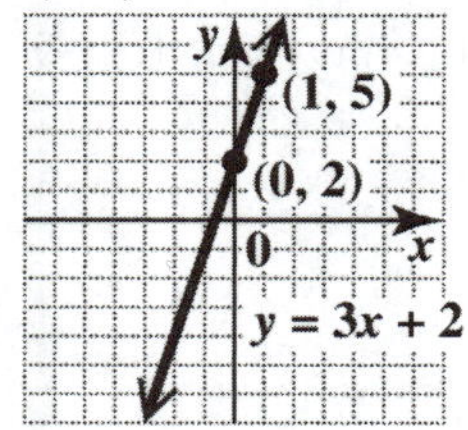

13. The slope of $y=-\frac{1}{3}x+4$ is the coefficient of x, $-\frac{1}{3}$. By definition, $m=\frac{\text{change in } y}{\text{change in } x}=\frac{-1}{3}$.

The y-intercept is $(0, 4)$. Plot that point and count down 1 unit and right 3 units to get to the point $(3, 3)$. Draw a line through the points.

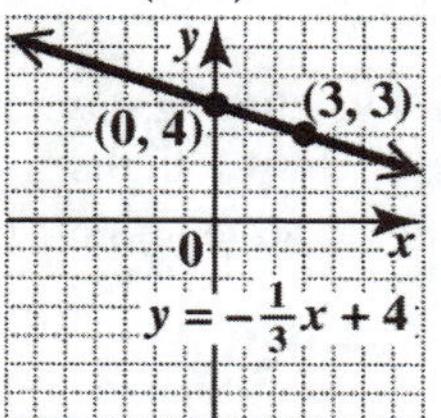

15. Solve the equation for y.

$2x+y=-5$

$y=-2x-5$

The slope is the coefficient of x, -2. By definition, $m=\frac{\text{change in } y}{\text{change in } x}=\frac{-2}{1}$. The y-intercept is $(0, -5)$. Plot that point and count down 2 units and right 1 unit to get to the point $(1, -7)$. Draw a line through the points.

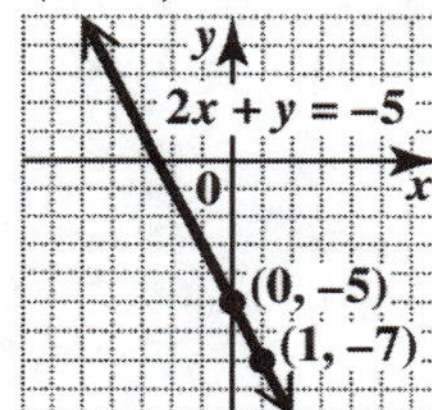

17. Solve the equation for y.

$4x-5y=20$	Given equation
$-5y=-4x+20$	Subtract $4x$.
$y=\frac{4}{5}x-4$	Divide by -5.

The slope is the coefficient of x, $\frac{4}{5}$. By definition, $m=\frac{\text{change in } y}{\text{change in } x}=\frac{4}{5}$. The y-intercept is $(0, -4)$. Plot that point and count up 4 units and right 5 units to get to the point $(5, 0)$. Draw a line through the points

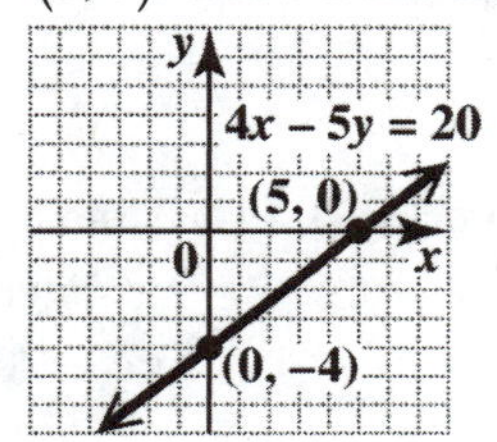

19. First, locate the point $(0, 1)$, which is the y-intercept of the line to be graphed. Write the slope as $m = \frac{\text{rise}}{\text{run}} = \frac{4}{1}$.
Locate another point by counting 4 units up and then 1 unit to the right. Draw a line through this new point, $(1, 5)$, and the given point, $(0, 1)$.

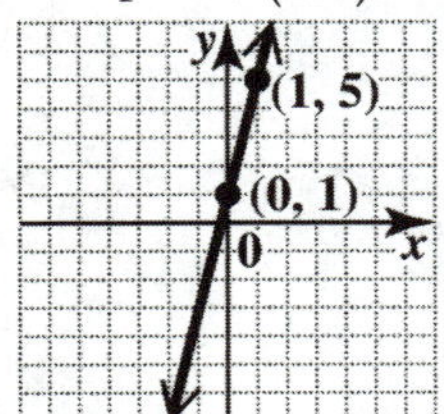

21. First, locate the point $(1, -5)$. Write the slope as $m = \frac{\text{rise}}{\text{run}} = \frac{-2}{5}$.
Locate another point by counting 2 units down (because of the negative sign) and then 5 units to the right. Draw a line through this new point, $(6, -7)$, and the given point, $(1, -5)$.

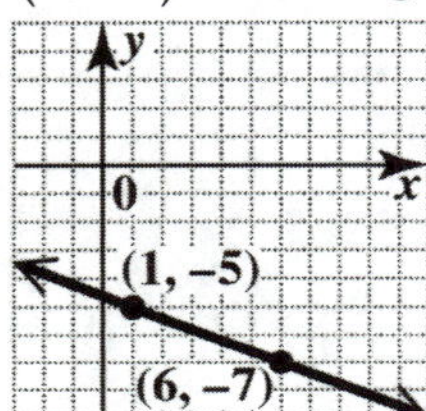

23. First, locate the point $(-1, 4)$. The slope is $m = \frac{\text{rise}}{\text{run}} = \frac{2}{5}$.
Locate another point by counting 2 units up and then 5 units to the right. Draw a line through this new point, $(4, 6)$, and the given point, $(-1, 4)$.

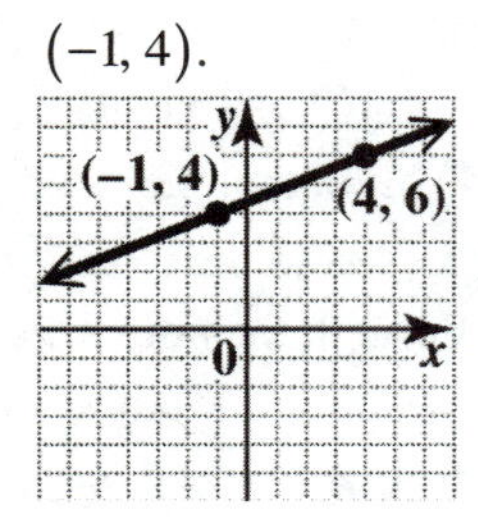

25. First, locate the point $(0, 0)$. The slope is $m = \frac{\text{rise}}{\text{run}} = \frac{-2}{1}$.

Locate another point by counting 2 units down and then 1 unit to the right. Draw a line through this new point, $(1, -2)$, and the given point, $(0, 0)$.

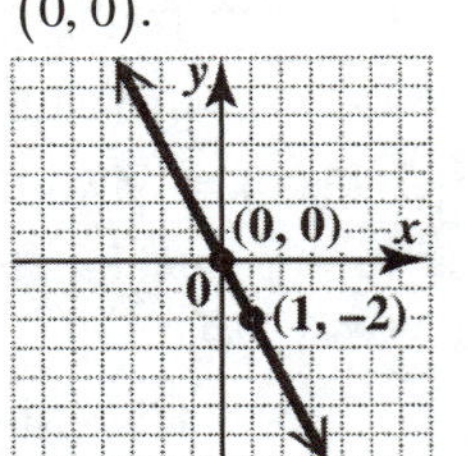

27. First, locate the point $(-2, 3)$. Since the slope is 0, the line will be horizontal. Draw a horizontal line through the point $(-2, 3)$.

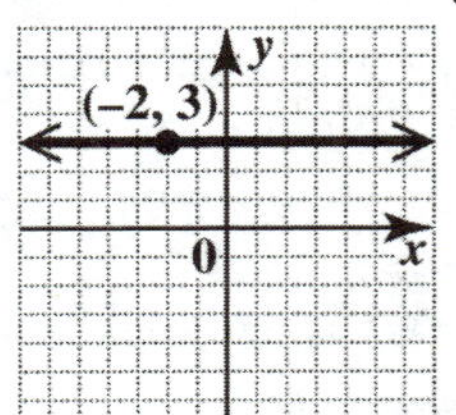

29. First, locate the point $(2, 4)$. Since the slope is undefined, the line will be vertical. Draw the vertical line through the point $(2, 4)$.

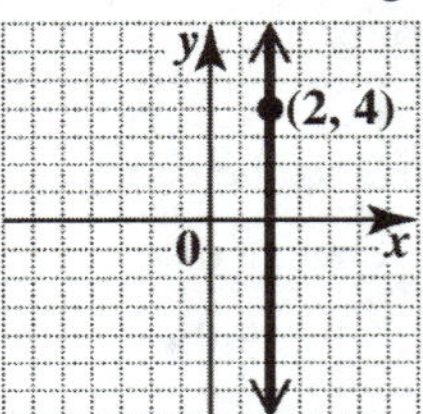

31. First, locate the point $(5, -5)$. Since the slope is 0, the line will be horizontal. Draw a horizontal line through the point $(5, -5)$.

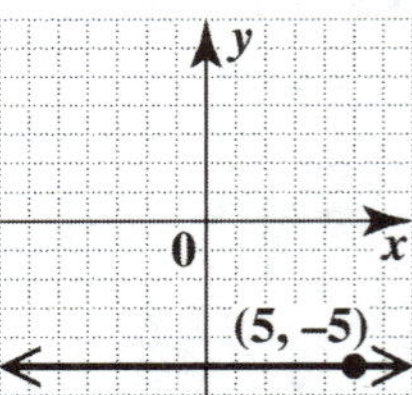

33. The common name given to a vertical line whose x-intercept is the origin is the y-axis.

35. The rise is 3 and the run is 1, so the slope is given by $m = \frac{\text{rise}}{\text{run}} = \frac{3}{1} = 3$.
The y-intercept is $(0, -3)$, so $b = -3$. The equation of the line, written in slope-intercept form, is $y = 3x - 3$.

37. Since the line falls from left to right, the "rise" is negative. For this line, the rise is -3 and the run is 3, so the slope is $m = \frac{\text{rise}}{\text{run}} = \frac{-3}{3} = -1.$
The y-intercept is $(0, 3)$, so $b = 3$. The slope-intercept form of the equation of the line is
$y = -1x + 3$
$y = -x + 3.$

39. Since the line falls from left to right, the "rise" is negative. For this line, the rise is -2 and the run is 4, so the slope is
$m = \frac{\text{rise}}{\text{run}} = \frac{-2}{4} = -\frac{1}{2}.$
The y-intercept is $(0, 2)$, so $b = 2$. The slope-intercept form of the equation of the line is
$y = -\frac{1}{2}x + 2.$

41. Since the y-intercept is $(0, -3)$, we have $b = -3$. Use the slope-intercept form and substitute the values for b and m.
$y = mx + b$
$y = 4x + (-3)$
$y = 4x - 3$
Thus, the equation of the line in slope-intercept form is $y = 4x - 3$.

43. Since the y-intercept is $(0, -7)$, we have $b = -7$. Use the slope-intercept form and substitute the values for b and m.
$y = mx + b$
$y = -1x - 7$
$y = -x - 7$
Thus, the equation of the line in slope-intercept form is $y = -x - 7$.

45. To write the equation in the form $y = mx + b$ find b by substituting the given point and slope into the slope-intercept form. Solve for b.

$y = mx + b$	Slope-intercept form
$1 = 2(4) + b$	Substitute.
$1 = 8 + b$	Multiply.
$-7 = b$	Subtract.

Use the slope-intercept form and substitute the values for b and m.
$y = mx + b$
$y = 2x - 7$
Thus, the equation of the line in slope-intercept form is $y = 2x - 7$.

47. To write the equation in the form $y = mx + b$ find b by substituting the given point and slope into the slope-intercept form. Solve for b.

$y = mx + b$	Slope-intercept form
$3 = -4(-1) + b$	Substitute.
$3 = 4 + b$	Multiply.
$-1 = b$	Subtract.

Use the slope-intercept form and substitute the values for b and m.
$y = mx + b$
$y = -4x - 1$
Thus, the equation of the line in slope-intercept form is $y = -4x - 1$.

49. To write the equation in the form $y = mx + b$ find b by substituting the given point and slope into the slope-intercept form. Solve for b.

$y = mx + b$	Slope-intercept form
$3 = 1(9) + b$	Substitute.
$3 = 9 + b$	Multiply.
$-6 = b$	Subtract.

Use the slope-intercept form and substitute the values for b and m.
$y = mx + b$
$y = x - 6$
Thus, the equation of the line in slope-intercept form is $y = x - 6$.

51. To write the equation in the form $y = mx + b$ find b by substituting the given point and slope into the slope-intercept form. Solve for b.

$y = mx + b$	Slope-intercept form
$1 = \frac{3}{4}(-4) + b$	Substitute.
$1 = -3 + b$	Multiply.
$4 = b$	Add.

Use the slope-intercept form and substitute the values for b and m.
$y = mx + b$
$y = \frac{3}{4}x + 4$
Thus, the equation of the line in slope-intercept form is $y = \frac{3}{4}x + 4$.

53. Since the y-intercept is $(0, 3)$, we have $b = 3$. Use the slope-intercept form.

$y = mx + b$

$y = 0x + 3$

$y = 3$

Thus, the equation of the line in slope-intercept form is $y = 3$.

55. Since the slope is undefined, the line is vertical and has equation $x = k$. Because the line goes through a point $(x, y) = (2, -6)$, we must have $x = 2$.

57. Since the slope is undefined, the line is vertical and has equation $x = k$. Because the line goes through a point $(x, y) = (0, -2)$, we must have $x = 0$.

59. Since the slope is 0, the line will be horizontal and with the equation $y = k$. Because the line goes through a point $(x, y) = (6, -6)$, we must have $y = -6$.

61. **(a)** Use the slope formula with the points $(0, -1)$ and $(3, 5)$. Note that other points can be used.

$$\text{slope } m = \frac{\text{change in } y}{\text{change in } x} = \frac{y_2 - y_1}{x_2 - x_1} = \frac{5-(-1)}{3-0} = \frac{6}{3} = 2$$

(b) The y-intercept, occurs when $x = 0$ in the given table. This occurs when $y = -1$. The y-intercept is the point $(0, -1)$.

(c) Note that because the y-intercept is the point $(0, -1)$, the value of b is $b = -1$.

Use the values of m and b from the previous steps to write the slope-intercept form for the line.

$y = mx + b$

$y = 2x - 1$

(d)

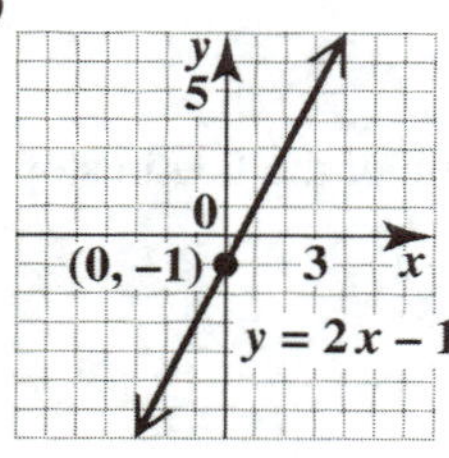

63. **(a)** Use the slope formula with the points $(-6, 0)$ and $(0, -2)$. Note that other points can be used.

$$\text{slope } m = \frac{\text{change in } y}{\text{change in } x} = \frac{y_2 - y_1}{x_2 - x_1} = \frac{-2-0}{0-(-6)} = \frac{-2}{6} = -\frac{1}{3}$$

(b) The y-intercept occurs when $x = 0$ in the given table. This occurs when $y = -2$. The y-intercept is the point $(0, -2)$.

(c) Note that because the y-intercept is the point $(0, -2)$, the value of b is $b = -2$.

Use the values of m and b from the previous steps to write the slope-intercept form for the line.

$y = mx + b$

$y = -\frac{1}{3}x - 2$

(d)

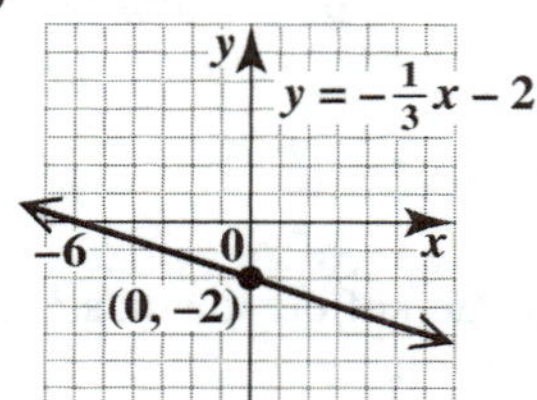

65. **(a)** The slope of the line $y = 0.05x + 2000$ is 0.05. This corresponds to the commission rate.

(b) The y-intercept, when $x = 0$, is the point $(0, 2000)$. The point corresponds to the base salary per month.

(c) Evaluate y when $x = 10{,}000$.

$y = 0.05x + 2000$

$y = 0.05(10{,}000) + 2000$ Let $x = 10{,}000$.

$y = 500 + 2000$ Multiply.

$y = 2500$ Add.

According to the equation, Andrew's salary is \$2500 when his sales are \$10,000. This corresponds to the ordered pair $(2500, 10000)$ which is present on the graph. The equation and the graph match.

(d) Solve for x when $y = 3500$.

$$y = 0.05x + 2000$$
$$3500 = 0.05x + 2000 \quad \text{Let } y = 3500.$$
$$1500 = 0.05x \quad \text{Subtract 2000.}$$
$$30{,}000 = x \quad \text{Divide by 0.05.}$$

According to the equation, Andrew's salary is \$3500 when his sales are \$30,000. This corresponds to the ordered pair (3500, 30,000) which is present on the graph. The equation and the graph match.

67. **(a)** The fixed cost is \$400.

(b) The variable cost is \$0.25.

(c) Substitute $m = 0.25$ and $b = 400$ into $y = mx + b$ to get the cost equation $y = 0.25x + 400$.

(d) Let $x = 100$ in the cost equation.

$$y = 0.25(100) + 400$$
$$y = 25 + 400$$
$$y = 425$$

The cost to produce 100 snow cones will be \$425.

(e) Let $y = 775$ in the cost equation.

$$775 = 0.25x + 400$$
$$375 = 0.25x \quad \text{Subtract 400.}$$
$$x = \frac{375}{0.25} = 1500 \quad \text{Divide by 0.25.}$$

If the total cost is \$775, then 1500 snow cones will be produced.

69. Start with the standard form equation of a line and solve for y to find the slope-intercept form.

$$Ax + By = C \quad \text{standard-form}$$
$$By = C - Ax \quad \text{Subtract } Ax.$$
$$y = \frac{C}{B} - \frac{A}{B}x \quad \text{Divide by } B.$$
$$y = -\frac{A}{B}x + \frac{C}{B} \quad \text{slope-intercept form}$$

Thus, the slope is $m = \frac{A}{B}$, where $B \neq 0$.

71. The y-intercept of the line $y = -\frac{A}{B}x + \frac{C}{B}$ is $\left(0, \frac{C}{B}\right)$.

3.5 Point-Slope Form of a Linear Equation and Modeling

Now Try Exercises

N1. Use the point-slope form of the equation of a line with $x_1 = 3$, $y_1 = -1$, and $m = -\frac{2}{5}$.

$$y - y_1 = m(x - x_1)$$
$$y - (-1) = -\frac{2}{5}(x - 3)$$
$$y + 1 = -\frac{2}{5}x + \frac{6}{5}$$
$$y = -\frac{2}{5}x + \frac{6}{5} - \frac{5}{5}$$
$$y = -\frac{2}{5}x + \frac{1}{5}$$

N2. **(a)** First, find the slope of the line.

$$m = \frac{y_2 - y_1}{x_2 - x_1} = \frac{-2 - 1}{6 - 4} = \frac{-3}{2} = -\frac{3}{2}$$

Now use the point $(4, 1)$ for (x_1, y_1) and $m = -\frac{3}{2}$ in the point-slope form.

$$y - y_1 = m(x - x_1)$$
$$y - 1 = -\frac{3}{2}(x - 4)$$
$$y - 1 = -\frac{3}{2}x + 6$$
$$y = -\frac{3}{2}x + 7$$

(b) Rewrite the equation in standard form $Ax + By = C$.

$$y = -\frac{3}{2}x + 7$$
$$2y = -3x + 14 \quad \text{Multiply by 2.}$$
$$3x + 2y = 14 \quad \text{Add } x.$$

N3. Let $(x_1, y_1) = (3, 4645)$ and $(x_2, y_2) = (5, 5491)$.

$$m = \frac{y_2 - y_1}{x_2 - x_1} = \frac{5491 - 4645}{5 - 3} = \frac{846}{2} = 423$$

Now use this slope and the point $(3, 4645)$ in the point-slope form to find an equation of the line.

$$y - y_1 = m(x - x_1)$$
$$y - 4645 = 423(x - 3)$$
$$y - 4645 = 423x - 1269$$
$$y = 423x + 3376$$

For 2011, let $x = 11$.

$$y = 423(11) + 3376$$
$$y = 8029$$

The equation gives $y = 8029$ when $x = 7$, which approximates the cost given in the table, 8244, reasonable well.

Exercises

1. (a) The point-slope form of the equation of a line with slope m and passing through the point (x_1, y_1) is $y - y_1 = m(x - x_1)$. This corresponds to Choice D.

(b) A horizontal line has a slope of 0. Writing its equation in slope-intercept form and simplifying yields the equation $y = b$.

$$y = mx + b$$
$$y = 0x + b$$
$$y = b$$

This corresponds to Choice C.

(c) The slope-intercept form of a line with slope m and y-intercept b is the equation $y = mx + b$. This corresponds to Choice B.

(d) The standard form of a line is the equation $Ax + By = C$. This corresponds to Choice E.

(e) A vertical line has undefined slope. All of its points have the same x coordinate of a and thus have the equation $x = a$. This corresponds to Choice A.

3. Write each equation in standard form to check if it is equivalent or not.

$y = \frac{2}{3}x - 2$ Given

$3(y) = 3\left(\frac{2}{3}x - 2\right)$ Multiply by 3.

$3y = 3\left(\frac{2}{3}x\right) - 3(2)$ Distributive prop.

$3y = 2x - 6$ Multiply.

$6 = 2x - 3y$ Standard form

This equation is equivalent to $2x - 3y = 6$.

$-2x + 3y = -6$ Given

$-1(-2x + 3y) = -1(-6)$ Mult. by -1.

$-(-2x) - 1(3y) = -1(-6)$ Dist. property

$2x - 3y = 6$ Multiply.

This equation is equivalent to $2x - 3y = 6$.

$y = -\frac{3}{2}x + 3$ Given

$2(y) = 2\left(-\frac{3}{2}x + 3\right)$ Multiply by 2.

$2y = 2\left(-\frac{3}{2}x\right) + 2(3)$ Dist. property

$2y = -3x + 6$ Multiply.

$3x + 2y = 6$ Standard form

This equation is not equivalent to $2x - 3y = 6$.

$y - 2 = \frac{2}{3}(x - 6)$ Given

$3(y - 2) = 3\left[\frac{2}{3}(x - 6)\right]$ Multiply by 3.

$3y - 6 = 2(x - 6)$ Distributive prop. Associative prop.

$3y - 6 = 2x - 12$ Distributive prop.

$6 = 2x - 3y$ Standard form

This equation is equivalent to $2x - 3y = 6$.

So A, B, and D are equivalent to $2x - 3y = 6$.

5. Use the values $x_1 = 1$, $y_1 = 7$, and $m = 5$ in the point-slope form.

$$y - y_1 = m(x - x_1)$$
$$y - 7 = 5(x - 1)$$
$$y - 7 = 5x - 5$$
$$y = 5x - 5 + 7$$
$$y = 5x + 2$$

7. Use the values $x_1 = 6$, $y_1 = -3$, and $m = 1$ in the point-slope form.

$$\begin{aligned} y - y_1 &= m(x - x_1) \\ y - (-3) &= 1(x - 6) \\ y + 3 &= x - 6 \\ y &= x - 6 - 3 \\ y &= x - 9 \end{aligned}$$

9. Use the values $x_1 = 1$, $y_1 = -7$, and $m = -3$ in the point-slope form.

$$\begin{aligned} y - y_1 &= m(x - x_1) \\ y - (-7) &= -3(x - 1) \\ y + 7 &= -3x + 3 \\ y &= -3x + 3 - 7 \\ y &= -3x - 4 \end{aligned}$$

11. Use the values $x_1 = 3$, $y_1 = -2$, and $m = -1$ in the point-slope form.

$$\begin{aligned} y - y_1 &= m(x - x_1) \\ y - (-2) &= -1(x - 3) \\ y + 2 &= -x + 3 \\ y &= -x + 3 - 2 \\ y &= -x + 1 \end{aligned}$$

13. Use the values $x_1 = -2$, $y_1 = 5$, and $m = \frac{2}{3}$ in the point-slope form.

$$\begin{aligned} y - y_1 &= m(x - x_1) \\ y - 5 &= \frac{2}{3}[x - (-2)] \\ y - 5 &= \frac{2}{3}(x + 2) \\ y - 5 &= \frac{2}{3}x + \frac{4}{3} \\ y &= \frac{2}{3}x + \frac{19}{3} \end{aligned}$$

15. Use the values $x_1 = 6$, $y_1 = -3$, and $m = -\frac{4}{5}$ in the point-slope form.

$$\begin{aligned} y - y_1 &= m(x - x_1) \\ y - (-3) &= -\frac{4}{5}(x - 6) \\ y + 3 &= -\frac{4}{5}x + \frac{24}{5} \\ y &= -\frac{4}{5}x + \frac{9}{5} \end{aligned}$$

17. **(a)** First, find the slope of the line.

$$m = \frac{12 - 10}{6 - 4} = \frac{2}{2} = 1$$

Now use the point $(4, 10)$ for (x_1, y_1) and $m = 1$ in the point-slope form.

$$\begin{aligned} y - y_1 &= m(x - x_1) \\ y - 10 &= 1(x - 4) \\ y - 10 &= x - 4 \\ y &= x + 6 \end{aligned}$$

The same result would be found by using $(6, 12)$ for (x_1, y_1).

(b) Write the equation in standard form.

$$\begin{aligned} y &= x + 6 && \\ -x + y &= 6 && \text{Subtract } x. \\ x - y &= -6 && \text{Multiply by } -1. \end{aligned}$$

19. **(a)** $m = \frac{2 - 0}{0 - (-4)} = \frac{2}{4} = \frac{1}{2}$

Use the point-slope form with $(x_1, y_1) = (-4, 0)$ and $m = \frac{1}{2}$.

$$\begin{aligned} y - y_1 &= m(x - x_1) \\ y - 0 &= \frac{1}{2}[x - (-4)] \\ y &= \frac{1}{2}(x + 4) \\ y &= \frac{1}{2}x + 2 \end{aligned}$$

(b) Write the equation in standard form.

$$\begin{aligned} y &= \frac{1}{2}x + 2 && \\ 2y &= x + 4 && \text{Multiply by 2.} \\ -x + 2y &= 4 && \text{Subtract } x. \\ x - 2y &= -4 && \text{Multiply by } -1. \end{aligned}$$

21. **(a)** First, find the slope of the line.

$$m=\frac{-4-(-1)}{3-(-2)}=\frac{-3}{5}=-\frac{3}{5}$$

Use the point-slope form with $(x_1, y_1)=(3,-4)$ and $m=-\frac{3}{5}$.

$$y-y_1=m(x-x_1)$$
$$y-(-4)=-\frac{3}{5}(x-3)$$
$$y+4=-\frac{3}{5}(x-3)$$
$$y+4=-\frac{3}{5}x+\frac{9}{5}$$
$$y=-\frac{3}{5}x-\frac{11}{5}$$

(b) Write the equation in standard form.

$$y=-\frac{3}{5}x-\frac{11}{5}$$

$5y=-3x-11$ Multiply by 5.

$3x+5y=-11$ Add $3x$.

23. **(a)** First, find the slope of the line.

$$m=\frac{\frac{7}{3}-\frac{8}{3}}{\frac{1}{3}-\left(-\frac{2}{3}\right)}=\frac{-\frac{1}{3}}{\frac{3}{3}}=\frac{-\frac{1}{3}}{1}=-\frac{1}{3}$$

Use the point-slope form with $(x_1, y_1)=\left(\frac{1}{3},\frac{7}{3}\right)$ and $m=-\frac{1}{3}$.

$$y-y_1=m(x-x_1)$$
$$y-\frac{7}{3}=-\frac{1}{3}(x-\frac{1}{3})$$
$$y-\frac{7}{3}=-\frac{1}{3}x+\frac{1}{9}$$
$$y=-\frac{1}{3}x+\frac{22}{9}$$

(b) Write the equation in standard form.

$$y=-\frac{1}{3}x+\frac{22}{9}$$

$9y=-3x+22$ Multiply by 9.

$3x+9y=22$ Add $3x$.

25. **(a)** First, find the slope of the line.

$$m=\frac{3-(-3)}{4-2}=\frac{6}{2}=3$$

Now use the point $(2,-3)$ for (x_1, y_1) and $m=3$ in the point-slope form.

$$y-y_1=m(x-x_1)$$
$$y-(-3)=3(x-2)$$
$$y+3=3x-6$$
$$y=3x-6-3$$
$$y=3x-9$$

(b) Write the equation in standard form.

$y=3x-9$

$y+9=3x$ Add 9.

$9=3x-y$ Subtract y.

Thus, the standard form is $3x-y=9$.

27. **(a)** First, find the slope of the line.

$$m=\frac{-2-0}{5-2}=-\frac{2}{3}$$

Now use the point $(2,0)$ for (x_1, y_1) and $m=-\frac{2}{3}$ in the point-slope form.

$$y-y_1=m(x-x_1)$$
$$y-0=-\frac{2}{3}(x-2)$$
$$y=-\frac{2}{3}x+\frac{4}{3}$$

(b) Write the equation in standard form.

$$y=-\frac{2}{3}x+\frac{4}{3}$$

$3y=-2x+4$ Multiply by 3.

$2x+3y=4$ Add $2x$.

29. Solve the equation for y.

$$x-2y=7$$
$$-2y=-x+7$$
$$y=\frac{1}{2}x-\frac{7}{2}$$

The slope is $\frac{1}{2}$.

A line perpendicular to this line has slope -2 $\left(\text{the negative reciprocal of } \frac{1}{2}\right)$.

Now use the slope-intercept form with $m = -2$ and y-intercept $(0, -3)$.

$y = mx + b$

$y = -2x - 3$

31. Solve the equation for y.

$4x - y = -2$

$4x + 2 = y$

The slope is 4. A line parallel to this line has the same slope. Now use the point-slope form with $m = 4$ and $(x_1, y_1) = (2, 3)$.

$y - y_1 = m(x - x_1)$

$y - 3 = 4(x - 2)$

$y - 3 = 4x - 8$

$y = 4x - 5$

33. Solve the equation for y.

$3x = 4y + 5$

$-4y = -3x + 5$

$y = \frac{3}{4}x - \frac{5}{4}$

The slope is $\frac{3}{4}$. A line parallel to this line has the same slope. Now use the point-slope form with $m = \frac{3}{4}$ and $(x_1, y_1) = (2, -3)$.

$y - y_1 = m(x - x_1)$

$y - (-3) = \frac{3}{4}(x - 2)$

$y + 3 = \frac{3}{4}x - \frac{3}{2}$

$y = \frac{3}{4}x - \frac{3}{2} - \frac{6}{2}$

$y = \frac{3}{4}x - \frac{9}{2}$

35. **(a)** x represents the year and y represents the cost in the ordered pairs (x, y). The ordered pairs are $(1, 2530)$, $(2, 2790)$, $(3, 2940)$, $(4, 3070)$, and $(5, 3220)$.

(b)

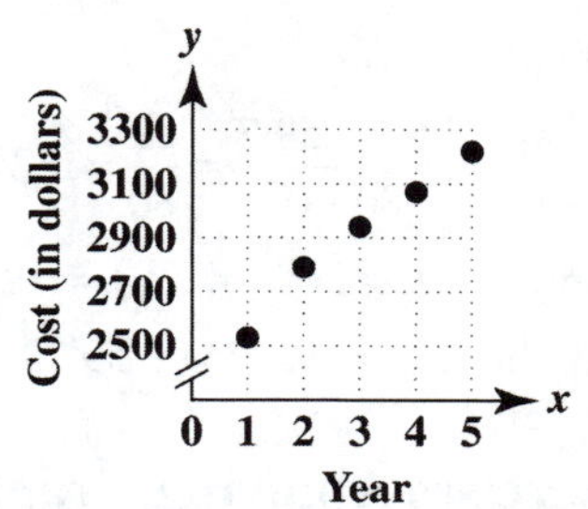

Yes, the points lie approximately in a straight line.

(c) Find the slope using $(x_1, y_1) = (1, 2530)$ and $(x_2, y_2) = (4, 3070)$.

$$m = \frac{y_2 - y_1}{x_2 - x_1} = \frac{3070 - 2530}{4 - 1} = \frac{540}{3} = 180$$

Now use the point-slope form with $m = 180$ and $(x_1, y_1) = (1, 2530)$.

$y - y_1 = m(x - x_1)$

$y - 2530 = 180(x - 1)$

$y - 2530 = 180x - 180$

$y = 180x + 2350$

(d) Since year 1 represents 2008, year 0 represents 2007.
For 2013, $x = 2013 - 2007 = 6$.

$y = 180x + 2350$

$y = 180(6) + 2350$

$y = 3430$

In 2013, the estimate of the average annual cost at 2-year colleges is \$3430.

37. Find the slope using $(x_1, y_1) = (1, 52)$ and $(x_2, y_2) = (6, 127)$.

$$m = \frac{y_2 - y_1}{x_2 - x_1} = \frac{127 - 52}{6 - 1} = \frac{75}{5} = 15$$

Now use the point-slope form with $m = 15$ and $(x_2, y_2) = (6, 127)$.

$y - y_1 = m(x - x_1)$

$y - 127 = 15(x - 6)$

$y - 127 = 15x - 90$

$y = 15x + 37$

39. Find the slope using $(x_1, y_1) = (0, 59.7)$ and $(x_2, y_2) = (80, 78.7)$.

$$m = \frac{y_2 - y_1}{x_2 - x_1} = \frac{78.7 - 59.7}{80 - 0} = \frac{19}{80} = 0.2375$$

Using the y-intercept of $(x_1, y_1) = (0, 59.7)$, $b = 59.7$. Thus, $y = 0.2375x + 59.7$.

Summary Exercises Applying Graphing and Equation-Writing Techniques for Lines

1. (a) "Slope $= -0.5$" indicates that $m = -\frac{1}{2}$ and "$b = -2$" indicates that the y-intercept is $(0, -2)$. Given the slope and y-intercept, we know that one form of the equation of the line is $y = -\frac{1}{2}x - 2$, which is choice B.

(b) The slope of the line passing through the x-intercept of $(4, 0)$ and the y-intercept of $(0, 2)$ is $m = \frac{2-0}{0-4} = \frac{2}{-4} = -\frac{1}{2}$.

The slope-intercept form is $y = -\frac{1}{2}x + 2$.
None of the choices has this equation, so change the form of $y = -\frac{1}{2}x + 2$ by multiplying each side by 2 to clear fractions.

$$2(y) = 2\left(-\frac{1}{2}x + 2\right)$$
$$2y = -x + 4$$

Add x to both sides to get $x + 2y = 4$, which is choice D.

(c) The slope of the line passing through $(4, -2)$ and $(0, 0)$ is

$$m = \frac{0-(-2)}{0-4} = \frac{2}{-4} = -\frac{1}{2}.$$

The slope-intercept form is $y = -\frac{1}{2}x$, which is choice A.

(d) Substitute $x = -2$, $y = -2$, and $m = \frac{1}{2}$ into the slope-intercept form.

$y = mx + b$	Slope-intercept form
$-2 = \frac{1}{2}(-2) + b$	Substitute.
$-2 = -1 + b$	Multiply.
$-1 = b$	Add 1.

The line has equation $y = \frac{1}{2}x - 1$.
Multiply each side by 2 to get $2y = x - 2$, which is the same as $2 = x - 2y$, which is choice C.

2. Solve the given equation for y.

$$2x + 5y = 20$$
$$5y = -2x + 20$$
$$y = -\frac{2}{5}x + 4$$

Choice A, $y = -\frac{2}{5}x + 4$, is already solved for y.
The equations are equivalent.
Solve choice B for y.

$$y - 2 = -\frac{2}{5}(x - 5)$$
$$y - 2 = -\frac{2}{5}x + 2$$
$$y = -\frac{2}{5}x + 4$$

The equations are equivalent.

Choice C, $y = \frac{5}{2}x - 4$, is already solved for y.
This is not equivalent to the given equation.
Solve choice D for y.

$$2x = 5y - 20$$
$$2x + 20 = 5y$$
$$\frac{2}{5}x + 4 = y$$

This is not equivalent to the given equation.
Thus, A and B are equivalent.

3. Solve the equation for *y*.

$x - 2y = -4$ Given equation

$-2y = -x - 4$ Subtract *x*.

$y = \frac{1}{2}x + 2$ Divide by -2.

The slope is the coefficient of $x, \frac{1}{2}$.

The slope is $\frac{1}{2} = \frac{\text{change in } y}{\text{change in } x}$, and the *y*-intercept is $(0, 2)$. Plot that point and count up 1 unit and right 2 units to get to the point $(2, 3)$. Draw a line through the points.

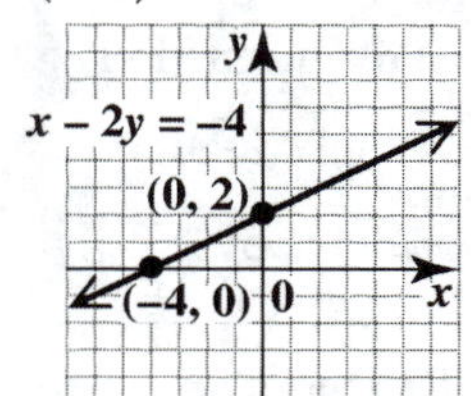

4. Solve the equation for *y*.

$2x + 3y = 12$ Given equation

$3y = -2x + 12$ Subtract $2x$.

$y = -\frac{2}{3}x + 4$ Divide by 3.

The slope is the coefficient of $x, -\frac{2}{3}$.

The slope is $-\frac{2}{3} = \frac{-2}{3} = \frac{\text{change in } y}{\text{change in } x}$, and the *y*-intercept is $(0, 4)$. Plot that point and count down 2 units and right 3 units to get to the point $(3, 2)$. Draw a line through the points.

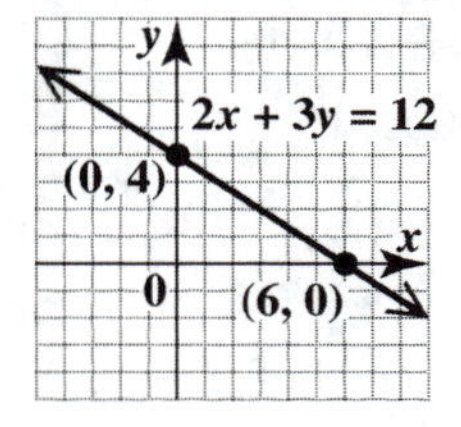

5. The slope is $1 = \frac{1}{1} = \frac{\text{change in } y}{\text{change in } x}$, and the *y*-intercept is $(0, -2)$. Plot that point and count up 1 unit and right 1 unit to get to the point $(1, -1)$. Draw a line through the points.

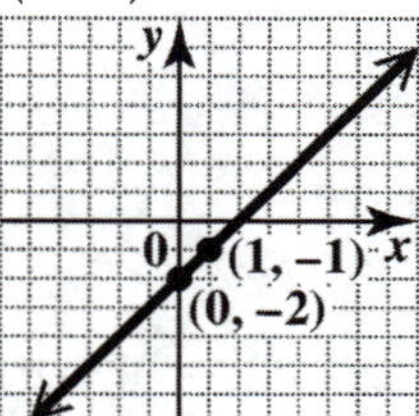

6. Solve the equation for *y*.

$y - 4 = -9$

$y = -5$ Add 4.

This is an equation of the horizontal line with *y*-intercept $(0, -5)$. The slope of the line is 0.

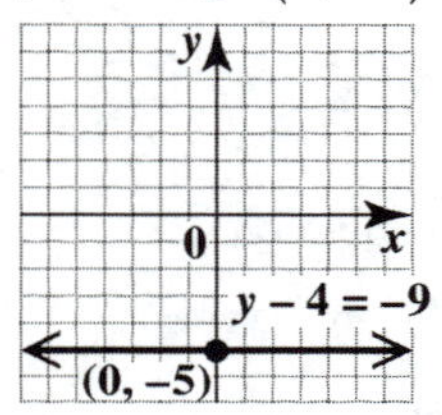

7. The slope is $-\frac{2}{3} = \frac{2}{-3} = \frac{\text{change in } y}{\text{change in } x}$. Plot the point $(3, -4)$ and count up 2 units and left 3 units to get to the point $(0, -2)$. Draw a line through the points.

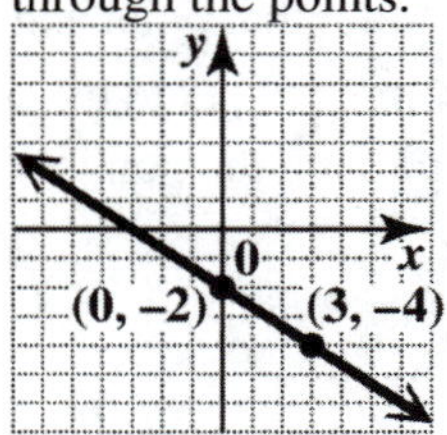

8. Find the *x*-intercept.

$8x = 6y + 24$

$8x = 6(0) + 24$ Let $y = 0$.

$8x = 24$

$x = 3$

The *x*-intercept is $(3, 0)$.

Find the *y*-intercept.

$8x = 6y + 24$

$8(0) = 6y + 24$ Let $x = 0$.

$-24 = 6y$

$-4 = y$

The y-intercept is $(0, -4)$.

Find a third point. Set $x = 2$.

$$8(2) = 6y + 24$$
$$16 = 6y + 24$$
$$-8 = 6y$$
$$-\frac{4}{3} = y$$

The third point is $\left(2, -\frac{4}{3}\right)$.

Plot $(3, 0), (0, -4)$, and $\left(2, -\frac{4}{3}\right)$ and draw a line through these points.

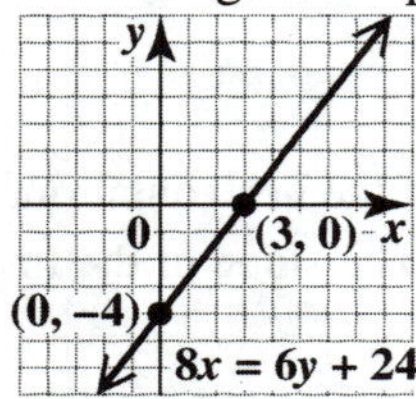

9. Solve the equation for y.

$x - 4y = 0$ Given equation

$-4y = -x$ Subtract x.

$y = \frac{1}{4}x$ Divide by -4.

The slope is the coefficient of x, $\frac{1}{4}$.

The slope is $\frac{1}{4} = \frac{\text{change in } y}{\text{change in } x}$, and the y-intercept is $(0, 0)$. Plot that point and count up 1 unit and right 4 units to get to the point $(4, 1)$. Draw a line through the points.

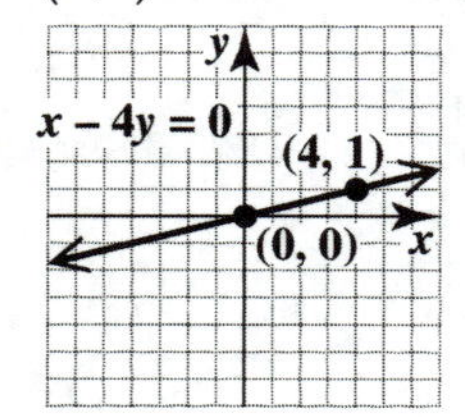

10. The slope is $-\frac{3}{4} = \frac{3}{-4} = \frac{\text{change in } y}{\text{change in } x}$. Plot the point $(4, -4)$ and count up 3 units and left 4 units to get to the point $(0, -1)$. Draw a line through the points.

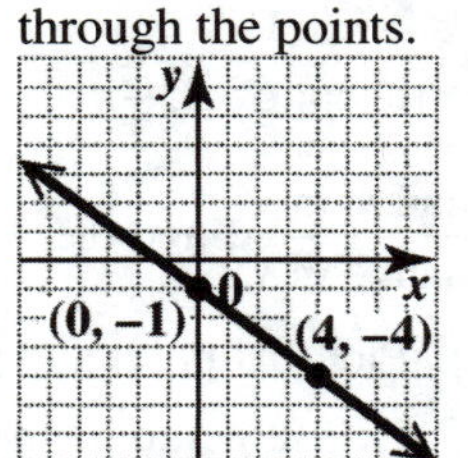

11. Solve the equation for y.

$5x + 2y = 10$ Given equation

$2y = -5x + 10$ Subtract $5x$.

$y = -\frac{5}{2}x + 5$ Divide by 2.

The slope is the coefficient of x, $-\frac{5}{2}$.

The slope is $-\frac{5}{2} = \frac{-5}{2} = \frac{\text{change in } y}{\text{change in } x}$, and the y-intercept is $(0, 5)$. Plot that point and count down 5 units and right 2 units to get to the point $(2, 0)$. Draw a line through the points.

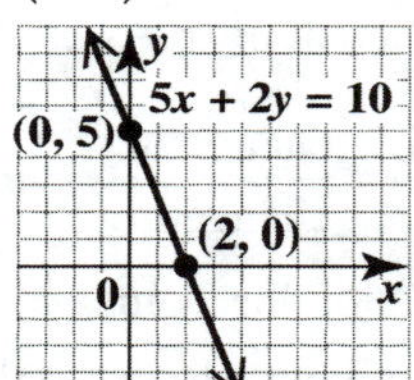

12. Solve the equation for y.

$x + 5y = 0$ Given equation

$5y = -x$ Subtract x.

$y = -\frac{1}{5}x$ Divide by 5.

The slope is the coefficient of x, $-\frac{1}{5}$.

The slope is $-\frac{1}{5}=\frac{-1}{5}=\frac{\text{change in } y}{\text{change in } x}$, and the y-intercept is $(0, 0)$. Plot that point and count down 1 unit and right 5 units to get to the point $(5, -1)$. Draw a line through the points.

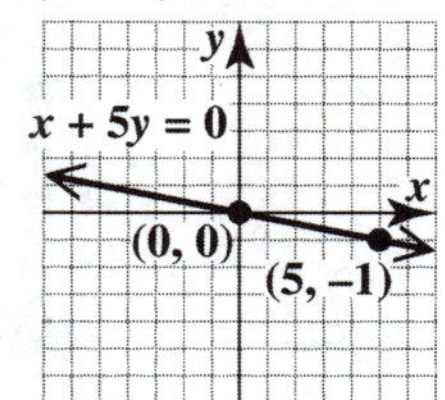

13. Solve the equation for x.

$x+4=0$

$x=-4$ Subtract 4.

This is an equation of the vertical line with x-intercept $(-4, 0)$. The slope of the line is undefined.

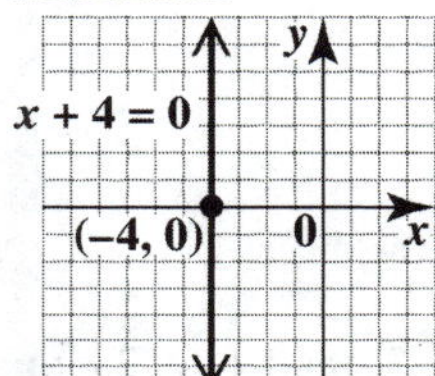

14. $y=-x+6$

The slope is $-1=\frac{-1}{1}=\frac{\text{change in } y}{\text{change in } x}$, and the y-intercept is $(0, 6)$. Plot that point and count down 1 unit and right 1 unit to get to the point $(1, 5)$. Draw a line through the points.

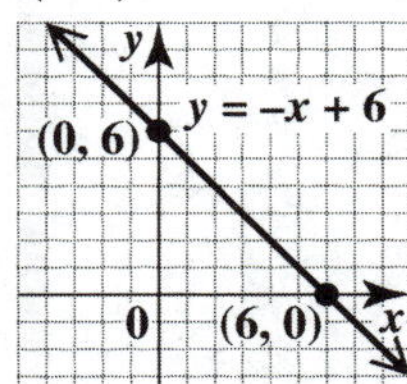

15. Use the points $(3, 0)$ and $(1, 4)$ to find the slope of the line.

$$m=\frac{4-0}{1-3}=\frac{4}{-2}=-2$$

Now use the point $(3, 0)$ for (x_1, y_1) and $m=-2$ in the point-slope form and solve for y.

$$y-y_1=m(x-x_1)$$
$$y-0=-2(x-3)$$
$$y=-2x+6$$

16. Use the points $(-6, 0)$ and $(0, 8)$ to find the slope of the line.

$$m=\frac{8-0}{0-(-6)}=\frac{8}{6}=\frac{4}{3}$$

Note that the y-intercept is $(0, 8)$ and thus $b=8$. Find the slope-intercept form for the line by substituting for m and b.

$$y=mx+b$$
$$y=\frac{4}{3}x+8$$

17. Use the points $(0, -2)$ and $(4, 0)$ to find the slope of the line.

$$m=\frac{0-(-2)}{4-0}=\frac{2}{4}=\frac{1}{2}$$

Note that the y-intercept is $(0, -2)$ and thus $b=-2$. Find the slope-intercept form for the line by substituting for m and b.

$$y=mx+b$$
$$y=\frac{1}{2}x-2$$

18. Use the points $(3, 3)$ and $(6, 1)$ to find the slope of the line.

$$m=\frac{1-3}{6-3}=-\frac{2}{3}$$

Now use the point $(3, 3)$ for (x_1, y_1) and $m=-2$ in the point-slope form and solve for y.

$$y-y_1=m(x-x_1)$$
$$y-3=-\frac{2}{3}(x-3)$$
$$y-3=-\frac{2}{3}x+\frac{2}{3}(3)$$
$$y=-\frac{2}{3}x+2+3$$
$$y=-\frac{2}{3}x+5$$

19. Substitute the slope and y-intercept into the slope-intercept form.

$$y=mx+b$$
$$y=-3x-6$$

20. Substitute the slope and point $(-4, 6)$ into the point-slope form and solve for y.

$$y - y_1 = m(x - x_1)$$
$$y - 6 = \frac{3}{2}[x - (-4)]$$
$$y - 6 = \frac{3}{2}x + 6$$
$$y = \frac{3}{2}x + 12$$

21. Use the points $(1, -7)$ and $(-2, 5)$ to find the slope of the line.

$$m = \frac{5 - (-7)}{-2 - 1} = \frac{12}{-3} = -4$$

Substitute the slope and point $(1, -7)$ into the point-slope form and solve for y.

$$y - y_1 = m(x - x_1)$$
$$y - (-7) = -4(x - 1)$$
$$y + 7 = -4x + 4$$
$$y = -4x - 3$$

22. A line with undefined slope has equation $x = a$ where $(a, 0)$ is the x-intercept. In this case $x = 0$.

23. Use the points $(0, 0)$ and $(3, 2)$ to find the slope of the line.

$$m = \frac{2 - 0}{3 - 0} = \frac{2}{3}$$

Substitute the slope and y-intercept into the slope-intercept form.

$$y = mx + b$$
$$y = \frac{2}{3}x + 0$$
$$y = \frac{2}{3}x$$

24. Substitute the slope and y-intercept into the slope-intercept form.

$$y = mx + b$$
$$y = -1x - 4$$
$$y = -x - 4$$

25. Use the points $(5, 0)$ and $(0, -5)$ to find the slope of the line.

$$m = \frac{-5 - 0}{0 - 5} = \frac{-5}{-5} = 1$$

Substitute the slope and y-intercept into the slope-intercept form.

$$y = mx + b$$
$$y = 1x - 5$$
$$y = x - 5$$

26. A line with slope 0 has equation $y = b$ where $(0, b)$ is the y-intercept. In this case, $y = 0$.

27. Substitute the slope and point $(-3, 0)$ into the point-slope form and solve for y.

$$y - y_1 = m(x - x_1)$$
$$y - 0 = \frac{5}{3}[x - (-3)]$$
$$y = \frac{5}{3}(x + 3)$$
$$y = \frac{5}{3}x + 5$$

28. Use the points $(1, -13)$ and $(-2, 2)$ to find the slope of the line.

$$m = \frac{2 - (-13)}{-2 - 1} = \frac{15}{-3} = -5$$

Substitute the slope and point $(-2, 2)$ into the point-slope form and solve for y.

$$y - y_1 = m(x - x_1)$$
$$y - 2 = -5[x - (-2)]$$
$$y - 2 = -5(x + 2)$$
$$y - 2 = -5x - 10$$
$$y = -5x - 8$$

3.6 Graphing Linear Inequalities in Two Variables

Now Try Exercises

N1. Start by graphing the equation $x + 3y = 6$. The intercepts are $(6, 0)$ and $(0, 2)$. Draw a solid line through these points to show that the points on the line are solutions to the inequality $x + 3y \le 6$.

Choose a test point not on the line, such as $(0, 0)$.

$$x+3y \le 6$$

$$(0)+3(0) \overset{?}{\le} 6 \qquad \text{Let } x=0 \text{ and } y=0.$$

$$0+0 \overset{?}{\le} 6$$

$$0 \overset{?}{\le} 6 \qquad \text{True}$$

Since the last statement is true, shade the region that includes the test point $(0, 0)$, that is, the region below the line. The shaded region, along with the boundary, is the desired graph.

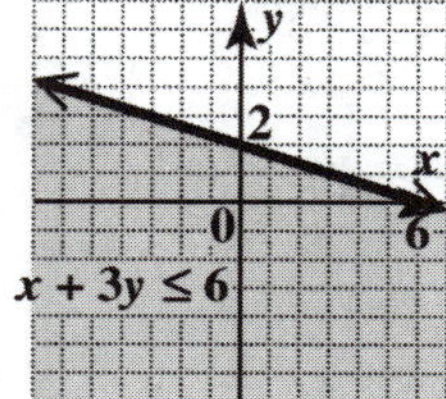

N2. Start by graphing the equation $2x-4y=8$. The intercepts are $(4, 0)$ and $(0, -2)$. Draw a dashed line through these points to show that the points on the line are not solutions to the inequality $2x-4y>8$.

Choose a test point not on the line, such as $(0, 0)$.

$$2x-4y>8$$

$$2(0)-4(0) \overset{?}{>} 8 \qquad \text{Let } x=0 \text{ and } y=0.$$

$$0>8 \qquad \text{False}$$

Since the last statement is false, shade the region that does *not* include the test point, $(0, 0)$, that is, the region below the line. The dashed line shows that the boundary is not part of the graph.

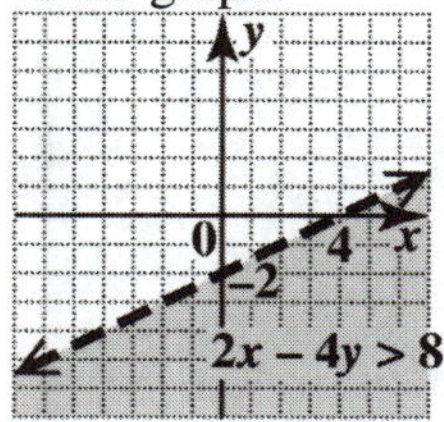

N3. First graph $x=2$, a vertical line through the point $(2, 0)$. Use a dashed line because of the > symbol. Choose $(0, 0)$ as a test point.

$$x>2$$

$$0>2 \qquad \text{Let } x=0. \text{ False}$$

Since the last statement is false, shade the region that does not include the test point $(0, 0)$, that is, the region to the right of line. The dashed line shows that the boundary is not part of the graph.

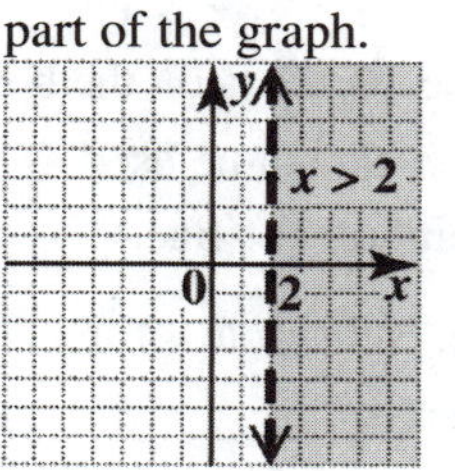

N4. Graph $y=-2x$ as a solid line through $(0, 0)$ and $(1, -2)$. We cannot use $(0, 0)$ as a test point because $(0, 0)$ is on the line $y=-2x$. Instead, we choose a test point off the line, $(1, 0)$.

$$y \le -2x$$

$$0 \overset{?}{\le} -2(1) \qquad \text{Let } x=1, y=0.$$

$$0 \le -2 \qquad \text{False}$$

Since the last statement is false, shade the region that does *not* include the test point $(1, 0)$, that is, the region below the line. The shaded region, along with the boundary, is the desired graph.

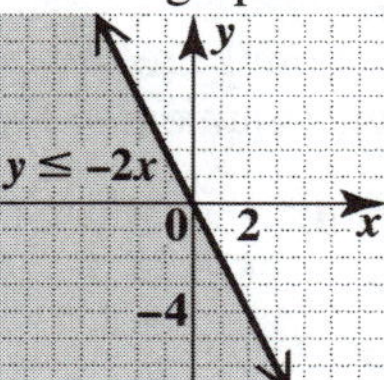

Exercises

1. The key phrase is "more than" (occurs twice), so use the symbol > twice.

3. The key phrase is "at most," so use the symbol ≤.

5. The key phrase is "less than," so use the symbol <.

7. For the point $(4, 0)$, substitute 4 for x and *0* for y in the inequality.

$$3x - 4y < 12$$
$$3(4) - 4(0) \overset{?}{<} 12$$
$$12 - 0 \overset{?}{<} 12$$
$$12 < 12 \quad \text{False}$$

The false result shows that $(4, 0)$ is not a solution of the given inequality.

9. This is false because $(0, 0)$ is on the boundary line $x + 4y = 0$, so it cannot be used as a test point. Use a test point off the line.

11. For the point $(4, 1)$, substitute 4 for x and 1 for y in the given inequality.

$$3x - 2y \geq 0$$
$$3(4) - 2(1) \overset{?}{\geq} 0$$
$$12 - 2 \overset{?}{\geq} 0$$
$$10 \geq 0 \quad \text{True}$$

The *true* result shows that $(4, 1)$ *is* a solution of the given inequality.
For the point $(0, 0)$, substitute 0 for x and 0 for y in the given inequality.

$$3x - 2y \geq 0$$
$$3(0) - 2(0) \overset{?}{\geq} 0$$
$$0 \geq 0 \quad \text{True}$$

The *true* result shows that $(0, 0)$ *is* a solution of the given inequality.
Since both of the given points are solutions to the inequality, the given statement is *true*.

13. Use $(0, 0)$ as a test point.

$$x + 2y \geq 7$$
$$0 + 2(0) \overset{?}{\geq} 7 \quad \text{Let } x = 0 \text{ and } y = 0.$$
$$0 \geq 7 \quad \text{False}$$

Because the last statement is false, we shade the region that does not include the test point $(0, 0)$. This is the region above the line.

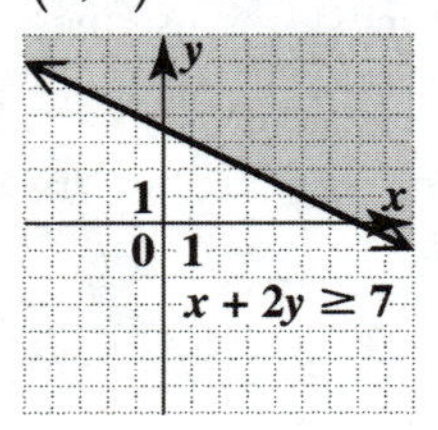

15. Use $(0, 0)$ as a test point.

$$-3x + 4y > 12$$
$$-3(0) + 4(0) \overset{?}{>} 12 \quad \text{Let } x = 0 \text{ and } y = 0.$$
$$0 > 12 \quad \text{False}$$

Because the last statement is false, we shade the region that does not include the test point $(0, 0)$. This is the region above the line.

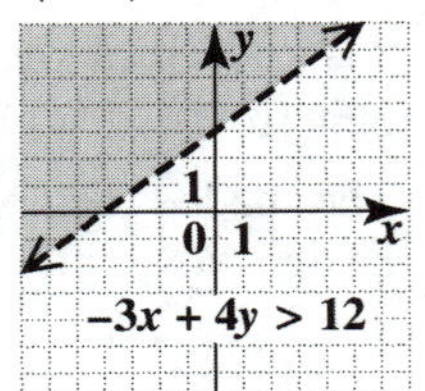

17. Use $(1, 0)$ as a test point.

$$x \leq -y$$
$$1 \overset{?}{\leq} -(0) \quad \text{Let } x = 1 \text{ and } y = 1.$$
$$1 \leq 0 \quad \text{False}$$

Because the last statement is false, we shade the region that does not include the test point $(1, 0)$. This is the region below the line.

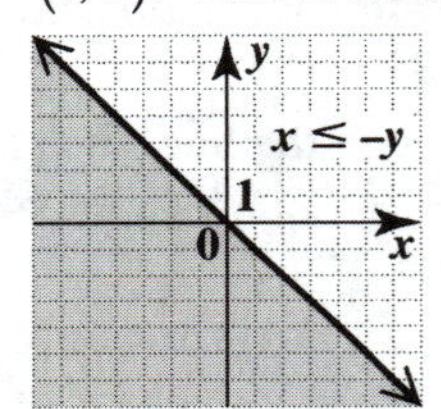

19. Use $(0, 0)$ as a test point.

$$y < -1$$
$$0 < -1 \quad \text{False}$$

Because $0 < -1$ is false, shade the region not containing $(0, 0)$. This is the region below the line.

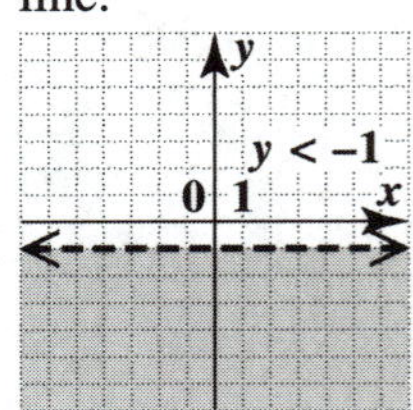

21. Use a *dashed* line if the symbol is < or >. Use a *solid* line if the symbol is ≤ or ≥.

23. *Step 1*
Graph the boundary of the region, the line with equation $x + y = 5$.

If $y = 0$, $x = 5$, so the x-intercept is $(5, 0)$.

If $x = 0$, $y = 5$, so the y-intercept is $(0, 5)$.

Draw the line through these intercepts. Make the line solid because of the $\leq$ sign.
Step 2
Choose the point $(0, 0)$ as a test point.

$x + y \leq 5$

$0 + 0 \overset{?}{\leq} 5$ Let $x = 0$ and $y = 0$.

$0 \leq 5$ True

Because $0 \leq 5$ is true, shade the region containing the origin. The shaded region, along with the boundary, is the desired graph.

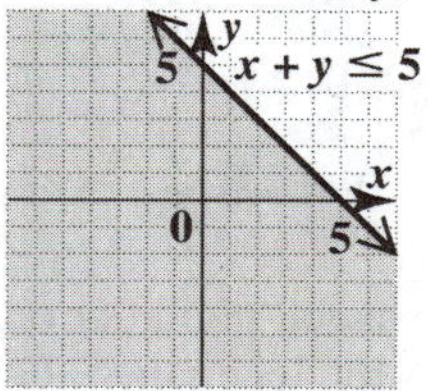

25. The boundary is the line with equation $2x + 3y = -6$. Draw this line through its intercepts, $(-3, 0)$ and $(0, -2)$. The line should be dashed because of the > sign. Choose $(0, 0)$ as a test point. Since $2(0) + 3(0) > -6$ is true, shade the region containing the origin. The dashed line shows that the boundary is not part of the graph.

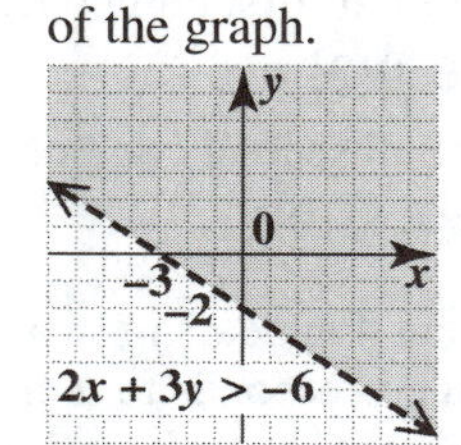

27. The boundary is the line with equation $y = 2x + 1$. This line has slope 2 and y-intercept $(0, 1)$. It may be graphed by starting at $(0, 1)$ and going 2 units up and then 1 unit to the right to reach the point $(1, 3)$. Draw a solid line through $(0, 1)$ and $(1, 3)$.

Using $(0, 0)$ as a test point will result in the inequality $0 \geq 1$, which is false. Shade the region *not* containing the origin, that is, the region above the line. The solid line shows that the boundary is part of the graph.

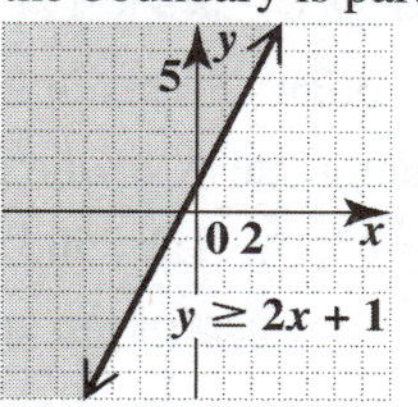

29. The boundary is the line with equation $x = -2$. This is a vertical line through $(-2, 0)$. Make this line dashed because of the < sign.
Using $(0, 0)$ as a test point will result in the inequality $0 < -2$, which is false. Shade the region *not* containing the origin. This is the region to the left of the boundary. The dashed line shows that the boundary is not part of the graph.

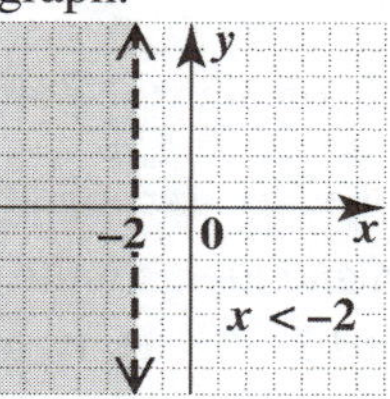

31. The boundary is the line with equation $y = 5$. This is the horizontal line through $(0, 5)$. Make this line solid because of the < sign.
Using $(0, 0)$ as a test point will result in the inequality $0 \leq 5$, which is true. Shade the region containing the origin, that is, the region below the line. The solid line shows that the boundary is part of the graph.

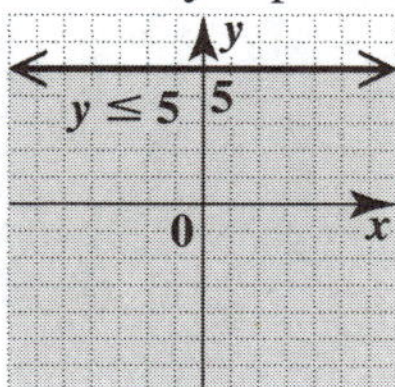

33. The boundary has the equation $y = 4x$. This line goes through the points $(0, 0)$ and $(1, 4)$. Make the line solid because of the $\geq$ sign. Because the boundary passes through the origin, we cannot use $(0, 0)$ as a test point. Using $(2, 0)$ as a test point will result in the inequality $0 \geq 8$, which is false. Shade the region *not* containing $(2, 0)$. The solid line shows that the boundary is part of the graph.

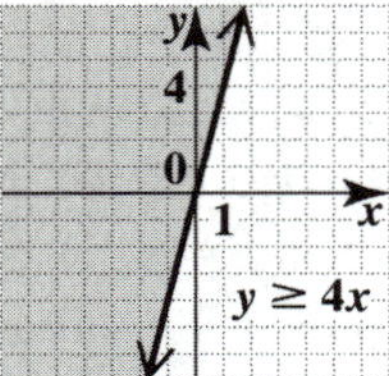

35. The boundary is the line with equation $x = -2y$. This can be rewritten as the line $-\frac{x}{2} = y$ by dividing both sides by -2. This line has slope $-\frac{1}{2}$ and contains the point $(0, 0)$. It may be graphed by starting at $(0, 0)$, and going 1 unit down and then 2 units to the right to reach the point $(2, -1)$. Draw a dashed line through $(0, 0)$ and $(2, -1)$.
Using $(1, 0)$ as a test point will result in the inequality $1 < 0$, which is false. Shade the region below the line that does not contain $(1, 0)$. The dashed line shows that the boundary is not part of the graph.

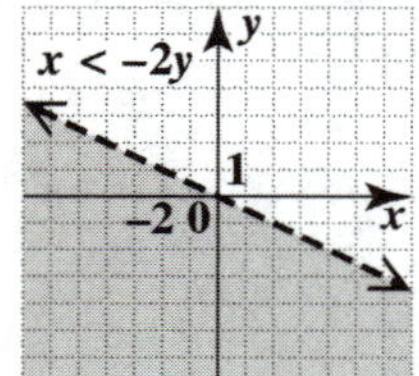

37. **(a)** Graph the inequality.
Step 1
Graph the line $x + y = 500$.
If $x = 0$, then $y = 500$, so the y-intercept is $(0, 500)$.
If $y = 0$, then $x = 500$, so the x-intercept is $(500, 0)$.
Graph the line with these intercepts.
The line is solid because of the $\geq$ sign.

Step 2
Use $(0, 0)$ as a test point.

$x + y \geq 500$ Original inequality

$0 + 0 \overset{?}{\geq} 500$ Let $x = 0$, $y = 0$.

$0 \geq 500$ False

Since $0 \geq 500$ is false, shade the side of the boundary not containing $(0, 0)$. Because of the restrictions $x \geq 0$ and $y \geq 0$ in this applied problem, only the portion of the graph that lies in quadrant I is included.

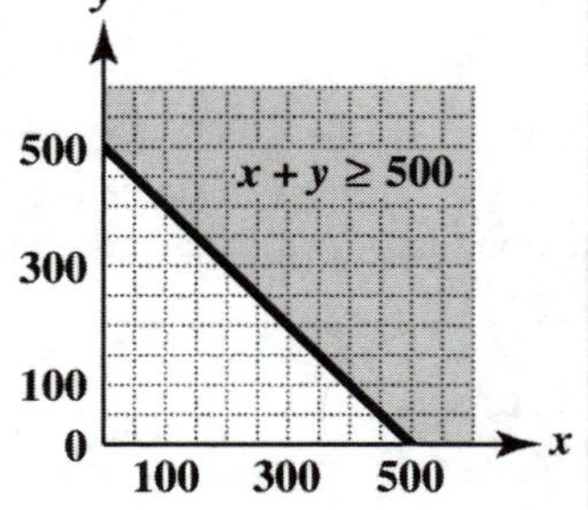

(b) Any point in the shaded region or on the boundary satisfies the inequality. Two ordered pairs are $(500, 0)$ and $(200, 400)$. There are many other ordered pairs that will also satisfy the inequality.

3.7 Introduction to Functions

Now Try Exercises

N1. The domain is the set of all first components in the ordered pairs, $\{-2, 0, 2\}$.
The range is the set of all second components in the ordered pairs, $\{3, 7, 8, 10\}$.

N2. **(a)** The first component 4 appears in two ordered pairs, and corresponds to more than one second component. Therefore, this relation is not a function.

(b) Notice that each first component appears once and only once. Because of this, the relation *is a function.*

N3. **(a)** This linear equation is in the form $y = mx + b$. Since the graph of this equation is a line that is not vertical, the equation defines a function.

(b) The equation $x = -5$ refers to a vertical line which is not a function.

(c) Use the vertical line test. A vertical line could intersect the graph twice, so this graph *is not* the graph of a function.

N4. (a) Any number may be input for x, so the domain is the set of all real numbers. Also, any number may be output for y, so the range is also the set of all real numbers. The graph of the equation is a straight line that extends infinitely in both directions. Both the domain and range are $(-\infty, \infty)$.

(b) Any number may be used for x, so the domain is the set of all real numbers, written $(-\infty, \infty)$.

The second power of a real number cannot be negative, and since $y = x^2 - 2$, the values of y cannot be less than -2. The range is the set of all real numbers greater than or equal to -2, written $[-2, \infty)$.

N5. Find $f(-2)$ by substituting -2 in for x.

$f(x) = x^3 - 7$

$f(-2) = (-2)^3 - 7$ Let $x = -2$.

$f(-2) = -8 - 7$

$f(-2) = -15$

Find $f(0)$ by substituting 0 in for x.

$f(x) = x^3 - 7$

$f(0) = (0)^3 - 7$ Let $x = 0$.

$f(0) = 0 - 7$

$f(0) = -7$

N6. (a) Since 2006 corresponds with 14.9, $f(2006) = 14.9$.

The population of Asian-Americans was 14.9 million in 2006.

(b) $f(x)$ is equal to 17.3 for the x-value 2010.

Exercises

1. Any set of ordered pairs is a(n) relation. The set of first components of the ordered pairs of a relation is the domain. The set of second components of the ordered pairs of a relation is the range.

3. If $x = 1$, then $x + 2 = 3$. Since $f(x) = x + 2$, $f(x)$ is also equal to 3. The ordered pair (x, y) is $(1, 3)$.

5. If $x = 3$, then $x + 2 = 5$. Since $f(x) = x + 2$, $f(x)$ is also equal to 5. The ordered pair (x, y) is $(3, 5)$.

7. If the domain of the function f in Exercises 3–6 is $\{0, 1, 2, 3, 4\}$, then the graph of f consists of the five points $(0, 2), (1, 3), (2, 4), (3, 5)$, and $(4, 6)$.

9. The domain is the set of all first components in the ordered pairs, $\{3, 1, 0, -1\}$.

The range is the set of all second components in the ordered pairs, $\{7, 4, -2, -1\}$.

Since each first component of the ordered pairs corresponds to exactly one second component, the relation *is a function.*

11. The domain is the set of all first components in the ordered pairs, $\{1, 2, 3\}$.

The range is the set of all second components in the ordered pairs, $\{-1, -2\}$.

Since each first component of the ordered pairs corresponds to exactly one second component, the relation is a function.

13. The domain is the set of all first components in the ordered pairs, $\{-4, -2, 0\}$.

The range is the set of all second components in the ordered pairs, $\{3, 1, 5, -8\}$.

This relation is not a function since one value of x, namely -2, corresponds to two values of y, namely 1 and -8.

15. The domain is $\{1, 3, 5, 7, 9\}$.

The range is $\{2, 3, 6, 4\}$.

The relation is a function since each of the first components 1, 3, 5, 7, and 9 corresponds to exactly one second component.

17. The graph consists of the following set of six ordered pairs:

$\{(-4, 1), (-2, 0), (-2, 2), (0, -2), (2, 1), (3, 3)\}$

The domain is the set of all first components in the ordered pairs, $\{-4, -2, 0, 2, 3\}$.

The range is the set of all second components in the ordered pairs, $\{-2, 0, 1, 2, 3\}$.

This relation *is not a function* since one value of x, namely -2, corresponds to two values of y, namely 0 and 2.

19. Any vertical line will intersect the graph in only one point. The graph passes the vertical line test, so this is the graph of a function. The domain is the set of all real numbers, written $(-\infty, \infty)$. The lowest y value on the graph is -1, so the range is $[-1, \infty)$.

21. A vertical line can cross the graph twice, so this is not the graph of a function. The lowest x value on the graph is 0, so the domain is $[0, \infty)$. The range is the set of all real numbers, written $(-\infty, \infty)$.

23. Every value of x will give one and only one value of y, so the equation defines a function.

25. The equation refers to a vertical line. Every value of y corresponds to an x value of -7. The equation is not a function.

27. The function is a horizontal line. Every value of x will give one and only one value of y, of 1, so the equation defines a function. The function is a horizontal line.

29. Every value of x will give one and only one value of y, so the equation defines a function.

31. Every positive value of x will give two values of y. For example, if $x = 16, 16 = y^2$ and $y = +4$ or -4. Therefore, the equation does not define a function.

33. Any number may be used for x, so the domain is the set of all real numbers, written $(-\infty, \infty)$. In $y = 3x - 2$, any number may be used for y, so the range is also $(-\infty, \infty)$.

35. Any number may be used for x, so the domain is the set of all real numbers, written $(-\infty, \infty)$. In $y = x$, any number may be used for y, so the range is also $(-\infty, \infty)$.

37. Any number may be used for x, so the domain is the set of all real numbers, written $(-\infty, \infty)$. The second power of a real number cannot be negative, and since $y = x^2 + 2$, the values of y cannot be less than 2. The range is the set of all real numbers greater than or equal to 2, written $[2, \infty)$.

39. Any number may be used for x, so the domain is the set of all real numbers, written $(-\infty, \infty)$. The second power of a real number, x^2, is zero or positive, so $y = -x^2$, is 0 or negative. The range is the set of all real numbers less than or equal to 0, written $(-\infty, 0]$.

41. **(a)** Substitute 2 for x in the function.
$f(2) = 4(2) + 3 = 8 + 3 = 11$

(b) Substitute 0 for x in the function.
$f(0) = 4(0) + 3 = 0 + 3 = 3$

(c) Substitute -3 for x in the function.
$f(-3) = 4(-3) + 3 = -12 + 3 = -9$

(d) Substitute $\frac{1}{2}$ for x in the function.

$$f\left(\frac{1}{2}\right) = 4(-3) + 3 = -12 + 3 = -9$$

(e) Substitute $-\frac{1}{3}$ for x in the function.

$$f\left(-\frac{1}{3}\right) = 4(-3) + 3 = -12 + 3 = -9$$

43. **(a)** Substitute 2 for x in the function.
$f(2) = -(2) - 2 = -2 - 2 = -4$

(b) Substitute 0 for x in the function.
$f(0) = -(0) - 2 = 0 - 2 = -2$

(c) Substitute -3 for x in the function.
$f(-3) = -(-3) - 2 = 3 - 2 = 1$

(d) Substitute $\frac{1}{2}$ for x in the function.

$$f\left(\frac{1}{2}\right) = -\left(\frac{1}{2}\right) - 2 = -\frac{1}{2} - \frac{4}{2} = -\frac{5}{2}$$

(e) Substitute $-\frac{1}{3}$ for x in the function.

$$f\left(-\frac{1}{3}\right) = -\left(-\frac{1}{3}\right) - 2 = \frac{1}{3} - \frac{6}{3} = -\frac{5}{3}$$

45. **(a)** Substitute 2 for x in the function.
$f(2) = (2)^2 - (2) + 2 = 4 - 2 + 2 = 4$

(b) Substitute 0 for x in the function.
$f(0) = (0)^2 - (0) + 2 = 0 - 0 + 2 = 2$

(c) Substitute -3 for x in the function.

$$f(-3)=(-3)^2-(-3)+2$$
$$=9+3+2=14$$

(d) Substitute $\frac{1}{2}$ for x in the function.

$$f\left(\frac{1}{2}\right)=\left(\frac{1}{2}\right)^2-\left(\frac{1}{2}\right)+2=\frac{1}{4}-\frac{2}{4}+\frac{8}{4}=\frac{7}{4}$$

(e) Substitute $-\frac{1}{3}$ for x in the function.

$$f\left(-\frac{1}{3}\right)=\left(-\frac{1}{3}\right)^2-\left(-\frac{1}{3}\right)+2$$
$$=\frac{1}{9}+\frac{3}{9}+\frac{18}{9}=\frac{22}{9}$$

47. **(a)** Substitute 2 for x in the function.

$$f(2)=|2|=2$$

(b) Substitute 0 for x in the function.

$$f(0)=|0|=0$$

(c) Substitute -3 for x in the function.

$$f(-3)=|-3|=-(-3)=3$$

(d) Substitute $\frac{1}{2}$ for x in the function.

$$f\left(\frac{1}{2}\right)=\left|\frac{1}{2}\right|=\frac{1}{2}$$

(e) Substitute $-\frac{1}{3}$ for x in the function.

$$f\left(-\frac{1}{3}\right)=\left|-\frac{1}{3}\right|=-\left(-\frac{1}{3}\right)=\frac{1}{3}$$

49. **(a)** Substitute 2 for x in the function.

$$f(2)=|2+7|=|9|=9$$

(b) Substitute 0 for x in the function.

$$f(0)=|0+7|=|7|=7$$

(c) Substitute -3 for x in the function.

$$f(-3)=|-3+7|=|4|=4$$

(d) Substitute $\frac{1}{2}$ for x in the function.

$$f\left(\frac{1}{2}\right)=\left|\frac{1}{2}+7\right|=\left|\frac{1}{2}+\frac{14}{2}\right|$$
$$=\left|\frac{15}{2}\right|=\frac{15}{2}$$

(e Substitute $-\frac{1}{3}$ for x in the function.

$$f\left(-\frac{1}{3}\right)=\left|-\frac{1}{3}+7\right|=\left|-\frac{1}{3}+\frac{21}{3}\right|$$
$$=\left|\frac{20}{3}\right|=\frac{20}{3}$$

51. Write the information in the graph as a set of ordered pairs of the form (year, population). The set is $\{(1970, 9.6), (1980, 14.1), (1990, 19.8), (2000, 28.4), (2010, 40.0)\}$. Since each year corresponds to exactly one number, the set defines a function.

53. Use the points from the table.
$g(1980)=14.1$; $g(2000)=28.4$

55. From the table, $g(2010)=40.0$, so $x=2010$.

57. **(a)** The domain value is 4.

(b) The range value is $\sqrt{4}$, which is 2.

59. If $f(2)=4$, one point on the line has coordinates $(2, 4)$.

61. Use the points $(2, 4)$ and $(-1, -4)$ in the slope formula.

$$m=\frac{-4-4}{-1-2}=\frac{-8}{-3}=\frac{8}{3}$$

Chapter 3 Review Exercises

1. The graph rises from 2010 to 2011, so the number of real trees purchased increased from 2010 to 2011.
2. The graph stays the same from 2008 to 2009, so the number of real trees purchased stayed the same from 2008 to 2009.
3. Locate 2011 on the horizontal scale and follow the line up to the line graph. Then move across to read the value on the vertical scale. The number of real trees purchased in 2011 is about 31 million. Similarly, we get about 24 million for 2012.
4. The change from 2011 to 2012 was
27 million − 31 million = −7 million.
This represents a decrease of about 7 million.

5. Find y when $x = -1$.

$y = 3x + 2$

$y = 3(-1) + 2$ Let $x = -1$.

$y = -3 + 2$

$y = -1$

Find y when $x = 0$.

$y = 3x + 2$

$y = 3(0) + 2$ Let $x = 0$.

$y = 0 + 2$

$y = 2$

Find x when $y = 5$.

$y = 3x + 2$

$5 = 3x + 2$ Let $y = 5$.

$3 = 3x$

$1 = x$

The ordered pairs are $(-1, \underline{-1}), (0, \underline{2})$, and $(\underline{1}, 5)$.

6. Find y when $x = 0$.

$4x + 3y = 6$

$4(0) + 3y = 6$ Let $x = 0$.

$3y = 6$

$y = 2$

Find x when $y = 0$.

$4x + 3y = 6$

$4x + 3(0) = 6$ Let $y = 0$.

$4x = 6$

$x = \frac{6}{4} = \frac{3}{2}$

Find y when $x = -2$.

$4x + 3y = 6$

$4(-2) + 3y = 6$ Let $x = -2$.

$-8 + 3y = 6$

$3y = 14$

$y = \frac{14}{3}$

The ordered pairs are $(0, \underline{2}), \left(\underline{\frac{3}{2}}, 0\right)$, and $\left(-2, \underline{\frac{14}{3}}\right)$.

7. Find y when $x = 0$.

$x = 3y$

$0 = 3y$ Let $x = 0$.

$0 = y$

Find y when $x = 8$.

$x = 3y$

$8 = 3y$ Let $x = 8$.

$\frac{8}{3} = y$

Find x when $y = -3$.

$x = 3y$

$x = 3(-3)$ Let $y = -3$.

$x = -9$

The ordered pairs are $(0, \underline{0}), \left(8, \underline{\frac{8}{3}}\right)$, and $(\underline{-9}, -3)$.

8. The given equation may be written $x = 7$. For any value of y, the value of x will always be 7. The ordered pairs are $(\underline{7}, -3), (\underline{7}, 0)$, and $(\underline{7}, 5)$.

9. Substitute 2 for x and 5 for y in the given equation.

$x + y = 7$

$2 + 5 \stackrel{?}{=} 7$

$7 = 7$ True

Yes, $(2, 5)$ is a solution of the given equation.

10. Substitute -1 for x and 3 for y in the given equation.

$2x + y = 5$

$2(-1) + 3 \stackrel{?}{=} 5$

$-2 + 3 \stackrel{?}{=} 5$

$1 = 5$ False

No, $(-1, 3)$ is not a solution of the given equation.

11. Substitute $\frac{1}{3}$ for x and -3 for y in the given equation.

$3x - y = 4$

$3\left(\frac{1}{3}\right) - (-3) \stackrel{?}{=} 4$

$1 + 3 \stackrel{?}{=} 4$

$4 = 4$ True

Yes, $\left(\frac{1}{3}, -3\right)$ is a solution of the given equation.

12. Substitute 0 for x and in the given equation.

$x = -1$

$0 \stackrel{?}{=} -1$

$0 = -1$ False

No, $(0, -1)$ is not a solution of the given equation.

13. To plot $(2, 3)$, start at the origin, go 2 units to the right, and then go up 3 units. The point lies in quadrant I.

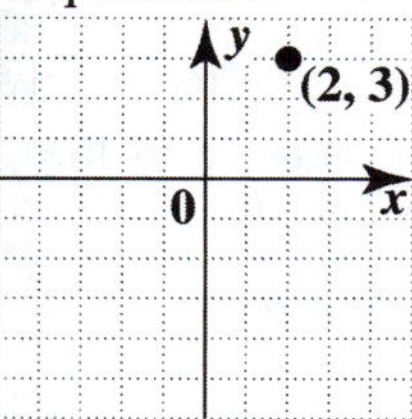

14. To plot $(-4, 2)$, start at the origin, go 4 units to the left, and then go up 2 units. The point lies in quadrant II.

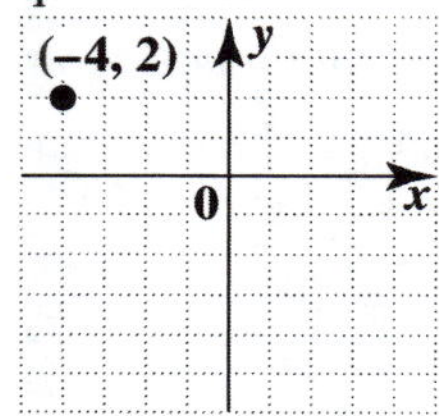

15. To plot $(3, 0)$, start at the origin, go 3 units to the right. The point lies on the x-axis (not in any quadrant).

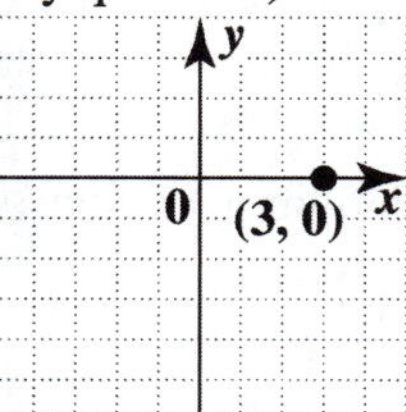

16. To plot $(0, -6)$, start at the origin, go down 6 units. The point lies on the y-axis (not in any quadrant).

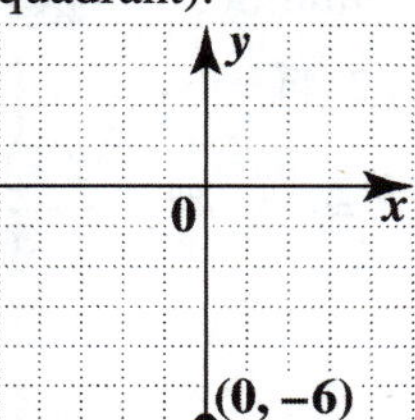

17. To find the x-intercept, let $y = 0$.

$y = 2x + 5$

$0 = 2x + 5$ Let $y = 0$.

$-2x = 5$

$x = -\frac{5}{2}$

The x-intercept is $\left(-\frac{5}{2}, 0\right)$.

To find the y-intercept, let $x = 0$.

$y = 2x + 5$

$y = 2(0) + 5$ Let $x = 0$.

$y = 5$

The y-intercept is $(0, 5)$.

To find a third point, choose $x = -1$.

$y = 2x + 5$

$y = 2(-1) + 5$ Let $x = -1$.

$y = 3$

This gives the ordered pair $(-1, 3)$. Plot $\left(-\frac{5}{2}, 0\right)$, $(0, 5)$, and $(-1, 3)$ and draw a line through them.

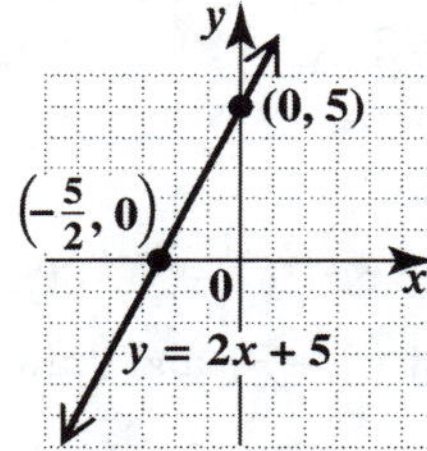

18. To find the x-intercept, let $y = 0$.

$3x + 2y = 8$

$3x + 2(0) = 8$ Let $y = 0$.

$3x = 8$

$x = \frac{8}{3}$

The x-intercept is $\left(\frac{8}{3}, 0\right)$.

To find the y-intercept, let $x = 0$.

$3x + 2y = 8$

$3(0) + 2y = 8$ Let $x = 0$.

$2y = 8$

$y = 4$

The y-intercept is $(0, 4)$.

To find a third point, choose $x = 1$.

$$3x + 2y = 8$$
$$3(1) + 2y = 8 \quad \text{Let } x = 1.$$
$$2y = 5$$
$$y = \frac{5}{2}$$

This gives the ordered pair $\left(1, \frac{5}{2}\right)$.

Plot $\left(\frac{8}{3}, 0\right)$, $(0, 4)$, and $\left(1, \frac{5}{2}\right)$ and draw a line through them.

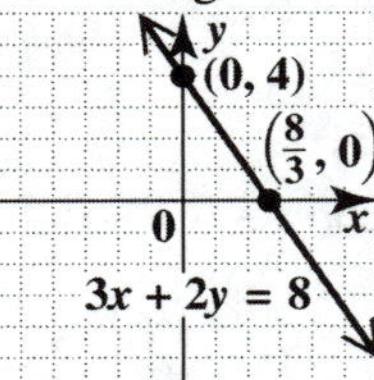

19. Find the intercepts.
If $y = 0$, $x = -4$, so the x-intercept is $(-4, 0)$.
If $x = 0$, $y = -2$, so the y-intercept is $(0, -2)$.
To find a third point, choose $x = 2$.

$$x + 2y = -4$$
$$2 + 2y = -4 \quad \text{Let } x = 2.$$
$$2y = -6$$
$$y = -3$$

This gives the ordered pair $(2, -3)$. Plot $(-4, 0)$, $(0, -2)$, and $(2, -3)$ and draw a line through them.

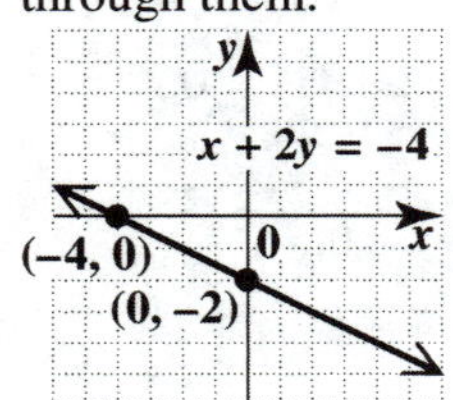

20. Find the intercepts.
In this equation, x is always -6, including when $y = 0$, so the x-intercept is $(-6, 0)$. There is no y-intercept as x is never 0. The graph is a horizontal line through $(-6, 0)$.

21. Let $(2, 3) = (x_1, y_1)$ and $(-4, 6) = (x_2, y_2)$.

$$\text{slope } m = \frac{\text{change in } y}{\text{change in } x} = \frac{y_2 - y_1}{x_2 - x_1} = \frac{6-3}{-4-2} = \frac{3}{-6} = -\frac{1}{2}$$

22. Let $(2, 5) = (x_1, y_1)$ and $(2, 8) = (x_2, y_2)$.

$$\text{slope } m = \frac{\text{change in } y}{\text{change in } x} = \frac{y_2 - y_1}{x_2 - x_1} = \frac{8-5}{2-2} = \frac{3}{0}$$

The slope is *undefined*.

23. The equation $y = 3x - 4$ is already solved for y, so the slope of the line is given by the coefficient of x. Thus, the slope is 3.

24. $y = 5$ is an equation of a horizontal line. Its slope is 0.

25. $x = -7$ is an equation of a vertical line. Its slope is undefined.

26. The indicated points have coordinates $(0, -2)$ and $(2, 1)$. Use the definition of slope with $(0, -2) = (x_1, y_1)$ and $(2, 1) = (x_2, y_2)$.

$$m = \frac{\text{change in } y}{\text{change in } x} = \frac{y_2 - y_1}{x_2 - x_1} = \frac{1-(-2)}{2-0} = \frac{3}{2}$$

Note: We could also simply count grid marks to get $m = \frac{\text{rise}}{\text{run}} = \frac{+3}{+2} = \frac{3}{2}$.

27. The indicated points have coordinates $(0, 1)$ and $(3, 0)$. Use the definition of slope with $(0, 1) = (x_1, y_1)$ and $(3, 0) = (x_2, y_2)$.

$$m = \frac{\text{change in } y}{\text{change in } x} = \frac{y_2 - y_1}{x_2 - x_1} = \frac{0-1}{3-0} = \frac{-1}{3} = -\frac{1}{3}$$

28. From the table, we choose the two points $(0, 1)$ and $(2, 4)$. Use the definition of slope with $(0, 1) = (x_1, y_1)$ and $(2, 4) = (x_2, y_2)$.

$$\text{slope } m = \frac{\text{change in } y}{\text{change in } x} = \frac{y_2 - y_1}{x_2 - x_1} = \frac{4-1}{2-0} = \frac{3}{2}$$

29. **(a)** Because parallel lines have equal slopes and the slope of the graph of $y = 2x + 3$ is 2, the slope of a line parallel to it will also be 2.

(b) Perpendicular lines have slopes which are negative reciprocals of each other. The slope of the graph of $y = -3x + 3$ is -3, so the slope of a line perpendicular to it will be $-\frac{1}{-3} = \frac{1}{3}$.

30. Find the slope of each line by solving the equations for y.

$3x + 2y = 6$

$2y = -3x + 6$ Subtract $3x$.

$y = -\frac{3}{2}x + 3$ Divide by 2.

The slope of the first line is $-\frac{3}{2}$.

$6x + 4y = 8$

$4y = -6x + 8$ Subtract $6x$.

$y = -\frac{6}{4}x + 2$ Divide by 4.

$y = -\frac{3}{2}x + 2$ Lowest terms

The slope of the second line is $-\frac{3}{2}$. The slopes are equal so the lines are *parallel*.

31. Find the slope of each line by solving the equations for y.

$x - 3y = 1$

$-3y = -x + 1$ Subtract x.

$y = \frac{1}{3}x - \frac{1}{3}$ Divide by -3.

The slope of the first line is $\frac{1}{3}$.

$3x + y = 4$

$y = -3x + 4$ Subtract $3x$.

The slope of the second line is -3.

The product of the slopes is $\frac{1}{3}(-3) = -1$, so the lines are *perpendicular*.

32. Find the slope of each line by solving the equations for y.

$x - 2y = 8$

$-2y = -x + 8$ Subtract x.

$y = \frac{1}{2}x - 4$ Divide by -2.

The slope of the first line is $\frac{1}{2}$.

$x + 2y = 8$

$2y = -x + 8$ Subtract x.

$y = -\frac{1}{2}x + 4$ Divide by 2.

The slope of the second line is $-\frac{1}{2}$.

The slopes are not equal and their product is $\left(\frac{1}{2}\right)\left(-\frac{1}{2}\right) = -\frac{1}{4} \neq -1$.

So the lines are *neither* parallel nor perpendicular.

33. Use the slope-intercept form.

$y = mx + b$

$y = -1 \cdot x + \frac{2}{3}$

$y = -x + \frac{2}{3}$

34. Use the point-slope form and solve for y.

$y - y_1 = m(x - x_1)$

$y - 3 = -\frac{1}{2}(x - 2)$

$y - 3 = -\frac{1}{2}x + 1$

$y = -\frac{1}{2}x + 4$

35. Use the point-slope form and solve for y.

$y - y_1 = m(x - x_1)$

$y - (-3) = 1(x - 4)$

$y + 3 = x - 4$

$y = x - 7$

36. Use the point-slope form and solve for y.

$y - y_1 = m(x - x_1)$

$y - 4 = \frac{2}{3}[x - (-1)]$

$y - 4 = \frac{2}{3}(x + 1)$

$y - 4 = \frac{2}{3}x + \frac{2}{3}$

$y = \frac{2}{3}x + \frac{14}{3}$

37. Use the point-slope form and solve for y.

$$y-(-1)=-\frac{3}{4}(x-1)$$
$$y+1=-\frac{3}{4}x+\frac{3}{4}$$
$$y=-\frac{3}{4}x-\frac{1}{4}$$

38. Use the slope-intercept form.
$y=mx+b.$

$$y=-\frac{1}{4}x+\frac{3}{2}$$

39. Horizontal lines have 0 slope and equations of the form $y=k$. In this case, k must equal 1 since the line goes through $(-4, 1)$, so the equation is $y=1$.

40. Vertical lines have undefined slope and equations of the form $x=k$. In this case, k must equal $\frac{1}{3}$ since the line goes through $\left(\frac{1}{3}, -\frac{5}{4}\right)$, so the equation is $x=\frac{1}{3}$.

41. First calculate the slope. The indicated points in the graph provided in Exercise 26 have coordinates $(0, -2)$ and $(2, 1)$. Use the definition of slope with $(0, -2)=(x_1, y_1)$ and $(2, 1)=(x_2, y_2)$.

$$m=\frac{\text{change in } y}{\text{change in } x}=\frac{y_2-y_1}{x_2-x_1}$$
$$=\frac{1-(-2)}{2-0}=\frac{3}{2}$$

Now, observe that the y-intercept in the graph provided in Exercise 26 is $(0, -2)$ and therefore $b=-2$. Use m and b to write the equation of the line in slope-intercept form.
$y=mx+b$

$$y=\frac{3}{2}x-2$$

42. First calculate the slope. The indicated points in the graph provided in Exercise 27 have coordinates $(0, 1)$ and $(3, 0)$. Use the definition of slope with $(0, 1)=(x_1, y_1)$ and $(3, 0)=(x_2, y_2)$.

$$m=\frac{\text{change in } y}{\text{change in } x}=\frac{y_2-y_1}{x_2-x_1}=\frac{0-1}{3-0}=\frac{-1}{3}=-\frac{1}{3}$$

Now, observe that the y-intercept in the graph provided in Exercise 27 is $(0, 1)$ and therefore $b=1$. Use m and b to write the equation of the line in slope-intercept form.
$y=mx+b$

$$y=-\frac{1}{3}x+1$$

43. To graph the boundary, which is the line $3x+5y=9$, find its intercepts.

Find x when $y=0$.

$$3x+5y=9$$
$$3x+5(0)=9 \quad \text{Let } y=0.$$
$$3x+0=9$$
$$3x=9$$
$$x=3$$

Find y when $x=0$.

$$3x+5y=9$$
$$3(0)+5y=9 \quad \text{Let } x=0.$$
$$0+5y=9$$
$$5y=9$$
$$y=\frac{9}{5}$$

The x-intercept is $(3, 0)$ and the y-intercept is $\left(0, \frac{9}{5}\right)$. (A third point may be used as a check.)

Plot these points. Draw a dashed line through these points. In order to determine which side of the line should be shaded, use $(0, 0)$ as a test point. Substituting 0 for x and 0 for y will result in the inequality $0>9$, which is false. Shade the region *not* containing the origin. This is the region above the line. The dashed line shows that the boundary is not part of the graph.

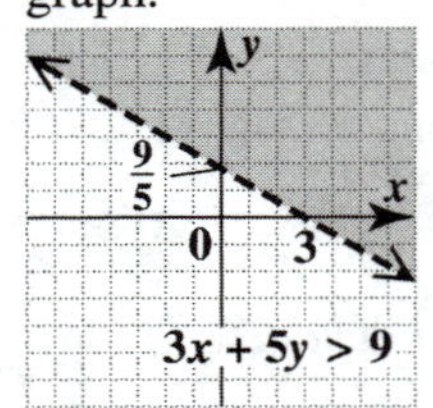

44. Use intercepts to graph the boundary, $2x-3y=-6$.

If $y=0$, $x=-3$, so the x-intercept is $(-3, 0)$.

If $x=0$, $y=2$, so the y-intercept is $(0, 2)$.

Plot these points. Draw a dashed line through $(-3, 0)$ and $(0, 2)$.

Using $(0, 0)$ as a test point will result in the inequality $0 > -6$, which is true. Shade the region containing the origin. This is the region below the line. The dashed line shows that the boundary is not part of the graph.

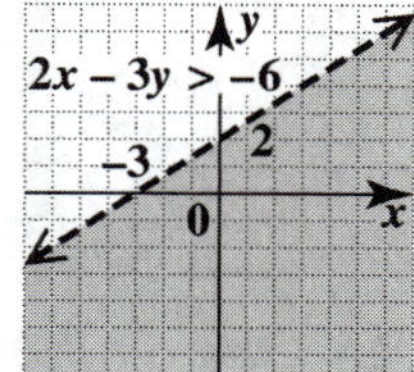

45. The equation of the boundary is $x - 2y = 0$. This line goes through the origin, so both intercepts are $(0, 0)$. Two other points on this line are $(2, 1)$ and $(-2, -1)$. Plot these points. Draw a solid line through $(0, 0), (2, 1)$, and $(-2, -1)$.

Because $(0, 0)$ lies on the boundary, we must choose another point as the test point. Using $(0, 3)$ results in the inequality $-6 \geq 0$, which is false. Shade the region *not* containing the test point $(0, 3)$.

This is the region below the line.
The solid line shows that the boundary is part of the graph.

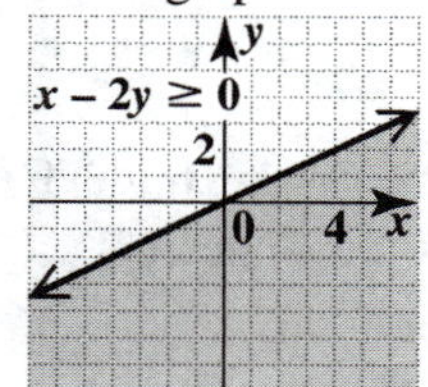

46. The equation of the boundary is $y = -1$. This is a horizontal line through the point $(0, -1)$, the x-intercept. It does not have a y-intercept. Plot these points. Draw a solid horizontal line through $(0, -1)$.

Using $(0, 0)$ as a test point will result in the inequality $0 \geq -1$, which is true. Shade the region containing the origin. This is the region above the line. The solid line shows that the boundary is part of the graph.

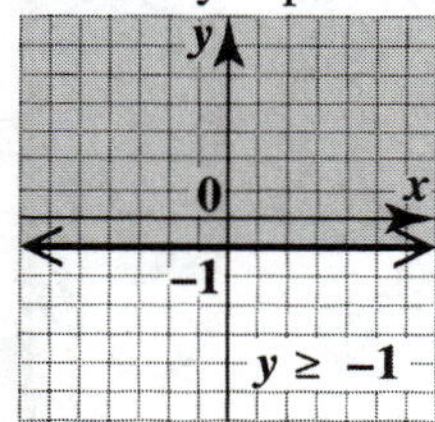

47. Since $x = 2$ appears in two ordered pairs, one value of x yields more than one value of y. Hence, this relation is not a function.
The domain is the set of first components of the ordered pairs, $\{-2, 0, 2\}$.

The range is the set of second components of the ordered pairs, $\{4, 8, 5, 3\}$.

48. Since each first component of the ordered pairs corresponds to exactly one second component, the relation is a function.
The domain is $\{8, 7, 6, 5, 4\}$.

The range is $\{3, 4, 5, 6, 7\}$.

49. Since a vertical line may cross the graph twice, this is not the graph of a function.
The highest x value on the graph is 3. So, the domain is $(-\infty, 3]$.

The range is all real numbers, written $(-\infty, \infty)$.

50. Any vertical line will cross this graph at exactly one point, so it is the graph of a function.
The domain is all real numbers, written $(-\infty, \infty)$.

The highest y value on the graph is 3. So, the range is $(-\infty, 3]$.

51. Solve the equation for y.

$$3y = -2x + 12$$

$$y = -\frac{2}{3}x + 4$$

Since one value of x will lead to only one value of y, $2x + 3y = 12$ is a function.

The domain is all real numbers, written $(-\infty, \infty)$.

The range is all real numbers, written $(-\infty, \infty)$.

52. Each value of x will lead to only one value of y, so $y = x^2$ is a function.
The domain is all real numbers, written $(-\infty, \infty)$.

The range is real numbers greater than or equal to 0, written $[0, \infty)$.

53. **(a)** Substitute 2 for x in the function.

$$f(2) = 3(2) + 2 = 6 + 2 = 8$$

(b) Substitute -1 for x in the function.

$$f(-1) = 3(-1) + 2 = -3 + 2 = -1$$

54. **(a)** Substitute 2 for x in the function.

$f(2) = 2(2)^2 - 1 = 2(4) - 1 = 8 - 1 = 7$

(b) Substitute -1 for x in the function.

$f(-1) = 2(-1)^2 - 1 = 2(1) - 1 = 2 - 1 = 1$

55. **(a)** Substitute 2 for x in the function.

$f(2) = |2+3| = |5| = 5$

(b) Substitute -1 for x in the function.

$f(-1) = |-1+3| = |2| = 2$

Chapter 3 Mixed Review Exercises

1. Vertical lines have undefined slopes. The answer is A.

2. Two graphs pass through $(0, -3)$. C and D are the answers.

3. Three graphs pass through the point $(-3, 0)$. A, B, and D are the answers.

4. Lines that fall from left to right have negative slope. The answer is D.

5. $y = -3$ is a horizontal line passing through $(0, -3)$. C is the answer.

6. B is the only graph that has a positive slope, so it is the only one we need to investigate. B passes through the points $(0, 3)$ and $(-3, 0)$.

Find the slope.

$m = \frac{0-3}{-3-0} = \frac{-3}{-3} = 1$

Since the slope is 1, B is the answer.

7. The equation is in the slope-intercept form, so the slope is -2 and the y-intercept is $(0, -5)$.

To find the x-intercept, let $y = 0$.

$0 = -2x - 5$ Let $y = 0$.

$2x = -5$

$x = -\frac{5}{2}$

The x-intercept is $\left(-\frac{5}{2}, 0\right)$.

Graph the line using the intercepts.

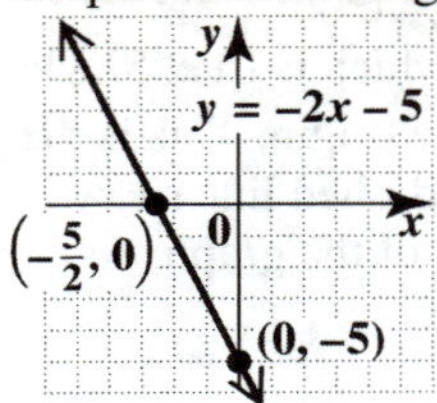

8. Solve the equation for y.

$x + 3y = 0$

$3y = -x$ Subtract x.

$y = -\frac{1}{3}x$ Divide by 3.

From this slope-intercept form, we see that the slope is $-\frac{1}{3}$ and the y-intercept is $(0, 0)$, which is also the x-intercept. To find another point, let $y = 1$.

$x + 3(-1) = 0$ Let $y = -1$.

$x = 3$

So the point $(3, -1)$ is on the graph. Graph the line through $(0,0)$ and $(3, -1)$.

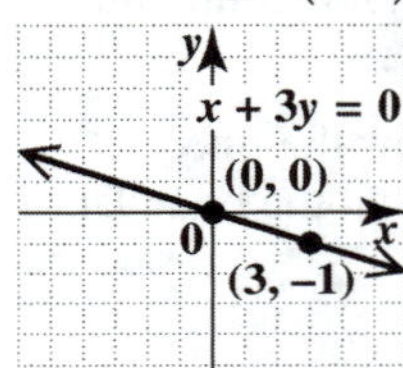

9. This is a horizontal line passing through the point $(0, 5)$, which is the y-intercept.

There is no x-intercept.

Horizontal lines have slopes of 0.

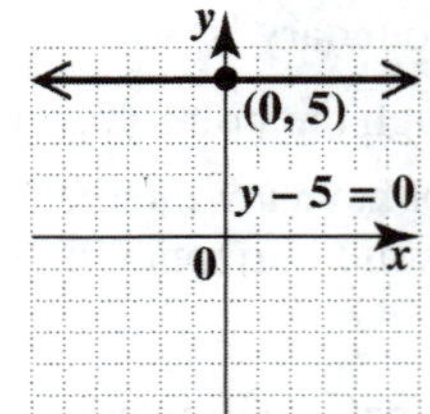

10. This is a vertical line passing through the point $(-5, 0)$, which is the x-intercept.
There is no y-intercept.
Vertical lines have undefined slopes.

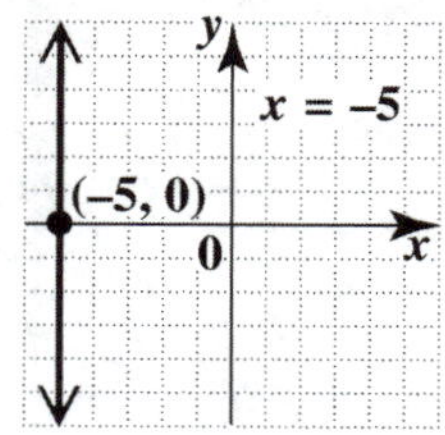

11. **(a)** Substitute the values $m = -\frac{1}{4}$ and $b = -\frac{5}{4}$ into the slope-intercept form.

$$y = mx + b$$
$$y = -\frac{1}{4}x - \frac{5}{4}$$

(b) Move all of the variables to one side of the equation to write the equation in standard form.

$y = -\frac{1}{4}x - \frac{5}{4}$	Slope-intercept form
$4y = -x - 5$	Multiply by 4.
$x + 4y = -5$	Add x.

Thus, the equation of the line in standard form is $x + 4y = -5$.

12. **(a)** Use the point-slope form with $(x_1, y_1) = (8, 6)$ and $m = -3$.

$$y - y_1 = m(x - x_1)$$
$$y - 6 = -3(x - 8)$$
$$y - 6 = -3x + 24$$
$$y = -3x + 30$$

(b) Move all of the variables to one side of the equation to write the equation in standard form.

$$y = -3x + 30$$
$$3x + y = 30$$

Thus, the equation of the line in standard form is $3x + y = 30$.

13. **(a)** First, find the slope of the line.

$$m = \frac{-1-(-5)}{-4-3} = \frac{4}{-7} = -\frac{4}{7}$$

Now use either point and the slope in the point-slope form. If we use $(3, -5)$, we get the following.

$$y - y_1 = m(x - x_1)$$
$$y - (-5) = -\frac{4}{7}(x - 3)$$
$$y + 5 = -\frac{4}{7}x + \frac{12}{7}$$
$$y = -\frac{4}{7}x - \frac{23}{7}$$

(b) Move all of the variables to one side of the equation to write the equation in standard form.

$y = -\frac{4}{7}x - \frac{23}{7}$	Slope-intercept form
$7y = -4x - 23$	Multiply by 7.
$4x + 7y = -23$	Add $4x$.

Thus, the equation of the line in standard form is $4x + 7y = -23$.

14. **(a)** Since the slope is 0, we have a horizontal line. Knowing that one y-value is -5 tells us that all y-values are -5, so the equation is $y = -5$.

(b) The equation in standard form is also $y = -5$.

15. This is a linear inequality, so its graph will be a shaded region. Graph the boundary line, $x - 2y = 6$, as a solid line through the intercepts $(0, -3)$ and $(6, 0)$.
Using $(0, 0)$ as a test point results in the true statement $0 \le 6$, so shade the region containing the origin. This is the region above the line. The solid line shows that the boundary is part of the graph.

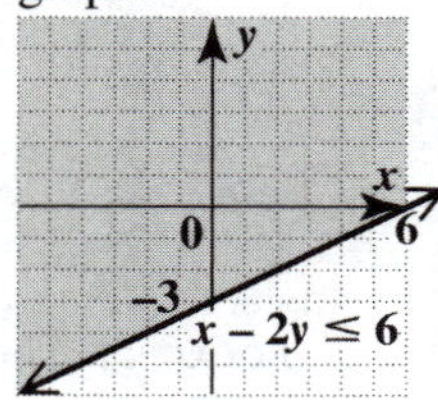

16. This is a linear inequality, so its graph will be a shaded region. Graph the boundary line, $y=-4x$, as a dashed line through $(0, 0)$, $(1, -4)$, and $(-1, 4)$.
Choose a test point that does not lie on the line. Using $(2, 0)$ results in the statement $0<-8$, which is false, so shade the region *not* containing $(2, 0)$. This is the region below the line. The dashed line shows that the boundary is not part of the graph.

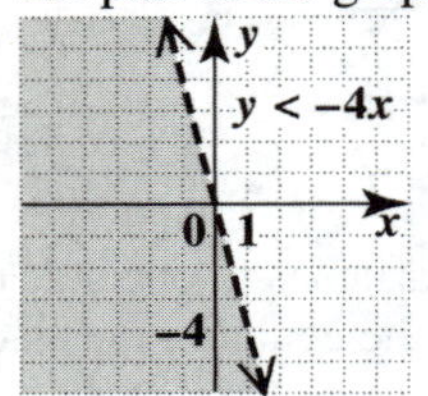

17. Since the graph falls from left to right, the slope is negative.

18. Let *x* represent the year and *y* represent the percent in the ordered pairs (x, y). Recall that $x=0$ represents the year 2010 and thus $x=3$ represents the year 2013. The ordered pairs are $(0, 39.6)$ and $(3, 36.0)$, for the years 2010 and 2013 respectively.

19. Find the slope using $(x_1, y_1)=(0, 39.6)$ and $(x_2, y_2)=(3, 36.0)$.

$$m=\frac{y_2-y_1}{x_2-x_1}=\frac{36.0-39.6}{3-0}=\frac{-3.6}{3}=-1.2$$

Observe that the *y*-intercept is $(0, 39.6)$ and thus $b=39.6$. Using *m* and *b* write the equation for the line in slope-intercept form.

$y=mx+b$

$y=-1.2x+39.6$

20. The year 2012 corresponds to $x=2$ because $2012-2010=2$. Substitute the value 2 for *x* $y=-1.2x+39.6$.

$y=-1.2x+39.6$

$y=-1.2(2)+39.6$

$y=-2.4+39.6$

$y=37.2$

It is a little high, as we might expect. The actual data point lies slightly below the graph of the line.

Chapter 3 Test

1. Find **y** when $x=0$.

$$\begin{aligned}3x+5y&=-30\\3(0)+5y&=-30 &&\text{Let } x=0.\\5y&=-30\\y&=-6\end{aligned}$$

Find *x* when $y=0$.

$$\begin{aligned}3x+5y&=-30\\3x+5(0)&=-30 &&\text{Let } y=0.\\3x&=-30\\x&=-10\end{aligned}$$

Find *x* when $y=-3$.

$$\begin{aligned}3x+5y&=-30\\3x+5(-3)&=-30 &&\text{Let } y=-3.\\3x-15&=-30\\3x&=-15\\x&=-5\end{aligned}$$

The ordered pairs are $(0, \underline{-6})$, $(\underline{-10}, 0)$, and $(\underline{-5}, -3)$.

2. Substitute the given point into the equation.

$$\begin{aligned}4x-7y&=9\\4(4)-7(-1)&\overset{?}{=}9 &&\text{Let } x=4 \text{ and } y=-1.\\16+7&\overset{?}{=}9\\23&\overset{?}{=}9 &&\text{No}\end{aligned}$$

So $(4, -1)$ is not a solution of $4x-7y=9$.

3. To find the *x*-intercept, let $y=0$ and solve for *x*. To find the *y*-intercept, let $x=0$ and solve for *y*.

4. If $y=0$, $x=2$, so the *x*-intercept is $(2, 0)$. If $x=0$, $y=6$, so the *y*-intercept is $(0, 6)$. A third point, such as $(1, 3)$, can be used as a check. Plot these points. Draw a line through $(0, 6)$, $(1, 3)$ and $(2, 0)$.

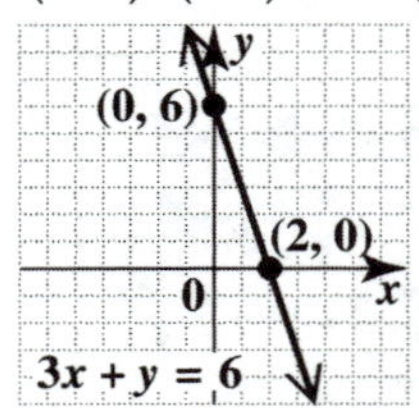

5. Solving for y gives us the slope-intercept form of the line, $y = 2x$. We see that the y-intercept is $(0, 0)$ and so the x-intercept is also $(0, 0)$. The slope is 2 and can be written as
$m = \frac{\text{rise}}{\text{run}} = \frac{2}{1}$.
Starting at the origin and moving to the right 1 unit and then up 2 units gives us the point $(1, 2)$. Draw a line through $(0, 0)$ and $(1, 2)$.

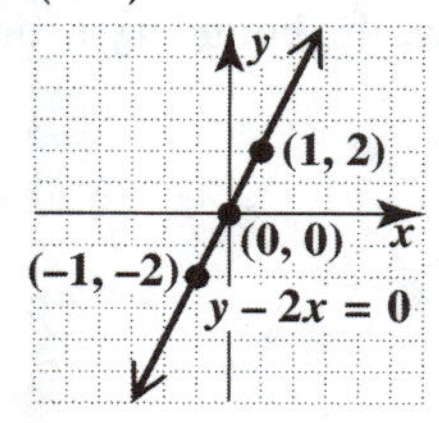

6. $x + 3 = 0$ can also be written as $x = -3$. Its graph is a vertical line with x-intercept $(-3, 0)$. There is no y-intercept.

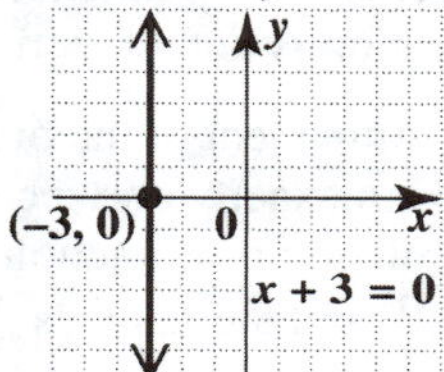

7. $y = 1$ is the graph of a horizontal line with y-intercept $(0, 1)$. There is no x-intercept.

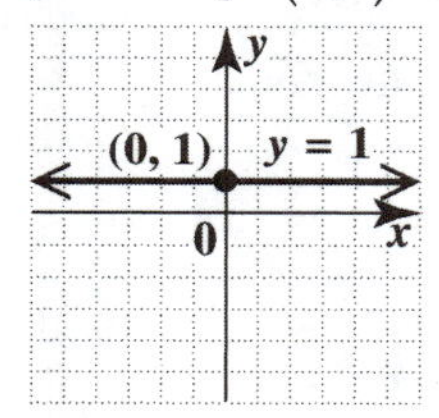

8. If $y = 0$, $x = 4$, so the x-intercept is $(4, 0)$. If $x = 0$, $y = -4$, so the y-intercept is $(0, -4)$. A third point, such as $(2, -2)$, can be used as a check. Draw a line through $(0, -4)$, $(2, -2)$, and $(4, 0)$.

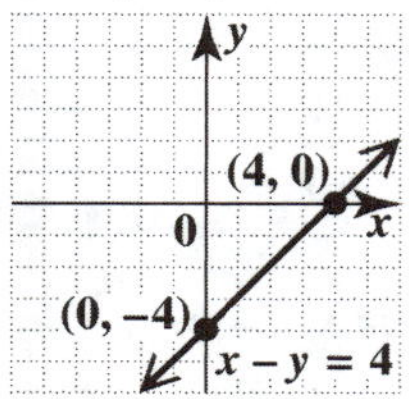

9. Use the definition of slope with $(x_1, y_1) = (-4, 6)$ and $(x_2, y_2) = (-1, -2)$.

$$\text{slope } m = \frac{y_2 - y_1}{x_2 - x_1} = \frac{-2-6}{-1-(-4)} = \frac{-8}{3} = -\frac{8}{3}$$

10. To find the slope, solve the given equation for y.

$$2x + y = 10$$
$$y = -2x + 10$$

The equation is now written in $y = mx + b$ form, so the slope is given by the coefficient of x, which is -2.

11. $x + 12 = 0$ can also be written as $x = -12$. Its graph is a vertical line with x-intercept $(-12, 0)$. The slope is undefined.

12. $y - 4 = 6$ can also be written as $y = 10$. Its graph is a horizontal line with y-intercept $(0, 10)$. Its slope is 0, as is the slope of any line parallel to it.

13. The indicated points are $(0, -4)$ and $(2, 1)$. Use the definition of slope with $(x_1, y_1) = (0, -4)$ and $(x_2, y_2) = (2, 1)$.

$$\text{slope } m = \frac{y_2 - y_1}{x_2 - x_1} = \frac{1-(-4)}{2-0} = \frac{5}{2}$$

14. Let $x_1 = -1$, $y_1 = 4$, and $m = 2$ in the point-slope form.

$$y - y_1 = m(x - x_1)$$
$$y - 4 = 2[x - (-1)]$$
$$y - 4 = 2(x + 1)$$
$$y - 4 = 2x + 2$$
$$y = 2x + 6$$

15. Find the slope using the given points $(0, -4)$ and $(2, 1)$.

$$m = \frac{1-(-4)}{2-0} = \frac{5}{2}$$

The y-intercept is $(0, -4)$, so the slope-intercept form is $y = \frac{5}{2}x - 4$.

16. Find the slope using the given points $(2, -6)$ and $(1, 3)$.

$$m = \frac{3-(-6)}{1-2} = \frac{9}{-1} = -9$$

Use the point-slope form of a line with $(1, 3) = (x_1, y_1)$ and $m = -9$.

$$y - y_1 = m(x - x_1)$$
$$y - 3 = -9(x - 1)$$
$$y - 3 = -9x + 9$$
$$y = -9x + 12$$

17. The slope is negative since worldwide snowmobile sales are decreasing, indicated by the line which falls from left to right.

18. **(a)** Two ordered pairs are $(0, 209)$ and $(13, 145)$. Use these points to find the slope.

$$m = \frac{y_2 - y_1}{x_2 - x_1} = \frac{145 - 209}{13 - 0} = -\frac{64}{13} \approx -4.9$$

The slope is approximately -4.9.

(b) Using the ordered pair $(0, 209)$, which is the y-intercept, and the slope of -4.9, we get the slope-intercept form $y = -4.9x + 209$.

19. For 2005, $x = 2005 - 2000 = 5$. Substitute this value for x.

$$y = -4.9x + 209$$
$$y = -4.9(5) + 209$$
$$y = 184.5$$

This represents worldwide snowmobile sales of 184.5 thousand in 2005. The equation gives an approximation that is a little high.

20. For 2013, $x = 2013 - 2000 = 13$. The ordered pair $(13, 145)$ is one of the two shown on the graph. In 2013, worldwide snowmobile sales were 145 thousand.

21. Graph the boundary, $x + y = 3$, as a solid line through the intercepts $(3, 0)$ and $(0, 3)$. Using $(0, 0)$ as a test point results in the true statement $0 \le 3$, so shade the region containing the origin. This is the region below the line. The solid line shows that the boundary is part of the graph.

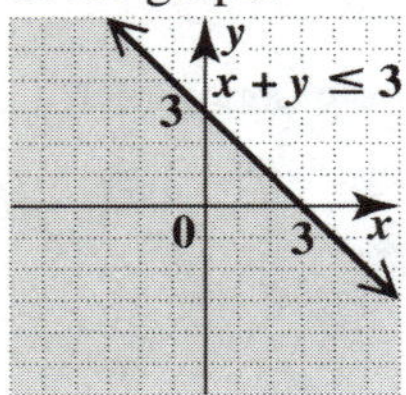

22. The boundary, $3x - y = 0$, goes through the origin, so both intercepts are $(0, 0)$. Two other points on this line are $(1, 3)$ and $(-1, -3)$. Draw the boundary as a dashed line. Choose a test point which is not on the boundary. Using $(3, 0)$ results in the true statement $9 > 0$, so shade the region containing $(3, 0)$. This is the region below the line. The dashed line shows that the boundary is not part of the graph.

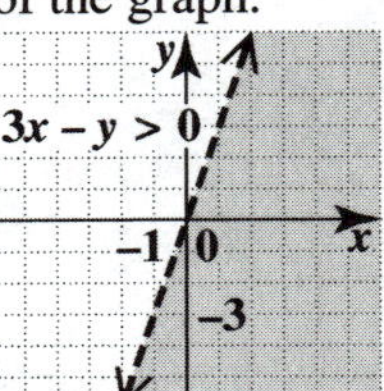

23. **(a)** Since $x = 2$ appears in two ordered pairs, one value of x yields more than one value of y. Hence, this relation is not a function.

(b) Since each first component of the ordered pairs corresponds to exactly one second component, the relation is a function. The domain is $\{0, 1, 2\}$ and the range is $\{2\}$.

24. The vertical line test shows that this graph is not the graph of a function. A vertical line could cross the graph twice.

25. Substitute -2 for x in the equation.

$$f(x) = 3x + 7$$
$$f(-2) = 3(-2) + 7$$
$$= -6 + 7$$
$$= 1$$

Chapters R–3 Cumulative Review Exercises

1. $10\frac{5}{8} - 3\frac{1}{10} = \frac{85}{8} - \frac{31}{10}$
$= \frac{425}{40} - \frac{124}{40}$
$= \frac{301}{40}$, or $7\frac{21}{40}$

2. $\frac{3}{4} \div \frac{1}{8} = \frac{3}{4} \cdot \frac{8}{1} = \frac{3 \cdot 2 \cdot 4}{4 \cdot 1} = 3 \cdot 2 = 6$

3. $5 - (-4) + (-2) = 9 + (-2) = 7$

4. $\dfrac{(-3)^2-(-4)(2^4)}{5(2)-(-2)^3}$

$=\dfrac{9-(-4)(16)}{10-(-8)}$ Evaluate exponents first.

$=\dfrac{9-(-64)}{10-(-8)}$ Multiply.

$=\dfrac{9+64}{10+8}=\dfrac{73}{18}$, or $4\dfrac{1}{18}$

5. $\dfrac{4(3-9)}{2-6}\overset{?}{\geq}6$

$\dfrac{4(-6)}{-4}\overset{?}{\geq}6$

$\dfrac{-24}{-4}\overset{?}{\geq}6$

$6\geq 6$

The statement is *true* since $6=6$.

6. Let $x=-2$, $y=-3$, and $z=-1$.

$xz^3-5y^2=(-2)(-1)^3-5(-3)^2$

$=(-2)(-1)+(-5)(9)$

$=2+(-45)$

$=-43$

7. $3(-2+x)=3\cdot(-2)+3(x)$

$=-6+3x$

This illustrates the *distributive property.*

8. $-4p-6+3p+8=(-4p+3p)+(-6+8)$

$=-p+2$

9. Solve for h.

$V=\frac{1}{3}\pi r^2h$

$3V=\pi r^2h$ Multiply by 3.

$\dfrac{3V}{\pi r^2}=h$ Divide by πr^2.

10. $6-3(1+x)=2(x+5)-2$

$6-3-3x=2x+10-2$

$3-3x=2x+8$

$-5x=5$

$x=-1$

The solution set is $\{-1\}$.

11. $-(m-3)=5-2m$

$-m+3=5-2m$ Distributive property

$m+3=5$ Add $2m$.

$m=2$ Subtract 3.

The solution set is $\{2\}$.

12. $\dfrac{x-2}{3}=\dfrac{2x+1}{5}$

$(x-2)(5)=(3)(2x+1)$ Cross products

$5x-10=6x+3$

$-10=x+3$

$-13=x$

The solution set is $\{-13\}$.

13. $-2.5x<6.5$

$\dfrac{-2.5x}{-2.5}>\dfrac{6.5}{-2.5}$

$x>-2.6$

Thus, the solution set is $(-2.6,\infty)$.

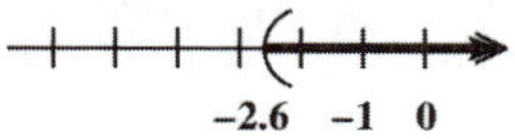

14. $4(x+3)-5x<12$

$4x+12-5x<12$ Distributive property

$-x+12<12$ Combine like terms.

$-x<0$ Subtract 12.

$x>0$ Divide by -1.

Thus, the solution set is $(0,\infty)$.

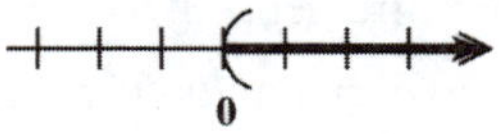

15. $\dfrac{2}{3}x-\dfrac{1}{6}x\leq -2$

$6\left(\dfrac{2}{3}x-\dfrac{1}{6}x\right)\leq 6(-2)$ Multiply by 6.

$6\left(\dfrac{2}{3}x\right)-6\left(\dfrac{1}{6}x\right)\leq 6(-2)$ Dist. property

$4x-x\leq -12$

$3x\leq -12$

$x\leq -4$

Thus, the solution set is $(-\infty,-4]$.

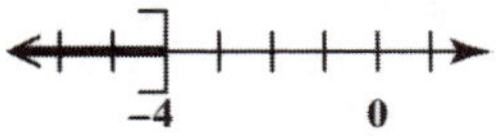

16. Mount Mayon can be modeled by a cone. The height of the cone, Mount Mayon, is known to be $h = 8100$ ft, but r is unknown.
The circumference of the circular base of the cone is given, $C = 80$ miles, and can be used to find r using the formula for the circumference of a circle, $C = 2\pi r$.

$$C = 2\pi r$$
$$80 = 2\pi r \quad \text{Let } C = 80 \text{ miles}$$
$$\frac{80}{2\pi} = r \quad \text{Divide by } 2\pi$$
$$\frac{40}{\pi} = r \quad \text{Use } \pi \approx 3.14.$$
$$13 \approx r$$

The radius of the base is about 13 miles.

17. **(a)** Multiply 14% (or 0.14) by the total, \$50,000.

$$0.14(50,000) = 7000$$

\$7000 is expected to go toward home purchase.

(b) Multiply 20% (or 0.20) by the total, \$50,000.

$$0.20(50,000) = 10,000$$

\$10,000 is expected to go toward retirement.

18. Since the sector for paying off debt or funding children's education is about three times larger than the sector for retirement, 3(\$10,000) or about \$30,000 is expected to go toward paying off debt or funding children's education.

19. To find the x-intercept, let $y = 0$.

$$-3x + 4y = 12$$
$$-3x + 4(0) = 12$$
$$-3x = 12$$
$$x = -4$$

The x-intercept is $(-4, 0)$.

To find the y-intercept, let $x = 0$.

$$-3x + 4y = 12$$
$$-3(0) + 4y = 12$$
$$4y = 12$$
$$y = 3$$

The y-intercept is $(0, 3)$.

20. To find the slope of the line, solve the equation for y.

$$-3x + 4y = 12$$
$$4y = 3x + 12$$
$$y = \frac{3}{4}x + 3$$

The slope is the coefficient of x, $\frac{3}{4}$.

21. To find a third point, let $x = 4$.

$$-3x + 4y = 12$$
$$-3(4) + 4y = 12$$
$$4y = 12 + 12 = 24$$
$$y = 6$$

Plot the points $(-4, 0)$, $(0, 3)$, and $(4, 6)$ and draw a line through them.

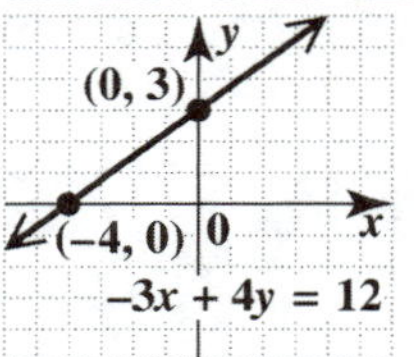

22. Solve the equation for y.

$$x + 5y = -6$$
$$5y = -x - 6$$
$$y = -\frac{1}{5}x - \frac{6}{5}$$

The slope of the first line is $-\frac{1}{5}$.

The slope of the second line, $y = 5x - 8$, is 5.

Since $-\frac{1}{5}$ is the negative reciprocal of 5, the lines are *perpendicular*.

23. Use the point-slope form of a line.

$$y - y_1 = m(x - x_1)$$
$$y - (-5) = 3(x - 2)$$
$$y + 5 = 3x - 6$$
$$y = 3x - 11$$

24. Find the slope using the given points.

$$\text{slope } m = \frac{\text{change in } y}{\text{change in } x} = \frac{4-4}{2-0} = \frac{0}{2} = 0$$

Since the slope is 0, the line is horizontal. Horizontal lines have equations of the form $y = k$. An equation of the line is $y = 4$.

Chapter 4 Systems of Linear Equations and Inequalities

4.1 Solving Systems of Linear Equations by Graphing

Now Try Exercises

N1. **(a)** To decide whether (5, 2) is a solution, substitute 5 for x and 2 for y in each equation.

$2x+5y=20$

$2(5)+5(2)\stackrel{?}{=}20$

$10+10\stackrel{?}{=}20$

$20=20$ True

$x-y=7$

$5-2\stackrel{?}{=}7$

$3=7$ False

Since (5, 2) does not satisfy the second equation, it is not a solution of the system.

(b) To decide whether (5, 2) is a solution, substitute 5 for x and 2 for y in each equation.

$3x-y=13$

$3(5)-2\stackrel{?}{=}13$

$15-2\stackrel{?}{=}13$ Multiply.

$13=13$ True

$2x+y=12$

$2(5)+2\stackrel{?}{=}12$

$10+2\stackrel{?}{=}12$ Multiply.

$12=12$ True

Since (5, 2) satisfies both equations, it is a solution of the system.

N2. $x-2y=4$

$2x+y=3$

Graph each equation by plotting several points for each line. The intercepts are good choices.

Create a table of values for $x-2y=4$.

x	y
0	−2
2	−1
4	0

Create a table of values for $2x+y=3$.

x	y
0	3
1	1
$\frac{3}{2}$	0

Graph $x-2y=4$ and $2x+y=3$.

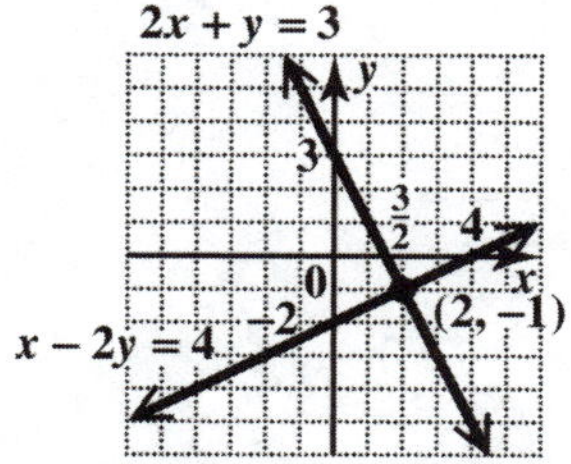

As suggested by the figure, the point at which the graphs of the two lines intersect is $(2, -1)$.

N3. **(a)** $5x-3y=2$

$10x-6y=4$

Notice that if we multiply each side of the first equation by 2, we obtain the second equation. Thus, the graphs of these two equations are the same line.

Create a table of values for $5x-3y=2$.

x	y
0	$-\frac{2}{3}$
$\frac{2}{5}$	0
1	1

Graph $5x-3y=2$.

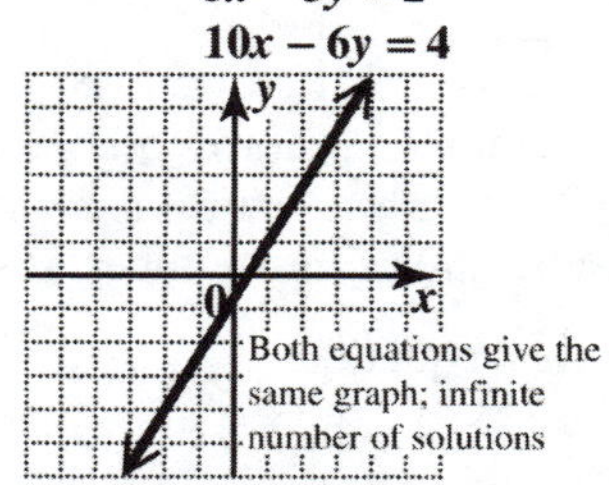

In this case, every point on the line is a solution of the system, and the solution set contains an infinite number of ordered pairs, each of which satisfies both equations of the system. We write the solution set as $\{(x, y)|5x-3y=2\}$.

(b) $4x + y = 7$

$12x + 3y = 10$

Create a table of values for $4x + y = 7$.

x	y
$\frac{7}{4}$	0
0	7
1	3

Create a table of values for $12x + 3y = 10$.

x	y
$\frac{5}{6}$	0
0	$\frac{10}{3}$
1	$-\frac{2}{3}$

Graph $4x + y = 7$ and $12x + 3y = 10$.

12x + 3y = 10 4x + y = 7

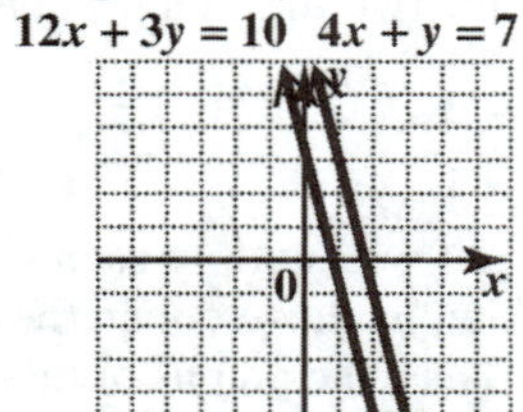

The lines each have slope -4 and hence, are parallel. Since they do not intersect, there is *no solution.* We write the solution set as $\varnothing$.

N4. **(a)** Solve each equation for y.

$$5x - 8y = 4 \qquad x - \frac{8}{5}y = \frac{4}{5}$$

$$-8y = -5x + 4 \qquad -\frac{8}{5}y = -x + \frac{4}{5}$$

$$y = \frac{5}{8}x - \frac{1}{2} \qquad y = \frac{5}{8}x - \frac{1}{2}$$

The slope-intercept forms are the same, so the equations represent the same line. The system has an infinite number of solutions.

(b) Solve each equation for y.

$$2x + y = 7 \qquad 3y = -6x - 12$$

$$y = -2x + 7 \qquad y = -2x - 4$$

The equations have the same slope, -2, but different y-intercepts, 7 and -4. The equations represent parallel lines. The system has no solution.

(c) Solve each equation for y.

$$y - 3x = 7 \qquad 3y - x = 0$$

$$y = 3x + 7 \qquad 3y = x$$

$$y = \frac{1}{3}x$$

The slopes, 3 and $\frac{1}{3}$, are different, so the graphs of these equations are neither parallel nor the same line. The system has exactly one solution.

Exercises

1. A system of linear equations consists of two or more linear equations with the *same* variables.

3. The equations of two parallel lines form an inconsistent system that has *no* solutions. The equations are independent because their graphs are different.

5. If two equations of a linear system have the same graph, the equations are dependent. The system is consistent and has infinitely many solutions.

7. The ordered pair $(1, -2)$ does make the first equation true, but if we replace x with 1 and y with -2 in the second equation, $2x + y = 4$, we get $2(1) + (-2) = 4$, or $0 = 4$, which is false. Because the ordered pair does not satisfy *both* equations in the system, it is not a solution of the system.

9. From the graph, the ordered pair that is a solution of the system is in the second quadrant. Choice A, $(-4, -4)$, is the only ordered pair given that is not in quadrant II, so it is the only valid choice.

11. **(a)** $(3, 4)$ is in quadrant I—choice B is correct.

    (b) $(-2, 3)$ is in quadrant II—choice C is correct since choice D is the only other choice with an intersection point in quadrant II, but its x-coordinate of the intersection point is -3, not -2.

    (c) $(-3, 2)$ is in quadrant II—choice D is correct.

    (d) $(5, -2)$ is in quadrant IV—choice A is correct.

13. To decide whether $(2, -3)$ is a solution of the system, substitute 2 for x and -3 for y in each equation.

$$x + y = -1$$
$$(2) + (-3) \stackrel{?}{=} -1$$
$$-1 = -1 \quad \text{True}$$
$$2x + 5y = 19$$
$$2(2) + 5(-3) \stackrel{?}{=} 19$$
$$4 + (-15) \stackrel{?}{=} 19$$
$$-11 = 19 \quad \text{False}$$

The ordered pair $(2, -3)$ satisfies the first equation but not the second. Because it does not satisfy *both* equations, it is not a solution of the system.

15. Substitute -1 for x and -3 for y in each equation.

$$3x + 5y = -18$$
$$3(-1) + 5(-3) \stackrel{?}{=} -18$$
$$-3 - 15 \stackrel{?}{=} -18$$
$$-18 = -18 \quad \text{True}$$
$$4x + 2y = -10$$
$$4(-1) + 2(-3) \stackrel{?}{=} -10$$
$$-4 - 6 \stackrel{?}{=} -10$$
$$-10 = -10 \quad \text{True}$$

Since $(-1, -3)$ satisfies both equations, it is a solution of the system.

17. Substitute 7 for x and -2 for y in each equation.

$$4x = 26 - y$$
$$4(7) \stackrel{?}{=} 26 - (-2)$$
$$28 \stackrel{?}{=} 26 + 2$$
$$28 = 28 \quad \text{True}$$
$$3x = 29 + 4y$$
$$3(7) \stackrel{?}{=} 29 + 4(-2)$$
$$21 \stackrel{?}{=} 29 - 8$$
$$21 = 21 \quad \text{True}$$

Since $(7, -2)$ satisfies both equations, it is a solution of the system.

19. Substitute 6 for x and -8 for y in each equation.

$$-2y = x + 10$$
$$-2(-8) \stackrel{?}{=} 6 + 10$$
$$16 = 16 \quad \text{True}$$
$$3y = 2x + 30$$
$$3(-8) \stackrel{?}{=} 2(6) + 30$$
$$-24 \stackrel{?}{=} 12 + 30$$
$$-24 = 42 \quad \text{False}$$

The ordered pair $(6, -8)$ satisfies the first equation but not the second. Because it does not satisfy *both* equations, it is not a solution of the system.

21. Substitute 0 for x and 0 for y in each equation.

$$4x + 2y = 0$$
$$4(0) + 2(0) \stackrel{?}{=} 0$$
$$0 = 0 \quad \text{True}$$
$$x + y = 0$$
$$0 + 0 \stackrel{?}{=} 0$$
$$0 = 0 \quad \text{True}$$

Since $(0, 0)$ satisfies both equations, it is a solution of the system.

23. To graph the equations, find the intercepts.

$x - y = 2$: Let $y = 0$; then $x = 2$.
Let $x = 0$; then $y = -2$.

Plot the intercepts, $(2, 0)$ and $(0, -2)$, and draw the line through them.

$x + y = 6$: Let $y = 0$; then $x = 6$.
Let $x = 0$; then $y = 6$.

Plot the intercepts, $(6, 0)$ and $(0, 6)$, and draw the line through them.

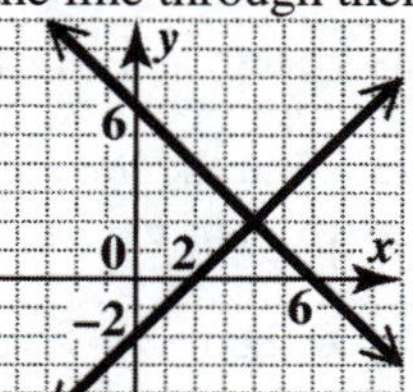

It appears that the lines intersect at the point $(4, 2)$. Check this by substituting 4 for x and 2 for y in both equations. Since $(4, 2)$ satisfies both equations, the solution set of this system is $\{(4, 2)\}$.

25. To graph the equations, find the intercepts.
$x+y=4$: Let $y=0$; then $x=4$.
Let $x=0$; then $y=4$.
Plot the intercepts, $(4, 0)$ and $(0, 4)$, and draw the line through them.
$y-x=4$: Let $y=0$; then $x=-4$.
Let $x=0$; then $y=4$.
Plot the intercepts, $(-4, 0)$ and $(0, 4)$, and draw the line through them.

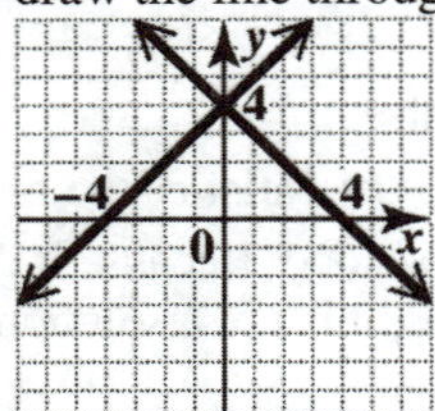

The lines intersect at their common y-intercept, $(0, 4)$, so $\{(0, 4)\}$ is the solution set of the system.

27. To graph the equations, find the intercepts.
$x-2y=6$: Let $y=0$; then $x=6$.
Let $x=0$; then $y=-3$.
Plot the intercepts, $(6, 0)$ and $(0, -3)$, and draw the line through them.
$x+2y=2$: Let $y=0$; then $x=2$.
Let $x=0$; then $y=1$.
Plot the intercepts, $(2, 0)$ and $(0, 1)$, and draw the line through them.

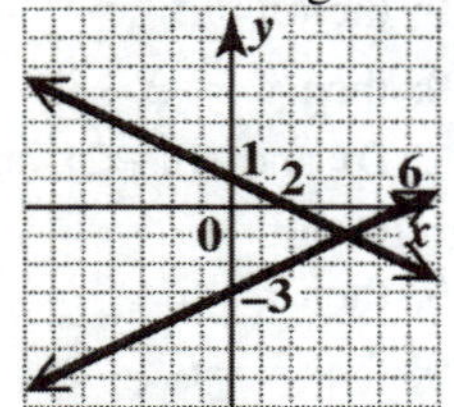

It appears that the lines intersect at the point $(4, -1)$. Since $(4, -1)$ satisfies both equations, the solution set of this system is $\{(4, -1)\}$.

29. To graph the equations, find the intercepts.
$3x-2y=-3$: Let $y=0$; then $x=-1$.
Let $x=0$; then $y=\frac{3}{2}$.
Plot the intercepts, $(-1, 0)$ and $\left(0, \frac{3}{2}\right)$, and draw the line through them.
$-3x-y=-6$: Let $y=0$; then $x=2$.
Let $x=0$; then $y=6$.
Plot the intercepts, $(2, 0)$ and $(0, 6)$, and draw the line through them.

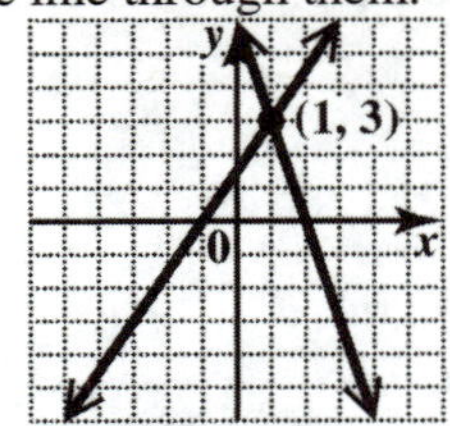

It appears that the lines intersect at the point $(1, 3)$. Since $(1, 3)$ satisfies both equations, the solution set of this system is $\{(1, 3)\}$.

31. To graph the equations, find the intercepts.
$3x+y=5$: Let $y=0$; then $x=\frac{5}{3}$.
Let $x=0$; then $y=5$.
Plot these intercepts, $\left(\frac{5}{3}, 0\right)$ and $(0, 5)$, and draw the line through them.
$6x+2y=10$: Let $y=0$; then $x=\frac{5}{3}$.
Let $x=0$; then $y=5$.
Plot these intercepts, $\left(\frac{5}{3}, 0\right)$ and $(0, 5)$, and draw the line through them.
Since both equations have the same intercepts, they are equations of the same line.

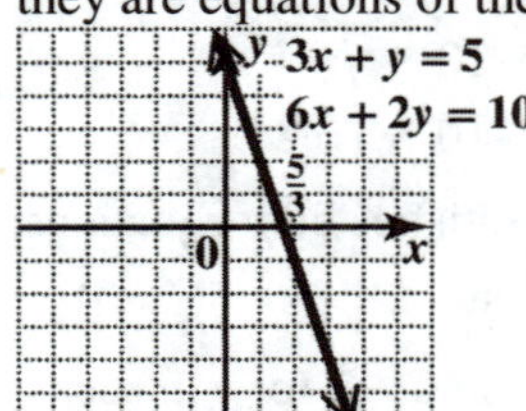

There is an infinite number of solutions. The equations are dependent equations and the solution set contains an infinite number of ordered pairs. The solution set is $\{(x, y) \mid 3x+y=5\}$.

33. To graph the first line, find the intercepts.
$2x-3y=-6$: Let $y=0$; then $x=-3$.
Let $x=0$; then $y=2$.
Plot the intercepts, $(-3, 0)$ and $(0, 2)$, and draw the line through them.
To graph the second line, start by plotting the y-intercept, $(0, 2)$. From this point, go 3 units down and 1 unit to the right (because the slope is -3) to reach the point $(1, -1)$. Draw the line through $(0, 2)$ and $(1, -1)$.

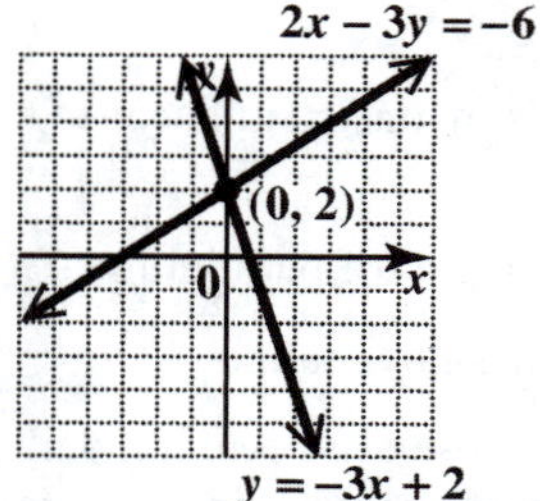

The lines intersect at their common y-intercept, $(0, 2)$, so $\{(0, 2)\}$ is the solution set of the system.

35. Graph the line $2x-y=6$ through its intercepts, $(3, 0)$ and $(0, -6)$.
Graph the line $4x-2y=8$ through its intercepts, $(2, 0)$ and $(0, -4)$.

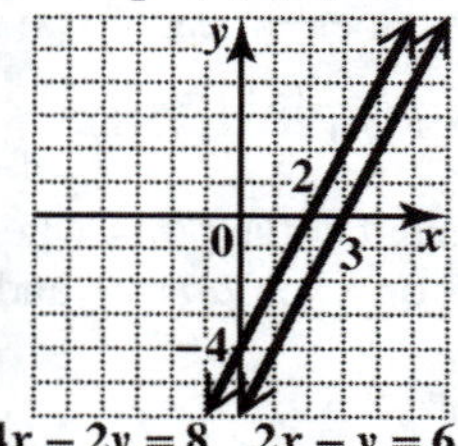

Solving each equation for y gives us $y=2x-6$ and $y=2x-4$, so the lines each have slope 2, and hence, are parallel. Since they do not intersect, there is no solution. This is an inconsistent system.

37. Graph the line $2y-6x=12$ through its intercepts, $(-2, 0)$ and $(0, 6)$.
Graph the line $3x-y=-6$ through its intercepts, $(-2, 0)$ and $(0, 6)$.
Since both equations have the same intercepts, they are equations of the same line.

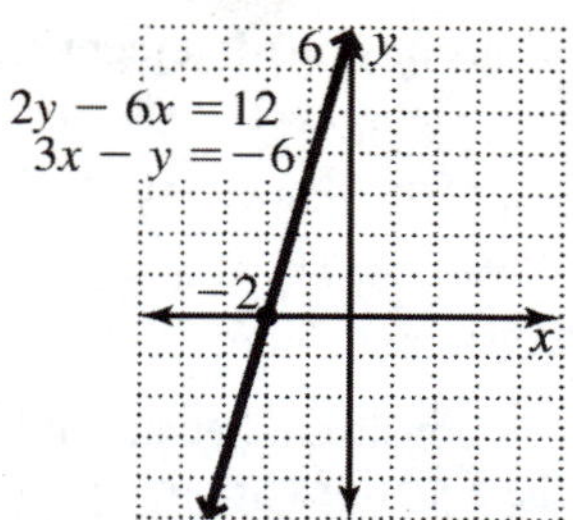

There is an infinite number of solutions. The equations are dependent equations, and the solution set contains an infinite number of ordered pairs. The solution set is $\{(x, y) \mid 3x-y=-6\}$.

39. Graph the line $3x-4y=24$ through its intercepts, $(8, 0)$ and $(0, -6)$.

To graph the line $y=-\frac{3}{2}x+3$, plot the y-intercept $(0, 3)$, and then go 3 units down and 2 units to the right (because the slope is $-\frac{3}{2}$) to reach the point $(2, 0)$. Draw the line through $(0, 3)$ and $(2, 0)$.

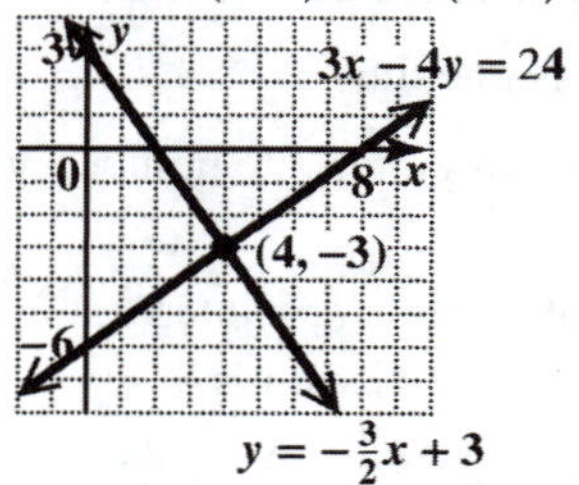

It appears that the lines intersect at the point $(4, -3)$. Since $(4, -3)$ satisfies both equations, the solution set of this system is $\{(4, -3)\}$.

41. Graph the line $2x=y-4$ through its intercepts, $(-2, 0)$ and $(0, 4)$.
Graph the line $4x+4=2y$ through its intercepts, $(-1, 0)$ and $(0, 2)$.

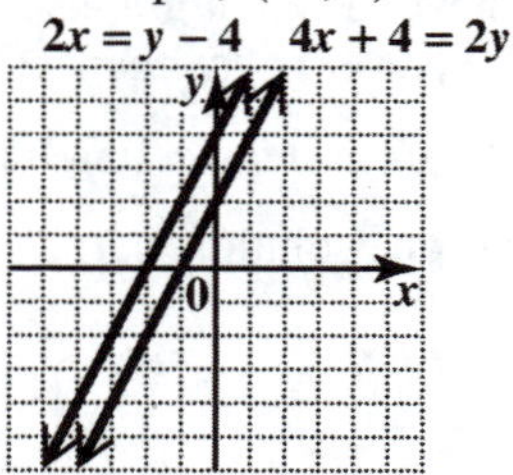

Solving each equation for y gives us $y=2x+4$ and $y=2x+2$, so the lines each have slope 2, and hence, are parallel. Since they do not intersect, there is no solution. This is an inconsistent system and the solution set is $\varnothing$.

43. Write the equations in slope-intercept form.

$y - x = -5$	$x + y = 1$
$y = x - 5$	$y = -x + 1$
$m = 1$	$m = -1$

The lines have different slopes.

(a) The system is consistent because it has a solution. The equations are independent because they have different graphs. Therefore, the answer is "neither."

(b) The graph is a pair of intersecting lines.

(c) The system has one solution.

45. Write the equations in slope-intercept form.

$x + 2y = 0$	$4y = -2x$
$2y = -x$	$y = -\frac{1}{2}x$
$y = -\frac{1}{2}x$	

For both lines, $m = -\frac{1}{2}$ and $b = 0$.

(a) Since the equations have the same slope and y-intercept, they are dependent.

(b) The graph is one line.

(c) The system has an infinite number of solutions.

47. Write the equations in slope-intercept form.

$x - 3y = 5$	$2x + y = 8$
$-3y = -x + 5$	$y = -2x + 8$
$y = \frac{1}{3}x - \frac{5}{3}$	$m = -2$
$m = \frac{1}{3}$	

The lines have different slopes.

(a) The system is consistent because it has a solution. The equations are independent because they have different graphs. Therefore, the answer is "neither."

(b) The graph is a pair of intersecting lines.

(c) The system has one solution.

49. Write the equations in slope-intercept form.

$5x + 4y = 7$	$10x + 8y = 4$
$4y = -5x + 7$	$8y = -10x + 4$
$y = -\frac{5}{4}x + \frac{7}{4}$	$y = -\frac{10}{8}x + \frac{4}{8}$
$m = -\frac{5}{4}, b = \frac{7}{4}$	$y = -\frac{5}{4}x + \frac{1}{2}$
	$m = -\frac{5}{4}, b = \frac{1}{2}$

The lines have the same slope but different y-intercepts.

(a) The system is inconsistent because it has no solution.

(b) The graph is a pair of parallel lines.

(c) The system has no solution.

51. **(a)** The graph representing evening dailies is above the graph representing morning dailies for the years 1980 to 2000, so there were more evening dailies than morning dailies during the years 1980–2000.

(b) The graphs intersect when the year is about 2001. The intersection point is about $\frac{3}{4}$ of the way between 600 and 800, so there were about 750 newspapers of each type in 2001.

(c) The ordered pair is $(2001, 750)$.

53. Supply equals demand at the point where the two lines intersect, or when $x = 40$ and $p = 30$.

55. When $x > 40$, the graph shows that supply exceeds demand.

4.2 Solving Systems of Linear Equations by Substitution

Now Try Exercises

N1. The second equation is already solved for y. Substitute $-3x$ for x in the first equation.

$$\begin{aligned} 2x - 4y &= 28 & \\ 2x - 4(-3x) &= 28 & \text{Let } y = -3x. \\ 2x + 12x &= 28 & \text{Multiply.} \\ 14x &= 28 & \text{Combine like terms.} \\ x &= 2 & \text{Divide by 2.} \end{aligned}$$

Now substitute 2 for x in the second equation.

$y = -3x = -3(2) = -6$

Check $x = 2,\ y = -6$:

$$2x - 4y = 28$$
$$2(2) - 4(-6) \stackrel{?}{=} 28$$
$$4 + 24 \stackrel{?}{=} 28$$
$$28 = 28 \quad \text{True}$$

Check $x = 2,\ y = -6$:

$$y = -3x$$
$$-6 \stackrel{?}{=} -3(2)$$
$$-6 = -6 \quad \text{True}$$

Since $(2, -6)$ satisfies both equations, the solution set is $\{(2, -6)\}$.

N2. Substitute $y - 3$ for x in the first equation, and solve for y.

$$4x + 9y = 1$$
$$4(y-3) + 9y = 1 \quad \text{Let } x = y - 3.$$
$$4y - 12 + 9y = 1$$
$$13y - 12 = 1$$
$$13y = 13$$
$$y = 1$$

To find x, use $x = y - 3$ and $y = 1$.

$$x = y - 3 = 1 - 3 = -2$$

The solution set is $\{(-2, 1)\}$.

N3.
$$2y = x - 2 \quad (1)$$
$$4x - 5y = -4 \quad (2)$$

Solve (1) for x.

$$2y = x - 2$$
$$2y + 2 = x \quad (3)$$

Substitute $2y + 2$ for x in equation (2) and solve for y.

$$4x - 5y = -4 \quad \text{Let } x = 2y + 2.$$
$$4(2y+2) - 5y = -4$$
$$8y + 8 - 5y = -4$$
$$3y + 8 = -4$$
$$3y = -12$$
$$y = -4$$

To find x, let $y = -4$ in equation (3).

$$x = 2y + 2$$
$$= 2(-4) + 2$$
$$= -8 + 2$$
$$= -6$$

Check that $(-6, -4)$ is the solution.

$$2y = x - 2 \quad (1)$$
$$2(-4) \stackrel{?}{=} -6 - 2$$
$$-8 = -8 \quad \text{True}$$
$$4x - 5y = -4 \quad (2)$$
$$4(-6) - 5(-4) \stackrel{?}{=} -4$$
$$-4 = -4 \quad \text{True}$$

The solution set of the system is $\{(-6, -4)\}$.

N4.
$$8x - 2y = 1 \quad (1)$$
$$y = 4x - 8 \quad (2)$$

Substitute $4x - 8$ for y in equation (1).

$$8x - 2y = 1$$
$$8x - 2(4x - 8) = 1 \quad \text{Let } y = 4x - 8.$$
$$8x - 8x + 16 = 8$$
$$16 = 8 \quad \text{False}$$

This false result indicates that the system has no solution, so the solution set is $\varnothing$.

N5.
$$5x - y = 6 \quad (1)$$
$$-10x + 2y = -12 \quad (2)$$

Solve equation (1) for y.

$$5x - y = 6$$
$$5x - 6 = y$$

Substitute $5x - 6$ for y in equation (2) and solve the resulting equation.

$$-10x + 2y = -12 \quad (2)$$
$$-10x + 2(5x - 6) = -12 \quad \text{Let } y = 5x - 6.$$
$$-10x + 10x - 12 = -12 \quad \text{Distributive property}$$
$$0 = 0 \quad \text{Simplify.}$$

This true result means that every solution of one equation is also a solution of the other, so the system has an infinite number of solutions. The solution set is $\{(x, y) \mid 5x - y = 6\}$.

N6.
$$x + \frac{1}{2}y = \frac{1}{2} \quad (1)$$
$$\frac{1}{6}x - \frac{1}{3}y = \frac{4}{3} \quad (2)$$

First, clear all fractions.

Equation (1):

$$2\left(x + \frac{1}{2}y\right) = 2\left(\frac{1}{2}\right) \quad \text{Mult. by the LCD.}$$
$$2(x) + 2\left(\frac{1}{2}y\right) = 1 \quad \text{Distributive property}$$
$$2x + y = 1 \quad (3)$$

Equation (2):

$6\left(\frac{1}{6}x-\frac{1}{3}y\right)=6\left(\frac{4}{3}\right)$ Mult. by the LCD.

$6\left(\frac{1}{6}x\right)-6\left(\frac{1}{3}y\right)=8$ Distributive property

$x-2y=8$ (4)

The system has been simplified to

$2x+y=1$ (3)

$x-2y=8$ (4)

Solve this system by the substitution method.

$x=2y+8$ (5) Solve (4) for x.

$2(2y+8)\ y=1$ Substitute for x in (3).

$4y+16+y=1$

$5y+16=1$

$5y=-15$

$y=-3$

To find x, let $y=-3$ in equation (5).

$x=2(-3)+8=-6+8=2$

The solution set is $\{(2,-3)\}$.

N7. $0.2x+0.3y=0.5$ (1)

$0.3x-0.1y=1.3$ (2)

First, clear all decimals.

Equation (1):

$10(0.2x+0.3y)=10(0.5)$ Multiply by 10.

$10(0.2x)+10(0.3y)=10(0.5)$ Dist. property

$2x+3y=5$ (3)

Equation (2):

$10(0.3x-0.1y)=10(1.3)$ Multiply by 10.

$10(0.3x)-10(0.1y)=10(1.3)$ Dist. property

$3x-y=13$ (4)

The system has been simplified to

$2x+3y=5$ (3)

$3x-y=13$ (4)

Solve this system by the substitution method.

$y=3x-13$ (5) Solve (4) for y.

$2x+3(3x-13)=5$ Substitute for x in (3).

$2x+9x-39=5$

$11x-39=5$

$11x=44$

$x=4$

To find y, let $x=4$ in equation (5).

$y=3(4)-13=12-13=-1$

The solution set is $\{(4,-1)\}$.

Exercises

1. The student must find the value of y and write the solution as an ordered pair. Substituting 3 for x in the first equation, $5x-y=15$, gives us $15-y=15$, so $y=0$. The correct solution set is $\{(3,0)\}$.

3. A false statement, such as $0=3$, occurs.

5. $x+y=12$ (1)

$y=3x$ (2)

Equation (2) is already solved for y. Substitute $3x$ for y in equation (1) and solve the resulting equation for x.

$x+y=12$

$x+3x=12$ Let $y=3x$.

$4x=12$

$x=3$

To find the y-value of the solution, substitute 3 for x in equation (2).

$y=3x$

$y=3(3)$ Let $x=3$.

$=9$

The solution set is $\{(3,9)\}$.

To check this solution, substitute 3 for x and 9 for y in both equations of the given system.

7. $3x+2y=27$ (1)

$x=y+4$ (2)

Equation (2) is already solved for x. Substitute $y+4$ for x in equation (1).

$3x+2y=27$

$3(y+4)+2y=27$

$3y+12+2y=27$

$5y=15$

$y=3$

To find x, substitute 3 for y in equation (2).

$x=y+4$

$x=(3)+4=7$

The solution set is $\{(7,3)\}$.

9. $3x+4=-y$ (1)
$2x+y=0$ (2)
Solve equation (1) for y.
$3x+4=-y$
$y=-3x-4$ (3)
Substitute $-3x-4$ for y in equation (2) and solve for x.
$$2x+y=0$$
$$2x+(-3x-4)=0$$
$$-x-4=0$$
$$-x=4$$
$$x=-4$$
To find y, substitute -4 for x in equation (3).
$y=-3x-4$
$=-3(-4)-4=8$
The solution set is $\{(-4, 8)\}$.

11. $7x+4y=13$ (1)
$x+y=1$ (2)
Solve equation (2) for y.
$x+y=1$
$y=1-x$ (3)
Substitute $1-x$ for y in equation (1).
$$7x+4y=13$$
$$7x+4(1-x)=13$$
$$7x+4-4x=13$$
$$3x+4=13$$
$$3x=9$$
$$x=3$$
To find y, substitute 3 for x in equation (3).
$y=1-x$
$y=1-(3)=-2$
The solution set is $\{(3, -2)\}$.

13. $3x+5y=25$ (1)
$x-2y=-10$ (2)
Solve equation (2) for x since its coefficient is 1.
$x-2y=-10$
$x=2y-10$ (3)
Substitute $2y-10$ for x in equation (1), and solve for y.
$$3x+5y=25$$
$$3(2y-10)+5y=25$$
$$6y-30+5y=25$$
$$11y=55$$
$$y=5$$
To find x, substitute 5 for y in equation (3).
$x=2y-10$
$x=2(5)-10=0$
The solution set is $\{(0, 5)\}$.

15. $3x-y=5$ (1)
$y=3x-5$ (2)
Equation (2) is already solved for y, so we substitute $3x-5$ for y in equation (1).
$$3x-(3x-5)=5$$
$$3x-3x+5=5$$
$$5=5 \quad \text{True}$$
This true result means that every solution of one equation is also a solution of the other, so the system has an infinite number of solutions. The solution set is $\{(x, y) \mid 3x-y=5\}$.

17. $2x+y=0$ (1)
$4x-2y=2$ (2)
Solve equation (1) for y.
$2x+y=0$
$y=-2x$ (3)
Substitute $-2x$ for y in equation (2) and solve for x.
$$4x-2(-2x)=2$$
$$4x+4x=2$$
$$8x=2$$
$$x=\frac{1}{4}$$
To find y, let $x=\frac{1}{4}$ in equation (3).
$$y=-2x$$
$$y=-2\left(\frac{1}{4}\right)$$
$$=-\frac{1}{2}$$
The solution set is $\left\{\left(\frac{1}{4}, -\frac{1}{2}\right)\right\}$.

19. $2x+8y=3$ (1)

$x=8-4y$ (2)

Equation (2) is already solved for x, so substitute $8-4y$ for x in equation (1).

$2(8-4y)+8y=3$

$16-8y+8y=3$

$16=3$ False

This false result means that the system is inconsistent and its solution set is $\varnothing$.

21. $2y=4x+24$ (1)

$2x-y=-12$ (2)

Solve equation (2) for y.

$2x-y=-12$

$2x=y-12$

$2x+12=y$

Substitute $2x+12$ for y in (1) and solve for x.

$2(2x+12)=4x+24$

$4x+24=4x+24$ True

This true result means that every solution of one equation is also a solution of the other, so the system has an infinite number of solutions. The solution set is $\{(x, y) \mid 2x-y=-12\}$.

23. $y=6-x$ (1)

$y=2x+3$ (2)

Equation (1) is already solved for y, so substitute $6-x$ for y in equation (2).

$y=2x+3$

$(6-x)=2x+3$

$6=3x+3$

$3=3x$

$1=x$

To find y, let $x=1$ in equation (1).

$y=6-(1)=5$

The solution set is $\{(1, 5)\}$.

25. $x=y-4$ (1)

$x-y=1$ (2)

Equation (1) is already solved for x, so substitute $y-4$ for x in equation (2).

$x-y=1$

$(y-4)-y=1$

$-4=1$ False

This false result means that the system is inconsistent and its solution set is $\varnothing$.

27. $x+y=0$ (1)

$3x-3y=0$ (2)

Solve equation (1) for y.

$y=-x$ (3)

$3x-3y=0$

Substitute $-x$ for y in (2), and solve for x.

$3x-3y=0$

$3x-3(-x)=0$

$6x=0$

$x=0$

To find y, let $x=0$ in equation (3).

$y=-(0)=0$

The solution set is $\{(0, 0)\}$.

29. $\frac{1}{2}x+\frac{1}{3}y=3$ (1)

$y=3x$ (2)

First, clear all fractions in equation (1).

$$6\left(\frac{1}{2}x+\frac{1}{3}x\right)=6(3)$$

$$6\left(\frac{1}{2}x\right)+6\left(\frac{1}{3}y\right)=18 \quad \text{Dist. property}$$

$$3x+2y=18 \quad (3)$$

From equation (2), substitute $3x$ for y in equation (3).

$3x+2(3x)=18$

$3x+6x=18$

$9x=18$

$x=2$

To find y, let $x=2$ in equation (2).

$y=3(2)=6$

The solution set is $\{(2, 6)\}$.

31. $\frac{1}{2}x+\frac{1}{3}y=-\frac{1}{3}$ (1)

$\frac{1}{2}x+2y=-7$ (2)

First, clear all fractions.

Equation (1):

$$6\left(\frac{1}{2}x+\frac{1}{3}y\right)=6\left(-\frac{1}{3}\right) \quad \text{Multiply by LCD.}$$

$$6\left(\frac{1}{2}x\right)+6\left(\frac{1}{3}y\right)=-2 \quad \text{Dist. property}$$

$$3x+2y=-2 \quad (3)$$

Equation (2):

$2\left(\frac{1}{2}x+2y\right)=2(-7)$ Multiply by 2.

$x+4y=-14$ (4)

The system has been simplified to

$3x+2y=-2$ (3)

$x+4y=-14$ (4)

Solve this system by the substitution method.

$x=-4y-14$ (5) Solve (4) for x.

$3(-4y-14)+2y=-2$ Substitute for x in (3).

$-12y-42+2y=-2$

$-10y-42=-2$

$-10y=40$

$y=-4$

To find x, let $y=-4$ in equation (5).

$x=-4(-4)-14=16-14=2$

The solution set is $\{(2,-4)\}$.

33. $\frac{x}{5}+2y=\frac{8}{5}$ (1)

$\frac{3x}{5}+\frac{y}{2}=-\frac{7}{10}$ (2)

First, clear all fractions.

Equation (1):

$5\left(\frac{x}{5}+2y\right)=5\left(\frac{8}{5}\right)$ Multiply by 5.

$x+10y=8$ (3)

Equation (2):

$10\left(\frac{3x}{5}+\frac{y}{2}\right)=10\left(-\frac{7}{10}\right)$ Multiply by 10.

$6x+5y=-7$ (4)

The system has been simplified to

$x+10y=8$ (3)

$6x+5y=-7.$ (4)

Solve this system by the substitution method.
Solve equation (3) for x.

$x=8-10y$ (5)

Substitute $8-10y$ for x in equation (4).

$6(8-10y)+5y=-7$

$48-60y+5y=-7$

$-55y=-55$

$y=1$

To find x, let $y=1$ in equation (5).

$x=8-10(1)=-2$

The solution set is $\{(-2,1)\}$.

35. $\frac{1}{6}x+\frac{1}{3}y=8$ (1)

$\frac{1}{4}x+\frac{1}{2}y=12$ (2)

Multiply equation (1) by 6.

$6\left(\frac{1}{6}x+\frac{1}{3}y\right)=6(8)$

$x+2y=48$ (3)

Multiply equation (2) by 4.

$4\left(\frac{1}{4}x+\frac{1}{2}y\right)=4(12)$

$x+2y=48$ (4)

Equations (3) and (4) are identical.
This means that every solution of one equation is also a solution of the other, so the system has an infinite number of solutions. The solution set is $\{(x, y) \mid x+2y=48\}$.

37. $0.2x-1.3y=-3.2$ (1)

$-0.1x+2.7y=9.8$ (2)

Multiply equation (1) and equation (2) by 10 to eliminate the decimals. Then solve the resulting equations by the substitution method.

$2x-13y=-32$ (3)

$-x+27y=98$ (4)

Solve equation (4) for x.

$27y-98=x$ (5)

Substitute $27y-98$ for x in equation (3).

$2x-13y=-32$

$2(27y-98)-13y=-32$ Let $x=27y-98$.

$54y-196-13y=-32$

$41y=164$

$y=4$

To find x, let $y=4$ in equation (5).

$x=27(4)-98=10$

The solution set is $\{(10, 4)\}$. Check in both of the original equations.

39. $0.3x-0.1y=2.1$ (1)

$0.6x+0.3y=-0.3$ (2)

Multiply equation (1) and equation (2) by 10 to eliminate the decimals. Then solve the resulting equations by the substitution method.

$3x-y=21$ (3)

$6x+3y=-3$ (4)

Solve equation (3) for y.

$y=3x-21$ (5)

Substitute $3x-21$ for y in equation (4).

$6x+3(3x-21)=-3$

$6x+9x-63=-3$ Let $y=3x-21$.

$15x-63=-3$

$15x=60$

$x=4$

To find y, let $x=4$ in equation (5).

$y=3(4)-21=-9$

The solution set is $\{(4,-9)\}$. Check in both of the original equations.

41. To find the total cost, multiply the number of bicycles (x) by the cost per bicycle ($400), and add the fixed cost ($5000). Thus, $y_1=400x+5000$ gives the total cost (in dollars).

43. $y_1=400x+5000$ (1)

$y_2=600x$ (2)

To solve this system by the substitution method, substitute $600x$ for y_1 in equation (1).

$600x=400x+5000$

$200x=5000$

$x=25$

If $x=25$, $y_2=600(25)=15{,}000$.

The solution set is $\{(25, 15{,}000)\}$.

4.3 Solving Systems of Linear Equations by Elimination

Now Try Exercises

N1. $x-y=4$ (1)

$3x+y=8$ (2)

To eliminate y, add equations (1) and (2).

$$\begin{array}{rcl} x - y &=& 4 \quad (1) \\ 3x + y &=& 8 \quad (2) \\ \hline 4x &=& 12 \quad \text{Add (1) and (2).} \\ x &=& 3 \end{array}$$

This result gives the x-value of the solution. To find the y-value of the solution, substitute 2 for x in either equation. We will use equation (2).

$3x+y=8$ (2)

$3(3)+y=8$

$9+y=8$

$y=-1$

Check by substituting 3 for x and -1 for y in both equations of the original system.

Check Equation (1).

$x-y=4$

$(3)-(-1)\stackrel{?}{=}4$

$3+1\stackrel{?}{=}4$

$4=4$ True

Check Equation (2).

$3x+y=8$

$3(3)+(-1)\stackrel{?}{=}8$

$9-1\stackrel{?}{=}8$

$8=8$ True

Since both results are true, the solution set of the system is $\{(3,-1)\}$.

N2. $2x-6=-3y$ (1)

$5x-3y=-27$ (2)

First, rewrite the first equation in the standard form, $Ax+By=C$.

$$\begin{array}{rcl} 2x + 3y &=& 6 \quad (1) \\ 5x - 3y &=& -27 \quad (2) \\ \hline 7x &=& -21 \quad \text{Add (1) and (2).} \\ x &=& -3 \end{array}$$

Substitute -3 for x in equation (1).

$2x+3y=6$

$2(-3)+3y=6$

$-6+3y=6$

$3y=12$

$y=4$

The solution set is $\{(-3,4)\}$.

N3. $3x-5y=25$ (1)

$2x+8y=-6$ (2)

If we simply add the equations, we will not eliminate either variable. To eliminate y, multiply equation (1) by 8, equation (2) by 5, and add the resulting equations.

$$\begin{array}{rcl} 24x - 40y &=& 200 \quad (3)\ 8\times \text{Eq.}(1) \\ 10x + 40y &=& -30 \quad (4)\ 5\times \text{Eq.}(2) \\ \hline 34x &=& 170 \quad \text{Add (3) and (4).} \\ x &=& 5 \end{array}$$

Substitute 5 for x in equation (1).

$$3x-5y=25$$
$$3(5)-5y=25$$
$$15-5y=25$$
$$-5y=10$$
$$y=-2$$

Check $x=5$, $y=-2$:

Eq. (1): $15-(-10)=25$ True

Eq. (2): $10+(-16)=-6$ True

The solution set is $\{(5,-2)\}$.

N4. $4x+9y=3$ (A)

$5y=6-3x$ (B)

Rewrite in standard form.

$4x+9y=3$ (1)

$3x+5y=6$ (2)

Solve for y by using elimination. The least common multiple of 4 and 3 is 12.

$$\begin{array}{rl} -12x-27y=-9 & (3)\ -3\times \text{Eq.}(1) \\ 12x+20y=24 & (4)\ 4\times \text{Eq.}(2) \\ \hline -7y=15 & \text{Add (3) and (4).} \\ y=-\frac{15}{7} & \end{array}$$

Solve for x by using elimination. The least common multiple of 9 and 5 is 45.

$$\begin{array}{rl} 20x+45y=15 & (5)\ 5\times \text{Eq.}(1) \\ -27x-45y=-54 & (6)\ -9\times \text{Eq.}(2) \\ \hline -7x=-39 & \text{Add (5) and (6).} \\ x=\frac{39}{7} & \end{array}$$

Check $x=\frac{39}{7}$, $y=-\frac{15}{7}$:

Eq. (A): $\frac{156}{7}-\frac{135}{7}=\frac{21}{7}$ True

Eq. (B): $-\frac{75}{7}=\frac{42}{7}-\frac{117}{7}$ True

The solution set is $\left\{\left(\frac{39}{7},-\frac{15}{7}\right)\right\}$.

N5. **(a)** $x-y=2$ (1)

$5x-5y=10$ (2)

$$\begin{array}{rl} -5x+5y=-10 & (3)\ -5\times \text{Eq. (1)} \\ 5x-5y=10 & (2) \\ \hline 0=0 & \text{Add (3) and (2).} \end{array}$$

The true statement, $0=0$, indicates that there are an infinite number of solutions, and the solution set is $\{(x, y)|x-y=2\}$.

(b)

$$\begin{array}{rl} 4x+3y=0 & (1) \\ -4x-3y=-1 & (2) \\ \hline 0=-1 & \text{Add (1) and (2).} \end{array}$$

This false statement, $0=-1$, indicates that the given system has solution set $\varnothing$.

Exercises

1. The statement is *false*; if the elimination method leads to the equation $0=-1$, the solution set of the system is $\varnothing$.

3. To eliminate y, add equations (1) and (2).

$$\begin{array}{rl} x-y=-2 & (1) \\ x+y=10 & (2) \\ \hline 2x=8 & \text{Add (1) and (2).} \\ x=4 & \end{array}$$

This result gives the x-value of the solution. To find the y-value of the solution, substitute 4 for x in either equation. We will use equation (2).

$$x+y=10$$
$$(4)+y=10 \quad \text{Let } x=4.$$
$$y=6$$

Check by substituting 4 for x and 6 for y in both equations of the original system.

Check Equation (1).

$$x-y=-2$$
$$(4)-(6)\stackrel{?}{=}-2$$
$$-2=-2 \quad \text{True}$$

Check Equation (2).

$$x+y=10$$
$$4+6\stackrel{?}{=}10$$
$$10=10 \quad \text{True}$$

The solution set is $\{(4,6)\}$.

5.
$$\begin{array}{rcll} 2x + y &=& -5 & (1) \\ x - y &=& 2 & (2) \\ \hline 3x \quad &=& -3 & \text{Add (1) and (2).} \\ x &=& -1 & \end{array}$$

Substitute -1 for x in equation (1) to find the y-value of the solution.

$$\begin{aligned} 2x + y &= -5 \\ 2(-1) + y &= -5 \quad \text{Let } x = -1. \\ -2 + y &= -5 \\ y &= -3 \end{aligned}$$

The solution set is $\{(-1, -3)\}$.

7. First, rewrite both equations in the standard form, $Ax + By = C$. We can write the first equation, $2y = -3x$, as $3x + 2y = 0$ by adding $3x$ to each side.

$$\begin{array}{rcll} 3x + 2y &=& 0 & (1) \\ -3x - y &=& 3 & (2) \\ \hline y &=& 3 & \text{Add (1) and (2).} \end{array}$$

Substitute 3 for y in equation (1).

$$\begin{aligned} 3x + 2y &= 0 \\ 3x + 2(3) &= 0 \\ 3x + 6 &= 0 \\ 3x &= -6 \\ x &= -2 \end{aligned}$$

The solution set is $\{(-2, 3)\}$.

9.
$$\begin{array}{ll} 6x - y = -1 & (1) \\ 5y = 17 + 6x & (2) \end{array}$$

Rewrite in standard form.

$$\begin{array}{rcll} 6x - y &=& -1 & (1) \\ -6x + 5y &=& 17 & (2) \\ \hline 4y &=& 16 & \text{Add (1) and (2).} \\ y &=& 4 & \text{Solve for } y. \end{array}$$

Substitute 4 for y in equation (1).

$$\begin{aligned} 6x - y &= -1 \\ 6x - (4) &= -1 \\ 6x &= 3 \\ x &= \frac{3}{6} = \frac{1}{2} \end{aligned}$$

The solution set is $\left\{\left(\frac{1}{2}, 4\right)\right\}$.

11.
$$\begin{array}{ll} 2x - y = 12 & (1) \\ 3x + 2y = -3 & (2) \end{array}$$

If we simply add the equations, we will not eliminate either variable. To eliminate y, multiply equation (1) by 2 and add the result to equation (2).

$$\begin{array}{rcll} 4x - 2y &=& 24 & (3) \\ 3x + 2y &=& -3 & (2) \\ \hline 7x \quad &=& 21 & \text{Add (3) and (2).} \\ x &=& 3 & \end{array}$$

Substitute 3 for x in equation (1).

$$\begin{aligned} 2x - y &= 12 \\ 2(3) - y &= 12 \\ -y &= 6 \\ y &= -6 \end{aligned}$$

The solution set is $\{(3, -6)\}$.

13.
$$\begin{array}{ll} x + 4y = 16 & (1) \\ 3x + 5y = 20 & (2) \end{array}$$

To eliminate x, multiply equation (1) by -3, and add the result to equation (2).

$$\begin{array}{rcll} -3x - 12y &=& -48 & (3) \\ 3x + 5y &=& 20 & (2) \\ \hline -7y &=& -28 & \text{Add (3) and (2).} \\ y &=& 4 & \end{array}$$

Substitute 4 for y in equation (1).

$$\begin{aligned} x + 4y &= 16 \\ x + 4(4) &= 16 \\ x + 16 &= 16 \\ x &= 0 \end{aligned}$$

The solution set is $\{(0, 4)\}$.

15.
$$\begin{array}{ll} 2x - 8y = 0 & (1) \\ 4x + 5y = 0 & (2) \end{array}$$

To eliminate x, multiply equation (1) by -2, and add the result to equation (2).

$$\begin{array}{rcll} -4x + 16y &=& 0 & (3) \\ 4x + 5y &=& 0 & (2) \\ \hline 21y &=& 0 & \text{Add (3) and (2).} \\ y &=& 0 & \end{array}$$

Substitute 0 for y in equation (1).

$$2x-8y=0$$
$$2x-8(0)=0$$
$$2x=0$$
$$x=0$$

The solution set is $\{(0, 0)\}$.

17. $3x+3y=33$ (1)

$5x-2y=27$ (2)

To eliminate y, multiply equation (1) by 2, and equation (2) by 3.

$$\begin{array}{rcrl} 6x + 6y &=& 66 & (3) \\ 15x - 6y &=& 81 & (4) \\ \hline 21x &=& 147 & \text{Add (3) and (4).} \\ x &=& 7 & \end{array}$$

Substitute 7 for x in equation (2).

$$5x-2y=27$$
$$5(7)-2y=27$$
$$35-2y=27$$
$$-2y=-8$$
$$y=4$$

The solution set is $\{(7, 4)\}$.

19. $5x+4y=12$ (1)

$3x+5y=15$ (2)

To eliminate x, we could multiply equation (1) by $-\frac{3}{5}$, but that would introduce fractions and make the solution more complicated. Instead, we'll work with the least common multiple of the coefficients of x, which is 15, and choose suitable multipliers of these coefficients so that the new coefficients are opposites. In this case, we could pick -3 times equation (1) and 5 times equation (2), or, 3 times equation (1) and -5 times equation (2). If we wanted to eliminate y, we could multiply equation (1) by -5 and equation (2) by 4, or equation (1) by 5 and equation (2) by -4.

$$\begin{array}{rcrl} -15x - 12y &=& -36 & (3)\ -3\times\text{Eq. }(1) \\ 15x + 25y &=& 75 & (4)\ \ 5\times\text{Eq. }(2) \\ \hline 13y &=& 39 & \text{Add (3) and (4).} \\ y &=& 3 & \end{array}$$

Substitute 3 for y in equation (1).

$$5x+4y=12$$
$$5x+4(3)=12$$
$$5x+12=12$$
$$5x=0$$
$$x=0$$

The solution set is $\{(0, 3)\}$.

21. $5x-4y=15$ (1)

$-3x+6y=-9$ (2)

$$\begin{array}{rcrl} 15x - 12y &=& 45 & (3)\ \ 3\times\text{Eq. }(1) \\ -15x + 30y &=& -45 & (4)\ \ 5\times\text{Eq. }(2) \\ \hline 18y &=& 0 & \text{Add (3) and (4).} \\ y &=& 0 & \end{array}$$

Substitute 0 for y in equation (1).

$$5x-4y=15$$
$$5x-4(0)=15$$
$$5x=15$$
$$x=3$$

The solution set is $\{(3, 0)\}$.

23. $-x+3y=4$ (1)

$-2x+6y=8$ (2)

$$\begin{array}{rcrl} 2x - 6y &=& -8 & (3)\ -2\times\text{Eq. }(1) \\ -2x + 6y &=& 8 & (2) \\ \hline 0 &=& 0 & \text{Add (3) and (2).} \end{array}$$

Since $0=0$ is a *true* statement, the equations are equivalent. This result indicates that every solution of one equation is also a solution of the other; there are an *infinite number of solutions.* The solution set is $\{(x, y)|x-3y=-4\}$.

25. $5x-2y=3$ (1)

$10x-4y=5$ (2)

$$\begin{array}{rcrl} -10x + 4y &=& -6 & (3)\ -2\times\text{Eq. }(1) \\ 10x - 4y &=& 5 & (2) \\ \hline 0 &=& -1 & \text{Add (3) and (2).} \end{array}$$

Since $0=-1$ is a *false* statement, there are no solutions of the system and the solution set is $\varnothing$.

27. $6x-2y=-22$ (1)
$-3x+4y=17$ (2)

$$\begin{array}{rcrl} 6x - 2y &=& -22 & (1) \\ -6x + 8y &=& 34 & (3) \quad 2\times\text{Eq.}(2) \\ \hline 6y &=& 12 & \text{Add}(1)\text{ and }(3). \\ y &=& 2 & \end{array}$$

Substitute 2 for y in equation (1).

$$\begin{aligned} 6x-2y&=-22 \\ 6x-2(2)&=-22 \\ 6x-4&=-22 \\ 6x&=-18 \\ x&=-3 \end{aligned}$$

The solution set is $\{(-3, 2)\}$.

29. $3x=3+2y$ (1)
$-\frac{4}{3}x+y=\frac{1}{3}$ (2)

Rewrite equation (1) in standard form and multiply equation (2) by 3 to clear fractions.

$3x-2y=3$ (3)
$-4x+3y=1$ (4)

$$\begin{array}{rcrl} 12x - 8y &=& 12 & (5) \quad 4\times\text{Eq.}(3) \\ -12x + 9y &=& 3 & (6) \quad 3\times\text{Eq.}(4) \\ \hline y &=& 15 & \text{Add }(5)\text{ and }(6). \end{array}$$

$$\begin{array}{rcrl} 9x - 6y &=& 9 & (7) \quad 3\times\text{Eq.}(3) \\ -8x + 6y &=& 2 & (8) \quad 2\times\text{Eq.}(4) \\ \hline x &=& 11 & \text{Add }(7)\text{ and }(8). \end{array}$$

The solution set is $\{(11, 15)\}$.

31. $\frac{1}{5}x+y=\frac{6}{5}$ (1)
$\frac{1}{10}x+\frac{1}{3}y=\frac{5}{6}$ (2)

First, clear all fractions.

Equation (1):

$5\left(\frac{1}{5}x+y\right)=5\left(\frac{6}{5}\right)$ Multiply by LCD.

$x+5y=6$ (3)

Equation (2):

$30\left(\frac{1}{10}x+\frac{1}{3}y\right)=30\left(\frac{5}{6}\right)$ Multiply by LCD.

$3x+10y=25$ (4)

The system has been simplified to

$x+5y=6$ (3)
$3x+10y=25$ (4)

To eliminate y, add -2 times equation (3) to equation (4).

$$\begin{array}{rcrl} -2x - 10y &=& -12 & (5) \quad -2\times\text{Eq.}(3) \\ 3x + 10y &=& 25 & (4) \\ \hline x &=& 13 & \text{Add }(5)\text{ and }(4). \end{array}$$

Substitute 13 for x in equation (3).

$$\begin{aligned} x+5y&=6 \quad (3) \\ (13)+5y&=6 \\ 5y&=-7 \\ y&=-\frac{7}{5} \end{aligned}$$

The solution set is $\left\{\left(13, -\frac{7}{5}\right)\right\}$.

33. $2.4x+1.7y=7.6$ (1)
$1.2x-0.5y=9.2$ (2)

Multiply equation (1) and equation (2) by 10 to eliminate the decimals. Then solve the resulting equations by the elimination method.

$24x+17y=76$ (3)
$12x-5y=92$ (4)

Multiply equation (2) by -2.

$$\begin{array}{rcrl} 24x + 17y &=& 76 & (3) \\ -24x + 10y &=& -184 & (4) \\ \hline 27y &=& -108 & \text{Add.} \\ y &=& -4 & \end{array}$$

Substitute -4 for y in equation (4).

$$\begin{aligned} 12x-5y&=92 \\ 12x-5(-4)&=92 \quad \text{Let } y=-4. \\ 12x+20&=92 \\ 12x&=72 \\ x&=6 \quad \text{Divide by 12.} \end{aligned}$$

The solution set is $\{(6, -4)\}$.

35. $x+3y=6 \quad (1)$

$-2x+12=6y \quad (2)$

Rewrite in standard form.

$x+3y=6 \quad (1)$

$-2x-6y=-12 \quad (2)$

To eliminate x, multiply equation (1) by 2 and add the result to equation (2).

$$\begin{array}{rcrl} 2x + 6y &=& 12 & (3) \\ -2x - 6y &=& -12 & (2) \\ \hline 0 &=& 0 & \text{Add (3) and (2).} \end{array}$$

The true statement, $0=0$, indicates that there are an infinite number of solutions and the solution set is $\{(x, y)|x+3y=6\}$.

37. $4x-3y=1 \quad (1)$

$8x=3+6y \quad (2)$

Rewrite in standard form.

$4x-3y=1 \quad (1)$

$8x-6y=3 \quad (2)$

To eliminate x, multiply equation (1) by -2 and add the result to equation (2).

$$\begin{array}{rcrl} -8x + 6y &=& -2 & (3) \\ 8x - 6y &=& 3 & (2) \\ \hline 0 &=& 1 & \text{Add (3) and (2).} \end{array}$$

The false statement, $0=1$, indicates that the given system has solution set $\varnothing$.

39. $4x=3y-2 \quad (1)$

$5x+3=2y \quad (2)$

Rewrite in standard form.

$-4x+3y=2 \quad (1)$

$5x-2y=-3 \quad (2)$

$$\begin{array}{rcrl} -8x + 6y &=& 4 & (3) \quad 2\times\text{Eq.}(1) \\ 15x - 6y &=& -9 & (4) \quad 3\times\text{Eq.}(2) \\ \hline 7x &=& -5 & \text{Add (3) and (4).} \end{array}$$

$$x = -\frac{5}{7}$$

Rather than substitute $-\frac{5}{7}$ for x in (1) or (2), we will eliminate x by multiplying equation (1) by 5, equation (2) by 4, and adding the results.

$$\begin{array}{rcrl} -20x + 15y &=& 10 & (5) \quad 5\times\text{Eq.}(1) \\ 20x - 8y &=& -12 & (6) \quad 4\times\text{Eq.}(2) \\ \hline 7y &=& -2 & \text{Add (5) and (6).} \end{array}$$

$$y = -\frac{2}{7}$$

Solving the system in this fashion reduces the chance of making an arithmetic error.

The solution set is $\left\{\left(-\frac{5}{7}, -\frac{2}{7}\right)\right\}$.

When you get a solution that has non-integer components, it is sometimes more difficult to check the problem than it is to solve it. A graphing calculator can be very helpful in this case. Just store the values for x and y in their respective memory locations, and then type the expressions as shown in the following screen. The results 2 and -3 (the right sides of equations (1) and (2)) indicate that we have found the correct solution.

```
-5/7→X:-2/7→Y
         -.2857142857
-4X+3Y
                    2
5X-2Y
                   -3
```

41. $24x+12y=-7 \quad (1)$

$16x-18y=17 \quad (2)$

$$\begin{array}{rcrl} 48x + 24y &=& -14 & (3) \quad 2\times\text{Eq. }(1) \\ -48x + 54y &=& -51 & (4) \quad -3\times\text{Eq. }(2) \\ \hline 78y &=& -65 & \text{Add (3) and (4).} \end{array}$$

$$y = \frac{-65}{78} = -\frac{5}{6}$$

$$\begin{array}{rcrl} 72x + 36y &=& -21 & (5) \quad 3\times\text{Eq.}(1) \\ 32x - 36y &=& 34 & (6) \quad 2\times\text{Eq.}(2) \\ \hline 104x &=& 13 & \text{Add (5) and (6).} \end{array}$$

$$x = \frac{13}{104} = \frac{1}{8}$$

The solution set is $\left\{\left(\frac{1}{8}, -\frac{5}{6}\right)\right\}$.

43. $y=ax+b$

$6.21=a(2004)+b \quad$ Let $x=2004,\ y=6.21.$

$6.21=2004a+b$

45. $2004a+b=6.21 \quad (1)$

$2012a+b=7.96 \quad (2)$

Multiply equation (1) by -1 and add the result to equation (2).

$$\begin{array}{rcrl} -2004a - b &=& -6.21 & (3) \quad -1\times\text{Eq. }(1) \\ 2012a + b &=& 7.96 & (2) \\ \hline 8a &=& 1.75 & \text{Add (3) and (2).} \end{array}$$

$$a = 0.21875$$

Substitute 0.21875 for a in equation (1).

$$2004(0.21875)+b=6.21$$
$$438.375+b=6.21$$
$$b=-432.165$$

The solution set is $\{(0.21875,-432.165)\}$.

Summary Exercises Applying Techniques for Solving Systems of Linear Equations

1. **(a)** $3x+5y=69$
$y=4x$
Use substitution because the second equation is solved for y.

(b) $3x+y=-7$
$x-y=-5$
Use elimination because the coefficients of the y-terms are opposites.

(c) $3x-2y=0$
$9x+8y=7$
Use elimination because the equations are in standard form with no coefficients of 1 or −1. Solving by substitution would involve fractions.

2. System A: $5x-3y=7$
$2x+8y=3$
System B: $7x+2y=4$
$y=-3x+1$
System B is easier to solve by substitution because the second equation is already solved for y.

3. $4x-3y=-8 \quad (1)$
$x+3y=13 \quad (2)$

(a) Solve the system by the elimination method.

$$\begin{array}{rcll} 4x-3y &=& -8 & (1) \\ x+3y &=& 13 & (2) \\ \hline 5x &=& 5 & \text{Add } (1) \text{ and } (2). \\ x &=& 1 & \end{array}$$

To find y, let $x=1$ in equation (2).

$$x+3y=13$$
$$1+3y=13$$
$$3y=12$$
$$y=4$$

The solution set is $\{(1,4)\}$.

(b) To solve this system by the substitution method, begin by solving equation (2) for x.

$$x+3y=13$$
$$x=-3y+13$$

Substitute $-3y+13$ for x in equation (1).

$$4(-3y+13)-3y=-8$$
$$-12y+52-3y=-8$$
$$-15y=-60$$
$$y=4$$

To find x, let $y=4$ in equation (2).

$$x+3y=13$$
$$x+3(4)=13$$
$$x+12=13$$
$$x=1$$

The solution set is $\{(1,4)\}$.

(c) Answers will vary. Sample Answer: For this particular system, the elimination method is preferable because both equations are already written in the form $Ax+By=C$, and the equations can be added without multiplying either by a constant. Comparing solutions by the two methods, we see that the elimination method requires fewer steps than the substitution method for this system.

4. $2x+5y=0 \quad (1)$
$x=-3y+1 \quad (2)$

(a) To solve this system by the elimination method, begin by rewriting equation (2) in the form $Ax+By=C$.

$x+3y=1 \quad (3)$

We now have the system

$2x+5y=0 \quad (1)$
$x+3y=1. \quad (3)$

Multiply equation (3) by -2 and add the result to equation (1).

$$\begin{array}{rcll} 2x+5y &=& 0 & (1) \\ -2x-6y &=& -2 & (4) \quad -2\times\text{Eq. } (3) \\ \hline -y &=& -2 & \text{Add } (1) \text{ and } (4). \\ y &=& 2 & \end{array}$$

To find x, let $y = 2$ in equation (1).

$$2x + 5y = 0$$
$$2x + 5(2) = 0$$
$$2x + 10 = 0$$
$$2x = -10$$
$$x = -5$$

The solution set is $\{(-5, 2)\}$.

(b) To solve this system by the substitution method, begin by substituting $-3y + 1$ for x in equation (1).

$$2x + 5y = 0$$
$$2(-3y + 1) + 5y = 0$$
$$-6y + 2 + 5y = 0$$
$$-y = -2$$
$$y = 2$$

To find x, let $y = 2$ in equation (2).

$$x = -3y + 1$$
$$x = -3(2) + 1 = -5$$

The solution set is $\{(-5, 2)\}$.

(c) Answers will vary. Sample Answer: For this particular system, the substitution method is probably preferable because equation (2) is already solved for x.

5. $3x + 5y = 69 \quad (1)$

$y = 4x \quad (2)$

Equation (2) is already solved for y, so we'll use the substitution method. Substitute $4x$ for y in equation (1) and solve the resulting equation for x.

$$3x + 5y = 69$$
$$3x + 5(4x) = 69 \qquad \text{Let } y = 4x.$$
$$3x + 20x = 69$$
$$23x = 69$$
$$x = \frac{69}{23} = 3$$

To find the y-value of the solution, substitute 3 for x in equation (2).

$$y = 4x$$
$$y = 4(3) \quad \text{Let } x = 3.$$
$$= 12$$

The solution set is $\{(3, 12)\}$.

6. Adding the equations will eliminate y, so we'll use the elimination method.

$$3x + y = -7 \qquad (1)$$
$$x - y = -5 \qquad (2)$$
$$4x = -12 \qquad \text{Add } (1) \text{ and } (2).$$
$$x = -\frac{12}{4} = -3$$

Substitute -3 for x in equation (2).

$$-3 - y = -5$$
$$-y = -2$$
$$y = 2$$

The solution set is $\{(-3, 2)\}$.

7. $3x - 2y = 0 \quad (1)$

$9x + 8y = 7 \quad (2)$

$$12x - 8y = 0 \qquad (3) \quad 4 \times \text{Eq. } (1)$$
$$9x + 8y = 7 \qquad (2)$$
$$21x = 7 \qquad \text{Add } (3) \text{ and } (2).$$
$$x = \frac{7}{21} = \frac{1}{3}$$

$$-9x + 6y = 0 \qquad (4) \quad -3 \times \text{Eq. } (1)$$
$$9x + 8y = 7 \qquad (2)$$
$$14y = 7 \qquad \text{Add } (4) \text{ and } (2).$$
$$y = \frac{7}{14} = \frac{1}{2}$$

The solution set is $\left\{\left(\frac{1}{3}, \frac{1}{2}\right)\right\}$.

8. $x + y = 7 \quad (1)$

$x = -3 - y \quad (2)$

Substitute $-3 - y$ for x in equation (1).

$$(-3 - y) + y = 7$$
$$-3 = 7 \quad \text{False}$$

The solution set is $\varnothing$.

9. $6x + 7y = 4 \quad (1)$

$5x + 8y = -1 \quad (2)$

$$-30x - 35y = -20 \quad (3) \quad -5 \times \text{Eq. } (1)$$
$$30x + 48y = -6 \quad (4) \quad 6 \times \text{Eq. } (2)$$
$$13y = -26 \quad \text{Add } (3) \text{ and } (4).$$
$$y = -2$$

Substitute -2 for y in equation (1).

$$6x+7y=4$$
$$6x+7(-2)=4 \quad \text{Let } y=-2.$$
$$6x-14=4$$
$$6x=18$$
$$x=3$$

The solution set is $\{(3,-2)\}$.

10. $6x-y=5 \quad (1)$

$y=11x \quad (2)$

Substitute $11x$ for y in equation (1).

$$6x-(11x)=5 \quad \text{Let } y=11x.$$
$$-5x=5$$
$$x=-1$$

Substitute -1 for x in equation (2).

$$y=11(-1) \quad \text{Let } x=-1.$$
$$=-11$$

The solution set is $\{(-1,-11)\}$.

11. $4x-6y=10 \quad (1)$

$-10x+15y=-25 \quad (2)$

Divide equation (1) by 2 and equation (2) by 5.

$$\begin{array}{rcl} 2x-3y&=&5 \quad (3)\\ -2x+3y&=&-5 \quad (4)\\ \hline 0&=&0 \quad \text{Add (3) and (4).}\end{array}$$

Since $0=0$ is a *true* statement, there are an *infinite number of solutions*. The solution set is $\{(x,y)\,|\,2x-3y=5\}$.

12. $3x-5y=7 \quad (1)$

$2x+3y=30 \quad (2)$

$$\begin{array}{rcll} 9x-15y&=&21 & (3)\ \ 3\times\text{Eq.}(1)\\ 10x+15y&=&150 & (4)\ \ 5\times\text{Eq.}(2)\\ \hline 19x&=&171 & \text{Add (3) and (4).}\end{array}$$

$$x=\frac{171}{19}=9$$

Substitute 9 for x in equation (2).

$$2x+3y=30$$
$$2(9)+3y=30 \quad \text{Let } x=9.$$
$$18+3y=30$$
$$3y=12$$
$$y=4$$

The solution set is $\{(9,4)\}$.

13. $5x=7+2y$

$5y=5-3x$

Rewrite in standard form.

$5x-2y=7 \quad (1)$

$3x+5y=5 \quad (2)$

$$\begin{array}{rcll} 25x-10y&=&35 & (3)\ \ 5\times\text{Eq.}(1)\\ 6x+10y&=&10 & (4)\ \ 2\times\text{Eq.}(2)\\ \hline 31x&=&45 & \text{Add (3) and (4).}\end{array}$$

$$x=\frac{45}{31}$$

$$\begin{array}{rcll} -15x+6y&=&-21 & (5)\ \ -3\times\text{Eq.}(1)\\ 15x+25y&=&25 & (6)\ \ 5\times\text{Eq.}(2)\\ \hline 31y&=&4 & \text{Add (5) and (6).}\end{array}$$

$$x=\frac{4}{31}$$

The solution set is $\left\{\left(\frac{45}{31},\frac{4}{31}\right)\right\}$.

14. $4x+3y=1 \quad (1)$

$3x+2y=2 \quad (2)$

$$\begin{array}{rcll} -12x-9y&=&-3 & (3)\ \ -3\times\text{Eq.}(1)\\ 12x+8y&=&8 & (4)\ \ 4\times\text{Eq.}(2)\\ \hline -y&=&5 & \text{Add (3) and (4).}\\ y&=&-5 & \end{array}$$

Substitute -5 for y in equation (1).

$$4x+3(-5)=1 \quad \text{Let } y=-5.$$
$$4x-15=1$$
$$4x=16$$
$$x=4$$

The solution set is $\{(4,-5)\}$.

15. $2x-3y=7 \quad (1)$

$-4x+6y=14 \quad (2)$

$$\begin{array}{rcll} 4x - 6y &=& 14 & (3) \quad 2\times \text{Eq. }(1) \\ -4x + 6y &=& 14 & (2) \\ \hline 0 &=& 28 & \text{Add (3) and (2).} \end{array}$$

Since $0=28$ is *false*, the solution set is $\varnothing$.

16. $2x+3y=10 \quad (1)$

$-3x+y=18 \quad (2)$

Solve equation (2) for *y*.

$y=3x+18 \quad (3)$

Substitute $3x+18$ for *y* in equation (1).

$2x+3(3x+18)=10$ Let $y=3x+18$.

$2x+9x+54=10$

$11x+54=10$

$11x=-44$

$x=-4$

Substitute -4 for *x* in equation (3).

$y=3(-4)+18$ Let $x=-4$.

$=-12+18$

$=6$

The solution set is $\{(-4,6)\}$.

17. $7x-4y=0$

$3x=2y$

Rewrite in standard form.

$7x-4y=0 \quad (1)$

$3x-2y=0 \quad (2)$

$$\begin{array}{rcll} 7x - 4y &=& 0 & (1) \\ -6x + 4y &=& 0 & (3) \quad -2\times \text{Eq.}(2) \\ \hline x &=& 0 & \text{Add (1) and (3).} \end{array}$$

Substitute 0 for *x* in equation (1).

$7x-4y=0$

$7(0)-4y=0$

$-4y=0$

$y=0$

The solution set is $\{(0,0)\}$.

18. $x-3y=7 \quad (1)$

$4x+y=5 \quad (2)$

Solve equation (1) for *x*.

$x=3y+7 \quad (3)$

Substitute $3y+7$ for *x* in equation (2).

$4(3y+7)+y=5$ Let $x=3y+7$.

$12y+28+y=5$

$13y+28=5$

$13y=-23$

$y=-\frac{23}{13}$

Substitute $-\frac{23}{13}$ for *y* in equation (3).

$x=3\left(-\frac{23}{13}\right)+7$ Let $y=-\frac{23}{13}$.

$=-\frac{69}{13}+\frac{91}{13}$

$=\frac{22}{13}$

The solution set is $\left\{\left(\frac{22}{13},-\frac{23}{13}\right)\right\}$.

19. $\frac{1}{5}x+\frac{2}{3}y=-\frac{8}{5} \quad (1)$

$3x-y=9 \quad (2)$

Multiply each side of equation (1) by 15 to clear fractions.

$15\left(\frac{1}{5}x+\frac{2}{3}y\right)=15\left(-\frac{8}{5}\right)$

$15\left(\frac{1}{5}x\right)+15\left(\frac{2}{3}y\right)=-24$

$3x+10y=-24 \quad (3)$

Now use the elimination method with equations (2) and (3).

$$\begin{array}{rcll} 3x + 10y &=& -24 & (3) \\ -3x + y &=& -9 & (4) \quad -1\times \text{Eq.}(2) \\ \hline 11y &=& -33 & \text{Add (3) and (4).} \\ y &=& -3 & \end{array}$$

Substitute -3 for *y* in equation (2).

$3x-(-3)=9$ Let $y=-3$.

$3x+3=9$

$3x=6$

$x=2$

The solution set is $\{(2,-3)\}$.

20. $\frac{1}{6}x+\frac{1}{6}y=2 \quad (1)$

$-\frac{1}{2}x-\frac{1}{3}y=-8 \quad (2)$

Multiply each side of equation (1) by 6 to clear fractions.

$$6\left(\frac{1}{6}x+\frac{1}{6}y\right)=6(2)$$

$$6\left(\frac{1}{6}x\right)+6\left(\frac{1}{6}y\right)=6(2)$$

$$x+y=12 \quad (3)$$

Multiply each side of equation (2) by the LCD, 6, to clear fractions

$$6\left(-\frac{1}{2}x-\frac{1}{3}y\right)=6(-8)$$

$$6\left(-\frac{1}{2}x\right)+6\left(-\frac{1}{3}y\right)=6(-8)$$

$$-3x-2y=-48 \quad (4)$$

The given system of equations has been simplified as follows.

$x+y=12 \quad (3)$

$-3x-2y=-48 \quad (4)$

Multiply equation (3) by 3 and add the result to equation (4).

$$\begin{aligned} 3x + 3y &= 36 \\ -3x - 2y &= -48 \\ \hline y &= -12 \end{aligned}$$

To find x, let $y=-12$ in equation (3).

$$\begin{aligned} x+y&=12 \\ x+(-12)&=12 \\ x-12&=12 \\ x&=24 \end{aligned}$$

The solution set is $\{(24, -12)\}$.

21. $\frac{x}{3}-\frac{3y}{4}=-\frac{1}{2} \quad (1)$

$\frac{x}{6}+\frac{y}{8}=\frac{3}{4} \quad (2)$

Multiply each side of equation (1) by 12 to clear fractions.

$$12\left(\frac{x}{3}-\frac{3y}{4}\right)=12\left(-\frac{1}{2}\right)$$

$$4x-9y=-6 \quad (3)$$

Multiply each side of equation (2) by 24 to clear fractions.

$$24\left(\frac{x}{6}+\frac{y}{8}\right)=24\left(\frac{3}{4}\right)$$

$$4x+3y=18 \quad (4)$$

The given system of equations has been simplified as follows.

$4x-9y=-6 \quad (3)$

$4x+3y=18 \quad (4)$

Multiply equation (3) by -1 and add the result to equation (4).

$$\begin{aligned} -4x + 9y &= 6 \\ 4x + 3y &= 18 \\ \hline 12y &= 24 \\ y &= 2 \end{aligned}$$

To find x, let $y=2$ in equation (4).

$$\begin{aligned} 4x+3y&=18 \\ 4x+3(2)&=18 \\ 4x+6&=18 \\ 4x&=12 \\ x&=3 \end{aligned}$$

The solution set is $\{(3, 2)\}$.

22. $\frac{x}{2}-\frac{y}{3}=9 \quad (1)$

$\frac{x}{5}-\frac{y}{4}=5 \quad (2)$

To clear fractions, multiply each side of equation (1) by the LCD, 6.

$$6\left(\frac{x}{2}-\frac{y}{3}\right)=6(9)$$

$$3x-2y=54 \quad (3)$$

To clear fractions, multiply each side of equation (2) by the LCD, 20.

$$20\left(\frac{x}{5}-\frac{y}{4}\right)=20(5)$$

$$4x-5y=100 \quad (4)$$

We now have the simplified system

$3x-2y=54 \quad (3)$

$4x-5y=100. \quad (4)$

To solve this system by the elimination method, multiply equation (3) by 5 and equation (4) by -2; then add the results.

$$\begin{aligned} 15x - 10y &= 270 \\ -8x + 10y &= -200 \\ \hline 7x &= 70 \\ x &= 10 \end{aligned}$$

To find y, let $x = 10$ in equation (3).

$3x - 2y = 54$
$3(10) - 2y = 54$
$30 - 2y = 54$
$-2y = 24$
$y = -12$

The solution set is $\{(10, -12)\}$.

23. $0.1x + y = 1.6$ (1)
$0.6x + 0.5y = -1.4$ (2)

Multiply equation (1) and equation (2) by 10 to eliminate the decimals. Then solve the resulting equations by the elimination method.

$x + 10y = 16$ (3)
$6x + 5y = -14$ (4)

Multiply equation (2) by -2.

$x + 10y = 16$ (3)
$-12x - 10y = 28$ (5)
$-11x = 44$ Add.
$x = -4$

Substitute -4 for x in equation (3).

$x + 10y = 16$
$-4 + 10y = 16$ Let $x = -4$.
$10y = 20$
$y = 2$ Divide by 10.

The solution set is $\{(-4, 2)\}$.

24. $0.2x - 0.3y = 0.1$ (1)
$0.3x - 0.2y = 0.9$ (2)

Multiply equation (1) and equation (2) by 10 to eliminate the decimals. Then solve the resulting equations by the elimination method.

$2x - 3y = 1$ (3)
$3x - 2y = 9$ (4)

Multiply equation (3) by -3 and equation (4) by 2.

$-6x + 9y = -3$ (5)
$6x - 4y = 18$ (6)
$5y = 15$ Add.
$y = 3$

Substitute 3 for y in equation (4).

$3x - 2y = 9$
$3x - 2(3) = 9$ Let $y = 3$.
$3x - 6 = 9$
$3x = 15$
$x = 15$ Divide by 3.

The solution set is $\{(5, 3)\}$.

4.4 Applications of Linear Systems

Now Try Exercises

N1. Let x = the amount per month that was spent on electricity and y = the amount per month that was spent on rent. The total was $1150, so the following equation applies.

$x + y = 1150$ (1)

Marina spent $650 more on rent than on electricity, so the following applies.

$y = x + 650$ (2)

Substitute $x + 650$ for y in equation (1).

$x + y = 1150$
$x + (x + 650) = 1150$
$2x + 650 = 1150$
$2x = 500$
$x = 250$

$250 was spent on electricity per month and $250 + $650 = $900 was spent on rent per month.

N2. *Step 2*
Let x = the number of adult tickets sold and y = the number of children's tickets sold.

Kind of Ticket	Number Sold	Cost of Each (in dollars)	Total Value (in dollars)
Adult	x	19	$19x$
Child	y	16	$16y$
Total	27	XXXXXX	$462

Step 3
The total number of tickets sold was 27 and is represented by the following.

$x + y = 27$ (1)

Since the total value was $462, the final column leads to the following.

$19x + 16y = 462$ (2)

Step 4
Multiply both sides of equation (1) by -16 and add this result to equation (2).

$$\begin{aligned} -16x - 16y &= -432 \\ 19x + 16y &= 462 \\ \hline 3x \quad &= 30 \\ x &= 10 \end{aligned}$$

From equation (1), the following is calculated.
$y = 27 - (10) = 17$

Step 5
There were 10 adults and 17 children at the game.

Step 6
Check: The total number of tickets sold was $10 + 17 = 27$. Since 10 adults paid \$19 each and 17 children paid \$16 each, the value of tickets sold should be $10(19) + 17(16) = 462$, or \$462. The result agrees with the given information.

N3. Let $x =$ the number of liters of 10% saline solution needed and $y =$ the number of liters of 35% saline solution needed.
Summarize the information in a table.

Liters	Percent	Liters of Pure Alcohol
x	0.10	$0.10x$
y	0.35	$0.35y$
80	0.30	0.30(80)

$x + y = 80$ From first column
$0.10x + 0.35y = 0.30(80)$ From third column

To eliminate the x-terms, multiply the first equation by -0.10. Then add the result to the second equation.

$$\begin{aligned} -0.10x - 0.10y &= -8 \\ 0.10x + 0.35y &= 24 \\ \hline 0.25y &= 16 \quad \text{Add.} \\ y &= 64 \end{aligned}$$

To find x, substitute 64 for y in the first equation of the original system.

$x + y = 80$
$x + 64 = 80$ Let $y = 64$.
$x = 16$

The solution is $x = 16$, $y = 64$. Mix 16 L of 10% solution with 64 L of 35% solution.

N4. Let $x =$ the rate of the faster truck, and $y =$ the rate of the slower truck.
Summarize the information in a table.

	r	t	d
Faster Truck	x	3	$3x$
Slower Truck	y	3	$3y$

Write a system of equations.
$3x + 3y = 405$ Total distance
$x = y + 5$ Faster truck is 5 mph faster.

Substitute $y + 5$ for x in the first equation, and solve for y.

$3x + 3y = 405$
$3(y + 5) + 3y = 405$ Let $x = y + 5$.
$3y + 15 + 3y = 405$
$6y + 15 = 405$
$6y = 390$
$y = 65$

To find x, use $x = y + 5$ and $y = 65$.
$x = y + 5 = 65 + 5 = 70$

The faster truck's rate is 70 miles per hour and the slower truck's rate is 65 miles per hour.

N5. Let $x =$ the rate that Gigi can row in still water and $y =$ the rate of the current. Summarize the information in a table.

	r	t	d
Rate of Gigi Rowing	x	1	x
Rate of the Current	y	1	y

Write a system of equations.
$x + y = 10$ Travel distance with current
$x - y = 2$ Travel distance against current

Solve the first equation for y.
$y = 10 - x$

Substitute $10 - x$ for y in the first equation, and solve for x.

$x - y = 2$
$x - (10 - x) = 2$ Let $y = 15 - x$.
$x - 10 + x = 2$
$2x = 12$
$x = 6$

To find y, use $y = 10 - x$ and $x = 6$.

$y = 10 - (6) = 4$

Gigi's rate is 6 mph and the rate of the current is 4 mph.

Exercises

1. To represent the monetary value of x 5-dollar bills, multiply 5 times x. The answer is D, $5x$ dollars.

3. The amount of interest earned on d dollars at an interest rate of 3% (0.03) is choice B, $0.03d$ dollars.

5. If a cheetah's rate is 70 mph and it runs at that rate for x hours, then the distance covered is choice D, $70x$ miles.

7. Since the plane is traveling *with* the wind, add the rate of the plane, 650 miles per hour, to the rate of the wind, r miles per hour. The answer is C, $(650 + r)$ mph.

9. If there are x liters of the 40% acid solution and y liters of the 35% acid solution, then the amount of pure acid in the two solutions is represented by the following.

$0.40x + 0.35y$

The amount of pure acid is also represented by 0.38(100). The correct equation is the following.

$0.40x + 0.35y = 0.38(100)$

So the correct choice is B.

11. *Step 2*
Let $x =$ the first number and let $y =$ <u>the second number</u>.
Step 3
First equation: $x + y = 98$

Second equation: $x - y = 48$

Step 4
Add the two equations.

$$\begin{aligned} x + y &= 98 \\ x - y &= 48 \\ \hline 2x \quad &= 146 \\ x &= 73 \end{aligned}$$

Substitute 73 for x in either equation to find $y = 25$.
Step 5
The two numbers are 73 and 25.
Step 6
The sum of 73 and 25 is 98. The difference between 73 and 25 is 48. The solution satisfies the conditions of the problem.

13. *Step 2*
Let $x =$ the number of performances of *Cats* and $y =$ the number of performances of *Phantom of the Opera.*
Step 3
The total number of performances was 18,188, so one equation is the following.

$x + y = 18{,}188 \quad (1)$

Phantom of the Opera had 3218 more performances than *Cats*, so another equation is the following.

$y = x + 3218 \quad (2)$

Step 4
Substitute $x + 3218$ for y in equation (1).

$$\begin{aligned} x + y &= 18{,}188 \\ x + (x + 3218) &= 18{,}188 \\ 2x + 3218 &= 18{,}188 \\ 2x &= 14{,}970 \\ x &= 7485 \end{aligned}$$

Substitute 7485 for x in (2) to find y.

$y = 7485 + 3218 = 10{,}703$

Step 5
Cats had 7485 performances, and *Phantom of the Opera* had 10,703 performances.
Step 6
The sum of 7485 and 10,703 is 18,188, and 10,703 is 3218 more than 7485.

15. *Step 2*
Let $x =$ the gross earnings in millions of *Marvel's The Avengers* and $y =$ the gross earnings in millions of *The Dark Knight Rises.*
Step 3
The total gross earnings of both movies in 2012 was \$1071 million, so one equation is represented by the following.

$x + y = 1071 \quad (1)$

The Dark Knight Rises grossed \$175 million less than *Marvel's The Avengers*, so another equation is the following.

$y = x - 175 \quad (2)$

Step 4
Substitute $x - 175$ for y in equation (1).

$$\begin{aligned} x + y &= 1071 \\ x + (x - 175) &= 1071 \\ 2x - 175 &= 1071 \\ 2x &= 1246 \\ x &= 623 \end{aligned}$$

Substitute 623 for x in (2) to find y.

$y = (623) - 175 = 448$

Step 5
Marvel's The Avengers grossed \$623 million, and *The Dark Knight Rises* grossed \$448 million.
Step 6
The sum of \$623 million and \$448 million is \$1071 million, and \$448 million is \$175 million less than \$623 million.

17. *Step 2*
Let $x =$ the height of the Terminal Tower and let $y =$ the height of the Key Tower.
Step 3
The total height of the two buildings is 1655 ft, so one equation is represented by the following.
$x + y = 1655 \quad (1)$
The Terminal Tower is 239 ft shorter than the Key Tower, so another equation is the following.
$x = y - 239 \quad (2)$
Step 4
Substitute $y - 239$ for x in equation (1).

$$\begin{aligned} x + y &= 1655 \\ (y - 239) + y &= 1655 \\ 2y - 239 &= 1655 \\ 2y &= 1894 \\ y &= 947 \end{aligned}$$

Substitute 947 for y in (2) to find x.
$x = (947) - 239 = 708$
Step 5
The Terminal Tower 708 ft tall, and the Key Tower is 947 ft tall.
Step 6
The sum of 708 ft and 947 ft is 1655 ft, and 708 ft is 239 ft shorter than 947 ft.

19. *Step 1*
Read the problem carefully, and assign variables.
Step 2
Let $x =$ the length (in yards) and let $y =$ the width (in yards).
Step 3
For the first equation, express the length in terms of the width.
$x = 38 + y \qquad (1)$
For the second equation, the perimeter of a rectangle equals twice the length plus twice the width.
$2x + 2y = 188 \qquad (2)$
Step 4
Substitute $38 + y$ for x in equation (2).

$$\begin{aligned} 2x + 2y &= 188 \\ 2(38 + y) + 2y &= 188 \\ 76 + 2y + 2y &= 188 \\ 76 + 4y &= 188 \\ 4y &= 112 \\ y &= 28 \end{aligned}$$

Substitute 28 for y in (1) to find x.
$x = 38 + (28) = 66$
Step 5
The length of the playing field is 66 yd, and the width is 28 yd.
Step 6
The sum of twice the length and twice the width is 188 yd, and the length is 38 yd longer than the width. The solution satisfies the conditions of the problem.

21. **(a)** $C = 85x + 900$; $R = 105x$; No more than 38 units can be sold. To find the break-even quantity, let $C = R$.

$$\begin{aligned} 85x + 900 &= 105x \\ 900 &= 20x \\ x &= \frac{900}{20} = 45 \end{aligned}$$

The break-even quantity is 45 units.

(b) Since no more than 38 units can be sold, do not produce the product (since $38 < 45$). The product will lead to a loss.

23. Multiply the Number of Coins by the Value per Coin to get the Total Value in the table.

Number of Coins	Value per Coin	Total Value
x	\$0.25	\$0.25x
y	\$0.10	\$0.10y
39	XXXXXX	\$7.50

Step 3
From the first and third columns of the table, we obtain the following system.

$$\begin{aligned} x + y &= 39 \quad (1) \\ 0.25x + 0.10y &= 7.50 \quad (2) \end{aligned}$$

Step 4
To solve this system by the elimination method, multiply equation (1) by -10, equation (2) by 100, and add the resulting equations.

$$\begin{aligned} -10x - 10y &= -390 \\ 25x + 10y &= 750 \\ \hline 15x &= 360 \\ x &= 24 \end{aligned}$$

To find y, let $x = 24$ for y in equation (1).

$$\begin{aligned} x + y &= 39 \\ 24 + y &= 39 \\ y &= 15 \end{aligned}$$

Step 5
Jonathan had 24 quarters and 15 dimes.
Step 6
24 quarters and 15 dimes give us 39 coins worth $7.50.

25. *Step 2*
Let $x =$ the number of *Iron Man 3* DVDs and $y =$ the number of *The Hunger Games: Catching Fire* Blue-ray discs.

Type of Gift	Number Bought	Cost of Each (in dollars)	Total Value
DVD	x	$14.95	$14.95x$
Blue-ray Disc	y	$16.88	$16.88y$
Totals	7	XXXXXX	$114.30

Step 3
From the second and fourth columns of the table, we obtain the following system.

$$\begin{aligned} x + y &= 7 \quad (1) \\ 14.95x + 16.88y &= 114.30 \quad (2) \end{aligned}$$

Step 4
Multiply equation (1) by -14.95, and add the result to equation (2).

$$\begin{aligned} -14.95x - 14.95y &= -104.65 \\ 14.95x + 16.88y &= 114.30 \\ \hline 1.93y &= 9.65 \\ y &= 5 \end{aligned}$$

From equation (1) with $y = 5$, $x = 2$.
Step 5
She bought 2 *Iron Man 3* DVDs and 5 *The Hunger Games: Catching Fire* Blue-ray discs.
Step 6
Two $14.95 DVDs and five $16.88 Blue-ray discs give us 7 gifts worth $114.30.

27. *Step 2*
Let $x =$ the amount invested at 5%; let $y =$ the amount invested at 4%.

Amount Invested	Interest Rate (as a decimal)	Interest Income (yearly)
x	0.05	$0.05x$
y	0.04	$0.04y$
XXXXX	XXXXXX	$350

Step 3
Karen has invested twice as much at 5% as at 4%, represented by the following.

$x = 2y \quad (1)$

Her total interest income is $350, represented by the following.

$0.05x + 0.04y = 350 \quad (2)$

Step 4
Solve the system by substitution. Substitute $2y$ for x in equation (2).

$$\begin{aligned} 0.05(2y) + 0.04y &= 350 \\ 0.14y &= 350 \\ y &= \frac{350}{0.14} = 2500 \end{aligned}$$

Substitute 2500 for y in equation (1).

$$\begin{aligned} x &= 2y \\ x &= 2(2500) = 5000 \end{aligned}$$

Step 5
Karen has $5000 invested at 5% and $2500 invested at 4%.
Step 6
$5000 is twice as much as $2500. 5% of $5000 is $250 and 4% of $2500 is $100, which is a total of $350 in interest.

29. *Step 2*
Let $x =$ the average ticket cost for Madonna; let $y =$ the average ticket cost for Bruce Springsteen and the E Street Band.

Average Ticket Cost	Number of Tickets Bought	Number of Tickets Bought
x	6	$6x$
y	5	5y
	XXXXXX	1300

Step 3
Six tickets for Madonna plus five tickets for Springsteen cost \$1300, so the following equation applies.
$6x+5y=1300 \quad (1)$
Three tickets for Madonna plus four tickets for Springsteen cost \$788, shown in the following equation.
$3x+4y=788 \quad (2)$
Step 4
Multiply (2) by -2, and add the results.

$$\begin{array}{rcr} 6x + 5y &=& 1300 \\ -6x - 8y &=& -1576 \\ \hline -3y &=& -276 \\ y &=& 92 \end{array}$$

Substitute $y=92$ in (1).

$$\begin{aligned} 6x+5(92)&=1300 \\ 6x+460&=1300 \\ 6x&=840 \\ x&=140 \end{aligned}$$

Step 5
The average ticket cost for Madonna is \$140 and the average ticket cost for Springsteen is \$92.
Step 6
Six tickets (on average) for Madonna cost \$840, and five tickets for Springsteen cost \$460, a sum of \$1300. Three tickets for Madonna cost \$420, and four tickets for Springsteen cost \$368, a sum of \$788.

30

31. *Step 2*
Let $x=$ the amount of 40% dye solution and $y=$ the amount of 70% dye solution.

Liters of Solution	Percent (as a decimal)	Liters of Pure Dye
x	0.40	$0.40x$
y	0.70	$0.70y$
120	0.50	$0.50(120)=60$

Step 3
The total number of liters in the final mixture is 120, so the following is true.
$x+y=120 \quad (1)$
The amount of pure dye in the 40% solution added to the amount of pure dye in the 70% solution is equal to the amount of pure dye in the 50% mixture, so $0.40+0.70y=60. \quad (2)$
Multiply equation (2) by 10 to clear decimals.
$4x+7y=600 \quad (3)$
We now have the following system.

$$\begin{aligned} x+y&=120 \quad (1) \\ 4x+7y&=600 \quad (3) \end{aligned}$$

Step 4
Solve this system by the elimination method.

$$\begin{array}{rcrl} -4x - 4y &=& -480 & \text{Multiply (1) by } -4. \\ 4x + 7y &=& 600 & \\ \hline 3y &=& 120 & \\ y &=& 40 & \end{array}$$

$x+40=120 \qquad$ Let $y=40$ in (1).
$x=120-40=80$

Step 5
80 liters of 40% solution should be mixed with 40 liters of 70% solution.
Step 6
$80+40=120$
$0.40(80)+0.70(40)=60$
Therefore, this mixture will give the 120 liters of 50 percent solution, as required in the original problem.

33. *Step 2*
Let $x=$ the number of pounds of coffee worth \$6 per pound; let $y=$ the number of pounds of coffee worth \$3 per pound. Complete the table given in the textbook.

Pounds	Dollars per Pound	Cost
x	6	$6x$
y	3	$3y$
90	4	$90(4)=\$360$

Step 3
The mixture contains 90 pounds, so the following equation applies.
$x+y=90 \quad (1)$
The cost of the mixture is \$360, so the following equation also applies.
$6x+3y=360 \quad (2)$
Equation (2) may be simplified by dividing each side by 3.
$2x+y=120 \quad (3)$
We now have the following system.

$$\begin{aligned} x+y&=90 \quad (1) \\ 2x+y&=120 \quad (3) \end{aligned}$$

Step 4
To solve this system by the elimination method, multiply equation (1) by -1 and add the result to equation (3).

$$\begin{array}{rcrcl} -x & - & y & = & -90 \\ 2x & + & y & = & 120 \\ \hline x & & & = & 30 \end{array}$$

From equation (1), the following is known.
$y = 60$

Step 5
Deoraj will need to mix 30 pounds of coffee at \$6 per pound with 60 pounds at \$3 per pound.

Step 6
$30 + 60 = 90$
$\$6(30) + \$3(60) = \$360$

Therefore, this mixture will give the 90 pounds worth \$4 per pound, as required in the original problem.

35. *Step 2*
Let $x =$ the number of pounds of nuts selling for \$6 per pound; let $y =$ the number of pounds of raisins selling for \$3 per pound. Complete a table.

Pounds	Dollars per Pound	Cost
x	6	$6x$
y	3	$3y$
60	5	$60(5) = \$300$

Step 3
The mixture contains 60 pounds, so the following equation applies.
$x + y = 60 \quad (1)$

The cost of the mixture is \$300, so the following also applies.
$6x + 3y = 300 \quad (2)$

Equation (2) may be simplified by dividing each side by 3.
$2x + y = 100 \quad (3)$

We now have the following system.
$x + y = 60 \quad (1)$
$2x + y = 100 \quad (3)$

Step 4
To solve this system by the elimination method, multiply equation (1) by -1 and add the result to equation (3).

$$\begin{array}{rcrcl} -x & - & y & = & -60 \\ 2x & + & y & = & 100 \\ \hline x & & & = & 40 \end{array}$$

From equation (1), the following is known.
$y = 20$

Step 5
Theresa will need to mix 40 pounds of nuts at \$6 per pound with 20 pounds of raisins at \$3 per pound.

Step 6
$40 + 20 = 60$
$\$6(40) + \$3(20) = \$300,$

Therefore, this mixture will give the 60 pounds worth \$5 per pound, as required in the original problem.

37. *Step 2*
Let $x =$ the average rate of the slower train; let $y =$ the average rate of the faster train.

	r	t	d
Slower Train	x	4.5	$4.5x$
Faster Train	y	4.5	$4.5y$

Step 3
The total distance is 495 miles, so the following equation applies.
$4.5x + 4.5y = 495 \quad (1)$

Or, dividing equation (1) by 4.5, it can be represented by the following.
$x + y = 110 \quad (2)$

The faster train traveled 10 mph faster than the slower train, so the following also applies.
$y = x + 10 \quad (3)$

Step 4
From equation (3), substitute $x + 10$ for y in (2).

$$\begin{aligned} x + (x + 10) &= 110 \\ 2x + 10 &= 110 \\ 2x &= 100 \\ x &= 50 \end{aligned}$$

From equation (3) with $x = 50$,
$y = 50 + 10 = 60.$

Step 5
The average rate of the slower train was 50 mph, and the average rate of the faster train was 60 mph.

Step 6
60 mph is 10 mph faster than 50 mph. The slower train traveled $4.5(50) = 225$ miles, and the faster train traveled $4.5(60) = 270$ miles. The sum is $225 + 270 = 495,$ as required.

39. *Step 2*
Let $x =$ the average rate of the slower car; let $y =$ the average rate of the faster car.

	r	t	d
Slower Car	x	6	$6x$
Faster Car	y	6	$6y$

Step 3
The total distance is 600 miles, so the following equation applies.
$6x + 6y = 600$ (1)
Or, dividing equation (1) by 6, it can be represented by the following.
$x + y = 100$ (2)
The faster car travels 30 miles per hour faster than the slower car, so the following also applies.
$y = x + 30$ (3)
Step 4
From equation (3), substitute $x + 30$ for y in (2).

$$x + (x + 30) = 100$$
$$2x = 70$$
$$x = 35$$

From equation (3), $y = 65.$
Step 5
The faster car travels at 65 miles per hour, and the slower car travels at 35 miles per hour.
Step 6
65 miles per hour is 30 miles per hour faster than 35 miles per hour. The slower car travels $35(6) = 210$ miles, and the faster car travels $65(6) = 390$ miles. The total distance traveled is $210 + 390 = 600,$ as required.

41. *Step 2*
Let $x =$ the average rate of the bicycle; let $y =$ the average rate of the car.

	r	t	d
Bicycle	x	6	$6x$
Car	y	6	$6y$

Step 3
The total distance is 405 miles, so the following equation applies.
$6x + 6y = 405$ (1)
Or, dividing equation (1) by 6, it can be represented by the following.
$x + y = 67.5$ (2)
The car traveled 40 mph faster than the bicycle, so the following also applies.
$y = x + 40$ (3)
Step 4
From equation (3), substitute $x + 40$ for y in (2).

$$x + (x + 40) = 67.5$$
$$2x + 40 = 67.5$$
$$2x = 27.5$$
$$x = 13.75$$

From equation (3) with $x = 13.75,$
$y = (13.75) + 40 = 53.75.$
Step 5
The average rate of the bicycle was 13.75 mph, and the average rate of the car was 53.75 mph.
Step 6
53.75 mph is 40 mph faster than 13.75 mph. The bicycle traveled $6(13.75) = 82.5$ miles, and the car traveled $6(53.75) = 322.5$ miles. The sum is $82.5 + 333.5 = 405,$ as required.

43. *Step 2*
Let $x =$ the rate of the plane in still air; let $y =$ the wind speed.

	r	t	d
Rate of the Plane	x	1	x
Wind Speed	y	1	y

Step 3
The distance traveled by the the plane with the wind after an hour is represented by the following.
$x + y = 500$ (1)

The distance traveled by the plane into the wind after an hour is represented by the following.
$x - y = 440$ (2)

Step 4
To solve the system by the elimination method, add equations (1) and (2).

$$\begin{aligned} x + y &= 500 \\ x - y &= 440 \\ \hline 2x \quad &= 940 \\ x &= 470 \end{aligned}$$

From equation (1), the following is true.
$y = 30$

Step 5
The wind speed is 30 miles per hour; the rate of the plane in still air is 470 miles per hour.

Step 6
The plane travels $470 + 30 = 500$ miles per hour with the wind and $470 - 30 = 440$ miles per hour into the wind, as required.

45. *Step 2*
Let $x =$ the rate of the boat in still water; let $y =$ the rate of the current.

	r	***t***	***d***
Downstream	$x+y$	3	$3(x+y)$
Upstream	$x-y$	3	$3(x-y)$

Step 3
Use the formula $d = rt$ and the completed table to write the system of equations.
$3(x+y) = 36$ (1) Distance downstream
$3(x-y) = 24$ (2) Distance upstream

Step 4
Divide equations (1) and (2) by 3.
$x + y = 12$ (3)
$x - y = 8$ (4)
Now add equations (3) and (4).

$$\begin{aligned} x + y &= 12 \quad (3) \\ x - y &= 8 \quad (4) \\ \hline 2x \quad &= 20 \quad \text{Add.} \\ x &= 10 \end{aligned}$$

From equation (3), $y = 2$.

Step 5
The rate of the current is 2 miles per hour; the rate of the boat in still water is 10 miles per hour.

Step 6
Traveling downstream for 3 hours at $10 + 2 = 12$ miles per hour gives us a 36-mile trip. Traveling upstream for 3 hours at $10 - 2 = 8$ miles per hour gives us a 24-mile trip, as required.

47. *Step 2*
Let $x =$ Yady's rate; let $y =$ Dane's rate.

Step 3
Use the formula $d = rt$ to complete two tables.

(Riding in same direction)	***r***	***t***	***d***
Yady	x	6	$6x$
Dane	y	6	$6y$

Yady rode 30 miles farther than Dane.
$6x = 6y + 30$
Or, after dividing the equation by 6, the following is also true.
$x - y = 5$ (1)

(Riding toward each other)	***r***	***t***	***d***
Yady	x	1	$1x$
Dane	y	1	$1y$

Yady and Dane rode a total of 30 miles, so the following applies.
$x + y = 30$ (2)
We have the following system.
$x - y = 5$ (1)
$x + y = 30$ (2)

Step 4
To solve the system by the elimination method, add equations (1) and (2).

$$\begin{aligned} x - y &= 5 \\ x + y &= 30 \\ \hline 2x \quad &= 35 \\ x &= 17.5 \end{aligned}$$

From equation (2) with $x = 17.5$, $y = 12.5$.

Step 5
Yady's rate is 17.5 miles per hour, and Dane's rate is 12.5 miles per hour.

Step 6
Riding in the same direction, Yady rides $6(17.5) = 105$ miles, and Dane rides $6(12.5) = 75$ miles. The difference is 30 miles, as required. Riding toward each other, Yady rides $1(17.5) = 17.5$ miles and Dane rides $1(12.5) = 12.5$ miles. The sum is 30 miles, as required.

4.5 Solving Systems of Linear Inequalities

Now Try Exercises

N1. The graph of $4x-2y=8$ has intercepts (2, 0) and $(0,-4)$. The graph of $x+3y=3$ has intercepts (3, 0) and (0, 1). Both are graphed as solid lines because of the ≤ and ≥ signs. Use (0, 0) as a test point in each case.

$4x-2y\le 8$

$4(0)-2(0)\le 8$ Let $x=0$ and $y=0$.

$0\le 8$ True

Shade the side of the graph for $4x-2y=8$ that contains (0, 0).

$x-3y\ge 3$

$0+3(0)\ge 3$ Let $x=0$ and $y=0$.

$0\ge 3$ False

Shade the side of the graph for $x+3y=3$ that does not contain (0, 0).

The solution set of this system is the intersection (overlap) of the two shaded regions, and includes the portions of the boundary lines that bound this region.

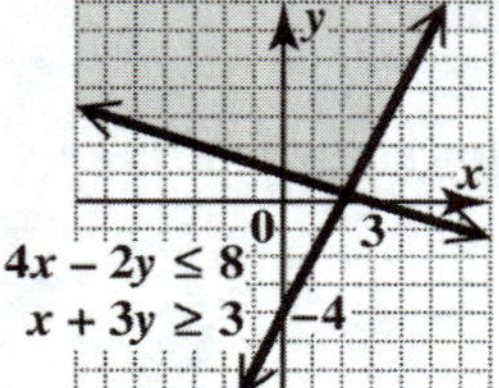

N2. Graph $2x+5y=10$ as a dashed line through its intercepts, $(5,0)$ and $(0,2)$. Use $(0,0)$ as a test point.

$2x+5y>10$

$2(0)+5(0)>10$ Let $x=0$ and $y=0$.

$0>10$ False

The solution is the region that does not include (0, 0).

Then graph $x-2y=0$ as a dashed line through the points $(0,0)$ and $(2,1)$. Use $(0,3)$ as a test point.

$x-2y<0$

$0-2(3)<0$ Let $x=0$ and $y=3$.

$-6<0$ True

The solution is the region that includes $(0,3)$.

The solution set of the system is the intersection of the two shaded regions. Because the inequality signs are $>$ and $<$, the solution set does not include the boundary lines.

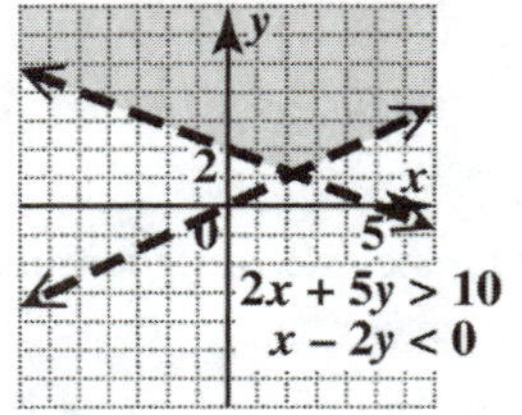

N3. Graph $x-y=2$ as a dashed line through its intercepts, $(2,0)$ and $(0,-2)$. Use $(0,0)$ as a test point to get $0<2$, a true statement, so shade above the line.

Recall that $x=-2$ is a vertical line through the point $(-2,0)$, and $y=4$ is a horizontal line through the point $(0,4)$. Graph these as solid lines. Shade to the right of $x=-2$ and below $y=4$.

The solution set of this system is the intersection (overlap) of the three shaded regions, and includes the portions of the horizontal and vertical boundary lines that bound this region.

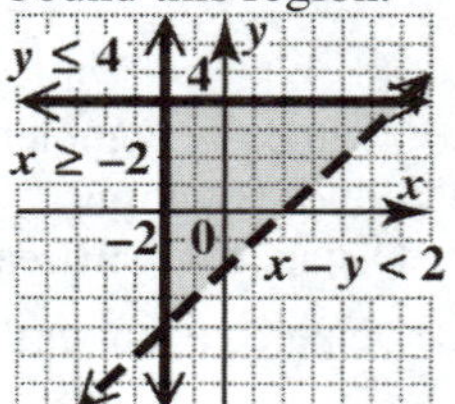

Exercises

1. $x\ge 5$ is the region to the right of the vertical line $x=5$ and includes the line. $y\le -3$ is the region below the horizontal line $y=-3$ and includes the line. The correct choice is C.

3. $x>5$ is the region to the right of the vertical line $x=5$. $y<-3$ is the region below the horizontal line $y=-3$. The correct choice is B.

5. **(a)** $x-3y \le 6$ $\quad$ $x \ge -4$

$(-5)-3(-4) \overset{?}{\le} 6$ $\quad$ $-5 \ge -4$ False

$7 \le 6$ False

So, the ordered pair $(-5,-4)$ *is not* a solution to the system.

(b) $x-3y \le 6$ $\quad$ $x \ge -4$

$(0)-3(-4) \overset{?}{\le} 6$ $\quad$ $0 \ge -4$ True

$12 \le 6$ False

So, the ordered pair $(0,-4)$ *is not* a solution to the system.

(c) $x-3y \le 6$ $\quad$ $x \ge -4$

$(0)-3(0) \overset{?}{\le} 6$ $\quad$ $0 \ge -4$ True

$0 \le 6$ True

So, the ordered pair $(0,0)$ *is* a solution to the system.

The shaded region of the system is shown.

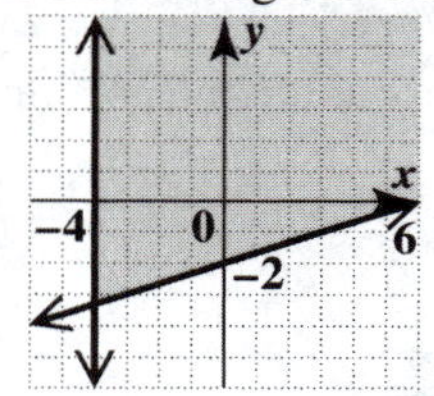

7. **(a)** $x+y>4$ $\quad$ $5x-3y<15$

$0+0 \overset{?}{>} 4$ $\quad$ $5(0)-3(0) \overset{?}{<} 15$

$0>4$ False $\quad$ $0 \le 15$ True

The ordered pair $(0,0)$ *is not* a solution to the system.

(b) $x+y>4$ $\quad$ $5x-3y<15$

$3+3 \overset{?}{>} 4$ $\quad$ $5(3)-3(3) \overset{?}{<} 15$

$6>4$ True $\quad$ $6 \le 15$ True

The ordered pair $(3,3)$ *is* a solution to the system.

(c) $x+y>4$ $\quad$ $5x-3y<15$

$5+0 \overset{?}{>} 4$ $\quad$ $5(5)-3(0) \overset{?}{<} 15$

$5>4$ True $\quad$ $25 \le 15$ False

The ordered pair $(5,0)$ *is not* a solution to the system.

The shaded region of the system is shown.

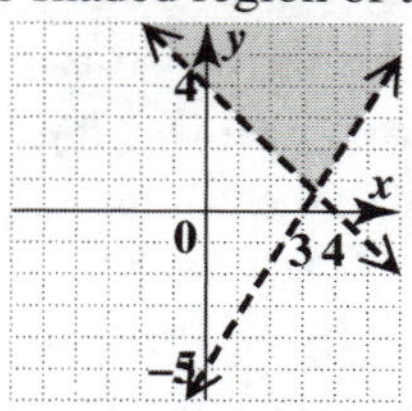

9. Graph the boundary $x+y=6$ as a solid line through its intercepts, $(6, 0)$ and $(0, 6)$. Using $(0, 0)$ as a test point will result in the true statement $0 \le 6$, so shade the region containing the origin.
Graph the boundary $x-y=1$ as a solid line through its intercepts, $(1, 0)$ and $(0, -1)$. Using $(0, 0)$ as a test point will result in the false statement $0 \ge 1$, so shade the region *not* containing the origin.
The solution set of this system is the intersection (overlap) of the two shaded regions, and includes the portions of the boundary lines that bound this region.

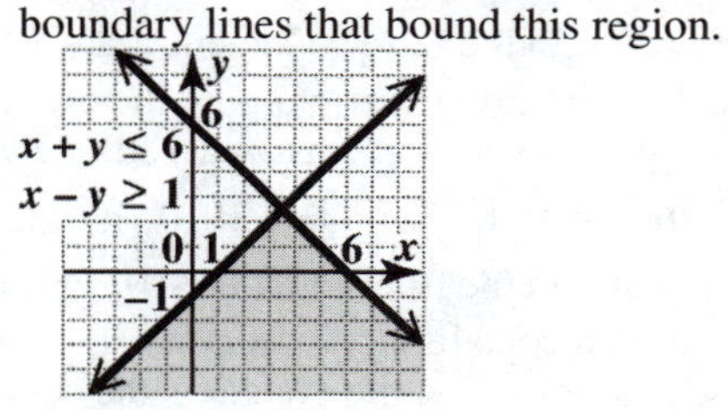

11. Graph the boundary $4x+5y=20$ as a solid line through its intercepts, $(5, 0)$ and $(0, 4)$. Using $(0, 0)$ as a test point will result in the false statement $0 \ge 20$, so shade the region *not* containing the origin.
Graph the boundary $x-2y=5$ as a solid line through $(5, 0)$ and $(1, -2)$. Using $(0, 0)$ as a test point will result in the true statement $0 \le 5$, so shade the region containing the origin.
The solution set of this system is the intersection of the two shaded regions, and includes the portions of the boundary lines that bound the region.

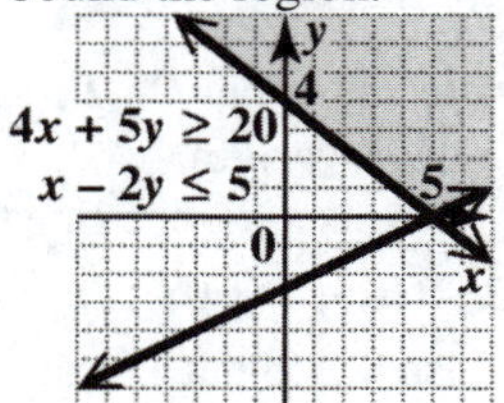

13. Graph $2x+3y=6$ as a dashed line through $(3, 0)$ and $(0, 2)$. Using $(0, 0)$ as a test point will result in the true statement $0<6$, so shade the region containing the origin.
Now graph $x-y=5$ as a dashed line through $(5, 0)$ and $(0, -5)$. Using $(0, 0)$ as a test point will result in the true statement $0<5$, so shade the region containing the origin.
The solution set of the system is the intersection of the two shaded regions. Because the inequality signs are both <, the solution set does not include the boundary lines.

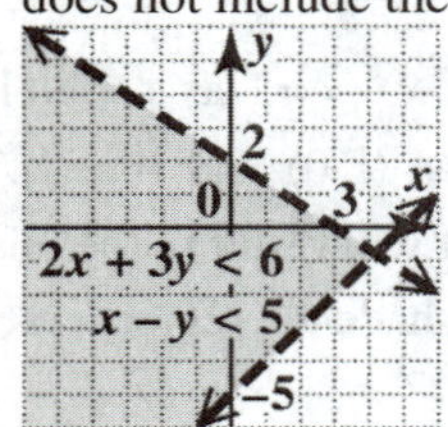

15. Graph $y=2x-5$ as a solid line through $(0, -5)$ and $(3, 1)$. Using $(0, 0)$ as a test point will result in the false statement $0 \le -5$, so shade the region *not* containing the origin.
Now graph $x=3y+2$ as a dashed line through $(2, 0)$ and $(-1, -1)$. Using $(0, 0)$ as a test point will result in the true statement $0<2$, so shade the region containing the origin.
The solution set of the system is the intersection of the two shaded regions. It includes the portion of the line $y=2x-5$ that bounds the region, but not the portion of the line $x=3y+2$.

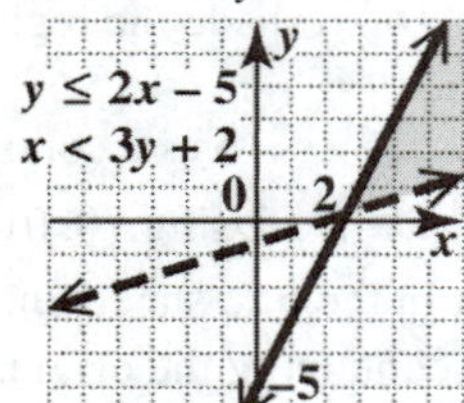

17. Graph $4x+3y=6$ as a dashed line through $\left(\frac{3}{2}, 0\right)$ and $(0, 2)$. Using $(0, 0)$ as a test point will result in the true statement $0<6$, so shade the region containing the origin.
Now graph $x-2y=4$ as a dashed line through $(4, 0)$ and $(0, -2)$. Using $(0, 0)$ as a test point will result in the false statement $0>4$, so shade the region *not* containing the origin.

The solution set of the system is the intersection of the two shaded regions. It does not include the boundary lines.

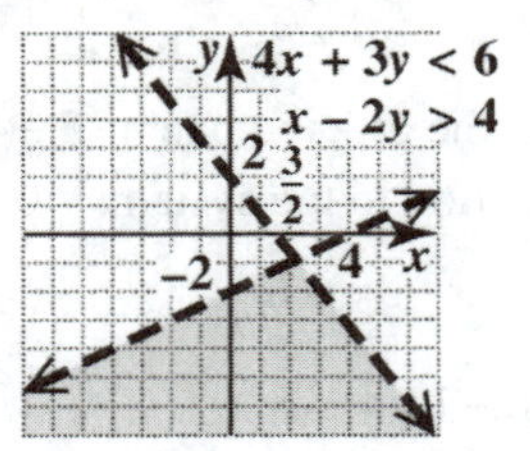

19. Graph $x=2y+3$ as a solid line through $(3, 0)$ and $(7, 2)$. Using $(0, 0)$ as a test point will result in the true statement $0 \le 3$, so shade the region containing the origin.
Now graph $x+y=0$ as a dashed line through $(0, 0)$ and $(1, -1)$. Using $(1, 0)$ as a test point will result in the false statement $1<0$, so shade the region *not* containing $(1, 0)$.
The solution set of the system is the intersection of the two shaded regions. It includes the portion of the line $x=2y+3$ that bounds the region, but not the portion of the line $x+y=0$.

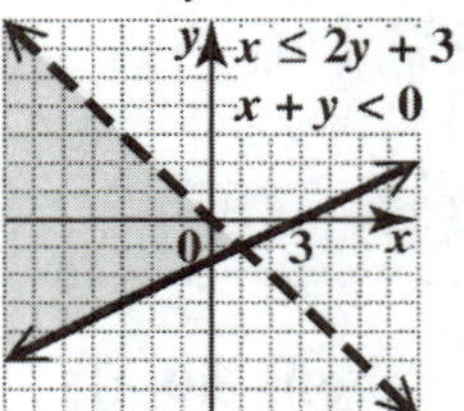

21. Graph the boundary $x-3y=6$ as a solid line through its intercepts, $(6, 0)$ and $(0, -2)$. Using $(0, 0)$ as a test point will result in the true statement $0 \le 6$, so shade the region containing the origin.
Graph the boundary $x=-5$ as a solid, vertical line through $(-5, 0)$. Using $(0, 0)$ as a test point will result in the true statement $0 \ge -5$, so shade the region containing the origin.
The solution set of this system is the intersection of the two shaded regions, and includes the portions of the boundary lines that bound the region.

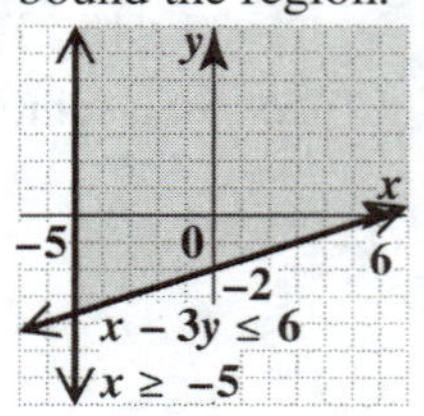

23. Graph $-3x + y = 1$ as a solid line through $\left(-\frac{1}{3}, 0\right)$ and $(0, 1)$. Using $(0, 0)$ as a test point will result in the false statement $0 \geq 1$, so shade the region *not* containing the origin. This is the region above the line.
Now graph $6x - 2y = -10$ as a solid line through $\left(-\frac{5}{3}, 0\right)$ and $(0, 5)$. Using $(0, 0)$ as a test point will result in the true statement $0 \geq -10$, so shade the region containing the origin. This is the region below the line.
The solution set of the system is the intersection of the two shaded regions. This is the region between the two parallel lines (both lines have slope 3). These boundary lines are included in the solution set.

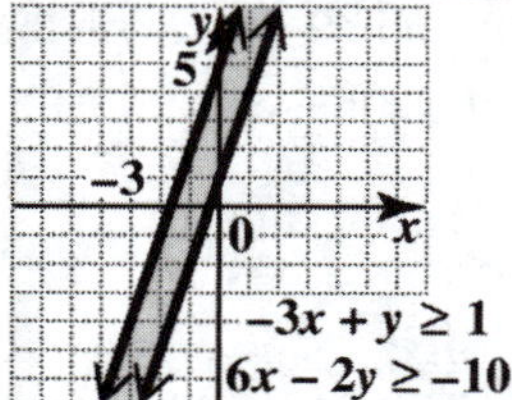

25. Graph $2x + 3y = 6$ as a dashed line through $(3, 0)$ and $(0, 2)$. Using $(0, 0)$ as a test point will result in the true statement $0 < 6$, so shade the region containing the origin. This is the region below the line.
Now graph $4x + 6y = 18$ as a dashed line through $\left(\frac{9}{2}, 0\right)$ and $(0, 3)$. Using $(0, 0)$ as a test point will result in the false statement $0 > 18$, so shade the region *not* containing the origin. This is the region above the line.
Note that the two boundary lines are parallel (both lines have slope $-\frac{2}{3}$) and that the shaded regions do not overlap. Therefore, the system of inequalities has no solution set.

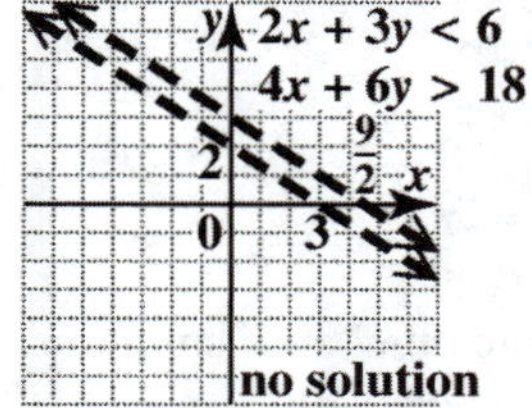

27. Graph $4x + 5y = 8$, $y = -2$, and $x = -4$ as dashed lines. All three inequalities are true for $(0, 0)$. Shade the region bounded by the three lines, which contains the test point $(0, 0)$.

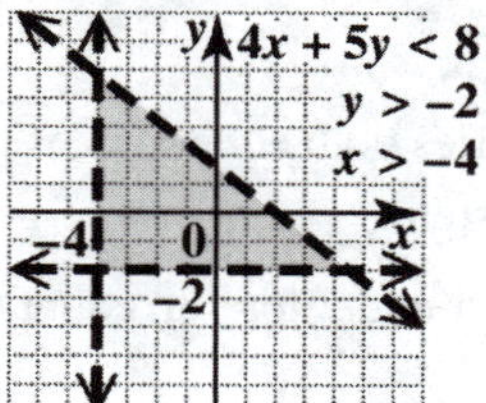

29. Graph $x + y = -3$, $x - y = 3$, and $y = 3$ as solid lines. All three inequalities are true for $(0, 0)$. Shade the region bounded by the three lines, which contains the test point $(0, 0)$.

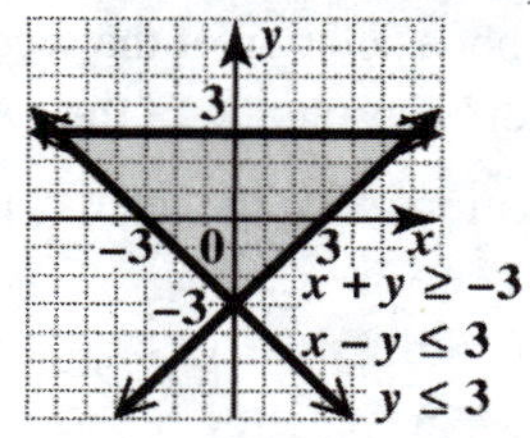

31. Graph $3x - 2y = 6$, $x + y = 4$, $x = 0$, and $y = -4$ as solid lines. All four inequalities are true for $(2, -2)$. Shade the region bounded by the four lines, which contains the test point $(2, -2)$.

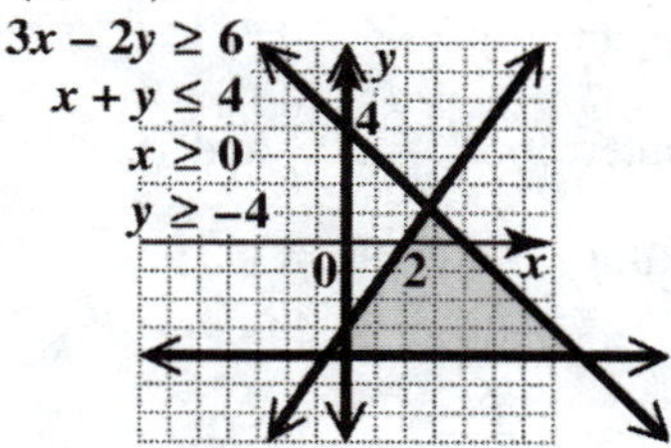

Chapter 4 Review Exercises

1. To decide whether $(3, 4)$ is a solution of the system, substitute 3 for *x* and 4 for *y* in each equation.

$$4x - 2y = 4$$
$$4(3) - 2(4) \stackrel{?}{=} 4$$
$$12 - 8 \stackrel{?}{=} 4$$
$$4 = 4 \quad \text{True}$$

$$5x + y = 19$$
$$5(3) + 4 \stackrel{?}{=} 19$$
$$15 + 4 \stackrel{?}{=} 19$$
$$19 = 19 \quad \text{True}$$

Since $(3, 4)$ satisfies both equations, it is a solution of the system.

2. Substitute -5 for x and 2 for y in equation (2).

$$2x + 3y = 4$$
$$2(-5) + 3(2) \stackrel{?}{=} 4$$
$$-10 + 6 \stackrel{?}{=} 4$$
$$-4 = 4 \quad \text{False}$$

Since $(-5, 2)$ is not a solution of the second equation, it cannot be a solution of the system.

3. To graph the equations, find the intercepts of $x + y = 4$.

Let $y = 0$; then $x = 4$.

Let $x = 0$; then $y = 4$.

Plot the intercepts, $(4, 0)$ and $(0, 4)$, and draw the line through them.

Find the intercepts of $2x - y = 5$.

Let $y = 0$; then $x = \frac{5}{2}$.

Let $x = 0$; then $y = -5$.

Plot the intercepts, $\left(\frac{5}{2}, 0\right)$ and $(0, -5)$, and draw the line through them.

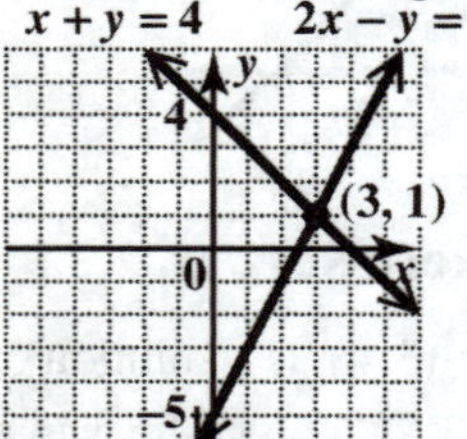

It appears that the lines intersect at the point $(3, 1)$. Check this by substituting 3 for x and 1 for y in both equations. Since $(3, 1)$ satisfies both equations, the solution set of this system is $\{(3, 1)\}$.

4. To graph the equations, find the intercepts of $x - 2y = 4$.

Let $y = 0$; then $x = 4$.

Let $x = 0$; then $y = -2$.

Plot the intercepts, $(4, 0)$ and $(0, -2)$, and draw the line through them.

Find the intercepts of $2x + y = -2$.

Let $y = 0$; then $x = -1$.

Let $x = 0$; then $y = -2$.

Plot the intercepts, $(-1, 0)$ and $(0, -2)$, and draw the line through them.

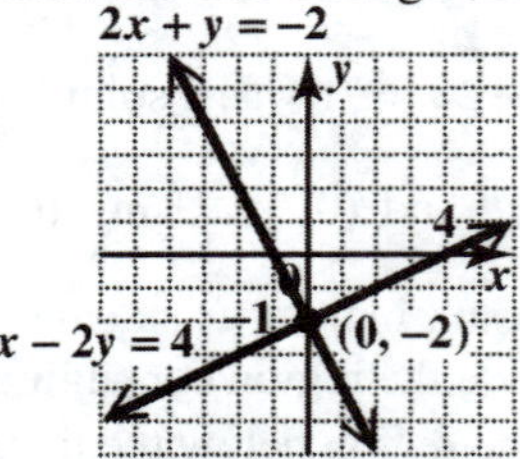

The lines intersect at their common y-intercept, $(0, -2)$, so $\{(0, -2)\}$ is the solution set of the system.

5. Graph the line $2x + 4 = 2y$ through its intercepts, $(-2, 0)$ and $(0, 2)$. Graph the line $y - x = -3$ through its intercepts, $(3, 0)$ and $(0, -3)$.

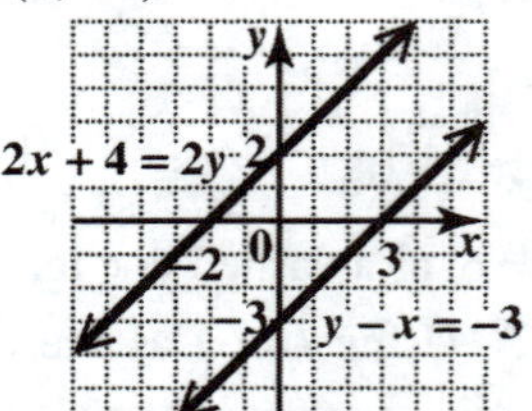

Solving each equation for y gives us $y = x + 2$ and $y = x - 3$, so the lines are parallel (both lines have slope 1). Since they do not intersect, there is no solution. This is an inconsistent system and the solution set is $\varnothing$.

6. Graph the line $x - 2 = 2y$ through its intercepts, $(2, 0)$ and $(0, -1)$. Graph the line $2x - 4y = 4$ through its intercepts, $(2, 0)$ and $(0, -1)$.

Since both equations have the same intercepts, they are equations of the same line.

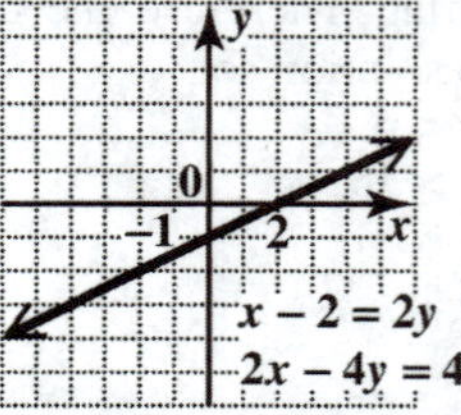

There is an infinite number of solutions. The equations are dependent equations and the solution set contains an infinite number of ordered pairs. The solution set is $\{(x, y) \mid x - 2y = 2\}$.

7. It would be easiest to solve for x in the second equation because its coefficient is -1. No fractions would be involved.

8. The true statement $0 = 0$ is an indication that the system has an infinite number of solutions. Write the solution set using set-builder notation and the equation of the system that is in standard form with integer coefficients having greatest common factor 1.

9. $3x + y = 7 \quad (1)$

$x = 2y \quad (2)$

Substitute $2y$ for x in equation (1) and solve the resulting equation for y.

$$\begin{aligned} 3x + y &= 7 \\ 3(2y) + y &= 7 \\ 6y + y &= 7 \\ 7y &= 7 \\ y &= 1 \end{aligned}$$

To find x, let $y = 1$ in equation (2).

$x = 2y$

$x = 2(1) = 2$

The solution set is $\{(2, 1)\}$.

10. $2x - 5y = -19 \quad (1)$

$y = x + 2 \quad (2)$

Substitute $x + 2$ for y in equation (1).

$$\begin{aligned} 2x - 5y &= -19 \\ 2x - 5(x+2) &= -19 \\ 2x - 5x - 10 &= -19 \\ -3x - 10 &= -19 \\ -3x &= -9 \\ x &= 3 \end{aligned}$$

To find y, let $x = 3$ in equation (2).

$y = x + 2$

$y = 3 + 2 = 5$

The solution set is $\{(3, 5)\}$.

11. $4x + 5y = 44 \quad (1)$

$x + 2 = 2y \quad (2)$

Solve equation (2) for x.

$x + 2 = 2y$

$x = 2y - 2 \quad (3)$

Substitute $2y - 2$ for x in equation (1) and solve the resulting equation for y.

$$\begin{aligned} 4x + 5y &= 44 \\ 4(2y - 2) + 5y &= 44 \\ 8y - 8 + 5y &= 44 \\ 13y - 8 &= 44 \\ 13y &= 52 \\ y &= 4 \end{aligned}$$

To find x, let $y = 4$ in equation (3).

$x = 2y - 2$

$x = 2(4) - 2 = 6$

The solution set is $\{(6, 4)\}$.

12. $5x + 15y = 30 \quad (1)$

$x + 3y = 6 \quad (2)$

Solve equation (2) for x.

$x + 3y = 6$

$x = 6 - 3y \quad (3)$

Substitute $6 - 3y$ for x in equation (1).

$$\begin{aligned} 5x + 15y &= 30 \\ 5(6 - 3y) + 15y &= 30 \\ 30 - 15y + 15y &= 30 \\ 30 &= 30 \quad \text{True} \end{aligned}$$

This true result means that every solution of one equation is also a solution of the other, so the system has an infinite number of solutions. The solution set is $\{(x, y) \mid x + 3y = 6\}$.

13. If we simply add the given equations without first multiplying one or both equations by a constant, choice C is the only system in which a variable will be eliminated. If we add the equations in C we get $3x = 17$. (The variable y was eliminated.)

14. $2x + 12y = 7 \quad (1)$

$3x + 4y = 1 \quad (2)$

(a) If we multiply equation (1) by -3, the first term will become $-6x$. To eliminate x, we need to change the first term on the left side of equation (2) from $3x$ to $6x$. In order to do this, we must multiply equation (2) by 2.

(b) If we multiply equation (1) by -3, the second term will become $-36y$. To eliminate y, we need to change the second term on the left side of equation (2) from $4y$ to $36y$. In order to do this, we must multiply equation (2) by 9.

15. $2x - y = 13$ (1)

$x + y = 8$ (2)

$3x = 21$ Add (1) and (2).

$x = 7$

From equation (2), $y = 1$.

The solution set is $\{(7, 1)\}$.

16. $-4x + 3y = 25$ (1)

$6x - 5y = -39$ (2)

Multiply equation (1) by 3 and equation (2) by 2; then add the results.

$$\begin{aligned} -12x + 9y &= 75 \\ 12x - 10y &= -78 \\ \hline -y &= -3 \\ y &= 3 \end{aligned}$$

To find x, let $y = 3$ in equation (1).

$$\begin{aligned} -4x + 3y &= 25 \\ -4x + 3(3) &= 25 \\ -4x + 9 &= 25 \\ -4x &= 16 \\ x &= -4 \end{aligned}$$

The solution set is $\{(-4, 3)\}$.

17. $3x - 4y = 9$ (1)

$6x - 8y = 18$ (2)

Multiply equation (1) by –2, and add the result to equation (2).

$$\begin{aligned} -6x + 8y &= -18 \\ 6x - 8y &= 18 \\ \hline 0 &= 0 \quad \text{True} \end{aligned}$$

This result indicates that all solutions of equation (1) are also solutions of equation (2). The given system has an infinite number of solutions. The solution set is $\{(x, y) \mid 3x - 4y = 9\}$.

18. $2x + y = 3$ (1)

$-4x - 2y = 6$ (2)

Multiply equation (1) by 2 and add the result to equation (2).

$$\begin{aligned} 4x + 2y &= 6 \\ -4x - 2y &= 6 \\ \hline 0 &= 12 \quad \text{False} \end{aligned}$$

This result indicates that the given system has solution set $\varnothing$.

19. $2x + 3y = -5$ (1)

$3x + 4y = -8$ (2)

Multiply equation (1) by -3 and equation (2) by 2; then add the results.

$$\begin{aligned} -6x - 9y &= 15 \\ 6x + 8y &= -16 \\ \hline -y &= -1 \\ y &= 1 \end{aligned}$$

To find x, let $y = 1$ in equation (1).

$$\begin{aligned} 2x + 3y &= -5 \\ 2x + 3(1) &= -5 \\ 2x + 3 &= -5 \\ 2x &= -8 \\ x &= -4 \end{aligned}$$

The solution set is $\{(-4, 1)\}$.

20. $6x - 9y = 0$ (1)

$2x - 3y = 0$ (2)

Multiply equation (2) by -3 and add the result to equation (1).

$$\begin{aligned} 6x - 9y &= 0 \\ -6x + 9y &= 0 \\ \hline 0 &= 0 \quad \text{True} \end{aligned}$$

This result indicates that the system has an infinite number of solutions. The solution set is $\{(x, y) \mid 2x - 3y = 0\}$.

21. $x - 2y = 5$ (1)

$y = x - 7$ (2)

From equation (2), substitute $x - 7$ for y in equation (1).

$$\begin{aligned} x - 2y &= 5 \\ x - 2(x - 7) &= 5 \\ x - 2x + 14 &= 5 \\ -x &= -9 \\ x &= 9 \end{aligned}$$

Let $x = 9$ in equation (2) to find y.

The solution set is $\{(9, 2)\}$.

22. $\frac{x}{2} + \frac{y}{3} = 7$ (1)

$\frac{x}{4} + \frac{2y}{3} = 8$ (2)

Multiply equation (1) by 6 to clear fractions.

$$6\left(\frac{x}{2} + \frac{y}{3}\right) = 6(7)$$

$3x + 2y = 42$ (3)

Multiply equation (2) by 12 to clear fractions.

$$12\left(\frac{x}{4}+\frac{2y}{3}\right)=12(8)$$

$$3x+8y=96 \quad (4)$$

To solve this system by the elimination method, multiply equation (3) by -1 and add the result to equation (4).

$$\begin{aligned} -3x - 2y &= -42 \\ 3x + 8y &= 96 \\ \hline 6y &= 54 \\ y &= 9 \end{aligned}$$

To find x, let $y=9$ in equation (3).

$$\begin{aligned} 3x+2y&=42 \\ 3x+2(9)&=42 \\ 3x+18&=42 \\ 3x&=24 \\ x&=8 \end{aligned}$$

The solution set is $\{(8, 9)\}$.

23. $\frac{3}{4}x-\frac{1}{3}y=\frac{7}{6}$ (1)

$\frac{1}{2}x+\frac{2}{3}y=\frac{5}{3}$ (2)

Multiply equation (1) by 12 to clear fractions.

$$12\left(\frac{3}{4}x\right)-12\left(\frac{1}{3}y\right)=12\left(\frac{7}{6}\right)$$

$$9x-4y=14 \quad (3)$$

Multiply equation (2) by 6 to clear fractions.

$$6\left(\frac{1}{2}x\right)+6\left(\frac{2}{3}y\right)=6\left(\frac{5}{3}\right)$$

$$3x+4y=10 \quad (4)$$

Add equations (3) and (4) to eliminate y.

$$\begin{aligned} 9x - 4y &= 14 \\ 3x + 4y &= 10 \\ \hline 12x &= 24 \\ x &= 2 \end{aligned}$$

To find y, let $x=2$ in equation (4).

$$\begin{aligned} 3(2)+4y&=10 \\ 6+4y&=10 \\ 4y&=4 \\ y&=1 \end{aligned}$$

The solution set is $\{(2, 1)\}$.

24. $0.4x-0.5y=-2.2$ (1)

$0.3x+0.2y=-0.5$ (2)

Multiply equation (1) and equation (2) by 10 to eliminate the decimals. Then solve the resulting equations by the substitution method.

$4x-5y=-22$ (3)

$3x+2y=-5$ (4)

To solve this system by the elimination method, multiply equation (3) by 2, multiply equation (4) by 5, and add the resulting equations.

$$\begin{aligned} 8x - 10y &= -44 \\ 15x + 10y &= -25 \\ \hline 23x &= -69 \\ x &= -3 \end{aligned}$$

To find y, let $x=-3$ in equation (4).

$$\begin{aligned} 3x+2y&=-5 \\ 3(-3)+2y&=-5 \\ -9+2y&=-5 \\ 2y&=4 \\ y&=2 \end{aligned}$$

The solution set is $\{(-3, 2)\}$.

25. *Step 2*

Let $x=$ the number of Domino's restaurants; let $y=$ the number of Pizza Hut restaurants.

Step 3

Pizza Hut had 1192 more locations than Domino's, so the following equation applies.

$y=x+1192$ (1)

Together, the two chains had 11,044 locations, so the following also applies.

$x+y=11{,}044$ (2)

Step 4

Substitute $x+1192$ for y in (2).

$$\begin{aligned} x+(x+1192)&=11{,}044 \\ 2x+1192&=11{,}044 \\ 2x&=9852 \\ x&=4926 \end{aligned}$$

From equation (1), the following is calculated.

$y=(4926)+1192=6118$

Step 5

In September 2013, Pizza Hut had 6118 locations, and Domino's had 4926 locations.

Step 6

6118 is 1192 more than 4926, and the sum of 6118 and 4926 is 11,044.

26. *Step 2*
Let $x =$ the circulation figure for *Reader's Digest;* let $y =$ the circulation figure for *People.*
Step 3
The total circulation was 8.7 million, so the following equation applies.
$x + y = 8.7 \quad (1)$
People's circulation was 1.7 million less than that of *Reader's Digest*, so the following is also true.
$y = x - 1.7 \quad (2)$
Step 4
Substitute $x - 1.7$ for y in equation (1).
$$\begin{aligned} x + (x - 1.7) &= 8.7 \\ 2x - 1.7 &= 8.7 \\ 2x &= 10.4 \\ x &= 5.2 \end{aligned}$$
From equation (2), the following is calculated.
$y = (5.2) - 1.7 = 3.5.$
Step 5
The average circulation for *Reader's Digest* was 5.2 million, and for *People* it was 3.5 million.
Step 6
3.5 is 1.7 less than 5.2, and the sum of 3.5 and 5.2 is 8.7.

27. *Step 2*
Let $x =$ the length of the rectangle; let $y =$ the width of the rectangle.
Step 3
The perimeter is 90 meters, so the following equation applies.
$2x + 2y = 90 \quad (1)$
The length is $1\frac{1}{2}\left(\text{or } \frac{3}{2}\right)$ times the width, so the following also applies.
$x = \frac{3}{2}y \quad (2)$
Step 4
Substitute $\frac{3}{2}y$ for x in equation (1).
$$\begin{aligned} 2x + 2y &= 90 \\ 2\left(\frac{3}{2}y\right) + 2y &= 90 \\ 3y + 2y &= 90 \\ 5y &= 90 \\ y &= 18 \end{aligned}$$
From equation (2), the following is calculated.
$x = \frac{3}{2}(18) = 27$
Step 5
The length is 27 meters and the width is 18 meters.
Step 6
27 is $1\frac{1}{2}$ times 18 and the perimeter is $2(27) + 2(18) = 90$ meters.

28. *Step 2*
Let $x =$ the number of \$20 bills; let $y =$ the number of \$10 bills.
Step 3
The total number of bills is 20, so the following equation applies.
$x + y = 20 \quad (1)$
The total value of the money is \$330, so the following also applies.
$20x + 10y = 330 \quad (2)$
We may simplify equation (2) by dividing each side by 10.
$2x + y = 33 \quad (3)$
Step 4
To solve this equation by the elimination method, multiply equation (1) by -1 and add the result to equation (3).
$$\begin{array}{rcrcr} -x & - & y & = & -20 \\ 2x & + & y & = & 33 \\ \hline x & & & = & 13 \end{array}$$
From equation (1), $y = 7$.
Step 5
She has 13 twenties and 7 tens.
Step 6
There are $13 + 7 = 20$ bills worth $13(\$20) + 7(\$10) = \$330$.

29. *Step 2*
Let $x =$ the number of pounds of \$1.30 per-pound candy; let $y =$ the number of pounds of \$0.90 per-pound candy.

Number of Pounds	**Cost per Pound (in dollars)**	**Total Value (in dollars)**
x	1.30	$1.30x$
y	0.90	$0.90y$
100	1.00	$100(1) = \$100$

Step 3
From the first and third columns of the table, we obtain the following system.

$$x + y = 100 \quad (1)$$
$$1.30x + 0.90y = 100 \quad (2)$$

Step 4
Multiply equation (2) by 10 (to clear decimals) and equation (1) by −9.

$$\begin{aligned} -9x - 9y &= -900 \\ 13x + 9y &= 1000 \\ \hline 4x &= 100 \\ x &= 25 \end{aligned}$$

From equation (1), $y = 75$.

Step 5
25 pounds of candy at $1.30 per pound and 75 pounds of candy at $0.90 per pound should be used.

Step 6
The value of the mixture is $25(1.30) + 75(0.90) = \$100$, giving us 100 pounds of candy that can sell for $1 per pound.

30. *Step 2*
Let $x =$ the number of liters of 40% antifreeze solution; let $y =$ the number of liters of 70% antifreeze solution.

Liters of Solution	Percent (as a decimal)	Amount of Pure Antifreeze
x	0.40	$0.40x$
y	0.70	$0.70y$
90	0.50	$0.50(90) = 45$

Step 3
From the first and third columns of the table, we obtain the following system.

$$x + y = 90 \quad (1)$$
$$0.40x + 0.70y = 45 \quad (2)$$

Step 4
Multiply equation (2) by 10 (to clear decimals) and equation (1) by −4.

$$\begin{aligned} -4x - 4y &= -360 \\ 4x + 7y &= 450 \\ \hline 3y &= 90 \\ y &= 30 \end{aligned}$$

From equation (1), $x = 60$.

Step 5
In order to make 90 liters of 50% antifreeze solution, 60 liters of 40% solution and 30 liters of 70% solution will be needed.

Step 6
60 liters added to 30 liters will give us the desired 90 liters. 40% of 60 liters gives us 24 liters of pure antifreeze and 70% of 30 liters gives us 21 liters of pure antifreeze. This is a total of 45 liters of pure antifreeze, as required.

31. *Step 2*
Let $x =$ the amount invested at 3%; let $y =$ the amount invested at 4%.

Amount of Principal	Rate	Interest
x	0.03	$0.03x$
y	0.04	$0.04y$
$18,000	XXXX	$650

Step 3
From the chart, we obtain the following equations.

$$x + y = 18{,}000 \quad (1)$$
$$0.03x + 0.04y = 650 \quad (2)$$

Step 4
To clear decimals, multiply each side of equation (2) by 100.

$$100(0.03x + 0.04y) = 100(650)$$
$$3x + 4y = 65{,}000 \quad (3)$$

Multiply equation (1) by −3 and add the result to equation (3).

$$\begin{aligned} -3x - 3y &= -54{,}000 \\ 3x + 4y &= 65{,}000 \\ \hline y &= 11{,}000 \end{aligned}$$

From equation (1), $x = 7000$.

Step 5
She invested $7000 at 3% and $11,000 at 4%.

Step 6
The sum of $7000 and $11,000 is $18,000. 3% of $7000 is $210, and 4% of $11,000 is $440. This gives us $\$210 + \$440 = \$650$ in interest, as required.

32. *Step 2*
Let $x =$ the rate of the plane in still air; let $y =$ the rate of the wind.

	d	*r*	*t*
With Wind	540	$x+y$	2
Against Wind	690	$x-y$	3

Step 3
Use the formula $d = rt$.

$540 = (x+y)(2)$

$270 = x+y$ (1) Divide by 2.

$690 = (x-y)(3)$

$230 = x-y$ (2) Divide by 3.

Step 4
Solve the system by the elimination method.

$270 = x + y$ (1)

$230 = x - y$ (2)

$500 = 2x$ Add (1) and (2).

$250 = x$

From equation (1), $y = 20$.

Step 5
The rate of the wind is 20 miles per hour; the rate of the plane in still air is 250 miles per hour.

Step 6
Flying with the wind at $250+20 = 270$ miles per hour for 2 hours results in a trip of 540 miles. Flying against the wind at $250-20 = 230$ miles per hour for 3 hours results in a trip of 690 miles.

33. Graph $x+y=2$ as a solid line through its intercepts, $(2, 0)$ and $(0, 2)$. Using $(0, 0)$ as a test point will result in the false statement $0 \geq 2$, so shade the region *not* containing the origin.
Graph $x-y=4$ as a solid line through its intercepts, $(4, 0)$ and $(0, -4)$. Using $(0, 0)$ as a test point will result in the true statement $0 \leq 4$, so shade the region containing the origin.
The solution set of this system is the intersection of the two shaded regions, and includes the portions of the two lines that bound this region.

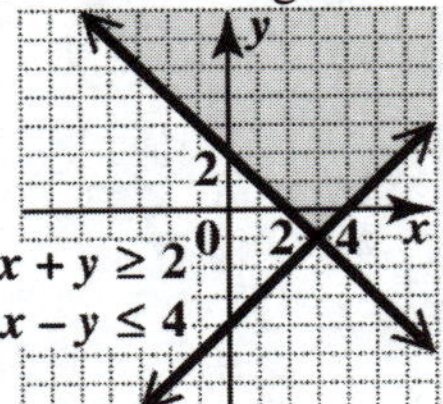

34. Graph $y = 2x$ as a solid line through $(0, 0)$ and $(1, 2)$. This line goes through the origin, so a different test point must be used.
Choosing $(-4, 0)$ as a test point will result in the true statement $0 \geq -8$, so shade the region containing $(-4, 0)$.
Graph $2x+3y=6$ as a solid line through its intercepts, $(3, 0)$ and $(0, 2)$. Choosing $(0, 0)$ as a test point will result in the true statement $0 \leq 6$, so shade the region containing the origin.
The solution set of this system is the intersection of the two shaded regions, and includes the portions of the two lines that bound this region.

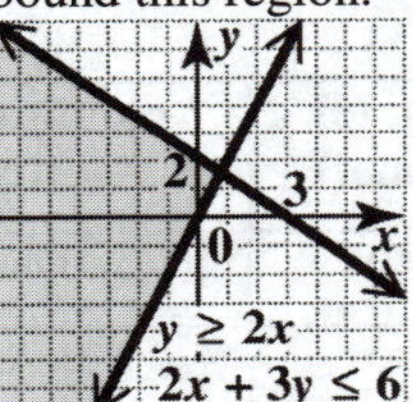

35. Graph $x+y=3$ as a dashed line through $(3, 0)$ and $(0, 3)$. Using $(0, 0)$ as a test point will result in the true statement $0 < 3$, so shade the region containing the origin.
Graph $2x = y$ as a dashed line through $(0, 0)$ and $(1, 2)$. Choosing $(0, -3)$ as a test point will result in the true statement $0 \geq -3$, so shade the region containing $(0, -3)$. The solution set of this system is the intersection of the two shaded regions. It does not contain the boundary lines.

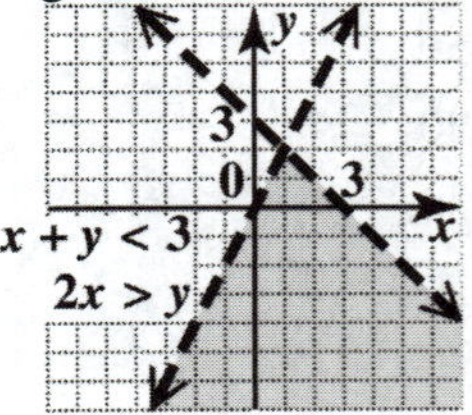

36. Graph the line $3x - y = 3$ through its intercepts $(1, 0)$ and $(0, -3)$, the vertical line $x = -1$, and the horizontal line $y = 2$. All of these lines should be solid because of the $\le$ and $\ge$ signs. All three inequalities are true for $(0, 0)$. Shade the triangular region bounded by the three lines, which contains the test point $(0, 0)$. The solution includes the portions of the three lines that bound the region and form the sides of the triangle.

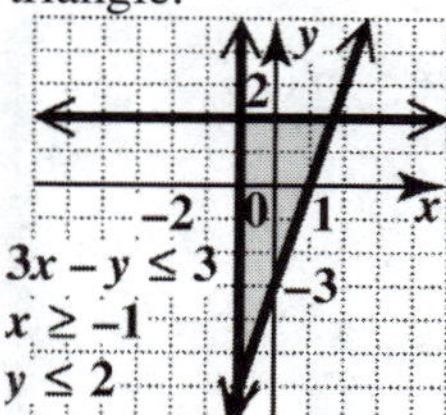

Chapter 4 Mixed Review Exercises

1. $\frac{2x}{3} + \frac{y}{4} = \frac{14}{3}$ (1)

$\frac{x}{2} + \frac{y}{12} = \frac{8}{3}$ (2)

To clear fractions, multiply both sides of each equation by 12.

$12\left(\frac{2x}{3} + \frac{y}{4}\right) = 12\left(\frac{14}{3}\right)$

$8x + 3y = 56$ (3)

$12\left(\frac{x}{2} + \frac{y}{12}\right) = 12\left(\frac{8}{3}\right)$

$6x + y = 32$ (4)

To solve this system by the elimination method, multiply both sides of equation (4) by -3, and add the result to equation (3).

$$\begin{aligned} 8x + 3y &= 56 \\ -18x - 3y &= -96 \\ \hline -10x \quad &= -40 \\ x &= 4 \end{aligned}$$

To find y, let $x = 4$ in equation (4).

$$\begin{aligned} 6x + y &= 32 \\ 6(4) + y &= 32 \\ 24 + y &= 32 \\ y &= 8 \end{aligned}$$

The solution set is $\{(4, 8)\}$.

2. $x = y + 6$ (1)

$2y - 2x = -12$ (2)

Rewrite the equations in the form $Ax + By = C$. Equation (1) becomes the following.

$x - y = 6$ (3)

After dividing by 2, equation (2) becomes the following.

$-x + y = -6$ (4)

To solve this system by the elimination method, add equations (3) and (4).

$$\begin{aligned} x - y &= 6 \\ -x + y &= -6 \\ \hline 0 &= 0 \text{ True} \end{aligned}$$

This result indicates that the system has an infinite number of solutions. The solution set is $\{(x, y) \mid x - y = 6\}$.

3. $3x + 4 = 6$ (1)

$4x - 5y = 8$ (2)

To solve this system by the elimination method, multiply equation (1) by -4 and equation (2) by 3; then add the results.

$$\begin{aligned} -12x - 16y &= -24 \\ 12x - 15y &= 24 \\ \hline -31y &= 0 \\ y &= 0 \end{aligned}$$

To find x, substitute 0 for y in equation (1).

$$\begin{aligned} 3x + 4(0) &= 6 \\ 3x &= 6 \\ x &= 2 \end{aligned}$$

The solution set is $\{(2, 0)\}$.

4. $0.4x - 0.9y = 0.7$ (1)

$0.3x + 0.2y = 1.4$ (2)

Multiply equation (1) and equation (2) by 10 to eliminate the decimals.

$4x - 9y = 7$ (3)

$3x + 2y = 14$ (4)

To solve this system by the elimination method, multiply equation (3) by 2, multiply equation (4) by 9, and add the resulting equations.

$$\begin{aligned} 8x - 18y &= 14 \\ 27x + 18y &= 126 \\ \hline 35x \quad &= 140 \\ x &= 4 \end{aligned}$$

To find y, let $x = 4$ in equation (4).

$$3x + 2y = 14$$
$$3(4) + 2y = 14$$
$$12 + 2y = 14$$
$$2y = 2$$
$$y = 1$$

The solution set is $\{(4, 1)\}$.

5. System B is easier to solve by the substitution method than system A because the bottom equation in system B is already solved for y. Solving system A would require our solving one of the equations for one of the variables before substituting, and the expression to be substituted would involve fractions.

6. **(a)** The graph of the fixed rate mortgage is above the graph of the variable rate mortgage for years 0 to 6, so that is when the monthly payment for the fixed rate mortgage is more than the monthly payment for the variable rate mortgage.

(b) The graphs intersect at year 6, so that's when the payments would be the same. The monthly payment at that time appears to be about $650.

7. Let $x =$ the length of each of the two equal sides; let $y =$ the length of the longer third side.
The perimeter is 29 inches, so the following is true.

$$x + x + y = 29$$
$$\text{or} \quad 2x + y = 29 \quad (1)$$

The third side is 5 inches longer than each of the two equal sides, so the following is also true.

$$y = x + 5 \quad (2)$$

Substitute $x + 5$ for y in (1).

$$2x + y = 29$$
$$2x + (x + 5) = 29$$
$$3x + 5 = 29$$
$$3x = 24$$
$$x = 8$$

From equation (2), $y = (8) + 5 = 13$.
The lengths of the sides of the triangle are 8 inches, 8 inches, and 13 inches.

8. Let $x =$ the number of people who visited the Lincoln Memorial; let $y =$ the number of people who visited the World War II Memorial.
The total was 10.6 million, so the following equation is true.

$$x + y = 10.6 \quad (1)$$

The World War II Memorial had 1.8 million fewer visitors than the Lincoln Memorial, so the following is also true.

$$y = x - 1.8 \quad (2)$$

Substitute $x - 1.8$ for y in (1).

$$(x) + (x - 1.8) = 10.6$$
$$2x - 1.8 = 10.6$$
$$2x = 12.4$$
$$y = 6.2$$

From equation (2), the following is calculated.

$$y = (6.2) - 1.8 = 4.4$$

In 2012, 4.4 million people visited the World War II Memorial, and 6.2 million people visited the Lincoln Memorial.

9. Let $x =$ the rate of the slower car; let $y =$ the rate of the faster car.
One car travels 30 miles per hour faster than the other, so the following equation is true.

$$y = x + 30 \quad (1)$$

In $2\frac{1}{2}$ $\left(\text{or } \frac{5}{2}\right)$ hours, the slower car travels $\left(\frac{5}{2}\right)(x)$ miles, and the faster car travels $\left(\frac{5}{2}\right)(y)$ miles. The cars will be 265 miles apart, so the following is also true.

$$\frac{5}{2}x + \frac{5}{2}y = 265 \quad (2)$$

Clear fractions from equation (2).

$$2\left(\frac{5}{2}x + \frac{5}{2}y\right) = 2(265) \quad \text{Multiply by 2.}$$
$$5x + 5y = 530$$
$$x + y = 106 \quad \text{Divide by 5.}$$

We now have the following system.

$$y = x + 30 \quad (1)$$
$$x + y = 106 \quad (3)$$

Substitute $x+30$ for y in equation (3).

$$x+y=106$$
$$x+(x+30)=106$$
$$2x+30=106$$
$$2x=76$$
$$x=38$$

From equation (1), the following is calculated.

$y=(38)+30=68$

The slower car went 38 miles per hour, and the faster car went 68 miles per hour.

10. The shaded region is to the left of the vertical line $x=3$, above the horizontal line $y=1$, and includes both lines. This is the graph of the following system.

$x \le 3$

$y \ge 1$

So, the answer is B.

11. Graph $x+y=5$ as a dashed line through its intercepts, $(5, 0)$ and $(0, 5)$. Using $(0, 0)$ as a test point will result in the true statement $0<5$, so shade the region containing the origin.
Graph $x-y=2$ as a solid line through its intercepts, $(2, 0)$ and $(0, -2)$. Using $(0, 0)$ as a test point will result in the false statement $0 \ge 2$, so shade the region *not* containing the origin.
The solution set of this system is the intersection of the two shaded regions. It includes the portion of the line $x-y=2$ that bounds this region, but not the line $x+y=5$.

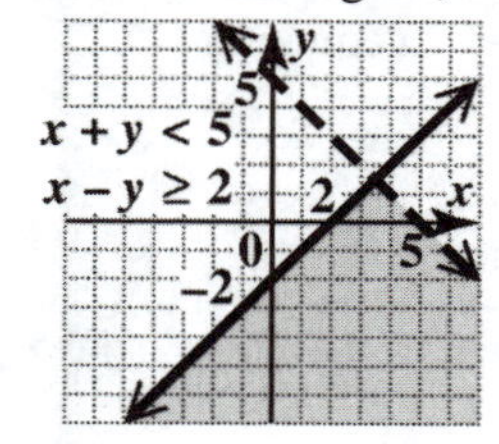

12. Graph $y=2x$ as a solid line through $(0, 0)$ and $(1, 2)$. Shade the region below the line.
Graph $x+2y=4$ as a dashed line through its intercepts, $(4, 0)$ and $(0, 2)$. Using $(0, 0)$ as a test point will result in the false statement $0>4$, so shade the region *not* containing the origin.
The solution set of this system is the intersection of the shaded regions. It includes the portion of the line $y=2x$ that bounds this region, but not the line $x+2y=4$.

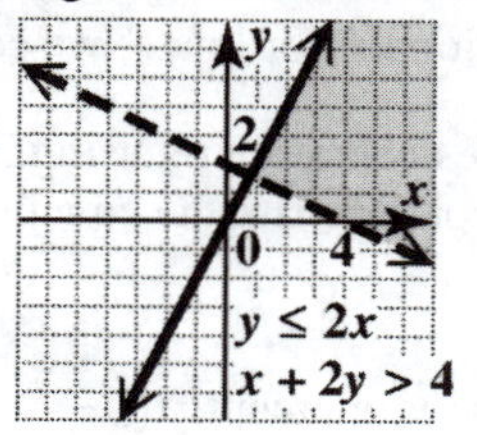

Chapter 4 Test

1. $2x+y=-3$ (1)
$x-y=-9$ (2)

(a) $(1, -5)$

Substitute 1 for x and -5 for y in (1) and (2).

(1) $2(1)+(-5) \stackrel{?}{=} -3$
$-3=-3$ True

(2) $(1)-(-5) \stackrel{?}{=} -9$
$6=-9$ False

Since $(1, -5)$ does not satisfy both equations, it *is not* a solution of the system.

(b) $(1, 10)$

(1) $2(1)+(10) \stackrel{?}{=} -3$
$12=-3$ False

The false result indicates that $(1, 10)$ *is not* a solution of the system.

(c) $(-4, 5)$

(1) $2(-4)+(5) \stackrel{?}{=} -3$
$-3=-3$ True

(2) $(-4)-(5) \stackrel{?}{=} -9$
$-9=-9$ True

Since $(-4, 5)$ satisfies both equations, it *is* a solution of the system.

2. Graph the line $x+2y=6$ through its intercepts, $(6, 0)$ and $(0, 3)$. Because the x-intercept of the line $-2x+y=-7$, which is $\left(\frac{7}{2}, 0\right)$ has a fractional coordinate, we can graph the line more accurately by using its slope and y-intercept. Rewrite the equation as the following.

$y=2x-7$

Start by plotting the y-intercept, $(0, -7)$. From this point, go 2 units up and 1 unit to the right to reach the point $(1, -5)$. Draw the line through $(0, -7)$ and $(1, -5)$.

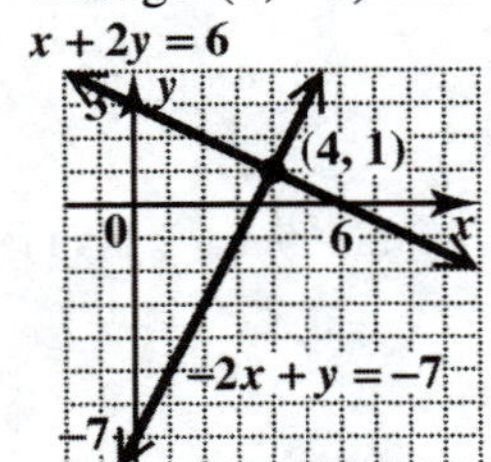

It appears that the lines intersect at the point $(4, 1)$. Since $(4, 1)$ satisfies both equations, the solution set of this system is $\{(4, 1)\}$.

3. Two lines with the same slope but different y-intercepts are parallel. Since the lines are parallel and have no points in common, the system has no solution.

4. $2x+y=-4 \quad (1)$

$x=y+7 \quad (2)$

Substitute $y+7$ for x in equation (1), and solve for y.

$$\begin{aligned} 2x+y&=-4 \\ 2(y+7)+y&=-4 \\ 2y+14+y&=-4 \\ 3y&=-18 \\ y&=-6 \end{aligned}$$

From equation (2), the following is calculated.

$x=(-6)+7=1$

The solution set is $\{(1, -6)\}$.

5. $4x+3y=-35 \quad (1)$

$x+y=0 \quad (2)$

Solve equation (2) for y.

$y=-x \quad (3)$

Substitute $-x$ for y in equation (1) and solve for x.

$$\begin{aligned} 4x+3y&=-35 \\ 4x+3(-x)&=-35 \\ 4x-3x&=-35 \\ x&=-35 \end{aligned}$$

From equation (3), the following is calculated.

$y=-(-35)=35$

The solution set is $\{(-35, 35)\}$.

6.
$$\begin{array}{rcrcrl} 2x & - & y & = & 4 & (1) \\ 3x & + & y & = & 21 & (2) \\ \hline 5x & & & = & 25 & \text{Add (1) and (2).} \\ & & x & = & 5 & \end{array}$$

To find y, let $x=5$ in equation (2).

$$\begin{aligned} 3x+y&=21 \\ 3(5)+y&=21 \\ 15+y&=21 \\ y&=6 \end{aligned}$$

The solution set is $\{(5, 6)\}$.

7. $4x+2y=2 \quad (1)$

$5x+4y=7 \quad (2)$

Multiply equation (1) by -2, and add the result to equation (2).

$$\begin{array}{rcrcr} -8x & - & 4y & = & -4 \\ 5x & + & 4y & = & 7 \\ \hline -3x & & & = & 3 \\ & & x & = & -1 \end{array}$$

To find y, let $x=-1$ in equation (1).

$$\begin{aligned} 4x+2y&=2 \\ 4(-1)+2y&=2 \\ -4+2y&=2 \\ 2y&=6 \\ y&=3 \end{aligned}$$

The solution set is $\{(-1, 3)\}$.

8. $3x+4y=9 \quad (1)$

$2x+5y=13 \quad (2)$

$$\begin{array}{rcll} 6x + 8y &=& 18 & (3)\ 2\times\text{Eq. }(1) \\ -6x - 15y &=& -39 & (4)\ -3\times\text{Eq. }(2) \\ \hline -7y &=& -21 & \text{Add (3) and (4).} \\ y &=& 3 & \end{array}$$

Substitute 3 for y in (1).

$$3x+4y=9$$
$$3x+4(3)=9$$
$$3x+12=9$$
$$3x=-3$$
$$x=-1$$

The solution set is $\{(-1, 3)\}$.

9. $4x+5y=2 \quad (1)$

$-8x-10y=6 \quad (2)$

Multiply equation (1) by 2, and add the result to equation (2).

$$\begin{array}{rcl} 8x + 10y &=& 4 \\ -8x - 10y &=& 6 \\ \hline 0 &=& 10 \quad \text{False} \end{array}$$

This result indicates that the system has solution set $\varnothing$.

10. $6x-5y=0 \quad (1)$

$-2x+3y=0 \quad (2)$

Multiply equation (2) by 3, and add the result to equation (1).

$$\begin{array}{rcl} 6x - 5y &=& 0 \\ -6x + 9y &=& 0 \\ \hline 4y &=& 0 \\ y &=& 0 \end{array}$$

To find x, let $y=0$ in equation (1).

$$6x-5y=0$$
$$6x-5(0)=0$$
$$6x=0$$
$$x=0$$

The solution set is $\{(0, 0)\}$.

11. $4y=-3x+5 \quad (1)$

$6x=-8y+10 \quad (2)$

Rewrite (1) and (2) in standard form.

$3x+4y=5 \quad (3)$

$6x+8y=10 \quad (4)$

Multiply equation (3) by -2, and add the result to equation (4).

$$\begin{array}{rcl} -6x - 8y &=& -10 \\ 6x + 8y &=& 10 \\ \hline 0 &=& 0 \quad \text{True} \end{array}$$

The true result indicates that this system has an infinite number of solutions.

The solution set is $\{(x, y) \mid 3x+4y=5\}$.

12. $\frac{6}{5}x-\frac{1}{3}y=-20 \quad (1)$

$-\frac{2}{3}x+\frac{1}{6}y=11 \quad (2)$

To clear fractions, multiply by the LCD for each equation. Multiply equation (1) by 15.

$$15\left(\frac{6}{5}x-\frac{1}{3}y\right)=15(-20)$$
$$18x-5y=-300 \qquad (3)$$

Multiply equation (2) by 6.

$$6\left(-\frac{2}{3}x+\frac{1}{6}y\right)=6(11)$$
$$-4x+y=66 \qquad (4)$$

To solve this system by the elimination method, multiply equation (4) by 5, and add the result to equation (3).

$$\begin{array}{rcl} 18x - 5y &=& -300 \\ -20x + 5y &=& 330 \\ \hline -2x &=& 30 \\ x &=& -15 \end{array}$$

To find y, let $x=-15$ in equation (4).

$$-4x+y=66$$
$$-4(-15)+y=66$$
$$60+y=66$$
$$y=6$$

The solution set is $\{(-15, 6)\}$.

13. $0.2x + 0.3y = 1.0$ (1)

$-0.3x + 0.1y = 1.8$ (2)

Multiply equation (1) and equation (2) by 10 to eliminate the decimals. Then solve the resulting equations by the elimination method.

$2x + 3y = 10$ (3)

$-3x + y = 18$ (4)

To solve this system by the elimination method, multiply equation (4) by -3, and add the result to equation (3).

$$\begin{array}{rcl} 2x + 3y &=& 10 \\ 9x - 3y &=& -54 \\ \hline 11x &=& -44 \\ x &=& -4 \end{array}$$

To find y, let $x = -4$ in equation (3).

$$\begin{aligned} 2x + 3y &= 10 \\ 2(-4) + 3y &= 10 \\ -8 + 3y &= 10 \\ 3y &= 18 \\ y &= 6 \end{aligned}$$

The solution set is $\{(-4, 6)\}$.

14. *Step 2*

Let $x =$ the distance between Memphis and Atlanta (in miles); let $y =$ the distance between Minneapolis and Houston (in miles).

Step 3

Since the distance between Memphis and Atlanta is 782 miles less than the distance between Minneapolis and Houston, this can be represented by the following.

$x = y - 782$ (1)

Together the two distances total 1570 miles, so the following is also true.

$x + y = 1570$ (2)

Step 4

Substitute $y - 782$ for x in equation (2).

$$\begin{aligned} (y - 782) + y &= 1570 \\ 2y - 782 &= 1570 \\ 2y &= 2352 \\ y &= 1176 \end{aligned}$$

From equation (1), the following is calculated.

$x = (1176) - 782 = 394$

Step 5

The distance between Memphis and Atlanta is 394 miles, while the distance between Minneapolis and Houston is 1176 miles.

Step 6

394 is 782 less than 1176, and the sum of 394 and 1176 is 1570, as required.

15. *Step 2*

Let $x =$ the number of visitors to the Magic Kingdom (in millions); let $y =$ the number of visitors to Disneyland (in millions).

Step 3

Disneyland had 1.5 million fewer visitors than the Magic Kingdom, so the following equation applies.

$y = x - 1.5$ (1)

Together they had 33.5 million visitors, so the following also applies.

$x + y = 33.5$ (2)

Step 4

Substitute $x - 1.5$ for y in (2).

$$\begin{aligned} x + (x - 1.5) &= 33.5 \\ 2x - 1.5 &= 33.5 \\ 2x &= 35 \\ x &= 17.5 \end{aligned}$$

From equation (1), the following is calculated.

$y = (17.5) - 1.5 = 16.0$

Step 5

In 2008, the Magic Kingdom had 17.5, million visitors, and Disneyland had 16.0 million visitors.

Step 6

16.0 is 1.5 million fewer than 17.5, and the sum of 16.0 million and 17.5 million is 33.5, as required.

16. *Step 2*

Let $x =$ the number of liters of 25% alcohol solution; let $y =$ the number of liters of 40% alcohol solution.

Liters of Solution	Percent (as a decimal)	Liters of Pure Alcohol
x	0.25	$0.25x$
y	0.40	$0.40y$
50	0.30	$0.30(50) = 15$

Step 3

From the first and third columns of the table, we obtain the following equations.

$x + y = 50$ (1)

$0.25x + 0.40y = 15$ (2)

To clear decimals, multiply both sides of equation (2) by 100.

$25x + 40y = 1500$ (3)

Step 4
To solve this system by the elimination method, multiply equation (1) by -25, and add the result to equation (3).

$$\begin{aligned} -25x - 25y &= -1250 \\ 25x + 40y &= 1500 \\ \hline 15y &= 250 \end{aligned}$$

$$y = \frac{250}{15} = \frac{50}{3} = 16\frac{2}{3}$$

From equation (1), $x = 33\frac{1}{3}$.

Step 5
To get 50 liters of a 30% alcohol solution, $33\frac{1}{3}$ liters of a 25% solution and $16\frac{2}{3}$ liters of a 40% solution should be used.

Step 6
Since $33\frac{1}{3} + 16\frac{2}{3} = 50$ and $0.25\left(33\frac{1}{3}\right) + 0.40\left(16\frac{2}{3}\right) = 15$, this mixture will give the 50 liters of 30% solution, as required.

17. *Step 2*
Let $x =$ the rate of the faster car; let $y =$ the rate of the slower car.
Use $d = rt$ in constructing the table.

	r	t	d
Faster Car	x	3	$3x$
Slower Car	y	3	$3y$

Step 3
The faster car travels $1\frac{1}{3}$ times as fast as the other car, so the following equations are true.

$$x = 1\frac{1}{3}y$$

or $x = \frac{4}{3}y$ (1)

After 3 hours, they are 45 miles apart, that is, the difference between their distances is 45.
$3x - 3y = 45$ (2)

Step 4
To solve the system by substitution, substitute $\frac{4}{3}y$ for x in equation (2).

$$3\left(\frac{4}{3}y\right) - 3y = 45$$
$$4y - 3y = 45$$
$$y = 45$$

From equation (1), the following is calculated.

$$x = \frac{4}{3}(45) = 60$$

Step 5
The rate of the faster car is 60 miles per hour, and the rate of the slower car is 45 miles per hour.

Step 6
60 is $1\frac{1}{3}$ times 45. In three hours, the faster car travels $3(60) = 180$ miles and the slower car travels $3(45) = 135$ miles. The cars are $180 - 135 = 45$ miles apart, as required.

18. Graph $2x + 7y = 14$ as a solid line through its intercepts, $(7, 0)$ and $(0, 2)$. Choosing $(0, 0)$ as a test point will result in the true statement $0 \le 14$, so shade the side of the line containing the origin.
Graph $x - y = 1$ as a solid line through its intercepts, $(1, 0)$ and $(0, -1)$. Choosing $(0, 0)$ as a test point will result in the false statement $0 \ge 1$, so shade the side of the line *not* containing the origin.
The solution set of the given system is the intersection of the two shaded regions, and includes the portions of the two lines that bound this region.

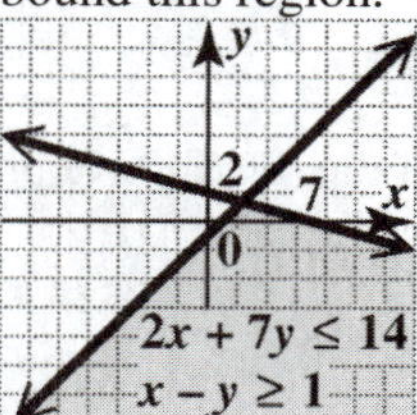

19. Graph $2x - y = 6$ as a dashed line through its intercepts, $(3, 0)$ and $(0, -6)$. Choosing $(0, 0)$ as a test point will result in the false statement $0 > 6$, so shade the side of the line *not* containing the origin.
Graph $4y + 12 = -3x$ as a solid line through its intercepts, $(-4, 0)$ and $(0, -3)$. Choosing $(0, 0)$ as a test point will result in the true statement $12 \ge 0$, so shade the side of the line containing the origin.

The solution set of the given system is the intersection of the two shaded regions. It includes the portion of the line $4y+12=-3x$ that bounds this region, but not the line $2x-y=6$.

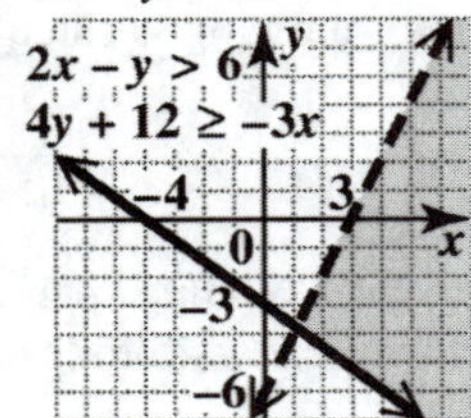

20. It is impossible for the sum of any two numbers to be both greater than 4 and less than 3. Therefore, system B has no solution.

Chapters R–4 Cumulative Review Exercises

1. The integer factors of 40 are $-1, 1, -2, 2, -4, 4, -5, 5, -8, 8, -10, 10, -20, 20, -40,$ and 40.

2. $$\begin{aligned} -2+6[3-(4-9)] &= -2+6[3-(-5)] \\ &= -2+6(8) \\ &= -2+48 \\ &= 46 \end{aligned}$$

3. $$\begin{aligned} \frac{3x^2+2y^2}{10y+3} &= \frac{3\cdot 1^2+2\cdot 5^2}{10(5)+3} \quad \text{Let } x=1,\ y=5. \\ &= \frac{3\cdot 1+2\cdot 25}{50+3} \\ &= \frac{3+50}{50+3} = \frac{53}{53} = 1 \end{aligned}$$

4. $r(s-k)=rs-rk$

This is an example of the distributive property.

5. $$\begin{aligned} 2-3(6x+2) &= 4(x+1)+18 \\ 2-18x-6 &= 4x+4+18 \\ -18x-4 &= 4x+22 \\ -22x &= 26 \\ x &= \frac{26}{-22} = -\frac{13}{11} \end{aligned}$$

The solution set is $\left\{-\frac{13}{11}\right\}$.

6. $\frac{3}{2}\left(\frac{1}{3}x+4\right)=6\left(\frac{1}{4}+x\right)$

Multiply each side by 2.

$$2\left[\frac{3}{2}\left(\frac{1}{3}x+4\right)\right]=2\left[6\left(\frac{1}{4}+x\right)\right]$$

$$3\left(\frac{1}{3}x+4\right)=12\left(\frac{1}{4}+x\right)$$

Use the distributive property to remove parentheses.

$$\begin{aligned} x+12 &= 3+12x \\ -11x &= -9 \\ x &= \frac{9}{11} \end{aligned}$$

The solution set is $\left\{\frac{9}{11}\right\}$.

7. $P=\frac{kT}{V}$ for T

$PV=kT$ Mutiply by V.

$\frac{PV}{k}=\frac{kT}{k}$ Divide by k.

$\frac{PV}{k}=T$ or $T=\frac{PV}{k}$

8. $-\frac{5}{6}x<15$

Multiply each side by the reciprocal of $-\frac{5}{6}$, which is $-\frac{6}{5}$, and reverse the direction of the inequality symbol.

$$-\frac{6}{5}\left(-\frac{5}{6}x\right)>-\frac{6}{5}(15)$$

$$x>-18$$

The solution set is $(-18, \infty)$.

9. $$\begin{aligned} -8 &< 2x+3 \\ -11 &< 2x \end{aligned}$$

$-\frac{11}{2}<x$ or $x>-\frac{11}{2}$

The solution set is $\left(-\frac{11}{2}, \infty\right)$.

10. 80.4% of 2500 is $0.804(2500) = 2010$

72.5% of 2500 is $0.725(2500) = 1812.5 \approx 1813$

$\frac{1570}{2500} = 0.628$ or 62.8%

$\frac{1430}{2500} = 0.572$ or 57.2%

Product or Company	Percent	Actual Number
Charmin	80.4%	2010
Wheaties	72.5%	1813
Budweiser	62.8%	1570
State Farm	57.2%	1430

11. Let $x =$ the number of points the Baltimore Ravens scored; let $y =$ the number of points the San Francisco 49ers scored.
The two teams scored 65 total points, and the Ravens defeated the 49ers by 3 points, so we have the following.

$$x + y = 65 \quad (1)$$
$$y = x - 3 \quad (2)$$

Substitute $x - 3$ for y in equation (1).

$$x + (x - 3) = 65$$
$$2x - 3 = 65$$
$$2x = 68$$
$$x = 34$$

From equation (2), the following is calculated.

$$y = (34) - 3 = 31$$

The Ravens scored 34 points, and the 49ers scored 31 points.

12. Let $x =$ the measure of the equal angles. Let $2x - 4 =$ the measure of the third angle.
The sum of the measures of the angles in a triangle is 180°.

$$x + x + (2x - 4) = 180$$
$$4x - 4 = 180$$
$$4x = 184$$
$$x = 46$$

The third angle has the following measure.

$$2x - 4 = 2(46) - 4 = 92 - 4 = 88$$

The measures of the angles are 46°, 46°, and 88°.

13. To graph this line, find the intercepts.
If $y = 0$, $x = 4$, so the x-intercept is $(4, 0)$.
If $x = 0$, $y = -4$, so the y-intercept is $(0, -4)$.
Graph the line through these intercepts. A third point, such as $(5, 1)$, may be used as a check.

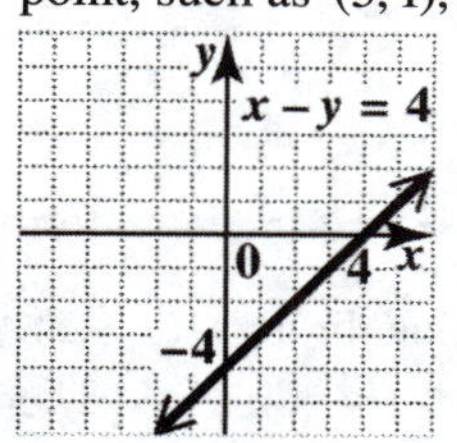

14. If $y = 0$, $x = 2$, so the x-intercept is $(2, 0)$.
If $x = 0$, $y = 6$, so the y-intercept is $(0, 6)$.
Graph the line through these intercepts. A third point, such as $(1, 3)$, may be used as a check.

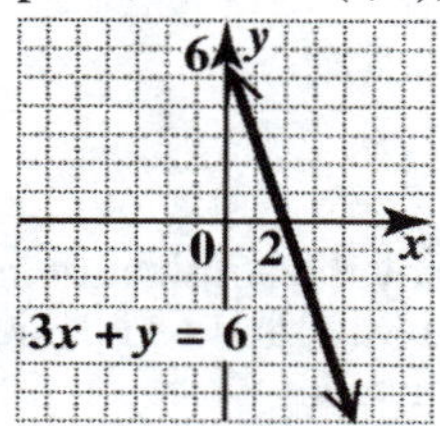

15. The slope m of the line passing through the points $(-5, 6)$ and $(1, -2)$ is

$$m = \frac{y_2 - y_1}{x_2 - x_1} = \frac{-2 - 6}{1 - (-5)} = \frac{-8}{6} = -\frac{4}{3}.$$

16. The slope of the line $y = 4x - 3$ is 4. The slope of the line whose graph is perpendicular to that of $y = 4x - 3$ is the negative reciprocal of 4, namely $-\frac{1}{4}$.

17. Use the point-slope form of a line.

$$y - y_1 = m(x - x_1)$$
$$y - 1 = \frac{1}{2}[x - (-4)]$$
$$y - 1 = \frac{1}{2}(x + 4)$$
$$y - 1 = \frac{1}{2}x + 2$$
$$y = \frac{1}{2}x + 3 \quad \text{Slope-intercept form}$$

18. Find the slope.

$$m = \frac{y_2 - y_1}{x_2 - x_1} = \frac{-3-3}{-2-1} = \frac{-6}{-3} = 2$$

Use the point-slope form of a line.

$$y - y_1 = m(x - x_1)$$
$$y - 3 = 2(x - 1)$$
$$y - 3 = 2x - 2$$
$$y = 2x + 1 \quad \text{Slope-intercept form}$$

19. **(a)** On the vertical line through $(9, -2)$, the x-coordinate of every point is 9. Therefore, an equation of this line is $x = 9$.

(b) On the horizontal line through $(4, -1)$, the y-coordinate of every point is -1. Therefore, an equation of this line is $y = -1$.

20. $2x - y = -8 \quad (1)$

$x + 2y = 11 \quad (2)$

To solve this system by the elimination method, multiply equation (1) by 2, and add the result to equation (2) to eliminate y.

$$\begin{array}{rcl} 4x - 2y &=& -16 \\ x + 2y &=& 11 \\ \hline 5x &=& -5 \\ x &=& -1 \end{array}$$

To find y, let $x = -1$ in equation (2).

$$-1 + 2y = 11$$
$$2y = 12$$
$$y = 6$$

The solution set is $\{(-1, 6)\}$.

21. $4x + 5y = -8 \quad (1)$

$3x + 4y = -7 \quad (2)$

Multiply equation (1) by -3 and equation (2) by 4. Add the resulting equations to eliminate x.

$$\begin{array}{rcl} -12x - 15y &=& 24 \\ 12x + 16y &=& -28 \\ \hline y &=& -4 \end{array}$$

To find x, let $y = -4$ in equation (1).

$$4x + 5y = -8$$
$$4x + 5(-4) = -8$$
$$4x - 20 = -8$$
$$4x = 12$$
$$x = 3$$

The solution set is $\{(3, -4)\}$.

22. $3x + 4y = 2 \quad (1)$

$6x + 8y = 1 \quad (2)$

Multiply equation (1) by -2, and add the result to equation (2).

$$\begin{array}{rcl} -6x - 8y &=& -4 \\ 6x + 8y &=& 1 \\ \hline 0 &=& -3 \quad \text{False} \end{array}$$

Since $0 = -3$ is a false statement, the solution set is $\emptyset$.

23. *Step 2*

Let $x =$ the number of adult tickets sold; let $y =$ the number of child tickets sold.

Kind of Ticket	Number Sold	Cost of Each (in dollars)	Total Value (in dollars)
Adult	x	6	$6x$
Child	y	2	$2y$
Total	454	XXXXXX	2528

Step 3

The total number of tickets sold was 454, so the following equation applies.

$x + y = 454 \quad (1)$

Since the total value was $2528, the final column leads to the following.

$6x + 2y = 2528 \quad (2)$

Step 4

Multiply both sides of equation (1) by -2, and add this result to equation (2).

$$\begin{array}{rcl} -2x - 2y &=& -908 \\ 6x + 2y &=& 2528 \\ \hline 4x &=& 1620 \\ x &=& 405 \end{array}$$

From equation (1), the following is calculated.

$y = 454 - (405) = 49.$

Step 5

There were 405 adult tickets and 49 child tickets sold.

Step 6

The total number of tickets sold was $405 + 49 = 454$. Since 405 adults paid $6 each, and 49 children paid $2 each, the value of tickets sold should be $405(6) + 49(2) = 2528$, or $2528. The result agrees with the given information.

24. Let $x =$ the number of liters of 20% alcohol solution; let $y =$ the number of liters of 50% alcohol solution.

Liters of Solution	Percent (as a decimal)	Liters of Pure Alcohol
x	0.20	$0.20x$
y	0.50	$0.50y$
12	0.40	$0.40(12) = 4.8$

Since 12 L of the mixture are needed, the following equation applies.

$x + y = 12 \quad (1)$

Since the amount of pure alcohol in the 20% solution plus the amount of pure alcohol in the 50% solution must equal the amount of pure alcohol in the mixture, the following also applies.

$0.20x + 0.50y = 4.8$

Multiply by 10 to clear the decimals.

$2x + 5y = 48 \quad (2)$

Multiply equation (1) by -2, and add the result to equation (2).

$$\begin{array}{rcrcr} -2x & - & 2y & = & -24 \\ 2x & + & 5y & = & 48 \\ \hline & & 3y & = & 24 \\ & & y & = & 8 \end{array}$$

From equation (1), the following is calculated.

$x = 12 - 8 = 4$

4 L of 20% alcohol solution and 8 L of 50% alcohol solution will be needed.

25. Graph the boundary with equation $x + 2y = 12$ as a solid line through its intercepts, $(12, 0)$ and $(0, 6)$. Using $(0, 0)$ as a test point results in a true statement, $0 \le 12$. Shade the region containing the origin.

Graph the boundary with equation $2x - y = 8$ as a solid line through its intercepts, $(4, 0)$ and $(0, -8)$. Using $(0, 0)$ as a test point results in a true statement, $0 \le 8$. Shade the region containing the origin.

The solution is the intersection of the two shaded regions and includes the portions of the lines that bound this region.

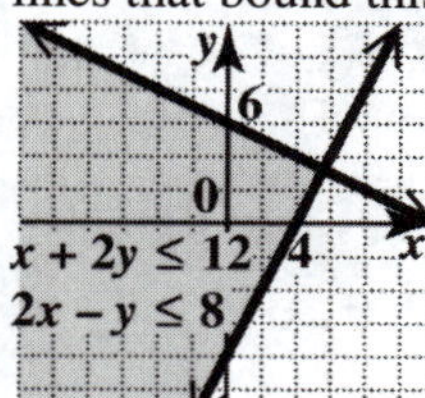

Chapter 5
Exponents and Polynomials

5.1 The Product Rule and Power Rules for Exponents

Now Try Exercises

N1. $4 \cdot 4 \cdot 4 = 4^3$
$= 64$
4 occurs as a factor 3 times.

N2. **(a)** $(-3)^4 = (-3)(-3)(-3)(-3)$
$= 81$
Base: −3; exponent: 4

(b) $-3^4 = -1 \cdot 3^4$
$= -1 \cdot (3 \cdot 3 \cdot 3 \cdot 3)$
$= -81$
Base: 3; exponent: 4

N3. **(a)** $(-5)^2(-5)^4 = (-5)^{2+4}$ Product rule
$= (-5)^6$

(b) $y^2 \cdot y \cdot y^5 = y^2 \cdot y^1 \cdot y^5 = y^{2+1+5} = y^8$

(c) $(2x^3)(4x^6) = (2 \cdot 4) \cdot (x^3 \cdot x^6)$
$= 8(x^{3+6})$
$= 8x^9$

(d) $2^4 \cdot 5^3$: the product rule does not apply because the bases are different.
$2^4 \cdot 5^3 = 16 \cdot 125 = 2000$

(e) $3^2 + 3^3$: the product rule does not apply because it is a sum, not a product.
$3^2 + 3^3 = 9 + 27 = 36$

N4. **(a)** $(4^7)^5 = 4^{7 \cdot 5}$ Power rule (a)
$= 4^{35}$

(b) $(y^4)^7 = y^{4 \cdot 7}$ Power rule (a)
$= y^{28}$

N5. **(a)** $(-5ab)^3$
$= (-1 \cdot 5 \cdot a \cdot b)^3$ $\quad -x = -1 \cdot x$
$= (-1)^3 \cdot 5^3 \cdot a^3 \cdot b^3$ Power rule (b)
$= -5^3 a^3 b^3$ or $-125a^3b^3$

(b) $(4t^3p^5)^2 = 4^2(t^3)^2(p^5)^2$ Power rule (b)
$= 16t^6p^{10}$ Power rule (a)

N6. **(a)** $\left(\frac{p}{q}\right)^5 = \frac{p^5}{q^5}$ Power rule (c), $q \neq 0$

(b) $\left(\frac{1}{4}\right)^3 = \frac{1^3}{4^3}$ Power rule (c)
$= \frac{1}{64}$

N7. **(a)** $\left(\frac{3}{5}\right)^3 \cdot 3^2 = \frac{3^3}{5^3} \cdot \frac{3^2}{1}$ Power rule (c)
$= \frac{3^{3+2}}{5^3}$ Power rule (b)
$= \frac{3^5}{5^3}$ or $\frac{243}{125}$

(b) $(8k)^5(8k)^4 = (8k)^{5+4}$ Product rule
$= (8k)^9$
$= 8^9k^9$ Power rule (b)

(c) $(x^4y)^5(-2x^2y^5)^3$
$= (x^4)^5 y^5 (-1)^3 2^3 (x^2)^3 (y^5)^3$
$= x^{4 \cdot 5} y^5 (-1) 2^3 x^{2 \cdot 3} y^{5 \cdot 3}$
$= -2^3 x^{20} y^5 x^6 y^{15}$
$= -2^3 x^{20+6} y^{5+15}$
$= -2^3 x^{26} y^{20}$ or $-8x^{26}y^{20}$

N8. Use the formula for the area of a triangle, $A = \frac{1}{2}bh$, with $b = 10x^8$ and $h = 3x^7$.
$A = \frac{1}{2}(10x^8)(3x^7)$ Area formula
$= \frac{1}{2} \cdot 10 \cdot 3 \cdot x^{8+7}$ Product rule
$= 15x^{15}$

Exercises

1. $3^3 = 3 \cdot 3 \cdot 3 = 27,$ so the statement $3^3 = 9$ is *false.*

3. $(-3)^4 = (-3)(-3)(-3)(-3) = 9 \cdot 9 = 81$ and $3^4 = 3 \cdot 3 \cdot 3 \cdot 3 = 9 \cdot 9 = 81,$ so the statement $(-3)^4 = 3^4$ is *true.*

5. $\underbrace{w \cdot w \cdot w \cdot w \cdot w \cdot w}_{6w\text{'s}} = w^6$

7. $\left(\frac{1}{2}\right)\left(\frac{1}{2}\right)\left(\frac{1}{2}\right)\left(\frac{1}{2}\right)\left(\frac{1}{2}\right)\left(\frac{1}{2}\right) = \left(\frac{1}{2}\right)^6$

9. $(-4)(-4)(-4)(-4) = (-4)^4$

11. $(-7y)(-7y)(-7y)(-7y) = (-7y)^4$

13. In $(-3)^4$, -3 is the base.
$(-3)^4 = (-3)(-3)(-3)(-3) = 81$
In -3^4, 3 is the base.
$-3^4 = -(3 \cdot 3 \cdot 3 \cdot 3) = -81$

15. In the exponential expression 3^5, the base is 3 and the exponent is 5.
$3^5 = 3 \cdot 3 \cdot 3 \cdot 3 \cdot 3 = 243$

17. In the expression $(-3)^5$, the base is -3 and the exponent is 5.
$(-3)^5 = (-3)(-3)(-3)(-3)(-3) = -243$

19. In the expression $(-6x)^4$, the base is $-6x$ and the exponent is 4.

21. In the expression $-6x^4, -6$ is not part of the base. The base is x and the exponent is 4.

23. $8^2 \cdot 8^5 = 8^{2+5}$ Product rule
$= \underline{8^7}$

25. $5^2 \cdot 5^6 = 5^{2+6} = 5^8$

27. $4^2 \cdot 4^7 \cdot 4^3 = 4^{2+7+3} = 4^{12}$

29. $(-7)^3(-7)^6 = (-7)^{3+6} = (-7)^9$

31. $t^3 \cdot t^8 \cdot t^{13} = t^{3+8+13} = t^{24}$

33.
$$\begin{aligned}(-8r^4)(7r^3) &= -8 \cdot 7 \cdot r^4 \cdot r^3 \\ &= -56r^{4+3} \\ &= -56r^7\end{aligned}$$

35.
$$\begin{aligned}(-6p^5)(-7p^5) &= (-6)(-7)\,p^5 \cdot p^5 \\ &= 42p^{5+5} \\ &= 42p^{10}\end{aligned}$$

37.
$$\begin{aligned}&(5x^2)(-2x^3)(3x^4) \\ &= (5)(-2)(3)x^2 \cdot x^3 \cdot x^4 \\ &= (-10)(3)x^{2+3+4} \\ &= -30x^9\end{aligned}$$

39. $3^8 + 3^9$ is a sum, so the product rule does not apply.

41. $5^8 \cdot 3^9$ is a product with different bases, so the product rule does not apply.

43. $(4^3)^2 = 4^{3 \cdot 2}$ Power rule (a)
$= 4^6$

45. $(t^4)^5 = t^{4 \cdot 5} = t^{20}$ Power rule (a)

47. $(7r)^3 = 7^3r^3$ Power rule (b)

49. $(5xy)^5 = 5^5x^5y^5$ Power rule (a)

51.
$$\begin{aligned}(-5^2)^6 &= (-1 \cdot 5^2)^6 \\ &= (-1)^6 \cdot (5^2)^6 && \text{Power rule (b)} \\ &= 1 \cdot 5^{2 \cdot 6} && \text{Power rule (a)} \\ &= 1 \cdot 5^{12} = 5^{12}\end{aligned}$$

53.
$$\begin{aligned}(-8^3)^5 &= (-1 \cdot 8^3)^5 \\ &= (-1)^5 \cdot (8^3)^5 && \text{Power rule (b)} \\ &= -1 \cdot 8^{3 \cdot 5} && \text{Power rule (a)} \\ &= -8^{15}\end{aligned}$$

55. $8(qr)^3 = 8q^3r^3$ Power rule (b)

57. $\left(\frac{9}{5}\right)^8 = \frac{9^8}{5^8}$ Power rule (c)

59. $\left(\frac{1}{2}\right)^3 = \frac{1^3}{2^3} = \frac{1}{2^3}$ Power rule (c)

61. $\left(\frac{a}{b}\right)^3 = \frac{a^3}{b^3}$ Power rule (c), $(b \neq 0)$

63. $\left(\frac{x}{2}\right)^3 = \frac{x^3}{2^3}$ Power rule (c)

65. $\left(-\frac{2x}{y}\right)^5 = \left(-1 \cdot \frac{2x}{y}\right)^5$

$= (-1)^5 \cdot \left(\frac{2x}{y}\right)^5$ Power rule (b)

$= -1 \cdot \frac{(2x)^5}{y^5}$ Power rule (c)

$= -\frac{2^5 x^5}{y^5}$ Power rule (b)

67. $\left(\frac{5}{2}\right)^3 \cdot \left(\frac{5}{2}\right)^2 = \left(\frac{5}{2}\right)^{3+2}$ Product rule

$= \left(\frac{5}{2}\right)^5$

$= \frac{5^5}{2^5}$ Power rule (c)

69. $\left(\frac{9}{8}\right)^3 \cdot 9^2 = \frac{9^3}{8^3} \cdot \frac{9^2}{1}$ Power rule (c)

$= \frac{9^3 \cdot 9^2}{8^3 \cdot 1}$ Multiply fractions

$= \frac{9^{3+2}}{8^3}$

Product rule

$= \frac{9^5}{8^3}$

71. $(2x)^9 (2x)^3 = (2x)^{9+3}$ Product rule

$= (2x)^{12}$

$= 2^{12} x^{12}$ Product rule (b)

73. $(-6p)^4 (-6p)$

$= (-6p)^4 (-6p)^1$

$= (-6p)^5$ Product rule

$= (-1)^5 6^5 p^5$ Power rule (b)

$= -6^5 p^5$

75. $\left(6x^2y^3\right)^5 = 6^5 \left(x^2\right)^5 \left(y^3\right)^5$ Power rule (b)

$= 6^5 x^{2 \cdot 5} y^{3 \cdot 5}$ Power rule (a)

$= 6^5 x^{10} y^{15}$

77. $\left(x^2\right)^3 \left(x^3\right)^5 = x^6 \cdot x^{15}$ Power rule (a)

$= x^{21}$ Power rule

79. $\left(2w^2x^3y\right)^2 \left(x^4y\right)^5$

$= \left[2^2 \left(w^2\right)^2 \left(x^3\right)^2 y^2\right]\left[\left(x^4\right)^5 y^5\right]$

$= \left(2^2 w^4 x^6 y^2\right)\left(x^{20} y^5\right)$

$= 2^2 w^4 \left(x^6 x^{20}\right)\left(y^2 y^5\right)$

$= 4w^4 x^{26} y^7$

81. $\left(-r^4 s\right)^2 \left(-r^2 s^3\right)^5$

$= \left[(-1) r^4 s\right]^2 \left[(-1) r^2 s^3\right]^5$

$= \left[(-1)^2 \left(r^4\right)^2 s^2\right]\left[(-1)^5 \left(r^2\right)^5 \left(s^3\right)^5\right]$

$= \left[(-1)^2 r^8 s^2\right]\left[(-1)^5 r^{10} s^{15}\right]$

$= (-1)^7 r^{18} s^{17}$

$= -r^{18} s^{17}$

83. $\left(\frac{5a^2b^5}{c^6}\right)^3 \quad (c \neq 0)$

$= \frac{\left(5a^2b^5\right)^3}{\left(c^6\right)^3}$ Power rule (c)

$= \frac{5^3 \left(a^2\right)^3 \left(b^5\right)^3}{\left(c^6\right)^3}$ Power rule (b)

$= \frac{125a^6b^{15}}{c^{18}}$ Power rule (a)

85. To simplify using $\left(10^2\right)^3$ as 1000^6 is not correct. Using power rule (a) to simplify $\left(10^2\right)^3$, we obtain

$$\begin{aligned}\left(10^2\right)^3 &= 10^{2\cdot 3}\\ &= 10^6\\ &= 10\cdot 10\cdot 10\cdot 10\cdot 10\cdot 10\\ &= 1{,}000{,}000.\end{aligned}$$

87. Use the formula for the area of a rectangle, $A = LW$, with $L = 4x^3$ and $W = 3x^2$.

$$\begin{aligned}A &= \left(4x^3\right)\left(3x^2\right)\\ &= 4\cdot 3\cdot x^3\cdot x^2\\ &= 12x^5\end{aligned}$$

89. Use the formula for the area of a parallelogram, $A = bh$, with $b = 2p^5$ and $h = 3p^2$.

$$\begin{aligned}A &= \left(2p^5\right)\left(3p^2\right)\\ &= 2\cdot 3\cdot p^5\cdot p^2\\ &= 6p^7\end{aligned}$$

91. Use the formula for the volume of a cube, $V = e^3$, with $e = 5x^2$.

$$\begin{aligned}V &= \left(5x^2\right)^3\\ &= 5^3\left(x^2\right)^3\\ &= 125x^6\end{aligned}$$

93. Use the formula $A = P(1+r)^n$ with $P = \$250$, $r = 0.04$, and $n = 5$.

$$\begin{aligned}A &= 250(1+0.04)^5\\ &= 250(1.04)^5\\ &\approx 304.16\end{aligned}$$

The amount of money in the account will be \$304.16.

95. Use the formula $A = P(1+r)^n$ with $P = \$1500$, $r = 0.015$, and $n = 6$.

$$\begin{aligned}A &= 1500(1+0.015)^6\\ &= 1500(1.015)^6\\ &\approx 1640.16\end{aligned}$$

The amount of money in the account will be \$1640.16.

5.2 Integer Exponents and the Quotient Rule

Now Try Exercises

N1. **(a)** $6^0 = 1$ Definition of zero exponent

(b) $-12^0 = -\left(12^0\right) = -1$

(c) $(-12x)^0 = 1\ (x \neq 0)$

(d) $14^0 - 12^0 = 1 - 1 = 0$

N2. **(a)** $2^{-3} = \frac{1}{2^3}$ Definition of negative exponent

$$= \frac{1}{8}$$

(b) $\left(\frac{1}{7}\right)^{-2} = \left(\frac{7}{1}\right)^2$ $\frac{1}{7}$ and 7 are reciprocals.

$$= 49$$

(c) $\left(\frac{3}{2}\right)^{-4} = \left(\frac{2}{3}\right)^4 = \frac{16}{81}$

(d)
$$\begin{aligned}3^{-2} + 4^{-2} &= \frac{1}{3^2} + \frac{1}{4^2}\\ &= \frac{1}{9} + \frac{1}{16}\\ &= \frac{16}{144} + \frac{9}{144}\\ &= \frac{25}{144}\end{aligned}$$

(e) $p^{-4} = \frac{1}{p^4},\ p \neq 0$

N3. **(a)** $\frac{5^{-3}}{6^{-2}} = \frac{6^2}{5^3}$, or $\frac{36}{125}$

(b) $m^2n^{-4} = \frac{m^2}{1}\cdot\frac{1}{n^4} = \frac{m^2}{n^4}$

(c) $\frac{x^2y^{-3}}{5z^{-4}} = \frac{x^2}{5}\cdot\frac{y^{-3}}{z^{-4}} = \frac{x^2}{5}\cdot\frac{z^4}{y^3} = \frac{x^2z^4}{5y^3}$

N4. **(a)** $\frac{6^3}{6^4} = 6^{3-4} = 6^{-1} = \frac{1}{6^1} = \frac{1}{6}$

(b) $\frac{t^4}{t^{-5}} = t^{4-(-5)} = t^{4+5} = t^9$

(c) $\frac{(p+q)^{-3}}{(p+q)^{-7}} = (p+q)^{-3-(-7)}$

$= (p+q)^{-3+7}$

$= (p+q)^4$

(d) $\frac{5^2 xy^{-3}}{3^{-1}x^{-2}y^2} = \frac{5^2}{3^{-1}} \cdot \frac{x}{x^{-2}} \cdot \frac{y^{-3}}{y^2}$

$= 5^2 3^1 \cdot x^{1-(-2)} + y^{-3-2}$

$= 25 \cdot 3 \cdot x^3 \cdot y^{-5}$

$= \frac{75}{1} \cdot \frac{x^3}{1} \cdot \frac{1}{y^5}$

$= \frac{75x^3}{y^5}$

N5. **(a)** $\frac{3^{15}}{(3^3)^4} = \frac{3^{15}}{3^{12}} = 3^{15-12} = 3^3$, or 27

(b) $(4t)^5(4t)^{-3} = (4t)^{5+(-3)}$

$= (4t)^2$

$= 4^2t^2$, or $16t^2$

(c) $\left(\frac{7y^4}{10}\right)^{-3} = \left(\frac{10}{7y^4}\right)^3$

$= \frac{10^3}{(7y^4)^3}$

$= \frac{10^3}{7^3(y^4)^3}$

$= \frac{10^3}{7^3y^{12}}$, or $\frac{1000}{343y^{12}}$

(d) $\frac{(a^2b^{-2}c)^{-3}}{(2ab^3c^{-4})^5} = \frac{(a^2)^{-3}(b^{-2})^{-3}c^{-3}}{2^5a^5(b^3)^5(c^{-4})^5}$

$= \frac{a^{-6}b^6c^{-3}}{2^5a^5b^{15}c^{-20}}$

$= \frac{b^6c^{20}}{2^5a^5a^6b^{15}c^3}$

$= \frac{c^{20-3}}{2^5a^{5+6}b^{15-6}}$

$= \frac{c^{17}}{2^5a^{11}b^9}$, or $\frac{c^{17}}{32a^{11}b^9}$

(e) $\frac{(5k)^{-6}(5k)^8}{(5k)^7(5k)^{-4}} = \frac{(5k)^{-6+8}}{(5k)^{7+(-4)}}$

$= \frac{(5k)^2}{(5k)^3}$

$= (5k)^{2-3}$

$= (5k)^{-1}$

$= \frac{1}{5k}$

Exercises

1. $(-2)^{-3} = \frac{1}{(-2)^3} = \frac{1}{-8} = -\frac{1}{8}$
The expression is negative.

3. $-2^4 = -1 \cdot 2 \cdot 2 \cdot 2 \cdot 2 = -16$
The expression is negative.

5. $\left(\frac{1}{4}\right)^{-2} = \left(\frac{4}{1}\right)^2 = 16$
The expression is positive.

7. $1 - 5^0 = 1 - 1 = 0$
The expression is zero.

9. By definition, $a^0 = 1\,(a \neq 0)$, so $9^0 = 1$.

11. $(-2)^0 = 1$ Definition of zero exponent

13. $-8^0 = -(8^0) = -(1) = -1$

15. $-(-6)^0 = -1 \cdot (-6)^0 = -1 \cdot 1 = -1$

17. $(-4)^0 - 4^0 = 1 - 1 = 0$

19. $\frac{0^{10}}{12^0} = \frac{0}{1} = 0$

21. $8^0 - 12^0 = 1 - 1 = 0$

23. $\dfrac{0^2}{2^0 + 0^0} = \dfrac{0}{1+0} = \dfrac{0}{1} = 0$

25. **(a)** $x^0 = 1$ (Choice B)

(b) $-x^0 = -1 \cdot x^0 = -1 \cdot 1 = -1$ (Choice C)

(c) $7x^0 = 7 \cdot x^0 = 7 \cdot 1 = 7$ (Choice D)

(d) $(7x)^0 = 1$ (Choice B)

(e) $-7x^0 = -7 \cdot x^0 = -7 \cdot 1 = -7$ (Choice E)

(f) $(-7x)^0 = 1$ (Choice B)

27. $6^0 + 8^0 = 1 + 1 = 2$

29. $4^{-3} = \dfrac{1}{4^3}$ Definition of negative exponent

$= \dfrac{1}{64}$

31. When we evaluate a fraction raised to a negative exponent, we can use a shortcut. Note that $\left(\dfrac{a}{b}\right)^{-n} = \dfrac{1}{\left(\dfrac{a}{b}\right)^n} = \dfrac{1}{\dfrac{a^n}{b^n}} = \dfrac{b^n}{a^n} = \left(\dfrac{b}{a}\right)^n$.

In words, a fraction raised to the negative of a number is equal to its reciprocal raised to the number. We will use the simple phrase "$\dfrac{a}{b}$ and $\dfrac{b}{a}$ are reciprocals" to indicate our use of this evaluation shortcut.

$\left(\dfrac{1}{2}\right)^{-4} = 2^4 = 16$ $\quad \dfrac{1}{2}$ and 2 are reciprocals.

33. $\left(\dfrac{6}{7}\right)^{-2} = \left(\dfrac{7}{6}\right)^2$ $\quad \dfrac{6}{7}$ and $\dfrac{7}{6}$ are reciprocals.

$= \dfrac{7^2}{6^2}$ Power rule (c)

$= \dfrac{49}{36}$

35. $(-3)^{-4} = \dfrac{1}{(-3)^4}$

$= \dfrac{1}{81}$

37. $5^{-1} + 3^{-1} = \dfrac{1}{5} + \dfrac{1}{3}$

$= \dfrac{3}{15} + \dfrac{5}{15} = \dfrac{8}{15}$

39. $3^{-2} - 2^{-1} = \dfrac{1}{3^2} - \dfrac{1}{2^1}$

$= \dfrac{1}{9} - \dfrac{1}{2}$

$= \dfrac{2}{18} - \dfrac{9}{18} = -\dfrac{7}{18}$

41. $\left(\dfrac{1}{2}\right)^{-1} + \left(\dfrac{2}{3}\right)^{-1} = \left(\dfrac{2}{1}\right)^1 + \left(\dfrac{3}{2}\right)^1$

$= \dfrac{4}{2} + \dfrac{3}{2}$

$= \dfrac{7}{2}$

43. $\dfrac{5^{11}}{5^8}$

$= 5^{\underline{11-8}}$

$= 5^{\underline{3}}$

$= \underline{125}$

45. $\dfrac{5^8}{5^5} = 5^{8-5} = 5^3$, or 125

47. $\dfrac{3^{-2}}{5^{-3}} = \dfrac{5^3}{3^2}$, or $\dfrac{125}{9}$

49. $\dfrac{5}{5^{-1}} = \dfrac{5^1}{5^{-1}} = 5^1 \cdot 5^1$

$= 5^{1+1} = 5^2$, or 25

51. $\dfrac{x^{12}}{x^{-3}} = x^{12} \cdot x^3$

$= x^{12+3} = x^{15}$

53. $\dfrac{1}{6^{-3}} = 6^3$, or 216

55. $\dfrac{2}{r^{-4}} = 2r^4$

57. $\dfrac{4^{-3}}{5^{-2}} = \dfrac{5^2}{4^3}$, or $\dfrac{25}{64}$

59. $p^5 q^{-8} = \dfrac{p^5}{q^8}$

61. $\dfrac{r^5}{r^{-4}} = r^5 \cdot r^4 = r^{5+4} = r^9$

Or we can use the quotient rule:

$\dfrac{r^5}{r^{-4}} = r^{5-(-4)} = r^{5+4} = r^9$

63. $\dfrac{x^{-3}y}{4z^{-2}} = \dfrac{yz^2}{4x^3}$

65. Treat the expression in parentheses as a single variable; that is, treat $(a+b)$ as you would treat x.

$$\begin{aligned}\frac{(a+b)^{-3}}{(a+b)^{-4}} &= (a+b)^{-3-(-4)}\\ &= (a+b)^{-3+4}\\ &= (a+b)^1 = a+b\end{aligned}$$

Another method:

$$\begin{aligned}\frac{(a+b)^{-3}}{(a+b)^{-4}} &= \frac{(a+b)^4}{(a+b)^3}\\ &= (a+b)^{4-3}\\ &= (a+b)^1 = a+b\end{aligned}$$

67. $$\begin{aligned}\frac{(x+2y)^{-3}}{(x+2y)^{-5}} &= (x+2y)^{-3-(-5)}\\ &= (x+2y)^{-3+5}\\ &= (x+2y)^2\end{aligned}$$

69. $$\begin{aligned}\frac{(7^4)^3}{7^9} &= \frac{7^{4\cdot 3}}{7^9} && \text{Power rule (a)}\\ &= \frac{7^{12}}{7^9}\\ &= 7^{12-9} && \text{Quotient rule}\\ &= 7^3,\ \text{or } 343\end{aligned}$$

71. $$\begin{aligned}&x^{-3}\cdot x^5\cdot x^{-4}\\ &= x^{-3+5+(-4)} && \text{Product rule}\\ &= x^{-2}\\ &= \frac{1}{x^2} && \text{Definition of negative exponent}\end{aligned}$$

73. $$\begin{aligned}\frac{(3x)^{-2}}{(4x)^{-3}} &= \frac{(4x)^3}{(3x)^2}\\ &= \frac{4^3x^3}{3^2x^2} && \text{Power rule (b)}\\ &= \frac{4^3x^{3-2}}{3^2} && \text{Quotient rule}\\ &= \frac{4^3x}{3^2}\\ &= \frac{64x}{9}\end{aligned}$$

75. $$\begin{aligned}\left(\frac{x^{-1}y}{z^2}\right)^{-2} &= \frac{(x^{-1}y)^{-2}}{(z^2)^{-2}} && \text{Power rule (c)}\\ &= \frac{(x^{-1})^{-2}y^{-2}}{(z^2)^{-2}} && \text{Power rule (b)}\\ &= \frac{x^2y^{-2}}{z^{-4}} && \text{Power rule (a)}\\ &= \frac{x^2z^4}{y^2}\end{aligned}$$

77. $$\begin{aligned}(6x)^4(6x)^{-3} &= (6x)^{4+(-3)} && \text{Product rule}\\ &= (6x)^1 = 6x\end{aligned}$$

79. $$\begin{aligned}\frac{(m^7n)^{-2}}{m^{-4}n^3} &= \frac{(m^7)^{-2}n^{-2}}{m^{-4}n^3}\\ &= \frac{m^{7(-2)}n^{-2}}{m^{-4}n^3}\\ &= \frac{m^{-14}n^{-2}}{m^{-4}n^3}\\ &= m^{-14-(-4)}n^{-2-3}\\ &= m^{-10}n^{-5}\\ &= \frac{1}{m^{10}n^5}\end{aligned}$$

81. $$\begin{aligned}\frac{(x^{-1}y^2z)^{-2}}{(x^{-3}y^3z)^{-1}} &= \frac{(x^{-1})^{-2}(y^2)^{-2}z^{-2}}{(x^{-3})^{-1}(y^3)^{-1}z^{-1}}\\ &= \frac{x^2y^{-4}z^{-2}}{x^3y^{-3}z^{-1}}\\ &= \frac{x^2y^3z^1}{x^3y^4z^2}\\ &= \frac{1}{xyz}\end{aligned}$$

83. $\left(\dfrac{xy^{-2}}{x^2y}\right)^{-3}=\dfrac{x^{-3}\left(y^{-2}\right)^{-3}}{\left(x^2\right)^{-3}y^{-3}}$
$=\dfrac{x^{-3}y^6}{x^{-6}y^{-3}}$
$=\dfrac{x^6y^6y^3}{x^3}$
$=x^3y^9$

85. $\dfrac{(2r)^{-4}(2r)^5}{(2r)^9(2r)^{-7}}=\dfrac{(2r)^{-4+5}}{(2r)^{9+(-7)}}$
$=\dfrac{(2r)^1}{(2r)^2}$
$=\dfrac{1}{2r}$

87. $\dfrac{(-4y)^8(-4y)^{-8}}{(-4y)^{-26}(-4y)^{27}}=\dfrac{(-4y)^{8+(-8)}}{(-4y)^{-26+27}}$
$=\dfrac{(-4y)^0}{(-4y)^1}$
$=\dfrac{1}{-4y}$

89. The student attempted to use the quotient rule with unequal bases. The correct way to simplify this expression is

$\dfrac{16^3}{2^2}=\dfrac{\left(2^4\right)^3}{2^2}=\dfrac{2^{12}}{2^2}=2^{10}=1024.$

91. $\dfrac{\left(4a^2b^3\right)^{-2}\left(2ab^{-1}\right)^3}{\left(a^3b\right)^{-4}}$
$=\dfrac{\left(a^3b\right)^4\left(2ab^{-1}\right)^3}{\left(4a^2b^3\right)^2}$
$=\dfrac{\left(a^3\right)^4b^42^3a^3\left(b^{-1}\right)^3}{4^2\left(a^2\right)^2\left(b^3\right)^2}$
$=\dfrac{a^{12}b^48a^3b^{-3}}{16a^4b^6}$
$=\dfrac{8a^{15}b^1}{16a^4b^6}$
$=\dfrac{a^{11}}{2b^5}$

93. $\dfrac{\left(2y^{-1}z^2\right)^2\left(3y^{-2}z^{-3}\right)^3}{\left(y^3z^2\right)^{-1}}$
$=\dfrac{2^2y^{-2}z^43^3y^{-6}z^{-9}}{y^{-3}z^{-2}}$
$=\dfrac{4\cdot 27y^{-8}z^{-5}}{y^{-3}z^{-2}}$
$=108y^{-5}z^{-3}$
$=\dfrac{108}{y^5z^3}$

95. $\dfrac{\left(9^{-1}z^{-2}x\right)^{-1}\left(4z^2x^4\right)^{-2}}{\left(5z^{-2}x^{-3}\right)^2}$
$=\dfrac{9^1z^2x^{-1}4^{-2}z^{-4}x^{-8}}{5^2z^{-4}x^{-6}}$
$=\dfrac{9z^{-2}x^{-9}}{25\cdot 4^2z^{-4}x^{-6}}$
$=\dfrac{9z^2x^{-3}}{400}$
$=\dfrac{9z^2}{400x^3}$

Summary Exercises Applying the Rules for Exponents

1. $(10x^2y^4)^2(10xy^2)^3$
$=10^2(x^2)^2(y^4)^2\cdot 10^3x^3(y^2)^3$
$=10^2x^4y^810^3x^3y^6$
$=10^5x^7y^{14}$

2. $(-2ab^3c)^4(-2a^2b)^3$
$=(-2)^4a^4(b^3)^4c^4(-2)^3(a^2)^3b^3$
$=16a^4b^{12}c^4(-8)a^6b^3$
$=-128a^{10}b^{15}c^4$

3. $\left(\dfrac{9wx^3}{y^4}\right)^3=\dfrac{\left(9wx^3\right)^3}{\left(y^4\right)^3}$
$=\dfrac{9^3w^3\left(x^3\right)^3}{y^{12}}$
$=\dfrac{729w^3x^9}{y^{12}}$

4. $(4x^{-2}y^{-3})^{-2} = 4^{-2}(x^{-2})^{-2}(y^{-3})^{-2}$
$= \dfrac{1}{4^2}x^{(-2)(-2)}y^{(-3)(-2)}$
$= \dfrac{x^4y^6}{16}$

5. $\dfrac{c^{11}(c^2)^4}{(c^3)^3(c^2)^{-6}} = \dfrac{c^{11}c^8}{c^9c^{-12}}$
$= \dfrac{c^{19}}{c^{-3}}$
$= c^{19-(-3)}$
$= c^{22}$

6. $\left(\dfrac{k^4t^2}{k^2t^{-4}}\right)^{-2} = (k^{4-2}t^{2-(-4)})^{-2}$
$= (k^2t^6)^{-2}$
$= (k^2)^{-2}(t^6)^{-2}$
$= k^{-4}t^{-12}$
$= \dfrac{1}{k^4t^{12}}$

7. $5^{-1} + 6^{-1} = \dfrac{1}{5^1} + \dfrac{1}{6^1}$
$= \dfrac{6}{30} + \dfrac{5}{30}$
$= \dfrac{11}{30}$

8. $\dfrac{(3y^{-1}z^3)^{-1}(3y^2)}{(y^3z^2)^{-3}}$
$= \dfrac{(y^3z^2)^3(3y^2)}{(3y^{-1}z^3)^1}$
$= \dfrac{(y^3)^3(z^2)^3(3y^2)}{3y^{-1}z^3}$
$= \dfrac{y^9z^6 3y^2}{3y^{-1}z^3}$
$= \dfrac{3y^{11}z^6}{3y^{-1}z^3}$
$= y^{12}z^3$

9. $\dfrac{(2xy^{-1})^3}{2^3x^{-3}y^2} = \dfrac{2^3x^3y^{-3}}{2^3x^{-3}y^2}$
$= x^6y^{-5}$
$= \dfrac{x^6}{y^5}$

10. $-4^0 + (-4)^0 = -1\cdot 4^0 + 1$
$= -1\cdot 1 + 1$
$= -1 + 1$
$= 0$

11. $(z^4)^{-3}(z^{-2})^{-5}$
$= z^{-12}z^{10} = z^{-2} = \dfrac{1}{z^2}$

12. $\left(\dfrac{r^2st^5}{3r}\right)^{-2} = \left(\dfrac{rst^5}{3}\right)^{-2}$
$= \left(\dfrac{3}{rst^5}\right)^2$
$= \dfrac{3^2}{(rst^5)^2}$
$= \dfrac{9}{r^2s^2t^{10}}$

13. $\dfrac{(3^{-1}x^{-3}y)^{-1}(2x^2y^{-3})^2}{(5x^{-2}y^2)^{-2}}$
$= \dfrac{(5x^{-2}y^2)^2(2x^2y^{-3})^2}{(3^{-1}x^{-3}y)^1}$
$= \dfrac{5^2x^{-4}y^4 2^2x^4y^{-6}}{3^{-1}x^{-3}y}$
$= \dfrac{25x^0y^{-2}\cdot 4}{3^{-1}x^{-3}y}$
$= 100x^3y^{-3}\cdot 3$
$= \dfrac{300x^3}{y^3}$

14. $\left(\dfrac{5x^2}{3x^{-4}}\right)^{-1} = \left(\dfrac{5x^6}{3}\right)^{-1}$
$= \left(\dfrac{3}{5x^6}\right)^1$
$= \dfrac{3}{5x^6}$

15. $\left(\dfrac{-9x^{-2}}{9x^2}\right)^{-2} = \left(\dfrac{9x^2}{-9x^{-2}}\right)^2$
$= \left(\dfrac{x^4}{-1}\right)^2 = \dfrac{(x^4)^2}{(-1)^2}$
$= \dfrac{x^8}{1}$
$= x^8$

16. $\dfrac{(x^{-4}y^2)^3(x^2y)^{-1}}{(xy^2)^{-3}}$

$= \dfrac{(x^{-4})^3(y^2)^3(x^2)^{-1}y^{-1}}{x^{-3}(y^2)^{-3}}$

$= \dfrac{x^{-12}y^6x^{-2}y^{-1}}{x^{-3}y^{-6}}$

$= \dfrac{x^{-14}y^5}{x^{-3}y^{-6}}$

$= x^{-11}y^{11}$

$= \dfrac{y^{11}}{x^{11}}$

17. $\dfrac{(a^{-2}b^3)^{-4}}{(a^{-3}b^2)^{-2}(ab)^{-4}}$

$= \dfrac{a^8b^{-12}}{a^6b^{-4}a^{-4}b^{-4}}$

$= \dfrac{a^8b^{-12}}{a^2b^{-8}}$

$= a^6b^{-4}$

$= \dfrac{a^6}{b^4}$

18. $(2a^{-30}b^{-29})(3a^{31}b^{30})$

$= (2\cdot 3)a^{-30+31}b^{-29+30}$

$= 6a^1b^1$

$= 6ab$

19. $5^{-2} + 6^{-2} = \dfrac{1}{5^2} + \dfrac{1}{6^2}$

$= \dfrac{1}{25} + \dfrac{1}{36}$

$= \dfrac{36}{25\cdot 36} + \dfrac{25}{25\cdot 36}$

$= \dfrac{36+25}{900}$

$= \dfrac{61}{900}$

20. $\left(\dfrac{(x^{43}y^{23})^2}{x^{-26}y^{-42}}\right)^0 = 1$ zero exponent

21. $\left(\dfrac{7a^2b^3}{2}\right)^3 = \dfrac{(7a^2b^3)^3}{2^3}$

$= \dfrac{7^3a^6b^9}{8}$

$= \dfrac{343a^6b^9}{8}$

22. $-(-19^0) = -(-1\cdot 19^0)$

$= -(-1\cdot 1)$

$= -(-1)$

$= 1$

23. $-(-13)^0 = -1$

24. $\dfrac{0^{13}}{13^0} = \dfrac{0}{1} = 0$

25. $\dfrac{(2xy^{-3})^{-2}}{(3x^{-2}y^4)^{-3}}$

$= \dfrac{(3x^{-2}y^4)^3}{(2xy^{-3})^2}$

$= \dfrac{3^3x^{-6}y^{12}}{2^2x^2y^{-6}}$

$= \dfrac{27x^{-8}y^{18}}{4}$

$= \dfrac{27y^{18}}{4x^8}$

26. $\left(\dfrac{a^2b^3c^4}{a^{-2}b^{-3}c^{-4}}\right)^{-2}$

$= (a^4b^6c^8)^{-2}$

$= \dfrac{1}{(a^4b^6c^8)^2}$

$= \dfrac{1}{a^8b^{12}c^{16}}$

27. $(6x^{-5}z^3)^{-3}$

$= 6^{-3}x^{15}z^{-9}$

$= \dfrac{x^{15}}{6^3z^9}$

$= \dfrac{x^{15}}{216z^9}$

28. $(2p^{-2}qr^{-3})(2p)^{-4}$
$= 2p^{-2}qr^{-3}2^{-4}p^{-4}$
$= 2^{-3}p^{-6}qr^{-3}$
$= \dfrac{q}{8p^6r^3}$

29. $\dfrac{(xy)^{-3}(xy)^5}{(xy)^{-4}} = (xy)^{-3+5-(-4)}$
$= (xy)^6$
$= x^6y^6$

30. $52^0 - (-8)^0 = 1 - 1 = 0$

31. $\dfrac{(7^{-1}x^{-3})^{-2}(x^4)^{-6}}{7^{-1}x^{-3}}$
$= \dfrac{7^2x^6x^{-24}}{7^{-1}x^{-3}}$
$= 7^{2-(-1)}x^{6-24-(-3)}$
$= 7^3x^{-15}$
$= \dfrac{343}{x^{15}}$

32. $\left(\dfrac{3^{-4}x^{-3}}{3^{-3}x^{-6}}\right)^{-2} = \left(\dfrac{x^3}{3}\right)^{-2}$
$= \left(\dfrac{3}{x^3}\right)^2$
$= \dfrac{9}{x^6}$

33. $(5p^{-2}q)^{-3}(5pq^3)^4$
$= 5^{-3}p^6q^{-3}5^4p^4q^{12}$
$= 5p^{10}q^9$

34. $8^{-1} + 6^{-1} = \dfrac{1}{8^1} + \dfrac{1}{6^1}$
$= \dfrac{3}{3\cdot 8} + \dfrac{4}{4\cdot 6}$
$= \dfrac{7}{24}$

35. $\left(\dfrac{4r^{-6}s^{-2}t}{2r^8s^{-4}t^2}\right)^{-1}$
$= \left(\dfrac{2s^2}{r^{14}t}\right)^{-1}$
$= \dfrac{r^{14}t}{2s^2}$

36. $(13x^{-6}y)(13x^{-6}y)^{-1}$
$= (13x^{-6}y)^{1+(-1)}$
$= (13x^{-6}y)^0$
$= 1$

37. $\dfrac{(8pq^{-2})^4}{(8p^{-2}q^{-3})^3}$
$= \dfrac{8^4p^4q^{-8}}{8^3p^{-6}q^{-9}}$
$= 8^{4-3}p^{4-(-6)}q^{-8-(-9)}$
$= 8p^{10}q$

38. $\left(\dfrac{mn^{-2}p}{m^2np^4}\right)^{-2}\left(\dfrac{mn^{-2}p}{m^2np^4}\right)^3$
$= \left(\dfrac{mn^{-2}p}{m^2np^4}\right)^{-2+3}$
$= \left(\dfrac{mn^{-2}p}{m^2np^4}\right)^1$
$= \dfrac{1}{m^{2-1}n^{1-(-2)}p^{4-1}}$
$= \dfrac{1}{mn^3p^3}$

39. $-(-8^0)^0 = -1$

40. **(a)** $2^0 + 2^0 = 1 + 1 = 2$ (D)

(b) $2^1 \cdot 2^0 = 2\cdot 1 = 2$ (D)

(c) $2^0 - 2^{-1} = 1 - \dfrac{1}{2} = \dfrac{1}{2}$ (E)

(d) $2^1 - 2^0 = 2 - 1 = 1$ (B)

(e) $2^0 \cdot 2^{-2} = 1\cdot\dfrac{1}{2^2} = 1\cdot\dfrac{1}{4} = \dfrac{1}{4}$ (J)

(f) $2^1 \cdot 2^1 = 2\cdot 2 = 4$ (F)

(g) $2^{-2} - 2^{-1} = \dfrac{1}{2^2} - \dfrac{1}{2^1}$
$= \dfrac{1}{4} - \dfrac{1}{2} = \dfrac{1}{4} - \dfrac{2}{4} = -\dfrac{1}{4}$ (I)

(h) $2^0 \cdot 2^0 = 1\cdot 1 = 1$ (B)

(i) $2^{-2} \div 2^{-1} = \frac{1}{2^2} \div \frac{1}{2^1}$

$= \frac{1}{4} \div \frac{1}{2} = \frac{1}{4} \cdot \frac{2}{1} = \frac{1}{2}$ (E)

(j) $2^0 \div 2^{-2} = 1 \div \frac{1}{2^2}$

$= 1 \div \frac{1}{4} = 1 \cdot \frac{4}{1} = 4$ (F)

5.3 Scientific Notation

Now Try Exercises

N1. **(a)** $12,600,000 = 1.26 \times 10^7$

The decimal point has been moved seven places to put it after the first nonzero digit. Since 1.26 is less than the original number, it must be multiplied by a number greater than 1 to get 12,600,000. Thus, the exponent on 10 must be positive.

(b) $0.00027 = 2.7 \times 10^{-4}$

The decimal point has been moved four places to put it after the first nonzero digit. Since 2.7 is greater than the original number, it must be multiplied by a number less than 1 to get 0.00027. Thus, the exponent on 10 must be negative.

(c) $-0.0000341 = -3.41 \times 10^{-5}$

Work with the absolute value of the number, and then apply the negative sign to the answer.

N2. **(a)** $5.71 \times 10^4 = 57,100$

Since the exponent is positive, move the decimal point four places to the *right*.

(b) $2.72 \times 10^{-5} = 0.0000272$

Since the exponent is negative, move the decimal point five places to the *left*.

(c) $-8.81 \times 10^{-4} = -0.000881$

Since the exponent is negative, move the decimal point five places to the *left*.

N3. **(a)** $(6 \times 10^7)(7 \times 10^{-4})$

$= (6 \times 7)(10^7 \times 10^{-4})$

$= 42 \times 10^3$

$= (4.2 \times 10^1) \times 10^3$

$= 4.2 \times (10^1 \times 10^3)$

$= 4.2 \times 10^4$, or 42,000

(b) $\frac{18 \times 10^{-3}}{6 \times 10^4} = \frac{18}{6} \times \frac{10^{-3}}{10^4}$

$= 3 \times 10^{-3-4}$

$= 3 \times 10^{-7}$, or 0.0000003

N4. $(8,000,000)(0.00000003937)$

$= (8 \times 10^6) \times (3.937 \times 10^{-8})$

$= (8 \times 3.937) \times (10^6 \times 10^{-8})$

$= 31.496 \times 10^{6+(-8)}$

$= 31.496 \times 10^{-2}$

$= 3.1496 \times 10^{-1}$, or 0.31496

Thus, 8,000,000 nanometers would measure 3.1496×10^{-1} in., or 0.31496 in.

N5. Divide the land area by the population.

$\frac{1.6 \times 10^5}{3.8 \times 10^7} = \frac{1.6}{3.8} \times \frac{10^5}{10^7}$

$= 0.42 \times 10^{5-7}$

$= 0.42 \times 10^{-2}$

$= 4.2 \times 10^{-3}$

$= 0.0042$

In 2008, the number of square miles per person living in California was about 4.2×10^{-3} mi^2, or 0.0042 mi^2.

Exercises

1. **(a)** Move the decimal point to the left four places due to the exponent, -4.

$4.6 \times 10^{-4} = 0.00046$

Choice C is correct.

(b) $4.6 \times 10^4 = 46,000$

Choice A is correct.

(c) Move the decimal point to the right five places due to the exponent, 5.
$4.6 \times 10^5 = 460{,}000$
Choice B is correct.

(d) $4.6 \times 10^{-5} = 0.000046$
Choice D is correct.

3. 4.56×10^4 is written in scientific notation because 4.56 is between 1 and 10, and 10^4 is a power of 10.

5. 5,600,000 is not in scientific notation. It can be written in scientific notation as 5.6×10^6.

7. 0.8×10^2 is not in scientific notation because $|0.8| = 0.8$ is not greater than or equal to 1 and less than 10. It can be written in scientific notation as 8×10^1.

9. 0.004 is not in scientific notation because $|0.004| = 0.004$ is not between 1 and 10. It can be written in scientific notation as 4×10^{-3}.

11. **(a)** 63,000
The first nonzero digit is 6. The decimal point should be moved four places.
$63{,}000 = 6.3 \times 10^4$

(b) 0.0571
The first nonzero digit is 5. The decimal point should be moved two places.
$0.0571 = 5.71 \times 10^{-2}$

13. 5,876,000,000
Move the decimal point to the right of the first nonzero digit and count the number of places the decimal point was moved.
The decimal point was moved nine places.
Because moving the decimal point to the *left* made the number *smaller*, we must multiply by a *positive* power of 10 so that the product 5.876×10^n will equal the larger number. Thus, $n = 9$, and $5{,}876{,}000{,}000 = 5.876 \times 10^9$.

15. 82,350
Move the decimal point left four places so it is to the right of the first nonzero digit.
The decimal point was moved four places.
Since the number got smaller, multiply by a positive power of 10.
$82{,}350 = 8.2350 \times 10^4 = 8.235 \times 10^4$
(Note that the final zero need not be written.)

17. 0.000007
Move the decimal point to the right of the first nonzero digit.
The decimal point was moved six places.
Since moving the decimal point to the *right* made the number *larger*, we must multiply by a *negative* power of 10 so that the product 7×10^n will equal the smaller number. Thus, $n = -6$, and $0.000007 = 7 \times 10^{-6}$.

19. 0.00203
To move the decimal point to the right of the first nonzero digit, we move it three places.
Since 2.03 is larger than 0.00203, the exponent on 10 must be negative.
$0.00203 = 2.03 \times 10^{-3}$

21. −13,000,000
Move the decimal point to the right of the first nonzero digit and count the number of places the decimal point was moved.
The decimal point was moved seven places.
Because moving the decimal point to the *left* made the number *smaller*, we must multiply by a *positive* power of 10 so that the product -1.3×10^n will equal the larger number. Thus, $n = 7$, and $-13{,}000{,}000 = -1.3 \times 10^7$.

23. −0.006
To move the decimal point to the right of the first nonzero digit, we move it three places.
Since 6 is larger than 0.006, the exponent on 10 must be negative. Remember to include the negative sign.
$-0.006 = -6 \times 10^{-3}$

25. 7.5×10^5
Since the exponent is positive, make 7.5 larger by moving the decimal point five places to the right.
$7.5 \times 10^5 = 750{,}000$

27. 5.677×10^{12}
Since the exponent is positive, make 5.677 larger by moving the decimal point 12 places to the right. We need to add 9 zeros.
$5.677 \times 10^{12} = 5{,}677{,}000{,}000{,}000$

29. 1×10^{12}
Since the exponent is positive, make 1 larger by moving the decimal point 12 places to the right.
We need to add 12 zeros.
$1 \times 10^{12} = 1{,}000{,}000{,}000{,}000$

31. 6.21×10^0

Because the exponent is 0, the decimal point should not be moved.

$6.21\times10^0 = 6.21$

We know this result is correct because $10^0 = 1$.

33. 7.8×10^{-4}

Since the exponent is negative, make 7.8 smaller by moving the decimal point four places to the left.

$7.8\times10^{-4} = 0.00078$

35. 5.134×10^{-9}

Since the exponent is negative, make 5.134 smaller by moving the decimal point nine places to the left.

$5.134\times10^{-9} = 0.000000005134$

37. -4×10^{-3}

Move the decimal point three places to the left.

$-4\times10^{-3} = -0.004$

39. -8.1×10^5

Move the decimal point five places to the right.

$-8.1\times10^5 = -810{,}000$

41. **(a)** $(2\times10^8)(3\times10^3)$

$= (2\times3)(10^8\times10^3)$

$= 6\times10^{11}$

(b) $6\times10^{11} = 600{,}000{,}000{,}000$

43. **(a)** $(5\times10^4)(3\times10^2)$

$= (5\times3)(10^4\times10^2)$

$= 15\times10^6$

$= 1.5\times10^7$

(b) $1.5\times10^7 = 15{,}000{,}000$

45. **(a)** $(3\times10^{-4})(-2\times10^8)$

$= (3\times(-2))(10^{-4}\times10^8)$

$= -6\times10^4$

(b) $-6\times10^4 = -60{,}000$

47. **(a)** $(6\times10^3)(4\times10^{-2})$

$= (6\times4)(10^3\times10^{-2})$

$= 24\times10^1$

$= 2.4\times10^2$

(b) $2.4\times10^2 = 240$

49. **(a)** $(9\times10^4)(7\times10^{-7})$

$= (9\times7)(10^4\times10^{-7})$

$= 63\times10^{-3}$

$= 6.3\times10^{-2}$

(b) $6.3\times10^{-2} = 0.063$

51. **(a)** $\frac{9\times10^{-5}}{3\times10^{-1}} = \frac{9}{3}\times\frac{10^{-5}}{10^{-1}}$

$= 3\times10^{-5-(-1)}$

$= 3\times10^{-4}$

(b) $3\times10^{-4} = 0.0003$

53. **(a)** $\frac{8\times10^3}{-2\times10^2} = \frac{8}{-2}\times\frac{10^3}{10^2}$

$= -4\times10^1 = -4\times10$

(b) $-4\times10^1 = -40$

55. **(a)** $\frac{2.6\times10^{-3}}{2\times10^2} = \frac{2.6}{2}\times\frac{10^{-3}}{10^2}$

$= 1.3\times10^{-5}$

(b) $1.3\times10^{-5} = 0.000013$

57. **(a)** $\frac{4\times10^5}{8\times10^2} = \frac{4}{8}\times\frac{10^5}{10^2}$

$= 0.5\times10^{5-2}$

$= 0.5\times10^3$

$= 5\times10^2$

(b) $5\times10^2 = 500$

59. **(a)** $\frac{-4.5\times10^4}{1.5\times10^{-2}} = \frac{-4.5}{1.5}\times\frac{10^4}{10^{-2}}$

$= -3\times10^{4-(-2)}$

$= -3\times10^6$

(b) $-3\times10^6 = -3{,}000{,}000$

61. **(a)** $\frac{-8\times10^{-4}}{-4\times10^{3}}=\frac{-8}{-4}\times\frac{10^{-4}}{10^{3}}$
$=2\times10^{-4-3}$
$=2\times10^{-7}$

(b) $2\times10^{-7}=0.0000002$

63. $0.00000047=4.7\times10^{-7}$

Prediction: $4.7\text{E}^{-}7$

65. $(8\text{E}5)/(4\text{E}^{-}2)=\frac{8\times10^{5}}{4\times10^{-2}}$
$=\frac{8}{4}\times10^{5-(-2)}$
$=2\times10^{7}$

Prediction: 2 E7

67. $(2\text{E}6)*(2\text{E}^{-}3)/(4\text{E}2)$
$=\frac{(2\times10^{6})(2\times10^{-3})}{4\times10^{2}}$
$=\frac{(2\times2)\times(10^{6}\times10^{-3})}{4\times10^{2}}$
$=\frac{4\times10^{3}}{4\times10^{2}}$
$=\frac{4}{4}\times10^{3-2}$
$=1\times10^{1}$

Prediction: 1 E1

69. $\frac{650{,}000{,}000(0.0000032)}{0.00002}$
$=\frac{(6.5\times10^{8})(3.2\times10^{-6})}{2\times10^{-5}}$
$=\frac{20.8\times10^{2}}{2\times10^{-5}}$
$=\frac{20.8}{2}\times\frac{10^{2}}{10^{-5}}$
$=10.4\times10^{7}$
$=1.04\times10^{8}$

71. $\frac{0.00000072(0.00023)}{0.000000018}$
$=\frac{(7.2\times10^{-7})(2.3\times10^{-4})}{1.8\times10^{-8}}$
$=\frac{16.56\times10^{-11}}{1.8\times10^{-8}}$
$=9.2\times10^{-3}$

73. $\frac{0.0000016(240{,}000{,}000)}{0.00002(0.0032)}$
$=\frac{(1.6\times10^{-6})(2.4\times10^{8})}{(2\times10^{-5})(3.2\times10^{-3})}$
$=\frac{3.84\times10^{2}}{6.4\times10^{-8}}$
$=0.6\times10^{10}$
$=6\times10^{9}$

75. $2\times10^{-6}=0.000002$

77. Move the decimal point to the right of the first nonzero digit and count the number of places the decimal point was moved. The decimal point was moved 42 places. Because moving the decimal point to the *left* made the number *smaller*, we must multiply by a *positive* power of 10. The electromagnetic attraction is written as 4.2×10^{42} in scientific notation.

79. To find the number of miles, multiply the number of light-years by the number of miles per light-year.
$(25{,}000)(6{,}000{,}000{,}000{,}000)$
$=(2.5\times10^{4})(6\times10^{12})$
$=(2.5\times6)(10^{4}\times10^{12})$
$=15\times10^{16}$
$=1.5\times10^{17}$

The nebula is about 1.5×10^{17} miles from Earth.

81. To find the amount each person would have to contribute, divide the amount by the population.

$$\frac{\$1,000,000,000,000}{313.9 \text{ million}} = \frac{\$1\times10^{12}}{313,900,000}$$
$$= \frac{\$1\times10^{12}}{3.139\times10^{8}}$$
$$= \frac{\$1}{3.139}\times\frac{10^{12}}{10^{8}}$$
$$= \$0.3186\times10^{4}$$
$$= \$3.186\times10^{3}$$
$$= \$3186$$

Each person would have to contribute about $3186.

83. To find the debt per person, divide the debt by the population.

$$\frac{\$1.67\times10^{13}}{314 \text{ million}} = \frac{\$1.67\times10^{13}}{314,000,000}$$
$$= \frac{\$1.67\times10^{13}}{3.14\times10^{8}}$$
$$= \$0.53185\times10^{5}$$
$$= \$5.3185\times10^{4}$$
$$= \$53,185$$

The debt will be about $53,185 for every person.

85. Use the formula $d = rt$.

$$\text{distance} = \text{rate}\times\text{time}$$
$$6.68\times10^{7} = \left(1.86\times10^{5}\right)\times\text{time}$$
$$\frac{6.68\times10^{7}}{1.86\times10^{5}} = \text{time}$$
$$\text{time} = \frac{6.68}{1.86}\times\frac{10^{7}}{10^{5}}$$
$$\approx 3.59\times10^{2}$$

It takes about 3.59×10^{2}, or 359, seconds for light to travel from the sun to Venus.

87. To get the average ticket price, divide the total gross by the attendance.

$$\frac{\$1.14\times10^{9}}{1.16\times10^{7}} = \frac{\$1.14}{1.16}\times\frac{10^{9}}{10^{7}}$$
$$= \frac{\$1.14}{1.16}\times10^{9-7}$$
$$= \$0.9828\times10^{2}$$
$$= \$98.28$$

The average ticket price was about $98.28.

89. First, write 10,000,000,000,000,000 in scientific notation.

$$10,000,000,000,000,000 = 10^{16}$$

Multiply by the number of seconds in a minute.

$$60\left(10^{16}\right) = \left(6\times10^{1}\right)\left(10^{16}\right)$$
$$= (6)\left(10^{1}\times10^{16}\right)$$
$$= 6\times10^{17}$$

The computer can perform 6×10^{17}, or 600,000,000,000,000,000, calculations in one minute.

Multiply the number of calculations in a minute by the number of minutes in an hour.

$$60\left(6\times10^{17}\right) = \left(6\times10^{1}\right)\left(6\times10^{17}\right)$$
$$= (6\times6)\left(10^{1}\times10^{17}\right)$$
$$= 36\times10^{18}$$
$$= 3.6\times10^{19}$$

The computer can perform 3.6×10^{19}, or 36,000,000,000,000,000,000, calculations in one hour.

91. Calculate the ratio of the earthquake intensities to compare them.

$$\frac{\text{intensity } 9.5}{\text{intensity } 8.5} = \frac{I_0\times10^{9.5}}{I_0\times10^{8.5}} = \frac{10^{9.5}}{10^{8.5}}$$
$$= 10^{9.5-8.5}$$
$$= 10^{1}$$

The 1960 Chile earthquake was 10 times as intense as the 2007 Southern Sumatra earthquake.

93. Calculate the ratio of the earthquake intensities to compare them.

$$\frac{\text{intensity } 6.9}{\text{intensity } 5.9} = \frac{I_0 \times 10^9}{I_0 \times 10^{8.5}} = \frac{10^9}{10^{8.5}}$$
$$= 10^{9-8.5}$$
$$= 10^{0.5}$$
$$= \sqrt{10}$$
$$\approx 3.16$$

The 1952 Karchatka earthquake was about 3.16 times as intense as the 2007 Southern Sumatra earthquake.

5.4 Adding, Subtracting, and Graphing Polynomials

Now Try Exercises

N1. The coefficient of t, or $1t$, is 1 and the coefficient of $-10t^2$, is -10. There are two terms.

N2. **(a)** $x - \frac{2}{5}x$

$= \frac{5}{5}x - \frac{2}{5}x$

$= \left(\frac{5}{5} - \frac{2}{5}\right)x$ Distributive property

$= \frac{3}{5}x$

(b) $3x^2 - x^2 + 2x$

$= \left(3x^2 - 1x^2\right) + 2x$ Associative property

$= (3-1)x^2 + 2x$ Distributive property

$= 2x^2 + 2x$

N3. **(a)** $3x^2 + 2x - 4$

The terms are unlike terms and the polynomial cannot be simplified. The largest exponent on the variable x is 2, so the degree of the polynomial is 2.
The polynomial has three terms, so it is a trinomial.

(b) $x^3 + 4x^3$

$= (1+4)x^3$ Distributive property

$= 5x^3$

The largest exponent on the variable x is 3, so the degree of the polynomial is 3.
The polynomial has one term, so it is a monomial.

(c) $x^8 - x^7 + 2x^8$

$= 1x^8 + 2x^8 - 1x^7$ Commutative property

$= \left(1x^8 + 2x^8\right) - 1x^7$ Associative property

$= (1+2)x^8 - 1x^7$ Distributive property

$= 3x^8 - x^7$

The largest exponent on the variable x is 8, so the degree of the polynomial is 8.
The polynomial has two terms, so it is a binomial.

N4. Replace t with -3.

$4t^3 - t^2 - t$

$= 4(-3)^3 - (-3)^2 - (-3)$

$= 4(-27) - (9) + 3$

$= -108 - 9 + 3$

$= -117 + 3$

$= -114$

N5. **(a)** Add, column by column.

$$\begin{array}{rrr} 7y^3 & -\ 4y^2 & +\ 2 \\ -6y^3 & +\ 5y^2 & -\ 3 \\ \hline y^3 & +\ y^2 & -\ 1 \end{array}$$

(b) Add, column by column.

$$\begin{array}{rrrr} -5x^4 & & -\ 2x & +\ 3 \\ & x^3 & -\ 5x & \\ \hline -5x^4 & +\ x^3 & -\ 7x & +\ 3 \end{array}$$

N6. Combine like terms.

$\left(10x^4 - 3x^2 - x\right) + \left(x^4 - 3x^2 + 5x\right)$

$= \left(10x^4 + x^4\right) + \left(-3x^2 - 3x^2\right) + (-x + 5x)$

$= 11x^4 - 6x^2 + 4x$

N7. **(a)** Change subtraction to addition.

$(3x-8) - (5x-9)$

$= (3x-8) + (-5x+9)$

Combine like terms.

$(3x-8) + (-5x+9)$

$= (3x + (-5x)) + (-8+9)$

$= -2x + 1$

Check by adding.

$5x - 9$	Second polynomial
$-2x + 1$	Answer
$3x - 8$	First polynomial

(b) Change subtraction to addition.

$(4t^4 - t^2 + 7) - (5t^4 - 3t^2 + 1)$

$= (4t^4 - t^2 + 7) + (-5t^4 + 3t^2 - 1)$

Combine like terms.

$\left(4t^4 - t^2 + 7\right) + \left(-5t^4 + 3t^2 - 1\right)$

$= \left(4t^4 - 5t^4\right) + \left(-t^2 + 3t^2\right) + (7 - 1)$

$= -t^4 + 2t^2 + 6$

Check by adding.

$5t^4 - 3t^2 + 1$	Second polynomial
$-t^4 + 2t^2 + 6$	Answer
$4t^4 - t^2 + 7$	First polynomial

N8. Arrange like terms in columns. Insert zeros for missing terms.

$12x^2 - 9x + 4$

$-10x^2 - 3x + 7$

Change all signs in the second row, and then add.

$12x^2 - 9x + 4$

$10x^2 + 3x - 7$

$22x^2 - 6x - 3$

N9. Change subtraction to addition.

$\left(6p^4 - 8p^3 + 2p - 1\right)$

$+\left(7p^4 - 6p^2 + 12\right)$

$+\left(p^4 - 3p + 8\right)$

Combine like terms. Arrange like terms in columns. Insert zeros for missing terms.

$6p^4 - 8p^3 + 0 + 2p - 1$

$7p^4 + 0 - 6p^2 + 0 + 12$

$p^4 + 0 + 0 - 3p + 8$

$14p^4 - 8p^3 - 6p^2 - p + 19$

N10. $(4x^2 - 2xy + y^2) - (6x^2 - 7xy + 2y^2)$

$= 4x^2 - 2xy + y^2 - 6x^2 + 7xy - 2y^2$

$= (4x^2 - 6x^2) + (-2xy + 7xy) + (y^2 - 2y^2)$

$= -2x^2 + 5xy - y^2$

N11. Graph $y = -x^2 - 1$.

Make a table of ordered pairs whose x-values are on either side of the vertex's x-value of $x = 0$.

x	$y = -x^2 - 1$
0	$-(0)^2 - 1 = 0 - 1 = -1$
± 1	$-(\pm 1)^2 - 1 = -1 - 1 = -2$
± 2	$-(\pm 2)^2 - 1 = -4 - 1 = -5$

Plot these seven ordered pairs and connect them with a smooth curve.

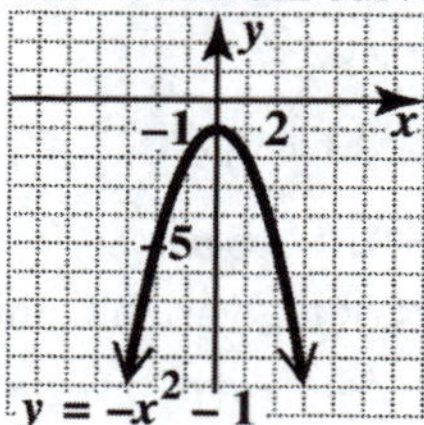

Exercises

1. In the term $4x^6$, the coefficient of x^6 is $\underline{4}$ and the exponent is $\underline{6}$.

3. The degree of the term $-3x^9$ is $\underline{9}$, which is the exponent.

5. When $x^2 + 10$ is evaluated for $x = 3$, the result is $3^2 + 10 = 9 + 10 = \underline{19}$.

7. $-3xy - 2xy + 5xy = (-3 - 2 + 5)xy$

$= (0)xy$

$= 0$

So, combining like terms in $-3xy - 2xy + 5xy$ gives $\underline{0}$.

9. The polynomial $6x^4$ has one term. The coefficient of this term is 6.

11. The polynomial t^4 has one term. Since $t^4 = 1 \cdot t^4$, the coefficient of this term is 1.

13. The polynomial $-19r^2 - r$ has two terms. The coefficient of r^2 is -19 and the coefficient of r is -1.

15. The polynomial $x + 8x^2 + 5x^3$ has three terms. The coefficient of x is 1, the coefficient of x^2 is 8, and the coefficient of x^3 is 5.

17. $-3m^5 + 5m^5 = (-3+5)m^5$
$= 2m^5$

19. $2r^5 + (-3r^5) = [2 + (-3)]r^5$
$= -1r^5 = -r^5$

21. The polynomial $0.2m^5 - 0.5m^2$ cannot be simplified. The two terms are unlike because the exponents on the variables are different, so they cannot be combined.

23. $-3x^5 + 3x^5 - 5x^5 = (-3+3-5)x^5$
$= -5x^5$

25. $-4p^7 + 8p^7 + 5p^9 = (-4+8)p^7 + 5p^9$
$= 4p^7 + 5p^9$

In descending powers of the variable, this polynomial is written $5p^9 + 4p^7$.

27. $-1.5x^2 + 5.3x^2 - 3.8x^2$
$= (-1.5 + 5.3 - 3.8)x^2$
$= 0x^2 = 0$

29. $-4xy^2 + 3xy^2 - 2xy^2 + xy^2$
$= (-4+3-2+1)xy^2$
$= -2xy^2$

31. $-\frac{1}{3}tu^7 + \frac{2}{5}tu^7 + \frac{1}{15}tu^7 - \frac{8}{5}tu^7$
$= \left(-\frac{1}{3} + \frac{2}{5} + \frac{1}{15} - \frac{8}{5}\right)tu^7$
$= \left(-\frac{5}{15} + \frac{6}{15} + \frac{1}{15} - \frac{24}{15}\right)tu^7$
$= -\frac{22}{15}tu^7$

33. This polynomial has no like terms, so it is already simplified. It is already written in descending powers of the variable *x*. The highest degree of any nonzero term is 4, so the degree of the polynomial is 4. There are two terms, so this is a *binomial.*

35. $5m^4 - 3m^2 + 6m^4 - 7m^3$
$= (5m^4 + 6m^4) + (-7m^3) + (-3m^2)$
$= 11m^4 - 7m^3 - 3m^2$

The resulting polynomial is a *trinomial* of degree 4.

37. $\frac{5}{3}x^4 - \frac{2}{3}x^4 = \left(\frac{5}{3} - \frac{2}{3}\right)x^4$
$= \frac{3}{3}x^4 = x^4$

The resulting polynomial is a *monomial* of degree 4.

39. $0.8x^4 - 0.3x^4 - 0.5x^4 + 7$
$= (0.8 - 0.3 - 0.5)x^4 + 7$
$= 0x^4 + 7 = 7$

Since 7 can be written as $7x^0$, the degree of the polynomial is 0. The simplified polynomial has one term, so it is a *monomial.*

41. $-11ab + 2ab - 4ab$
$= (-11 + 2 - 4)ab$
$= -13ab$

The variables each have an exponent of 1, so the degree of the polynomial is 2. The simplified polynomial has one term, so it is a *monomial.*

43. **(a)** $2x^2 - 3x - 5$
$= 2(2)^2 - 3(2) - 5$ Let $x = 2$.
$= 2(4) - 6 - 5$
$= 8 - 6 - 5$
$= 2 - 5$
$= -3$

(b) $2x^2 - 3x - 5$
$= 2(-1)^2 - 3(-1) - 5$ Let $x = -1$.
$= 2(1) + 3 - 5$
$= 2 + 3 - 5$
$= 5 - 5$
$= 0$

45. **(a)** $-3x^2 + 14x - 2$
$= -3(2)^2 + 14(2) - 2$ Let $x = 2$.
$= -3(4) + 28 - 2$
$= -12 + 28 - 2$
$= 16 - 2$
$= 14$

(b) $-3x^2 + 14x - 2$

$= -3(-1)^2 + 14(-1) - 2$ Let $x = -1$.

$= -3(1) - 14 - 2$

$= -3 - 14 - 2$

$= -17 - 2$

$= -19$

47. **(a)** $2x^5 - 4x^4 + 5x^3 - x^2$

$= 2(2)^5 - 4(2)^4 + 5(2)^3 - (2)^2$ Let $x = 2$.

$= 2(32) - 4(16) + 5(8) - 4$

$= 64 - 64 + 40 - 4$

$= 36$

(b) $2x^5 - 4x^4 + 5x^3 - x^2$

Let $x = -1$.

$2x^5 - 4x^4 + 5x^3 - x^2$

$= 2(-1)^5 - 4(-1)^4 + 5(-1)^3 - (-1)^2$

$= 2(-1) - 4(1) + 5(-1) - 1$

$= -2 - 4 - 5 - 1$

$= -12$

49. Add, column by column.

$$\begin{array}{rr} 2x^2 & -\ 4x \\ \underline{3x^2} & \underline{+\ 2x} \\ 5x^2 & -\ 2x \end{array}$$

51. Add, column by column.

$$\begin{array}{rrr} 3m^2 & +\ 5m & +\ 6 \\ \underline{2m^2} & \underline{-\ 2m} & \underline{-\ 4} \\ 5m^2 & +\ 3m & +\ 2 \end{array}$$

53. Add, column by column.

$$\begin{array}{rrr} \frac{2}{3}x^2 & +\ \frac{1}{5}x & +\ \frac{1}{6} \\ \underline{\frac{1}{2}x^2} & \underline{-\ \frac{1}{3}x} & \underline{+\ \frac{2}{3}} \end{array}$$

Rewrite the fractions so that the fractions in each column have a common denominator; then add column by column.

$$\begin{array}{rrr} \frac{4}{6}x^2 & +\ \frac{3}{15}x & +\ \frac{1}{6} \\ \underline{\frac{3}{6}x^2} & \underline{-\ \frac{5}{15}x} & \underline{+\ \frac{4}{6}} \\ \frac{7}{6}x^2 & -\ \frac{2}{15}x & +\ \frac{5}{6} \end{array}$$

55. $(9m^3 - 5m^2 + 4m - 8) + (-3m^3 + 6m^2 - 6)$

$= (9m^3 - 3m^3) + (-5m^2 + 6m^2) + 4m + (-8 - 6)$

$= 6m^3 + m^2 + 4m - 14$

57. Subtract, column by column.

$$\begin{array}{rr} 5y^3 & -\ 3y^2 \\ \underline{2y^3} & \underline{+\ 8y^2} \end{array}$$

Change all signs in the second row, and then add.

$$\begin{array}{rr} 5y^3 & -\ 3y^2 \\ \underline{-2y^3} & \underline{-\ 8y^2} \\ 3y^3 & -\ 11y^2 \end{array}$$

59. Subtract, column by column.

$$\begin{array}{rrr} 12x^4 & -\ x^2 & +\ x \\ \underline{8x^4} & \underline{+\ 3x^2} & \underline{-\ 3x} \end{array}$$

Change all signs in the second row, and then add.

$$\begin{array}{rrr} 12x^4 & -\ x^2 & +\ x \\ \underline{-8x^4} & \underline{-\ 3x^2} & \underline{+\ 3x} \\ 4x^4 & -\ 4x^2 & +\ 4x \end{array}$$

61. Subtract, column by column.

$$\begin{array}{rrrr} 12m^3 & -\ 8m^2 & +\ 6m & +\ 7 \\ \underline{-3m^3} & \underline{+\ 5m^2} & \underline{-\ 2m} & \underline{-\ 4} \end{array}$$

Change all signs in the second row, and then add.

$$\begin{array}{rrrr} 12m^3 & -\ 8m^2 & +\ 6m & +\ 7 \\ \underline{3m^3} & \underline{-\ 5m^2} & \underline{+\ 2m} & \underline{+\ 4} \\ 15m^3 & -\ 13m^2 & +\ 8m & +\ 11 \end{array}$$

63. $(8m^2 - 7m) - (3m^2 + 7m - 6)$

$= (8m^2 - 7m) + (-3m^2 - 7m + 6)$

$= (8 - 3)m^2 + (-7 - 7)m + 6$

$= 5m^2 - 14m + 6$

65. $(16x^3 - x^2 + 3x) + (-12x^3 + 3x^2 + 2x)$

$= 16x^3 - x^2 + 3x - 12x^3 + 3x^2 + 2x$

$= (16 - 12)x^3 + (-1 + 3)x^2 + (3 + 2)x$

$= 4x^3 + 2x^2 + 5x$

67. $(7y^4 + 3y^2 + 2y) - (18y^4 - 5y^2 + y)$
$= (7y^4 + 3y^2 + 2y) + (-18y^4 + 5y^2 - y)$
$= (7 - 18)y^4 + (3 + 5)y^2 + (2 - 1)y$
$= -11y^4 + 8y^2 + y$

69. $(9a^4 - 3a^2 + 2) + (4a^4 - 4a^2 + 2)$
$+ (-12a^4 + 6a^2 - 3)$
$= (9a^4 + 4a^4 - 12a^4) + (-3a^2 - 4a^2 + 6a^2)$
$+ (2 + 2 - 3)$
$= a^4 - a^2 + 1$

71. $[(8m^2 + 4m - 7) - (2m^2 - 5m + 2)]$
$- (m^2 + m + 1)$
$= (8m^2 + 4m - 7) + (-2m^2 + 5m - 2)$
$+ (-m^2 - m - 1)$
$= (8 - 2 - 1)m^2 + (4 + 5 - 1)m + (-7 - 2 - 1)$
$= 5m^2 + 8m - 10$

73. $[(3x^2 - 2x + 7) - (4x^2 + 2x - 3)]$
$- [(9x^2 + 4x - 6) + (-4x^2 + 4x + 4)]$
$= [(3 - 4)x^2 + (-2 - 2)x + (7 + 3)]$
$- [(9 - 4)x^2 + (4 + 4)x + (-6 + 4)]$
$= (-x^2 - 4x + 10) - (5x^2 + 8x - 2)$
$= -x^2 - 4x + 10 - 5x^2 - 8x + 2$
$= -6x^2 - 12x + 12$

75. The coefficients of the x^2 terms are $-4, -(-2)$, and -8. The sum of these numbers is $-4 + 2 - 8 = -10$.

77. $(6b + 3c) + (-2b - 8c)$
$= (6b - 2b) + (3c - 8c)$
$= 4b - 5c$

79 $(4x + 2xy - 3) - (-2x + 3xy + 4)$
$= (4x + 2xy - 3) + (2x - 3xy - 4)$
$= (4x + 2x) + (2xy - 3xy) + (-3 - 4)$
$= 6x - xy - 7$

81. $(5x^2y - 2xy + 9xy^2)$
$- (8x^2y + 13xy + 12xy^2)$
$= (5x^2y - 2xy + 9xy^2)$
$+ (-8x^2y - 13xy - 12xy^2)$
$= (5x^2y - 8x^2y) + (-2xy - 13xy)$
$+ (9xy^2 - 12xy^2)$
$= -3x^2y - 15xy - 3xy^2$

83. Use the formula for the perimeter of a rectangle, $P = 2L + 2W$, with length $L = 4x^2 + 3x + 1$ and width $W = x + 2$.
$P = 2L + 2W$
$= 2(4x^2 + 3x + 1) + 2(x + 2)$
$= 8x^2 + 6x + 2 + 2x + 4$
$= 8x^2 + 8x + 6$
A polynomial that represents the perimeter of the rectangle is $8x^2 + 8x + 6$.

85. Use the formula for the perimeter of a square, $P = 4s$, with $s = \frac{1}{2}x^2 + 2x$.
$P = 4s$
$= 4\left(\frac{1}{2}x^2 + 2x\right)$
$= 4\left(\frac{1}{2}x^2\right) + 4(2x)$
$= 2x^2 + 8x$
A polynomial that represents the perimeter of the square is $2x^2 + 8x$.

87. Use the formula for the perimeter of a triangle, $P = a + b + c$, with $a = 3t^2 + 2t + 7$, $b = 5t^2 + 2$, and $c = 6t + 4$.
$P = (3t^2 + 2t + 7) + (5t^2 + 2) + (6t + 4)$
$= (3t^2 + 5t^2) + (2t + 6t) + (7 + 2 + 4)$
$= 8t^2 + 8t + 13$
A polynomial that represents the perimeter of the triangle is $8t^2 + 8t + 13$.

89. **(a)** Use the formula for the perimeter of a triangle, $P = a + b + c$, with $a = 2y - 3t$, $b = 5y + 3t$, and $c = 16y + 5t$.

$$P = (2y - 3t) + (5y + 3t) + (16y + 5t)$$
$$= (2y + 5y + 16y) + (-3t + 3t + 5t)$$
$$= 23y + 5t$$

The perimeter of the triangle is $23y + 5t$.

(b) Use the fact that the sum of the angles of any triangle is 180°.

$$(7x - 3)° + (5x + 2)° + (2x - 1)° = 180°$$
$$(7x + 5x + 2x) + (-3 + 2 - 1) = 180$$
$$14x - 2 = 180$$
$$14x = 182$$
$$x = \frac{182}{14} = 13$$

Let $x = 13$.

$7x - 3 = 7(13) - 3 = 88$

$5x + 2 = 5(13) + 2 = 67$

$2x - 1 = 2(13) - 1 = 25$

The measures of the angles are 25°, 67°, and 88°.

91. First find the two sums and then find their difference.

$$[(5x^2 + 2x - 3) + (x^2 - 8x + 2)]$$
$$-[(7x^2 - 3x + 6) + (-x^2 + 4x - 6)]$$
$$= (6x^2 - 6x - 1) - (6x^2 + x)$$
$$= (6x^2 - 6x - 1) + (-6x^2 - x)$$
$$= (6x^2 - 6x^2) + (-6x - x) + (-1)$$
$$= -7x - 1$$

93. $y = x^2 - 4$

x	$y = x^2 - 4$
−2	$(-2)^2 - 4 = 4 - 4 = 0$
−1	$(-1)^2 - 4 = 1 - 4 = -3$
0	$(0)^2 - 4 = 0 - 4 = -4$
1	$(1)^2 - 4 = 1 - 4 = -3$
2	$(2)^2 - 4 = 4 - 4 = 0$

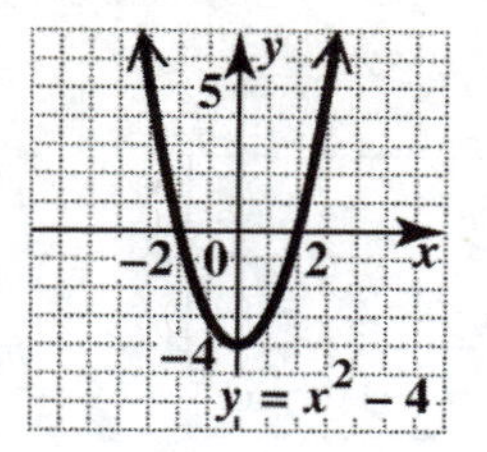

95. $y = 2x^2 - 1$

x	$y = 2x^2 - 1$
−2	$2(-2)^2 - 1 = 2 \cdot 4 - 1 = 7$
−1	$2(-1)^2 - 1 = 2 \cdot 1 - 1 = 1$
0	$2(0)^2 - 1 = 2 \cdot 0 - 1 = -1$
1	$2(1)^2 - 1 = 2 \cdot 1 - 1 = 1$
2	$2(2)^2 - 1 = 2 \cdot 4 - 1 = 7$

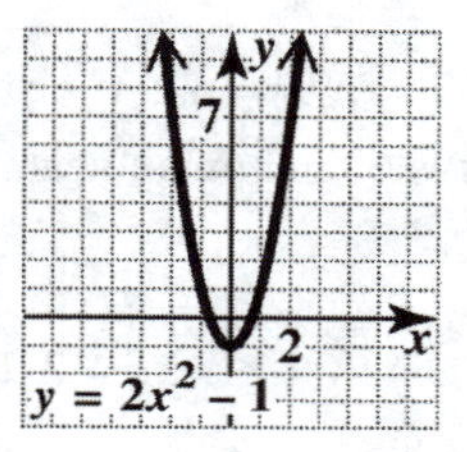

97. $y = -x^2 + 4$

x	$y = -x^2 + 4$
−2	$-(-2)^2 + 4 = -4 + 4 = 0$
−1	$-(-1)^2 + 4 = -1 + 4 = 3$
0	$-(0)^2 + 4 = -0 + 4 = 4$
1	$-(1)^2 + 4 = -1 + 4 = 3$
2	$-(2)^2 + 4 = -4 + 4 = 0$

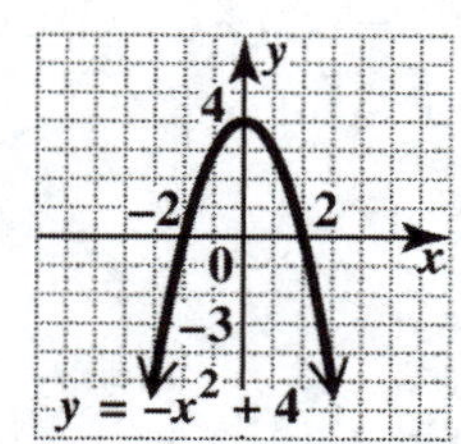

99. $y=(x+3)^2$

x	$y=(x+3)^2$
-5	$(-5+3)^2=(-2)^2=4$
-4	$(-4+3)^2=(-1)^2=1$
-3	$(-3+3)^2=(0)^2=0$
-2	$(-2+3)^2=(1)^2=1$
-1	$(-1+3)^2=(2)^2=4$

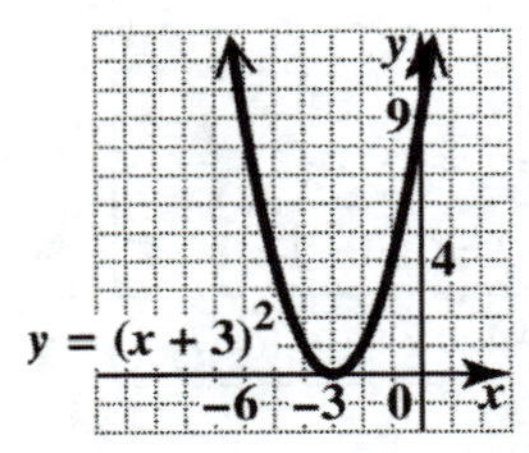

101. If $x=9$, then $y=7x=7(9)=63$. If a dog is 9 in dog years, then it is 63 in human years.

103. $-16x^2+60x+80$

$=-16(2.5)^2+60(2.5)+80$ Let $x=2.5$.

$=-16(6.25)+150+80$

$=-100+150+80$

$=130$

If 2.5 seconds have elapsed, the height of the object is 130 feet.

5.5 Multiplying Polynomials

Now Try Exercises

N1. $-3x^5(2x^3-5x^2+10)$

$=-3x^5(2x^3)+(-3x^5)(-5x^2)+(-3x^5)(10)$

$=-6x^8+15x^7+(-30x^5)$

$=-6x^8+15x^7-30x^5$

N2. $(x^2-4)(2x^2-5x+3)$

$=x^2(2x^2)+x^2(-5x)+x^2(3)$

$\quad -4(2x^2)-4(-5x)-4(3)$

$=2x^4-5x^3+3x^2-8x^2+20x-12$

$=2x^4-5x^3-5x^2+20x-12$

N3.

$$\begin{array}{rl} 5t^2-7t+4 & \\ 2t-6 & \\ \hline -30t^2+42t-24 & \leftarrow -6(5t^2-7t+4) \\ 10t^3-14t^2+8t \quad\quad & \leftarrow 2t(5t^2-7t+4) \\ \hline 10t^3-44t^2+50t-24 & \leftarrow \text{Add like terms.} \end{array}$$

N4.

$$\begin{array}{rl} 9x^3-12x^2+3 & \\ \frac{1}{3}x^2-\frac{2}{3} & \\ \hline -6x^3+8x^2-2 & \left(\text{times } -\frac{2}{3}\right) \\ 3x^5-4x^4 \quad\quad +x^2 \quad\quad & \left(\text{times } \frac{1}{3}x^2\right) \\ \hline 3x^5-4x^4-6x^3+9x^2-2 & \text{Add.} \end{array}$$

The product is $3x^5-4x^4-6x^3+9x^2-2$.

N5. $(t-6)(t+5)$

F: $t(t)=t^2$

O: $t(5)=5t$

I: $-6(t)=-6t$

L: $-6(5)=-30$

$(t-6)(t+5)=t^2+5t-6t-30$

$=t^2-t-30$

N6. $(7y-3)(2x+5)$

F: $7y(2x)=14yx$

O: $7y(5)=35y$

I: $-3(2x)=-6x$

L: $-3(5)=-15$

The product $(7y-3)(2x+5)$ is equal to $14yx+35y-6x-15$.

N7. **(a)** $(3p-5q)(4p-q)$

F: $3p(4p)=12p^2$

O: $3p(-q)=-3pq$

I: $-5q(4p)=-20pq$

L: $-5q(-q)=5q^2$

$(3p-5q)(4p-q)$

$=12p^2-3pq-20pq+5p^2$

$=12p^2-23pq+5q^2$

(b) $5x^2(3x+1)(x-5)$

First multiply the binomials and then multiply that result by $5x^2$.

F: $3x(x) = 3x^2$

O: $3x(-5) = -15x$

I: $1(x) = x$

L: $1(-5) = -5$

$$\begin{aligned}5x^2(3x+1)(x-5) &= 5x^2\left(3x^2-15x+x-5\right)\\ &= 5x^2\left(3x^2-14x-5\right)\\ &= 15x^4-70x^3-25x^2\end{aligned}$$

Exercises

1. **(a)** $5x^3\left(6x^7\right)$

$= 5\cdot 6x^{3+7}$

$= 30x^{10}$

Choice B is correct.

(b) $-5x^7\left(6x^3\right)$

$= -5\cdot 6x^{7+3}$

$= -30x^{10}$

Choice D is correct.

(c) $\left(5x^7\right)^3 = (5)^3\left(x^7\right)^3$

$= 125x^{7\cdot 3}$

$= 125x^{21}$

Choice A is correct.

(d) $\left(-6x^3\right)^3 = (-6)^3\left(x^3\right)^3$

$= -216x^{3\cdot 3}$

$= -216x^9$

Choice C is correct.

3. In multiplying a monomial by a polynomial, the first property that is used is the distributive property.

5. The product $2x^2\left(-3x^5\right)$ has exactly one term after the multiplication is performed.

7. $5y^4\left(3y^7\right) = 5(3)\,y^{4+7}$

$= 15y^{11}$

9. $-15a^4(-2a^5) = -15(-2)a^{4+5}$

$= 30a^9$

11. $5p(3q^2) = 5(3)pq^2$

$= 15pq^2$

13. $-6m^3(3n^2) = -6(3)m^3n^2$

$= -18m^3n^2$

15. $y^5\cdot 9y\cdot y^4 = 9y^{5+1+4}$

$= 9y^{10}$

17. $\left(4x^3\right)\left(2x^2\right)\left(-x^5\right) = (-4\cdot 2)x^{3+2+5}$

$= -8x^{10}$

19. $2m(3m+2) = 2m(3m)+2m(2)$

$= 6m^2+4m$

21. $3p\left(-2p^3+4p^2\right) = 3p\left(-2p^3\right)+3p\left(4p^2\right)$

$= -6p^4+12p^3$

23. $-8z\left(2z+3z^2+3z^3\right)$

$= -8z(2z)+(-8z)\left(3z^2\right)+(-8z)\left(3z^3\right)$

$= -16z^2-24z^3-24z^4$

25. $2y^3\left(3+2y+5y^4\right)$

$= 2y^3(3)+2y^3(2y)+2y^3\left(5y^4\right)$

$= 6y^3+4y^4+10y^7$

27. $-4r^3\left(-7r^2+8r-9\right)$

$= -4r^3\left(-7r^2\right)+\left(-4r^3\right)(8r)+\left(-4r^3\right)(-9)$

$= 28r^5-32r^4+36r^3$

29. $3a^2\left(2a^2-4ab+5b^2\right)$

$= 3a^2\left(2a^2\right)+3a^2(-4ab)+3a^2\left(5b^2\right)$

$= 6a^4-12a^3b+15a^2b^2$

31. $7m^3n^2\left(3m^2+2mn-n^3\right)$

$= 7m^3n^2\left(\underline{3m^2}\right)+7m^3n^2\left(\underline{2mn}\right)+7m^3n^2\left(\underline{-n^3}\right)$

$= \underline{21m^5n^2+14m^4n^3-7m^3n^5}$

33. $(6x+1)(2x^2+4x+1)$

$= 6x(2x^2)+6x(4x)+6x(1)$
$\quad +1(2x^2)+1(4x)+1(1)$
$= 12x^3+24x^2+6x+2x^2+4x+1$
$= 12x^3+26x^2+10x+1$

35. $(9y-2)(8y^2-6y+1)$

Multiply vertically.

$$\begin{array}{rrrr} & 8y^2 & -6y & +1 \\ & & 9y & -2 \\ \hline & -16y^2 & +12y & -2 \\ 72y^3 & -54y^2 & +9y & \\ \hline 72y^3 & -70y^2 & +21y & -2 \end{array}$$

37. $(4m+3)(5m^3-4m^2+m-5)$

Multiply vertically.

$$\begin{array}{rrrrr} & 5m^3 & -4m^2 & +m & -5 \\ & & & 4m & +3 \\ \hline & 15m^3 & -12m^2 & +3m & -15 \\ 20m^4 & -16m^3 & +4m^2 & -20m & \\ \hline 20m^4 & -m^3 & -8m^2 & -17m & -15 \end{array}$$

39. $(2x-1)(3x^5-2x^3+x^2-2x+3)$

Multiply vertically.

$$\begin{array}{rrrrrrr} & 3x^5 & & -2x^3 & +x^2 & -2x & +3 \\ & & & & & 2x & -1 \\ \hline & -3x^5 & & +2x^3 & -x^2 & +2x & -3 \\ 6x^6 & & -4x^4 & +2x^3 & -4x^2 & +6x & \\ \hline 6x^6 & -3x^5 & -4x^4 & +4x^3 & -5x^2 & +8x & -3 \end{array}$$

41. $(5x^2+2x+1)(x^2-3x+5)$

Multiply vertically.

$$\begin{array}{rrrrr} & & 5x^2 & +2x & +1 \\ & & x^2 & -3x & +5 \\ \hline & & 25x^2 & +10x & +5 \\ & -15x^3 & -6x^2 & -3x & \\ 5x^4 & +2x^3 & +x^2 & & \\ \hline 5x^4 & -13x^3 & +20x^2 & +7x & +5 \end{array}$$

43. $(6x^4-4x^2+8x)\left(\frac{1}{2}x+3\right)$

$= 6x^4\left(\frac{1}{2}x\right)+(-4x^2)\left(\frac{1}{2}x\right)+8x\left(\frac{1}{2}x\right)$
$\quad +6x^4(3)+(-4x^2)(3)+8x(3)$
$= 3x^5-2x^3+4x^2+18x^4-12x^2+24x$
$= 3x^5+18x^4-2x^3-8x^2+24x$

45. $(x+3)(x+4)$

	x	4
x	x^2	$4x$
3	$3x$	12

The product is the sum of the four monomial products.

$x^2+3x+4x+12 = x^2+7x+12$

Product: $\underline{x^2+7x+12}$

47 $(2x+1)(x^2+3x+2)$

	x^2	$3x$	2
$2x$	$2x^3$	$6x^2$	$4x$
1	x^2	$3x$	2

The product is the sum of the six monomial products.

$2x^3+6x^2+x^2+4x+3x+2$
$= 2x^3+7x^2+7x+2$

Product: $\underline{2x^3+7x^2+7x+2}$

49. **(a)** Product of first terms:

$\underline{2p}(\underline{3p}) = \underline{6p^2}$

(b) Outer product:

$\underline{2p}(\underline{7}) = \underline{14p}$

(c) Inner product:

$\underline{-5}(\underline{3p}) = \underline{-15p}$

(d) Product of last terms:

$\underline{-5}(\underline{7}) = \underline{-35}$

(e) The product is the sum of the four monomial products.
Complete product in simplified form:

$6p^2+14p+(-15p)+(-35)$
$= \underline{6p^2-p-35}$

51. $(m+7)(m+5)$
$= m(m)+m(5)+7(m)+7(5)$
$= m^2+5m+7m+35$
$= m^2+12m+35$

53. $(n-1)(n+4)$
$= n(n)+n(4)+(-1)(n)+(-1)(4)$
$= n^2+4n+(-1n)+(-4)$
$= n^2+3n-4$

55. $(2x+3)(6x-4)$
$= 2x(6x)+2x(-4)+3(6x)+3(-4)$
$= 12x^2-8x+18x-12$
$= 12x^2+10x-12$

57. $(9+t)(9-t)$
$= 9(9)+9(-t)+t(9)+t(-t)$
$= 81-9t+9t-t^2$
$= 81-t^2$

59. $(3x-2)(3x-2)$
$= 3x(3x)+3x(-2)+(-2)(3x)+(-2)(-2)$
$= 9x^2-6x-6x+4$
$= 9x^2-12x+4$

61. $(5a+1)(2a+7)$
$= 5a(2a)+5a(7)+1(2a)+1(7)$
$= 10a^2+35a+2a+7$
$= 10a^2+37a+7$

63. $(6-5m)(2+3m)$
$= 6(2)+6(3m)+(-5m)(2)+(-5m)(3m)$
$= 12+18m-10m-15m^2$
$= 12+8m-15m^2$

65. $(5-3x)(4+x)$
$= 5(4)+5(x)+(-3x)(4)+(-3x)(x)$
$= 20+5x-12x-3x^2$
$= 20-7x-3x^2$

67. $(3t-4s)(t+3s)$
$= 3t(t)+3t(3s)+(-4s)(t)+(-4s)(3s)$
$= 3t^2+9st-4st-12s^2$
$= 3t^2+5st-12s^2$

69. $(4x+3)(2y-1)$
$= 4x(2y)+4x(-1)+3(2y)+3(-1)$
$= 8xy-4x+6y-3$

71. $(3x+2y)(5x-3y)$
$= 3x(5x)+3x(-3y)+2y(5x)+2y(-3y)$
$= 15x^2-9xy+10xy-6y^2$
$= 15x^2+xy-6y^2$

73. $3y^3(2y+3)(y-5)$
$(2y+3)(y-5)$
$= 2y(y)+2y(-5)+3(y)+3(-5)$
$= 2y^2-10y+3y-15$
$= 2y^2-7y-15$

Now multiply this result by $3y^3$.
$3y^3(2y^2-7y-15)$
$= 3y^3(2y^2)+3y^3(-7y)+3y^3(-15)$
$= 6y^5-21y^4-45y^3$

75. $-8r^3(5r^2+2)(5r^2-2)$
$(5r^2+2)(5r^2-2)$
$= 5r^2(5r^2)+5r^2(-2)+2(5r^2)+2(-2)$
$= 25r^4-10r^2+10r^2-4$
$= 25r^4-4$

Now multiply this result by $-8r^3$.
$-8r^3(25r^4-4)$
$= -8r^3(25r^4)+(-8r^3)(-4)$
$= -200r^7+32r^3$

77. **(a)** Use the formula for the area of a rectangle, $A = LW$, with $L = 3y+7$ and $W = y+1$.
$A = (3y+7)(y+1)$
$= (3y+7)(y)+(3y+7)(1)$
$= 3y^2+7y+3y+7$
$= 3y^2+10y+7$

(b) Use the formula for the perimeter of a rectangle, $P = 2L + 2W$, with $L = 3y + 7$ and $W = y + 1$.

$$\begin{aligned} P &= 2(3y+7)+2(y+1) \\ &= 2(3y)+2(7)+2(y)+2(1) \\ &= 6y+14+2y+2 \\ &= 8y+16 \end{aligned}$$

79. $(x+7)^2 = (x+7)(x+7)$ Use FOIL.

$$\begin{aligned} &= x^2+7x+7x+49 \\ &= x^2+14x+49 \end{aligned}$$

81. $(a-4)(a+4)$ Use FOIL.

$$\begin{aligned} &= a^2+4a-4a-16 \\ &= a^2-16 \end{aligned}$$

83. $(2p-5)^2 = (2p-5)(2p-5)$ Use FOIL.

$$\begin{aligned} &= 4p^2-10p-10p+25 \\ &= 4p^2-20p+25 \end{aligned}$$

85. $(5k-3q)^2 = (5k+3q)(5k+3q)$ Use FOIL.

$$\begin{aligned} &= 25k^2+15kq+15kq+9q^2 \\ &= 25k^2+30kq+9q^2 \end{aligned}$$

87. Recall that a^3 means $(a)(a)(a)$, so $(m-5)^3 = (m-5)(m-5)(m-5)$.

We'll start by finding $(m-5)(m-5)$.

$$\begin{aligned} (m-5)(m-5) &= m^2-5m-5m+25 \\ &= m^2-10m+25 \end{aligned}$$

Now multiply that result by $m-5$.

$$\begin{array}{rrrr} & m^2 & -10m & +25 \\ & & m & -5 \\ \hline & -5m^2 & +50m & -125 \\ m^3 & -10m^2 & +25m & \\ \hline m^3 & -15m^2 & +75m & -125 \end{array}$$

89. $(2a+1)^3 = (2a+1)(2a+1)(2a+1)$

$$\begin{aligned} (2a+1)(2a+1) &= 4a^2+2a+2a+1 \\ &= 4a^2+4a+1 \end{aligned}$$

Now multiply vertically.

$$\begin{array}{rrrr} & 4a^2 & +4a & +1 \\ & & 2a & +1 \\ \hline & 4a^2 & +4a & +1 \\ 8a^3 & +8a^2 & +2a & \\ \hline 8a^3 & +12a^2 & +6a & +1 \end{array}$$

91. $-3a(3a+1)(a-4)$

$$\begin{aligned} &= -3a\left(3a^2-12a+a-4\right) \quad \text{FOIL} \\ &= -3a\left(3a^2-11a-4\right) \\ &= -9a^3+33a^2+12a \end{aligned}$$

93. $7(4m-3)(2m+1)$

$$\begin{aligned} &= 7\left(8m^2+4m-6m-3\right) \quad \text{FOIL} \\ &= 7\left(8m^2-2m-3\right) \\ &= 56m^2-14m-21 \end{aligned}$$

95. $(3r-2s)^4 = (3r-2s)^2(3r-2s)^2$

First we find $(3r-2s)^2$.

$$\begin{aligned} (3r-2s)^2 &= (3r-2s)(3r-2s) \\ &= 9r^2-6rs-6rs+4s^2 \\ &= 9r^2-12rs+4s^2 \end{aligned}$$

Now multiply this result by itself.

$$\begin{array}{rrrrr} & & 9r^2 & -12rs & +4s^2 \\ & & 9r^2 & -12rs & +4s^2 \\ \hline & & 36r^2s^2 & -48rs^3 & +16s^4 \\ & -108r^3s & +144r^2s^2 & -48rs^3 & \\ 81r^4 & -108r^3s & +36r^2s^2 & & \\ \hline 81r^4 & -216r^3s & +216r^2s^2 & -96rs^3 & +16s^4 \end{array}$$

97. $3p^3(2p^2+5p)(p^3+2p+1)$
$=[3p^3(2p^2)+3p^3(5p)](p^3+2p+1)$
$=(6p^5+15p^4)(p^3+2p+1)$

Now multiply vertically.

$$\begin{array}{rrrrr} & p^3 & +2p & & +1 \\ & & & 6p^5 & +15p^4 \\ \hline & 15p^7 & & +30p^5 & +15p^4 \\ 6p^8 & & +12p^6 & +6p^5 & \\ \hline 6p^8+15p^7 & +12p^6 & +36p^5 & & +15p^4 \end{array}$$

99. $-2x^5(3x^2+2x-5)(4x+2)$
$=[-2x^5(3x^2)+(-2x^5)(2x)$
$+(-2x^5)(-5)](4x+2)$
$=(-6x^7-4x^6+10x^5)(4x+2)$

Now multiply vertically.

$$\begin{array}{rrrr} & -6x^7 & -4x^6 & +10x^5 \\ & & 4x & +2 \\ \hline & -12x^7 & -8x^6 & +20x^5 \\ -24x^8 & -16x^7 & +40x^6 & \\ \hline -24x^8 & -28x^7 & +32x^6 & +20x^5 \end{array}$$

101. $\left(3p^2+\frac{5}{4}q\right)\left(2p^2-\frac{5}{3}q\right)$
$=3p^2(2p^2)+3p^2\left(-\frac{5}{3}q\right)+\frac{5}{4}q(2p^2)$
$+\frac{5}{4}q\left(-\frac{5}{3}q\right)$
$=6p^4-5p^2q+\frac{5}{2}p^2q-\frac{25}{12}q^2$
$=6p^4+\left(-\frac{10}{2}+\frac{5}{2}\right)p^2q-\frac{25}{12}q^2$
$=6p^4-\frac{5}{2}p^2q-\frac{25}{12}q^2$

103. The area A of the shaded region is the difference between the area of the larger square, which has sides of length $x+7$, and the area of the smaller square, which has sides of length x.
$A=(x+7)^2-(x)^2$
$=(x+7)(x+7)-x^2$
$=(x^2+7x+7x+49)-x^2$
$=x^2+14x+49-x^2$
$=14x+49$

105. The area A of the shaded region is the difference between the area of the circle, which has radius x, and the area of the square, which has sides of length 3.
$A=\pi r^2-s^2$
$=\pi(x)^2-(3)^2$
$=\pi x^2-9$

107. Use the formula for the area of a rectangle, $A=LW$, with $L=3x+6$ and $W=10$.
$A=(3x+6)(10)$
$=(3x)(10)+(6)(10)$
$=30x+60$
The area of the rectangle can be represented with the polynomial $30x+60$.

109. The length of the rectangle is $3x+6$, where $x=18$.
$3x+6=3(18)+6$
$=54+6$
$=60$
The length of the rectangle is 60 ft. The width of the rectangle is given as 10 ft. So, the dimensions of the rectangle are 10 ft by 60 ft.

111. Multiply the area of the rectangle by the cost per square foot.
$C=600\cdot\$0.75$
$=\$450$
It would cost $450 to sod the entire lawn.

5.6 Special Products

Now Try Exercises

N1. $(x+5)^2 = x^2 + 2(x)(5) + 5^2$
$= x^2 + 10x + 25$

N2. **(a)** $(3x-1)^2 = (3x)^2 - 2(3x)(1) + 1^2$
$= 9x^2 - 6x + 1$

(b) $(4p-5q)^2 = (4p)^2 - 2(4p)(5q) + (5q)^2$
$= 16p^2 - 40pq + 25q^2$

(c) $\left(6t - \frac{1}{3}\right)^2 = (6t)^2 - 2(6t)^2\left(\frac{1}{3}\right) + \left(\frac{1}{3}\right)^2$
$= 36t^2 - 4t + \frac{1}{9}$

(d) $-(3y+2)^2 = -(9y^2 + 12y + 4)$
$= -9y^2 - 12y - 4$

(e) $m(2m+3)^2 = m(4m^2 + 12m + 9)$
$= 4m^3 + 12m^2 + 9m$

N3. $(t+10)(t-10) = t^2 - 10^2$
$= t^2 - 100$

N4. **(a)** $(4x-6)(4x+6) = (4x)^2 - 6^2$
$= 16x^2 - 36$

(b) $\left(5r - \frac{4}{5}\right)\left(5r + \frac{4}{5}\right) = (5r)^2 - \left(\frac{4}{5}\right)^2$
$= 25r^2 - \frac{16}{25}$

(c) $y(3y+1)(3y-1)$

First use the rule for the product of the sum and difference of two terms.

$(3y+1)(3y-1) = (3y)^2 - 1^2$
$= 9y^2 - 1$

Now multiply this result by y.

$y(9y^2 - 1) = y(9y^2) + y(-1)$
$= 9y^3 - y$

(d) $-5(p+q^2)(p-q^2)$

First use the rule for the product of the sum and difference of two terms.

$(p+q^2)(p-q^2) = (p)^2 - (q^2)^2$
$= p^2 - q^4$

Now multiply this result by -5.

$-5(p^2 - q^4) = -5(p^2) + (-5)(-q^4)$
$= -5p^2 + 5q^4$

N5. $(2m-1)^3$

Since $(2m-1)^3 = (2m-1)^2(2m-1)$, the first step is to find the product $(2m-1)^2$.

$(2m-1)^2 = (2m)^2 - 2(2m)(1) + 1^2$
$= 4m^2 - 4m + 1$

Now multiply this result by $2m-1$.

$(2m-1)^3 = (2m-1)(4m^2 - 4m + 1)$
$= 8m^3 - 8m^2 + 2m - 4m^2 + 4m - 1$
$= 8m^3 - 12m^2 + 6m - 1$

Exercises

1. **(a)** The square of the first term is
$(4x)^2 = (4x)(4x) = 16x^2.$

(b) Twice the product of the two terms is
$2(\underline{4x})(\underline{3}) = \underline{24x}.$

(c) The square of the last term is $3^2 = 9$.

(d) The final product is the trinomial
$16x^2 + 24x + 9.$

3. $(m+2)^2 = m^2 + 2(m)(2) + 2^2$
$= m^2 + 4m + 4$

5. $(r-3)^2 = r^2 - 2(r)(3) + 3^2$
$= r^2 - 6r + 9$

7. $(x+2y)^2 = x^2 + 2(x)(2y) + (2y)^2$
$= x^2 + 4xy + 4y^2$

9. $(5p+2q)^2 = (5p)^2 + 2(5p)(2q) + (2q)^2$
$= 25p^2 + 20pq + 4q^2$

11. $(4x-3)^2=(4x)^2+2(4x)(-3)+(-3)^2$
$=16x^2-24x+9$

13. $(4a+5b)^2=(4a)^2+2(4a)(5b)+(5b)^2$
$=16a^2+40ab+25b^2$

15. $\left(6m-\frac{4}{5}n\right)^2=(6m)^2-2(6m)\left(\frac{4}{5}n\right)+\left(\frac{4}{5}n\right)^2$
$=36m^2-\frac{48}{5}mn+\frac{16}{25}n^2$

17. $\left(\frac{1}{2}x+\frac{1}{3}\right)^2=\left(\frac{1}{2}x\right)^2+2\left(\frac{1}{2}x\right)\left(\frac{1}{3}\right)+\left(\frac{1}{3}\right)^2$
$=\frac{1}{4}x^2+\frac{1}{3}x+\frac{1}{9}$

19. $2(x+6)^2=2\left[x^2+2(x)(6)+6^2\right]$
$=2\left(x^2+12x+36\right)$
$=2x^2+24x+72$

21. $t(3t-1)^2$
$(3t-1)^2=(3t)^2-2(3t)(1)+1^2$
$=9t^2-6t+1$
Now multiply by t.
$t\left(9t^2-6t+1\right)=9t^3-6t^2+t$

23. $3t(4t+1)^2$
$(4t+1)^2=(4t)^2+2(4t)(1)+1^2$
$=16t^2+8t+1$
Now multiply by $3t$.
$3t\left(16t^2+8t+1\right)=48t^3+24t^2+3t$

25. $-(4r-2)^2$
$(4r-2)^2=(4r)^2-2(4r)(2)+2^2$
$=16r^2-16r+4$
Now multiply by -1.
$-1\left(16r^2-16r+4\right)=-16r^2+16r-4$

27. $(k+5)(k-5)=k^2-5^2$
$=k^2-25$

29. $(4-3t)(4+3t)=4^2-(3t)^2$
$=16-9t^2$

31. $(5x+2)(5x-2)=(5x)^2-2^2$
$=25x^2-4$

33. $(5y+3x)(5y-3x)=(5y)^2-(3x)^2$
$=25y^2-9x^2$

35. $(10x+3y)(10x-3y)=(10x)^2-(3y)^2$
$=100x^2-9y^2$

37. $\left(2x^2-5\right)\left(2x^2+5\right)=\left(2x^2\right)^2-5^2$
$=4x^4-25$

39. $\left(\frac{3}{4}-x\right)\left(\frac{3}{4}+x\right)=\left(\frac{3}{4}\right)^2-x^2$
$=\frac{9}{16}-x^2$

41. $\left(9y+\frac{2}{3}\right)\left(9y-\frac{2}{3}\right)=\left(9y^2\right)-\left(\frac{2}{3}\right)^2$
$=81y^2-\frac{4}{9}$

43. $q(5q-1)(5q+1)$
$(5q-1)(5q+1)=(5q)^2-1^2$
$=25q^2-1$
Now multiply by q.
$q\left(25q^2-1\right)=25q^3-q$

45. $-5\left(a-b^3\right)\left(a+b^3\right)$
$\left(a-b^3\right)\left(a+b^3\right)=(a)^2-\left(b^3\right)^2$
$=a^2-b^6$
Now multiply by -5.
$-5\left(a^2-b^6\right)=-5\left(a^2\right)-(-5)\left(b^6\right)$
$=-5a^2+5b^6$

47. $\frac{1}{2}(2k-1)(2k+1)$

$(2k-1)(2k+1)=(2k)^2-(1)^2$
$=4k^2-1$

Now multiply by $\frac{1}{2}$.

$\frac{1}{2}(4k^2-1)=\frac{1}{2}(4k^2)-\left(\frac{1}{2}\right)(1)$
$=2k^2-\frac{1}{2}$

49. $-\frac{1}{100}(10x+10)(10x-10)$

$(10x+10)(10x-10)=(10x)^2-(10)^2$
$=100x^2-100$

Now multiply by $-\frac{1}{100}$.

$-\frac{1}{100}(100x^2-100)$
$=-\frac{1}{100}(100x^2)-\left(-\frac{1}{100}\right)(100)$
$=-x^2+1$

51. $(x+1)^3$
$=(x+1)^2(x+1)$
$=(x^2+2x+1)(x+1)$
$=x^3+2x^2+x+x^2+2x+1$
$=x^3+3x^2+3x+1$

53. $(t-3)^3$
$=(t-3)^2(t-3)$
$=(t^2-6t+9)(t-3)$
$=t^3-6t^2+9t-3t^2+18t-27$
$=t^3-9t^2+27t-27$

55. $(r+5)^3$
$=(r+5)^2(r+5)$
$=(r^2+10r+25)(r+5)$
$=r^3+10r^2+25r+5r^2+50r+125$
$=r^3+15r^2+75r+125$

57. $(2a+1)^3$
$=(2a+1)^2(2a+1)$
$=(4a^2+4a+1)(2a+1)$
$=8a^3+8a^2+2a+4a^2+4a+1$
$=8a^3+12a^2+6a+1$

59. $(4x-1)^4$
$=(4x-1)^2(4x-1)^2$
$=(16x^2-8x+1)(16x^2-8x+1)$
$=256x^4-128x^3+16x^2-128x^3+64x^2-8x$
$+16x^2-8x+1$
$=256x^4-256x^3+96x^2-16x+1$

61. $(3r-2t)^4$
$=(3r-2t)^2(3r-2t)^2$
$=(9r^2-12rt+4t^2)(9r^2-12rt+4t^2)$
$=81r^4-108r^3t+36r^2t^2-108r^3t+144r^2t^2$
$-48rt^3+36r^2t^2-48rt^3+16t^4$
$=81r^4-216r^3t+216r^2t^2-96rt^3+16t^4$

63. $2x(x+1)^3$
$=2x(x+1)(x+1)^2$
$=2x(x+1)(x^2+2x+1)$
$=2x(x^3+2x^2+x+x^2+2x+1)$
$=2x(x^3+3x^2+3x+1)$
$=2x^4+6x^3+6x^2+2x$

65. $-4t(t+3)^3$
$=-4t(t+3)(t+3)^2$
$=-4t(t+3)(t^2+6t+9)$
$=-4t(t^3+6t^2+9t+3t^2+18t+27)$
$=-4t(t^3+9t^2+27t+27)$
$=-4t^4-36t^3-108t^2-108t$

67. $(x+y)^2(x-y)^2$
$=[(x+y)(x-y)]^2$
$=(x^2-y^2)^2$
$=(x^2)^2-2x^2y^2+(y^2)^2$
$=x^4-2x^2y^2+y^4$

69. No. For example, $(a+b)^2$ equals $a^2+2ab+b^2$, which is not equivalent to a^2+b^2.

71. $101\times 99=(100+1)(100-1)$
$=100^2-1^2$
$=10{,}000-1$
$=9999$

73. $201\times 199=(200+1)(200-1)$
$=200^2-1^2$
$=40{,}000-1$
$=39{,}999$

75. $20\frac{1}{2}\times 19\frac{1}{2}=\left(20+\frac{1}{2}\right)\left(20-\frac{1}{2}\right)$
$=20^2-\left(\frac{1}{2}\right)^2$
$=400-\frac{1}{4}$
$=399\frac{3}{4}$

77. Use the formula for the area of a triangle, $A=\frac{1}{2}bh$, with $b=m+2n$ and $h=m-2n$.
$A=\frac{1}{2}(m+2n)(m-2n)$
$=\frac{1}{2}\left[m^2-(2n)^2\right]$
$=\frac{1}{2}(m^2-4n^2)$
$=\frac{1}{2}m^2-2n^2$

79. Use the formula for the area of a parallelogram, $A=bh$, with $b=3a+2$ and $h=3a-2$.
$A=(3a+2)(3a-2)$
$=(3a)^2-2^2$
$=9a^2-4$

81. Use the formula for the area of a circle, $A=\pi r^2$, with $r=x+2$.
$A=\pi(x+2)^2$
$=\pi(x^2+4x+4)$
$=\pi x^2+4\pi x+4\pi$

83. Use the formula for the volume of a cube, $V=e^3$, with $e=x+2$.
$V=(x+2)^3$
$=(x+2)^2(x+2)$
$=(x^2+4x+4)(x+2)$
$=x^3+4x^2+4x+2x^2+8x+8$
$=x^3+6x^2+12x+8$

85. The large square has sides of length $a+b$, so its area is $(a+b)^2$.

87. Each blue rectangle has length a and width b, so each has an area of ab. Thus, the sum of the areas of the blue rectangles is $ab+ab=2ab$.

89. $\text{Sum}=a^2+2ab+b^2$

91. $35^2=(35)(35)$

$$\begin{array}{r} 35 \\ \times\ 35 \\ \hline 175 \\ +\ 105 \\ \hline 1225 \end{array}$$

93. $30^2+2(30)(5)+5^2$
$=900+60(5)+25$
$=900+300+25$
$=1225$

5.7 Dividing Polynomials

Now Try Exercises

N1. $\dfrac{16a^6 - 12a^4}{4a^2}$

$= \dfrac{16a^6}{4a^2} - \dfrac{12a^4}{4a^2}$

$= 4a^4 - 3a^2$

N2. $\dfrac{36x^5 + 24x^4 - 12x^3}{6x^4}$

$= \dfrac{36x^5}{6x^4} + \dfrac{24x^4}{6x^4} - \dfrac{12x^3}{6x^4}$

$= 6x + 4 - \dfrac{2}{x}$

N3. Divide $7y^4 - 40y^5 + 100y^2$ by $-5y^2$.

$\dfrac{7y^4 - 40y^5 + 100y^2}{-5y^2}$

$= \dfrac{7y^4}{-5y^2} - \dfrac{40y^5}{-5y^2} + \dfrac{100y^2}{-5y^2}$

$= -\dfrac{7y^2}{5} + 8y^3 - 20$ or $8y^3 - \dfrac{7y^2}{5} - 20$

N4. Divide $35m^5n^4 - 49m^2n^3 + 12mn$ by $7m^2n$.

$\dfrac{35m^5n^4 - 49m^2n^3 + 12mn}{7m^2n}$

$= \dfrac{35m^5n^4}{7m^2n} - \dfrac{49m^2n^3}{7m^2n} + \dfrac{12mn}{7m^2n}$

$= 5m^3n^3 - 7n^2 + \dfrac{12}{7m}$

N5. Divide. $\dfrac{4x^2 + x - 18}{x - 2}$

Step 1 (see the division following Step 4)
$4x^2$ divided by x is $4x$;
$4x(x-2) = 4x^2 - 8x.$

Step 2
Subtract $4x^2 - 8x$ from $4x^2 + x.$
Bring down $-18.$

Step 3
$9x$ divided by x is 9;
$9(x-2) = 9x - 18.$

Step 4
Subtract $9x - 18$ from $9x - 18.$
The remainder is 0.
The answer is the quotient, $4x + 9.$

$$
\begin{array}{r}
4x + 9 \\
x-2\overline{)\,4x^2 + x - 18} \\
\underline{4x^2 - 8x} \\
9x - 18 \\
\underline{9x - 18} \\
0
\end{array}
$$

Check: Multiply the divisor, $x - 2,$ by the quotient, $4x + 9.$ The product must be the original dividend, $4x^2 + x - 18.$

$(x-2)(4x+9)$

$= 4x^2 - 8x + 9x - 18$

$= 4x^2 + x - 18$

N6. Divide $6k^3 - 20k - k^2 + 1$ by $2k - 3.$
Write the numerator in descending powers and divide.

$$
\begin{array}{r}
3k^2 + 4k - 4 \\
2k-3\overline{)\,6k^3 - k^2 - 20k + 1} \\
\underline{6k^3 - 9k^2} \\
8k^2 - 20k \\
\underline{8k^2 - 12k} \\
-8k + 1 \\
\underline{-8k + 12} \\
-11
\end{array}
$$

The remainder is $-11.$

The answer is $3k^2 + 4k - 4 + \dfrac{-11}{2k-3}.$

N7. Divide $m^3 - 1000$ by $m - 10$.

Use 0 as the coefficient of the missing m^2- and m-terms.

$$\begin{array}{r|l}
 & m^2 + 10m + 100 \\
\hline
m - 10 & m^3 + 0m^2 + 0m - 1000 \\
 & \underline{m^3 - 10m^2} \\
 & \quad 10m^2 + 0m \\
 & \quad \underline{10m^2 - 100m} \\
 & \qquad 100m - 1000 \\
 & \qquad \underline{100m - 1000} \\
 & \qquad\qquad 0
\end{array}$$

$(m^3 - 1000)$ divided by $(m - 10)$ is $m^2 + 10m + 100$.

N8. Divide $y^4 - 5y^3 + 6y^2 + y - 4$ by $y^2 + 2$.

$$\begin{array}{r|l}
 & y^2 - 5y + 4 \\
\hline
y^2 + 0y + 2 & y^4 - 5y^3 + 6y^2 + y - 4 \\
 & \underline{y^4 + 0y^3 + 2y^2} \\
 & \quad -5y^3 + 4y^2 + y \\
 & \quad \underline{-5y^3 - 0y^2 - 10y} \\
 & \qquad 4y^2 + 11y - 4 \\
 & \qquad \underline{4y^2 + 0y + 8} \\
 & \qquad\qquad 11y - 12
\end{array}$$

$(y^4 - 5y^3 + 6y^2 + y - 4)$ divided by $(y^2 + 2)$ is $y^2 - 5y + 4 + \dfrac{11y - 12}{y^2 + 2}$.

N9. Divide $10x^3 + 21x^2 + 5x - 8$ by $2x + 4$.

$$\begin{array}{r|l}
 & 5x^2 + \frac{1}{2}x + \frac{3}{2} \\
\hline
2x + 4 & 10x^3 + 21x^2 + 5x - 8 \\
 & \underline{10x^3 + 20x^2} \\
 & \quad x^2 + 5x \\
 & \quad \underline{x^2 + 2x} \\
 & \qquad 3x - 8 \\
 & \qquad \underline{3x + 6} \\
 & \qquad -14
\end{array}$$

$(10x^3 + 21x^2 + 5x - 8)$ divided by $(2x + 4)$ is $5x^2 + \dfrac{1}{2}x + \dfrac{3}{2} + \dfrac{-14}{2x + 4}$.

N10. Divide $x^3 + 7x^2 + 17x + 20$ by $x + 4$.

$$\begin{array}{r|l}
 & x^2 + 3x + 5 \\
\hline
x + 4 & x^3 + 7x^2 + 17x + 20 \\
 & \underline{x^3 + 4x^2} \\
 & \quad 3x^2 + 17x \\
 & \quad \underline{3x^2 + 12x} \\
 & \qquad 5x + 20 \\
 & \qquad \underline{5x + 20} \\
 & \qquad\quad 0
\end{array}$$

The length of the rectangle is $x^2 + 3x + 5$ units.

Exercises

1. In the statement $\dfrac{10x^2 + 8}{2} = 5x^2 + 4$, $\underline{10x^2 + 8}$ is the dividend, $\underline{2}$ is the divisor, and $\underline{5x^2 + 4}$ is the quotient.

3. To check the division shown in Exercise 1, multiply $\underline{5x^2 + 4}$ by $\underline{2}$ (or $\underline{2}$ by $\underline{5x^2 + 4}$) and show that the product is $\underline{10x^2 + 8}$.

5. $\dfrac{6p^4 + 18p^7}{3p^2}$

$= \dfrac{6p^4}{3p^2} + \dfrac{18p^7}{3p^2}$

$= \dfrac{6}{3}p^{4-2} + \dfrac{18}{3}p^{7-2}$

$= \underline{2p^2 + 6p^5}$

7. $\dfrac{60x^4 - 20x^2 + 10x}{2x}$

$= \dfrac{60x^4}{2x} - \dfrac{20x^2}{2x} + \dfrac{10x}{2x}$

$= \dfrac{60}{2}x^{4-1} - \dfrac{20}{2}x^{2-1} + \dfrac{10}{2}x^{1-1}$

$= 30x^3 - 10x + 5$

9. $\dfrac{20m^5 - 10m^4 + 5m^2}{5m^2}$

$= \dfrac{20m^5}{5m^2} - \dfrac{10m^4}{5m^2} + \dfrac{5m^2}{5m^2}$

$= \dfrac{20}{5}m^{5-2} - \dfrac{10}{5}m^{4-2} + \dfrac{5}{5}m^{2-2}$

$= 4m^3 - 2m^2 + 1$

11. $\dfrac{8t^5 - 4t^3 + 4t^2}{2t}$

$= \dfrac{8t^5}{2t} - \dfrac{4t^3}{2t} + \dfrac{4t^2}{2t}$

$= 4t^4 - 2t^2 + 2t$

13. $\dfrac{4a^5 - 4a^2 + 8}{4a}$

$= \dfrac{4a^5}{4a} - \dfrac{4a^2}{4a} + \dfrac{8}{4a}$

$= a^4 - a + \dfrac{2}{a}$

15. $\dfrac{18p^5 + 12p^3 - 6p^2}{-6p^3}$

$= \dfrac{18p^5}{-6p^3} + \dfrac{12p^3}{-6p^3} - \dfrac{6p^2}{-6p^3}$

$= -3p^2 - 2 + \dfrac{1}{p}$

17. $\dfrac{-7r^7 + 6r^5 - r^4}{-r^5}$

$= \dfrac{-7r^7}{-r^5} + \dfrac{6r^5}{-r^5} - \dfrac{r^4}{-r^5}$

$= 7r^2 - 6 + \dfrac{1}{r}$

19. $\dfrac{12x^5 - 9x^4 + 6x^3}{3x^2}$

$= \dfrac{12x^5}{3x^2} - \dfrac{9x^4}{3x^2} + \dfrac{6x^3}{3x^2}$

$= 4x^3 - 3x^2 + 2x$

21. $\dfrac{3x^2 + 15x^3 - 27x^4}{3x^2}$

$= \dfrac{3x^2}{3x^2} + \dfrac{15x^3}{3x^2} - \dfrac{27x^4}{3x^2}$

$= 1 + 5x - 9x^2$ or $-9x^2 + 5x + 1$

23. $\dfrac{36x + 24x^2 + 6x^3}{3x^2}$

$= \dfrac{36x}{3x^2} + \dfrac{24x^2}{3x^2} + \dfrac{6x^3}{3x^2}$

$= \dfrac{12}{x} + 8 + 2x$ or $2x + 8 + \dfrac{12}{x}$

25. $\dfrac{4x^4 + 3x^3 + 2x}{3x^2}$

$= \dfrac{4x^4}{3x^2} + \dfrac{3x^3}{3x^2} + \dfrac{2x}{3x^2}$

$= \dfrac{4x^2}{3} + x + \dfrac{2}{3x}$

27. $\dfrac{-81x^5 + 30x^4 + 12x^2}{3x^2}$

$= \dfrac{-81x^5}{3x^2} + \dfrac{30x^4}{3x^2} + \dfrac{12x^2}{3x^2}$

$= -27x^3 + 10x^2 + 4$

29. $\dfrac{-27r^4 + 36r^3 - 6r^2 - 26r + 2}{-3r}$

$= \dfrac{-27r^4}{-3r} + \dfrac{36r^3}{-3r} - \dfrac{6r^2}{-3r} - \dfrac{26r}{-3r} + \dfrac{2}{-3r}$

$= 9r^3 - 12r^2 + 2r + \dfrac{26}{3} - \dfrac{2}{3r}$

31. $\dfrac{2m^5 - 6m^4 + 8m^2}{-2m^3}$

$= \dfrac{2m^5}{-2m^3} - \dfrac{6m^4}{-2m^3} + \dfrac{8m^2}{-2m^3}$

$= -m^2 + 3m - \dfrac{4}{m}$

33. $\left(20a^4 - 15a^5 + 25a^3\right) \div \left(5a^4\right)$

$= \dfrac{20a^4 - 15a^5 + 25a^3}{5a^4}$

$= \dfrac{20a^4}{5a^4} - \dfrac{15a^5}{5a^4} + \dfrac{25a^3}{5a^4}$

$= 4 - 3a + \dfrac{5}{a}$ or $-3a + 4 + \dfrac{5}{a}$

35. $\left(120x^{11}-60x^{10}+140x^{9}-100x^{8}\right)\div\left(10x^{12}\right)$

$$=\frac{120x^{11}-60x^{10}+140x^{9}-100x^{8}}{10x^{12}}$$

$$=\frac{120x^{11}}{10x^{12}}-\frac{60x^{10}}{10x^{12}}+\frac{140x^{9}}{10x^{12}}-\frac{100x^{8}}{10x^{12}}$$

$$=\frac{12}{x}-\frac{6}{x^{2}}+\frac{14}{x^{3}}-\frac{10}{x^{4}}$$

37. $\left(120x^{5}y^{4}-80x^{2}y^{3}+40x^{2}y^{4}-20x^{5}y^{3}\right)$ $\div\left(20xy^{2}\right)$

$$=\frac{120x^{5}y^{4}-80x^{2}y^{3}+40x^{2}y^{4}-20x^{5}y^{3}}{20xy^{2}}$$

$$=\frac{120x^{5}y^{4}}{20xy^{2}}-\frac{80x^{2}y^{3}}{20xy^{2}}+\frac{40x^{2}y^{4}}{20xy^{2}}-\frac{20x^{5}y^{3}}{20xy^{2}}$$

$$=6x^{4}y^{2}-4xy+2xy^{2}-x^{4}y$$

39. Divide: $\dfrac{x^{2}-x-6}{x-3}$

$$\begin{array}{r}
x+2\\
x-3\overline{)x^{2}-x-6}\\
\underline{x^{2}-3x}\\
2x-6\\
\underline{2x-6}\\
0
\end{array}$$

The remainder is 0. The answer is the quotient, $x+2$.

41. Divide: $\dfrac{2y^{2}+9y-35}{y+7}$

$$\begin{array}{r}
2y-5\\
y+7\overline{)2y^{2}+9y-35}\\
\underline{2y^{2}+14y}\\
-5y-35\\
\underline{-5y-35}\\
0
\end{array}$$

The remainder is 0. The answer is the quotient, $2y-5$.

43. Divide: $\dfrac{p^{2}+2p+20}{p+6}$

$$\begin{array}{r}
p-4\\
p+6\overline{)p^{2}+2p+20}\\
\underline{p^{2}+6p}\\
-4p+20\\
\underline{-4p-24}\\
44
\end{array}$$

The remainder is 44. Write the remainder as the numerator of a fraction that has the divisor $p+6$ as its denominator. The answer is

$$p-4+\frac{44}{p+6}.$$

45. Divide: $\dfrac{12m^{2}-20m+3}{2m-3}$

$$\begin{array}{r}
6m-1\\
2m-3\overline{)12m^{2}-20m+3}\\
\underline{12m^{2}-18m}\\
-2m+3\\
\underline{-2m+3}\\
0
\end{array}$$

The remainder is 0. The answer is the quotient, $6m-1$.

47. Divide: $\dfrac{4a^{2}-22a+32}{2a+3}$

$$\begin{array}{r}
2a-14\\
2a+3\overline{)4a^{2}-22a+32}\\
\underline{4a^{2}+6a}\\
-28a+32\\
\underline{-28a-42}\\
74
\end{array}$$

The remainder is 74. The answer is

$$2a-14+\frac{74}{2a+3}.$$

49. Divide: $\dfrac{8x^3 - 10x^2 - x + 3}{2x+1}$

$$
\begin{array}{r}
4x^2 - 7x + 3 \\
2x+1\overline{\big)\,8x^3 - 10x^2 - x + 3} \\
\underline{8x^3 + 4x^2} \quad\quad\quad\quad \\
-14x^2 - x \quad\quad \\
\underline{-14x^2 - 7x} \quad\quad \\
6x + 3 \\
\underline{6x + 3} \\
0
\end{array}
$$

The remainder is 0. The answer is the quotient, $4x^2 - 7x + 3$.

51. Divide: $\dfrac{8k^4 - 12k^3 - 2k^2 + 7k - 6}{2k-3}$

$$
\begin{array}{r}
4k^3 \quad\quad - k + 2 \\
2k-3\overline{\big)\,8k^4 - 12k^3 - 2k^2 + 7k - 6} \\
\underline{8k^4 - 12k^3} \quad\quad\quad\quad\quad\quad \\
-2k^2 + 7k \quad\quad \\
\underline{-2k^2 + 3k} \quad\quad \\
4k - 6 \\
\underline{4k - 6} \\
0
\end{array}
$$

The remainder is 0. The answer is the quotient, $4k^3 - k + 2$.

53. Divide: $\dfrac{5y^4 + 5y^3 + 2y^2 - y - 8}{y+1}$

$$
\begin{array}{r}
5y^3 \quad\quad + 2y - 3 \\
y+1\overline{\big)\,5y^4 + 5y^3 + 2y^2 - y - 8} \\
\underline{5y^4 + 5y^3} \quad\quad\quad\quad\quad\quad \\
2y^2 - y \quad\quad \\
\underline{2y^2 + 2y} \quad\quad \\
-3y - 8 \\
\underline{-3y - 3} \\
-5
\end{array}
$$

The remainder is -5.

The quotient is $5y^3 + 2y - 3$.

The answer is $5y^3 + 2y - 3 + \dfrac{-5}{y+1}$.

55. Divide: $\dfrac{3k^3 - 4k^2 - 6k + 10}{k-2}$

$$
\begin{array}{r}
3k^2 + 2k - 2 \\
k-2\overline{\big)\,3k^3 - 4k^2 - 6k + 10} \\
\underline{3k^3 - 6k^2} \quad\quad\quad\quad\quad \\
2k^2 - 6k \quad\quad \\
\underline{2k^2 - 4k} \quad\quad \\
-2k + 10 \\
\underline{-2k + 4} \\
6
\end{array}
$$

The remainder is 6.

The quotient is $3k^2 + 2k - 2$.

The answer is $3k^2 + 2k - 2 + \dfrac{6}{k-2}$.

57. Divide: $\dfrac{6p^4 - 16p^3 + 15p^2 - 5p + 10}{3p+1}$

$$
\begin{array}{r}
2p^3 - 6p^2 + 7p - 4 \\
3p+1\overline{\big)\,6p^4 - 16p^3 + 15p^2 - 5p + 10} \\
\underline{6p^4 + 2p^3} \quad\quad\quad\quad\quad\quad\quad\quad \\
-18p^3 + 15p^2 \quad\quad\quad\quad \\
\underline{-18p^3 - 6p^2} \quad\quad\quad\quad \\
21p^2 - 5p \quad\quad \\
\underline{21p^2 + 7p} \quad\quad \\
-12p + 10 \\
\underline{-12p - 4} \\
14
\end{array}
$$

The remainder is 14.

The quotient is $2p^3 - 6p^2 + 7p - 4$.

The answer is $2p^3 - 6p^2 + 7p - 4 + \dfrac{14}{3p+1}$.

59. Divide: $\left(x^3+2x^2-3\right)\div(x-1)$

Use 0 as the coefficient of the missing x-term.

$$\begin{array}{r}
x^2+3x+3 \\
x-1\overline{)x^3+2x^2+0x-3} \\
\underline{x^3-\ x^2\qquad\qquad} \\
3x^2+0x\quad \\
\underline{3x^2-3x\quad} \\
3x-3 \\
\underline{3x-3} \\
0
\end{array}$$

$$\left(x^3+2x^2-3\right)\div(x-1)=x^2+3x+3$$

61. Divide: $\left(2x^3+x+2\right)\div(x+1)$

Use 0 as the coefficient of the missing x^2-term.

$$\begin{array}{r}
2x^2-2x+3 \\
x+1\overline{)2x^3+0x^2+\ x+2} \\
\underline{2x^3+2x^2\qquad\qquad} \\
-2x^2+\ x\quad \\
\underline{-2x^2-2x\quad} \\
3x+2 \\
\underline{3x+3} \\
-1
\end{array}$$

$$\left(2x^3+x+2\right)\div(x+1)=2x^2-2x+3+\frac{-1}{x+1}$$

63. Divide: $\dfrac{5-2r^2+r^4}{r^2-1}$

Use 0 as the coefficient for the missing terms. Rearrange terms of the dividend in descending powers.

$$\begin{array}{r}
r^2\qquad -1 \\
r^2+0r-1\overline{)r^4+0r^3-2r^2+0r+5} \\
\underline{r^4+0r^3-\ r^2\qquad\qquad} \\
-r^2+0r+5 \\
\underline{-r^2+0r+1} \\
4
\end{array}$$

The remainder is 4.

The quotient is r^2-1.

The answer is $r^2-1+\dfrac{4}{r^2-1}$.

65. Divide: $\dfrac{-4x+3x^3+2}{x-1}$

Use 0 as the coefficient for the missing terms. Rearrange terms of the dividend in descending powers.

$$\begin{array}{r}
3x^2+3x-1 \\
x-1\overline{)3x^3+0x^2-4x+2} \\
\underline{3x^3-3x^2\qquad\qquad} \\
3x^2-4x\quad \\
\underline{3x^2-3x\quad} \\
-x+2 \\
\underline{-x+1} \\
1
\end{array}$$

$$\frac{-4x+3x^3+2}{x-1}=3x^2+3x-1+\frac{1}{x-1}$$

67. Divide: $\dfrac{y^3+27}{y+3}$

$$\begin{array}{r}
y^2-3y+\ 9 \\
y+3\overline{)y^3+0y^2+0y+27} \\
\underline{y^3+3y^2\qquad\qquad} \\
-3y^2+0y\quad \\
\underline{-3y^2-9y\quad} \\
9y+27 \\
\underline{9y+27} \\
0
\end{array}$$

The remainder is 0. The answer is the quotient, y^2-3y+9.

69. Divide: $\dfrac{a^4 - 25}{a^2 - 5}$

$$\begin{array}{r}
a^2 \qquad + 5 \\
a^2 + 0a - 5\overline{\big)a^4 + 0a^3 + 0a^2 + 0a - 25} \\
\underline{a^4 + 0a^3 - 5a^2} \qquad\qquad \\
5a^2 + 0a - 25 \\
\underline{5a^2 + 0a - 25} \\
0
\end{array}$$

The remainder is 0. The answer is the quotient, $a^2 + 5$.

71. Divide: $\dfrac{x^4 - 4x^3 + 5x^2 - 3x + 2}{x^2 + 3}$

$$\begin{array}{r}
x^2 - 4x + 2 \\
x^2 + 0x + 3\overline{\big)x^4 - 4x^3 + 5x^2 - 3x + 2} \\
\underline{x^4 + 0x^3 + 3x^2} \qquad\qquad \\
-4x^3 + 2x^2 - 3x \qquad \\
\underline{-4x^3 + 0x^2 - 12x} \qquad \\
2x^2 + 9x + 2 \\
\underline{2x^2 + 0x + 6} \\
9x - 4
\end{array}$$

The remainder is $9x - 4$. The answer is

$$\frac{x^4 - 4x^3 + 5x^2 - 3x + 2}{x^2 + 3} = x^2 - 4x + 2 + \frac{9x - 4}{x^2 + 3}.$$

73. Divide: $\dfrac{2x^5 + 9x^4 + 8x^3 + 10x^2 + 14x + 5}{2x^2 + 3x + 1}$

$$\begin{array}{r}
x^3 + 3x^2 - x + 5 \\
2x^2 + 3x + 1\overline{\big)2x^5 + 9x^4 + 8x^3 + 10x^2 + 14x + 5} \\
\underline{2x^5 + 3x^4 + x^3} \qquad\qquad\qquad\qquad \\
6x^4 + 7x^3 + 10x^2 \qquad\qquad \\
\underline{6x^4 + 9x^3 + 3x^2} \qquad\qquad \\
-2x^3 + 7x^2 + 14x \qquad \\
\underline{-2x^3 - 3x^2 - x} \qquad \\
10x^2 + 15x + 5 \\
\underline{10x^2 + 15x + 5} \\
0
\end{array}$$

The remainder is 0. The answer is the quotient, $x^3 + 3x^2 - x + 5$.

75. Divide: $(3a^2 - 11a + 17) \div (2a + 6)$

$$\begin{array}{r}
\frac{3}{2}a - 10 \\
2a + 6\overline{\big)3a^2 - 11a + 17} \\
\underline{3a^2 + 9a} \qquad \\
-20a + 17 \\
\underline{-20a - 60} \\
77
\end{array}$$

The remainder is 77.

The quotient is $\dfrac{3}{2}a - 10$.

The answer is $\dfrac{3}{2}a - 10 + \dfrac{77}{2a + 6}$.

77. Divide: $\dfrac{3x^3 + 5x^2 - 9x + 5}{3x - 3}$

$$\begin{array}{r}
x^2 + \frac{8}{3}x - \frac{1}{3} \\
3x - 3\overline{\big)3x^3 + 5x^2 - 9x + 5} \\
\underline{3x^3 - 3x^2} \qquad\qquad \\
8x^2 - 9x \qquad \\
\underline{8x^2 - 8x} \qquad \\
-x + 5 \\
\underline{-x + 1} \\
4
\end{array}$$

$(3x^3 + 5x^2 - 9x + 5)$ divided by $(3x - 3)$ is $x^2 + \dfrac{8}{3}x - \dfrac{1}{3} + \dfrac{4}{3x - 3}$.

79. Use the formula for the area of a rectangle, $A = LW$, with $A = 5x^3 + 7x^2 - 13x - 6$ and $W = 5x + 2$.

$$5x^3 + 7x^2 - 13x - 6 = L(5x + 2)$$

$$\frac{5x^3 + 7x^2 - 13x - 6}{5x + 2} = L$$

$$\begin{array}{r} x^2 + x - 3 \\ 5x + 2\overline{)5x^3 + 7x^2 - 13x - 6} \\ \underline{5x^3 + 2x^2} \\ 5x^2 - 13x \\ \underline{5x^2 + 2x} \\ -15x - 6 \\ \underline{-15x - 6} \\ 0 \end{array}$$

The length L is $(x^2 + x - 3)$ units.

81. Use the formula for the area of a triangle, $A = \frac{1}{2}bh$, with $A = 24m^3 + 48m^2 + 12m$ and $h = m$.

$$24m^3 + 48m^2 + 12m = \frac{1}{2}(b)m$$

$$48m^3 + 96m^2 + 24m = bm$$

$$\frac{48m^3 + 96m^2 + 24m}{m} = b$$

$$\frac{48m^3}{m} + \frac{96m^2}{m} + \frac{24m}{m} = b$$

$$48m^2 + 96m + 24 = b$$

83. Use the distance formula, $d = rt$, with $d = (5x^3 - 6x^2 + 3x + 14)$ miles and $r = (x + 1)$ miles per hour.

$$(5x^3 - 6x^2 + 3x + 14) = (x + 1)t$$

$$\frac{(5x^3 - 6x^2 + 3x + 14)}{(x + 1)} = t$$

$$\begin{array}{r} 5x^2 - 11x + 14 \\ x + 1\overline{)5x^3 - 6x^2 + 3x + 14} \\ \underline{5x^3 + 5x^2} \\ -11x^2 + 3x \\ \underline{-11x^2 - 11x} \\ 14x + 14 \\ \underline{14x + 14} \\ 0 \end{array}$$

The time t is $(5x^2 - 11x + 14)$ hours.

Chapter 5 Review Exercises

1. $4^3 \cdot 4^8 = 4^{3+8} = 4^{11}$

2. $(-5)^6(-5)^5 = (-5)^{6+5} = (-5)^{11}$

3. $(-8x^4)(9x^3) = -8(9)(x^4)(x^3)$
$= -72x^{4+3} = -72x^7$

4. $(2x^2)(5x^3)(x^9) = 2(5)(x^2)(x^3)(x^9)$
$= 10x^{2+3+9} = 10x^{14}$

5. $(19x)^5 = 19^5 x^5$

6. $(-4y)^7 = (-4)^7 y^7$

7. $5(pt)^4 = 5p^4t^4$

8. $\left(\frac{7}{5}\right)^6 = \frac{7^6}{5^6}$

9. $(3x^2y^3)^3 = 3^3(x^2)^3(y^3)^3$
$= 3^3 x^{2\cdot 3} y^{3\cdot 3}$
$= 3^3 x^6 y^9$

10. $(t^4)^8(t^2)^5 = t^{4\cdot 8}t^{2\cdot 5}$
$= t^{32}t^{10}$
$= t^{32+10}$
$= t^{42}$

11. $(6x^2z^4)^2(x^3yz^2)^4$
$= 6^2(x^2)^2(z^4)^2(x^3)^4(y)^4(z^2)^4$
$= 6^2 x^4 z^8 x^{12} y^4 z^8$
$= 6^2 x^{4+12} y^4 z^{8+8}$
$= 6^2 x^{16} y^4 z^{16}$

12. $\left(\dfrac{2m^3n}{p^2}\right)^3 = \dfrac{2^3(m^3)^3n^3}{(p^2)^3}$
$= \dfrac{2^3m^9n^3}{p^6}$

13. $-10^0 = -(10^0) = -(1) = -1$

14. $-(-23)^0 = -1\cdot(-23)^0 = -1\cdot 1 = -1$

15. $6^0 + (-6)^0 = 1 + 1 = 2$

16. $-3^0 - 2^0 = -1 - 1 = -2$

17. $-7^{-2} = -\dfrac{1}{7^2} = -\dfrac{1}{49}$

18. $\left(\dfrac{5}{8}\right)^{-2} = \left(\dfrac{8}{5}\right)^2 = \dfrac{64}{25}$

19. $(2^{-2})^{-3} = 2^{(-2)(-3)}$
$= 2^6 = 64$

20. $9^3\cdot 9^{-5} = 9^{3+(-5)} = 9^{-2} = \dfrac{1}{9^2} = \dfrac{1}{81}$

21. $2^{-1} + 4^{-1} = \dfrac{1}{2^1} + \dfrac{1}{4^1}$
$= \dfrac{1}{2} + \dfrac{1}{4}$
$= \dfrac{2}{4} + \dfrac{1}{4} = \dfrac{3}{4}$

22. $\dfrac{6^{-5}}{6^{-3}} = \dfrac{6^3}{6^5} = \dfrac{1}{6^2} = \dfrac{1}{36}$

23. $\dfrac{x^{-7}}{x^{-9}} = \dfrac{x^9}{x^7} = x^{9-7} = x^2$

24. $\dfrac{y^4\cdot y^{-2}}{y^{-5}} = \dfrac{y^4\cdot y^5}{y^2} = \dfrac{y^9}{y^2} = y^7$

25. $(3r^{-2})^{-4} = (3)^{-4}(r^{-2})^{-4}$
$= (3^{-4})(r^{-2(-4)})$
$= \dfrac{1}{3^4}r^8$
$= \dfrac{r^8}{81}$

26. $(3p)^4(3p^{-7}) = (3^4p^4)(3p^{-7})$
$= 3^{4+1}\cdot p^{4+(-7)}$
$= 3^5p^{-3}$
$= \dfrac{3^5}{p^3}$

27. $\dfrac{ab^{-3}}{a^4b^2} = \dfrac{a}{a^4b^2b^3} = \dfrac{1}{a^3b^5}$

28. $\dfrac{(6r^{-1})^2(2r^{-4})}{r^{-5}(r^2)^{-3}} = \dfrac{(6^2r^{-2})(2r^{-4})}{r^{-5}r^{-6}}$
$= \dfrac{72r^{-6}}{r^{-11}}$
$= \dfrac{72r^{11}}{r^6}$
$= 72r^5$

29. $48{,}000{,}000 = 4.8\times 10^7$
Move the decimal point left seven places so it is to the right of the first nonzero digit. 48,000,000 is *greater* than 4.8, so the power is *positive.*

30. $28{,}988{,}000{,}000 = 2.8988\times 10^{10}$
Move the decimal point left 10 places so it is to the right of the first nonzero digit. 28,988,000,000 is *greater* than 2.8988, so the power is *positive.*

31. $0.0000000824 = 8.24\times 10^{-8}$
Move the decimal point right eight places so it is to the right of the first nonzero digit. 0.0000000824 is *less* than 8.24, so the power is *negative.*

32. $-4{,}820{,}000 = -4.82\times 10^6$
Move the decimal point left six places so it is to the right of the first nonzero digit. $|-4{,}820{,}000|$ is *greater* than $|-4.82|$, so the power is *positive.*

33. $2.4\times 10^4 = 24{,}000$
Move the decimal point four places to the right.

34. $7.83\times 10^7 = 78{,}300{,}000$
Move the decimal point seven places to the right.

35. $8.97\times 10^{-7} = 0.000000897$
Move the decimal point seven places to the left.

36. $-7.6\times10^{-4}=-0.00076$

Move the decimal point four places to the left.

37. $(2\times10^{-3})(4\times10^{5})$

$=(2\times4)(10^{-3}\times10^{5})$

$=8\times10^{-3+5}=8\times10^{2}=800$

38. $(2.5\times10^{-51})(2.0\times10^{51})$

$=(2.5\times2.0)(10^{-51}\times10^{51})$

$=5.0\times10^{-51+51}=5.0\times10^{0}=5$

39. $\dfrac{8\times10^{4}}{2\times10^{-2}}=\dfrac{8}{2}\times\dfrac{10^{4}}{10^{-2}}=4\times10^{4-(-2)}$

$=4\times10^{6}=4{,}000{,}000$

40. $\dfrac{60\times10^{-1}}{24\times10}=\dfrac{60}{24}\times\dfrac{10^{-1}}{10^{1}}=\dfrac{5}{2}\times10^{-1-1}$

$=2.5\times10^{-2}=0.025$

41. $8.1887\times10^{16}=81{,}887{,}000{,}000{,}000{,}000$

Move the decimal point 16 places to the right.

42. $3.72174\times10^{7}=37{,}217{,}400$

Move the decimal point seven places to the right.

43. $1{,}008{,}600{,}000{,}000{,}000=1.0086\times10^{15}$

Move the decimal point left 15 places so it is to the right of the first nonzero digit. $1{,}008{,}600{,}000{,}000{,}000$ is *greater* than 1.0086, so the power is *positive*.

44. $67{,}800{,}000{,}000{,}000=6.78\times10^{13}$

Move the decimal point left 13 places so it is to the right of the first nonzero digit. $67{,}800{,}000{,}000{,}000$ is *greater* than 6.78, so the power is *positive*.

45. $9m^2+11m^2=(9+11)m^2$

$=20m^2$

The degree is 2. There is one term, so this is a *monomial.*

46. $-4p+p^3-p^2=p^3-p^2-4p$

The degree is 3.There are three terms, so this is a *trinomial.*

47. $-7y^5-8y^4-y^5+y^4$

$=-7y^5-1y^5-8y^4+1y^4$

$=(-7-1)y^5+(-8+1)y^4$

$=-8y^5-7y^4$

The degree is 5. There are two terms, so the polynomial is a *binomial.*

48. Change all signs in the second polynomial and then add.

$(12r^4-7r^3+2r^2)-(5r^4-3r^3+2r^2-1)$

$-(7r^4-3r^3+2r-6)$

$=(12r^4-7r^3+2r^2)+(-5r^4+3r^3-2r^2+1)$

$+(-7r^4+3r^3-2r+6)$

$=12r^4-7r^3+2r^2-5r^4+3r^3-2r^2+1$

$-7r^4+3r^3-2r+6$

$=-r^3-2r+7$

49. $(5x^3y^2-3xy^5+12x^2)$

$-(-9x^2-8x^3y^2+2xy^5)$

$=(5x^3y^2-3xy^5+12x^2)$

$+(9x^2+8x^3y^2-2xy^5)$

$=(5x^3y^2+8x^3y^2)+(-3xy^5-2xy^5)$

$+(12x^2+9x^2)$

$=13x^3y^2-5xy^5+21x^2$

50. Add.

$$\begin{array}{rcr} -2a^3 & + & 5a^2 \\ 3a^3 & - & a^2 \\ \hline a^3 & + & 4a^2 \end{array}$$

51. Subtract.

$$\begin{array}{rcrcr} 6y^2 & - & 8y & + & 2 \\ \underline{5y^2} & \underline{+} & \underline{2y} & \underline{-} & \underline{7} \end{array}$$

Change all signs in the second row and then add.

$$\begin{array}{rcrcr} 6y^2 & - & 8y & + & 2 \\ -5y^2 & - & 2y & + & 7 \\ \hline y^2 & - & 10y & + & 9 \end{array}$$

52. Subtract.

$$\begin{array}{r} -12k^4 - 8k^2 + 7k \\ \underline{k^4 + 7k^2 - 11k} \end{array}$$

Change all signs in the second row and then add.

$$\begin{array}{r} -12k^4 - 8k^2 + 7k \\ \underline{-k^4 - 7k^2 + 11k} \\ -13k^4 - 15k^2 + 18k \end{array}$$

53. $y = -x^2 + 5$

$x = -2: y = -(-2)^2 + 5 = -4 + 5 = 1$

$x = -1: y = -(-1)^2 + 5 = -1 + 5 = 4$

$x = 0: y = -(0)^2 + 5 = 0 + 5 = 5$

$x = 1: y = -(1)^2 + 5 = -1 + 5 = 4$

$x = 2: y = -(2)^2 + 5 = -4 + 5 = 1$

x	-2	-1	0	1	2
y	1	4	5	4	1

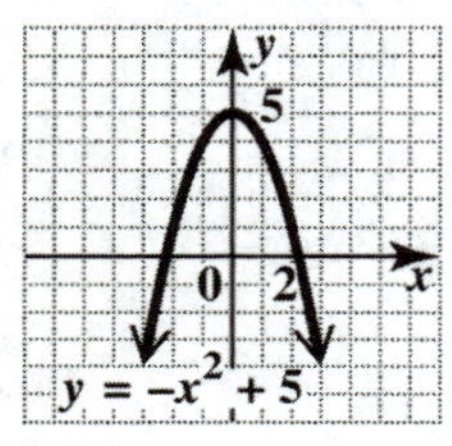

54. $y = 3x^2 - 2$

$x = -2: y = 3(-2)^2 - 2 = 3 \cdot 4 - 2 = 10$

$x = -1: y = 3(-1)^2 - 2 = 3 \cdot 1 - 2 = 1$

$x = 0: y = 3(0)^2 - 2 = 3 \cdot 0 - 2 = -2$

$x = 1: y = 3(1)^2 - 2 = 3 \cdot 1 - 2 = 1$

$x = 2: y = 3(2)^2 - 2 = 3 \cdot 4 - 2 = 10$

x	-2	-1	0	1	2
y	10	1	-2	1	10

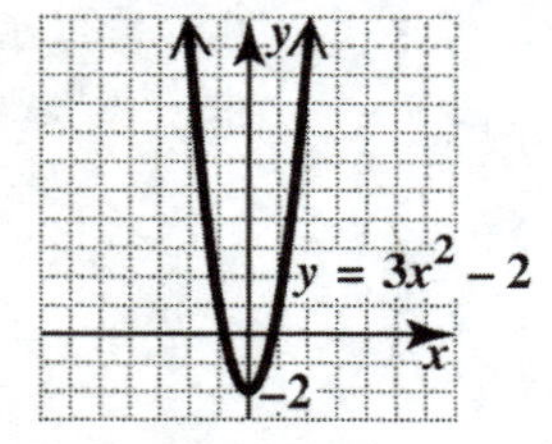

55. Multiply vertically.

$$\begin{array}{r} a^2 - 4a + 1 \\ \underline{a + 2} \\ 2a^2 - 8a + 2 \\ \underline{a^3 - 4a^2 + a \quad\quad} \\ a^3 - 2a^2 - 7a + 2 \end{array}$$

56. Multiply vertically.

$$\begin{array}{r} 2r^2 + 4r - 3 \\ \underline{3r - 2} \\ -4r^2 - 8r + 6 \\ \underline{6r^3 + 12r^2 - 9r \quad\quad} \\ 6r^3 + 8r^2 - 17a + 6 \end{array}$$

57. $(5p^2 + 3p)(p^3 - p^2 + 5)$

$= 5p^2(p^3) + 5p^2(-p^2) + 5p^2(5)$
$\quad + 3p(p^3) + 3p(-p^2) + 3p(5)$

$= 5p^5 - 5p^4 + 25p^2 + 3p^4 - 3p^3 + 15p$

$= 5p^5 - 2p^4 - 3p^3 + 25p^2 + 15p$

58. $(m - 9)(m + 2)$

$= m(m) + m(2) + (-9)(m) + (-9)(2)$

$= m^2 + 2m - 9m - 18$

$= m^2 - 7m - 18$

59. $(3k - 6)(2k + 1)$

$= 3k(2k) + 3k(1) + (-6)(2k) + (-6)(1)$

$= 6k^2 + 3k - 12k - 6$

$= 6k^2 - 9k - 6$

60. $(a + 3b)(2a - b)$

$= a(2a) + a(-b) + 3b(2a) + 3b(-b)$

$= 2a^2 - ab + 6ab - 3b^2$

$= 2a^2 + 5ab - 3b^2$

61. $(6k + 5q)(2k - 7q)$

$= 6k(2k) + 6k(-7q) + 5q(2k) + 5q(-7q)$

$= 12k^2 - 42kq + 10kq - 35q^2$

$= 12k^2 - 32kq - 35q^2$

62. $(s-1)^3 = (s-1)^2(s-1)$
$= (s^2 - 2s + 1)(s - 1)$

Now, use vertical multiplication.

$$\begin{array}{r} s^2 - 2s + 1 \\ s - 1 \\ \hline -s^2 + 2s - 1 \\ s^3 - 2s^2 + s \quad\quad \\ \hline s^3 - 3s^2 + 3s - 1 \end{array}$$

$(s-1)^3 = s^3 - 3s^2 + 3s - 1$

63. $(a+4)^2 = a^2 + 2(a)(4) + 4^2$
$= a^2 + 8a + 16$

64. $(2r+5t)^2 = (2r)^2 + 2(2r)(5t) + (5t)^2$
$= 4r^2 + 20rt + 25t^2$

65. $(6m-5)(6m+5) = (6m)^2 - 5^2$
$= 36m^2 - 25$

66. $(5a+6b)(5a-6b) = (5a)^2 - (6b)^2$
$= 25a^2 - 36b^2$

67. $(r+2)^3 = (r+2)^2(r+2)$
$= (r^2 + 4r + 4)(r + 2)$
$= r^3 + 4r^2 + 4r + 2r^2 + 8r + 8$
$= r^3 + 6r^2 + 12r + 8$

68. $t(5t-3)^2 = t(25t^2 - 30t + 9)$
$= 25t^3 - 30t^2 + 9r$

69. Answers will vary. One example is given here.

(a) $(x+y)^2 \neq x^2 + y^2$
Let $x = 2$ and $y = 3$.
$(x+y)^2 = (2+3)^2 = 5^2 = 25$
$x^2 + y^2 = 2^2 + 3^2 = 4 + 9 = 13$
Since $25 \neq 13$,
$(x+y)^2 \neq x^2 + y^2$.

(b) $(x+y)^3 \neq x^3 + y^3$
Let $x = 2$ and $y = 3$.
$(x+y)^3 = (2+3)^3 = 5^3 = 125$
$x^3 + y^3 = 2^3 + 3^3 = 8 + 27 = 35$
Since $125 \neq 35$,
$(x+y)^3 \neq x^3 + y^3$.

70. To find the third power of a binomial, such as $(a+b)^3$, first square the binomial and then multiply that result by the binomial:
$(a+b)^3 = (a+b)^2(a+b)$
$= (a^2 + 2ab + b^2)(a+b)$
$= (a^3 + 2a^2b + ab^2)$
$+ (a^2b + 2ab^2 + b^3)$
$= a^3 + 3a^2b + 3ab^2 + b^3$

71. Use the formula for the volume of a cube, $V = e^3$, with $e = x^2 + 2$ centimeters.
$V = (x^2 + 2)^3$
$= (x^2 + 2)^2(x^2 + 2)$
$= (x^4 + 4x^2 + 4)(x^2 + 2)$

Now use vertical multiplication.

$$\begin{array}{r} x^2 + 4x^2 + 4 \\ x^2 + 2 \\ \hline 2x^4 + 8x^2 + 8 \\ x^6 + 4x^4 + 4x^2 \quad\quad \\ \hline x^6 + 6x^4 + 12x^2 + 8 \end{array}$$

The volume of the cube is $x^6 + 6x^4 + 12x^2 + 8$ cubic centimeters.

72. Use the formula for the volume of a sphere, $V = \frac{4}{3}\pi r^3$, with $r = x + 1$ inches.

$V = \frac{4}{3}\pi(x+1)^3$
$= \frac{4}{3}\pi(x+1)^2(x+1)$
$= \frac{4}{3}\pi(x^2 + 2x + 1)(x+1)$

Now use vertical multiplication.

$$\begin{array}{r} x^2 + 2x + 1 \\ x + 1 \\ \hline x^2 + 2x + 1 \\ x^3 + 2x^2 + x \quad\quad \\ \hline x^3 + 3x^2 + 3x + 1 \end{array}$$

$V = \frac{4}{3}\pi(x^3 + 3x^2 + 3x + 1)$
$= \frac{4}{3}\pi x^3 + 4\pi x^2 + 4\pi x + \frac{4}{3}\pi$

The volume of the sphere is $\frac{4}{3}\pi x^3 + 4\pi x^2 + 4\pi x + \frac{4}{3}\pi$ cubic inches.

73. $\dfrac{-15y^4}{9y^2} = \dfrac{-15y^{4-2}}{9} = \dfrac{-5y^2}{3}$

74 $\left(-10m^4n^2 + 5m^3n^2 + 6m^2n^4\right) \div \left(5m^2n\right)$

$$= \frac{-10m^4n^2 + 5m^3n^2 + 6m^2n^4}{5m^2n}$$

$$= \frac{-10m^4n^2}{5m^2n} + \frac{5m^3n^2}{5m^2n} + \frac{6m^2n^4}{5m^2n}$$

$$= -2m^2n + mn + \frac{6n^3}{5}$$

75. $\dfrac{6y^4 - 12y^2 + 18y}{6y} = \dfrac{6y^4}{6y} - \dfrac{12y^2}{6y} + \dfrac{18y}{6y}$

$$= y^3 - 2y + 3$$

76. $\dfrac{24r^8s^6 + 12r^7s^5 - 8r}{-4r^3s^5}$

$$= \frac{24r^8s^6}{-4r^3s^5} + \frac{12r^7s^5}{-4r^3s^5} - \frac{8r}{-4r^3s^5}$$

$$= -6r^5s - 3r^4 + \frac{2}{r^2s^5}$$

77. Let P be the polynomial that when multiplied by $6m^2n$ gives the product $12m^3n^2 + 18m^6n^3 - 24m^2n^2$.

$$(P)\left(6m^2n\right) = 12m^3n^2 + 18m^6n^3 - 24m^2n^2$$

$$P = \frac{12m^3n^2 + 18m^6n^3 - 24m^2n^2}{6m^2n}$$

$$= \frac{12m^3n^2}{6m^2n} + \frac{18m^6n^3}{6m^2n} - \frac{24m^2n^2}{6m^2n}$$

$$= 2mn + 3m^4n^2 - 4n$$

78. The error made was not dividing both terms in the numerator by 6. The correct method is as follows:

$$\frac{6x^2 - 12x}{6} = \frac{6x^2}{6} - \frac{12x}{6} = x^2 - 2x.$$

79. Divide: $\dfrac{2r^2 + 3r - 14}{r - 2}$

$$\begin{array}{r|l} & 2r + 7 \\ r - 2 & 2r^2 + 3r - 14 \\ & \underline{2r^2 - 4r} \\ & \quad 7r - 14 \\ & \quad \underline{7r - 14} \\ & \qquad 0 \end{array}$$

The remainder is 0.
The answer is the quotient, $2r + 7$.

80. Divide: $\dfrac{10a^3 + 9a^2 - 14a + 9}{5a - 3}$

$$\begin{array}{r|l} & 2a^2 + 3a - 1 \\ 5a - 3 & 10a^3 + 9a^2 - 14a + 9 \\ & \underline{10a^3 - 6a^2} \\ & \quad 15a^2 - 14a \\ & \quad \underline{15a^2 - 9a} \\ & \qquad -5a + 9 \\ & \qquad \underline{-5a + 3} \\ & \qquad\qquad 6 \end{array}$$

The answer is $2a^2 + 3a - 1 + \dfrac{6}{5a - 3}$.

81. Divide: $\dfrac{x^4 - 5x^2 + 3x^3 - 3x + 4}{x^2 - 1}$

Write the dividend in descending powers and use 0 as the coefficient of the missing term.

$$\begin{array}{r|l} & x^2 + 3x - 4 \\ x^2 + 0x - 1 & x^4 + 3x^3 - 5x^2 - 3x + 4 \\ & \underline{x^4 + 0x^3 - x^2} \\ & \quad 3x^3 - 4x^2 - 3x \\ & \quad \underline{3x^3 + 0x^2 - 3x} \\ & \qquad -4x^2 + 0x + 4 \\ & \qquad \underline{-4x^2 + 0x + 4} \\ & \qquad\qquad 0 \end{array}$$

The remainder is 0. The answer is the quotient, $x^2 + 3x - 4$.

82. Divide: $\dfrac{m^4+4m^3-12m-5m^2+6}{m^2-3}$

Write the dividend in descending powers and use 0 as the coefficient of the missing term.

$$\begin{array}{r}
m^2+4m-2 \\
m^2+0m-3\overline{\smash{)}m^4+4m^3-5m^2-12m+6} \\
\underline{m^4+0m^3-3m^2} \\
4m^3-2m^2-12m \\
\underline{4m^3+0m^2-12m} \\
-2m^2+0m+6 \\
\underline{-2m^2+0m+6} \\
0
\end{array}$$

The remainder is 0. The answer is the quotient, m^2+4m-2.

83. Divide: $\dfrac{16x^2-25}{4x+5}$

$$\begin{array}{r}
4x-5 \\
4x+5\overline{\smash{)}16x^2+0x-25} \\
\underline{16x^2+20x} \\
-20x-25 \\
\underline{-20x-25} \\
0
\end{array}$$

The remainder is 0.
The answer is the quotient, $4x-5$.

84. Divide: $\dfrac{25y^2-100}{5y+10}$

$$\begin{array}{r}
5y-10 \\
5y+10\overline{\smash{)}25y^2+0y-100} \\
\underline{25y^2+50y} \\
-50y-100 \\
\underline{-50y-100} \\
0
\end{array}$$

The remainder is 0.
The answer is the quotient, $5y-10$.

85. Divide: $\dfrac{y^3-8}{y-2}$

$$\begin{array}{r}
y^2+2y+4 \\
y-2\overline{\smash{)}y^3+0y^2+0y-8} \\
\underline{y^3-2y^2} \\
2y^2+0y \\
\underline{2y^2-4y} \\
4y-8 \\
\underline{4y-8} \\
0
\end{array}$$

The remainder is 0.
The answer is the quotient, y^2+2y+4.

86. Divide: $\dfrac{1000x^6+1}{10x^2+1}$

$$\begin{array}{r}
100x^4-10x^2+1 \\
10x^2+1\overline{\smash{)}1000x^6+0x^4+0x^2+1} \\
\underline{1000x^6+100x^4} \\
-100x^4+0x^2 \\
\underline{-100x^4-10x^2} \\
10x^2+1 \\
\underline{10x^2+1} \\
0
\end{array}$$

The remainder is 0.
The answer is the quotient, $100x^4-10x^2+1$.

87. Divide: $\dfrac{6y^4-15y^3+14y^2-5y-1}{3y^2+1}$

$$\begin{array}{r}
2y^2-5y+4 \\
3y^2+1\overline{\smash{)}6y^4-15y^3+14y^2-5y-1} \\
\underline{6y^4+0y^3+2y^2} \\
-15y^3+12y^2-5y \\
\underline{-15y^3+0y^2-5y} \\
12y^2+0y-1 \\
\underline{12y^2+0y+4} \\
-5
\end{array}$$

The answer is $2y^2-5y+4+\dfrac{-5}{3y^2+1}$.

88. Divide: $\dfrac{4x^5-8x^4-3x^3+22x^2-15}{4x^2-3}$

$$\begin{array}{r} x^3-2x^2\qquad\;\;+4 \\ 4x^2-3\overline{)4x^5-8x^4-3x^3+22x^2+0x-15} \\ \underline{4x^5+0x^4-3x^3}\qquad\qquad\qquad\quad \\ -8x^4+0x^3+22x^2\qquad\qquad \\ \underline{-8x^4+0x^3+\;\;6x^2}\qquad\qquad \\ 16x^2+0x-15 \\ \underline{16x^2+0x-12} \\ -3 \end{array}$$

The answer is $x^3-2x^2+4+\dfrac{-3}{4x^2-3}$.

Chapter 5 Mixed Review Exercises

1. $5^0+7^0=1+1=2$

2. $\left(\dfrac{6r^2p}{5}\right)^3=\dfrac{6^3\left(r^2\right)^3p^3}{5^3}$
$=\dfrac{216r^6p^3}{5^3}$

3. $(12a+1)(12a-1)=(12a)^2-1^2$
$=144a^2-1$

4. $2^{-4}=\dfrac{1}{2^4}=\dfrac{1}{16}$

5. $\left(4^{-2}\right)^2=\left(\dfrac{1}{4^2}\right)^2=\left(\dfrac{1}{16}\right)^2=\dfrac{1}{256}$

6. $\dfrac{2p^3-6p^2+5p}{2p^2}=\dfrac{2p^3}{2p^2}-\dfrac{6p^2}{2p^2}+\dfrac{5p}{2p^2}$
$=\dfrac{2}{2}p^{3-2}-\dfrac{6}{2}p^{2-2}+\dfrac{5}{2}p^{1-2}$
$=1p^1-3p^0+\dfrac{5}{2}p^{-1}$
$=p-3+\dfrac{5}{2p}$

7. $\dfrac{\left(2m^{-5}\right)\left(3m^2\right)^{-1}}{m^{-2}\left(m^{-1}\right)^2}$
$=\dfrac{2\cdot m^{-5}\cdot(3)^{-1}\cdot\left(m^2\right)^{-1}}{m^{-2}\cdot\left(m^{-1}\right)^2}$
$=\dfrac{2\cdot m^{-5}\cdot 3^{-1}\cdot m^{-2}}{m^{-2}\cdot m^{-2}}$
$=\dfrac{2\cdot m^2\cdot m^2}{m^5\cdot m^2\cdot 3^1}$
$=\dfrac{2m^4}{3m^7}$
$=\dfrac{2}{3}m^{4-7}$
$=\dfrac{2}{3}m^{-3}=\dfrac{2}{3m^3}$

8. $(3k-6)\left(2k^2+4k+1\right)$

Multiply vertically.

$$\begin{array}{rrrr} & 2k^2 & +\;4k & +\;1 \\ & & 3k & -\;6 \\ \hline & -12k^2 & -\;24k & -\;6 \\ 6k^3 & +\;12k^2 & +\;3k & \\ \hline 6k^3 & & -\;21k & -\;6 \end{array}$$

9. $\dfrac{r^9\cdot r^{-5}}{r^{-2}\cdot r^{-7}}=\dfrac{r^{9+(-5)}}{r^{-2+(-7)}}$
$=\dfrac{r^4}{r^{-9}}$
$=r^{4-(-9)}$
$=r^{13}$

10. Use the following formula for the square of a binomial: $(x+y)^2=x^2+2xy+y^2$

$(2r+5s)^2=(2r)^2+2(2r)(5s)+(5s)^2$
$=4r^2+20rs+25s^2$

11. $\dfrac{2y^3+17y^2+37y+7}{2y+7}$

$$\begin{array}{r} y^2+5y+1 \\ 2y+7\overline{)2y^3+17y^2+37y+7} \\ \underline{2y^3+7y^2} \\ 10y^2+37y \\ \underline{10y^2+35y} \\ 2y+7 \\ \underline{2y+7} \\ 0 \end{array}$$

The answer is y^2+5y+1.

12. $(2r+5)(5r-2)$

F: $2r(5r)=10r^2$
O: $2r(-2)=-4r$
I: $5(5r)=25r$
L: $5(-2)=-10$

$$(2r+5)(5r-2)=10r^2-4r+25r-10$$
$$=10r^2+21r-10$$

13. Add, column by column.

$$\begin{array}{rrr} -5y^2 & +3y & -11 \\ 4y^2 & -7y & +15 \\ \hline -y^2 & -4y & +4 \end{array}$$

14. $\left(25x^2y^3-8xy^2+15x^3y\right)\div\left(10x^2y^3\right)$

$$=\frac{25x^2y^3-8xy^2+15x^3y}{10x^2y^3}$$
$$=\frac{25x^2y^3}{10x^2y^3}-\frac{8xy^2}{10x^2y^3}+\frac{15x^3y}{10x^2y^3}$$
$$=\frac{25}{10}x^0y^0-\frac{8}{10}x^{-1}y^{-1}+\frac{15}{10}x^1y^{-2}$$
$$=\frac{5}{2}-\frac{4}{5xy}+\frac{3x}{2y^2}$$

15. $\left(6p^2-p-8\right)-\left(-4p^2+2p-3\right)$

$$=\left(6p^2-p-8\right)+\left(4p^2-2p+3\right)$$
$$=\left(6p^2+4p^2\right)+[-p+(-2p)]+[(-8)+3]$$
$$=10p^2-3p-5$$

16. Use 0 as the coefficient of the missing x-term.

$$\begin{array}{r} 3x^2+9x+25 \\ x-3\overline{)3x^3+0x^2-2x+5} \\ \underline{3x^3-9x^2} \\ 9x^2-2x \\ \underline{9x^2-27x} \\ 25x+5 \\ \underline{25x-75} \\ 80 \end{array}$$

$$\frac{3x^3-2x+5}{x-3}=3x^2+9x+25+\frac{80}{x-3}$$

17. Use the following formula for the square of a binomial: $(x+y)^2=x^2+2xy+y^2$

$$(-7+2k)^2=(-7)^2+2(-7)(2k)+(2k)^2$$
$$=49-28k+4k^2$$

18. $$\left(\frac{x}{y^{-3}}\right)^{-4}=\frac{x^{-4}}{\left(y^{-3}\right)^{-4}}$$
$$=\frac{1}{x^4y^{12}}$$

19. **(a)** Use the formula for the perimeter of a rectangle, $P=2L+2W$, with $L=2x-3$ and $W=x+2$.

$$P=2(2x-3)+2(x+2)$$
$$=(4x-6)+(2x+4)$$
$$=(4x+2x)+(-6+4)$$
$$=6x-2$$

The perimeter of the rectangle can be expressed as $6x-2$.

(b) Use the formula for the area of a rectangle, $A=LW$, with $L=2x-3$ and $W=x+2$.

$$A=(2x-3)(x+2)$$
$$=2x(x)+2x(2)+(-3)(x)+(-3)(2)$$
$$=2x^2+4x-3x-6$$
$$=2x^2+x-6$$

The area of the rectangle can be expressed as $2x^2+x-6$.

20 **(a)** Use the formula for the perimeter of a square, $P = 4s$, with $s = 5x^4 + 2x^2$.

$$P = 4(5x^4 + 2x^2)$$
$$= 4(5x^4) + 4(2x^2)$$
$$= 20x^4 + 8x^2$$

(b) Use the formula for the area of a square, $A = s^2$, with $s = 5x^4 + 2x^2$.

$$A = (5x^4 + 2x^2)^2$$

Use the following formula for the square of a binomial: $(x+y)^2 = x^2 + 2xy + y^2$

$$(5x^4 + 2x^2)^2$$
$$= (5x^4)^2 + 2(5x^4)(2x^2) + (2x^2)^2$$
$$= 25x^8 + 20x^6 + 4x^4$$

Chapter 5 Test

1. $5^{-4} = \frac{1}{5^4} = \frac{1}{625}$

2. $(-3)^0 + 4^0 = 1 + 1 = 2$

3. $4^{-1} + 3^{-1} = \frac{1}{4^1} + \frac{1}{3^1} = \frac{3}{12} + \frac{4}{12} = \frac{7}{12}$

4.
$$\frac{(3x^2y)^2(xy^3)^2}{(xy)^3} = \frac{3^2(x^2)^2 y^2x^2(y^3)^2}{x^3y^3}$$
$$= \frac{9x^4y^2x^2y^6}{x^3y^3}$$
$$= \frac{9x^6y^8}{x^3y^3}$$
$$= 9x^3y^5$$

5. $\frac{8^{-1}\cdot 8^4}{8^{-2}} = \frac{8^{(-1)+4}}{8^{-2}} = \frac{8^3}{8^{-2}} = 8^{3-(-2)} = 8^5$

6.
$$\frac{(x^{-3})^{-2}(x^{-1}y)^2}{(xy^{-2})^2} = \frac{(x^{-3})^{-2}(x^{-1})^2(y)^2}{(x)^2(y^{-2})^2}$$
$$= \frac{x^6x^{-2}y^2}{x^2y^{-4}}$$
$$= \frac{x^4y^2}{x^2y^{-4}}$$
$$= x^{4-2}y^{2-(-4)}$$
$$= x^2y^6$$

7. **(a)** $3^{-4} = \frac{1}{3^4} = \frac{1}{81}$, which is *positive.*

A negative exponent indicates a reciprocal, not a negative number.

(b) $(-3)^4 = 81$, which is *positive.*

(c) $-3^4 = -1\cdot 3^4 = -81$, which is *negative.*

(d) $3^0 = 1$, which is *positive.*

(e) $(-3)^0 - 3^0 = 1 - 1 = 0$ *(zero)*

(f) $(-3)^{-3} = \frac{1}{(-3)^3} = \frac{1}{-27}$, which is *negative.*

8. **(a)** $45{,}000{,}000{,}000 = 4.5\times 10^{10}$

Move the decimal point left 10 places so it is to the right of the first nonzero digit. 45,000,000,000 is *greater* than 4.5, so the power is *positive.*

(b) $3.6\times 10^{-6} = 0.0000036$

Move the decimal point six places to the left.

(c)
$$\frac{9.5\times 10^{-1}}{5\times 10^3} = \frac{9.5}{5}\times\frac{10^{-1}}{10^3}$$
$$= 1.9\times 10^{-1-3}$$
$$= 1.9\times 10^{-4}$$
$$= 0.00019$$

9. **(a)** $1000 = 1 \times 10^3$

$5{,}890{,}000{,}000{,}000 = 5.89 \times 10^{12}$

(b) $(1 \times 10^3)(5.89 \times 10^{12})$
$= 5.89 \times 10^{3+12}$
$= 5.89 \times 10^{15}$

The Large Magellanic Cloud is 5.89×10^{15} miles across.

10. $5x^2 + 8x - 12x^2 = 5x^2 - 12x^2 + 8x$
$= -7x^2 + 8x$

degree 2; binomial (2 terms)

11. $13n^3 - n^2 + n^4 + 3n^4 - 9n^2$
$= n^4 + 3n^4 + 13n^3 - n^2 - 9n^2$
$= 4n^4 + 13n^3 - 10n^2$

degree 4; trinomial (3 terms)

12. $y = 2x^2 - 4$

$x = -2: y = 2(-2)^2 - 4 = 2 \cdot 4 - 4 = 4$

$x = -1: y = 2(-1)^2 - 4 = 2 \cdot 1 - 4 = -2$

$x = 0: y = 2(0)^2 - 4 = 2 \cdot 0 - 4 = -4$

$x = 1: y = 2(1)^2 - 4 = 2 \cdot 1 - 4 = -2$

$x = 2: y = 2(2)^2 - 4 = 2 \cdot 4 - 4 = 4$

x	−2	−1	0	1	2
y	4	−2	−4	−2	4

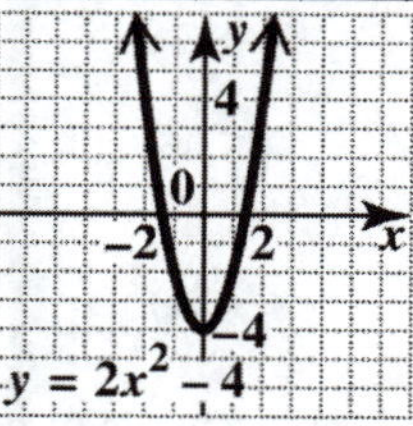

13. $(2y^2 - 8y + 8) + (-3y^2 + 2y + 3)$
$-(y^2 + 3y - 6)$
$= (2y^2 - 8y + 8) + (-3y^2 + 2y + 3)$
$+(-y^2 - 3y + 6)$
$= (2y^2 - 3y^2 - y^2) + (-8y + 2y - 3y)$
$+(8 + 3 + 6)$
$= -2y^2 - 9y + 17$

14. $(-9a^3b^2 + 13ab^5 + 5a^2b^2)$
$-(6ab^5 + 12a^3b^2 + 10a^2b^2)$
$= (-9a^3b^2 + 13ab^5 + 5a^2b^2)$
$+(-6ab^5 - 12a^3b^2 - 10a^2b^2)$
$= (-9a^3b^2 - 12a^3b^2) + (13ab^5 - 6ab^5)$
$+(5a^2b^2 - 10a^2b^2)$
$= -21a^3b^2 + 7ab^5 - 5a^2b^2$

15. Add.

$$\begin{array}{rrr} -6r^5 & +4r^2 & -3 \\ 6r^5 & +12r^2 & -16 \\ \hline & 16r^2 & -19 \end{array}$$

16. Subtract.

$$\begin{array}{l} 9t^3 - 4t^2 + 2t + 2 \\ \underline{9t^3 + 8t^2 - 3t - 6} \end{array}$$

Change all signs in the second row and then add.

$$\begin{array}{rrrr} 9t^3 & -\ 4t^2 & +\ 2t & +\ 2 \\ -9t^3 & -\ 8t^2 & +\ 3t & +\ 6 \\ \hline & -12t^2 & +\ 5t & +\ 8 \end{array}$$

17. $3x^2(-9x^3 + 6x^2 - 2x + 1)$
$= 3x^2(-9x^3) + 3x^2(6x^2) + 3x^2(-2x) + 3x^2(1)$
$= -27x^5 + 18x^4 - 6x^3 + 3x^2$

18. $(t - 8)(t + 3)$
$= t^2 + 3t - 8t - 24$
$= t^2 - 5t - 24$

19. $(4x + 3y)(2x - y)$
$= 8x^2 - 4xy + 6xy - 3y^2$
$= 8x^2 + 2xy - 3y^2$

20. $(5x - 2y)^2 = (5x)^2 - 2(5x)(2y) + (2y)^2$
$= 25x^2 - 20xy + 4y^2$

21. $(10v + 3w)(10v - 3w) = (10v)^2 - (3w)^2$
$= 100v^2 - 9w^2$

22. $(2r-3)\left(r^2+2r-5\right)$

Multiply vertically.

$$\begin{array}{rrrr} & r^2 & +\ 2r & -\ 5 \\ & & 2r & -\ 3 \\ \hline & -3r^2 & -\ 6r & +\ 15 \\ 2r^3\ + & 4r^2 & -\ 10r & \\ \hline 2r^3\ + & r^2 & -\ 16r & +\ 15 \end{array}$$

23. Use the formula for the perimeter of a square, $P=4s$, with $s=3x+9$.

$$\begin{aligned} P &= 4s \\ &= 4(3x+9) \\ &= 4(3x)+4(9) \\ &= 12x+36 \end{aligned}$$

The perimeter of the square is $(12x+36)$ units.

24. Use the formula for the area of a square, $A=s^2$, with $s=3x+9$.

$$\begin{aligned} A &= s^2 \\ &= (3x+9)^2 \\ &= (3x)^2+2(3x)(9)+9^2 \\ &= 9x^2+54x+81 \end{aligned}$$

The area of the square is $\left(9x^2+54x+81\right)$ square units.

25. $\dfrac{8y^3-6y^2+4y+10}{2y}$

$$\begin{aligned} &= \frac{8y^3}{2y}-\frac{6y^2}{2y}+\frac{4y}{2y}+\frac{10}{2y} \\ &= 4y^2-3y+2+\frac{5}{y} \end{aligned}$$

26. $\left(-9x^2y^3+6x^4y^3+12xy^3\right)\div(3xy)$

$$\begin{aligned} &= \frac{-9x^2y^3+6x^4y^3+12xy^3}{3xy} \\ &= \frac{-9x^2y^3}{3xy}+\frac{6x^4y^3}{3xy}+\frac{12xy^3}{3xy} \\ &= -3xy^2+2x^3y^2+4y^2 \end{aligned}$$

27. Divide: $\dfrac{5x^2-x-18}{5x+9}$

$$\begin{array}{r} x-\ 2 \\ 5x+9\overline{\big)5x^2-\ x-18} \\ \underline{5x^2+9x} \\ -10x-18 \\ \underline{-10x-18} \\ 0 \end{array}$$

The remainder is 0. The answer is the quotient, $x-2$.

28. Divide: $\left(3x^3-x+4\right)\div(x-2)$

$$\begin{array}{r} 3x^2+\ 6x+\ 11 \\ x-2\overline{\big)3x^3+0x^2-\ x+\ 4} \\ \underline{3x^3-6x^2} \\ 6x^2-\ x \\ \underline{6x^2-12x} \\ 11x+\ 4 \\ \underline{11x-22} \\ 26 \end{array}$$

The answer is $3x^2+6x+11+\dfrac{26}{x-2}$.

Chapters R–5 Cumulative Review Exercises

1. $\dfrac{28}{16}=\dfrac{7\cdot4}{4\cdot4}=\dfrac{7}{4}$

2. $\dfrac{55}{11}=\dfrac{5\cdot11}{1\cdot11}=\dfrac{5}{1}=5$

3. Each shed requires $1\frac{1}{4}$ cubic yards of concrete, so the total amount of concrete needed for 25 sheds would be

$$\begin{aligned} 25\times1\frac{1}{4} &= 25\times\frac{5}{4} \\ &= \frac{125}{4} \\ &= 31\frac{1}{4}\text{ cubic yards.} \end{aligned}$$

4. Use the formula for simple interest, $I = Prt$, with $P = \$34{,}000$, $r = 5.4\%$, and $t = 1$.

$$I = Prt$$
$$= (34{,}000)(0.054)(1)$$
$$= 1836$$

She earned $1836 in interest.

5. The positive integer factors of 45 are $1, 3, 5, 9, 15,$ and 45.

6. $$\frac{4x-2y}{x+y} = \frac{4(-2)-2(4)}{(-2)+4} = \frac{-8-8}{2} = \frac{-16}{2} = -8$$

7. $$\frac{(-13+15)-(3+2)}{6-12} = \frac{2-5}{-6} = \frac{-3}{-6} = \frac{1}{2}$$

8. $$-7-3[2+(5-8)] = -7-3[2+(-3)]$$
$$= -7-3[-1]$$
$$= -7+3 = -4$$

9. $(9+2)+3 = 9+(2+3)$

The numbers are in the same order but grouped differently, so this is an example of the associative property of addition.

10. $6(4+2) = 6(4)+6(2)$

The number 6 outside the parentheses is "distributed" over the 4 and the 2. This is an example of the distributive property.

11. $$-3(2x^2-8x+9)-(4x^2+3x+2)$$
$$= -6x^2+24x-27-4x^2-3x-2$$
$$= -10x^2+21x-29$$

12. $$2-3(t-5) = 4+t$$
$$2-3t+15 = 4+t$$
$$-3t+17 = 4+t$$
$$-4t+17 = 4$$
$$-4t = -13$$
$$t = \frac{-13}{-4} = \frac{13}{4}$$

The solution set is $\left\{\frac{13}{4}\right\}$.

13. $$2(5x+1) = 10x+4$$
$$10x+2 = 10x+4$$
$$2 = 4 \quad \text{False}$$

The false statement indicates that the equation has no solution, symbolized by $\varnothing$.

14. Solve $d = rt$ for r.

$$\frac{d}{t} = \frac{rt}{t} \quad \text{Divide by t.}$$
$$\frac{d}{t} = r$$

15. $$\frac{x}{5} = \frac{x-2}{7}$$
$$7x = 5(x-2) \quad \text{Cross products are equal.}$$
$$7x = 5x-10$$
$$2x = -10$$
$$x = -5$$

The solution set is $\{-5\}$.

16. $3x-(4+2x) = -4$

Write without parentheses.

$$3x-4-2x = -4 \quad \text{Write without parentheses}$$
$$x-4 = -4 \quad \text{Combine like terms.}$$
$$x = 0 \quad \text{Add 4 to both sides}$$

The solution set is $\{0\}$.

17. $0.05x+0.15(50-x) = 5.50$

To clear decimals, multiply both sides of the equation by 100.

$$100[0.05x+0.15(50-x)] = 100(5.50)$$
$$100(0.05x)+100[0.15(50-x)] = 100(5.50)$$
$$5x+15(50-x) = 550$$
$$5x+750-15x = 550$$
$$-10x+750 = 550$$
$$-10x = -200$$
$$x = 20$$

The solution set is $\{20\}$.

18. $\frac{1}{3}p - \frac{1}{6}p = -2$

To clear fractions, multiply both sides of the equation by the least common denominator, which is 6.

$$6\left(\frac{1}{3}p - \frac{1}{6}p\right) = (6)(-2)$$
$$6\left(\frac{1}{3}p\right) - 6\left(\frac{1}{6}p\right) = -12$$
$$2p - p = -12$$
$$p = -12$$

The solution set is $\{-12\}$.

19. $4 - (3x + 12) = -7 - (3x + 1)$

$$4 - 3x - 12 = -7 - 3x - 1$$
$$-3x - 8 = -3x - 8 \quad \text{True}$$

The true statement indicates that the solution set is {all real numbers}.

20. Let $x =$ the number of calories burned in thermoregulation.

Then $5\frac{3}{8}x = \frac{43}{8}x =$ the number of calories burned in exertion.

$$x + \frac{43}{8}x = 11{,}200$$
$$8\left(x + \frac{43}{8}x\right) = 8(11{,}200)$$
$$8x + 43x = 89{,}600$$
$$51x = 89{,}600$$
$$x = \frac{89{,}600}{51} \approx 1757$$

A husky burns approximately 1757 calories for thermoregulation and $\frac{43}{8}\left(\frac{89{,}600}{51}\right) \approx 9443$ calories for exertion.

21. Let $x =$ one side of the triangle.

Then $2x =$ the other (unknown) side. The third side is 17 feet. The perimeter of the triangle cannot be more than 50 feet. This is equivalent to stating that the sum of the lengths of the sides must be less than or equal to 50 feet. Write this statement as an inequality and solve.

$$x + 2x + 17 \le 50$$
$$3x + 17 \le 50$$
$$3x \le 33$$
$$x \le 11$$

One side cannot be more than 11 feet. The other side cannot be more than $2 \cdot 11 = 22$ feet.

22. $-2(x + 4) > 3x + 6$

$$-2x - 8 > 3x + 6$$
$$-2x > 3x + 14$$
$$-5x > 14$$
$$\frac{-5x}{-5} < \frac{14}{-5}$$
$$x < -\frac{14}{5}$$

The solution set is $\left(-\infty, -\frac{14}{5}\right)$.

23. $-3 \le 2x + 5 < 9$

$$-8 \le 2x < 4 \quad \text{Subtract 5.}$$
$$\frac{-8}{2} \le \frac{2x}{2} < \frac{4}{2} \quad \text{Divide by 2.}$$
$$-4 \le x < 2$$

The solution set is $[-4, 2)$.

24. We recognize $y = -3x + 6$ as the equation of a line with y-intercept $(0, 6)$ and slope -3. From the point (0, 6), we can move right 1 unit and down 3 units to the point $(1, 3)$ to get another point on the graph.

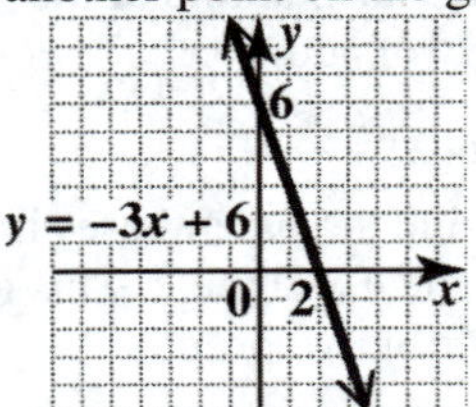

25. **(a)** Use the definition of slope with $(x_1, y_1) = (-1, 5)$ and $(x_2, y_2) = (2, 8)$.

$$m = \frac{\text{change in } y}{\text{change in } x} = \frac{y_2 - y_1}{x_2 - x_1}$$
$$= \frac{8 - 5}{2 - (-1)} = \frac{3}{3} = 1$$

(b) Use the point-slope form of the equation of a line with $m = 1$ and $(x_1, y_1) = (-1, 5)$.

$$y - y_1 = m(x - x_1)$$
$$y - 5 = 1[x - (-1)]$$
$$y - 5 = x + 1$$
$$y = x + 6$$

26. $f(x) = x + 7$

$$f(-8) = -8 + 7 = -1$$

27. $y = 2x + 5 \quad (1)$
$x + y = -4 \quad (2)$

To solve the system by the substitution method, let $y = 2x + 5$ in equation (2).

$$\begin{aligned} x + y &= -4 \\ x + (2x + 5) &= -4 \\ 3x + 5 &= -4 \\ 3x &= -9 \\ x &= -3 \end{aligned}$$

From (1), $y = 2(-3) + 5 = -1$.

The solution set is $\{(-3, -1)\}$.

28. $3x + 2y = 2 \quad (1)$
$2x + 3y = -7 \quad (2)$

We solve this system by the elimination method. To eliminate x, multiply equation (1) by 2, equation (2) by -3, and then add.

$$\begin{array}{rcrcr} 6x & + & 4y & = & 4 \\ -6x & - & 9y & = & 21 \\ \hline & & -5y & = & 25 \\ & & y & = & -5 \end{array}$$

To eliminate y, multiply equation (1) by 3, equation (2) by -2, and then add.

$$\begin{array}{rcrcr} 9x & + & 6y & = & 6 \\ -4x & - & 6y & = & 14 \\ \hline 5x & & & = & 20 \\ & & x & = & 4 \end{array}$$

The solution set is $\{(4, -5)\}$.

29. $4^{-1} + 3^0 = \dfrac{1}{4^1} + 1 = 1\dfrac{1}{4}$, or $\dfrac{5}{4}$

30. $\dfrac{8^{-5} \cdot 8^7}{8^2} = \dfrac{8^{-5+7}}{8^2} = \dfrac{8^2}{8^2} = 1$

31.
$$\begin{aligned} \frac{\left(a^{-3}b^2\right)^2}{\left(2a^{-4}b^{-3}\right)^{-1}} &= \frac{\left(a^{-3}\right)^2\left(b^2\right)^2}{2^{-1}\left(a^{-4}\right)^{-1}\left(b^{-3}\right)^{-1}} \\ &= \frac{a^{-6}b^4}{2^{-1}a^4b^3} \\ &= \frac{2b^4}{a^6a^4b^3} = \frac{2b}{a^{10}} \end{aligned}$$

32.
$$\begin{aligned} &\left(3.6\times10^1\right)\left(3.0\times10^5\right) \\ &= 3.6\times3.0\times10^{1+5} \\ &= 10.8\times10^6 \\ &= 1.08\times10^7 \end{aligned}$$

Venus is about 10,800,000 km from the sun.

33. $y = (x + 4)^2$

$x = -6: y = (-6 + 4)^2 = (-2)^2 = 4$

$x = -5: y = (-5 + 4)^2 = (-1)^2 = 1$

$x = -4: y = (-4 + 4)^2 = (0)^2 = 0$

$x = -3: y = (-3 + 4)^2 = (1)^2 = 1$

$x = -2: y = (-2 + 4)^2 = (2)^2 = 4$

x	-6	-5	-4	-3	-2
y	4	1	0	1	4

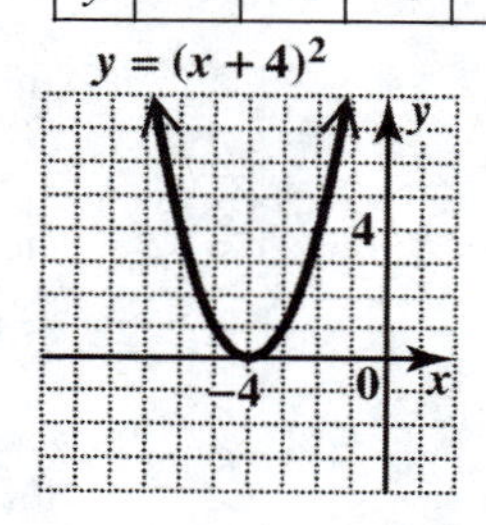

34.
$$\begin{aligned} &\left(7x^3 - 12x^2 - 3x + 8\right) + \left(6x^2 + 4\right) \\ &\quad - \left(-4x^3 + 8x^2 - 2x - 2\right) \\ &= \left(7x^3 - 12x^2 - 3x + 8\right) + \left(6x^2 + 4\right) \\ &\quad + \left(4x^3 - 8x^2 + 2x + 2\right) \\ &= (7 + 4)x^3 + (-12 + 6 - 8)x^2 \\ &\quad + (-3 + 2)x + (8 + 4 + 2) \\ &= 11x^3 - 14x^2 - x + 14 \end{aligned}$$

35.
$$\begin{aligned} &(7x + 4)(9x + 3) \\ &= 63x^2 + 21x + 36x + 12 \quad \text{FOIL} \\ &= 63x^2 + 57x + 12 \end{aligned}$$

36. Divide: $\dfrac{y^3 - 3y^2 + 8y - 6}{y - 1}$

$$
\begin{array}{r}
y^2 - 2y + 6 \\
y-1 \overline{\big) y^3 - 3y^2 + 8y - 6} \\
\underline{y^3 - y^2 } \\
-2y^2 + 8y \\
\underline{-2y^2 + 2y} \\
6y - 6 \\
\underline{6y - 6} \\
0
\end{array}
$$

The remainder is 0. The answer is the quotient, $y^2 - 2y + 6$.

Chapter 6 Factoring and Applications

6.1 The Greatest Common Factor; Factoring by Grouping

Now Try Exercises

N1. **(a)** Write each number in prime factored form.
$24 = 2 \cdot 2 \cdot 2 \cdot 3, \quad 36 = 2 \cdot 2 \cdot 3 \cdot 3$
Use each prime the *least* number of times it appears in *all* the factored forms. The least number of times 2 appears in all the factored forms is 2, and the least number of times 3 appears is 1. The GCF is $2 \cdot 2 \cdot 3 = 12$.

(b) Write each number in prime factored form.
$54 = 2 \cdot 3 \cdot 3 \cdot 3, \qquad 90 = 2 \cdot 3 \cdot 3 \cdot 5,$
$108 = 2 \cdot 2 \cdot 3 \cdot 3 \cdot 3$
Use each prime the least number of times it appears. The greatest common factor is $2 \cdot 3 \cdot 3 = 18$.

(c) Write each number in prime factored form.
$15 = 3 \cdot 5, \quad 19 = 19, \quad 25 = 5 \cdot 5$
Use each prime the least number of times it appears. Since no prime appears in all the factored forms, the GCF is 1.

N2. **(a)** Write each number in prime factored form.
$25k^3 = 5^2 \cdot k^3, \quad 15k^2 = 3 \cdot 5 \cdot k^2,$
$35k^5 = 5 \cdot 7 \cdot k^5$
The greatest common factor of the coefficients 25, 15, and 35 is 5. The greatest common factor of the terms k^3, k^2, and k^5 is k^2 since 2 is the least exponent on k. Thus, the GCF of these terms is the product of 5 and k^2, that is, $5k^2$.

(b) The least exponent on m is 3, and the least exponent on n is 2. Thus, the GCF is m^3n^2.

N3. **(a)** $7t^4 - 14t^3$
7 is the greatest common factor of 7 and 14, and 3 is the least exponent on t. Hence, $7t^3$ is the GCF.
$$7t^4 - 14t^3 = 7t^3(t) - 7t^3(2)$$
$$= 7t^3(t-2)$$

(b) $8x^6 - 20x^5 + 28x^4$
4 is the greatest common factor of 8, 20, and 28, so 4 can be factored out. Since x occurs in every term and the least exponent on x is 4, x^4 can be factored out. Hence, $4x^4$ is the GCF.
$$8x^6 - 20x^5 + 28x^4$$
$$= 4x^4(2x^2) - 4x^4(5x) + 4x^4(7)$$
$$= 4x^4(2x^2 - 5x + 7)$$

(c) $30m^4n^3 - 42m^2n^2$
6 is the greatest common factor of 30 and 42. The least exponents on m and n are 2 and 2, respectively. Hence, $6m^2n^2$ is the GCF.
$$30m^4n^3 - 42m^2n^2$$
$$= (6m^2n^2)(5m^2n) - (6m^2n^2)(7)$$
$$= 6m^2n^2(5m^2n - 7)$$

N4. $-14b^2 - 21b^3 + 7b$
$$= -7b(2b) - 7b(3b^2) - 7b(-1)$$
$$= -7b(2b + 3b^2 - 1)$$
The monomial $-7b$ is the greatest common factor.

N5. **(a)** $x(x+2) + 5(x+2)$
The binomial $x+2$ is the greatest common factor here.
$$x(x+2) + 5(x+2) = (x+2)(x+5)$$

(b) $a(t+10) - b(t+10)$
The binomial $t+10$ is the greatest common factor.
$$a(t+10) - b(t+10) = (t+10)(a-b)$$

N6. **(a)** $ab + 3a + 5b + 15$
$= (ab + 3a) + (5b + 15)$ Group terms.
$= a(b+3) + 5(b+3)$ Factor each group.
$= (b+3)(a+5)$ Factor out $b+3$.

(b) $12xy + 3x + 4y + 1$
$$= (12xy + 3x) + (4y + 1)$$
$$= 3x(4y+1) + 1(4y+1)$$
$$= (4y+1)(3x+1)$$

(c) $x^3 + 5x^2 - 8x - 40$
$$= (x^3 + 5x^2) + (-8x - 40)$$
$$= x^2(x+5) - 8(x+5)$$
$$= (x+5)(x^2 - 8)$$

N7. **(a)** $12p^2-28q-16pq+21p$

Factoring out the common factor 4 from the first two terms and the common factor p from the last two terms gives

$12p^2-28q-16pq+21p$
$=4(3p^2-7q)+p(-16q+21).$

This does not lead to a common factor, so we try rearranging the terms.

$=(12p^2-16pq)+(21p-28q)$
$=4p(3p-4q)+7(3p-4q)$
$=(3p-4q)(4p+7)$

Here's another rearrangement.

$=(12p^2+21p)+(-16pq-28q)$
$=3p(4p+7)-4q(4p+7)$
$=(4p+7)(3p-4q)$

This is an equivalent answer.

(b) $5xy-6-15x+2y$
$=(5xy+2y)+(-15x-6)$
$=y(5x+2)-3(5x+2)$
$=(5x+2)(y-3)$

Exercises

1. To factor a number or quantity means to write it as a product. Factoring is the opposite, or inverse, process of multiplying.

3. Find the prime factored form of each number.
$12=2\cdot2\cdot3$
$16=2\cdot2\cdot2\cdot2$
The least number of times 2 appears in all the factored forms is 2. There is no 3 in the prime factored form of 16, so the
$\text{GCF}=2\cdot2=4$

5. Find the prime factored form of each number.
$40=2\cdot2\cdot2\cdot5$
$20=2\cdot2\cdot5$
$4=2\cdot2$
The least number of times 2 appears in all the factored forms is 2. There is no 5 in the prime factored form of 4, so the
$\text{GCF}=2\cdot2=4.$

7. Find the prime factored form of each number.
$18=2\cdot3\cdot3$
$24=2\cdot2\cdot2\cdot3$
$36=2\cdot2\cdot3\cdot3$
$48=2\cdot2\cdot2\cdot2\cdot3$
The least number of times the primes 2 and 3 appear in all four factored forms is once, so the $\text{GCF}=2\cdot3=6.$

9. 6, 8, 9
Find the prime factored form of each number.
$6=2\cdot3$
$8=2\cdot2\cdot2$
$9=3\cdot3$
There are no primes common to all three numbers, so the GCF is 1.

11. Write each term in prime factored form.
$16y=2^4\cdot y$
$24=2^3\cdot3$
There is no y in the second term, so y will not appear in the GCF. Thus, the GCF of $16y$ and 24 is $2^3=8$.

13. $30x^3=2\cdot3\cdot5\cdot x^3$
$40x^6=2^3\cdot5\cdot x^6$
$50x^7=2\cdot5^2\cdot x^7$
The GCF of the coefficients, 30, 40, and 50, is $2^1\cdot5^1=10$. The smallest exponent on the variable x is 3. Thus, the GCF of the given terms is $10x^3$.

15. $x^4y^3=x^4\cdot y^3;\quad xy^2=x\cdot y^2$
The GCF is xy^2.

17. $42ab^3=2\cdot3\cdot7\cdot a\cdot b\cdot b\cdot b$
$36a=2\cdot2\cdot3\cdot3\cdot a$
$90b=2\cdot3\cdot3\cdot5\cdot b$
$48ab=2\cdot2\cdot2\cdot2\cdot3\cdot a\cdot b$
The GCF is $2\cdot3=6$.

19. $12m^3n^2=2^2\cdot3\cdot m^3\cdot n^2$
$18m^5n^4=2\cdot3^2\cdot m^5\cdot n^4$
$36m^8n^3=2^2\cdot3^2\cdot m^8\cdot n^3$
The GCF is $2\cdot3\cdot m^3\cdot n^2=6m^3n^2$.

21. $2k^2(5k)$ is written as a product of $2k^2$ and $5k$, and hence, it is *factored.*

23. $2k^2+(5k+1)$ is written as a sum of $2k^2$ and $(5k+1)$, and hence, it is *not factored.*

25. The correct factored form is
$18x^3y^2 + 9xy = 9xy(2x^2y + 1)$.
If a polynomial has two terms, the product of the factors must have two terms.
$9xy(2x^2y) = 18x^3y^2$ has just one term.

27. $9m^4 = 3m^2(3m^2)$
Factor out $3m^2$ from $9m^4$ to obtain $3m^2$.

29. $-8z^9 = -4z^5(2z^4)$
Factor out $-4z^5$ from $-8z^9$ to obtain $2z^4$.

31. $6m^4n^5 = 3m^3n(2mn^4)$
Factor out $3m^3n$ from $6m^4n^5$ to obtain $2mn^4$.

33. $12y + 24 = 12 \cdot y + 12 \cdot 2$
$= 12(y + 2)$

35. $10a^2 - 20a = 10a(a) - 10a(2)$
$= 10a(a - 2)$

37. $8x^2y + 12x^3y^2 = 4x^2y(2) + 4x^2y(3xy)$
$= 4x^2y(2 + 3xy)$

39. The greatest common factor is x.
$x^2 - 4x = x(x) + x(-4)$
$= x(x - 4)$

41. The greatest common factor is $3t$.
$6t^2 + 15t = 3t(2t) + 3t(5)$
$= 3t(2t + 5)$

43. The GCF is $9m$.
$27m^3 - 9m = 9m(3m^2) + 9m(-1)$
$= 9m(3m^2 - 1)$

45. The GCF is m^2.
$m^3 - m^2 = m^2(m) - m^2(1)$
$= m^2(m - 1)$

47. The GCF is $8z^2$.
$16z^4 + 24z^2 = 8z^2(2z^2) + 8z^2(3)$
$= 8z^2(2z^2 + 3)$

49. The GCF is $-6x^2$.
$-12x^3 - 6x^2 = -6x^2(2x) - 6x^2(1)$
$= -6x^2(2x + 1)$

51. The GCF is $5y^6$.
$65y^{10} + 35y^6 = (5y^6)(13y^4) + (5y^6)(7)$
$= 5y^6(13y^4 + 7)$

53. The two terms of this expression have no common factor (except 1), so $11w^3 - 100$ is in factored form.

55. The GCF is $8mn^3$.
$8mn^3 + 24m^2n^3$
$= (8mn^3)(1) + (8mn^3)(3m)$
$= 8mn^3(1 + 3m)$

57. The GCF is $13y^2$.
$13y^8 + 26y^4 - 39y^2$
$= 13y^2(y^6) + 13y^2(2y^2) + 13y^2(-3)$
$= 13y^2(y^6 + 2y^2 - 3)$

59. The GCF is $-2x$.
$-4x^3 + 10x^2 - 6x$
$= -2x(2x^2) - 2x(-5x) - 2x(3)$
$= -2x(2x^2 - 5x + 3)$

61. The GCF is $9p^3q$.
$36p^6q + 45p^5q^4 + 81p^3q^2$
$= 9p^3q(4p^3) + 9p^3q(5p^2q^3) + 9p^3q(9q)$
$= 9p^3q(4p^3 + 5p^2q^3 + 9q)$

63. The GCF is a^3.
$a^5 + 2a^3b^2 - 3a^5b^2 + 4a^4b^3$
$= a^3(a^2) + a^3(2b^2) + a^3(-3a^2b^2) + a^3(4ab^3)$
$= a^3(a^2 + 2b^2 - 3a^2b^2 + 4ab^3)$

65. The GCF is the binomial $x + 2$.
$c(x + 2) - d(x + 2)$
$= (x + 2)(c) + (x + 2)(-d)$
$= (x + 2)(c - d)$

67. The GCF is the binomial $m + 2n$.
$m(m + 2n) + n(m + 2n)$
$= (m + 2n)(m) + (m + 2n)(n)$
$= (m + 2n)(m + n)$

69. The GCF is $p-4$.

$q^2(p-4)+1(p-4)=(p-4)(q^2+1)$

71. This expression is the *sum* of two terms, $8(7t+4)$ and $x(7t+4)$, so it is not in factored form. We can factor out $7t+4$.

$8(7t+4)+x(7t+4)$
$=(7t+4)(8)+(7t+4)(x)$
$=(7t+4)(8+x)$

73. This expression is the *product* of two factors, $8+x$ and $7t+4$, so it is in factored form.

75. This expression is the *sum* of two terms, $18x^2(y+4)$ and $7(y-4)$, so it is not in factored form.

77. The student should factor out -2, instead of 2, in the second step to obtain $x^2(x+4)-2(x+4)$, which can be factored as $(x+4)(x^2-2)$.

79. The first two terms have a common factor of p, and the last two terms have a common factor of q. Thus,

$p^2+4p+pq+4q$
$=(p^2+4p)+(pq+4q)$
$=p(p+4)+q(p+4)$.

Now we have two terms that have a common binomial factor of $p+4$. Thus,

$p^2+4p+pq+4q$
$=p(p+4)+q(p+4)$
$=(p+4)(p+q)$.

81. $a^2-2a+ab-2b$

$=(a^2-2a)+(ab-2b)$ Group the terms.
$=a(a-2)+b(a-2)$ Factor each group.
$=(a-2)(a+b)$ Factor out $a-2$.

83. $7z^2+14z-az-2a$

$=(7z^2+14z)+(-az-2a)$ Group the terms.
$=7z(z+2)-a(z+2)$ Factor each group.
$=(z+2)(7z-a)$ Factor out $z+2$.

85. $18r^2+12ry-3xr-2xy$

$=(18r^2+12ry)+(-3xr-2xy)$
$=6r(3r+2y)-x(3r+2y)$
$=(3r+2y)(6r-x)$

87. $3a^3+3ab^2+2a^2b+2b^3$

$=(3a^3+3ab^2)+(2a^2b+2b^3)$
$=3a(a^2+b^2)+2b(a^2+b^2)$
$=(a^2+b^2)(3a+2b)$

89. $12-4a-3b+ab$

$=(12-4a)+(-3b+ab)$ Group the terms.
$=4(3-a)-b(3-a)$ Factor each group.
$=(3-a)(4-b)$ Factor out $3-a$.

91. $16m^3-4m^2p^2-4mp+p^3$

$=(16m^3-4m^2p^2)+(-4mp+p^3)$
$=4m^2(4m-p^2)-p(4m-p^2)$
$=(4m-p^2)(4m^2-p)$

93. $y^2+3x+3y+xy$

$=y^2+3y+xy+3x$ Rearrange.
$=(y^2+3y)+(xy+3x)$
$=y(y+3)+x(y+3)$
$=(y+3)(y+x)$

95. $5m-6p-2mp+15$

We need to rearrange these terms to get two groups that each have a common factor. We could group $5m$ with either $-2mp$ or 15.

$5m+15-2mp-6p$ Rearrange.
$=(5m+15)+(-2mp-6p)$ Group the terms.
$=5(m+3)-2p(m+3)$ Factor each group.
$=(m+3)(5-2p)$ Factor out $m+3$.

97. $18r^2-2ty+12ry-3rt$

We'll rearrange the terms so that $18r^2$ is grouped with another term containing r.

$18r^2+12ry-3rt-2ty$
$=(18r^2+12ry)+(-3rt-2ty)$
$=6r(3r+2y)-t(3r+2y)$
$=(3r+2y)(6r-t)$

99. $a^5 - 3 + 2a^5b - 6b$

$= a^5 + 2a^5b - 3 - 6b$ Rearrange.

$= (a^5 + 2a^5b) + (-3 - 6b)$ Group the terms.

$= a^5(1 + 2b) - 3(1 + 2b)$ Factor each group.

$= (1 + 2b)(a^5 - 3)$ Factor out $1 + 2b$.

101. $16a^2 + 40ab^2 + 16ab + 40b^3$

$= 8(2a^2 + 5ab^2 + 2ab + 5b^3)$

$= 8(a(2a + 5b^2) + b(2a + 5b^2))$

$= 8(2a + 5b^2)(a + b)$

103. $2p^2q^2 - 2p^2q + 2p^3 - 2pq^3$

$= 2p(pq^2 - pq + p^2 - q^3)$

$= 2(q^2(p - q) + p(p - q))$

$= 2p(p + q^2)(p - q)$

6.2 Factoring Trinomials

Now Try Exercises

N1. Find factors of 10 and the sum of the factors.

Factors of 10	Sums of Factors
10, 1	$10 + 1 = 11$
5, 2	$5 + 2 = 7 \leftarrow$

The pair of integers whose product is 10 and whose sum is 7 is 5 and 2. Thus, $p^2 + 7p + 10 = (p + 5)(p + 2)$.

N2. Find factors of 18 and the sum of the factors.

Factors of 18	Sums of Factors
−18, −1	$-18 + (-1) = -19$
−9, −2	$-9 + (-2) = -11$
−6, −3	$-6 + (-3) = -9 \leftarrow$

The pair of integers whose product is 18 and whose sum is −9 is −6 and −3. Thus, $t^2 - 9t + 18 = (t - 3)(t - 6)$.

N3. Find the two integers whose product is −42 and whose sum is 1. Because the last term is negative, the pair must include one positive and one negative integer.

Factors of −42	Sums of Factors
42, −1	$42 + (-1) = 41$
21, −2	$21 + (-2) = 19$
14, −3	$14 + (-3) = 11$
7, −6	$7 + (-6) = 1 \leftarrow$
−42, 1	$-42 + 1 = -41$
−21, 2	$-21 + 2 = -19$
−14, 3	$-14 + 3 = -11$
−7, 6	$-7 + 6 = -1$

The required integers are −6 and 7, so $x^2 + x - 42 = (x - 6)(x + 7)$.

N4. Find the two integers whose product is −21 and whose sum is −4. Because the last term is negative, the pair must include one positive and one negative integer.

Factors of −21	Sums of Factors
21, −1	$21 + (-1) = 20$
7, −3	$7 + (-3) = 4$
−21, 1	$-21 + 1 = -20$
−7, 3	$-7 + 3 = -4 \leftarrow$

The required integers are −7 and 3, so $x^2 - 4x - 21 = (x - 7)(x + 3)$.

N5. **(a)** There is no pair of integers whose product is 8 and whose sum is 5, so $m^2 + 5m + 8$ is a prime polynomial.

(b) There is no pair of integers whose product is −24 and whose sum is 11, so $t^2 + 11t - 24$ is a prime polynomial.

N6. Two expressions whose product is $-15b^2$ and whose sum is $2b$ are $5b$ and $-3b$, so
$a^2+2ab-15b^2=(a+5b)(a-3b).$

N7. $3y^4-27y^3+60y^2=3y^2\left(y^2-9y+20\right)$

Factor $y^2-9y+20$. The integers -5 and -4 have a product of 20 and a sum of -9, so
$y^2-9y+20=(y-5)(y-4).$
The completely factored form is
$3y^4-27y^3+60y^2=3y^2(y-5)(y-4).$

Exercises

1. If the coefficient of the last term of the trinomial is negative, then a and b must have different signs, one positive and one negative.

3. Factor $x^2-12x+32$.

Factors of 32	Sums of Factors
$-32,\ -1$	$-32+(-1)=-33$
$-16,-2$	$-16+(-2)=(-18)$
$-8,-4$	$-8+(-4)=(-12)$

The pair of integers whose product is 32 and whose sum is -12 is -8 and -4. Therefore, the correct factored form is choice C, $(x-8)(x-4)$.

5. Multiply the factors using FOIL to determine the polynomial.
$(a+9)(a+4)$
F O I L
$=a(a)+a(4)+9(a)+9(4)$
$=a^2+4a+9a+36$
$=a^2+13a+36$

7. List all pairs of integers whose product is 48, and then find the sum of each pair.

Factors of 48	Sums of Factors
$1,\ 48$	$1+48=49$
$-1,-48$	$-1+(-48)=-49$
$2,\ 24$	$2+24=26$
$-2,-24$	$-2+(-24)=-26$
3, 16	$3+16=19$
$-3,-16$	$-3+(-16)=-19\leftarrow$
4, 12	$4+12=16$
$-4,-12$	$-4+(-12)=-16$
$6,\ 8$	$6+8=14$
$-6,-8$	$-6+(-8)=-14$

The pair of integers whose product is 48 and whose sum is -19 is -3 and -16.

9. List all pairs of integers whose product is -24, and then find the sum of each pair.

Factors of -24	Sums of Factors
$1,-24$	$1+(-24)=-23$
$-1, 24$	$-1+24=23$
$2,-12$	$2+(-12)=-10$
$-2, 12$	$-2+12=10$
$3,-8$	$3+(-8)=-5\leftarrow$
$-3, 8$	$-3+8=5$
$4,-6$	$4+(-6)=-2$
$-4, 6$	$-4+6=2$

The pair of integers whose product is -24 and whose sum is -5 is 3 and -8.

11. To factor $y^2 + 12y + 20$, find two integers whose product is 20 and whose sum is 12.

Factors of 20	Sums of Factors
1, 20	$1 + 20 = 21$
10, 2	$10 + 2 = 12$
5, 4	$5 + 4 = 9$

The pair of factors whose product is 20 and whose sum is 12 is 10 and 2.

$y^2 + 12y + 20 = (y + 10)(y + 2)$

13. Look for an integer whose product with 5 is 30 and whose sum with 5 is 11. That integer is 6.

$p^2 + 11p + 30 = (p + 5)(p + 6)$

15. Look for an integer whose product with 4 is 44 and whose sum with 4 is 15. That integer is 11.

$x^2 + 15x + 44 = (x + 4)(x + 11)$

17. Look for an integer whose product with -1 is 8 and whose sum with -1 is -9. That integer is -8.

$x^2 - 9x + 8 = (x - 1)(x - 8)$

19. Look for an integer whose product with 3 is -15 and whose sum with 3 is -2. That integer is -5.

$y^2 - 2y - 15 = (y + 3)(y - 5)$

21. Look for an integer whose product with -2 is -22 and whose sum with -2 is 9. That integer is 11.

$x^2 + 9x - 22 = (x - 2)(x + 11)$

23. Look for an integer whose product with 2 is -18 and whose sum with 2 is -7. That integer is -9.

$y^2 - 7y - 18 = (y + 2)(y - 9)$

25. Look for two integers whose product is 8 and whose sum is 9. Both integers must be positive because b and c are both positive.

Factors of 8	Sums of Factors
1, 8	9 ←
2, 4	6

Thus, $y^2 + 9y + 8 = (y + 8)(y + 1)$.

27. Look for two integers whose product is 15 and whose sum is 8. Both integers must be positive because b and c are both positive.

Factors of 15	Sums of Factors
1, 15	16
3, 5	8 ←

Thus, $b^2 + 8b + 15 = (b + 3)(b + 5)$.

29. Look for two integers whose product is -20 and whose sum is 1. Since c is negative, one integer must be positive and one must be negative.

Factors of -20	Sums of Factors
−1, 20	19
1, −20	−19
−2, 10	8
2, −10	−8
−4, 5	1 ←
4, −5	−1

Thus, $m^2 + m - 20 = (m - 4)(m + 5)$.

31. Find two integers whose product is 15 and whose sum is -8. Since c is positive and b is negative, both integers must be negative.

Factors of 15	Sums of Factors
−1, −15	−16
−3, −5	−8 ←

Thus, $y^2 - 8y + 15 = (y - 5)(y - 3)$.

33. Look for two integers whose product is 5 and whose sum is 4. Both integers must be positive since b and c are both positive.

Factors of 5	Sum of Factors
$5 \cdot 1 = 5$	$5 + 1 = 6$

There is no other pair of positive integers whose product is 5. Since there is no pair of integers whose product is 5 and whose sum is 4, $x^2 + 4x + 5$ is a prime polynomial.

35. Find two integers whose product is 56 and whose sum is −15. Since *c* is positive and *b* is negative, both integers must be negative.

Factors of 56	Sums of Factors
−1, −56	−57
−2, −28	−30
−4, −14	−18
−7, −8	−15 ←

Thus, $z^2 - 15z + 56 = (z-7)(z-8)$.

37. Look for two integers whose product is −30 and whose sum is −1. Because *c* is negative, one integer must be positive and the other must be negative.

Factors of −30	Sums of Factors
−1, 30	29
1, −30	−29
−2, 15	13
2, −15	−13
−3, 10	7
3, −10	−7
−5, 6	1
5, −6	−1 ←

Thus, $r^2 - r - 30 = (r+5)(r-6)$.

39. Find two integers whose product is −48 and whose sum is −8. Since *c* is negative, one integer must be positive and one must be negative.

Factors of −48	Sums of Factors
−1, 48	47
1, −48	−47
−2, 24	22
2, −24	−22
−3, 16	13
3, −16	−13
−4, 12	8
4, −12	−8 ←
−6, 8	2
6, −8	−2

Thus, $a^2 - 8a - 48 = (a+4)(a-12)$.

41. Look for two integers whose product is −39 and whose sum is 3. Because *c* is negative, one integer must be positive and one must be negative.

Factors of −39	Sums of Factors
−1, 39	38
1, −39	−38
−3, 13	10
3, −13	−10

This list does not produce the required integers, and there are no other possibilities to try. Therefore, $x^2 + 3x - 39$ is prime.

43. Look for two integers whose product is −32 and whose sum is 14. Since c is negative, one integer must be positive and one must be negative.

Factors of −32	Sums of Factors
−1, 32	31
1, −32	−31
−2, 16	14 ←
2, −16	−14
−4, 8	4
4, −8	−4

Thus, $-32+14x+x^2=(x-2)(x+16)$.

45. Look for two expressions whose product is $2a^2$ and whose sum is $3a$. They are $2a$ and a, so $r^2+3ra+2a^2=(r+2a)(r+a)$.

47. Look for two expressions whose product is $3y^2$ and whose sum is $4y$. The expressions are $3y$ and y, so $x^2+4xy+3y^2=(x+3y)(x+y)$.

49. Look for two expressions whose product is $-6z^2$ and whose sum is $-z$. They are $2z$ and $-3z$, so $t^2-tz-6z^2=(t+2z)(t-3z)$.

51. Look for two integers whose product is $30w^2$ and whose sum is $-11w$.

Factors of $30w^2$	Sums of Factors
$-30w, -w$	$-31w$
$-15w, -2w$	$-17w$
$-10w, -3w$	$-13w$
$-5w, -6w$	$-11w$ ←

Therefore, $v^2-11vw+30w^2=(v-5w)(v-6w)$.

53. Look for two expressions whose product is $-12n^2$ and whose sum is $4n$.

Factors of $-12n^2$	Sums of Factors
$-12n, n$	$-13n$
$12n, -n$	$11n$
$-6n, 2n$	$-4n$
$6n, -2n$	$4n$ ←
$-4n, 3n$	$-n$
$4n, -3n$	n

Thus, $m^2+4mn-12n^2=(m+6n)(m-2n)$.

55. Look for two expressions whose product is $18b^2$ and whose sum is $-9b$.

Factors of $18b^2$	Sums of Factors
$-18b, -b$	$-19b$
$-9b, -2b$	$-11b$
$-6b, -3b$	$-9b$ ←

Thus, $a^2-9ab+18b^2=(a-6b)(a-3b)$.

57. First, factor out the GCF, 4.

$4x^2+12x-40=4(x^2+3x-10)$

Then factor $x^2+3x-10$. The integers −2 and 5 have a product of −10 and a sum of 3.

Thus, $x^2+3x-10=(x-2)(x+5)$.

The completely factored form is

$4x^2+12x-40=4(x-2)(x+5)$.

59. First, factor out the GCF, $2t$.

$2t^3+8t^2+6t=2t(t^2+4t+3)$

Then factor t^2+4t+3.

$t^2+4t+3=(t+1)(t+3)$

The completely factored form is

$2t^3+8t^2+6t=2t(t+1)(t+3)$.

61. First, factor out the GCF, $2x^4$.

$2x^6 + 8x^5 - 42x^4 = 2x^4(x^2 + 4x - 21)$

Now factor $x^2 + 4x - 21$. The integers -3 and 7 have a product of -21 and a sum of 4. Thus, $x^2 + 4x - 21 = (x-3)(x+7)$.

The completely factored form is

$2x^6 + 8x^5 - 42x^4 = 2x^4(x-3)(x+7)$.

63. Factor out the GCF, $6z^2$.

$6z^4 - 24z^3 + 18z^2 = 6z^2(z^2 - 4z + 3)$

The integers -3 and -1 have a product of 3 and a sum of -4. The completely factored form is $6z^4 - 24z^3 + 18z^2 = 6z^2(z-3)(z-1)$.

65. Factor out the GCF, $5m^2$.

$5m^5 - 25m^4 + 40m^2 = 5m^2(m^3 - 5m^2 + 8)$

67. Factor out the GCF, x.

$x^3 - 7x^2y + 12xy^2 = x(x^2 - 7xy + 12y^2)$

The integers $-4y$ and $-3y$ have a product of $12y^2$ and a sum of $-7y$. The completely factored form is

$x^3 - 7x^2y + 12xy^2 = x(x-4y)(x-3y)$.

69. The GCF is a^3.

$a^5 + 3a^4b - 4a^3b^2 = a^3(a^2 + 3ab - 4b^2)$

Now factor $a^2 + 3ab - 4b^2$. The expressions $4b$ and $-b$ have a product of $-4b^2$ and a sum of $3b$. The completely factored form is

$a^5 + 3a^4b - 4a^3b^2 = a^3(a+4b)(a-b)$.

71. First, factor out the GCF, z^8.

$z^{10} - 4z^9y - 21z^8y^2 = z^8(z^2 - 4zy - 21y^2)$

The expressions $-7y$ and $3y$ have a product of $-21y^2$ and a sum of $-4y$. The completely factored form is

$z^{10} - 4z^9y - 21z^8y^2 = z^8(z-7y)(z+3y)$.

73. First, factor out the GCF, mn.

$m^3n - 10m^2n^2 + 24mn^3$

$= mn(m^2 - 10mn + 24n^2)$

The expressions $-6n$ and $-4n$ have a product of $24n^2$ and a sum of $-10n$. The completely factored form is

$m^3n - 10m^2n^2 + 24mn^3 =$

$mn(m-6n)(m-4n)$.

75. The GCF is yz, so

$y^3z + y^2z^2 - 6yz^3 = yz(y^2 + yz - 6z^2)$.

Now factor $y^2 + yz - 6z^2$. The expressions $3z$ and $-2z$ have a product of $-6z^2$ and a sum of z. The completely factored form is

$y^3z + y^2z^2 - 6yz^3 = yz(y+3z)(y-2z)$.

77. The GCF is $(a+b)$, so

$(a+b)x^2 + (a+b)x - 12(a+b)$

$= (a+b)(x^2 + x - 12)$.

Now factor $x^2 + x - 12$.

$x^2 + x - 12 = (x+4)(x-3)$

The completely factored form is

$(a+b)x^2 + (a+b)x - 12(a+b)$

$= (a+b)(x+4)(x-3)$.

79. The GCF is $(2p+q)$, so

$(2p+q)r^2 - 12(2p+q)r + 27(2p+q)$

$= (2p+q)(r^2 - 12r + 27)$.

Now factor $r^2 - 12r + 27$.

$r^2 - 12r + 27 = (r-9)(r-3)$

The completely factored form is

$(2p+q)r^2 - 12(2p+q)r + 27(2p+q)$

$= (2p+q)(r-9)(r-3)$.

6.3 More on Factoring Trinomials

Now Try Exercises

N1. Find the two integers whose product is 6 and whose sum is 7. The integers are 6 and 1. Use these integers to write the middle term $7m$ as $6m+m$.

$$\begin{aligned} 2m^2+7m+3 &= 2m^2+6m+m+3 \\ &= (2m^2+6m)+(m+3) \\ &= 2m(m+3)+1(m+3) \\ &= (2m+1)(m+3) \end{aligned}$$

N2. **(a)** Find the two integers whose product is $2(3)=6$ and whose sum is 5. The integers are 2 and 3. Write the middle term, $5z$, as $2z+3z$.

$$\begin{aligned} 2z^2+5z+3 &= 2z^2+2z+3z+3 \\ &= (2z^2+2z)+(3z+3) \\ &= 2z(z+1)+3(z+1) \\ &= (z+1)(2z+3) \end{aligned}$$

(b) Find two integers whose product is $15(-2)=-30$ and whose sum is 1. The integers are 6 and -5.

$$\begin{aligned} 15m^2+m-2 &= 15m^2+6m-5m-2 \\ &= (15m^2+6m)+(-5m-2) \\ &= 3m(5m+2)-1(5m+2) \\ &= (5m+2)(3m-1) \end{aligned}$$

Note that if we had written $-5m+6m$ instead of $6m-5m$, the order of the factors in the final expression would be reversed, that is, $(3m-1)(5m+2)$.

(c) Find two integers whose product is $8(-3)=-24$ and whose sum is -2. The integers are -6 and 4.

$$\begin{aligned} &8x^2-2xy-3y^2 \\ &= 8x^2-6xy+4xy-3y^2 \\ &= (8x^2-6xy)+(4xy-3y^2) \\ &= 2x(4x-3y)+y(4x-3y) \\ &= (4x-3y)(2x+y) \end{aligned}$$

N3. First factor out the greatest common factor, $3z^4$.

$$15z^6+18z^5-24z^4=3z^4(5z^2+6z-8)$$

Find two integers whose product is $5(-8)=-40$ and whose sum is 6. The integers are -4 and 10.

$$\begin{aligned} 5z^2+6z-8 &= 5z^2-4z+10z-8 \\ &= (5z^2-4z)+(10z-8) \\ &= z(5z-4)+2(5z-4) \\ &= (5z-4)(z+2) \end{aligned}$$

The completely factored form is $15z^6+18z^5-24z^4=3z^4(5z-4)(z+2)$.

N4. Find two integers whose product is 18 and whose sum is 9. The integers are 6 and 3.

$$\begin{aligned} 2p^2+9p+9 &= 2p^2+6p+3p+9 \\ &= (2p^2+6p)+(3p+9) \\ &= 2p(p+3)+3(p+3) \\ &= (2p+3)(p+3) \end{aligned}$$

The completely factored form is $2p^2+9p+9=(2p+3)(p+3)$.

N5. The factors of $8y^2$ are $2y$ and $4y$, or $8y$ and y. The factors of 5 are 5 and 1. Try various combinations, checking to see if the middle term is $22y$ in each case.

$(4y+5)(2y+1)=8y^2+14y+5$ Incorrect

$(4y+1)(2y+5)=8y^2+22y+5$ Correct

Thus, $8y^2+22y+5=(4y+1)(2y+5)$.

N6. The factors of $10x^2$ are $10x$ and x, or $5x$ and $2x$. Try $5x$ and $2x$. Since the last term is positive and the coefficient of the middle term is negative, only negative factors of 2 should be considered. The factors of 2 are -2 and -1.

$(5x-1)(2x-2)=10x^2-12x+2$ Incorrect

$(5x-2)(2x-1)=10x^2-9m+2$ Correct

Thus, $10x^2-9x+2=(5x-2)(2x-1)$.

N7. The factors of $10a^2$ are $5a$ and $2a$, or $10a$ and a. Some factors of -14 are 14 and -1, or 7 and -2. Try various possibilities.

$(5a-14)(2a+1)=10a^2-23a-14$ Incorrect

$(5a-2)(2a+7)=10a^2+31a-14$ Correct

Thus, $10a^2+31a-14=(5a-2)(2a+7)$.

N8. Try various possibilities.

$(4z+5w)(2z-3w)=8z^2-2wz-15w^2$

This is incorrect.

The middle terms differ only in sign, so reverse the signs of the two factors.

$(4z-5w)(2z+3w)=8z^2+2wz-15w^2$

This is correct.

Thus, $8z^2+2wz-15w^2=(4z-5w)(2z+3w)$.

N9. The common factor could be $5x$ or $-5x$. If we factor out $-5x$, the first term of the trinomial factor will be positive, which makes it easier to factor.

$-10x^3-45x^2+90x$

$=-5x\left(2x^2+9x-18\right)$

Factor $2x^2+9x-18$ by trial and error to obtain $2x^2+9x-18=(2x-3)(x+6)$.

The completely factored form is

$-10x^3-45x^2+90x=-5x(2x-3)(x+6)$.

Exercises

1. $10t^2+5t+4t+2$

$=\left(10t^2+5t\right)+(4t+2)$ Group terms.

$=5t(2t+1)+2(2t+1)$ Factor each group.

$=(2t+1)(5t+2)$ Factor out $2t+1$.

3. $15z^2-10z-9z+6$

$=(15z^2-10z)+(-9z+6)$ Group terms.

$=5z(3z-2)-3(3z-2)$ Factor each group.

$=(3z-2)(5z-3)$ Factor out $3z-2$.

5. $8s^2-4st+6st-3t^2$

$=\left(8s^2-4st\right)+\left(6st-3t^2\right)$ Group terms.

$=4s(2s-t)+3t(2s-t)$ Factor each group.

$=(2s-t)(4s+3t)$ Factor out $2s-t$.

7. **(a)** Find two integers whose product is $\underline{2}\cdot\underline{12}=\underline{24}$ and whose sum is $\underline{11}$.

(b) The required integers are $\underline{3}$ and $\underline{8}$. (Order is irrelevant.)

(c) Write the middle term, $11m$, as $\underline{3m}+\underline{8m}$.

(d) Rewrite the given trinomial as $\underline{2m^2+3m+8m+12}$.

(e) $\left(2m^2+3m\right)+(8m+12)$ Group terms.

$=m(2m+3)+4(2m+3)$ Factor.

$=(2m+3)(m+4)$

(f) Use the FOIL method.

$(2m+3)(m+4)$

$=2m(m)+2m(4)+3(m)+3(4)$

$=2m^2+8m+3m+12$

$=2m^2+11m+12$

9. To factor $12y^2+5y-2$, we must find two integers with a product of $12(-2)=-24$ and a sum of 5. The only pair of integers satisfying those conditions is 8 and -3, choice B.

11. Multiply the factors in the choices together to see which ones give the correct product. Since

$$(2x-1)(x+1)=2x^2+x-1$$

and $(2x+1)(x-1)=2x^2-x-1$,

the correct factored form is choice B, $(2x+1)(x-1)$.

13. Multiply the factors in the choices together to see which ones give the correct product. Since

$$(y+5)(4y-3)=4y^2+17y-15$$

and $(2y-5)(2y+3)=4y^2-4y-15$,

the correct factored form is choice A, $(y+5)(4y-3)$.

15. $(4x+4)$ cannot be a factor because its terms have a common factor of 4, but those of the polynomial do not. The correct factored form is $(4x-3)(3x+4)$.

17. The first term in the missing expression must be $2a$ since $(3a)(2a) = 6a^2$. The second term in the missing expression must be $5b$ since $(-4b)(5b) = -20b^2$. Checking our answer by multiplying, we see that $(3a-4b)(\underline{2a+5b}) = 6a^2 + 7ab - 20b^2$, as desired.

19. $2x^2 + 6x - 8 = 2(\underline{x^2 + 3x - 4})$

To factor $x^2 + 3x - 4$, we look for two integers whose product is -4 and whose sum is 3. The integers are 4 and -1. Thus, $2x^2 + 6x - 8 = 2(\underline{x+4})(\underline{x-1})$.

21. $4z^3 - 10z^2 - 6z = 2z(\underline{2z^2 - 5z - 3})$
$= 2z(\underline{2z+1})(\underline{z-3})$

23. Factor by the grouping method. Look for two integers whose product is $3(7) = 21$ and whose sum is 10. The integers are 3 and 7. Use these integers to rewrite the middle term, $10a$, as $3a + 7a$, and then factor the resulting four-term polynomial by grouping.

$3a^2 + 10a + 7$
$= 3a^2 + 3a + 7a + 7$ $10a = 3a + 7a$
$= (3a^2 + 3a) + (7a + 7)$ Group the terms.
$= 3a(a+1) + 7(a+1)$ Factor each group.
$= (a+1)(3a+7)$ Factor out $a+1$.

25. Factor by the grouping method. Look for two integers whose product is $2(6) = 12$ and whose sum is 7. The integers are 3 and 4.

$2y^2 + 7y + 6$
$= 2y^2 + 3y + 4y + 6$ $7y = 3y + 4y$
$= (2y^2 + 3y) + (4y + 6)$ Group terms.
$= y(2y+3) + 2(2y+3)$ Factor each group.
$= (2y+3)(y+2)$ Factor out $2y+3$.

27. Factor by the grouping method. Look for two integers whose product is $15(-2) = -30$ and whose sum is 1. The integers are 6 and -5.

$15m^2 + m - 2$
$= 15m^2 + 6m - 5m - 2$ $m = 6m - 5m$
$= (15m^2 + 6m) + (-5m - 2)$ Group the terms.
$= 3m(5m+2) - 1(5m+2)$ Factor.
$= (5m+2)(3m-1)$ Factor out $5m+2$.

29. Factor by trial and error.
Possible factors of $12s^2$ are s and $12s$, $2s$ and $6s$, or $3s$ and $4s$. Factors of -5 are -1 and 5, or -5 and 1.

$(2s-1)(6s+5) = 12s^2 + 4s - 5$ Incorrect
$(2s+1)(6s-5) = 12s^2 - 4s - 5$ Incorrect
$(3s-1)(4s+5) = 12s^2 + 11s - 5$ Correct

31. Factor by the grouping method. Look for two integers whose product is $10(12) = 120$ and whose sum is -23. The integers are -8 and -15.

$10m^2 - 23m + 12 = 10m^2 - 8m - 15m + 12$
$= (10m^2 - 8m) + (-15m + 12)$
$= 2m(5m-4) - 3(5m-4)$
$= (5m-4)(2m-3)$

33. Factor by trial and error. Possible factors of $8w^2$ are w and $8w$, or $2w$ and $4w$.
Factors of 3 are -1 and -3 (since $b = -14$ is negative).

$(4w-3)(2w-1) = 8w^2 - 10w + 3$ Incorrect
$(4w-1)(2w-3) = 8w^2 - 14w + 3$ Correct

35. Factor by the grouping method. Look for two integers whose product is $20(-11) = -220$ and whose sum is -39. The integers are -44 and 5.

$20y^2 - 39y - 11 = 20y^2 - 44y + 5y - 11$
$= (20y^2 - 44y) + (5y - 11)$
$= 4y(5y-11) + 1(5y-11)$
$= (5y-11)(4y+1)$

37. Factor by the grouping method. Look for two integers whose product is $3(16)=48$ and whose sum is -15. The negative factors and their sums are:

$-1+(-48)=-49$
$-2+(-24)=-26$
$-3+(-16)=-19$
$-4+(-12)=-16$
$-6+(-8)=-14$

So there are no integers satisfying the conditions, and the polynomial is prime.

39. First, factor out the greatest common factor, 2.

$20x^2+22x+6=2\left(10x^2+11x+3\right)$

Now factor $10x^2+11x+3$ by trial and error to obtain $10x^2+11x+3=(5x+3)(2x+1)$. The complete factorization is

$20x^2+22x+6=2(5x+3)(2x+1)$.

41. First, factor out the GCF, 3.

$24x^2-42x+9=3\left(8x^2-14x+3\right)$

Use the grouping method to factor $8x^2-14x+3$. Look for two integers whose product is $8(3)=24$ and whose sum is -14. The integers are -12 and -2.

$$\begin{aligned}24x^2-42x+9&=3\left(8x^2-12x-2x+3\right)\\&=3\left[\left(8x^2-12x\right)+\left(-2x+3\right)\right]\\&=3\left[4x(2x-3)-1(2x-3)\right]\\&=3(2x-3)(4x-1)\end{aligned}$$

43. First, factor out the GCF, q.

$-40m^2q-mq+6q=-q\left(40m^2+m-6\right)$

Now factor $40m^2+m-6$ by trial and error to obtain

$40m^2+m-6=(5m+2)(8m-3)$.

The complete factorization is

$-40m^2q-mq+6q=-q(5m+2)(8m-3)$.

45. First, factor out the GCF, $3n^2$.

$15n^4-39n^3+18n^2=3n^2\left(5n^2-13n+6\right)$

Factor $5n^2-13n+6$ by the trial and error method. Possible factors of $5n^2$ are $5n$ and n. Possible factors of 6 are -6 and -1, or -3 and -2.

$(5n-6)(n-1)=5n^2-11n+6$ Incorrect

$(5n-3)(n-2)=5n^2-13n+6$ Correct

The completely factored form is

$15n^4-39n^3+18n^2=3n^2(5n-3)(n-2)$.

47. First, factor out the GCF, y^2.

$15x^2y^2-7xy^2-4y^2=y^2\left(15x^2-7x-4\right)$

Factor $15x^2-7x-4$ by the grouping method. Look for two integers whose product is $15(-4)=-60$ and whose sum is -7. The integers are -12 and 5.

$$\begin{aligned}&15x^2y^2-7xy^2-4y^2\\&=y^2(15x^2-12x+5x-4)\\&=y^2[3x(5x-4)+1(5x-4)]\\&=y^2(5x-4)(3x+1)\end{aligned}$$

49. Factor by the grouping method. Look for two integers whose product is $5(-6)=-30$ and whose sum is -7. The integers are -10 and 3.

$$\begin{aligned}&5a^2-7ab-6b^2\\&=5a^2-10ab+3ab-6b^2\\&=\left(5a^2-10ab\right)+\left(3ab-6b^2\right)\\&=5a(a-2b)+3b(a-2b)\\&=(a-2b)(5a+3b)\end{aligned}$$

51. Factor by the grouping method. Look for two integers whose product is $12(-5)=-60$ and whose sum is 11. The integers are 15 and -4.

$$\begin{aligned}&12s^2+11st-5t^2\\&=12s^2+15st-4st-5t^2\\&=\left(12s^2+15st\right)+\left(-4st-5t^2\right)\\&=3s(4s+5t)-t(4s+5t)\\&=(4s+5t)(3s-t)\end{aligned}$$

53. Factor out the greatest common factor.

$6m^6n+7m^5n^2+2m^4n^3$

$=m^4n\left(6m^2+7mn+2n^2\right)$ GCF $=m^4n$

Now factor $6m^2+7mn+2n^2$ by trial and error. Possible factors of $6m^2$ are 6m and m or 3m and 2m. Possible factors of $2n^2$ are 2n and n.

$(3m+2n)(2m+n)=6m^2+7mn+2n^2$ Correct

The completely factored form is
$6m^6n + 7m^5n^2 + 2m^4n^3$
$= m^4n(3m = 2n)(2m + n).$

55. The problem cannot be factored. It is prime.

57. $16 + 16x + 3x^2 = 3x^2 + 16x + 16$
Factor by the grouping method. Find two integers whose product is $(3)(16) = 48$ and whose sum is 16. The numbers are 4 and 12.
$$\begin{aligned} 3x^2 + 16x + 16 &= 3x^2 + 4x + 12x + 16 \\ &= (3x^2 + 4x) + (12x + 16) \\ &= x(3x + 4) + 4(3x + 4) \\ &= (3x + 4)(x + 4) \end{aligned}$$

59. First, factor out $-5x$; then complete the factoring by trial and error, using FOIL to test various possibilities until the correct one is found.
$$\begin{aligned} -10x^3 + 5x^2 + 140x &= -5x(2x^2 - x - 28) \\ &= -5x(2x + 7)(x - 4) \end{aligned}$$

61. The problem cannot be factored. It is prime.

63. Factor by using trial and error. Only positive factors of 24 should be considered: 1 and 24, 2 and 12, 3 and 8, or 4 and 6. The factors of -14 include 7 and -2. The correct factorization is
$24y^2 - 41xy - 14x^2 = (24y + 7x)(y - 2x).$

65. Factor by using trial and error. For $15y^2$, consider only $-y$ and $-15y$, and $-3y$ and $-5y$.
$36x^4 - 64x^2y + 15y^2 = (18x^2 - 5y)(2x^2 - 3y)$

67. $48a^2 - 94ab - 4b^2$
$= 2(24a^2 - 47ab - 2b^2) \quad \text{GCF} = 2$

Now factor $24a^2 - 47ab - 2b^2$ by the grouping method. Look for two integers whose product is $24(-2) = -48$ and whose sum is -47. The integers are 1 and -48.
$24a^2 - 47ab - 2b^2$
$= 24a^2 + ab - 48ab - 2b^2$
$= (24a^2 + ab) + (-48ab - 2b^2)$
$= a(24a + b) - 2b(24a + b)$
$= (24a + b)(a - 2b)$

The completely factored form is
$48a^2 - 94ab - 4b^2 = 2(24a + b)(a - 2b).$

69. $10x^4y^5 + 39x^3y^5 - 4x^2y^5$
$= x^2y^5(10x^2 + 39x - 4) \quad \text{GCF} = x^2y^5$

Now factor $10x^2 + 39x - 4$ by the grouping method. Look for two integers whose product is $10(-4) = -40$ and whose sum is 39. The integers are -1 and 40.
$10x^2 + 39x - 4$
$= 10x^2 - x + 40x - 4$
$= (10x^2 - x) + (40x - 4)$
$= x(10x - 1) + 4(10x - 1)$
$= (10x - 1)(x + 4)$
The completely factored form is
$10x^4y^5 + 39x^3y^5 - 4x^2y^5$
$= x^2y^5(10x - 1)(x + 4).$

71. $36a^3b^2 - 104a^2b^2 - 12ab^2$
$= 4ab^2(9a^2 - 26a - 3) \quad \text{GCF} = 4ab^2$

Now factor $9a^2 - 26a - 3$ by the grouping method. Look for two integers whose product is $9(-3) = -27$ and whose sum is -26. The integers are 1 and -27.
$$\begin{aligned} 9a^2 - 26a - 3 &= 9a^2 + a - 27a - 3 \\ &= (9a^2 + a) + (-27a - 3) \\ &= a(9a + 1) - 3(9a + 1) \\ &= (9a + 1)(a - 3) \end{aligned}$$
The completely factored form is
$36a^3b^2 - 104a^2b^2 - 12ab^2$
$= 4ab^2(9a + 1)(a - 3).$

73. Factor by the grouping method. Look for two integers whose product is $24(15) = 360$ and whose sum is -46. The integers are -10 and -36.
$24x^2 - 46x + 15$
$= 24x^2 - 10x - 36x + 15$
$= (24x^2 - 10x) + (-36x + 15)$
$= 2x(12x - 5) - 3(12x - 5)$
$= (12x - 5)(2x - 3)$

75. Factor by the grouping method. Look for two integers whose product is $24(-24) = -576$ and whose sum is 55. The integers are -9 and 64. Note the x^2-term.

$$\begin{aligned} &24x^4 + 55x^2 - 24 \\ &= 24x^4 - 9x^2 + 64x^2 - 24 \\ &= (24x^4 - 9x^2) + (64x^2 - 24) \\ &= 3x^2(8x^2 - 3) + 8(8x^2 - 3) \\ &= (8x^2 - 3)(3x^2 + 8) \end{aligned}$$

77. Factor by the grouping method. Look for two integers whose product is $24(15) = 360$ and whose sum is 38. The integers are 18 and 20. Note the *xy*-term.

$$\begin{aligned} &24x^2 + 38xy + 15y^2 \\ &= 24x^2 + 18xy + 20xy + 15y^2 \\ &= (24x^2 + 18xy) + (20xy + 15y^2) \\ &= 6x(4x + 3y) + 5y(4x + 3y) \\ &= (4x + 3y)(6x + 5y) \end{aligned}$$

79. $$\begin{aligned} -x^2 - 4x + 21 &= -1(x^2 + 4x - 21) \\ &= -1(x + 7)(x - 3) \end{aligned}$$

81. $$\begin{aligned} -3x^2 - x + 4 &= -1(3x^2 + x - 4) \\ &= -1(3x + 4)(x - 1) \end{aligned}$$

83. $$\begin{aligned} &-2a^2 - 5ab - 2b^2 \\ &= -1(2a^2 + 5ab + 2b^2) \\ &= -1(2a^2 + 4ab + ab + 2b^2) \\ &= -1[(2a^2 + 4ab) + (ab + 2b^2)] \\ &= -1[2a(a + 2b) + b(a + 2b)] \\ &= -1(a + 2b)(2a + b) \end{aligned}$$

85. First, factor out the GCF, $(m+1)^3$; then factor the resulting trinomial by trial and error.

$$\begin{aligned} &25q^2(m+1)^3 - 5q(m+1)^3 - 2(m+1)^3 \\ &= (m+1)^3(25q^2 - 5q - 2) \\ &= (m+1)^3(5q - 2)(5q + 1) \end{aligned}$$

87. $$\begin{aligned} &9x^2(r+3)^3 + 12xy(r+3)^3 + 4y^2(r+3)^3 \\ &= (r+3)^3(9x^2 + 12xy + 4y^2) \\ &= (r+3)^3(3x + 2y)(3x + 2y) \\ &= (r+3)^3(3x + 2y)^2 \end{aligned}$$

89. Look for two integers whose product is $5(-1) = -5$ and whose sum is 5.

Factors of -5	Sums of Factors
$-5, 1$	-4
$5, -1$	4

Thus, there are two possible integer values for k: -4 and 4.

91. Look for two integers whose product is $2(5) = 10$ and whose sum is *k*.

Factors of 10	Sums of Factors
$-10, -1$	-11
$10, 1$	11
$-5, -2$	-7
$5, 2$	7

Thus, there are four possible integer values for k: $-11, -7, 7,$ and 11.

6.4 Special Factoring Techniques

Now Try Exercises

N1. **(a)** $$\begin{aligned} x^2 - 100 &= x^2 - 10^2 \\ &= (x + 10)(x - 10) \end{aligned}$$

(b) This binomial is the sum of two squares and the terms have no common factor. Unlike the difference of two squares, it cannot be factored. It is a prime polynomial.

(c) The two terms do not have a common factor. Because 32 is not the square of an integer, this binomial is not a difference of squares. It is a prime polynomial.

N2. **(a)** $9t^2 - 100 = (3t)^2 - 10^2$
$= (3t+10)(3t-10)$

(b) $36a^2 - 49b^2 = (6a)^2 - (7b)^2$
$= (6a+7b)(6a-7b)$

N3. **(a)** $16k^2 - 64 = 16(k^2 - 4)$
$= 16(k^2 - 2^2)$
$= 16(k+2)(k-2)$

(b) $m^4 - 144 = (m^2)^2 - 12^2$
$= (m^2+12)(m^2-12)$

(c) $v^4 - 625 = (v^2)^2 - 25^2$
$= (v^2+25)(v^2-25)$
$= (v^2+25)(v^2-5^2)$
$= (v^2+25)(v+5)(v-5)$

N4. The term y^2 is a perfect square, and so is 49. Try to factor the trinomial as
$y^2 + 14y + 49 = (y+7)^2$.
To check, take twice the product of the two terms in the squared binomial.
$2 \cdot y \cdot 7 = 14y$
Since $14y$ is the middle term of the trinomial, the trinomial is a perfect square and can be factored as $(y+7)^2$. Thus,
$y^2 + 14y + 49 = (y+7)^2$.

N5. **(a)** $t^2 - 18t + 81 \stackrel{?}{=} (t-9)^2$
$2 \cdot t \cdot 9 = 18t$, so this is a perfect square trinomial, and $t^2 - 18t + 81 = (t-9)^2$.

(b) $4p^2 - 28p + 49 \stackrel{?}{=} (2p-7)^2$
$2 \cdot 2p \cdot 7 = 28p$, so this is a perfect square trinomial, and
$4p^2 - 28p + 49 = (2p-7)^2$.

(c) $9x^2 + 6x + 4 \stackrel{?}{=} (3x+2)^2$
$2 \cdot 3x \cdot 2 = 12x \neq 6x$, so this is not a perfect square trinomial. The trinomial cannot be factored even with the methods of the previous sections. It is prime.

(d) $80x^3 + 120x^2 + 45x$
Factor out the greatest common factor, $5x$.
$80x^3 + 120x^2 + 45x = 5x(16x^2 + 24x + 9)$
$\stackrel{?}{=} 5x(4x+3)^2$
$2 \cdot 4x \cdot 3 = 24x$, so this is a perfect square trinomial, and
$5x(16x^2 + 24x + 9) = 5x(4x+3)^2$.

N6. **(a)** $a^3 - 27 = a^3 - 3^3$
Let $x = a$ and $y = 3$ in the pattern for the difference of cubes.
$x^3 - y^3 = (x-y)(x^2 + xy + y^2)$
$a^3 - 3^3 = (a-3)(a^2 + a \cdot 3 + 3^2)$
$= (a-3)(a^2 + 3a + 9)$

(b) $8t^3 - 125 = (2t)^3 - 5^3$
Let $x = 2t$ and $y = 5$ in the pattern for the difference of cubes.
$x^3 - y^3 = (x-y)(x^2 + xy + y^2)$
$(2t)^3 - 5^3 = (2t-5)[(2t)^2 + 2t \cdot 5 + 5^2]$
$= (2t-5)(4t^2 + 10t + 25)$

(c) $3k^3 - 192 = 3(k^3 - 64)$
$= 3(k^3 - 4^3)$
$= 3(k-4)(k^2 + k \cdot 4 + 4^2)$
$= 3(k-4)(k^2 + 4k + 16)$

(d)
$125x^3 - 343y^6$
$= (5x)^3 - (7y^2)^3$
$= (5x - 7y^2)\left[(5x)^2 + 5x(7y^2) + (7y^2)^2\right]$
$= (5x - 7y^2)(25x^2 + 35xy^2 + 49y^4)$

N7. **(a)** $x^3+125=x^3+5^3$

Let $x=x$ and $y=5$ in the pattern for the sum of cubes.

$$x^3+y^3=(x+y)(x^2-xy+y^2)$$
$$x^3+5^3=(x+5)(x^2-x\cdot 5+5^2)$$
$$=(x+5)(x^2-5x+25)$$

(b) $27a^3+8b^3$

$$=(3a)^3+(2b)^3$$
$$=(3a+2b)\left[(3a)^2-3a(2b)+(2b)^2\right]$$
$$=(3a+2b)(9a^2-6ab+4b^2)$$

Section Exercises

1.

$1^2=\underline{1}$	$2^2=\underline{4}$	$3^2=\underline{9}$
$4^2=\underline{16}$	$5^2=\underline{25}$	$6^2=\underline{36}$
$7^2=\underline{49}$	$8^2=\underline{64}$	$9^2=\underline{81}$
$10^2=\underline{100}$	$11^2=\underline{121}$	$12^2=\underline{144}$
$13^2=\underline{169}$	$14^2=\underline{196}$	$15^2=\underline{225}$
$16^2=\underline{256}$	$17^2=\underline{289}$	$18^2=\underline{324}$
$19^2=\underline{361}$	$20^2=\underline{400}$	

2. The following powers of x are all perfect squares: $x^2, x^4, x^6, x^8, x^{10}$. On the basis of this observation, we may make a conjecture (an educated guess) that if the power of a variable is divisible by $\underline{2}$ (with 0 remainder), then we have a perfect square.

3. The binomial in A is a difference of squares.

$$x^2-4=x^2-2^2$$
$$=(x+2)(x-2)$$

The binomial in B is the sum of two squares and the terms have no common factor. Unlike the difference of two squares, it cannot be factored. It is a prime polynomial.
The two terms in the binomial in C do not have a common factor. This binomial is not a difference of squares. It is a prime polynomial.
The binomial in D is a difference of squares.

$$9m^2-1=9m^2-1^2$$
$$=(3x+1)(3x-1)$$

4. Choice A cannot be factored since it is the sum of two squares. It is a prime polynomial.
B can be factored.

$$x^3+x=x(x^2+1)$$

C can be factored.

$$3x^2+12=3(x^2+4)$$

D cannot be factored since the two terms are the sum of squares and have no common factors.

5. The student forgot that the terms both have a factor of 4 that can be pulled out. The binomial can be factored as $4x^2+16=4(x^2+4)$. After the common factor is removed, it is then a sum of squares and cannot be factored further.

6. The student's answer is not a complete factorization because x^2-9 can be factored further. The correct complete factorization is $k^4-81=(k^2+9)(k+3)(k-3)$.

7. To factor this binomial, use the rule for factoring a difference of squares.

$$a^2-b^2=(a+b)(a-b)$$
$$\downarrow \quad \downarrow \qquad \downarrow \ \downarrow \ \downarrow \ \downarrow$$
$$y^2-25=y^2-5^2=(y+5)(y-5)$$

8. To factor this binomial, use the rule for factoring a difference of squares.

$$a^2-b^2=(a+b)(a-b)$$
$$\downarrow \quad \downarrow \qquad \downarrow \ \downarrow \ \downarrow \ \downarrow$$
$$t^2-36=t^2-6^2=(t+6)(t-6)$$

9. To factor this binomial, use the rule for factoring a difference of squares.

$$x^2-144=x^2-12^2$$
$$=(x+12)(x-12)$$

10. To factor this binomial, use the rule for factoring a difference of squares.

$$x^2-400=x^2-20^2$$
$$=(x+20)(x-20)$$

11. The two terms do not have a common factor. Because 12 is not the square of an integer, this binomial is not a difference of squares. It is a prime polynomial.

12. The two terms do not have a common factor. Because 18 is not the square of an integer, this binomial is not a difference of squares. It is a prime polynomial.

13. This binomial is the sum of squares and the terms have no common factor. Unlike the difference of squares, it cannot be factored. It is a prime polynomial.

14. This binomial is the sum of squares and the terms have no common factor. Unlike the difference of squares, it cannot be factored. It is a prime polynomial.

15. Factor out the GCF, 4. The resulting sum of squares cannot be factored further.
$$4m^2 + 16 = 4(m^2 + 4)$$

16. Factor out the GCF, 9. The resulting sum of squares cannot be factored further.
$$9x^2 + 81 = 9(x^2 + 9)$$

17. To factor this binomial, use the rule for factoring a difference of squares.
$$\begin{aligned} 9r^2 - 4 &= (3r)^2 - 2^2 \\ &= (3r + 2)(3r - 2) \end{aligned}$$

18. To factor this binomial, use the rule for factoring a difference of squares.
$$\begin{aligned} 4x^2 - 9 &= (2x)^2 - 3^2 \\ &= (2x + 3)(2x - 3) \end{aligned}$$

19. First factor out the GCF, 4; then use the rule for factoring the difference of squares.
$$\begin{aligned} 36x^2 - 16 &= 4(9x^2 - 4) \\ &= 4[(3x^2) - 2^2] \\ &= 4(3x + 2)(3x - 2) \end{aligned}$$

20. First factor out the GCF, 8; then use the rule for factoring the difference of squares.
$$\begin{aligned} 32a^2 - 8 &= 8(4a^2 - 1) \\ &= 8[(2a)^2 - 1^2] \\ &= 8(2a + 1)(2a - 1) \end{aligned}$$

21. To factor this binomial, use the rule for factoring a difference of squares.
$$\begin{aligned} 196p^2 - 225 &= (14p)^2 - 15^2 \\ &= (14p + 15)(14p - 15) \end{aligned}$$

22. To factor this binomial, use the rule for factoring a difference of squares.
$$\begin{aligned} 361q^2 - 400 &= (19q)^2 - 20^2 \\ &= (19q + 20)(19q - 20) \end{aligned}$$

23. To factor this binomial, use the rule for factoring a difference of squares.
$$\begin{aligned} 16r^2 - 25a^2 &= (4r)^2 - (5a)^2 \\ &= (4r + 5a)(4r - 5a) \end{aligned}$$

24. To factor this binomial, use the rule for factoring a difference of squares.
$$\begin{aligned} 49m^2 - 100p^2 &= (7m)^2 - (10p)^2 \\ &= (7m + 10p)(7m - 10p) \end{aligned}$$

25. To factor this binomial, use the rule for factoring a difference of squares.
$$\begin{aligned} 81x^2 - 49y^2 &= (9x)^2 - (7y)^2 \\ &= (9x + 7y)(9x - 7y) \end{aligned}$$

26. To factor this binomial, use the rule for factoring a difference of squares.
$$\begin{aligned} 36y^2 - 121z^2 &= (6y)^2 - (11z)^2 \\ &= (6y + 11z)(6y - 11z) \end{aligned}$$

27. To factor this binomial, use the rule for factoring a difference of squares.
$$\begin{aligned} 54x^2 - 6y^2 &= 6(9x^2 - y^2) \\ &= 6\left[(3x)^2 - (1y)^2\right] \\ &= 6(3x + y)(3x - y) \end{aligned}$$

28. To factor this binomial, use the rule for factoring a difference of squares.
$$\begin{aligned} 48m^2 - 75n^2 &= 3(16m^2 - 25n^2) \\ &= 3\left[(4m)^2 - (5n)^2\right] \\ &= 3(4m + 5n)(4m - 5n) \end{aligned}$$

29. This binomial is the sum of squares and the terms have no common factor. Unlike the difference of squares, it cannot be factored. It is a prime polynomial.

30. This binomial is the sum of squares and the terms have no common factor. Unlike the difference of squares, it cannot be factored. It is a prime polynomial.

31. To factor this binomial, use the rule for factoring a difference of squares.
$$\begin{aligned} 4 - x^2 &= (2)^2 - (x)^2 \\ &= (2 + x)(2 - x) \end{aligned}$$

32. To factor this binomial, use the rule for factoring a difference of squares.

$$\begin{aligned}25-x^2&=(5)^2-(x)^2\\&=(5+x)(5-x)\end{aligned}$$

33. To factor this binomial, use the rule for factoring a difference of squares.

$$\begin{aligned}36-25t^2&=(6)^2-(5t)^2\\&=(6+5t)(6-5t)\end{aligned}$$

34. To factor this binomial, use the rule for factoring a difference of squares.

$$\begin{aligned}16-49p^2&=(4)^2-(7p)^2\\&=(4+7p)(4-7p)\end{aligned}$$

35. First factor out the GCF, x; then use the rule for factoring the difference of squares.

$$x^3+4x=x\left(x^2+4\right)$$

36. First factor out the GCF, z. The resulting sum of squares cannot be factored further.

$$z^3+25z=z\left(z^2+25\right)$$

37. First factor out the GCF, x^2, and then use the rule for factoring the difference of squares.

$$\begin{aligned}x^4-x^2&=x^2\left(x^2-1\right)\\&=x^2\left[(x)^2-(1)^2\right]\\&=x^2(x+1)(x-1)\end{aligned}$$

38. First factor out the GCF, y^2, and then use the rule for factoring the difference of squares.

$$\begin{aligned}y^4-9y^2&=y^2\left(y^2-9\right)\\&=y^2\left[(y)^2-(3)^2\right]\\&=y^2(y+3)(y-3)\end{aligned}$$

39. To factor this binomial, use the rule for factoring a difference of squares.

$$\begin{aligned}p^4-49&=(p^2)^2-7^2\\&=(p^2+7)(p^2-7)\end{aligned}$$

40. To factor this binomial, use the rule for factoring a difference of squares.

$$\begin{aligned}r^4-25&=(r^2)^2-5^2\\&=(r^2+5)(r^2-5)\end{aligned}$$

41. To factor this binomial completely, factor the difference of squares twice.

$$\begin{aligned}x^4-1&=\left(x^2\right)^2-1^2\\&=\left(x^2+1\right)\left(x^2-1\right)\\&=\left(x^2+1\right)\left(x^2-1^2\right)\\&=\left(x^2+1\right)(x+1)(x-1)\end{aligned}$$

42. To factor this binomial completely, factor the difference of squares twice.

$$\begin{aligned}y^4-10{,}000&=\left(y^2\right)^2-100^2\\&=\left(y^2+100\right)\left(y^2-100\right)\\&=\left(y^2+100\right)\left(y^2-10^2\right)\\&=\left(y^2+100\right)(y+10)(y-10)\end{aligned}$$

43. To factor this binomial completely, factor the difference of squares twice.

$$\begin{aligned}p^4-256&=\left(p^2\right)^2-16^2\\&=\left(p^2+16\right)\left(p^2-16\right)\\&=\left(p^2+16\right)\left(p^2-4^2\right)\\&=\left(p^2+16\right)(p+4)(p-4)\end{aligned}$$

44. To factor this binomial completely, factor the difference of squares twice.

$$\begin{aligned}k^4-81&=\left(k^2\right)^2-9^2\\&=\left(k^2+9\right)\left(k^2-9\right)\\&=\left(k^2+9\right)\left(k^2-3^2\right)\\&=\left(k^2+9\right)(k+3)(k-3)\end{aligned}$$

45. $y^2-13y+36\stackrel{?}{=}(y+6)^2$

$2\cdot y\cdot 6=12y\neq -13y,$ so A is not a perfect square trinomial. The trinomial cannot be factored. It is prime.

B is a perfect square trinomial.

$x^2+6x+9=(x+3)(x+3)$ since

$$\begin{aligned}(x+3)(x+3)&=x^2+3x+3x+9\\&=x^2+6x+9\end{aligned}$$

C is a perfect square trinomial.

$4z^2-4z+1=(2z-1)(2z-1)$ since
$(2z-1)(2z-1)=4z^2-2z-2z+1$
$=4z^2-4z+1$

$16m^2+10m+1\stackrel{?}{=}(4m+1)^2$
$2\cdot m\cdot 4=8m\neq 10m,$ so D is not a perfect square trinomial. The trinomial cannot be factored. It is prime.

46. No, it is not a perfect square trinomial because the middle term would have to be $30y$.

47. Find b so that $x^2+bx+25=(x+5)^2$.
Since $(x+5)^2=x^2+10x+25,\ b=10$.

48. Find c so that $4m^2-12m+c=(2m-3)^2$.
Since $(2m-3)^2=4m^2-12m+9,\ c=9$.

49. Find a so that $ay^2-12y+4=(3y-2)^2$.
Since $(3y-2)^2=9y^2-12y+4,\ a=9$.

50. Find b so that $100a^2+ba+9=(10a+3)^2$.
Since $(10a+3)^2=100a^2+60a+9,\ b=60$.

51. The first and last terms are perfect squares, w^2 and 1^2. This trinomial is a perfect square, since the middle term is twice the product of w and 1, or $2\cdot w\cdot 1=2w$. Therefore,
$w^2+2w+1=(w+1)^2$.

52. The first and last terms are perfect squares, p^2 and 2^2. This trinomial is a perfect square, since the middle term is twice the product of p and 2, or $2\cdot p\cdot 2=4p$. Therefore,
$p^2+4p+4=(p+2)^2$.

53. The first and last terms are perfect squares, x^2 and $(-4)^2$. This trinomial is a perfect square, since the middle term is twice the product of x and -4, or $2\cdot x\cdot(-4)=-8x$. Therefore,
$x^2-8x+16=(x-4)^2$.

54. The first and last terms are perfect squares, x^2 and $(-5)^2$. This trinomial is a perfect square, since the middle term is twice the product of x and -5, or $2\cdot x\cdot(-5)=-10x$. Therefore,
$x^2-10x+25=(x-5)^2$.

55. $x^2-10x+100\stackrel{?}{=}(x+10)^2$
$2\cdot x\cdot 10=20x\neq -10x,$ so this is not a perfect square trinomial. The trinomial cannot be factored even with the methods of the previous sections. It is prime.

56. $x^2-18x+36\stackrel{?}{=}(x+6)^2$
$2\cdot x\cdot 6=12x\neq -18x,$ so this is not a perfect square trinomial. The trinomial cannot be factored even with the methods of the previous sections. It is prime.

57. First, factor out the GCF, 2.
$2x^2+24x+72=2(x^2+12x+36)$
Now factor $x^2+12x+36$ as a perfect square trinomial.
$x^2+12x+36=(x+6)^2$
The final factored form is
$2x^2+24x+72=2(x+6)^2$.

58 First, factor out the GCF, 3.
$3y^2+48y+192=3(y^2+16y+64)$
Now factor $y^2+16y+64$ as a perfect square trinomial.
$y^2+16y+64=(y+8)^2$
The final factored form is
$3y^2+48y+192=3(y+8)^2$.

59. The first and last terms are perfect squares, $(2x)^2$ and $(3)^2$. The middle term is $2(2x)(3)=12x$. Therefore, the final factored form is $4x^2+12x+9=(2x+3)^2$.

60. The first and last terms are perfect squares, $(5x)^2$ and $(1)^2$. The middle term is $2(5x)(1)=10x$. Therefore, the final factored form is $25x^2+10x+1=(5x+1)^2$.

61. The first and last terms are perfect squares, $(4x)^2$ and $(-5)^2$. The middle term is $2(4x)(-5)=-40x$. Therefore,
$16x^2-40x+25=(4x-5)^2$.

62. The first and last terms are perfect squares, $(6y)^2$ and $(-5)^2$. The middle term is $2(6y)(-5) = -60y$. Therefore, $36y^2 - 60y + 25 = (6y-5)^2$.

63. The first and last terms are perfect squares, $(7x)^2$ and $(-2y)^2$. The middle term is $2(7x)(-2y) = -28xy$. Therefore, $49x^2 - 28xy + 4y^2 = (7x-2y)^2$.

64. $4z^2 - 12zw + 9w^2 = (2z)^2 - 2(2z)(3w) + (3w)^2$
$= (2z-3w)^2$

65. $64x^2 + 48xy + 9y^2$
$= (8x)^2 + 2(8x)(3y) + (3y)^2$
$= (8x+3y)^2$

66. $9t^2 + 24tr + 16r^2 = (3t)^2 + 2(3t)(4r) + (4r)^2$
$= (3t+4r)^2$

67. $50h^2 - 40hy + 8y^2$
$= 2(25h^2 - 20hy + 4y^2)$
$= 2[(5h)^2 - 2(5h)(2y) + (2y)^2]$
$= 2(5h-2y)^2$

68. $18x^2 - 48xy + 32y^2$
$= 2(9x^2 - 24xy + 16y^2)$
$= 2[(3x)^2 - 2(3x)(4y) + (4y)^2]$
$= 2(3x-4y)^2$

69. First, factor out the GCF, k.
$4k^3 - 4k^2 + 9k = k(4k^2 - 4k + 9)$
Since $4k^2 - 4k + 9$ cannot be factored, $k(4k^2 - 4k + 9)$ is the final factored form.

70. First, factor out the GCF, r.
$9r^3 - 6r^2 + 16r = r(9r^2 - 6r + 16)$
Since $9r^2 - 6r + 16$ cannot be factored, $r(9r^2 - 6r + 16)$ is the final factored form.

71. First, factor out the GCF, z^2.
$25z^4 + 5z^3 + z^2 = z^2(25z^2 + 5z + 1)$
Since $25z^2 + 5z + 1$ cannot be factored, $z^2(25z^2 + 5z + 1)$ is the final factored form.

72. First, factor out the GCF, x^2.
$4x^4 + 2x^3 + x^2 = x^2(4x^2 + 2x + 1)$
Since $4x^2 + 2x + 1$ cannot be factored, $x^2(4x^2 + 2x + 1)$ is the final factored form.

73. $1^3 = \underline{1}$ $\quad 2^3 = \underline{8}$ $\quad 3^3 = \underline{27}$
$4^3 = \underline{64}$ $\quad 5^3 = \underline{125}$ $\quad 6^3 = \underline{216}$
$7^3 = \underline{343}$ $\quad 8^3 = \underline{512}$ $\quad 9^3 = \underline{729}$
$10^3 = \underline{1000}$

74. The following powers of x are all perfect cubes: $x^3, x^6, x^9, x^{12}, x^{15}$. On the basis of this observation, we may make a conjecture that if the power of a variable is divisible by $\underline{3}$ (with 0 remainder), then we have a perfect cube.

75. The two terms in the binomial in A do not have a common factor. This binomial is not a difference of cubes. It is a prime polynomial. The two terms in the binomial in B do not have a common factor. This binomial is not a difference of cubes. It is a prime polynomial. C is a difference of cubes.
$x^3 - 1 = (x)^3 - (1)^3$
$= (x-1)(x^2 + x + 1)$
D is a difference of cubes.
$8x^3 - 27y^4 = (2x)^3 - (3y)^3$
$= (2x-3y)(4x^2 + 6xy + 9y^2)$

76. A is a sum of cubes.
$x^3 + 1 = (x)^3 + (1)^3$
$= (x+1)(x^2 - x + 1)$
B does not have any common factors and 36 is not a perfect cube. It is not a sum of cubes. The binomial C can be factored into the form of $12x^3 + 27 = 3(4x^3 + 9)$, but cannot be factored anymore. 4 and 9 are not perfect cubes. It is not a sum of cubes.

D is a sum of cubes.

$64x^3 + 216y^3$

$= (4x)^3 + (6y)^3$

$= (4x + 6y)(16x^2 - 24xy + 36y^2)$

77. **(a)** Since x^3 is not a perfect square, $4x^3$ is not a perfect square. Since 4 is not a perfect cube, $4x^3$ is not a perfect cube.

(b) $8y^6 = (2y^2)^3$, so $8y^6$ is a perfect cube.

Since 8 is not a perfect square, $8y^6$ is not a perfect square.

(c) $49x^{12} = (7x^6)^2$, so $49x^{12}$ is a perfect square. Since 49 is not a perfect cube, $49x^{12}$ is not a perfect cube.

(d) $81r^{10} = (9r^5)^2$, so $81r^{10}$ is a perfect square. It is not a perfect cube.

(e) $64x^6y^{12} = (8x^3y^6)^2$, so $64x^6y^{12}$ is a perfect square. $64x^6y^{12} = (4x^2y^4)^3$, so $64x^6y^{12}$ is also a perfect cube. Therefore, the answer is "both of these."

(f) $125t^6 = (5t^2)^3$, so $125t^6$ is a perfect cube.

Since 125 is not a perfect square, $125t^6$ is not a perfect square.

78. From the results of Exercises 2 and 4, we see that for x^n to be both a perfect square and a perfect cube, *n* must be divisible by 2 and by 3 (with 0 remainder). This means that the exponent *n* must be divisible by 6 (with 0 remainder).

79. Let $x = a$ and $y = 1$ in the pattern for the difference of cubes.

$x^3 - y^3 = (x - y)(x^2 + xy + y^2)$

$a^3 - 1 = a^3 - 1^3$

$= (a - 1)(a^2 + a \cdot 1 + 1^2)$

$= (a - 1)(a^2 + a + 1)$

80. Let $x = m$ and $y = 2$ in the pattern for the difference of cubes.

$x^3 - y^3 = (x - y)(x^2 + xy + y^2)$

$m^3 - 8 = m^3 - 2^3$

$= (m - 2)(m^2 + m \cdot 2 + 2^2)$

$= (m - 2)(m^2 + 2m + 4)$

81. Let $x = m$ and $y = 2$ in the pattern for the sum of cubes.

$x^3 + y^3 = (x + y)(x^2 - xy + y^2)$

$m^3 + 8 = m^3 + 2^3$

$= (m + 2)(m^2 - m \cdot 2 + 2^2)$

$= (m + 2)(m^2 - 2m + 4)$

82. Let $x = b$ and $y = 1$ in the pattern for the sum of cubes.

$x^3 + y^3 = (x + y)(x^2 - xy + y^2)$

$b^3 + 1 = b^3 + 1^3$

$= (b + 1)(b^2 - b \cdot 1 + 1^2)$

$= (b + 1)(b^2 - b + 1)$

83. Factor $y^3 - 216$ as the difference of cubes.

$y^3 - 216 = y^3 - 6^3$

$= (y - 6)(y^2 + 6y + 6^2)$

$= (y - 6)(y^2 + 6y + 36)$

84. Factor $x^3 - 343$ as the difference of cubes.

$x^3 - 343 = x^3 - 7^3$

$= (x - 7)(x^2 + 7x + 7^2)$

$= (x - 7)(x^2 + 7x + 49)$

85. Factor $k^3 + 1000$ as the sum of cubes.

$k^3 + 1000 = k^3 + 10^3$

$= (k + 10)(k^2 - 10k + 10^2)$

$= (k + 10)(k^2 - 10k + 100)$

86. Factor p^3+512 as the sum of cubes.

$$\begin{aligned} p^3+512 &= p^3+8^3 \\ &= (p+8)(p^2-8p+8)^2 \\ &= (p+8)(p^2-8p+64) \end{aligned}$$

87. Factor $27x^3-64$ as the difference of cubes.

$$\begin{aligned} 27x^3-64 &= (3x)^3-4^3 \\ &= (3x-4)\left[(3x)^2+3x\cdot 4+4^2\right] \\ &= (3x-4)(9x^2+12x+16) \end{aligned}$$

88. Factor $64y^3-27$ as the difference of cubes.

$$\begin{aligned} 64y^3-27 &= (4y)^3-3^3 \\ &= (4y-3)\left[(4y)^2+4y\cdot 3+3^2\right] \\ &= (4y-3)(16y^2+12y+9) \end{aligned}$$

89. Factor out the GCF, 6, and then factor as the sum of cubes.

$$\begin{aligned} 6p^3+6 &= 6(p^3+1) \\ &= 6(p^3+1^3) \\ &= 6(p+1)(p^2-p\cdot 1+1^2) \\ &= 6(p+1)(p^2-p+1) \end{aligned}$$

90. Factor out the GCF, 3, and then factor as the sum of cubes.

$$\begin{aligned} 81x^3+3 &= 3(27x^3+1) \\ &= 3\left[(3x)^3+1^3\right] \\ &= 3(3x+1)\left[(3x)^2-(3x)\cdot 1+1^3\right] \\ &= 3(3x+1)(9x^2-3x+1) \end{aligned}$$

91. Factor out the GCF, 5, and then factor as the sum of cubes.

$$\begin{aligned} 5x^3+40 &= 5(x^3+8) \\ &= 5(x^3+2^3) \\ &= 5(x+2)(x^2-x\cdot 2+2^2) \\ &= 5(x+2)(x^2-2x+4) \end{aligned}$$

92. Factor out the GCF, 2, and then factor as the sum of cubes.

$$\begin{aligned} 128y^3+54 &= 2(64y^3+27) \\ &= 2\left[(4y)^3+3^3\right] \\ &= 2(4y+3)\left[(4y)^2-4y\cdot 3+3^2\right] \\ &= 2(4y+3)(16y^2-12y+9) \end{aligned}$$

93. Factor y^3-8x^3 as the difference of cubes.

$$\begin{aligned} y^3-8x^3 &= y^3-(2x)^3 \\ &= (y-2x)\left[y^2+y(2x)+(2x)^2\right] \\ &= (y-2x)(y^2+2yx+4x^2) \end{aligned}$$

94. Factor w^3-216z^3 as the difference of cubes.

$$\begin{aligned} w^3-216z^3 &= w^3-(6z)^3 \\ &= (w-6z)\left[w^2+w(6z)+(6z)^2\right] \\ &= (w-6z)(w^2+6wz+36z^2) \end{aligned}$$

95. Factor out the GCF, 2, and then factor as the difference of cubes.

$$\begin{aligned} 2x^3-16y^3 &= 2(x^3-8y^3) \\ &= 2\left[x^3-(2y)^3\right] \\ &= 2(x-2y)\left[x^2+x\cdot 2y+(2y)^2\right] \\ &= 2(x-2y)(x^2+2xy+4y^2) \end{aligned}$$

96. Factor out the GCF, 27, and then factor as the difference of cubes.

$$\begin{aligned} &27w^3-216z^3 \\ &= 27(w^3-8z^3) \\ &= 27\left[w^3-(2z)^3\right] \\ &= 27(w-2z)\left[w^2+w\cdot 2z+(2z)^2\right] \\ &= 27(w-2z)(w^2+2wz+4z^2) \end{aligned}$$

97. Factor $8p^3+729q^3$ as the sum of cubes.

$$\begin{aligned} &8p^3+729q^3 \\ &= (2p)^3+(9q)^3 \\ &= (2p+9q)\left[(2p)^2-2p\cdot 9q+(9q)^2\right] \\ &= (2p+9q)(4p^2-18pq+81q^2) \end{aligned}$$

98. Factor $64x^3 + 125y^3$ as the sum of cubes.

$$64x^3 + 125y^3$$
$$= (4x)^3 + (5y)^3$$
$$= (4x+5y)\left[(4x)^2 - 4x\cdot 5y + (5y)^2\right]$$
$$= (4x+5y)(16x^2 - 20xy + 25y^2)$$

99. Factor $27a^3 + 64b^3$ as the sum of cubes.

$$27a^3 + 64b^3$$
$$= (3a)^3 + (4b)^3$$
$$= (3a+4b)\left[(3a)^2 - 3a\cdot 4b + (4b)^2\right]$$
$$= (3a+4b)(9a^2 - 12ab + 16b^2)$$

100. Factor $125m^3 + 8p^3$ as the sum of cubes.

$$125m^3 + 8p^3$$
$$= (5m)^3 + (2p)^3$$
$$= (5m+2p)\left[(5m)^2 - 5m\cdot 2p + (2p)^2\right]$$
$$= (5m+2p)(25m^2 - 10mp + 4p^2)$$

101. Factor $125t^3 + 8s^3$ as the sum of cubes.

$$125t^3 + 8s^3$$
$$= (5t)^3 + (2s)^3$$
$$= (5t+2s)\left[(5t)^2 - 5t\cdot 2s + (2s)^2\right]$$
$$= (5t+2s)(25t^2 - 10ts + 4s^2)$$

102. Factor $27r^3 + 1000s^3$ as the sum of cubes.

$$27r^3 + 1000s^3$$
$$= (3r)^3 + (10s)^3$$
$$= (3r+10s)\left[(3r)^2 - 3r\cdot 10s + (10s)^2\right]$$
$$= (3r+10s)(9r^2 - 30rs + 100s^2)$$

103. Factor $8x^3 - 125y^6$ as the difference of cubes.

$$8x^3 - 125y^6$$
$$= (2x)^3 - (5y^2)^3$$
$$= (2x - 5y^2)\left[(2x)^2 + 2x(5y)^2 + (5y^2)^2\right]$$
$$= (2x - 5y^2)(4x^2 + 10xy^2 + 25y^4)$$

104. Factor $27t^3 - 64s^6$ as the difference of cubes.

$$27t^3 - 64s^6$$
$$= (3t)^3 - 4s^{23}$$
$$= (3t - 4s^2)\left[(3t)^2 + 3t(4s)^2 + (4s^2)^2\right]$$
$$= (3t - 4s^2)(9t^2 + 12ts^2 + 16s^4)$$

105. Factor $27m^6 + 8n^3$ as the sum of cubes.

$$27m^6 + 8n^3$$
$$= (3m^2)^3 + (2n)^3$$
$$= (3m^2 + 2n)\left[(3m^2)^2 - 3m^2(2n) + (2n)^2\right]$$
$$= (3m^2 + 2n)(9m^4 - 6m^2n + 4n^2)$$

106. Factor $1000r^6 + 27s^3$ as the sum of cubes.

$$1000r^6 + 27s^3$$
$$= (10r^2)^3 + (3s)^3$$
$$= (10r^2 + 3s)\left[(10r^2)^2 - 10r^2(3s) + (3s)^2\right]$$
$$= (10r^2 + 3s)(100r^4 - 30r^2s + 9s^2)$$

107. Factor $x^9 + y^9$ as the sum of cubes.

$$x^9 + y^9$$
$$= (x^3)^3 + (y^3)^3$$
$$= (x^3 + y^3)\left[(x^3)^2 - x^3(y^3) + (y^3)^2\right]$$
$$= (x+y)(x^2 - xy + y^2)(x^6 - x^3y^3 + y^6)$$

108. Factor $x^9 - y^9$ as the difference of cubes.

$$x^9 - y^9$$
$$= (x^3)^3 - (y^3)^3$$
$$= (x^3 - y^3)\left[(x^3)^2 + x^3(y^3) + (y^3)^2\right]$$
$$= (x-y)(x^2 + xy + y^2)(x^6 + x^3y^3 + y^6)$$

109. $p^2 - \frac{1}{9} = p^2 - \left(\frac{1}{3}\right)^2$

$$= \left(p + \frac{1}{3}\right)\left(p - \frac{1}{3}\right)$$

110. $q^2 - \frac{1}{4} = q^2 - \left(\frac{1}{2}\right)^2$
$= \left(q + \frac{1}{2}\right)\left(q - \frac{1}{2}\right)$

111. $36m^2 - \frac{16}{25} = (6m)^2 - \left(\frac{4}{5}\right)^2$
$= \left(6m + \frac{4}{5}\right)\left(6m - \frac{4}{5}\right)$

112. $100b^2 - \frac{4}{49} = (10b)^2 - \left(\frac{2}{7}\right)^2$
$= \left(10b + \frac{2}{7}\right)\left(10b - \frac{2}{7}\right)$

113. $x^2 - 0.64 = x^2 - (0.8)^2$
$= (x + 0.8)(x - 0.8)$

114. $y^2 - 0.36 = y^2 - (0.6)^2$
$= (y + 0.6)(y - 0.6)$

115. t^2 is a perfect square, and $\frac{1}{4}$ is a perfect square since $\frac{1}{2} \cdot \frac{1}{2} = \frac{1}{4}$. The middle term is twice the product of t and $\frac{1}{2}$, or $t = 2(t)\left(\frac{1}{2}\right)$.

Therefore, $t^2 + t + \frac{1}{4} = \left(t + \frac{1}{2}\right)^2$.

116. The first and last terms are perfect squares, m^2 and $\left(\frac{1}{3}\right)^2$. The trinomial is a perfect square, since the middle term is $2 \cdot m \cdot \left(\frac{1}{3}\right) = \frac{2}{3}m.$

Therefore, $m^2 + \frac{2}{3}m + \frac{1}{9} = \left(m + \frac{1}{3}\right)^2$.

117. The first and last terms are perfect squares, x^2 and $(-0.9)^2$. The trinomial is a perfect square, since the middle term is $2 \cdot x \cdot (-0.9) = -1.8x$. Therefore, $x^2 - 1.8x + 0.81 = (x - 0.9)^2$.

118. The first and last terms are perfect squares, y^2 and $(-0.7)^2$. The trinomial is a perfect square, since the middle term is $2(y)(-0.7) = -1.4y$. Therefore, $y^2 - 1.4y + 0.49 = (y - 0.7)^2$.

119. Factor $x^3 + \frac{1}{8}$ as the sum of cubes.

$x^3 + \frac{1}{8} = x^3 + \left(\frac{1}{2}\right)^3$
$= \left(x + \frac{1}{2}\right)\left[x^2 - \frac{1}{2}x + \left(\frac{1}{2}\right)^2\right]$
$= \left(x + \frac{1}{2}\right)\left(x^2 - \frac{1}{2}x + \frac{1}{4}\right)$

120. Factor $x^3 + \frac{1}{64}$ as the sum of cubes.

$x^3 + \frac{1}{64} = x^3 + \left(\frac{1}{4}\right)^3$
$= \left(x + \frac{1}{4}\right)\left[x^2 - \frac{1}{4}x + \left(\frac{1}{4}\right)^2\right]$
$= \left(x + \frac{1}{4}\right)\left(x^2 - \frac{1}{4}x + \frac{1}{16}\right)$

121. Factor as the difference of squares. Substitute into the rule using $x = m + n$ and $y = m - n$.

$(m + n)^2 - (m - n)^2$
$= [(m + n) + (m - n)] \cdot [(m + n) - (m - n)]$
$= (2m)(2n)$
$= 4mn$

122. Factor as the difference of two cubes. Substitute into the rule using $x = a - b$ and $y = a + b$.

$(a-b)^3 - (a+b)^3$
$= [(a-b) - (a+b)][(a-b)^2$
$+ (a-b)(a+b) + (a+b)^2]$
$= (a - b - a - b)[(a^2 - 2ab + b^2)$
$+ (a^2 - b^2) + (a^2 + 2ab + b^2)]$
$= -2b(3a^2 + b^2)$

123. This expression can be factored by grouping.

$m^2 - p^2 + 2m + 2p$
$= (m + p)(m - p) + 2(m + p)$
$= (m + p)(m - p + 2)$

124. First factor out the GCF, 3.

$3r-3k+3r^2-3k^2=3\left(r-k+r^2-k^2\right)$

Now factor $r-k+r^2-k^2$ by grouping, noting that r^2-k^2 is the difference of squares.

$$\begin{aligned} r-k+r^2-k^2 &= 1(r-k)+(r+k)(r-k) \\ &= (r-k)[1+(r+k)] \\ &= (r-k)(1+r+k) \end{aligned}$$

Therefore,

$3r-3k+3r^2-3r^2=3(r-k)(1+r+k).$

Summary Exercises Recognizing and Applying Factoring Strategies

Now Try Exercises

N1. Factor out the GCF, 3. Then factor by grouping.

$$\begin{aligned} 24m^2-42my+9y^2 &= 3\left(8m^2-13my+3y^2\right) \\ &= 3(4m-y)(2m-3y) \end{aligned}$$

Exercises

1. $$\begin{aligned} 12x^2+20x+8 &= 4\left(3x^2+5x+2\right) \\ &= 4(3x+2)(x+1) \end{aligned}$$

Match with choice G. [Factor out the GCF; then factor a trinomial by grouping or trial and error.]

3. $16m^2n+24mn-40mn^2=8mn(2m+3-5n)$

Match with choice A. [Factor out the GCF; no further factoring is possible.]

5. The first and last terms are perfect squares, $(6p)^2$ and $(-5q)^2$. The middle term is $2(6p)(-5q)=-60pq$. Therefore, $36p^2-60pq+25q^2=(6p-5q)^2$. Match with choice E. [Factor a perfect square trinomial.]

7. Factor $8r^3-125$ as the difference of cubes.

$$\begin{aligned} 8r^3-125 &= (2r)^3-5^3 \\ &= (2r-5)\left[(2r)^2+2r(5)+5^2\right] \\ &= (2r-5)\left(4r^2+10r+25\right) \end{aligned}$$

Match with choice C. [Factor a difference of cubes.]

9. Match with choice I. [The polynomial is prime.]

11. $a^2-4a-12=(a-6)(a+2)$

13. $$\begin{aligned} 6y^2-6y-12 &= 6\left(y^2-y-2\right) \\ &= 6(y-2)(y+1) \end{aligned}$$

15. Factor out the GCF.

$6a+12b+18c=6(a+2b+3c)$

17. $p^2-17p+66=(p-11)(p-6)$

19. Use the grouping method. Look for two integers whose product is $10(-6)=-60$ and whose sum is -7. The integers are -12 and 5.

$$\begin{aligned} 10z^2-7z-6 &= 10z^2-12z+5z-6 \\ &= 2z(5z-6)+1(5z-6) \\ &= (5z-6)(2z+1) \end{aligned}$$

21. $17x^3y^2+51xy=17xy\left(x^2y+3\right)$

23. $$\begin{aligned} 8a^5-8a^4-48a^3 &= 8a^3\left(a^2-a-6\right) \\ &= 8a^3(a-3)(a+2) \end{aligned}$$

25. $z^2-3za-10a^2=(z-5a)(z+2a)$

27. $$\begin{aligned} x^2-4x-5x+20 &= x(x-4)-5(x-4) \\ &= x^2-4x-5x+20 \end{aligned}$$

29. $6n^2-19n+10=(3n-2)(2n-5)$

31. $16x+20=4(4x+5)$

33. Factor by grouping. Find two integers whose product is $6(-4)=-24$ and whose sum is -5. The integers are -8 and 3.

$$\begin{aligned} 6y^2-5y-4 &= 6y^2-8y+3y-4 \\ &= 2y(3y-4)+1(3y-4) \\ &= (3y-4)(2y+1) \end{aligned}$$

35. $6z^2+31z+5=(6z+1)(z+5)$

37. $$\begin{aligned} 4k^2-2k+9 &= (2k)^2-2\cdot 2k\cdot 3+3^2 \\ &= (2k-3)^2 \end{aligned}$$

39. $$\begin{aligned} 54m^2-24z^2 &= 6\left(9m^2-4z^2\right) \\ &= 6\left[(3m)^2-(2z)^2\right] \\ &= 6(3m+2z)(3m-2z) \end{aligned}$$

41. $3k^2 + 4k - 4 = (3k - 2)(k + 2)$

43. $14k^3 + 7k^2 - 70k = 7k\left(2k^2 + k - 10\right)$
$= 7k(2k + 5)(k - 2)$

45. $y^4 - 16 = \left(y^2\right)^2 - 4^2$
$= \left(y^2 + 4\right)\left(y^2 - 4\right)$
$= \left(y^2 + 4\right)(y + 2)(y - 2)$

47. $8m - 16m^2 = 8m(1 - 2m)$

49. Factor $z^3 - 8$ as the difference of cubes.
$z^3 - 8 = z^3 - 2^3$
$= (z - 2)\left(z^2 + z \cdot 2 + 2^2\right)$
$= (z - 2)\left(z^2 + 2z + 4\right)$

51. $k^2 + 9$ cannot be factored because it is the sum of squares with no GCF. The expression is prime.

53. $32m^9 + 16m^5 + 24m^3 = 8m^3\left(4m^6 + 2m^2 + 3\right)$

55. $16r^2 + 24rm + 9m^2 = (4r)^2 + 2 \cdot 4r \cdot 3m + (3m)^2$
$= (4r + 3m)^2$

57. Factor by grouping. Look for two integers whose product is $15(-14) = -210$ and whose sum is 11. The integers are 21 and -10.
$15h^2 + 11hg - 14g^2$
$= 15h^2 + 21hg - 10hg - 14g^2$
$= 3h(5h + 7g) - 2g(5h + 7g)$
$= (5h + 7g)(3h - 2g)$

59. $k^2 - 11k + 30 = (k - 5)(k - 6)$

61. $3k^3 - 12k^2 - 15k = 3k\left(k^2 - 4k - 5\right)$
$= 3k(k - 5)(k + 1)$

63. $1000p^3 + 27 = (10p)^3 + 3^3$
$= (10p + 3)\left[(10p)^2 - 10p \cdot 3 + 3^2\right]$
$= (10p + 3)\left(100p^2 - 30p + 9\right)$

65. $6 + 3m + 2p + mp = (6 + 3m) + (2p + mp)$
$= 3(2 + m) + p(2 + m)$
$= (2 + m)(3 + p)$

67. $16z^2 - 8z + 1 = (4z)^2 - 2 \cdot 4z \cdot 1 + 1^2$
$= (4z - 1)^2$

69. $108m^2 - 36m + 3 = 3\left(36m^2 - 12m + 1\right)$
$= 3(6m - 1)^2$

71. $x^2 - xy + y^2$ is prime. The middle term would have to be $+2xy$ or $-2xy$ in order to make this a perfect square trinomial.

73. $32z^3 + 56z^2 - 16z = 8z\left(4z^2 + 7z - 2\right)$
$= 8z(4z - 1)(z + 2)$

75. $64m^2 - 80mn + 25n^2 = (8n - 5n)^2$

77. $6a^2 + 10a - 4 = 2\left(3a^2 + 5a - 2\right)$
$= 2(3a - 1)(a + 2)$

79. The trinomial cannot be factored, it is prime.

81. $20 + 5m + 12n + 3mn = (4 + m)(5 + 3n)$

83. $8k^2 - 2kh - 3h^2 = 8k^2 - 6kh + 4kh - 3h^2$
$= 2k(4k - 3h) + h(4k - 3h)$
$= (4k - 3h)(2k + h)$

85. $2x^3 + 128 = 2\left(x^3 + 64\right)$
$= 2\left(x^3 + 4^3\right)$
$= 2(x + 4)\left(x^2 - x \cdot 4 + 4^2\right)$
$= 2(x + 4)\left(x^2 - 4x + 16\right)$

87. $10y^2 - 7yz - 6z^2 = 10y^2 - 12yz + 5yz - 6z^2$
$= 2y(5y - 6z) + z(5y - 6z)$
$= (5y - 6z)(2y + z)$

89. $8a^2 + 23ab - 3b^2 = 8a^2 + 24ab - ab - 3b^2$
$= 8a(a + 3b) - b(a + 3b)$
$= (a + 3b)(8a - b)$

6.5 Solving Quadratic Equations Using the Zero-Factor Property

Now Try Exercises

N1. **(a)** By the zero-factor property, either $x-4=0$ or $3x+1=0$. The equation $x-4=0$ gives the solution $x=4$; $3x+1=0$ gives $x=-\frac{1}{3}$. Check both solutions by substituting first 4 and then $-\frac{1}{3}$ for x in the original equation. The solution set is $\left\{-\frac{1}{3},4\right\}$.

(b) By the zero-factor property, either $y=0$ or $4y-5=0$, so the solutions are $y=0$ and $y=\frac{5}{4}$. Check each solution in the original equation. The solution set is $\left\{0,\frac{5}{4}\right\}$.

N2. Rewrite in standard form by adding $3t$ to each side and subtracting 18 from each side.

$$t^2+3t-18=0$$
$$(t-3)(t+6)=0$$
$$t-3=0 \quad \text{or} \quad t+6=0$$
$$t=3 \quad \text{or} \quad t=-6$$

The solution set is $\{-6, 3\}$.

N3.

$$10p^2+65p=35$$
$$10p^2+65p-35=0 \quad \text{Standard form}$$
$$5(2p^2+13p-7)=0 \quad \text{Factor out 5.}$$
$$2p^2+13p-7=0 \quad \text{Divide by 5.}$$
$$(2p-1)(p+7)=0 \quad \text{Factor.}$$
$$2p-1=0 \quad \text{or} \quad p+7=0$$
$$p=\frac{1}{2} \quad \text{or} \quad p=-7$$

Check the solutions by substituting in the original equation. The solution set is $\left\{-7,\frac{1}{2}\right\}$.

N4. **(a)**

$$9x^2-64=0$$
$$(3x+8)(3x-8)=0$$
$$3x+8=0 \quad \text{or} \quad 3x-8=0$$
$$3x=-8 \quad \text{or} \quad 3x=8$$
$$x=-\frac{8}{3} \quad \text{or} \quad x=\frac{8}{3}$$

Check the solutions by substituting in the original equation. The solution set is $\left\{-\frac{8}{3},\frac{8}{3}\right\}$.

(b)

$$m^2=5m$$
$$m^2-5m=0 \quad \text{Standard form}$$
$$m(m-5)=0 \quad \text{Factor.}$$

Apply the zero-factor property.

$$m=0 \quad \text{or} \quad m-5=0 \quad \text{Zero-factor prop.}$$
$$m=5 \quad \text{Solve.}$$

Check the solutions by substituting in the original equation. The solution set is $\{0, 5\}$.

(c) Write the equation in standard form.

$$6p^2-p=2$$
$$6p^2-p-2=0$$
$$(3p-2)(2p+1)=0$$
$$3p-2=0 \quad \text{or} \quad 2p+1=0$$
$$3p=2 \quad \text{or} \quad 2p=-1$$
$$p=\frac{2}{3} \qquad p=-\frac{1}{2}$$

Check the solutions by substituting in the original equation. The solution set is $\left\{-\frac{1}{2},\frac{2}{3}\right\}$.

N5. Factor $4x^2-4x+1$ as a perfect square trinomial.

$$(2x-1)^2=0$$

Set the factor $2x-1$ equal to 0 and solve.

$$2x-1=0$$
$$2x=1$$
$$x=\frac{1}{2}$$

Check this solution by substituting it in the original equation. The solution set is $\left\{\frac{1}{2}\right\}$.

N6. **(a)**

$$3x^3 - 27x = 0$$
$$3x(x^2 - 9) = 0$$
$$3x(x+3)(x-3) = 0$$
$$3x = 0 \quad \text{or} \quad x+3=0 \quad \text{or} \quad x-3=0$$
$$x = 0 \quad \text{or} \quad x=-3 \quad \text{or} \quad x=3$$

Check the solutions by substituting in the original equation. The solution set is $\{-3, 0, 3\}$.

(b) We know that $a = \frac{1}{3}$ is the solution of $3a - 1 = 0$, so we need to find the solutions of $2a^2 - 5a - 12 = 0$.

$$2a^2 - 5a - 12 = 0$$
$$(2a+3)(a-4) = 0 \quad \text{Factor.}$$

Apply the zero-factor property.

$$2a+3=0 \quad \text{or} \quad a-4=0$$
$$a = -\frac{3}{2} \qquad a = 4$$

The solutions of the original equation are $-\frac{3}{2}, \frac{1}{3}$, and 4. Check the solutions by substituting in the original equation. The solution set is $\left\{-\frac{3}{2}, \frac{1}{3}, 4\right\}$.

N7.

$$x(4x-9) = (x-2)^2 + 24$$
$$4x^2 - 9x = x^2 - 4x + 4 + 24 \quad \text{Multiply.}$$
$$3x^2 - 5x - 28 = 0 \quad \text{St. form}$$
$$(3x+7)(x-4) = 0 \quad \text{Factor.}$$
$$3x+7=0 \quad \text{or} \quad x-4=0 \quad \text{Zero-factor prop.}$$
$$x = -\frac{7}{3} \quad \text{or} \quad x = 4 \quad \text{Solve.}$$

Check the solutions by substituting in the original equation. The solution set is $\left\{-\frac{7}{3}, 4\right\}$.

Exercises

1. A quadratic equation is an equation that can be put in the form $\underline{ax^2 + bx + c} = 0$.

3. If the product of two numbers is 0, then at least one of the numbers is <u>0</u>. This is called the <u>zero-factor</u> property.

5. The equation $x^3 + x^2 + x = 0$ is not a quadratic equation, because <u>the term of greatest degree is greater than 2. (It is cubic.)</u>

7. **(a)** $2x - 5 = 6$ can be written as $2x - 11 = 0$, so it is a linear equation.

(b) $x^2 - 5 = -4$ can be written as $x^2 - 1 = 0$, so it is a quadratic equation.

(c) $x^2 + 2x - 3 = 2x^2 - 2$ can be written as $-x^2 + 2x - 1 = 0$, so it is a quadratic equation.

(d) $5^2 x + 2 = 0$ can be written as $25x + 2 = 0$, so it is a linear equation.

9. The variable x is another factor to set equal to 0, so the solution set is $\left\{0, \frac{1}{7}\right\}$.

Not including 0 as a solution is what went wrong.

11. By the zero-factor property, the only way that the product of these two factors can be zero is if at least one of the factors is zero.

$$x+5=0 \quad \text{or} \quad x-2=0$$

Solve each of these linear equations.

$$x=-5 \quad \text{or} \quad x=2$$

Check $x=-5$: $\quad 0(-7)=0$ True

Check $x=2$: $\quad 7(0)=0$ True

The solution set is $\{-5, 2\}$.

13. Set each factor equal to zero, and solve the resulting linear equations.

$$2m-7=0 \quad \text{or} \quad m-3=0$$
$$2m=7 \quad \text{or} \quad m=3$$
$$m=\frac{7}{2}$$

The solution set is $\left\{3, \frac{7}{2}\right\}$.

15. Set each factor equal to zero, and solve the resulting linear equations.

$$2x+1=0 \quad \text{or} \quad 6x-1=0$$
$$2x=-1 \quad \text{or} \quad 6x=1$$
$$x=-\frac{1}{2} \qquad x=\frac{1}{6}$$

The solution set is $\left\{-\frac{1}{2}, \frac{1}{6}\right\}$.

17. Set each factor equal to zero, and solve the resulting linear equations.
$t = 0$ or $6t + 5 = 0$
$6t = -5$
$t = -\frac{5}{6}$
The solution set is $\left\{-\frac{5}{6}, 0\right\}$.

19. Set each factor equal to zero, and solve the resulting linear equations.
$2x = 0$ or $3x - 4 = 0$
$x = 0$ or $3x = 4$
$x = \frac{4}{3}$
The solution set is $\left\{0, \frac{4}{3}\right\}$.

21. Set each factor equal to zero, and solve the resulting linear equations.
$x - 6 = 0$ or $x - 6 = 0$
$x = 6$ or $x = 6$
6 is called a double solution for $(x-6)^2 = 0$ because it occurs twice when the equation is solved. The solution set is $\{6\}$.

23. Factor the polynomial.
$(y+2)(y+1) = 0$
Set each factor equal to 0.
$y + 2 = 0$ or $y + 1 = 0$
Solve each equation.
$y = -2$ or $y = -1$
Check these solutions by substituting -2 for y and then -1 for y in the original equation.
$y^2 + 3y + 2 = 0$
$(-2)^2 + 3(-2) + 2 \stackrel{?}{=} 0$ Let $y = -2$.
$4 - 6 + 2 \stackrel{?}{=} 0$
$-2 + 2 = 0$ True
$y^2 + 3y + 2 = 0$
$(-1)^2 + 3(-1) + 2 \stackrel{?}{=} 0$ Let $y = -1$.
$1 - 3 + 2 \stackrel{?}{=} 0$
$-2 + 2 = 0$ True
The solution set is $\{-2, -1\}$.

25. Factor the polynomial.
$(y-1)(y-2) = 0$
Set each factor equal to 0.
$y - 1 = 0$ or $y - 2 = 0$
Solve each equation.
$y = 1$ or $y = 2$
The solution set is $\{1, 2\}$.

27. Write the equation in standard form.
$x^2 + 5x - 24 = 0$
Factor the polynomial.
$(x+8)(x-3) = 0$
Set each factor equal to 0.
$x + 8 = 0$ or $x - 3 = 0$
Solve each equation.
$x = -8$ or $x = 3$
The solution set is $\{-8, 3\}$.

29. Write the equation in standard form.
$x^2 - 2x - 3 = 0$
$(x+1)(x-3) = 0$
$x + 1 = 0$ or $x - 3 = 0$
$x = -1$ or $x = 3$
The solution set is $\{-1, 3\}$.

31. Write the equation in standard form.
$z^2 + 3z + 2 = 0$
Factor the polynomial.
$(z+2)(z+1) = 0$
Set each factor equal to 0.
$z + 2 = 0$ or $z + 1 = 0$
$z = -2$ or $z = -1$
The solution set is $\{-2, -1\}$.

33. Factor $m^2 + 8m + 16$ as a perfect square trinomial.
$(m+4)^2 = 0$
Set the factor $m + 4$ equal to 0 and solve.
$m + 4 = 0$
$m = -4$
The solution set is $\{-4\}$.

35. Factor the polynomial.

$(3x-1)(x+2)=0$

Set each factor equal to 0.

$3x-1=0 \quad \text{or} \quad x+2=0$

$3x=1 \quad \text{or} \quad x=-2$

$x=\frac{1}{3}$

The solution set is $\left\{-2, \frac{1}{3}\right\}$.

37. $12p^2=8-10p$

$12p^2+10p-8=0$ Standard form

$2(6p^2+5p-4)=0$ Factor out 2.

$2(3p+4)(2p-1)=0$

$3p+4=0 \quad \text{or} \quad 2p-1=0$

$3p=-4 \quad \text{or} \quad 2p=1$

$p=-\frac{4}{3} \quad \text{or} \quad p=\frac{1}{2}$

The solution set is $\left\{-\frac{4}{3}, \frac{1}{2}\right\}$.

39. $9s^2+12s=-4$

$9s^2+12s+4=0$

$(3s+2)^2=0$

Set the factor $3s+2$ equal to 0 and solve.

$3s+2=0$

$3s=-2$

$s=-\frac{2}{3}$

The solution set is $\left\{-\frac{2}{3}\right\}$.

41. $y^2-9=0$

$(y+3)(y-3)=0$

$y+3=0 \quad \text{or} \quad y-3=0$

$y=-3 \quad \text{or} \quad y=3$

The solution set is $\{-3, 3\}$.

43. $16x^2-49=0$

$(4x+7)(4x-7)=0$

$4x+7=0 \quad \text{or} \quad 4x-7=0$

$4x=-7 \quad \text{or} \quad 4x=7$

$x=-\frac{7}{4} \quad \text{or} \quad x=\frac{7}{4}$

The solution set is $\left\{-\frac{7}{4}, \frac{7}{4}\right\}$.

45. $n^2=121$

$n^2-121=0$

$(n+11)(n-11)=0$

$n+11=0 \quad \text{or} \quad n-11=0$

$n=-11 \quad \text{or} \quad n=11$

The solution set is $\{-11, 11\}$.

47. $x^2+6x=0$

$x(x+6)=0$

$x=0 \quad \text{or} \quad x+6=0$

$x=-6$

The solution set is $\{-6, 0\}$.

49. $x^2=7x$

$x^2-7x=0$

$x(x-7)=0$

$x=0 \quad \text{or} \quad x-7=0$

$x=0 \quad \text{or} \quad x=7$

Check $x=0$: $0=0$ True

Check $x=7$: $49=49$ True

The solution set is $\{0, 7\}$.

51. $6r^2=3r$

$6r^2-3r=0$

$3r(2r-1)=0$

$3r=0 \quad \text{or} \quad 2r-1=0$

$r=0 \quad \text{or} \quad 2r=1$

$r=\frac{1}{2}$

The solution set is $\left\{0, \frac{1}{2}\right\}$.

53.
$$x(x-7)=-10$$
$$x^2-7x=-10$$
$$x^2-7x+10=0$$
$$(x-2)(x-5)=0$$
$$x-2=0 \quad \text{or} \quad x-5=0$$
$$x=2 \quad \text{or} \quad x=5$$
The solution set is $\{2, 5\}$.

55.
$$3z(2z+7)=12$$
$$z(2z+7)=4 \quad \text{Divide by 3.}$$
$$2z^2+7z=4$$
$$2z^2+7z-4=0$$
$$(2z-1)(z+4)=0$$
$$2z-1=0 \quad \text{or} \quad z+4=0$$
$$2z=1 \quad \text{or} \quad z=-4$$
$$z=\frac{1}{2}$$
The solution set is $\left\{-4, \frac{1}{2}\right\}$.

57.
$$2y(y+13)=136$$
$$y(y+13)=68 \quad \text{Divide by 2.}$$
$$y^2+13y=68 \quad \text{Multiply.}$$
$$y^2+13y-68=0 \quad \text{Standard form}$$
$$(y+17)(y-4)=0 \quad \text{Factor.}$$
$$x+17=0 \quad \text{or} \quad y-4=0$$
$$y=-17 \quad \text{or} \quad y=4$$
Check $y=-17$:
$$2y(y+13)=136$$
$$2(-17)(-17+13)\stackrel{?}{=}136 \quad \text{Let } y=-17.$$
$$(-34)(-4)=136 \quad \text{True}$$
Check $y=4$:
$$2(4)(4+13)\stackrel{?}{=}136 \quad \text{Let } y=4.$$
$$8(17)=136 \quad \text{True}$$
The solution set is $\{-17, 4\}$.

59.
$$(x-8)(x+6)=6x$$
$$x^2-2x-48=6x$$
$$x^2-8x-48=0$$
$$(x-12)(x+4)=0$$
$$x-12=0 \quad \text{or} \quad x+4=0$$
$$x=12 \quad \text{or} \quad x=-4$$
The solution set is $\{-4, 12\}$.

61.
$$(x+4)(x+7)=10$$
$$x^2+11x+28=10$$
$$x^2+11x+18=0$$
$$(x+9)(x+2)=0$$
$$x+9=0 \quad \text{or} \quad x+2=0$$
$$x=-9 \quad \text{or} \quad x=-2$$
The solution set is $\{-9, -2\}$.

63. To factor the polynomial, begin by factoring out the greatest common factor.
$$y\left(9y^2-49\right)=0$$
Now factor $9y^2-49$ as the difference of squares.
$$y(3y+7)(3y-7)=0$$
Set each of the three factors equal to 0 and solve.
$$y=0 \quad \text{or} \quad 3y+7=0 \quad \text{or} \quad 3y-7=0$$
$$3y=-7 \qquad 3y=7$$
$$y=0 \quad \text{or} \quad y=-\frac{7}{3} \quad \text{or} \quad y=\frac{7}{3}$$
The solution set is $\left\{-\frac{7}{3}, 0, \frac{7}{3}\right\}$.

65.
$$r^3-2r^2-8r=0$$
$$r\left(r^2-2r-8\right)=0 \quad \text{Factor out } r.$$
$$r(r-4)(r+2)=0 \quad \text{Factor.}$$
Set each factor equal to zero and solve.
$$r=0 \quad \text{or} \quad r-4=0 \quad \text{or} \quad r+2=0$$
$$r=0 \quad \text{or} \quad r=4 \quad \text{or} \quad r=-2$$
The solution set is $\{-2, 0, 4\}$.

67. $x^3+x^2-20x=0$

$x(x^2+x-20)=0$ Factor out x.

$x(x+5)(x-4)=0$ Factor.

Set each factor equal to zero and solve.

$x=0$ or $x+5=0$ or $x-4=0$

$x=0$ or $x=-5$ or $x=4$

The solution set is $\{-5, 0, 4\}$.

69. $4x^3-18x^2+8x=0$

$2x(2x^2-9x+4)=0$

$2x(2x-1)(x-4)=0$

The solution set is $\left\{0, \frac{1}{2}, 4\right\}$.

71. Rewrite with all terms on the left side.

$r^4-2r^3-15r^2=0$

$r^2(r^2-2r-15)=0$ Factor out r^2.

$r^2(r-5)(r+3)=0$ Factor.

Set each factor equal to zero and solve.

$r^2=0$ or $r-5=0$ or $r+3=0$

$r=0$ or $r=5$ or $r=-3$

The solution set is $\{-3, 0, 5\}$.

73. Begin by factoring $3r^2-16r+5$.

$(2r+5)(3r-1)(r-5)=0$

Set each of the three factors equal to 0 and solve the resulting equations.

$2r+5=0$ or $3r-1=0$ or $r-5=0$

$2r=-5$ $\quad 3r=1$

$r=-\frac{5}{2}$ or $r=\frac{1}{3}$ or $r=5$

The solution set is $\left\{-\frac{5}{2}, \frac{1}{3}, 5\right\}$.

75. $(2x+7)(x^2+2x-3)=0$

$(2x+7)(x+3)(x-1)=0$

$2x+7=0$ or $x+3=0$ or $x-1=0$

$2x=-7$

$x=-\frac{7}{2}$ or $x=-3$ or $x=1$

The solution set is $\left\{-\frac{7}{2}, -3, 1\right\}$.

77. $3x(x+1)=(2x+3)(x+1)$

$3x^2+3x=2x^2+5x+3$

$x^2-2x-3=0$

$(x+1)(x-3)=0$

$x+1=0$ or $x-3=0$

$x=-1$ or $x=3$

The solution set is $\{-1, 3\}$.

Alternatively, we could begin by moving all the terms to the left side and then factoring out $x+1$.

$3x(x+1)-(2x+3)(x+1)=0$

$(x+1)[3x-(2x+3)]=0$

$(x+1)(x-3)=0$

The rest of the solution is the same.

79. $x^2+(x+1)^2=(x+2)^2$

$x^2+x^2+2x+1=x^2+4x+4$

$x^2-2x-3=0$

$(x+1)(x-3)=0$

$x+1=0$ or $x-3=0$

$x=-1$ or $x=3$

The solution set is $\{-1, 3\}$.

81. $(2x)^2=(2x+4)^2-(x+5)^2$

$4x^2=4x^2+16x+16-(x^2+10x+25)$

$4x^2=4x^2+16x+16-x^2-10x-25$

$4x^2=3x^2+6x-9$

$0=-x^2+6x-9$

$0=x^2-6x+9$

$0=(x-3)(x-3)$

Set the factor $x-3$ equal to 0 and solve.

$x-3=0$

$x=3$

The solution set is $\{3\}$.

83. First square the binomials.

$(x+3)^2-(2x-1)^2=0$

$(x^2+6x+9)-(4x^2-4x+1)=0$

Then combine like terms.

$x^2+6x+9-4x^2+4x-1=0$

$-3x^2+10x+8=0$

$-1(3x^2-10x-8)=0$ Factor out -1.

$-1(3x+2)(x-4)=0$ Factor.

Set each factor equal to zero and solve.

$3x+2=0$ or $x-4=0$

$3x=-2$

$x=-\frac{2}{3}$ or $x=4$

The solution set is $\left\{-\frac{2}{3}, 4\right\}$.

Alternatively we could begin by factoring the left side as the difference of two squares.

$(x+3)^2-(2x-1)^2=0$

$[(x+3)+(2x-1)][(x+3)-(2x-1)]=0$

$[3x+2][-x+4]=0$

The same solution set is obtained.

85. Rewrite with all terms on the right side.

$0=-6p^2(p+1)-5p(p+1)+4(p+1)$

$0=6p^2(p+1)+5p(p+1)-4(p+1)$

$0=(p+1)(6p^2+5p-4)$

$0=(p+1)(2p-1)(3p+4)$

Set each factor equal to zero and solve.

$p+1=0$ or $2p-1=0$ or $3p+4=0$

$2p=1$ $\quad$ $3p=-4$

$p=-1$ or $p=\frac{1}{2}$ or $p=-\frac{4}{3}$

The solution set is $\left\{-\frac{4}{3}, -1, \frac{1}{2}\right\}$.

87. **(a)** Use the equation $d=16t^2$.

$t=2$: $d=16(2)^2=16(4)=64$

$t=3$: $d=16(3)^2=16(9)=144$

$d=256$: $256=16t^2; 16=t^2; t=4$

$d=576$: $576=16t^2; 36=t^2; t=6$

t in seconds	0	1	2	3	4	6
d in feet	0	16	64	144	256	576

(b) When $t=0$, $d=0$. No time has elapsed, so the object hasn't fallen (been released) yet.

6.6 Applications of Quadratic Equations

Now Try Exercises

N1. *Step 2*

Let $x=$ the length of one leg.

Then $x-4=$ the length of the other leg.

Step 3 $A=\frac{1}{2}bh$

$6=\frac{1}{2}x(x-4)$

Step 4

Solve the equation.

$12=x(x-4)$ Multiply by 2.

$12=x^2-4x$ Distributive prop.

$x^2-4x-12=0$ Standard form

$(x+2)(x-6)=0$ Factor.

$x+2=0$ or $x-6=0$ Zero-factor prop.

$x=-2$ or $x=6$

Step 5

Because a triangle's side cannot be negative, $x=6$ and $x-4=2$. The lengths of the legs are 2 feet and 6 feet.

Step 6

The length of one leg is 4 meters less than the length of the other leg, and the area is $\frac{1}{2}(6)(2)=6\text{ m}^2$, as required.

N2. Let $x=$ the first integer.

Then $x+1=$ the second integer and $x+2=$ the third integer.

From the given information, we write the equation

$x(x+1)=8(x+2)+2$.

Solve this equation.

$x^2+x=8x+16+2$

$x^2-7x-18=0$

$(x+2)(x-9)=0$

$x+2=0$ or $x-9=0$

$x=-2$ or $x=9$

If $x=-2$, then $x+1=-1$ and $x+2=0$.

If $x=9$, then $x+1=10$ and $x+2=11$. There are two sets of integers that satisfy the statement of the problem: $-2, -1, 0$, and $9, 10, 11$.

N3. Let $x =$ the length of the shorter leg of the right triangle.
Then $x+7 =$ the length of the longer leg and $x+8 =$ the length of the hypotenuse.
Use the Pythagorean theorem, substituting x for a, $x+7$ for b, and $x+8$ for c.

$$a^2+b^2=c^2$$
$$x^2+(x+7)^2=(x+8)^2$$
$$x^2+x^2+14x+49=x^2+16x+64$$
$$2x^2+14x+49=x^2+16x+64$$
$$x^2-2x-15=0$$
$$(x+3)(x-5)=0$$
$$x+3=0 \quad \text{or} \quad x-5=0$$
$$x=-3 \quad \text{or} \quad x=5$$

Discard -3 since the length of a side of a triangle cannot be negative. The length of the shorter leg is 5 feet, the length of the longer leg is $x+7=5+7=12$ feet, and the length of the hypotenuse is $x+8=5+8=13$ feet.

N4. To find how long it will take for the ball to reach a height of 50 feet, let $h=50,$ and solve the resulting equation. (For convenience, we reverse the sides of the equation.)

$$-16t^2+180t+6=50$$
$$-16t^2+180t-44=0 \quad \text{Standard form}$$
$$4t^2-45t+11=0 \quad \text{Divide by } -4.$$
$$(4t-1)(t-11)=0 \quad \text{Factor.}$$
$$4t-1=0 \quad \text{or} \quad t-11=0$$
$$t=\frac{1}{4} \quad \text{or} \quad t=11$$

The ball will reach a height of 50 feet after $\frac{1}{4}$ second (on the way up) and 11 seconds (on the way down).

N5. For 2000, $x=2000-1930=70.$
Let $x=70.$

$$y=0.009665x^2-0.4942x+15.12$$
$$y=0.009665(70)^2-0.4942(70)+15.12$$
$$=27.8845\approx 27.9$$

According to the model, the foreign-born population of the United States in 2000, to the nearest tenth of a million, was about 27.9 million. The actual value from the table is 28.4 million, so the answer using the model is slightly high.

Exercises

1. *Step 1:* Read the problem carefully.
Step 2: Assign a variable to represent the unknown value.
Step 3: Write a(n) equation using the variable expressions (s).
Step 4: Solve the equation.
Step 5: State the answer.
Step 6: Check the answer in the words of the original problem.

3. Formula for the area of a parallelogram: $A=bh$

Step 3: $45=(2x+1)(x+1)$

Step 4: $x=4$ or $x=-\frac{11}{2}$

Step 5: base: 9 units; height: 5 units

Step 6: $9\cdot 5=45$

5. Formula for the area of a rectangle: $A=LW$

Step 3: $80=(x+8)(x-8)$

Step 4: $x=12$ or $x=-12$

Step 5: length: 20 units; width: 4 units

Step 6: $20\cdot 4=80$

7. *Step 2*
Let $x=$ the width of the case.
Then $x+2=$ the length of the case.
Step 3
$A=LW$
$168=(x+2)x$ Substitute.
Step 4

$$168=x^2+2x \quad \text{Multiply.}$$
$$x^2+2x-168=0 \quad \text{Standard form}$$
$$(x+14)(x-12)=0 \quad \text{Factor.}$$

Apply the zero-factor property and solve.

$$x+14=0 \quad \text{or} \quad x-12=0$$
$$x=-14 \quad \text{or} \quad x=12$$

Step 5
Because a width cannot be negative, $x=12$ and $x+2=14.$ The width is 12 cm, and the length is 14 cm.
Step 6
The length is 2 cm more than the width, and the area is $14(12)=168\ \text{cm}^2$, as required.

9. Let $h =$ the height of the triangle.
Then $2h+2 =$ the base of the triangle.
The area of the triangle is 30 square inches.

$$A = \frac{1}{2}bh$$
$$30 = \frac{1}{2}(2h+2)\cdot h$$
$$60 = (2h+2)h$$
$$60 = 2h^2 + 2h$$
$$0 = 2h^2 + 2h - 60$$
$$0 = 2(h^2 + h - 30)$$
$$0 = 2(h+6)(h-5)$$
$$h+6=0 \quad \text{or} \quad h-5=0$$
$$h=-6 \quad \text{or} \quad h=5$$

The solution $h = -6$ must be discarded since a triangle cannot have a negative height. Thus, $h = 5$ and $2h+2 = 2(5)+2 = 12.$

The height is 5 inches, and the base is 12 inches.

11. Let $x =$ the width of the aquarium.
Then $x+3 =$ the height of the aquarium.
Use the formula for the volume of a rectangular box.

$$V = LWH$$
$$2730 = 21x(x+3)$$
$$130 = x(x+3) \qquad \text{Divide by 21.}$$
$$130 = x^2 + 3x$$
$$0 = x^2 + 3x - 130$$
$$0 = (x+13)(x-10)$$
$$x+13=0 \quad \text{or} \quad x-10=0$$
$$x=-13 \quad \text{or} \quad x=10$$

We discard -13 because the width cannot be negative. The width is 10 inches. The height is $10+3=13$ inches.

13. *Step 2*
Let $x =$ the width of the monitor.
Then $x+3 =$ the the length of the monitor.
The area is $x(x+3)$.
Step 3
If the length were doubled $[2(x+3)]$, and if the width were decreased by 1 in. $[x-1]$, the area would be increased by 150 in.^{2} $[x(x+3)+150]$.

Write an equation.

$$LW = A$$
$$[2(x+3)](x-1) = x(x+3)+150$$

Step 4

$$(2x+6)(x-1) = x^2 + 3x + 150$$
$$2x^2 + 4x - 6 = x^2 + 3x + 150$$
$$x^2 + x - 156 = 0$$
$$(x+13)(x-12) = 0$$
$$x+13=0 \quad \text{or} \quad x-12=0$$
$$x=-13 \quad \text{or} \quad x=12$$

Step 5
Reject -13, so the width is 12 inches, and the length is $12+3=15$ inches.
Step 6
The length is 3 inches more than the width. The area of the monitor is $15(12)=180$ in.^{2}.
Doubling the length and decreasing the width by 1 inch gives us an area of $30(11)=330$ in.^{2}, which is 150 in.^{2} more than the area of the original monitor, as required.

15. Let $x =$ the length of a side of the square painting. Then $x-2 =$ the length of a side of the square mirror.
Since the formula for the area of a square is $A = s^2$, the area of the painting is x^2, and the area of the mirror is $(x-2)^2$. The difference between their areas is 32, so

$$x^2 - (x-2)^2 = 32$$
$$x^2 - (x^2 - 4x + 4) = 32$$
$$x^2 - x^2 + 4x - 4 = 32$$
$$4x - 4 = 32$$
$$4x = 36$$
$$x = 9.$$

The length of a side of the painting is 9 feet. The length of a side of the mirror is $9-2=7$ feet.
Check: $9^2 - 7^2 = 81 - 49 = 32$

17. Let $x =$ the first volume number.
Then $x+1 =$ the second volume number.
The product of the numbers is 420.

$$x(x+1)=420$$
$$x^2+x-420=0$$
$$(x-20)(x+21)=0$$
$$x-20=0 \quad \text{or} \quad x+21=0$$
$$x=20 \quad \text{or} \quad x=-21$$

The volume number cannot be negative, so we reject -21. The volume numbers are 20 and $x+1=20+1=21$.

19. Let $x =$ the first integer.
Then $x+1 =$ the second integer,
and $x+2 =$ the third integer.

The product of the second and third	is	2	more than	10 times the first.
↓	↓	↓	↓	↓
$(x+1)(x+2)$	$=$	2	$+$	$10x$

$$x^2+3x+2=2+10x$$
$$x^2-7x=0$$
$$x(x-7)=0$$
$$x=0 \quad \text{or} \quad x-7=0$$
$$x=7$$

If $x=0$, then $x+1=1$, and $x+2=2$.
If $x=7$, then $x+1=8$, and $x+2=9$.
So there are two sets of consecutive integers that satisfy the condition: 0, 1, 2 or 7, 8, 9.

Check 0, 1, 2: $1(2)\stackrel{?}{=}2+10(0)$
$2=2+0$ True

Check 7, 8, 9: $8(9)\stackrel{?}{=}2+10(7)$
$72=2+70$ True

21. Let $x =$ the first odd integer.
Then $x+2 =$ the second odd integer.

$$3(x+x+2)+15=x(x+2)$$
$$6x+6+15=x^2+2x$$
$$6x+21=x^2+2x$$
$$-x^2+4x+21=0$$
$$x^2-4x-21=0$$
$$(x-7)(x+3)=0$$
$$x-7=0 \quad \text{or} \quad x+3=0$$
$$x=7 \qquad x=-3$$

If $x=7$, $x+2=9$, and if $x=-3$, $x+2=-1$.
The two integers are -3 and -1, or 7 and 9.

23. Let $x =$ the first odd integer.
Then $x+2 =$ the second odd integer and $x+4 =$ the third odd integer.

$$3[x+(x+2)+(x+4)]=x(x+2)+18$$
$$3(3x+6)=x^2+2x+18$$
$$9x+18=x^2+2x+18$$
$$0=x^2-7x$$
$$0=x(x-7)$$
$$x=0 \quad \text{or} \quad x-7=0$$
$$x=7$$

We must discard 0 because it is even and the problem requires the integers to be odd. If $x=7$, $x+2=9$, and $x+4=11$. The three integers are 7, 9, and 11.

25. Let $x =$ the first even integer. Then $x+2$ and $x+4$ are the next two even integers.

$$x^2+(x+2)^2=(x+4)^2$$
$$x^2+x^2+4x+4=x^2+8x+16$$
$$x^2-4x-12=0$$
$$(x-6)(x+2)=0$$
$$x-6=0 \quad \text{or} \quad x+2=0$$
$$x=6 \quad \text{or} \quad x=-2$$

If $x=6$, $x+2=8$, and $x+4=10$.
If $x=-2$, $x+2=0$, and $x+4=2$.
The three integers are 6, 8, 10 or -2, 0, 2.

27. Let $x =$ the length of the longer leg of the right triangle. Then $x+1 =$ the length of the hypotenuse, and $x-7 =$ the length of the shorter leg.
Use the Pythagorean theorem with $a=x$, $b=x-7$, and $c=x+1$.

$$a^2+b^2=c^2$$
$$x^2+(x-7)^2=(x+1)^2$$
$$x^2+\left(x^2-14x+49\right)=x^2+2x+1$$
$$2x^2-14x+49=x^2+2x+1$$
$$x^2-16x+48=0$$
$$(x-12)(x-4)=0$$
$$x-12=0 \quad \text{or} \quad x-4=0$$
$$x=12 \quad \text{or} \quad x=4$$

Discard 4 because if the length of the longer leg is 4 centimeters, by the conditions of the problem, the length of the shorter leg would be $4-7=-3$ centimeters, which is impossible. The length of the longer leg is 12 centimeters.

Check: $12^2+5^2=13^2; 169=169$ True

29. Let $x=$ the width. Then $x+5=$ the length, and $2x-5=$ the diagonal.
Use the Pythagorean theorem with x for a, $x+5$ for b, and $2x-5$ for c.

$$a^2+b^2=c^2$$
$$x^2+(x+5)^2=(2x-5)^2$$
$$x^2+\left(x^2+10x+25\right)=4x^2-20x+25$$
$$2x^2+10x+25=4x^2-20x+25$$
$$-2x^2+30x=0$$
$$2x^2-30x=0$$
$$2x(x-15)=0$$
$$2x=0 \quad \text{or} \quad x-15=0$$
$$x=0 \quad \text{or} \quad x=15$$

Thus, the width is 15 inches. The length is $x+5=20$ inches, and the diagonal is $2x-5=25$ inches.

Check: $15^2+20^2=25^2; 625=625$ True

31. Let $x=$ Alan's distance from home. Then $x+1=$ the distance between Tram and Alan.
Refer to the diagram in the textbook.
Use the Pythagorean theorem.

$$a^2+b^2=c^2$$
$$x^2+5^2=(x+1)^2$$
$$x^2+25=x^2+2x+1$$
$$24=2x$$
$$12=x$$

Alan is 12 miles from home.

Check: $12^2+5^2=13^2; 169=169$ True

33. Let $x=$ the the length of the ladder. Then $x-4=$ the distance from the bottom of the ladder to the building, and $x-2=$ the distance on the side of the building to the top of the ladder.

Substitute into the Pythagorean theorem.

$$a^2+b^2=c^2$$
$$(x-2)^2+(x-4)^2=x^2$$
$$x^2-4x+4+x^2-8x+16=x^2$$
$$x^2-12x+20=0$$
$$(x-10)(x-2)=0$$
$$x-10=0 \quad \text{or} \quad x-2=0$$
$$x=10 \quad \text{or} \quad x=2$$

The solution cannot be 2 because then a negative distance results. Thus, $x=10$, and the top of the ladder reaches $x-2=10-2=8$ feet up the side of the building.

Check: $8^2+6^2=10^2; 100=100$ True

35. $h=-16t^2+128t$

$$h=-16(1)^2+128(1) \quad \text{Let } t=1.$$
$$=-16+128$$
$$=112$$

After 1 second, the height is 112 feet.

37. $h=-16t^2+128t$

$$h=-16(4)^2+128(4) \quad \text{Let } t=4.$$
$$=-16(16)+512$$
$$=-256+512$$
$$=256$$

After 4 seconds, the height is 256 feet.

39. **(a)** Let $h=64$ in the given formula and solve for t.

$$h=-16t^2+32t+48$$
$$64=-16t^2+32t+48$$
$$16t^2-32t+16=0$$
$$16(t^2-2t+1)=0$$
$$16(t-1)^2=0$$
$$t-1=0$$
$$t=1$$

The height of the object will be 64 feet after 1 second.

(b) To find the time when the height is 60 feet, let $h = 60$ in the given equation and solve for t.

$$h = -16t^2 + 32t + 48$$
$$60 = -16t^2 + 32t + 48$$
$$16t^2 - 32t + 12 = 0$$
$$4(4t^2 - 8t + 3) = 0$$
$$4(2t - 1)(2t - 3) = 0$$
$$2t - 1 = 0 \quad \text{or} \quad 2t - 3 = 0$$
$$2t = 1 \quad \text{or} \quad 2t = 3$$
$$t = \frac{1}{2} \quad \text{or} \quad t = \frac{3}{2}$$

The height of the object is 60 feet after $\frac{1}{2}$ second (on the way up) and after $\frac{3}{2}$ or $1\frac{1}{2}$ seconds (on the way down).

(c) To find the time when the object hits the ground, let $h = 0$ and solve for t.

$$h = -16t^2 + 32t + 48$$
$$0 = -16t^2 + 32t + 48$$
$$16t^2 - 32t - 48 = 0$$
$$16(t^2 - 2t - 3) = 0$$
$$16(t + 1)(t - 3) = 0$$
$$t + 1 = 0 \quad \text{or} \quad t - 3 = 0$$
$$t = -1 \quad \text{or} \quad t = 3$$

We discard -1 because time cannot be negative. The object will hit the ground after 3 seconds.

(d) The negative solution, -1, does not make sense, since t represents time, which cannot be negative.

41. **(a)** $x = 2000 - 1990 = 10$

$x = 10$ corresponds to 2000.

(b) $y = 0.339x^2 + 8.50x - 8.26$

$y = 0.339(10)^2 + 8.50(10) - 8.26$

$y = 110.64 \approx 111$

The model indicates there were about 111 million cellular phone subscribers in 2000. The result using the model is more than 109 million, the actual number for 2000.

(c) $x = 2012 - 1990 = 22$

$x = 22$ corresponds to 2012.

(d) $y = 0.339(22)^2 + 8.50(22) - 8.26$

$y = 342.816 \approx 343$

The model indicates there were about 343 million cellular phone subscribers in 2012. The result using the model is more than 326 million, the actual number for 2012.

(e) $x = 2014 - 1990 = 24$

$x = 24$ corresponds to 2014.

(f) $y = 0.339(24)^2 + 8.50(24) - 8.26$

$y = 391.004 \approx 391$

The model gives an estimate of 391 million cellular phone subscribers in 2014.

Chapter 6 Review Exercises

1. $7t + 14 = 7 \cdot t + 7 \cdot 2 = 7(t + 2)$

2. $60z^3 + 30z = 30z \cdot 2z^2 + 30z \cdot 1$
$= 30z(2z^2 + 1)$

3. $-3x^3 + 6x^2 + 3x = 3x \cdot (-x^2) + 3x \cdot 2x + 3x \cdot 1$
$= 3x(-x^2 + 2x + 1)$
$= -3x(x^2 - 2x - 1)$

4. $100m^2n^3 - 50m^3n^4 + 150m^2n^2$
$= 50m^2n^2 \cdot 2n + 50m^2n^2 \cdot (-mn^2) + 50m^2n^2 \cdot 3$
$= 50m^2n^2(2n - mn^2 + 3)$

5. $2xy - 8y + 3x - 12 = (2xy - 8y) + (3x - 12)$
$= 2y(x - 4) + 3(x - 4)$
$= (x - 4)(2y + 3)$

6. $6y^2 + 9y + 4xy + 6x = (6y^2 + 9y) + (4xy + 6x)$
$= 3y(2y + 3) + 2x(2y + 3)$
$= (2y + 3)(3y + 2x)$

7. Find two integers whose product is 6 and whose sum is 5. The integers are 3 and 2. Thus, $x^2 + 5x + 6 = (x + 3)(x + 2)$.

8. Find two integers whose product is 40 and whose sum is -13.

Factors of 40	Sums of Factors
$-1, -40$	-41
$-2, -20$	-22
$-4, -10$	-14
$-5, -8$	$-13 \leftarrow$

The integers are -5 and -8, so $y^2 - 13y + 40 = (y-5)(y-8)$.

9. Find two integers whose product is -27 and whose sum is 6. The integers are -3 and 9, so $q^2 + 6q - 27 = (q-3)(q+9)$.

10. Find two integers whose product is -56 and whose sum is -1. The integers are 7 and -8, so $r^2 - r - 56 = (r+7)(r-8)$.

11. The factors of 1 of 1 and 1 add up to 2. The factors of 1 of -1 and -1 add up to -2. It is not possible to find two integers whose product is 1 and whose sum is 1. Consequently, the polynomial is prime.

12. First, factor out the GCF, $8p$.
$8p^3 - 24p^2 - 80p = 8p\left(p^2 - 3p - 10\right)$
Now factor $p^2 - 3p - 10$.
$p^2 - 3p - 10 = (p+2)(p-5)$
The completely factored form is
$8p^3 - 24p^2 - 80p = 8p(p+2)(p-5)$.

13.
$$\begin{aligned} 3x^4 + 30x^3 + 48x^2 &= 3x^2\left(x^2 + 10x + 16\right) \\ &= 3x^2(x+2)(x+8) \end{aligned}$$

14. Find two expressions whose product is $-96s^2$ and whose sum is $-4s$. The expressions are $8s$ and $-12s$, so $r^2 - 4rs - 96s^2 = (r+8s)(r-12s)$.

15. Find two expressions whose product is $-120q^2$ and whose sum is $2q$. The expressions are $12q$ and $-10q$, so $p^2 + 2pq - 120q^2 = (p+12q)(p-10q)$.

16.
$$\begin{aligned} p^7 - p^6q - 2p^5q^2 &= p^5\left(p^2 - pq - 2q^2\right) \\ &= p^5(p+q)(p-2q) \end{aligned}$$

17.
$$\begin{aligned} 3r^5 - 6r^4s - 45r^3s^2 &= 3r^3\left(r^2 - 2rs - 15s^2\right) \\ &= 3r^3(r+3s)(r-5s) \end{aligned}$$

18.
$$\begin{aligned} 2x^7 + 2x^6y - 12x^5y^2 &= 2x^5\left(x^2 + xy - 6y^2\right) \\ &= 2x^5(x-2y)(x+3y) \end{aligned}$$

19. To begin factoring $6r^2 - 5r - 6$, the possible first terms of the two binomial factors are r and $6r$, or $2r$ and $3r$, if we consider only positive integer coefficients.

20. When factoring $2z^3 + 9z^2 - 5z$, the first step is to factor out the GCF, z.

21. Factor $2k^2 - 5k + 2$ by trial and error.
$2k^2 - 5k + 2 = (2k-1)(k-2)$

22. Factor $3r^2 + 11r - 4$ by grouping. Look for two integers whose product is $3(-4) = -12$ and whose sum is 11. The integers are 12 and -1.
$$\begin{aligned} 3r^2 + 11r - 4 &= 3r^2 + 12r - r - 4 \\ &= \left(3r^2 + 12r\right) + (-r - 4) \\ &= 3r(r+4) - 1(r+4) \\ &= (r+4)(3r-1) \end{aligned}$$

23. Factor $6r^2 - 5r - 6$ by grouping. Find two integers whose product is $6(-6) = -36$ and whose sum is -5. The integers are -9 and 4.
$$\begin{aligned} 6r^2 - 5r - 6 &= 6r^2 - 9r + 4r - 6 \\ &= \left(6r^2 - 9r\right) + (4r - 6) \\ &= 3r(2r-3) + 2(2r-3) \\ &= (2r-3)(3r+2) \end{aligned}$$

24. Factor $10z^2 - 3z - 1$ by trial and error.
$10z^2 - 3z - 1 = (5z+1)(2z-1)$

25. Factor $5t^2 - 11t + 12$. Find two integers whose product is 12 and whose sum is -11.

Factors of 12	Sums of Factors
$-1, -12$	-13
$-2, -6$	-8
$-3, -4$	-7

There are no factors of 12 whose sum is -11. The polynomial is prime.

26. Factor out the GCF, $4x^3$. Then complete the factoring by trial and error.

$$24x^5 - 20x^4 + 4x^3 = 4x^3(6x^2 - 5x + 1)$$
$$= 4x^3(3x-1)(2x-1)$$

27. $$-6x^2 + 3x + 30 = -3(2x^2 - x - 10)$$
$$= -3(2x-5)(x+2)$$

28. $$10r^3s + 17r^2s^2 + 6rs^3 = rs(10r^2 + 17rs + 6s^2)$$
$$= rs(5r+6s)(2r+s)$$

29. $$48x^4y + 4x^3y^2 - 4x^2y^3$$
$$= 4x^2y(12x^2 + xy - y^2)$$
$$= 4x^2y(3x+y)(4x-y)$$

30. The student stopped too soon. He needs to factor out the common factor $4x - 1$ to get $(4x-1)(4x-5)$ as the correct answer.

31. Only choice B, $4x^2y^2 - 25z^2$, is the difference of squares. In A, 32 is not a perfect square. In C, we have a sum, not a difference. In D, y^3 is not a square. The correct choice is B.

32. Only choice D, $x^2 - 20x + 100$, is a perfect square trinomial because $x^2 = x \cdot x$, $100 = 10 \cdot 10$, and $-20x = -2(x)(10)$.

33. Factor the difference of squares.

$$n^2 - 49 = n^2 - 7^2 = (n+7)(n-7)$$

34. Factor the difference of squares.

$$25b^2 - 121 = (5b)^2 - 11^2$$
$$= (5b+11)(5b-11)$$

35. Factor the difference of squares.

$$49y^2 - 25w^2 = (7y)^2 - (5w)^2$$
$$= (7y+5w)(7y-5w)$$

36. Use the rule for factoring a difference of squares.

$$144p^2 - 36q^2 = 36(4p^2 - q^2)$$
$$= 36\left[(2p)^2 - q^2\right]$$
$$= 36(2p+q)(2p-q)$$

37. This polynomial is prime because it is the sum of squares, and the two terms have no common factor.

38. Factor the perfect square trinomial.

$$r^2 - 12r + 36 = r^2 - 2(6)(r) + 6^2$$
$$= (r-6)^2$$

39. Factor the perfect square trinomial.

$$9t^2 - 42t + 49 = (3t)^2 - 2(3t)(7) + 7^2$$
$$= (3t-7)^2$$

40. Factor the sum of cubes.

$$m^3 + 1000 = m^3 + 10^3$$
$$= (m+10)(m^2 - 10 \cdot m + 10^2)$$
$$= (m+10)(m^2 - 10m + 100)$$

41. Factor the sum of cubes.

$$125k^3 + 64x^3$$
$$= (5k)^3 + (4x)^3$$
$$= (5k+4x)\left[(5k)^2 - 5k \cdot 4x + (4x)^2\right]$$
$$= (5k+4x)(25k^2 - 20kx + 16x^2)$$

42. Factor the difference of cubes.

$$343x^3 - 64 = (7x)^3 - 4^3$$
$$= (7x-4)\left[(7x)^2 + 7x \cdot 4 + 4^2\right]$$
$$= (7x-4)(49x^2 + 28x + 16)$$

43. Factor the difference of cubes.

$$1000 - 27x^6$$
$$= 10^3 - (3x^2)^3$$
$$= (10 - 3x^2)\left[10^2 + 10(3x^2) + (3x^2)^2\right]$$
$$= (10 - 3x^2)(100 + 30x^2 + 9x^4)$$

44. $x^6 - y^6 = (x^3)^2 - (y^3)^2 = (x^3 + y^3)(x^3 - y^3)$

Now factor as the sum and difference of cubes.

$(x+y)(x^2 - xy + y^2)(x-y)(x^2 + xy + y^2)$

45. $(4t+3)(t-1) = 0$

$4t + 3 = 0 \quad \text{or} \quad t - 1 = 0$

$4t = -3 \quad \text{or} \quad t = 1$

$t = -\frac{3}{4}$

The solution set is $\left\{-\frac{3}{4}, 1\right\}$.

46. $x(2x-5) = 0$

$x = 0 \quad \text{or} \quad 2x - 5 = 0$

$2x = 5$

$x = \frac{5}{2}$

The solution set is $\left\{0, \frac{5}{2}\right\}$.

47. $z^2 + 4z + 3 = 0$

$(z+3)(z+1) = 0$

$z + 3 = 0 \quad \text{or} \quad z + 1 = 0$

$z = -3 \quad \text{or} \quad z = -1$

The solution set is $\{-3, -1\}$.

48. $m^2 - 5m + 4 = 0$

$(m-1)(m-4) = 0$

$m - 1 = 0 \quad \text{or} \quad m - 4 = 0$

$m = 1 \quad \text{or} \quad m = 4$

The solution set is $\{1, 4\}$.

49. $x^2 = -15 + 8x$

$x^2 - 8x + 15 = 0$

$(x-3)(x-5) = 0$

$x - 3 = 0 \quad \text{or} \quad x - 5 = 0$

$x = 3 \quad \text{or} \quad x = 5$

The solution set is $\{3, 5\}$.

50. $3z^2 - 11z - 20 = 0$

$(3z+4)(z-5) = 0$

$3z + 4 = 0 \quad \text{or} \quad z - 5 = 0$

$3z = -4 \quad \text{or} \quad z = 5$

$z = -\frac{4}{3}$

The solution set is $\left\{-\frac{4}{3}, 5\right\}$.

51. $81t^2 - 64 = 0$

$(9t+8)(9t-8) = 0$

$9t + 8 = 0 \quad \text{or} \quad 9t - 8 = 0$

$9t = -8 \quad \text{or} \quad 9t = 8$

$t = -\frac{8}{9} \quad \text{or} \quad t = \frac{8}{9}$

The solution set is $\left\{-\frac{8}{9}, \frac{8}{9}\right\}$.

52. $y^2 = 8y$

$y^2 - 8y = 0$

$y(y-8) = 0$

$y = 0 \quad \text{or} \quad y - 8 = 0$

$y = 8$

The solution set is $\{0, 8\}$.

53. $n(n-5) = 6$

$n^2 - 5n = 6$

$n^2 - 5n - 6 = 0$

$(n+1)(n-6) = 0$

$n + 1 = 0 \quad \text{or} \quad n - 6 = 0$

$n = -1 \quad \text{or} \quad n = 6$

The solution set is $\{-1, 6\}$.

54. $t^2 - 14t + 49 = 0$

$(t-7)^2 = 0$

$t - 7 = 0$

$t = 7$

The solution set is $\{7\}$.

55.

$$t^2 = 12(t-3)$$
$$t^2 = 12t - 36$$
$$t^2 - 12t + 36 = 0$$
$$(t-6)^2 = 0$$
$$t - 6 = 0$$
$$t = 6$$

The solution set is $\{6\}$.

56.

$$x^2 = 9$$
$$x^2 - 9 = 0$$
$$(x+3)(x-3) = 0$$
$$x+3=0 \quad \text{or} \quad x-3=0$$
$$x=-3 \quad \text{or} \quad x=3$$

The solution set is $\{-3, 3\}$.

57.

$$(5z+2)(z^2+3z+2)=0$$
$$(5z+2)(z+2)(z+1)=0$$
$$5z+2=0 \quad \text{or} \quad z+2=0 \quad \text{or} \quad z+1=0$$
$$5z=-2$$
$$z=-\frac{2}{5} \quad \text{or} \quad z=-2 \quad \text{or} \quad z=-1$$

The solution set is $\left\{-2, -1, -\frac{2}{5}\right\}$.

58.

$$64x^3 - 9x = 0$$
$$x(64x^2-9)=0$$
$$x(8x+3)(8x-3)=0$$
$$x=0 \quad \text{or} \quad 8x+3=0 \quad \text{or} \quad 8x-3=0$$
$$8x=-3 \qquad 8x=3$$
$$x=0 \quad \text{or} \quad z=-\frac{3}{8} \quad \text{or} \quad z=\frac{3}{8}$$

The solution set is $\left\{-\frac{3}{8}, 0, \frac{3}{8}\right\}$.

59. Let x be the width of the rug. Then $x+6$ is the length of the rug.

$$A = LW$$
$$40 = (x+6)x$$
$$40 = x^2 + 6x$$
$$0 = x^2 + 6x - 40$$
$$0 = (x+10)(x-4)$$
$$x+10=0 \quad \text{or} \quad x-4=0$$
$$x=-10 \quad \text{or} \quad x=4$$

Reject -10 since the width cannot be negative. The width of the rug is 4 feet, and the length is $4+6$ or 10 feet.

60. The surface area S of a box is given by $S = 2WH + 2WL + 2LH$, where L is the length, W is the width, and H is the height of the box. From the figure, we have $L = 20$, $W = x$, and $H = x + 4$.

$$S = 2WH + 2WL + 2LH$$
$$650 = 2x(x+4) + 2x(20) + 2(20)(x+4)$$
$$650 = 2x^2 + 8x + 40x + 40(x+4)$$
$$650 = 2x^2 + 48x + 40x + 160$$
$$0 = 2x^2 + 88x - 490$$
$$0 = 2(x^2 + 44x - 245)$$
$$0 = 2(x+49)(x-5)$$
$$x+49=0 \quad \text{or} \quad x-5=0$$
$$x=-49 \quad \text{or} \quad x=5$$

Reject -49 because the width cannot be negative. The width of the chest is 5 feet.

61. Let x be the distance traveled west. Then $x-14$ is the distance traveled south, and $(x-14)+16 = x+2$ is the distance between cars. These three distances form a right triangle with x and $x-14$ representing the lengths of the legs and $x+2$ representing the length of the hypotenuse. Use the Pythagorean theorem.

$$a^2 + b^2 = c^2$$
$$x^2 + (x-14)^2 = (x+2)^2$$
$$x^2 + x^2 - 28x + 196 = x^2 + 4x + 4$$
$$x^2 - 32x + 192 = 0$$
$$(x-8)(x-24) = 0$$
$$x-8=0 \quad \text{or} \quad x-24=0$$
$$x=8 \quad \text{or} \quad x=24$$

If $x = 8$, then $x - 14 = -6$, which is not possible because a distance cannot be negative. If $x = 24$, then $x - 14 = 10$ and $x + 2 = 26$. The cars were 26 miles apart.

62. Let x be the width of the base.
Then $x+2$ is the length of the base.
The area of the base, B, is given by LW, so $B = x(x+2)$.
Use the formula for the volume of a pyramid,

$$V = \frac{1}{3} \cdot B \cdot h.$$
$$48 = \frac{1}{3}x(x+2)(6)$$
$$48 = 2x(x+2)$$
$$24 = x^2 + 2x$$
$$x^2 + 2x - 24 = 0$$
$$(x+6)(x-4) = 0$$
$$x+6=0 \quad \text{or} \quad x-4=0$$
$$x=-6 \quad \text{or} \quad x=4$$

Reject −6. The width of the base is 4 meters, and the length is $4+2$ or 6 meters.

63. Let $x =$ the first integer. Then $x+1$ and $x+2$ are the next two integers.
The product of the first two of three consecutive integers is equal to 23 plus the third.

$$x(x+1) = 23 + (x+2)$$
$$x^2 + x = 23 + x + 2$$
$$x^2 - 25 = 0$$
$$(x+5)(x-5) = 0$$
$$x+5=0 \quad \text{or} \quad x-5=0$$
$$x=-5 \quad \text{or} \quad x=5$$

If $x=-5$, then $x+1=-4$ and $x+2=-3$. If $x=5$, then $x+1=6$ and $x+2=7$. The integers are −5, −4, and −3, or 5, 6, and 7.

64. **(a)** $t = 4: d = 16(4)^2 = 256$
In 4 seconds, the object would fall 256 feet.

(b) $t = 8: d = 16(8)^2 = 1024$
In 8 seconds, the object would fall 1024 feet.

65. **(a)** In 2007, $x = 7$.

$$y = 7.02x^2 - 15.5x + 469$$
$$y = 7.02(7)^2 - 15.5(7) + 469$$
$$y = 343.98 - 108.5 + 469$$
$$y = 704.48 \approx 704$$

In 2007, $y = 704$. The equation predicts 704 thousand vehicles.

In 2011, $x = 11$.

$$y = 7.02x^2 - 15.5x + 469$$
$$y = 7.02(11)^2 - 15.5(11) + 469$$
$$y = 849.42 - 170.5 + 469$$
$$y = 1147.92 \approx 1148$$

In 2011 $y = 1148$. The equation predicts 1148 thousand vehicles.

(b) In 2007, the result of 704 thousand is slightly higher than the actual number. In 2011, the result of 1148 thousand is lower than the actual number.

Chapter 6 Mixed Review Exercises

1. D is not factored completely.
$3(7t+4) + x(7t+4) = (7t+4)(3+x)$

2. The factor $2x+8$ has a factor of 2. The completely factored form is $2(x+4)(3x-4)$.

3. Two integers with product $3(10) = 30$ and sum 11 are 5 and 6.

$$\begin{aligned} 3k^2 + 11k + 10 &= 3k^2 + 5k + 6k + 10 \\ &= (3k^2 + 5k) + (6k + 10) \\ &= k(3k+5) + 2(3k+5) \\ &= (3k+5)(k+2) \end{aligned}$$

4. $z^2 - 11zx + 10x^2 = (z-x)(z-10x)$

5.
$$\begin{aligned} y^4 - 625 &= (y^2)^2 - 25^2 \\ &= (y^2+25)(y^2-25) \\ &= (y^2+25)(y+5)(y-5) \end{aligned}$$

6.
$$\begin{aligned} &6m^3 - 21m^2 - 45m \\ &= 3m(2m^2 - 7m - 15) \\ &= 3m[(2m^2 - 10m) + (3m - 15)] \\ &= 3m[2m(m-5) + 3(m-5)] \\ &= 3m(m-5)(2m+3) \end{aligned}$$

7. $25a^2 + 15ab + 9b^2$ is a prime polynomial.

8.
$$\begin{aligned} 2a^5 - 8a^4 - 24a^3 &= 2a^3(a^2 - 4a - 12) \\ &= 2a^3(a-6)(a+2) \end{aligned}$$

9. $15m^2 + 20m - 12mp - 16p$
$= 5m(3m + 4) - 4p(3m + 4)$
$= (3m + 4)(5m - 4p)$

10. $24ab^3c^2 - 56a^2bc^3 + 72a^2b^2c$
$= 8abc(3b^2c - 7ac^2 + 9ab)$

11. $12x^2yz^3 + 12xy^2z - 30x^3y^2z^4$
$= 6xyz(2xz^2 + 2y - 5x^2yz^3)$

12. $12r^2 + 18rq - 10r - 15q$
$= 6r(2r + 3q) - 5(2r + 3q)$
$= (2r + 3q)(6r - 5)$

13. $49t^2 + 56t + 16 = (7t)^2 + 2(7t)(4) + 4^2$
$= (7t + 4)^2$

14. $1000a^3 + 27$
$= (10a)^3 + 3^3$
$= (10a + 3)\left[(10a)^2 - 10a \cdot 3 + 3^2\right]$
$= (10a + 3)(100a^2 - 30a + 9)$

15. $t(t - 7) = 0$
$t = 0$ or $t - 7 = 0$
$t = 7$
The solution set is $\{0, 7\}$.

16. $x^2 + 3x = 10$
$x^2 + 3x - 10 = 0$
$(x + 5)(x - 2) = 0$
$x + 5 = 0$ or $x - 2 = 0$
$x = -5$ or $x = 2$
The solution set is $\{-5, 2\}$.

17. $25x^2 + 20x + 4 = 0$
$(5x)^2 + 2(5x)(2) + 2^2 = 0$
$(5x + 2)^2 = 0$
$5x + 2 = 0$
$5x = -2$
$x = -\frac{2}{5}$
The solution set is $\left\{-\frac{2}{5}\right\}$.

18. Let x be the length of the shorter leg.
Then $2x + 6$ is the length of the longer leg, and $(2x + 6) + 3 = 2x + 9$ is the length of the hypotenuse.
Use the Pythagorean theorem, $a^2 + b^2 = c^2$.
$x^2 + (2x + 6)^2 = (2x + 9)^2$
$x^2 + 4x^2 + 24x + 36 = 4x^2 + 36x + 81$
$x^2 - 12x - 45 = 0$
$(x - 15)(x + 3) = 0$
$x - 15 = 0$ or $x + 3 = 0$
$x = 15$ or $x = -3$
Reject -3 because a length cannot be negative.
The sides of the lot are 15 meters,
$2(15) + 6 = 36$ meters, and
$36 + 3 = 39$ meters.

19. Let b be the base of the sail.
Then $b + 4$ is the height of the sail.
Use the formula for the area of a triangle.
$A = \frac{1}{2}bh$
$30 = \frac{1}{2}(b)(b + 4)$ Let $A = 30$.
$60 = b^2 + 4b$
$0 = b^2 + 4b - 60$
$0 = (b + 10)(b - 6)$
$b + 10 = 0$ or $b - 6 = 0$
$b = -10$ or $b = 6$
Discard -10 since the base of a triangle cannot be negative. The base of the triangular sail is 6 meters.

20. Let x be the width of the house.
Then $x + 7$ is the length of the house.
Use $A = LW$ with 170 for A, $x + 7$ for L, and x for W.
$170 = (x + 7)(x)$
$170 = x^2 + 7x$
$0 = x^2 + 7x - 170$
$0 = (x + 17)(x - 10)$
$x + 17 = 0$ or $x - 10 = 0$
$x = -17$ or $x = 10$
Discard -17 because the width cannot be negative. If $x = 10$, $x + 7 = 10 + 7 = 17$.
The width is 10 meters, and the length is 17 meters.

Chapter 6 Test

1. $2x^2 - 2x - 24 = 2(x^2 - x - 12)$
$= 2(x + 3)(x - 4)$

The correct completely factored form is choice D. Note that the factored forms A, $(2x + 6)(x - 4)$, and B, $(x + 3)(2x - 8)$, also can be multiplied to give a product of $2x^2 - 2x - 24$, but neither of these is completely factored because $2x + 6$ and $2x - 8$ both contain a common factor of 2.

2. Factor out the GCF.
$12x^2 - 30x = 6x(2x - 5)$

3. Factor out the GCF.
$2m^3n^2 + 3m^3n - 5m^2n^2$
$= m^2n(2mn + 3m - 5n)$

4. $2ax - 2bx + ay - by = 2x(a - b) + y(a - b)$
$= (a - b)(2x + y)$

5. Find two integers whose product is -24 and whose sum is -5. The integers are 3 and -8.
$x^2 - 5x - 24 = (x + 3)(x - 8)$

6. Factor $2x^2 + x - 3$ by trial and error.
$2x^2 + x - 3 = (2x + 3)(x - 1)$

7. Factor $10z^2 - 17z + 3$ by trial and error.
$10z^2 - 17z + 3 = (2z - 3)(5z - 1)$

8. $3x^2 - 12x - 15 = 3(x^2 - 4x - 5)$
$= 3(x + 1)(x - 5)$

9. We cannot find two integers whose product is 3 and whose sum is 2. This polynomial is prime.

10. This polynomial is prime because the sum of squares cannot be factored and the two terms have no common factor.

11. $12 - 6a + 2b - ab = (12 - 6a) + (2b - ab)$
$= 6(2 - a) + b(2 - a)$
$= (2 - a)(6 + b)$

12. $9y^2 - 64 = (3y)^2 - 8^2$
$= (3y + 8)(3y - 8)$

13. The polynomial is a difference of squares.
$81a^2 - 121b^2 = (9a + 11b)(9a - 11b)$

14. The polynomial is a perfect square trinomial.
$x^2 + 16x + 64 = (x + 8)^2$

15. $4x^2 - 28xy + 49y^2$
$= (2x)^2 - 2(2x)(7y) + (7y)^2$
$= (2x - 7y)^2$

16. $6t^4 + 3t^3 - 108t^2 = 3t^2(2t^2 + t - 36)$
$= 3t^2(2t + 9)(t - 4)$

17. $r^3 - 125 = r^3 - 5^3$
$= (r - 5)(r^2 + 5 \cdot r + 5^2)$
$= (r - 5)(r^2 + 5r + 25)$

18. $8k^3 + 64 = 8(k^3 + 8)$
$= 8(k^3 + 2^3)$
$= 8(k + 2)(k^2 - 2 \cdot k + 2^2)$
$= 8(k + 2)(k^2 - 2k + 4)$

19. $x^4 - 81 = (x^2)^2 - 9^2$
$= (x^2 + 9)(x^2 - 9)$
$= (x^2 + 9)(x + 3)(x - 3)$

20. $81x^4 - 16y^4$
$= (9x^2)^2 - (4y^2)^2$
$= (9x^2 + 4y^2)(9x^2 - 4y^2)$
$= (9x^2 + 4y^2)[(3x)^2 - (2y)^2]$
$= (9x^2 + 4y^2)[(3x + 2y)(3x - 2y)]$
$= (3x + 2y)(3x - 2y)(9x^2 + 4y^2)$

21. $(x + 3)(x - 9) = 0$
$x + 3 = 0$ or $x - 9 = 0$
$x = -3$ or $x = 9$
The solution set is $\{-3, 9\}$.

22. $2r^2 - 13r + 6 = 0$

$(2r-1)(r-6) = 0$

$2r - 1 = 0$ or $r - 6 = 0$

$2r = 1$ or $r = 6$

$r = \frac{1}{2}$

The solution set is $\left\{\frac{1}{2}, 6\right\}$.

23. $25x^2 - 4 = 0$

$(5x+2)(5x-2) = 0$

$5x + 2 = 0$ or $5x - 2 = 0$

$5x = -2$ or $5x = 2$

$x = -\frac{2}{5}$ or $x = \frac{2}{5}$

The solution set is $\left\{-\frac{2}{5}, \frac{2}{5}\right\}$.

24. $t^2 = 9t$

$t^2 - 9t = 0$

$t(t-9) = 0$

$t = 0$ or $t - 9 = 0$

$t = 0$ or $t = 9$

The solution set is $\{0, 9\}$.

25. $x(x-20) = -100$

$x^2 - 20x = -100$

$x^2 - 20x + 100 = 0$

$(x-10)^2 = 0$

$x - 10 = 0$

$x = 10$

The solution set is $\{10\}$.

26. $(s+8)(6s^2 + 13s - 5) = 0$

If $s + 8 = 0$, then $s = -8$.

$6s^2 + 13s - 5 = 0$

$(2s+5)(3s-1) = 0$

$2s + 5 = 0$ or $3s - 1 = 0$

$2s = -5$ or $3s = 1$

$s = -\frac{5}{2}$ $\quad$ $s = \frac{1}{3}$

The solution set is $\left\{-8, -\frac{5}{2}, \frac{1}{3}\right\}$.

27. Let $x =$ the width of the flower bed.
Then $2x - 3 =$ the length of the flower bed.
Use the formula $\mathcal{A} = LW$.

$x(2x-3) = 54$

$2x^2 - 3x = 54$

$2x^2 - 3x - 54 = 0$

$(2x+9)(x-6) = 0$

$2x + 9 = 0$ or $x - 6 = 0$

$2x = -9$ or $x = 6$

$x = -\frac{9}{2}$

Reject $-\frac{9}{2}$. If $x = 6, 2x - 3 = 2(6) - 3 = 9$.

The dimensions of the flower bed are 6 feet by 9 feet.

28. Let x be the first integer.
Then $x + 1$ is the second integer.
The square of the sum of the two integers is 11 more than the first integer.

$[x + (x+1)]^2 = x + 11$

$(2x+1)^2 = x + 11$

$4x^2 + 4x + 1 = x + 11$

$4x^2 + 3x - 10 = 0$

$(4x-5)(x+2) = 0$

$4x - 5 = 0$ or $x + 2 = 0$

$4x = 5$ or $x = -2$

$x = \frac{5}{4}$

Reject $\frac{5}{4}$ because it is not an integer. If $x = -2, x + 1 = -1$. The integers are -2 and -1.

29. Let x be the length of the stud.
Then $3x - 7$ is the length of the brace.
The figure shows that a right triangle is formed with the brace as the hypotenuse. Use the Pythagorean theorem, $a^2 + b^2 = c^2$.

$$x^2 + 15^2 = (3x - 7)^2$$
$$x^2 + 225 = 9x^2 - 42x + 49$$
$$0 = 8x^2 - 42x - 176$$
$$0 = 2(4x^2 - 21x - 88)$$
$$0 = 2(4x + 11)(x - 8)$$
$$4x + 11 = 0 \quad \text{or} \quad x - 8 = 0$$
$$4x = -11 \quad \text{or} \quad x = 8$$
$$x = -\frac{11}{4}$$

Reject $-\frac{11}{4}$. If $x = 8, 3x - 7 = 24 - 7 = 17,$ so the brace should be 17 feet long.

30. For 2014, $x = 2014 - 2000 = 14.$

$$y = 57.53x^2 - 72.93x + 3417$$
$$y = 57.53(14)^2 - 72.93(14) + 3417$$
$$y = 11,275.88 - 1021.02 + 3417$$
$$y = 13,671.86 \approx 13,672$$

In 2014, the model estimates that the public debt of the United States was $13,672 billion.

Chapters R–6 Cumulative Review Exercises

1. $3x + 2(x - 4) = 4(x - 2)$

$$3x + 2x - 8 = 4x - 8$$
$$5x - 8 = 4x - 8$$
$$x - 8 = -8$$
$$x = 0$$

The solution set is $\{0\}$.

2. Multiply both sides by 100 to clear decimals.

$$100(0.3x + 0.9x) = 100(0.06)$$
$$30x + 90x = 6$$
$$120x = 6$$
$$x = \frac{6}{129} = \frac{1}{20} = 0.05$$

The solution set is $\{0.05\}$.

3. To clear fractions, multiply both sides by the least common denominator, which is 6.

$$6\left[\frac{2}{3}m - \frac{1}{2}(m - 4)\right] = 6(3)$$
$$4m - 3(m - 4) = 18$$
$$4m - 3m + 12 = 18$$
$$m + 12 = 18$$
$$m = 6$$

The solution set is $\{6\}$.

4.

$$A = P + Prt$$
$$A = P(1 + rt) \quad \text{Factor out } P.$$
$$\frac{A}{1 + rt} = \frac{P(1 + rt)}{1 + rt} \quad \text{Divide by } 1 + rt.$$
$$\frac{A}{1 + rt} = P \quad \text{or} \quad P = \frac{A}{1 + rt}$$

5. The angles are supplementary, so the sum of the angles is $180°$.

$$(2x + 16) + (x + 23) = 180$$
$$3x + 39 = 180$$
$$3x = 141$$
$$x = 47$$

Since $x = 47, 2x + 16 = 2(47) + 16 = 110$ and $x + 23 = 47 + 23 = 70.$
The angles are $110°$ and $70°$.

6. Let x be number of bronze medals.
Then $x - 5$ is the number of silver medals, and $(x - 5) + 2 = x - 3$ is the number of gold medals.
The total number of medals was 28.

$$x + (x - 5) + (x - 3) = 28$$
$$3x - 8 = 28$$
$$3x = 36$$
$$x = 12$$

Since $x = 12, x - 5 = 7$ and $x - 3 = 9.$
The United States won 9 gold medals, 7 silver medals, and 12 bronze medals.

7. Find 46% of 500.

$$\frac{\text{part}}{\text{whole}} = \frac{p}{100}$$
$$\frac{a}{500} = \frac{46}{100}$$
$$100a = 46(500)$$
$$a = 230$$

Find 41% of 500.

$$\frac{\text{part}}{\text{whole}} = \frac{p}{100}$$
$$\frac{a}{500} = \frac{41}{100}$$
$$100a = 500(41)$$
$$a = 205$$

What percent of 500 is 190?

$$\frac{\text{part}}{\text{whole}} = \frac{p}{100}$$
$$\frac{190}{500} = \frac{p}{100}$$
$$500p = 190(100)$$
$$p = 38$$

What percent of 500 is 60?

$$\frac{\text{part}}{\text{whole}} = \frac{p}{100}$$
$$\frac{60}{500} = \frac{p}{100}$$
$$500p = 60(100)$$
$$p = 12$$

Item	Percent	Number
Personal Computer	46%	230
Cell Phone	41%	205
High-speed Internet	38%	190
MP3 player	12%	60

8. **(a)** quadrant II if a is *negative* and b is *positive*.

(b) quadrant III if a is *negative* and b is *negative*.

9. **(a)** The equation $y = -2x - 4$ is in slope-intercept form, so the y-intercept is $(0, -4)$.

Let $y = 0$ to find the x-intercept.

$$0 = -2x - 4$$
$$4 = -2x$$
$$-2 = x$$

The x-intercept is $(-2, 0)$.

(b) The equation $y = -2x - 4$ is in slope-intercept form, so the slope is the coefficient of x, that is, -2.

(c)

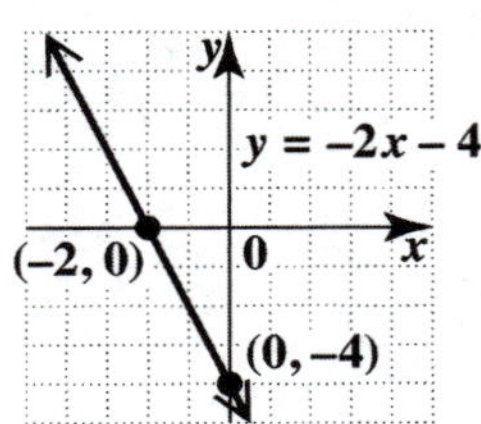

10. **(a)** $(2004, 217), (2010, 266)$

$$m = \frac{y_2 - y_1}{x_2 - x_1} = \frac{266 - 217}{2010 - 2004} = \frac{49}{6} \approx 8 \text{ (to the nearest whole number)}$$

A slope of (approximately) 8 means that retail sales of prescription drugs increased by about \$8 billion per year.

(b) The graph of the line will fall about 8 units from 2010 to 2009 and also from 2009 to 2008, so the y-value is $266 - 8 - 8 = 250$.

The ordered pair is $(2008, 250)$.

11.
$$4x - y = -6 \quad (1)$$
$$2x + 3y = 4 \quad (2)$$
$$12x - 3y = -18 \quad (3) \quad 3 \times \text{Eq. } (1)$$
$$2x + 3y = 4 \quad (2)$$
$$14x = -14 \quad \text{Add (3) and (2).}$$
$$x = -1$$

To find y, substitute -1 for x in equation (1).

$$4(-1) - y = -6$$
$$-4 - y = -6$$
$$-y = -2$$
$$y = 2$$

The solution set is $\{(-1, 2)\}$.

12.
$$5x + 3y = 10 \quad (1)$$
$$2x + \frac{6}{5}y = 5 \quad (2)$$
$$-10x - 6y = -20 \quad (3) \quad -2 \times \text{Eq. (1)}$$
$$10x + 6y = 20 \quad (4) \quad 5 \times \text{Eq. (2)}$$
$$0 = 5 \quad \text{Add (3) and (4).}$$

The system of equations has no solution, symbolized by $\varnothing$.

13. $\left(\frac{3}{4}\right)^{-2} = \left(\frac{4}{3}\right)^{2} = \frac{16}{9}$

14. $\left(\frac{4^{-3}\cdot 4^{4}}{4^{5}}\right)^{-1} = \left(\frac{4^{5}}{4^{-3}\cdot 4^{4}}\right)^{1} = \frac{4^5}{4^1} = 4^4 = 256$

15.
$$\frac{(p^2)^3 p^{-4}}{(p^{-3})^{-1} p} = \frac{p^{2\cdot 3} p^{-4}}{p^{(-3)(-1)} p}$$
$$= \frac{p^6 p^{-4}}{p^3 p^1}$$
$$= \frac{p^{6-4}}{p^{3+1}}$$
$$= \frac{p^2}{p^4} = \frac{1}{p^2}$$

16.
$$\frac{(m^{-2})^3 m}{m^5 m^{-4}} = \frac{m^{-2(3)} m^1}{m^{5+(-4)}}$$
$$= \frac{m^{-6+1}}{m^1}$$
$$= \frac{m^{-5}}{m^1} = \frac{1}{m^6}$$

17. $(2k^2 + 4k) - (5k^2 - 2) - (k^2 + 8k - 6)$
$= (2k^2 + 4k) + (-5k^2 + 2) + (-k^2 - 8k + 6)$
$= 2k^2 + 4k - 5k^2 + 2 - k^2 - 8k + 6$
$= -4k^2 - 4k + 8$

18. Use the FOIL method.
$(9x + 6)(5x - 3)$
$= 9x(5x) + 9x(-3) + 6(5x) + 6(-3)$
$= 45x^2 - 27x + 30x - 18$
$= 45x^2 + 3x - 18$

19. $(3p + 2)^2 = (3p)^2 + 2\cdot 3p \cdot 2 + 2^2$
$= 9p^2 + 12p + 4$

20.
$$\frac{8x^4 + 12x^3 - 6x^2 + 20x}{2x}$$
$$= \frac{8x^4}{2x} + \frac{12x^3}{2x} + \frac{6x^2}{2x} + \frac{20x}{2x}$$
$$= 4x^3 + 6x^2 - 3x + 10$$

21. $55{,}000 = 5.5\times 10^4$
Move the decimal point left 4 places so it is to the right of the first nonzero digit. 55,000 is greater than 5.5, so the power is positive.
$2{,}000{,}000 = 2.0\times 10^6$
Move the decimal point left 6 places so it is to the right of the first nonzero digit. 2,000,000 is greater than 2, so the power is positive.

22. Factor $2a^2 + 7a - 4$ by trial and error.
$2a^2 + 7a - 4 = (a + 4)(2a - 1)$

23. To factor by grouping, find two integers whose product is $10(6) = 60$ and whose sum is 19. The integers are 15 and 4.
$10m^2 + 19m + 6 = 10m^2 + 15m + 4m + 6$
$= 5m(2m + 3) + 2(2m + 3)$
$= (2m + 3)(5m + 2)$

24. Factor $8t^2 + 10tv + 3v^2$ by trial and error.
$8t^2 + 10tv + 3v^2 = (4t + 3v)(2t + v)$

25. $4p^2 - 12p + 9 = (2p - 3)(2p - 3)$
$= (2p - 3)^2$

26. $25r^2 - 81t^2 = (5r)^2 - (9t)^2$
$= (5r + 9t)(5r - 9t)$

27. $2pq + 6p^3q + 8p^2q = 2pq(1 + 3p^2 + 4p)$
$= 2pq(3p^2 + 4p + 1)$
$= 2pq(3p + 1)(p + 1)$

28. $6m^2 + m - 2 = 0$
$(3m + 2)(2m - 1) = 0$
$3m + 2 = 0$ or $2m - 1 = 0$
$3m = -2$ or $2m = 1$
$m = -\frac{2}{3}$ or $m = \frac{1}{2}$

The solution set is $\left\{-\frac{2}{3}, \frac{1}{2}\right\}$.

29. $8x^2 = 64x$

$8x^2 - 64x = 0$

$8x(x-8) = 0$

$8x = 0$ or $x - 8 = 0$

$x = 0$ or $x = 8$

The solution set is $\{0, 8\}$.

30. Let x be the length of the shorter leg.
Then $x + 7$ is the length of the longer leg, and $2x + 3$ is the length of the hypotenuse.
Use the Pythagorean theorem.

$$x^2 + (x+7)^2 = (2x+3)^2$$
$$x^2 + (x^2 + 14x + 49) = 4x^2 + 12x + 9$$
$$2x^2 + 14x + 49 = 4x^2 + 12x + 9$$
$$0 = 2x^2 - 2x - 40$$
$$0 = 2(x^2 - x - 20)$$
$$0 = (x-5)(x+4)$$

$x - 5 = 0$ or $x + 4 = 0$

$x = 5$ or $x = -4$

Reject -4 because the length of a leg cannot be negative. Since $x = 5$, $x + 7 = 12$ and $2x + 3 = 2(5) + 3 = 13$. The length of the sides are 5 meters, 12 meters, and 13 meters.

Chapter 7 Rational Expressions and Applications

7.1 The Fundamental Property of Rational Expressions

Now Try Exercises

N1. (a) $\frac{2x-1}{x+4}=\frac{2(-3)-1}{-3+4}$ Let $x=-3$.

$=\frac{-6-1}{-3+4}$

$=\frac{-7}{1}$

$=-7$

(b) $\frac{x+3}{4}=\frac{-3+3}{4}$ Let $x=-3$.

$=\frac{0}{4}$

$=0$

(c) $\frac{4}{x+3}=\frac{4}{-3+3}$ Let $x=-3$.

$=\frac{4}{0}$

The expression is undefined for $x=-3$.

N2. (a) $\frac{k-4}{2k-1}$

Solve $2k-1=0$.

Since $k=\frac{1}{2}$ will make the denominator zero, the expression is undefined for $k=\frac{1}{2}$.

We write the answer as $k\neq\frac{1}{2}$.

(b) $\frac{2x}{x^2+5x-14}$

Solve $x^2+5x-14=0$.

$(x+7)(x-2)=0$, so the denominator is zero when either $x+7=0$ or $x-2=0$, that is, when $x=-7$ or $x=2$. The expression is undefined for $x=-7$ and $x=2$. We write the answer as $x\neq-7$, $x\neq 2$.

(c) $\frac{y+10}{y^2+10}$

This denominator will not equal 0 for any value of y, because y^2 is always greater than or equal to 0, and adding 10 makes the sum greater than 0. Thus, there are no values for which this rational expression is undefined.

N3. (a) $\frac{20}{48}=\frac{2\cdot2\cdot5}{2\cdot2\cdot2\cdot2\cdot3}$ Factor.

$=\frac{5(2\cdot2)}{2\cdot2\cdot3(2\cdot2)}$ Group.

$=\frac{5}{2\cdot2\cdot3}$ Fundamental property

$=\frac{5}{12}$

(b) $\frac{21y^5}{7y^2}=\frac{3\cdot7\cdot y\cdot y\cdot y\cdot y\cdot y}{7\cdot y\cdot y}$ Factor.

$=\frac{3\cdot y\cdot y\cdot y(7\cdot y\cdot y)}{1(7\cdot y\cdot y)}$ Group.

$=\frac{3\cdot y\cdot y\cdot y}{1}$ Fund. prop.

$=3y^3$

N4. (a) $\frac{3x+15}{5x+25}=\frac{3(x+5)}{5(x+5)}$ Factor.

$=\frac{3}{5}$ Fundamental property

(b) $\frac{k^2-36}{k^2+8k+12}$

$=\frac{(k+6)(k-6)}{(k+6)(k+2)}$ Factor.

$=\frac{k-6}{k+2}$ Fundamental property

N5. $\frac{10-a^2}{a^2-10}=\frac{1(10-a^2)}{-1(10-a^2)}=\frac{1}{-1}=-1$

N6. **(a)** $\dfrac{p-4}{4-p}$

Since $p-4$ and $4-p$ are opposites, this expression equals -1.

(b) $$\frac{4m^2-n^2}{2n-4m}=\frac{(2m)^2-n^2}{2(n-2m)}=\frac{(2m+n)(2m-n)}{2(-1)(2m-n)}=\frac{2m+n}{2(-1)}=\frac{2m+n}{-2}\text{, or, }-\frac{2m+n}{2}$$

(c) $\dfrac{x+y}{x-y}$

$x-y=-1(-x+y)\neq-1(x+y)$

The expressions $x+y$ and $x-y$ are not opposites of each other. They do not have any common factors (other than 1), so the rational expression is already in lowest terms.

N7. To write four equivalent expressions for $-\dfrac{4k-9}{k+3}$, we will follow the outline in Example 7.

Applying the negative sign to the numerator, we have the following.

$$\frac{-(4k-9)}{k+3}$$

Distributing the negative sign gives us the following.

$$\frac{-4k+9}{k+3}$$

Applying the negative sign to the denominator yields the following.

$$\frac{4k-9}{-(k+3)}$$

Again, we distribute to get the following.

$$\frac{4k-9}{-k-3}$$

3. $$\frac{2x+3}{2x-3}\neq-1$$

$$\frac{2x-3}{3-2x}=\frac{-1(3-2x)}{3-2x}=-1$$

$$\frac{2x+3}{3+2x}=1\neq-1$$

$$\frac{2x+3}{-2x-3}=\frac{2x+3}{-1(2x+3)}=-1$$

B and D are equal to -1.

5. $$\frac{3-x}{x-4}=\frac{-1(3-x)}{-1(x-4)}=\frac{-3+x}{-x+4}=\frac{x-3}{4-x}$$

$$-\frac{3-x}{4-x}=\frac{-1(3-x)}{4-x}=\frac{-3+x}{4-x}=\frac{x-3}{4-x}$$

$$-\frac{x-3}{x-4}=\frac{x-3}{-1(x-4)}=\frac{x-3}{-x+4}=\frac{x-3}{4-x}$$

Since A, C, and D are equivalent to $\dfrac{x-3}{4-x}$, B is the one that is not.

7. Find the numerical value of the rational expression for $x=-3$.

$$\frac{x}{2x+1}$$

$$=\frac{\underline{-3}}{2(\underline{-3})+1}\quad\text{Let } x=-3.$$

$$=\frac{\underline{-3}}{\underline{-6}+1}$$

$$=\underline{\frac{3}{5}}$$

9. **(a)** $$\frac{3x+1}{5x}=\frac{3(2)+1}{5(2)}\quad\text{Let } x=2.$$

$$=\frac{7}{10}$$

(b) $$\frac{3x+1}{5x}=\frac{3(-3)+1}{5(-3)}\quad\text{Let } x=-3.$$

$$=\frac{-8}{-15}=\frac{8}{15}$$

Exercises

1. The rational expression $\dfrac{x+5}{x-3}$ is undefined when x is $\underline{3}$, so $x\neq\underline{3}$. This rational expression is equal to 0 when $x=\underline{-5}$.

11. (a) $\frac{x^2-4}{2x+1}=\frac{(2)^2-4}{2(2)+1}$ Let $x=2$.

$=\frac{0}{5}=0$

(b) $\frac{x^2-4}{2x+1}=\frac{(-3)^2-4}{2(-3)+1}$ Let $x=-3$.

$=\frac{5}{-5}=-1$

13. (a) $\frac{(-2x)^3}{3x+9}=\frac{(-2\cdot 2)^3}{3\cdot 2+9}$ Let $x=2$.

$=\frac{-64}{15}=-\frac{64}{15}$

(b) $\frac{(-2x)^3}{3x+9}=\frac{[-2(-3)]^3}{3(-3)+9}$ Let $x=-3$.

$=\frac{216}{0}$

Since substituting -3 for x makes the denominator zero, the given rational expression is undefined when $x=-3$.

15. (a) $\frac{7-3x}{3x^2-7x+2}$

$=\frac{7-3(2)}{3(2)^2-7(2)+2}$ Let $x=2$.

$=\frac{7-6}{12-14+2}=\frac{1}{0}$

Since substituting 2 for x makes the denominator zero, the given rational expression is undefined when $x=2$.

(b) $\frac{7-3x}{3x^2-7x+2}$

$=\frac{7-3(-3)}{3(-3)^2-7(-3)+2}$ Let $x=-3$.

$=\frac{7+9}{27+21+2}=\frac{16}{50}=\frac{8}{25}$

17. (a) $\frac{(x+3)(x-2)}{500x}$

$=\frac{(2+3)(2-2)}{500(2)}$ Let $x=2$.

$=\frac{5(0)}{1000}=\frac{0}{1000}=0$

(b) $\frac{(x+3)(x-2)}{500x}$

$=\frac{(-3+3)(-3-2)}{500(-3)}$ Let $x=-3$.

$=\frac{0(-5)}{-1500}=\frac{0}{-1500}=0$

19. (a) $\frac{x^2-4}{x^2-9}=\frac{2^2-4}{2^2-9}$ Let $x=2$.

$=\frac{4-4}{4-9}=\frac{0}{-5}=0$

(b) $\frac{x^2-4}{x^2-9}=\frac{(-3)^2-4}{(-3)^2-9}$ Let $x=-3$.

$=\frac{9-4}{9-9}=\frac{5}{0}$

Since substituting -3 for x makes the denominator zero, the given rational expression is undefined when $x=-3$.

21. $-\frac{5}{x}$

The denominator x will be zero when $x=0$, so the given expression is undefined for $x=0$. We write the answer as $x\neq 0$.

23. $\frac{12}{5y}$

The denominator $5y$ will be zero when $y=0$, so the given expression is undefined for $y=0$. We write the answer as $y\neq 0$.

25. $\frac{x+1}{x-6}$

To find the values for which this expression is undefined, set the denominator equal to zero and solve for x.

$x-6=0$

$x=6$

Because $x=6$ will make the denominator zero, the given expression is undefined for 6. We write the answer as $x\neq 6$.

27. $\dfrac{4x^2}{3x+5}$

To find the values for which this expression is undefined, set the denominator equal to zero and solve for x.

$$3x+5=0$$
$$3x=-5$$
$$x=-\frac{5}{3}$$

Because $x=-\dfrac{5}{3}$ will make the denominator zero, the given expression is undefined for $-\dfrac{5}{3}$.

We write the answer as $x \neq -\dfrac{5}{3}$.

29. $\dfrac{5m+2}{m^2+m-6}$

To find the numbers that make the denominator 0, we must solve.

$$m^2+m-6=0$$
$$(m+3)(m-2)=0$$
$$m+3=0 \quad \text{or} \quad m-2=0$$
$$m=-3 \quad \text{or} \quad m=2$$

The given expression is undefined for $m=-3$ and for $m=2$. We write the answer as $m \neq -3$ and $m \neq 2$.

31. $\dfrac{x^2+3x}{4}$ is never undefined since the denominator is never zero.

33. $\dfrac{3x-1}{x^2+2}$

This denominator cannot equal zero for any value of x because x^2 is always greater than or equal to zero, and adding 2 makes the sum greater than zero. Thus, the given rational expression is never undefined.

35. $\dfrac{x^2+4x}{x+4}$

The two terms in the numerator are x^2 and $4x$. The two terms in the denominator are x and 4.

37. $\dfrac{18r^3}{6r}=\dfrac{3r^2(6r)}{1(6r)}$ Factor.

$=3r^2$ Fundamental property

39. $\dfrac{4(y-2)}{10(y-2)}=\dfrac{2\cdot 2(y-2)}{5\cdot 2(y-2)}$ Factor.

$=\dfrac{2}{5}$ Fundamental prop.

41. $\dfrac{(x+1)(x-1)}{(x+1)^2}=\dfrac{(x+1)(x-1)}{(x+1)(x+1)}$

$=\dfrac{x-1}{x+1}$

43. $\dfrac{7m+14}{5m+10}=\dfrac{7(m+2)}{5(m+2)}$ Factor.

$=\dfrac{7}{5}$ Fundamental property

45. $\dfrac{6m-18}{7m-21}=\dfrac{6(m-3)}{7(m-3)}$ Factor.

$=\dfrac{6}{7}$ Fundamental property

47. $\dfrac{m^2-n^2}{m+n}=\dfrac{(m+n)(m-n)}{m+n}$

$=m-n$

49. $\dfrac{2t+6}{t^2-9}=\dfrac{2(t+3)}{(t+3)(t-3)}$

$=\dfrac{2}{t-3}$

51. $\dfrac{12m^2-3}{8m-4}=\dfrac{3(4m^2-1)}{4(2m-1)}$

$=\dfrac{3(2m+1)(2m-1)}{4(2m-1)}$

$=\dfrac{3(2m+1)}{4}$

53. $\dfrac{3m^3-3m}{5m-5}=\dfrac{3m(m-1)}{5(m-1)}$

$=\dfrac{3m}{5}$

55. $\dfrac{9r^2-4s^2}{9r+6s}=\dfrac{(3r+2s)(3r-2s)}{3(3r+2s)}$

$=\dfrac{3r-2s}{3}$

57. $\frac{x-6}{x^2-36}=\frac{x-6}{(x+6)(x-6)}$

$=\frac{1}{x+6}$

59. $\frac{x^2-9}{x^2-6x+9}=\frac{(x+3)(x-3)}{(x-3)(x-3)}$

$=\frac{x+3}{x-3}$

61. $\frac{13x^2-39x^3}{7x-21x^2}=\frac{13x^2(1-3x)}{7x(1-3x)}$

$=\frac{13x^2}{7x}$

$=\frac{13x}{7}$

63. $\frac{5k^2-13k-6}{5k+2}=\frac{(5k+2)(k-3)}{5k+2}$

$=k-3$

65. $\frac{x^2+2x-15}{x^2+6x+5}=\frac{(x+5)(x-3)}{(x+5)(x+1)}$

$=\frac{x-3}{x+1}$

67. $\frac{2x^2-3x-5}{2x^2-7x+5}=\frac{(2x-5)(x+1)}{(2x-5)(x-1)}$

$=\frac{x+1}{x-1}$

69. $\frac{3x^3+13x^2+14x}{3x^3-5x^2-28x}$

$=\frac{x(3x^2+13x+14)}{x(3x^2-5x-28)}$

$=\frac{x(3x+7)(x+2)}{x(3x+7)(x-4)}$

$=\frac{x+2}{x-4}$ Fundamental property

71. $\frac{-3t+6t^2-3t^3}{7t^2-14t^3+7t^4}$

$=\frac{-3t(1-2t+t^2)}{7t^2(1-2t+t^2)}$

$=-\frac{3}{7t}$ Fundamental property

73. $\frac{zw+4z-3w-12}{zw+4z+5w+20}$

$=\frac{z(w+4)-3(w+4)}{z(w+4)+5(w+4)}$ Factor by grouping.

$=\frac{(w+4)(z-3)}{(w+4)(z+5)}$

$=\frac{z-3}{z+5}$ Fundamental prop.

75. $\frac{pr+qr+ps+qs}{pr+qr-ps-qs}$

$=\frac{r(p+q)+s(p+q)}{r(p+q)-s(p+q)}$ Factor by grouping.

$=\frac{(p+q)(r+s)}{(p+q)(r-s)}$

$=\frac{r+s}{r-s}$ Fundamental prop.

77. $\frac{ac-ad+bc-bd}{ac-ad-bc+bd}$

$=\frac{a(c-d)+b(c-d)}{a(c-d)-b(c-d)}$ Factor by grouping.

$=\frac{(c-d)(a+b)}{(c-d)(a-b)}$

$=\frac{a+b}{a-b}$ Fundamental property

79. $\frac{m^2-n^2-4m-4n}{2m-2n-8}$

$=\frac{(m+n)(m-n)-4(m+n)}{2(m-n-4)}$

$=\frac{(m+n)(m-n-4)}{2(m-n-4)}$

$=\frac{m+n}{2}$

81. $\frac{x^2y+y+x^2z+z}{xy+xz}$

$=\frac{y(x^2+1)+z(x^2+1)}{x(y+z)}$ Factor by grouping.

$=\frac{(x^2+1)(y+z)}{x(y+z)}$

$=\frac{x^2+1}{x}$ Fundamental prop.

83. The numerator is the sum of cubes.

$$\frac{1+p^3}{1+p}=\frac{1^3+p^3}{1+p}$$
$$=\frac{(1+p)(1-p+p^2)}{1+p}$$
$$=1-p+p^2$$

85. The numerator is the difference of cubes.

$$\frac{x^3-27}{x-3}=\frac{x^3-3^3}{x-3}$$
$$=\frac{(x-3)(x^2+3x+9)}{x-3}$$
$$=x^2+3x+9$$

87. The numerator is the difference of cubes, and the denominator is the difference of squares.

$$\frac{b^3-a^3}{a^2-b^2}$$
$$=\frac{(b-a)(b^2+ba+a^2)}{(a-b)(a+b)}$$
$$=(-1)\cdot\frac{(b^2+ba+a^2)}{(a+b)}$$
$$=-\frac{b^2+ba+a^2}{a+b}$$

89. The numerator is the sum of cubes, and the denominator is the difference of squares.

$$\frac{k^3+8}{k^2-4}=\frac{k^3+2^3}{(k+2)(k-2)}$$
$$=\frac{(k+2)(k^2-2k+4)}{(k+2)(k-2)}$$
$$=\frac{k^2-2k+4}{k-2}$$

91. The numerator is the sum of cubes. The denominator has a common factor of z.

$$\frac{z^3+27}{z^3-3z^2+9z}=\frac{z^3+3^3}{z(z^2-3z+9)}$$
$$=\frac{(z+3)(z^2-3z+9)}{z(z^2-3z+9)}$$
$$=\frac{z+3}{z}$$

93. The numerator is the difference of cubes. The denominator has a common factor of 2.

$$\frac{1-8r^3}{8r^2+4r+2}=\frac{1^3-(2r)^3}{2\left(4r^2+2r+1\right)}$$
$$=\frac{(1-2r)\left(1+2r+4r^2\right)}{2\left(4r^2+2r+1\right)}$$
$$=\frac{1-2r}{2}$$

95. $\frac{6-t}{t-6}=\frac{-1(t-6)}{1(t-6)}=\frac{-1}{1}=-1$

Note that $6-t$ and $t-6$ are opposites, so we know that their quotient will be -1.

97. $\frac{m^2-1}{1-m}=\frac{(m+1)(m-1)}{-1(m-1)}$

$$=\frac{m+1}{-1}$$
$$=-(m+1) \quad\text{or}\quad -m-1$$

99. $\frac{q^2-4q}{4q-q^2}=\frac{q(q-4)}{q(4-q)}$

$$=\frac{q-4}{4-q}=-1$$

$q-4$ and $4-q$ are opposites.

101. In the expression $\frac{p+6}{p-6}$, neither numerator nor denominator can be factored. It is already in lowest terms. *Note:* $(p+6)$ and $(p-6)$ are not opposites.

103. $\frac{-2m+2n}{m-n}=\frac{-2(m-n)}{m-n}$

$$=-2$$

105. Answers may vary. To write four equivalent expressions for $-\frac{x+4}{x-3}$, we will follow the outline in Example 7.
Applying the negative sign to the numerator, we have $\frac{-(x+4)}{x-3}$.

Distributing the negative sign gives us $\frac{-x-4}{x-3}$.

Applying the negative sign to the denominator yields $\frac{x+4}{-(x-3)}$.

Again, we distribute to get $\frac{x+4}{-x+3}$.

107. Answers may vary. $-\frac{2x-3}{x+3}$ is equivalent to each of the following:

$\frac{-(2x-3)}{x+3}, \frac{-2x+3}{x+3},$

$\frac{2x-3}{-(x+3)}, \frac{2x-3}{-x-3}$

109. Answers may vary. $-\frac{3x-1}{5x-6}$ is equivalent to each of the following:

$\frac{-(3x-1)}{5x-6}, \frac{-3x+1}{5x-6},$

$\frac{3x-1}{-(5x-6)}, \frac{3x-1}{-5x+6}$

111.

$$L \cdot W = \mathcal{A}$$

$$W = \frac{\mathcal{A}}{L}$$

$$W = \frac{x^4 + 10x^2 + 21}{x^2 + 7}$$

$$= \frac{(x^2 + 7)(x^2 + 3)}{x^2 + 7}$$

$$= x^2 + 3$$

The width of the rectangle is $x^2 + 3$.

113. Let $w = \frac{x^2}{2(1-x)}$.

(a) $w = \frac{(0.1)^2}{2(1-0.1)}$ Let $x = 0.1$.

$$= \frac{0.01}{2(0.9)} \approx 0.006$$

To the nearest tenth, w is 0.

(b) $w = \frac{(0.8)^2}{2(1-0.8)}$ Let $x = 0.8$.

$$= \frac{0.64}{2(0.2)} = 1.6$$

(c) $w = \frac{(0.9)^2}{2(1-0.9)}$ Let $x = 0.9$.

$$= \frac{0.81}{2(0.1)} = 4.05 \approx 4.1$$

(d) Based on the answers in (a), (b), and (c), we see that as the traffic intensity increases, the waiting time also increases.

115. First, perform long division.

$$\begin{array}{r} 2x+3 \\ 4x+7 \overline{) 8x^2+26x+21} \\ \underline{8x^2+14x} \\ 12x+21 \\ \underline{12x+21} \\ 0 \end{array}$$

Then, simplify the rational expression.

$$\frac{8x^2+26x+21}{4x+7}$$

$$= \frac{(2x+3)(4x+7)}{4x+7}$$

$$= 2x+3$$

Both yield the expression $2x+3$.

117. First, perform long division.

$$\begin{array}{r} x^2 \quad +1 \\ x+1 \overline{) x^3+x^2+x+1} \\ \underline{x^3+x^2} \\ 0 \; +x+1 \\ \underline{x+1} \\ 0 \end{array}$$

Then, simplify the rational expression.

$$\frac{x^3+x^2+x+1}{x+1}$$

$$= \frac{(x^2+1)(x+1)}{x+1}$$

$$= x^2+1$$

Both yield the expression x^2+1.

7.2 Multiplying and Dividing Rational Expressions

Now Try Exercises

N1. **(a)** $\frac{7}{18}\cdot\frac{9}{14}=\frac{7\cdot 3\cdot 3}{2\cdot 3\cdot 3\cdot 2\cdot 7}$

$=\frac{1}{2\cdot 2}=\frac{1}{4}$

(b) $\frac{4k^2}{7}\cdot\frac{14}{11k}=\frac{2\cdot 2\cdot k\cdot k\cdot 2\cdot 7}{7\cdot 11\cdot k}$

$=\frac{2\cdot 2\cdot k\cdot 2}{11}=\frac{8k}{11}$

N2. $\frac{m-3}{3m}\cdot\frac{9m^2}{8(m-3)^2}$

$=\frac{(m-3)\cdot 9m^2}{3m\cdot 8(m-3)^2}$

$=\frac{(m-3)\cdot 3\cdot 3\cdot m\cdot m}{3\cdot m\cdot 2\cdot 2\cdot 2(m-3)^2}=\frac{3m}{8(m-3)}$

N3. $\frac{y^2-3y-28}{y^2-9y+14}\cdot\frac{y^2-7y+10}{y^2+4y}$

$=\frac{(y-7)(y+4)}{(y-7)(y-2)}\cdot\frac{(y-2)(y-5)}{y(y+4)}$ Factor.

$=\frac{(y-7)(y+4)(y-2)(y-5)}{(y-7)(y-2)y(y+4)}$ Multiply.

$=\frac{y-5}{y}$

N4. **(a)** $\frac{3}{10}\div\frac{11}{20}$

$=\frac{3}{10}\cdot\frac{20}{11}$ Multiply by reciprocal.

$=\frac{3}{2\cdot 5}\cdot\frac{2\cdot 2\cdot 5}{11}$ Factor.

$=\frac{3\cdot 2\cdot 2\cdot 5}{11\cdot 2\cdot 5}$ Multiply.

$=\frac{6}{11}$ Lowest terms

(b) $\frac{2x-5}{3x^2}\div\frac{2x-5}{12x}$

$=\frac{2x-5}{3x^2}\cdot\frac{12x}{2x-5}$ Multiply by reciprocal.

$=\frac{2x-5}{3\cdot x\cdot x}\cdot\frac{2\cdot 2\cdot 3\cdot x}{2x-5}$ Factor.

$=\frac{2\cdot 2\cdot 3\cdot x\cdot (2x-5)}{3\cdot x\cdot x\cdot (2x-5)}$ Multiply.

$=\frac{4}{x}$ Lowest terms

N5. $\frac{(3k)^3}{2j^4}\div\frac{9k^2}{6j}$

$=\frac{(3k)^3}{2j^4}\cdot\frac{6j}{9k^2}$ Multiply by reciprocal.

$=\frac{27k^3\cdot 6j}{2j^4\cdot 9k^2}$ Multiply.

$=\frac{9k}{j^3}$ Lowest terms

N6. $\frac{(t+2)(t-5)}{-4t}\div\frac{t^2-25}{(t+5)(t+2)}$

$=\frac{(t+2)(t-5)}{-4t}\cdot\frac{(t+5)(t+2)}{t^2-25}$ Multiply by reciprocal.

$=\frac{(t+2)(t-5)}{-4t}\cdot\frac{(t+5)(t+2)}{(t+5)(t-5)}$ Factor.

$=\frac{(t+2)(t-5)(t+5)(t+2)}{-4t(t+5)(t-5)}$ Multiply.

$=-\frac{(t+2)^2}{4t}$ Lowest terms

N7. $\frac{7-x}{2x+6}\div\frac{x^2-49}{x^2+6x+9}$

$=\frac{7-x}{2x+6}\cdot\frac{x^2+6x+9}{x^2-49}$

$=\frac{(7-x)\left(x^2+6x+9\right)}{(2x+6)\left(x^2-49\right)}$

$=\frac{(7-x)(x+3)(x+3)}{2(x+3)(x+7)(x-7)}$

$=\frac{(-1)(x+3)}{2(x+7)}$

$=-\frac{x+3}{2(x+7)}$

Exercises

1. (a) $\frac{5x^3}{10x^4} \cdot \frac{10x^7}{4x} = \frac{5 \cdot 10 \cdot x^3 \cdot x^7}{10 \cdot 4 \cdot x^4 \cdot x}$

$= \frac{5x^{10}}{4x^5}$

$= \frac{5x^5}{4}$ (B)

(b) $\frac{10x^4}{5x^3} \cdot \frac{10x^7}{4x} = \frac{5 \cdot 2 \cdot 5 \cdot 2 \cdot x^4 \cdot x^7}{5 \cdot 2 \cdot 2 \cdot x^3 \cdot x}$

$= \frac{5x^{11}}{1x^4}$

$= 5x^7$ (D)

(c) $\frac{5x^3}{10x^4} \cdot \frac{4x}{10x^7} = \frac{5 \cdot 2 \cdot 2 \cdot x^3 \cdot x}{5 \cdot 2 \cdot 5 \cdot 2 \cdot x^4 \cdot x^7}$

$= \frac{1x^4}{5x^{11}}$

$= \frac{1}{5x^7}$ (C)

(d) $\frac{10x^4}{5x^3} \cdot \frac{4x}{10x^7} = \frac{10 \cdot 4 \cdot x^4 \cdot x}{5 \cdot 10 \cdot x^3 \cdot x^7}$

$= \frac{4x^5}{5x^{10}}$

$= \frac{4}{5x^5}$ (A)

3. $\frac{15a^2}{14} \cdot \frac{7}{5a} = \frac{3 \cdot 5 \cdot a \cdot a \cdot 7}{2 \cdot 7 \cdot 5 \cdot a}$ Multiply and factor.

$= \frac{3 \cdot a(5 \cdot 7 \cdot a)}{2(5 \cdot 7 \cdot a)}$

$= \frac{3a}{2}$ Lowest terms

5. $\frac{12x^4}{18x^3} \cdot \frac{-8x^5}{4x^2} = \frac{-96x^9}{72x^5}$ Multiply.

$= \frac{-4x^4\left(24x^5\right)}{3\left(24x^5\right)}$ Group common factors.

$= -\frac{4x^4}{3}$ Lowest terms

7. $\frac{2(c+d)}{3} \cdot \frac{18}{6(c+d)^2}$

$= \frac{3 \cdot 3 \cdot 2 \cdot 2(c+d)}{3 \cdot 3 \cdot 2(c+d)(c+d)}$ Multiply and factor.

$= \frac{2}{c+d}$ Lowest terms

9. $\frac{(x-y)^2}{2} \cdot \frac{24}{3(x-y)}$

$= \frac{6 \cdot 4(x-y)(x-y)}{6(x-y)}$

$= 4(x-y)$

11. $\frac{t-4}{8} \cdot \frac{4t^2}{t-4}$

$= \frac{4t^2(t-4)}{2 \cdot 4(t-4)}$

$= \frac{t^2}{2}$

13. $\frac{3x}{x+3} \cdot \frac{(x+3)^2}{6x^2}$

$= \frac{3x(x+3)(x+3)}{2 \cdot 3 \cdot x \cdot x(x+3)}$

$= \frac{x+3}{2x}$

15. $\frac{5x-10}{6} \cdot \frac{9}{10x-20}$

$= \frac{5(x-2)}{6} \cdot \frac{3 \cdot 3}{10(x-2)}$

$= \frac{5(x-2) \cdot 3 \cdot 3}{2 \cdot 3 \cdot 2 \cdot 5 \cdot (x-2)}$

$= \frac{3}{4}$

17. $\frac{9z^4}{3z^5} \div \frac{3z^2}{5z^3} = \frac{9z^4}{3z^5} \cdot \frac{5z^3}{3z^2}$

$= \frac{9 \cdot 5z^7}{3 \cdot 3z^7}$

$= 5$

19. Rewrite the division as a multiplication problem and simplify.

$$\frac{4t^4}{2t^5} \div \frac{(2t)^3}{-6} = \frac{4t^4}{2t^5} \cdot \frac{-6}{(2t)^3}$$
$$= \frac{4t^4}{2t^5} \cdot \frac{-6}{8t^3}$$
$$= \frac{-24t^4}{16t^8}$$
$$= \frac{-3\left(8t^4\right)}{2t^4\left(8t^4\right)}$$
$$= \frac{-3}{2t^4} = -\frac{3}{2t^4}$$

21. $$\frac{3}{2y-6} \div \frac{6}{y-3} = \frac{3}{2y-6} \cdot \frac{y-3}{6}$$
$$= -\frac{3}{2(y-3)} \cdot \frac{y-3}{6}$$
$$= \frac{3(y-3)}{2 \cdot 2 \cdot 3(y-3)}$$
$$= \frac{1}{2 \cdot 2} = \frac{1}{4}$$

23. $$\frac{7t+7}{-6} \div \frac{4t+4}{15}$$
$$= \frac{7t+7}{-6} \cdot \frac{15}{4t+4}$$
$$= \frac{7(t+1)}{-2 \cdot 3} \cdot \frac{3 \cdot 5}{4(t+1)}$$
$$= \frac{3 \cdot 5 \cdot 7(t+1)}{-2 \cdot 3 \cdot 4(t+1)} = -\frac{35}{8}$$

25. $$\frac{2x}{x-1} \div \frac{x^2}{x+2}$$
$$= \frac{2x}{x-1} \cdot \frac{x+2}{x^2}$$
$$= \frac{2x(x+2)}{x \cdot x(x-1)} = \frac{2(x+2)}{x(x-1)}$$

27. $$\frac{(x-3)^2}{6x} \div \frac{x-3}{x^2}$$
$$= \frac{(x-3)^2}{6x} \cdot \frac{x^2}{x-3}$$
$$= \frac{x \cdot x(x-3)(x-3)}{6x(x-3)} = \frac{x(x-3)}{6}$$

29. $$\frac{5x^3}{x^2-16} \div \frac{x^5}{(x-4)^2}$$
$$= \frac{5x^3}{x^2-16} \cdot \frac{(x-4)^2}{x^5}$$
$$= \frac{5x^3(x-4)^2}{(x^2-16)x^5}$$
$$= \frac{5x^3(x-4)^2}{(x+4)(x-4)x^5}$$
$$= \frac{5(x-4)}{x^2(x+4)}$$

31. $$\frac{-4t^3}{t^2-1} \div \frac{t^2}{(t+1)^2}$$
$$= \frac{-4t^3}{t^2-1} \cdot \frac{(t+1)^2}{t^2}$$
$$= \frac{-4t^3(t+1)^2}{(t^2-1)t^2}$$
$$= \frac{-4t^3(t+1)^2}{(t+1)(t-1)t^2}$$
$$= \frac{-4t(t+1)}{t-1}$$

33. $$\frac{5x-15}{3x+9} \cdot \frac{4x+12}{6x-18}$$
$$= \frac{5(x-3)}{3(x+3)} \cdot \frac{4(x+3)}{6(x-3)}$$
$$= \frac{5 \cdot 4 \cdot (x-3)(x+3)}{3 \cdot 6 \cdot (x-3)(x+3)}$$
$$= \frac{10}{9}$$

35. $$\frac{2-t}{8} \div \frac{t-2}{6}$$
$$= \frac{2-t}{8} \cdot \frac{6}{t-2}$$ Multiply by reciprocal.
$$= \frac{6(2-t)}{8(t-2)}$$ Multiply.
$$= \frac{6(-1)}{8}$$ Factor.
$$= -\frac{3}{4}$$ Lowest terms

37. $\frac{27-3z}{4}\cdot\frac{12}{2z-18}$

$=\frac{3(9-z)}{4}\cdot\frac{3\cdot 4}{2(z-9)}$

$=\frac{3\cdot 3\cdot 4(9-z)}{4\cdot 2(z-9)}$

$=\frac{3\cdot 3\cdot(-1)}{2}$

$=-\frac{9}{2}$

39. $\frac{p^2+4p-5}{p^2+7p+10}\div\frac{p-1}{p+4}$

$=\frac{p^2+4p-5}{p^2+7p+10}\cdot\frac{p+4}{p-1}$

$=\frac{(p+5)(p-1)\cdot(p+4)}{(p+5)(p+2)\cdot(p-1)}$

$=\frac{p+4}{p+2}$

41. $\frac{m^2-4}{16-8m}\div\frac{m+2}{8}$

$=\frac{(m+2)(m-2)}{8(2-m)}\cdot\frac{8}{m+2}$

$=\frac{8(m+2)(m-2)}{8(m+2)(2-m)}$

$=-1$

43. $\frac{m^2-4}{16-8m}\div\frac{m^2+3m+2}{8m+16}$

$=\frac{(m+2)(m-2)}{8(2-m)}\cdot\frac{8(m+2)}{(m+2)(m+1)}$

$=\frac{8(m+2)^2(m-2)}{8(2-m)(m+2)(m+1)}$

$=-\frac{m+2}{m+1}$

45. $\frac{2x^2-7x+3}{x-3}\cdot\frac{x+2}{x-1}$

$=\frac{(2x-1)(x-3)}{x-3}\cdot\frac{x+2}{x-1}$

$=\frac{(2x-1)(x+2)}{x-1}$

47. $\frac{2k^2-k-1}{2k^2+5k+3}\div\frac{4k^2-1}{2k^2+k-3}$

$=\frac{2k^2-k-1}{2k^2+5k+3}\cdot\frac{2k^2+k-3}{4k^2-1}$

$=\frac{(2k+1)(k-1)(2k+3)(k-1)}{(2k+3)(k+1)(2k+1)(2k-1)}$

$=\frac{(k-1)(k-1)}{(k+1)(2k-1)}$

$=\frac{(k-1)^2}{(k+1)(2k-1)}$

49. $\frac{2k^2+3k-2}{6k^2-7k+2}\cdot\frac{4k^2-5k+1}{k^2+k-2}$

$=\frac{(2k-1)(k+2)}{(3k-2)(2k-1)}\cdot\frac{(4k-1)(k-1)}{(k+2)(k-1)}$

$=\frac{(2k-1)(k+2)(4k-1)(k-1)}{(3k-2)(2k-1)(k+2)(k-1)}$

$=\frac{4k-1}{3k-2}$

51. $\frac{m^2+2mp-3p^2}{m^2-3mp+2p^2}\div\frac{m^2+4mp+3p^2}{m^2+2mp-8p^2}$

$=\frac{m^2+2mp-3p^2}{m^2-3mp+2p^2}\cdot\frac{m^2+2mp-8p^2}{m^2+4mp+3p^2}$

$=\frac{(m+3p)(m-p)(m+4p)(m-2p)}{(m-2p)(m-p)(m+3p)(m+p)}$

$=\frac{m+4p}{m+p}$

53. $\frac{m^2+3m+2}{m^2+5m+4}\cdot\frac{m^2+10m+24}{m^2+5m+6}$

$=\frac{(m+2)(m+1)}{(m+4)(m+1)}\cdot\frac{(m+6)(m+4)}{(m+3)(m+2)}$

$=\frac{(m+2)(m+1)(m+6)(m+4)}{(m+4)(m+1)(m+3)(m+2)}$

$=\frac{m+6}{m+3}$

55. $\dfrac{y^2+y-2}{y^2+3y-4} \div \dfrac{y+2}{y+3}$

$= \dfrac{y^2+y-2}{y^2+3y-4} \cdot \dfrac{y+3}{y+2}$

$= \dfrac{(y+2)(y-1)}{(y+4)(y-1)} \cdot \dfrac{y+3}{y+2}$

$= \dfrac{y+3}{y+4}$

57. $\dfrac{2m^2+7m+3}{m^2-9} \cdot \dfrac{m^2+3m}{2m^2+11m+5}$

$= \dfrac{(2m+1)(m+3)}{(m-3)(m+3)} \cdot \dfrac{m(m-3)}{(2m+1)(m+5)}$

$= \dfrac{(2m+1)(m+3)m(m-3)}{(m-3)(m+3)(2m+1)(m+5)}$

$= \dfrac{m}{m+5}$

59. $\dfrac{r^2+rs-12s^2}{r^2-rs-20s^2} \div \dfrac{r^2-2rs+3s^2}{r^2+rs-30s^2}$

$= \dfrac{r^2+rs-12s^2}{r^2-rs-20s^2} \cdot \dfrac{r^2+rs-30s^2}{r^2+2rs-3s^2}$

$= \dfrac{(r-3s)(r+4s)(r+6s)(r-5s)}{(r-5s)(r+4s)(r-3s)(r+s)}$

$= \dfrac{r+6s}{r+s}$

61. $\dfrac{(q-3)^4(q+2)}{q^2+3q+2} \div \dfrac{q^2-6q+9}{q^2+4q+4}$

$= \dfrac{(q-3)^4(q+2)}{q^2+3q+2} \cdot \dfrac{q^2+4q+4}{q^2-6q+9}$

$= \dfrac{(q-3)^4(q+2)(q+2)^2}{(q+2)(q+1)(q-3)^2}$

$= \dfrac{(q-3)^2(q+2)^2}{q+1}$

63. $\dfrac{3a-3b-a^2+b^2}{4a^2-4ab+b^2} \cdot \dfrac{4a^2-b^2}{2a^2-ab-b^2}$

Factor $3a-3b-a^2+b^2$ by grouping.

$3a-3b-a^2+b^2$

$= 3(a-b)-(a^2-b^2)$

$= 3(a-b)-(a-b)(a+b)$

$= (a-b)[3-(a+b)]$

$= (a-b)(3-a-b)$

Thus,

$\dfrac{3a-3b-a^2+b^2}{4a^2-4ab+b^2} \cdot \dfrac{4a^2-b^2}{2a^2-ab-b^2}$

$= \dfrac{(a-b)(3-a-b)}{(2a-b)(2a-b)} \cdot \dfrac{(2a-b)(2a+b)}{(2a+b)(a-b)}$

$= \dfrac{(a-b)(3-a-b)(2a-b)(2a+b)}{(2a-b)(2a-b)(2a+b)(a-b)}$

$= \dfrac{3-a-b}{2a-b}.$

65. $\dfrac{-x^3-y^3}{x^2-2xy+y^2} \div \dfrac{3y^2-3xy}{x^2-y^2}$

$= \dfrac{-1(x^3+y^3)}{x^2-2xy+y^2} \cdot \dfrac{x^2-y^2}{3y^2-3xy}$

$= \dfrac{-1(x+y)(x^2-xy+y^2)}{(x-y)(x-y)} \cdot \dfrac{(x-y)(x+y)}{3y(y-x)}$

$= \dfrac{-1(x+y)(x^2-xy+y^2)(x-y)(x+y)}{-1(x-y)(x-y)(3y)(x-y)}$

$= \dfrac{(x+y)^2(x^2-xy+y^2)}{3y(x-y)^2}$

If we had not changed $y-x$ to $-1(x-y)$ in the denominator, we would have obtained an alternate form of the answer,

$-\dfrac{(x+y)^2(x^2-xy+y^2)}{3y(y-x)(x-y)}.$

67. $\dfrac{x+5}{x+10} \div \left(\dfrac{x^2+10x+25}{x^2+10x} \cdot \dfrac{10x}{x^2+15x+50}\right)$

$= \dfrac{x+5}{x+10} \div \left[\dfrac{(x+5)^2 \cdot 10x}{x(x+10)(x+5)(x+10)}\right]$

$= \dfrac{x+5}{x+10} \div \left[\dfrac{10(x+5)}{(x+10)^2}\right]$

$= \dfrac{x+5}{x+10} \cdot \dfrac{(x+10)^2}{10(x+5)}$

$= \dfrac{x+10}{10}$

69. Use the formula for the area of a rectangle with $A = \dfrac{5x^2y^3}{2pq}$ and $L = \dfrac{2xy}{p}$ to solve for W.

$$A = L \cdot W$$

$$\frac{5x^2y^3}{2pq} = \frac{2xy}{p} \cdot W$$

$$W = \frac{5x^2y^3}{2pq} \div \frac{2xy}{p}$$

$$= \frac{5x^2y^3}{2pq} \cdot \frac{p}{2xy}$$

$$= \frac{5x^2y^3p}{4pqxy}$$

$$= \frac{5xy^2}{4q}$$

Thus, the rational expression $\dfrac{5xy^2}{4q}$ represents the width of the rectangle.

7.3 Least Common Denominators

Now Try Exercises

N1. **(a)** $\dfrac{5}{48}, \dfrac{1}{30}$

Factor each denominator.

$48 = 2 \cdot 2 \cdot 2 \cdot 2 \cdot 3, \quad 30 = 2 \cdot 3 \cdot 5$

Take each different factor the *greatest* number of times it appears as a factor in any of the denominators, and use it to form the least common denominator (LCD).

$\text{LCD} = 2 \cdot 2 \cdot 2 \cdot 2 \cdot 3 \cdot 5 = 240$

(b) $\dfrac{3}{10y}, \dfrac{1}{6y}$

Factor each denominator.

$10y = 2 \cdot 5 \cdot y$

$6y = 2 \cdot 3 \cdot y$

Take each factor the greatest number of times it appears in any denominator; then multiply.

$\text{LCD} = 2 \cdot 3 \cdot 5 \cdot y = 30y$

N2. $\dfrac{5}{6x^4}, \dfrac{7}{8x^3}$

Factor each denominator.

$6x^4 = 2 \cdot 3 \cdot x^4$

$8x^3 = 2 \cdot 2 \cdot 2 \cdot x^3$

Take each factor the greatest number of times it appears in any denominator; then multiply.

$\text{LCD} = 2 \cdot 2 \cdot 2 \cdot 3 \cdot x^4 = 24x^4$

N3. **(a)** $\dfrac{3t}{2t^2 - 10t}, \dfrac{t+4}{t^2 - 25}$

Factor each denominator.

$2t^2 - 10t = 2t(t-5)$

$t^2 - 25 = (t+5)(t-5)$

Take each factor the greatest number of times it appears in any denominator; then multiply.

$\text{LCD} = 2t(t-5)(t+5)$

(b) $\dfrac{1}{x^2 + 7x + 12}, \dfrac{2}{x^2 + 6x + 9}, \dfrac{5}{x^2 + 2x - 8}$

Factor each denominator.

$x^2 + 7x + 12 = (x+3)(x+4)$

$x^2 + 6x + 9 = (x+3)(x+3)$

$x^2 + 2x - 8 = (x+4)(x-2)$

Take each factor the greatest number of times it appears in any denominator; then multiply.

$\text{LCD} = (x+3)^2(x+4)(x-2)$

(c) $\dfrac{2}{a-4}, \dfrac{1}{4-a}$

The expressions $a - 4$ and $4 - a$ are opposites of each other because of the following.

$-(a-4) = -a + 4 = 4 - a$

Therefore, either $a - 4$ or $4 - a$ can be used as the LCD.

N4. **(a)** $\frac{2}{9}=\frac{?}{27}$

First factor the denominator on the right. Then compare the denominator on the left with the one on the right to decide what factors are missing.

$\frac{2}{9}=\frac{?}{3\cdot 9}$

A factor of 3 is missing, so multiply $\frac{2}{9}$ by $\frac{3}{3}$, which is equal to 1.

$\frac{2}{9}=\frac{2}{9}\cdot\frac{3}{3}=\frac{6}{27}$

(b) $\frac{4t}{11}=\frac{?}{33t}$

Factor the denominator on the right; then compare it to the denominator on the left.

$\frac{4t}{11}=\frac{?}{3\cdot 11\cdot t}$

The factors 3 and t are missing on the left, so multiply by $\frac{3t}{3t}$.

$\frac{4t}{11}=\frac{4t}{11}\cdot\frac{3t}{3t}=\frac{12t^2}{33t}$

N5. **(a)** $\frac{8k}{5k-2}=\frac{?}{25k-10}$

Factor the denominator on the right.

$\frac{8k}{5k-2}=\frac{?}{5(5k-2)}$

The factor 5 is missing on the left, so multiply by $\frac{5}{5}$.

$\frac{8k}{5k-2}=\frac{8k}{5k-2}\cdot\frac{5}{5}=\frac{40k}{25k-10}$

(b) $\frac{2t-1}{t^2+4t}=\frac{?}{t^3+12t^2+32t}$

Factor and compare the denominators.

$\frac{2t-1}{t(t+4)}=\frac{?}{t(t+4)(t+8)}$

The factor $t+8$ is missing on the left, so multiply by $\frac{t+8}{t+8}$.

$$\frac{2t-1}{t^2+4t}=\frac{2t-1}{t(t+4)}\cdot\frac{t+8}{t+8}=\frac{(2t-1)(t+8)}{t(t+4)(t+8)}$$

Exercises

1. The factor x appears at most one time in any denominator as does the factor y. Thus, the LCD is the product of the two factors, xy. The correct response is C.

3. Since $20=2^2\cdot 5$, the LCD of $\frac{9}{20}$ and $\frac{1}{2}$ must have 5 as a factor and 2^2 as a factor. Because 2 appears twice in $2^2\cdot 5$, we don't have to include another 2 in the LCD for the number $\frac{1}{2}$. Thus, the LCD is just $2^2\cdot 5=20$. Note that this is a specific case of Exercise 2 since 2 is a factor of 20. The correct response is C.

5. Find the LCD for the pair of fractions.

$\frac{7}{10},\frac{1}{25}$

Step 1

$10=2\cdot\underline{5}$

$25=\underline{5}\cdot 5$

Step 2

The greatest number of times 2 appears is <u>one</u>.
The greatest number of times <u>5</u> appears is two.

Step 3

$\text{LCD}=\underline{2}\cdot\underline{5}\cdot 5$

$\text{LCD}=\underline{50}$

7. $\frac{7}{15},\frac{21}{20}$

Factor each denominator.

$15=3\cdot 5$

$20=2\cdot 2\cdot 5=2^2\cdot 5$

Take each factor the greatest number of times it appears as a factor in any one of the denominators.

$\text{LCD}=2^2\cdot 3\cdot 5=60$

9. $\frac{17}{100}, \frac{23}{120}, \frac{43}{180}$

Factor each denominator.

$100 = 2^2 \cdot 5^2$

$120 = 2^3 \cdot 3 \cdot 5$

$180 = 2^2 \cdot 3^2 \cdot 5$

Take each factor the greatest number of times it appears as a factor in any one of the denominators.

LCD $= 2^3 \cdot 3^2 \cdot 5^2 = 1800$

11. $\frac{9}{x^2}, \frac{8}{x^5}$

The greatest number of times x appears as a factor in any denominator is the greatest exponent on x, which is 5.

LCD $= x^5$

13. $\frac{-2}{5p}, \frac{13}{6p}$

Factor each denominator.

$5p = 5 \cdot p$

$6p = 2 \cdot 3 \cdot p$

Take each factor the greatest number of times it appears; then multiply.

LCD $= 2 \cdot 3 \cdot 5 \cdot p = 30p$

15. $\frac{17}{15y^2}, \frac{55}{36y^4}$

Factor each denominator.

$15y^2 = 3 \cdot 5 \cdot y^2$

$36y^4 = 2^2 \cdot 3^2 \cdot y^4$

Take each factor the greatest number of times it appears; then multiply.

LCD $= 2^2 \cdot 3^2 \cdot 5 \cdot y^4 = 180y^4$

17. $\frac{5}{21r^3}, \frac{7}{12r^5}$

Factor each denominator.

$21r^3 = 3 \cdot 7 \cdot r^3$

$12r^5 = 2^2 \cdot 3 \cdot r^5$

Take each factor the greatest number of times it appears; then multiply.

LCD $= 2^2 \cdot 3 \cdot 7 \cdot r^5 = 84r^5$

19. $\frac{13}{5a^2b^3}, \frac{29}{15a^5b}$

Factor each denominator.

$5a^2b^3 = 5 \cdot a^2 \cdot b^3$

$15a^5b = 3 \cdot 5 \cdot a^5 \cdot b$

Take each factor the greatest number of times it appears; then multiply.

LCD $= 3 \cdot 5 \cdot a^5 \cdot b^3 = 15a^5b^3$

21. $\frac{1}{r^2t^3}, \frac{1}{r^5t}, \frac{1}{r^9t^2}$

Factor each denominator.

$r^2t^3 = r^2 \cdot t^3$

$r^5t = r^5 \cdot t$

$r^9t^2 = r^9 \cdot t^2$

Take each factor the greatest number of times it appears; then multiply.

LCD $= r^9 \cdot t^3 = r^9t^3$

23. $\frac{7}{x+1}, \frac{9}{x-1}$

Since there is only one factor in each denominator, take each factor the greatest number of times it appears; then multiply.

LCD $= (x+1) \cdot (x-1) = (x+1)(x-1)$

25. $\frac{7}{6p}, \frac{15}{4p-8}$

Factor each denominator.

$6p = 2 \cdot 3 \cdot p$

$4p - 8 = 4(p-2) = 2^2(p-2)$

Take each factor the greatest number of times it appears; then multiply.

LCD$=2^2 \cdot 3 \cdot p(p-2) = 12p(p-2)$

27. $\frac{9}{28m^2}, \frac{3}{12m-20}$

Factor each denominator.

$28m^2 = 2^2 \cdot 7 \cdot m^2$

$12m - 20 = 4(3m-5) = 2^2(3m-5)$

Take each factor the greatest number of times it appears; then multiply.

LCD $= 2^2 \cdot 7m^2(3m-5) = 28m^2(3m-5)$

29. $\frac{7}{5b-10}, \frac{11}{6b-12}$

Factor each denominator.

$5b-10=5(b-2)$

$6b-12=6(b-2)=2\cdot 3(b-2)$

Take each factor the greatest number of times it appears; then multiply.

$\text{LCD}=2\cdot 3\cdot 5(b-2)=30(b-2)$

31. $\frac{37}{6r-12}, \frac{25}{9r-18}$

Factor each denominator.

$6r-12=6(r-2)=2\cdot 3(r-2)$

$9r-18=9(r-2)=3^2(r-2)$

Take each factor the greatest number of times it appears; then multiply.

$\text{LCD}=2\cdot 3^2(r-2)=18(r-2)$

33. $\frac{5}{c-d}, \frac{8}{d-c}$

The denominators, $c-d$ and $d-c$, are opposites of each other since the following is true.

$-(c-d)=-c+d=d-c$

Therefore, either $c-d$ or $d-c$ can be used as the LCD.

35. $\frac{12}{m-3}, \frac{-4}{3-m}$

The expression $3-m$ can be written as $-1(m-3)$, since the following is true.

$-1(m-3)=-m+3=3-m$

Because of this, either $m-3$ or $3-m$ can be used as the LCD.

37. $\frac{29}{p-q}, \frac{18}{q-p}$

The expression $q-p$ can be written as $-1(p-q)$, since the following is true.

$-1(p-q)=-p+q=q-p$

Because of this, either $p-q$ or $q-p$ can be used as the LCD.

39. $\frac{13}{x^2-1}, \frac{-5}{2x+2}$

Factor each denominator.

$x^2-1=(x+1)(x-1)$

$2x+2=2(x+1)$

$\text{LCD}=2(x+1)(x-1)$

41. $\frac{4x^2}{(x-4)^2}, \frac{17x}{3x-12}$

Factor each denominator.

$(x-4)^2=(x-4)(x-4)$

$3x-12=3(x-4)$

$\text{LCD}=3(x-4)^2$

43. $\frac{5}{12p+60}, \frac{-17}{p^2+5p}, \frac{-16}{p^2+10p+25}$

Factor each denominator.

$12p+60=12(p+5)=2^2\cdot 3(p+5)$

$p^2+5p=p(p+5)$

$p^2+10p+25=(p+5)(p+5)$

$\text{LCD}=2^2\cdot 3\cdot p(p+5)^2=12p(p+5)^2$

45. $\frac{-3}{8y+16}, \frac{22}{y^2+3y+2}$

Factor each denominator.

$8y+16=8(y+2)=2^3(y+2)$

$y^2+3y+2=(y+2)(y+1)$

$\text{LCD}=8(y+2)(y+1)$

47. $\frac{3}{k^2+5k}, \frac{2}{k^2+3k-10}$

Factor each denominator.

$k^2+5k=k(k-5)$

$k^2+3k-10=(k-5)(k-2)$

$\text{LCD}=k(k+5)(k-2)$

49. $\frac{6}{a^2+6a}, \frac{-5}{a^2+3a-18}$

Factor each denominator.

$a^2+6a=a(a+6)$

$a^2+3a-18=(a+6)(a-3)$

$\text{LCD}=a(a+6)(a-3)$

51. $\frac{5}{p^2+8p+15}, \frac{3}{p^2-3p-18}, \frac{12}{p^2-p-30}$

Factor each denominator.

$p^2+8p+15=(p+5)(p+3)$

$p^2-3p-18=(p-6)(p+3)$

$p^2-p-30=(p-6)(p+5)$

$\text{LCD}=(p+3)(p+5)(p-6)$

53. $\frac{-5}{k^2+2k-35}, \frac{-8}{k^2+3k-40}, \frac{19}{k^2-2k-15}$

Factor each denominator.

$k^2+2k-35=(k+7)(k-5)$

$k^2+3k-40=(k+8)(k-5)$

$k^2-2k-15=(k-5)(k+3)$

$\text{LCD}=(k+7)(k-5)(k+8)(k+3)$

55. $\frac{4}{11}=\frac{?}{55}$

First factor the denominator on the right. Then compare the denominator on the left with the one on the right to decide what factors are missing.

$\frac{4}{11}=\frac{?}{11\cdot 5}$

A factor of 5 is missing, so multiply $\frac{4}{11}$ by $\frac{5}{5}$, which is equal to 1.

$\frac{4}{11}\cdot\frac{5}{5}=\frac{20}{55}$

57. $\frac{-5}{k}=\frac{?}{9k}$

A factor of 9 is missing in the first fraction, so multiply the numerator and the denominator by 9.

$\frac{-5}{k}\cdot\frac{9}{9}=\frac{-45}{9k}$

59. $\frac{15m^2}{8k}=\frac{?}{32k^4}$

$32k^4=(8k)(4k^3)$, so we must multiply the numerator and the denominator by $4k^3$.

$\frac{15m^2}{8k}=\frac{15m^2}{8k}\cdot\frac{4k^3}{4k^3}$ Mult. identity prop.

$=\frac{60m^2k^3}{32k^4}$

61. $\frac{19}{2z-6}=\frac{?}{6z-18}$

Begin by factoring each denominator.

$2z-6=2(z-3)$

$6z-18=6(z-3)$

The fractions may now be written as follows.

$\frac{19z}{2(z-3)}=\frac{?}{6(z-3)}$

Comparing the two factored forms, we see that the denominator of the fraction on the left side must be multiplied by 3; the numerator must also be multiplied by 3.

$\frac{19z}{2z-6}$

$=\frac{19z}{2(z-3)}\cdot\frac{3}{3}$ Multiplicative identity property

$=\frac{19z(3)}{2(z-3)(3)}$ Multiply.

$=\frac{57z}{6z-18}$ Multiply the factors.

63. $\frac{-2a}{9a-18}=\frac{?}{18a-36}$

$\frac{-2a}{9(a-2)}=\frac{?}{18(a-2)}$ Factor denominators.

$\frac{-2a}{9a-18}=\frac{-2a}{9(a-2)}\cdot\frac{2}{2}$ Mult. identity prop.

$=\frac{-4a}{18a-36}$ Multiply.

65. $\frac{6}{k^2-4k}=\frac{?}{k(k-4)(k+1)}$

$\frac{6}{k(k-4)}=\frac{?}{k(k-4)(k+1)}$

$\frac{6}{k^2-4k}=\frac{6}{k(k-4)}\cdot\frac{(k+1)}{(k+1)}$

$=\frac{6(k+1)}{k(k-4)(k+1)}$

67. $\frac{4r-t}{r^2+rt+t^2}=\frac{?}{t^3-r^3}$

Factor the second denominator as the difference of cubes.

$t^3-r^3=(t-r)(t^2+rt+r^2)$

$\frac{4r-t}{r^2+rt+t^2}=\frac{(4r-t)}{(r^2+rt+t^2)}\cdot\frac{(t-r)}{(t-r)}$

$=\frac{(t-r)(4r-t)}{t^3-r^3}$

69. $\dfrac{2(z-y)}{y^2+yz+z^2}=\dfrac{?}{y^4-z^3y}$

Factor the second denominator.

$$y^4-z^3y=y\left(y^3-z^3\right)$$
$$=y(y-z)\left(y^2+yz+z^2\right)$$

$$\frac{2(z-y)}{y^2+yz+z^2}=\frac{?}{y(y-z)\left(y^2+yz+z^2\right)}$$

$$\frac{2(z-y)}{y^2+yz+z^2}=\frac{2(z-y)}{\left(y^2+yz+z^2\right)}\cdot\frac{y(y-z)}{y(y-z)}$$
$$=\frac{2y(z-y)(y-z)}{y(y-z)\left(y^2+yz+z^2\right)}$$
$$=\frac{2y(z-y)(y-z)}{y\left(y^3-z^3\right)}$$
$$=\frac{2y(z-y)(y-z)}{y^4-z^3y},$$
$$\text{or } \frac{-2y(y-z)^2}{y^4-z^3y}$$

71. $\dfrac{36}{r^2-r-6}=\dfrac{?}{(r-3)(r+2)(r+1)}$

$$\frac{36}{(r-3)(r+2)}=\frac{?}{(r-3)(r+2)(r+1)}$$
$$\frac{36r}{r^2-r-6}=\frac{36r}{(r-3)(r+2)}\cdot\frac{(r+1)}{(r+1)}$$
$$=\frac{36r(r+1)}{(r-3)(r+2)(r+1)}$$

73. $\dfrac{a+2b}{2a^2+ab-b^2}=\dfrac{?}{2a^3b+a^2b^2-ab^3}$

$$\frac{a+2b}{2a^2+ab-b^2}=\frac{?}{ab\left(2a^2+ab-b^2\right)}$$
$$\frac{a+2b}{2a^2+ab-b^2}=\frac{(a+2b)}{\left(2a^2+ab-b^2\right)}\cdot\frac{ab}{ab}$$
$$=\frac{ab(a+2b)}{2a^3b+a^2b^2-ab^3}$$

75. $\dfrac{3}{4}=\dfrac{?}{28}$

To change 4 into 28, multiply by 7. If you multiply the denominator by 7, you must multiply the numerator by 7.

77. Since $\frac{7}{7}$ has a value of 1, the multiplier is 1.

The *identity property of multiplication* is being used when we write a common fraction as an equivalent one with a larger denominator.

79. $\dfrac{2x+5}{x-4}=\dfrac{?}{7x-28}=\dfrac{?}{7(x-4)}$

To form the new denominator, 7 must be used as the multiplier for the denominator. To form an equivalent fraction, the same multiplier must be used for numerator and denominator. Thus, the multiplier is $\frac{7}{7}$, which is equal to 1.

7.4 Adding and Subtracting Rational Expressions

Now Try Exercises

N1. (a) $\dfrac{2}{7k}+\dfrac{4}{7k}=\dfrac{2+4}{7k}$

$$=\frac{6}{7k}$$

(b) $\dfrac{4y}{y+3}+\dfrac{12}{y+3}=\dfrac{4y+12}{y+3}$

$$=\frac{4(y+3)}{y+3}$$
$$=4$$

N2. (a) $\dfrac{5}{12}+\dfrac{3}{20}$

Step 1

$\text{LCD}=2\cdot2\cdot3\cdot5=60$

Step 2

$$\frac{5}{12}=\frac{5(5)}{12(5)}=\frac{25}{60}$$
$$\frac{3}{20}=\frac{3(3)}{20(3)}=\frac{9}{60}$$

Step 3

$$\frac{25}{60}+\frac{9}{60}=\frac{25+9}{60}$$
$$=\frac{34}{60}$$

Step 4

$$\frac{34}{60}=\frac{17(2)}{30(2)}=\frac{17}{30}$$

(b) $\frac{3}{5x}+\frac{2}{7x}$

Step 1

$\text{LCD}=5\cdot 7\cdot x=35x$

Step 2

$\frac{3}{5x}=\frac{3(7)}{5x(7)}=\frac{21}{35x}$

$\frac{2}{7x}=\frac{2(5)}{7x(5)}=\frac{10}{35x}$

Step 3

$\frac{21}{35x}+\frac{10}{35x}=\frac{31}{35x}$

N3. $\frac{6t}{t^2-9}+\frac{-3}{t+3}$

$=\frac{6t}{(t+3)(t-3)}+\frac{-3}{t+3}$ Factor.

$=\frac{6t}{(t+3)(t-3)}+\frac{-3(t-3)}{(t+3)(t-3)}$

$=\frac{6t+[-3(t-3)]}{(t+3)(t-3)}$ Add numerators.

$=\frac{6t-3t+9}{(t+3)(t-3)}$ Dist. property

$=\frac{3t+9}{(t+3)(t-3)}$ Combine terms.

$=\frac{3(t+3)}{(t+3)(t-3)}$ Factor numerator.

$=\frac{3}{t-3}$ Lowest terms

N4. $\frac{x-1}{x^2+6x+8}+\frac{4x}{x^2+x-12}$

$=\frac{x-1}{(x+2)(x+4)}+\frac{4x}{(x+4)(x-3)}$

$=\frac{(x-1)(x-3)}{(x-2)(x+4)(x-3)}+\frac{4x(x+2)}{(x+4)(x-3)(x+2)}$

$=\frac{(x-1)(x-3)+4x(x+2)}{(x+4)(x+2)(x-3)}$

$=\frac{x^2-4x+3+4x^2+8x}{(x+4)(x+2)(x-3)}$

$=\frac{5x^2+4x+3}{(x+4)(x+2)(x-3)}$

N5. $\frac{2k}{k-7}+\frac{5}{7-k}$

$=\frac{2k}{k-7}+\frac{5(-1)}{(7-k)(-1)}$

$=\frac{2k}{k-7}+\frac{-5}{k-7}$

$=\frac{2k-5}{k-7}$, or $\frac{5-2k}{7-k}$

N6. $\frac{2x}{x+5}-\frac{x+1}{x+5}$

$=\frac{2x-(x+1)}{x+5}$ Use parentheses.

$=\frac{2x-x-1}{x+5}$

$=\frac{x-1}{x+5}$

N7. $\frac{6}{y-6}-\frac{2}{y}$

$=\frac{6(y)}{(y-6)(y)}-\frac{2(y-6)}{y(y-6)}$ $\text{LCD}=y(y-6)$

$=\frac{6y-2(y-6)}{y(y-6)}$

$=\frac{6y-2y+12}{y(y-6)}$

$=\frac{4y+12}{y(y-6)}$

$=\frac{4(y+3)}{y(y-6)}$

N8. The denominators are opposites, so either may be used as the common denominator. We will choose $m-4$.

$\frac{2m}{m-4}-\frac{m-12}{4-m}$

$=\frac{2m}{m-4}-\frac{(m-12)(-1)}{(4-m)(-1)}$

$=\frac{2m}{m-4}-\frac{-m+12}{m-4}$

$=\frac{2m-(-m+12)}{m-4}$

$=\frac{2m+m-12}{m-4}$

$=\frac{3m-12}{m-4}$

$=\frac{3(m-4)}{m-4}=3$

N9. $\dfrac{5}{t^2-6t+9}-\dfrac{2t}{t^2-9}$

$=\dfrac{5}{(t-3)^2}-\dfrac{2t}{(t+3)(t-3)}$ Factor.

$=\dfrac{5(t+3)}{(t-3)^2(t+3)}-\dfrac{2t(t-3)}{(t-3)^2(t+3)}$

$=\dfrac{5(t+3)-2t(t-3)}{(t-3)^2(t+3)}$

$=\dfrac{5t+15-2t^2+6t}{(t-3)^2(t+3)}$

$=\dfrac{-2t^2+11t+15}{(t-3)^2(t+3)}$

N10. $\dfrac{q}{2q^2+5q-3}-\dfrac{3q+4}{3q^2+10q+3}$

$=\dfrac{q}{(2q-1)(q+3)}-\dfrac{3q+4}{(3q+1)(q+3)}$

$=\dfrac{q(3q+1)}{(2q-1)(q+3)(3q+1)}-\dfrac{(3q+4)(2q-1)}{(2q-1)(3q+1)(q+3)}$

$=\dfrac{q(3q+1)-(3q+4)(2q-1)}{(2q-1)(q+3)(3q+1)}$

$=\dfrac{3q^2+q-6q^2-5q+4}{(2q-1)(q+3)(3q+1)}$

$=\dfrac{-3q^2-4q+4}{(2q-1)(q+3)(3q+1)}$

Exercises

1. $\dfrac{x}{x+8}+\dfrac{8}{x+8}$

The denominators are the same, so the sum is found by adding the two numerators and keeping the same (common) denominator.

$\dfrac{x}{x+8}+\dfrac{8}{x+8}=\dfrac{x+8}{x+8}=1$

Choice E is correct.

3. $\dfrac{8}{x-8}-\dfrac{x}{x-8}$

The denominators are the same, so the difference is found by subtracting the two numerators and keeping the same (common) denominator.

$\dfrac{8}{x-8}-\dfrac{x}{x-8}=\dfrac{8-x}{x-8}$

$=\dfrac{-1(x-8)}{x-8}=1$

Choice C is correct.

5. $\dfrac{x}{x+8}-\dfrac{8}{x+8}$

The denominators are the same, so the difference is found by subtracting the two numerators and keeping the same (common) denominator.

$\dfrac{x}{x+8}-\dfrac{8}{x+8}=\dfrac{x-8}{x+8}$

Choice B is correct.

7. $\dfrac{1}{8}-\dfrac{1}{x}$

The LCD is $8x$. Now rewrite each rational expression as a fraction with the LCD as its denominator.

$\dfrac{1}{8}\cdot\dfrac{x}{x}=\dfrac{x}{8x}$

$\dfrac{1}{x}\cdot\dfrac{8}{8}=\dfrac{8}{8x}$

Since the fractions now have a common denominator, subtract the numerators and use the LCD as the denominator of the sum.

$\dfrac{1}{8}-\dfrac{1}{x}=\dfrac{x}{8x}-\dfrac{8}{8x}=\dfrac{x-8}{8x}$

Choice G is correct.

9. $\dfrac{6}{5x}+\dfrac{9}{2x}$

$=\dfrac{6(\underline{2})}{5x(\underline{2})}+\dfrac{9(\underline{5})}{2x(\underline{5})}$

$=\dfrac{12+45}{10x}$

$=\dfrac{57}{10x}$

11. The denominators are the same, so the sum is found by adding the two numerators and keeping the same (common) denominator.

$\dfrac{4}{m}+\dfrac{7}{m}=\dfrac{4+7}{m}=\dfrac{11}{m}$

13. The denominators are the same, so the difference is found by subtracting the two numerators and keeping the same (common) denominator.

$\dfrac{5}{y+4}-\dfrac{1}{y+4}=\dfrac{5-1}{y+4}=\dfrac{4}{y+4}$

15. The denominators are the same, so the sum is found by adding the two numerators and keeping the same (common) denominator.

$$\frac{x}{x+y}+\frac{y}{x+y}=\frac{x+y}{x+y}=1$$

17. The denominators are the same, so the difference is found by subtracting the two numerators and keeping the same (common) denominator. Don't forget the parentheses on the second numerator.

$$\frac{5m}{m+1}-\frac{1+4m}{m+1}=\frac{5m-(1+4m)}{m+1}$$
$$=\frac{5m-1-4m}{m+1}$$
$$=\frac{m-1}{m+1}$$

19. The denominators are the same, so the difference is found by subtracting the two numerators and keeping the same (common) denominator. Don't forget the parentheses on the second numerator.

$$\frac{a+b}{2}-\frac{a-b}{2}=\frac{(a+b)-(a-b)}{2}$$
$$=\frac{a+b-a+b}{2}$$
$$=\frac{2b}{2}=b$$

21. $\frac{x^2}{x+5}+\frac{5x}{x+5}=\frac{x^2+5x}{x+5}$ Add numerators.

$=\frac{x(x+5)}{x+5}$ Factor numerator.

$=x$ Lowest terms

23.
$$\frac{y^2-3y}{y+3}+\frac{-18}{y+3}=\frac{y^2-3y-18}{y+3}$$
$$=\frac{(y-6)(y+3)}{y+3}$$
$$=y-6$$

25. The denominators are the same, so the sum is found by adding the two numerators and keeping the same (common) denominator.

$$\frac{x}{x^2-9}-\frac{-3}{x^2-9}=\frac{x+3}{x^2-9}$$
$$=\frac{x+3}{(x+3)(x-3)}$$
$$=\frac{1}{x-3}$$

27. The denominators are the same, so the difference is found by subtracting the two numerators and keeping the same (common) denominator.

$$\frac{y^2+x^2}{x^2-y^2}-\frac{2x^2}{x^2-y^2}=\frac{y^2+x^2-2x^2}{x^2-y^2}$$
$$=\frac{y^2-x^2}{x^2-y^2}$$
$$=\frac{-1(x^2-y^2)}{x^2-y^2}$$
$$=-1$$

29. $\frac{z}{5}+\frac{1}{3}$

The LCD is 15. Now rewrite each rational expression as a fraction with the LCD as its denominator.

$$\frac{z}{5}\cdot\frac{3}{3}=\frac{3z}{15}$$
$$\frac{1}{3}\cdot\frac{5}{5}=\frac{5}{15}$$

Since the fractions now have a common denominator, add the numerators, and use the LCD as the denominator of the sum.

$$\frac{z}{5}+\frac{1}{3}=\frac{3z}{15}+\frac{5}{15}=\frac{3z+5}{15}$$

31. $\frac{5}{7}-\frac{r}{2}=\frac{5}{7}\cdot\frac{2}{2}-\frac{r}{2}\cdot\frac{7}{7}$ LCD = 14

$$=\frac{10}{14}-\frac{7r}{14}$$
$$=\frac{10-7r}{14}$$

33. $-\frac{3}{4}-\frac{1}{2x}=-\frac{3\cdot x}{4\cdot x}-\frac{1\cdot 2}{2x\cdot 2}$ LCD = 4x

$$=\frac{-3x-2}{4x}$$

35. $\frac{7}{4t}+\frac{3}{7t}=\frac{7}{4t}\cdot\frac{7}{7}+\frac{3}{7t}\cdot\frac{4}{4}$ LCD = $28t$

$=\frac{49+12}{28t}$

$=\frac{61}{28t}$

37. $\frac{x+1}{6}+\frac{3x+3}{9}$

First reduce the second fraction.

$\frac{3x+3}{9}=\frac{3(x+1)}{9}=\frac{x+1}{3}$

Now the LCD of $\frac{x+1}{6}$ and $\frac{x+1}{3}$ is 6. Thus,

$\frac{x+1}{6}+\frac{x+1}{3}=\frac{x+1}{6}+\frac{x+1}{3}\cdot\frac{2}{2}$

$=\frac{x+1+2x+2}{6}$

$=\frac{3x+3}{6}$

$=\frac{3(x+1)}{6}=\frac{x+1}{2}.$

39. $\frac{x+3}{3x}+\frac{2x+2}{4x}=\frac{x+3}{3x}+\frac{2(x+1)}{4x}$

$=\frac{x+3}{3x}+\frac{x+1}{2x}$

$=\frac{x+3}{3x}\cdot\frac{2}{2}+\frac{x+1}{2x}\cdot\frac{3}{3}$

$=\frac{2x+6+3x+3}{6x}$

$=\frac{5x+9}{6x}$

41. $\frac{7}{3p^2}-\frac{2}{p}=\frac{7}{3p^2}-\frac{2}{p}\cdot\frac{3p}{3p}$ LCD = $3p^2$

$=\frac{7-6p}{3p^2}$

43. $\frac{1}{k+4}-\frac{2}{k}=\frac{1}{k+4}\cdot\frac{k}{k}-\frac{2}{k}\cdot\frac{k+4}{k+4}$

$=\frac{k}{k(k+4)}-\frac{2(k+4)}{k(k+4)}$

$=\frac{k-2k-8}{k(k+4)}$

$=\frac{-k-8}{k(k+4)}$

45. $\frac{x}{x-2}+\frac{-8}{x^2-4}$

$=\frac{x}{x-2}+\frac{-8}{(x+2)(x-2)}$

$=\frac{x}{x-2}\cdot\frac{x+2}{x+2}+\frac{-8}{(x+2)(x-2)}$

$=\frac{x(x+2)-8}{(x+2)(x-2)}$

$=\frac{x^2+2x-8}{(x+2)(x-2)}$

$=\frac{(x+4)(x-2)}{(x+2)(x-2)}=\frac{x+4}{x+2}$

47. $\frac{4m}{m^2+3m+2}+\frac{2m-1}{m^2+6m+5}$

$=\frac{4m}{(m+2)(m+1)}+\frac{2m-1}{(m+1)(m+5)}$

$=\frac{4m(m+5)}{(m+2)(m+1)(m+5)}$

$+\frac{(2m-1)(m+2)}{(m+1)(m+5)(m+2)}$

$=\frac{\left(4m^2+20m\right)+\left(2m^2+3m-2\right)}{(m+2)(m+1)(m+5)}$

$=\frac{6m^2+23m-2}{(m+2)(m+1)(m+5)}$

49. $\frac{4y}{y^2-1}-\frac{5}{y^2+2y+1}$

$=\frac{4y}{(y+1)(y-1)}-\frac{5}{(y+1)(y+1)}$

$=\frac{4y(y+1)}{(y+1)^2(y-1)}-\frac{5(y-1)}{(y+1)^2(y-1)}$

$=\frac{\left(4y^2+4y\right)-(5y-5)}{(y+1)^2(y-1)}$

$=\frac{4y^2-y+5}{(y+1)^2(y-1)}$

51. $\dfrac{t}{t+2}+\dfrac{5-t}{t}-\dfrac{4}{t^2+2t}$

$=\dfrac{t}{t+2}+\dfrac{5-t}{t}-\dfrac{4}{t(t+2)}$

$=\dfrac{t}{t+2}\cdot\dfrac{t}{t}+\dfrac{5-t}{t}\cdot\dfrac{t+2}{t+2}-\dfrac{4}{t(t+2)}$

$=\dfrac{t\cdot t+(5-t)(t+2)-4}{t(t+2)}$

$=\dfrac{t^2+5t+10-t^2-2t-4}{t(t+2)}$

$=\dfrac{3t+6}{t(t+2)}$

$=\dfrac{3(t+2)}{t(t+2)}=\dfrac{3}{t}$

53. $\dfrac{10}{m-2}+\dfrac{5}{2-m}$

Consider the following.

$2-m=-1(m-2)$

Therefore, either $m-2$ or $2-m$ could be used as the LCD.

55. $\dfrac{4}{x-5}+\dfrac{6}{5-x}$

The two denominators, $x-5$ and $5-x$, are opposites of each other, so either one may be used as the common denominator. We will work the exercise both ways and compare the answers.

$$\dfrac{4}{x-5}+\dfrac{6}{5-x}=\dfrac{4}{x-5}+\dfrac{6(-1)}{(5-x)(-1)}=\dfrac{4}{x-5}+\dfrac{-6}{x-5}=\dfrac{-2}{x-5}$$

$$\dfrac{4}{x-5}+\dfrac{6}{5-x}=\dfrac{4(-1)}{(x-5)(-1)}+\dfrac{6}{5-x}=\dfrac{-4}{5-x}+\dfrac{6}{5-x}=\dfrac{2}{5-x}$$

The two answers are equivalent as shown.

$\dfrac{-2}{x-5}\cdot\dfrac{-1}{-1}=\dfrac{2}{5-x}$

57. $\dfrac{-1}{1-y}-\dfrac{4y-3}{y-1}$

The LCD is either $1-y$ or $y-1$.

We will use $y-1$.

$$\dfrac{-1}{1-y}-\dfrac{4y-3}{y-1}=\dfrac{-1\cdot-1}{-1\cdot(1-y)}-\dfrac{4y-3}{y-1}=\dfrac{1-(4y-3)}{y-1}=\dfrac{1-4y+3}{y-1}=\dfrac{4-4y}{y-1}=\dfrac{4(1-y)}{y-1}=-4$$

59. $\dfrac{2}{x-y^2}+\dfrac{7}{y^2-x}$

LCD $=x-y^2$ or y^2-x

We will use $x-y^2$.

$$\dfrac{2}{x-y^2}+\dfrac{7}{y^2-x}=\dfrac{2}{x-y^2}+\dfrac{-1(7)}{-1(y^2-x)}=\dfrac{2}{x-y^2}+\dfrac{-7}{-y^2+x}=\dfrac{2}{x-y^2}+\dfrac{-7}{x-y^2}=\dfrac{2+(-7)}{x-y^2}=\dfrac{-5}{x-y^2}$$

If y^2-x is used as the LCD, we will obtain the following equivalent answer.

$\dfrac{5}{y^2-x}$

61. $\dfrac{x}{5x-3y}-\dfrac{y}{3y-5x}$

LCD $= 5x-3y$ or $3y-5x$

We will use $5x-3y$.

$$\frac{x}{5x-3y}-\frac{y}{3y-5x}$$
$$=\frac{x}{5x-3y}-\frac{-1(y)}{-1(3y-5x)}$$
$$=\frac{x}{5x-3y}-\frac{-y}{-3y+5x}$$
$$=\frac{x}{5x-3y}-\frac{-y}{5x-3y}$$
$$=\frac{x-(-y)}{5x-3y}=\frac{x+y}{5x-3y}$$

If $3y-5x$ is used as the LCD, we will obtain the following equivalent answer.

$$\frac{-x-y}{3y-5x}$$

63. $\dfrac{3}{4p-5}+\dfrac{9}{5-4p}$

LCD $= 4p-5$ or $5-4p$

We will use $4p-5$.

$$\frac{3}{4p-5}+\frac{9}{5-4p}$$
$$=\frac{3}{4p-5}+\frac{-1(9)}{-1(5-4p)}$$
$$=\frac{3}{4p-5}+\frac{-9}{-5+4p}$$
$$=\frac{3}{4p-5}+\frac{-9}{4p-5}$$
$$=\frac{3+(-9)}{4p-5}=\frac{-6}{4p-5}$$

If $5-4p$ is used as the LCD, we will obtain the following equivalent answer.

$$\frac{6}{5-4p}$$

65. $\dfrac{15x}{5x-7}-\dfrac{-21}{7-5x}$

$$=\frac{15x}{5x-7}-\frac{-1(-21)}{-1(7-5x)}$$
$$=\frac{15x}{5x-7}-\frac{21}{5x-7}$$
$$=\frac{15x-21}{5x-7}$$
$$=\frac{3(5x-7)}{5x-7}$$
$$=3$$

67. $\dfrac{2m}{m-n}-\dfrac{5m+n}{2m-2n}$

$$=\frac{2m}{m-n}-\frac{5m+n}{2(m-n)} \quad \text{Factor.}$$
$$=\frac{2m}{m-n}\cdot\frac{2}{2}-\frac{5m+n}{2(m-n)} \quad \text{LCD} = 2(m-n)$$
$$=\frac{4m-(5m+n)}{2(m-n)}$$
$$=\frac{4m-5m-n}{2(m-n)}$$
$$=\frac{-m-n}{2(m-n)}$$

69. $\dfrac{5}{x^2-9}-\dfrac{x+2}{x^2+4x+3}$

To find the LCD, factor the denominators.

$x^2-9=(x+3)(x-3)$

$x^2+4x+3=(x+3)(x+1)$

The LCD is $(x+3)(x-3)(x+1)$.

$$\frac{5}{x^2-9}-\frac{x+2}{x^2+4x+3}$$
$$=\frac{5\cdot(x+1)}{(x+3)(x-3)\cdot(x+1)}-\frac{(x+2)\cdot(x-3)}{(x+3)(x+1)\cdot(x-3)}$$
$$=\frac{5x+5}{(x+3)(x-3)\cdot(x+3)}-\frac{x^2-x-6}{(x+3)(x+1)(x-3)}$$
$$=\frac{(5x-5)-(x^2-x-6)}{(x+3)(x-3)(x+1)}$$
$$=\frac{5x+5-x^2+x+6}{(x+3)(x-3)(x+1)}$$
$$=\frac{-x^2+6x+11}{(x+3)(x-3)(x+1)}$$

71. $\dfrac{2q+1}{3q^2+10q-8}-\dfrac{3q+5}{2q^2+5q-12}$

$$=\frac{2q+1}{(3q-2)(q+4)}-\frac{3q+5}{(2q-3)(q+4)}$$
$$=\frac{(2q+1)\cdot(2q-3)}{(3q-2)(q+4)\cdot(2q-3)}-\frac{(3q+5)\cdot(3q-2)}{(2q-3)(q+4)\cdot(3q-2)}$$
$$=\frac{(4q^2-4q-3)-(9q^2+9q-10)}{(3q-2)(q+4)(2q-3)}$$
$$=\frac{4q^2-4q-3-9q^2-9q+10}{(3q-2)(q+4)(2q-3)}$$
$$=\frac{-5q^2-13q+7}{(3q-2)(q+4)(2q-3)}$$

73. $$\frac{y^2}{y-2}-\frac{9y-14}{y-2}=\frac{y^2-(9y-14)}{y-2}$$
$$=\frac{y^2-9y+14}{y-2}$$
$$=\frac{(y-7)(y-2)}{y-2}$$
$$=y-7$$

75. $$\frac{3}{x+4}+7=\frac{3}{x+4}+\frac{7(x+4)}{x+4}$$
$$=\frac{3+7(x+4)}{x+4}$$
$$=\frac{3+7x+28}{x+4}$$
$$=\frac{7x+31}{x+4}$$

77. $$\frac{-x+2}{x}-\frac{x-5}{4x}=\frac{4(-x+2)}{4x}-\frac{x-5}{4x}$$
$$=\frac{4(-x+2)-(x-5)}{4x}$$
$$=\frac{-4x+8-x+5}{4x}$$
$$=\frac{-5x+13}{4x}$$

79. $$\frac{5x}{x-7}-\frac{3x}{x-3}=\frac{5x(x-3)}{(x-7)(x-3)}-\frac{3x(x-7)}{(x-7)(x-3)}$$
$$=\frac{5x(x-3)-3x(x-7)}{(x-7)(x-3)}$$
$$=\frac{5x^2-15x-3x^2+21x}{(x-7)(x-3)}$$
$$=\frac{2x^2+6x}{(x-7)(x-3)},\text{ or, }\frac{2x(x+3)}{(x-7)(x-3)}$$

81. $$\frac{5a}{3a-6}-\frac{a-7}{a-2}=\frac{5a}{3(a-2)}-\frac{3(a-7)}{3(a-2)}$$
$$=\frac{5a-3(a-7)}{3(a-2)}$$
$$=\frac{5a-3a+21}{3(a-2)}$$
$$=\frac{2a+21}{3(a-2)}$$

83. $\frac{4}{3-x}+\frac{x}{2x-6}=\frac{(-1)4}{(-1)(3-x)}+\frac{x}{2(x-3)}$

$=\frac{-4}{x-3}+\frac{x}{2(x-3)}$

$=\frac{(2)(-4)}{(2)(x-3)}+\frac{x}{2(x-3)}$

$=\frac{x-8}{2(x-3)}$, or $\frac{8-x}{2(3-x)}$

85. $\frac{5x+11}{x^2-11x+18}-\frac{4x+20}{x^2-11x+18}$

$=\frac{5x+11-(4x+20)}{x^2-11x+18}$

$=\frac{x-9}{x^2-11x+18}$

$=\frac{x-9}{(x-9)(x-2)}$

$=\frac{1}{x-2}$

87. $\frac{4}{r^2-r}+\frac{6}{r^2+2r}-\frac{1}{r^2+r-2}$

$=\frac{4}{r(r-1)}+\frac{6}{r(r+2)}-\frac{1}{(r+2)(r-1)}$

$=\frac{4\cdot(r+2)}{r(r-1)\cdot(r+2)}+\frac{6\cdot(r-1)}{r(r+2)\cdot(r-1)}$

$-\frac{1\cdot r}{r\cdot(r+2)(r-1)}$

$=\frac{4r+8+6r-6-r}{r(r+2)(r-1)}$

$=\frac{9r+2}{r(r+2)(r-1)}$

89. $\frac{x+3y}{x^2+2xy+y^2}+\frac{x-y}{x^2+4xy+3y^2}$

$=\frac{x+3y}{(x+y)(x+y)}+\frac{x-y}{(x+3y)(x+y)}$

$=\frac{(x+3y)\cdot(x+3y)}{(x+y)(x+y)\cdot(x+3y)}$

$+\frac{(x-y)\cdot(x+y)}{(x+3y)(x+y)\cdot(x+y)}$

$=\frac{\left(x^2+6xy+9y^2\right)+\left(x^2-y^2\right)}{(x+y)(x+y)(x+3y)}$

$=\frac{2x^2+6xy+8y^2}{(x+y)(x+y)(x+3y)}$

$=\frac{2\left(x^2+3xy+4y^2\right)}{(x+y)(x+y)(x+3y)}$,

or $\frac{2\left(x^2+3xy+4y^2\right)}{(x+y)^2(x+3y)}$

91. $\frac{r+y}{18r^2+9ry-2y^2}+\frac{3r-y}{36r^2-y^2}$

Factor the first denominator by grouping. Find integers whose product is $18(-2)=-36$ and whose sum is 9. The integers are 12 and -3.

$18r^2+9ry-2y^2$

$=18r^2+12ry-3ry-2y^2$

$=6r(3r+2y)-y(3r+2y)$

$=(3r+2y)(6r-y)$

Factor the second denominator.

$\frac{r+y}{(3r+2y)(6r-y)}+\frac{3r-y}{(6r-y)(6r+y)}$

Rewrite fractions with the following LCD.

$$(3r+2y)(6r-y)(6r+y)$$

$$=\frac{(r+y)\cdot(6r+y)}{(3r+2y)(6r-y)\cdot(6r+y)}+\frac{(3r-y)\cdot(3r+2y)}{(6r-y)(6r+y)\cdot(3r+2y)}$$

$$=\frac{6r^2+7ry+y^2}{(3r+2y)(6r-y)(6r+y)}+\frac{9r^2+3ry-2y^2}{(3r+2y)(6r-y)(6r+y)}$$

$$=\frac{6r^2+7ry+y^2+9r^2+3ry-2y^2}{(3r+2y)(6r-y)(6r+y)}$$

$$=\frac{15r^2+10ry-y^2}{(3r+2y)(6r-y)(6r+y)}$$

93. **(a)** $P=2L+2W$

$$=2\left(\frac{3k+1}{10}\right)+2\left(\frac{5}{6k+2}\right)$$

$$=2\left(\frac{3k+1}{2\cdot 5}\right)+2\left(\frac{5}{2(3k+1)}\right)$$

$$=\frac{3k+1}{5}+\frac{5}{3k+1}$$

To add the two fractions on the right, use $5(3k+1)$ as the LCD.

$$P=\frac{(3k+1)(3k+1)}{5(3k+1)}+\frac{(5)(5)}{5(3k+1)}$$

$$=\frac{(3k+1)(3k+1)+(5)(5)}{5(3k+1)}$$

$$=\frac{9k^2+6k+1+25}{5(3k+1)}$$

$$=\frac{9k^2+6k+26}{5(3k+1)}$$

(b) $\mathcal{A}=L\cdot W$

$$\mathcal{A}=\frac{3k+1}{10}\cdot\frac{5}{6k+2}$$

$$=\frac{3k+1}{5\cdot 2}\cdot\frac{5}{2(3k+1)}$$

$$=\frac{1}{2\cdot 2}=\frac{1}{4}$$

95. $$\frac{1010}{49(101-x)}-\frac{10}{49}$$

$$=\frac{1010}{49(101-x)}-\frac{10(101-x)}{49(101-x)}$$

$$=\frac{1010-1010+10x}{49(101-x)}$$

$$=\frac{10x}{49(101-x)}$$

7.5 Complex Fractions

Now Try Exercises

N1. **(a)** $$\frac{\frac{2}{5}+\frac{1}{4}}{\frac{1}{6}+\frac{3}{8}}$$

$$=\frac{\frac{2(4)}{5(4)}+\frac{1(5)}{4(5)}}{\frac{1(4)}{6(4)}+\frac{3(3)}{8(3)}}$$

$$=\frac{\frac{8+5}{20}}{\frac{4+9}{24}}$$

$$=\frac{13}{20}\div\frac{13}{24}$$

$$=\frac{13}{20}\cdot\frac{24}{13}$$ Multiply by reciprocal.

$$=\frac{13\cdot 4\cdot 6}{4\cdot 5\cdot 13}$$

$$=\frac{6}{5}$$

(b) $$\frac{2+\frac{4}{x}}{\frac{5}{6}+\frac{5x}{12}}=\frac{\frac{2}{1}+\frac{4}{x}}{\frac{5}{6}+\frac{5x}{12}}=\frac{\frac{2(x)}{1(x)}+\frac{4}{x}}{\frac{5(2)}{6(2)}+\frac{5x}{12}}$$

$$=\frac{\frac{2x+4}{x}}{\frac{10+5x}{12}}=\frac{2x+4}{x}\div\frac{10+5x}{12}$$

$$=\frac{2x+4}{x}\cdot\frac{12}{10+5x}$$

$$=\frac{2(x+2)\cdot 12}{x\cdot 5(2+x)}$$

$$=\frac{24}{5x}$$

N2. $$\frac{\frac{a^2b}{c}}{\frac{ab^2}{c^3}} = \frac{a^2b}{c} \div \frac{ab^2}{c^3}$$

$$= \frac{a^2b}{c} \cdot \frac{c^3}{ab^2} = \frac{ac^2}{b}$$

N3. $$\frac{5 + \frac{2}{a-3}}{\frac{1}{a-3} - 2}$$

$$= \frac{\frac{5(a-3)}{1(a-3)} + \frac{2}{a-3}}{\frac{1}{a-3} - \frac{2(a-3)}{1(a-3)}}$$

$$= \frac{\frac{5(a-3)+2}{a-3}}{\frac{1-2(a-3)}{a-3}}$$

$$= \frac{\frac{5a-15+2}{a-3}}{\frac{1-2a+6}{a-3}}$$

$$= \frac{\frac{5a-13}{a-3}}{\frac{7-2a}{a-3}}$$

$$= \frac{5a-13}{a-3} \div \frac{7-2a}{a-3}$$

$$= \frac{5a-13}{a-3} \cdot \frac{a-3}{7-2a}$$

$$= \frac{5a-13}{7-2a}$$

N4. **(a)** $$\frac{\frac{3}{5} - \frac{1}{4}}{\frac{1}{8} + \frac{3}{20}}$$

$$= \frac{40\left(\frac{3}{5} - \frac{1}{4}\right)}{40\left(\frac{1}{8} + \frac{3}{20}\right)} \qquad \text{LCD} = 2^3 \times 5 = 40$$

$$= \frac{40\left(\frac{3}{5}\right) - 40\left(\frac{1}{4}\right)}{40\left(\frac{1}{8}\right) + 40\left(\frac{3}{20}\right)}$$

$$= \frac{24-10}{5+6}$$

$$= \frac{14}{11}$$

(b) $$\frac{\frac{2}{x} - 3}{7 + \frac{x}{5}} \qquad \text{LCD} = 5x$$

$$= \frac{5x\left(\frac{2}{x} - 3\right)}{5x\left(7 + \frac{x}{5}\right)} \qquad \text{Multiply by } 5x.$$

$$= \frac{5x\left(\frac{2}{x}\right) - 5x(3)}{5x(7) + 5x\left(\frac{x}{5}\right)}$$

$$= \frac{10-15x}{35x + x^2}, \text{ or } \frac{10-15x}{x^2 + 35x}$$

N5. $$\frac{\frac{1}{y} + \frac{2}{3y^2}}{\frac{5}{4y^2} - \frac{3}{2y^3}} \qquad \text{LCD} = 12y^3$$

$$= \frac{12y^3\left(\frac{1}{y} + \frac{2}{3y^2}\right)}{12y^3\left(\frac{5}{4y^2} - \frac{3}{2y^3}\right)}$$

$$= \frac{12y^3\left(\frac{1}{y}\right) + 12y^3\left(\frac{2}{3y^2}\right)}{12y^3\left(\frac{5}{4y^2}\right) - 12y^3\left(\frac{3}{2y^3}\right)}$$

$$= \frac{12y^2 + 8y}{15y - 18}$$

N6. (a) $\dfrac{1-\frac{2}{x}-\frac{15}{x^2}}{1+\frac{5}{x}+\frac{6}{x^2}}$ LCD $= x^2$

$$= \frac{x^2\left(1-\frac{2}{x}-\frac{15}{x^2}\right)}{x^2\left(1+\frac{5}{x}+\frac{6}{x^2}\right)}$$

$$= \frac{x^2(1)-x^2\left(\frac{2}{x}\right)-x^2\left(\frac{15}{x^2}\right)}{x^2(1)+x^2\left(\frac{5}{x}\right)+x^2\left(\frac{6}{x^2}\right)}$$

$$= \frac{x^2-2x-15}{x^2+5x+6}$$

$$= \frac{(x-5)(x+3)}{(x+2)(x+3)} = \frac{x-5}{x+2}$$

(b) $\dfrac{\frac{9y^2-16}{y^2-100}}{\frac{3y-4}{y+10}} = \dfrac{9y^2-16}{y^2-100} \div \dfrac{3y-4}{y+10}$

$$= \frac{9y^2-16}{y^2-100} \cdot \frac{y+10}{3y-4}$$

$$= \frac{(3y+4)(3y-4)}{(y+10)(y-10)} \cdot \frac{y+10}{3y-4}$$

$$= \frac{3y+4}{y-10}$$

Exercises

1. (a) The LCD of $\frac{1}{2}$ and $\frac{1}{3}$ is $2 \cdot 3 = 6$. The simplified form of the numerator is the following.

$$\frac{1}{2}-\frac{1}{3}=\frac{3}{6}-\frac{2}{6}=\frac{1}{6}$$

(b) The LCD of $\frac{5}{6}$ and $\frac{1}{12}$ is 12 since 12 is a multiple of 6. The simplified form of the denominator is the following.

$$\frac{5}{6}-\frac{1}{12}=\frac{10}{12}-\frac{1}{12}=\frac{9}{12}=\frac{3}{4}$$

(c) $\dfrac{\frac{1}{6}}{\frac{3}{4}} = \dfrac{1}{6} \div \dfrac{3}{4}$

(d) $\dfrac{1}{6} \div \dfrac{3}{4} = \dfrac{1}{6} \cdot \dfrac{4}{3}$

$$= \frac{4}{18} = \frac{2}{9}$$

3. $\dfrac{2-\frac{1}{4}}{3-\frac{1}{2}} = \dfrac{-2+\frac{1}{4}}{-3+\frac{1}{2}}$

Choice D is equivalent to the given fraction. Each term of the numerator and denominator has been multiplied by -1. Since $\frac{-1}{-1}=1$, the fraction has been multiplied by the identity element, so its value is unchanged.

5. Simplify the complex fraction using Method 1.

$$\frac{\frac{2}{5}+\frac{1}{4}}{\frac{1}{2}+\frac{1}{3}}$$

Step 1

Write the numerator as a single fraction.

$$\frac{2}{5}+\frac{1}{4}=\frac{2}{5}\cdot\frac{4}{4}+\frac{1}{4}\cdot\frac{5}{5}$$

$$=\frac{8}{20}+\frac{5}{20}$$

$$=\underline{\frac{13}{20}}$$

Write the denominator as a single fraction.

$$\frac{1}{2}+\frac{1}{3}=\frac{1}{2}\cdot\frac{3}{3}+\frac{1}{3}\cdot\frac{2}{2}$$

$$=\frac{3}{6}+\frac{2}{6}$$

$$=\underline{\frac{5}{6}}$$

Step 2

Write the equivalent fraction as a division problem.

$$\frac{\frac{13}{20}}{\frac{5}{6}} = \underline{\frac{13}{20} \div \frac{5}{6}}$$

Step 3

Write the division problem as a multiplication problem.

$$\frac{13}{20} \div \frac{5}{6} = \underline{\frac{13}{20} \cdot \frac{6}{5}}$$

Multiply and write the answer in lowest terms.

$$\frac{13}{20}\cdot\frac{6}{5}=\frac{78}{100}$$
$$=\frac{39}{50}$$

7. To use Method 1, divide the numerator of the complex fraction by the denominator.

$$\frac{-\frac{4}{3}}{\frac{2}{9}}=-\frac{4}{3}\div\frac{2}{9}$$
$$=-\frac{4}{3}\cdot\frac{9}{2}=-\frac{36}{6}=-6$$

9. To use Method 2, multiply the numerator and denominator of the complex fraction by the LCD, y^2.

$$\frac{\frac{x}{y^2}}{\frac{x^2}{y}}=\frac{y^2\left(\frac{x}{y^2}\right)}{y^2\left(\frac{x^2}{y}\right)}$$
$$=\frac{x}{yx^2}=\frac{1}{xy}$$

11. $$\frac{\frac{4a^4b^3}{3a}}{\frac{2ab^4}{b^2}}=\frac{4a^4b^3}{3a}\div\frac{2ab^4}{b^2}\quad\text{Method 1}$$
$$=\frac{4a^4b^3}{3a}\cdot\frac{b^2}{2ab^4}$$
$$=\frac{4a^4b^3\cdot b^2}{3a\cdot 2ab^4}$$
$$=\frac{4a^4b^5}{6a^2b^4}$$
$$=\frac{2a^2b}{3}$$

13. To use Method 2, multiply the numerator and denominator of the complex fraction by the LCD, $3m$.

$$\frac{\frac{m+2}{3}}{\frac{m-4}{m}}=\frac{3m\left(\frac{m+2}{3}\right)}{3m\left(\frac{m-4}{m}\right)}$$
$$=\frac{m(m+2)}{3(m-4)}$$

15. $$\frac{\frac{2}{x}-3}{\frac{2-3x}{2}}=\frac{2x\left(\frac{2}{x}-3\right)}{2x\left(\frac{2-3x}{2}\right)}\quad\text{Method 2}$$
$$=\frac{2x\left(\frac{2}{x}\right)-2x(3)}{x(2-3x)}$$
$$=\frac{4-6x}{x(2-3x)}$$
$$=\frac{2(2-3x)}{x(2-3x)}\quad\text{Factor.}$$
$$=\frac{2}{x}\quad\text{Lowest terms}$$

17. $$\frac{\frac{1}{x}+x}{\frac{x^2+1}{8}}=\frac{8x\left(\frac{1}{x}+x\right)}{8x\left(\frac{x^2+1}{8}\right)}\quad\text{Method 2; LCD} = 8x$$
$$=\frac{8+8x^2}{x(x^2+1)}\quad\text{Distributive property}$$
$$=\frac{8(1+x^2)}{x(x^2+1)}\quad\text{Factor.}$$
$$=\frac{8}{x}\quad\text{Lowest terms}$$

19. $$\frac{a-\frac{5}{a}}{a+\frac{1}{a}}=\frac{a\left(a-\frac{5}{a}\right)}{a\left(a+\frac{1}{a}\right)}\quad\text{Method 2; LCD} = a$$
$$=\frac{a^2-5}{a^2+1}$$

21. $$\frac{\frac{5}{8}+\frac{2}{3}}{\frac{7}{3}-\frac{1}{4}}=\frac{24\left(\frac{5}{8}+\frac{2}{3}\right)}{24\left(\frac{7}{3}-\frac{1}{4}\right)}\quad\text{Method 2; LCD} = 24$$
$$=\frac{24\left(\frac{5}{8}\right)+24\left(\frac{2}{3}\right)}{24\left(\frac{7}{3}\right)-24\left(\frac{1}{4}\right)}$$
$$=\frac{15+16}{56-6}=\frac{31}{50}$$

23. $\dfrac{\frac{1}{x^2}+\frac{1}{y^2}}{\frac{1}{x}-\frac{1}{y}}$

$=\dfrac{x^2y^2\left(\frac{1}{x^2}+\frac{1}{y^2}\right)}{x^2y^2\left(\frac{1}{x}-\frac{1}{y}\right)}$ Method 2; LCD $= x^2y^2$

$=\dfrac{x^2y^2\left(\frac{1}{x^2}\right)+x^2y^2\left(\frac{1}{y^2}\right)}{x^2y^2\left(\frac{1}{x}\right)-x^2y^2\left(\frac{1}{y}\right)}$

$=\dfrac{y^2+x^2}{xy^2-x^2y}=\dfrac{y^2+x^2}{xy(y-x)}$

25. $\dfrac{\frac{2}{p^2}-\frac{3}{5p}}{\frac{4}{p}+\frac{1}{4p}}=\dfrac{20p^2\left(\frac{2}{p^2}-\frac{3}{5p}\right)}{20p^2\left(\frac{4}{p}+\frac{1}{4p}\right)}$ Method 2

$=\dfrac{20p^2\left(\frac{2}{p^2}\right)-20p^2\left(\frac{3}{5p}\right)}{20p^2\left(\frac{4}{p}\right)+20p^2\left(\frac{1}{4p}\right)}$

$=\dfrac{40-12p}{80p+5p}$

$=\dfrac{40-12p}{85p}$

27. $\dfrac{\frac{5}{x^2y}-\frac{2}{xy^2}}{\frac{3}{x^2y^2}+\frac{4}{xy}}$

$=\dfrac{x^2y^2\left(\frac{5}{x^2y}-\frac{2}{xy^2}\right)}{x^2y^2\left(\frac{3}{x^2y^2}+\frac{4}{xy}\right)}$ Method 2; LCD $= x^2y^2$

$=\dfrac{x^2y^2\left(\frac{5}{x^2y}\right)-x^2y^2\left(\frac{2}{xy^2}\right)}{x^2y^2\left(\frac{3}{x^2y^2}\right)+x^2y^2\left(\frac{4}{xy}\right)}$

$=\dfrac{5y-2x}{3+4xy}$

29. $\dfrac{\frac{1}{4}-\frac{1}{a^2}}{\frac{1}{2}+\frac{1}{a}}$

$=\dfrac{4a^2\left(\frac{1}{4}-\frac{1}{a^2}\right)}{4a^2\left(\frac{1}{2}+\frac{1}{a}\right)}$ Method 2; LCD $= 4a^2$

$=\dfrac{a^2-4}{2a^2+4a}$ Distributive property

$=\dfrac{(a-2)(a+2)}{2a(a+2)}$ Factor.

$=\dfrac{a-2}{2a}$ Fundamental property

31. $\dfrac{\frac{1}{z+5}}{\frac{4}{z^2-25}}$

$=\dfrac{1}{z+5}\div\dfrac{4}{z^2-25}$ Method 1

$=\dfrac{1}{z+5}\cdot\dfrac{z^2-25}{4}$ Multiply by reciprocal.

$=\dfrac{1\cdot(z^2-25)}{(z+5)\cdot 4}$ Multiply.

$=\dfrac{(z+5)(z-5)}{(z+5)\cdot 4}$ Factor numerator.

$=\dfrac{z-5}{4}$ Fundamental property

33. $\dfrac{\frac{1}{m+1}-1}{\frac{1}{m+1}+1}$

$=\dfrac{(m+1)\left(\frac{1}{m+1}-1\right)}{(m+1)\left(\frac{1}{m+1}+1\right)}$ Method 2; LCD $= m+1$

$=\dfrac{1-1(m+1)}{1+1(m+1)}$ Distributive property

$=\dfrac{1-m-1}{1+m+1}$ Distributive property

$=\dfrac{-m}{m+2}$

35. $\dfrac{\dfrac{12}{x+2}+2}{\dfrac{18}{x+2}-2}$

$$=\frac{(x+2)\left(\dfrac{12}{x+2}+2\right)}{(x+2)\left(\dfrac{18}{x+2}-2\right)}$$
$$=\frac{12+2(x+2)}{18-2(x+2)}$$
$$=\frac{12+2x+4}{18-2x-4}$$
$$=\frac{2x+16}{-2x+14}$$
$$=\frac{2(x+8)}{2(-x+7)}$$
$$=\frac{x+8}{-x+7}$$

37. $\dfrac{\dfrac{x}{y}+\dfrac{y}{x}}{\dfrac{x}{y}-\dfrac{y}{x}}=\dfrac{xy\left(\dfrac{x}{y}+\dfrac{y}{x}\right)}{xy\left(\dfrac{x}{y}-\dfrac{y}{x}\right)}$

$$=\frac{x^2+y^2}{x^2-y^2}, \text{ or } \frac{x^2+y^2}{(x+y)(x-y)}$$

39. $\dfrac{1}{\dfrac{1}{a}+\dfrac{1}{b}}=\dfrac{ab\cdot 1}{ab\left(\dfrac{1}{a}+\dfrac{1}{b}\right)}$

$$=\frac{ab}{b+a}$$

41. $\dfrac{\dfrac{1}{m-1}+\dfrac{2}{m+2}}{\dfrac{2}{m+2}-\dfrac{1}{m-3}}$

$$=\frac{(m-1)(m+2)(m-3)\left(\dfrac{1}{m-1}+\dfrac{2}{m+2}\right)}{(m-1)(m+2)(m-3)\left(\dfrac{2}{m+2}-\dfrac{1}{m-3}\right)}$$
$$=\frac{(m+2)(m-3)+2(m-1)(m-3)}{2(m-1)(m-3)-(m-1)(m+2)}$$
$$=\frac{(m-3)[(m+2)+2(m-1)]}{(m-1)[2(m-3)-(m+2)]}$$
$$=\frac{(m-3)[m+2+2m-2]}{(m-1)[2m-6-m-2]}$$
$$=\frac{3m(m-3)}{(m-1)(m-8)}$$

43. $\dfrac{2+\dfrac{1}{x}-\dfrac{28}{x^2}}{3+\dfrac{13}{x}+\dfrac{4}{x^2}}$ LCD $= x^2$

$$=\frac{x^2\left(2+\dfrac{1}{x}-\dfrac{28}{x^2}\right)}{x^2\left(3+\dfrac{13}{x}+\dfrac{4}{x^2}\right)} \quad \text{Method 2}$$
$$=\frac{2x^2+x-28}{3x^2+13x+4}$$
$$=\frac{(2x-7)(x+4)}{(3x+1)(x+4)}$$
$$=\frac{2x-7}{3x+1}$$

45. $\dfrac{\dfrac{y+8}{y-4}}{\dfrac{y^2-64}{y^2-16}}$

$$=\frac{y+8}{y-4}\div\frac{y^2-64}{y^2-16} \quad \text{Method 1}$$
$$=\frac{y+8}{y-4}\cdot\frac{y^2-16}{y^2-64}$$
$$=\frac{(y+8)\cdot(y^2-16)}{(y-4)\cdot(y^2-64)} \quad \text{Multiply.}$$
$$=\frac{(y+8)(y+4)(y-4)}{(y-4)(y+8)(y-8)} \quad \text{Factor.}$$
$$=\frac{y+4}{y-8} \quad \text{Fundamental prop.}$$

47. $\dfrac{\frac{15a^2+15b^2}{5}}{\frac{a^4-b^4}{10}}$

$$=\frac{10\left(\frac{15a^2+15b^2}{5}\right)}{10\left(\frac{a^4-b^4}{10}\right)}$$

$$=\frac{2\left(15a^2+15b^2\right)}{a^4-b^4}$$

$$=\frac{30\left(a^2+b^2\right)}{\left(a^2+b^2\right)\left(a^2-b^2\right)}$$

$$=\frac{30\left(a^2+b^2\right)}{\left(a^2+b^2\right)(a+b)(a-b)}$$

$$=\frac{30}{(a+b)(a-b)}$$

49. $\dfrac{\frac{1}{x^3-y^3}}{\frac{1}{x^2-y^2}}=\dfrac{1}{x^3-y^3}\div\dfrac{1}{x^2-y^2}$

$$=\frac{1}{x^3-y^3}\cdot\frac{x^2-y^2}{1}$$

$$=\frac{x^2-y^2}{x^3-y^3}$$

$$=\frac{(x+y)(x-y)}{(x-y)(x^2+xy+y^2)}$$

$$=\frac{x+y}{x^2+xy+y^2}$$

51. $\dfrac{1+x^{-1}-12x^{-2}}{1-x^{-1}-20x^{-2}}$

$$=\frac{1+\frac{1}{x}-\frac{12}{x^2}}{1-\frac{1}{x}-\frac{20}{x^2}} \quad \text{LCD} = x^2$$

$$=\frac{x^2\left(1+\frac{1}{x}-\frac{12}{x^2}\right)}{x^2\left(1-\frac{1}{x}-\frac{20}{x^2}\right)} \quad \text{Method 2}$$

$$=\frac{x^2+x-12}{x^2-x-20} \quad \text{Multiply.}$$

$$=\frac{(x+4)(x-3)}{(x-5)(x+4)} \quad \text{Factor.}$$

$$=\frac{x-3}{x-5} \quad \text{Fundamental property}$$

53. $1+\dfrac{1}{1+\frac{1}{1+1}}=1+\dfrac{1}{1+\frac{1}{2}}$

$$=1+\frac{1}{\frac{2}{2}+\frac{1}{2}}$$

$$=1+\frac{1}{\frac{3}{2}}$$

$$=1+1\cdot\frac{2}{3}$$

$$=1+\frac{2}{3}$$

$$=\frac{3}{3}+\frac{2}{3}=\frac{5}{3}$$

55. $7-\dfrac{3}{5+\frac{2}{4-2}}=7-\dfrac{3}{5+\frac{2}{2}}$

$$=7-\frac{3}{5+1}$$

$$=7-\frac{3}{6}$$

$$=7-\frac{1}{2}$$

$$=\frac{14}{2}-\frac{1}{2}=\frac{13}{2}$$

57.
$$r+\frac{r}{4-\frac{2}{6+2}}=r+\frac{r}{4-\frac{2}{8}}$$
$$=r+\frac{r}{4-\frac{1}{4}}$$
$$=r+\frac{r}{\frac{16}{4}-\frac{1}{4}}$$
$$=r+\frac{r}{\frac{15}{4}}$$
$$=r+r\cdot\frac{4}{15}$$
$$=r+\frac{4r}{15}$$
$$=\frac{15r}{15}+\frac{4r}{15}$$
$$=\frac{19r}{15}$$

59. "The sum of $\frac{3}{8}$ and $\frac{5}{6}$, divided by 2" is written as follows.
$$\frac{\frac{3}{8}+\frac{5}{6}}{2}$$

61.
$$\frac{\frac{3}{8}+\frac{5}{6}}{2}=\frac{24\left(\frac{3}{8}+\frac{5}{6}\right)}{24(2)} \quad \text{Method 2; LCD = 24}$$
$$=\frac{24\left(\frac{3}{8}\right)+24\left(\frac{5}{6}\right)}{24(2)}$$
$$=\frac{9+20}{48}=\frac{29}{48}$$

7.6 Solving Equations with Rational Expressions

Now Try Exercises

N1. **(a)** $\frac{3}{2}t-\frac{5}{7}t=\frac{11}{7}$ has an equals sign, so this is an *equation* to be solved. Use the multiplication property of equality to clear fractions. The LCD is 14.

$$\frac{3}{2}t-\frac{5}{7}t=\frac{11}{7}$$
$$14\left(\frac{3}{2}t-\frac{5}{7}t\right)=14\left(\frac{11}{7}\right)$$
$$14\left(\frac{3}{2}t\right)-14\left(\frac{5}{7}t\right)=14\left(\frac{11}{7}\right) \quad \text{Dist. prop.}$$
$$21t-10t=22 \quad \text{Multiply.}$$
$$11t=22$$
$$t=2 \quad \text{Divide by 11.}$$

Check $t=2$: $\frac{21}{7}-\frac{10}{7}=\frac{11}{7}$ True

The solution set is $\{2\}$.

(b) $\frac{3}{2}t-\frac{5}{7}t$ is the difference of two terms, so it is an *expression* to be simplified. Simplify by finding the LCD, writing each coefficient with this LCD, and combining like terms.

$$\frac{3}{2}t-\frac{5}{7}t=\frac{3\cdot 7}{2\cdot 7}t-\frac{5\cdot 2}{7\cdot 2}t \quad \text{LCD}=14$$
$$=\frac{21}{14}t-\frac{10}{14}t \quad \text{Multiply.}$$
$$=\frac{11}{14}t \quad \text{Combine terms.}$$

N2.
$$\frac{x}{6}+\frac{x}{3}=6+x$$
$$6\left(\frac{x}{6}+\frac{x}{3}\right)=6(6+x)$$
$$x+2x=36+6x$$
$$3x=36+6x$$
$$-3x=36$$
$$x=-12$$

Check $x=-12: -6=-6$ True

The solution set is $\{-12\}$.

N3.

$$\frac{x}{7}-\frac{x+5}{5}=-\frac{3}{7}$$
$$35\left(\frac{x}{7}-\frac{x+5}{5}\right)=35\left(-\frac{3}{7}\right)$$
$$5x-7(x+5)=5(-3)$$
$$5x-7x-35=-15$$
$$-2x-35=-15$$
$$-2x=20$$
$$x=-10$$

Check $x=-10: -\frac{10}{7}+1=-\frac{3}{7}$ True

The solution set is $\{-10\}$.

N4.

$$4+\frac{6}{x-3}=\frac{2x}{x-3}$$
$$(x-3)\left(4+\frac{6}{x-3}\right)=(x-3)\frac{2x}{x-3}$$
$$4(x-3)+6=2x$$
$$4x-12+6=2x$$
$$4x-6=2x$$
$$2x=6$$
$$x=3$$

Check $x=3$: $4+\frac{6}{0}=\frac{6}{0}$

The fractions are undefined, so the equation has no solution. The solution set is $\varnothing$.

N5.

$$\frac{3}{2x^2-8x}=\frac{1}{x^2-16}$$
$$\frac{3}{2x(x-4)}=\frac{1}{(x+4)(x-4)}$$
$$\left(\frac{2x(x+4)(x-4)(3)}{2x(x-4)}\right)=\left(\frac{2x(x+4)(x-4)(1)}{(x+4)(x-4)}\right)$$
$$3(x+4)=2x(1)$$
$$3x+12=2x$$
$$x=-12$$

Check $x=-12$: $\frac{3}{384}=\frac{1}{128}$ True

The solution set is $\{-12\}$.

N6.

$$\frac{2y}{y^2-25}=\frac{8}{y+5}-\frac{1}{y-5}$$
$$\frac{2y}{(y+5)(y-5)}=\frac{8}{y+5}-\frac{1}{y-5}$$
$$\left(\frac{(y+5)(y-5)(2y)}{(y+5)(y-5)}\right)$$
$$=(y+5)(y-5)\left(\frac{8}{y+5}-\frac{1}{y-5}\right)$$
$$2y=8(y-5)-1(y+5)$$
$$2y=8y-40-y-5$$
$$2y=7y-45$$
$$45=5y$$
$$9=y$$

Check $y=9: \frac{18}{56}=\frac{8}{14}-\frac{1}{4}$ True

The solution set is $\{9\}$.

N7.

$$\frac{3}{m^2-9}=\frac{1}{2(m-3)}-\frac{1}{4}$$
$$\frac{3}{(m+3)(m-3)}=\frac{1}{2(m-3)}-\frac{1}{4}$$
$$\left(\frac{4(m-3)(m+3)(3)}{(m+3)(m-3)}\right)$$
$$=4(m-3)(m+3)\left(\frac{1}{2(m-3)}-\frac{1}{4}\right)$$
$$4(3)=2(m+3)-(m-3)(m+3)$$
$$12=2m+6-(m^2-9)$$
$$12=2m+6-m^2+9$$
$$0=-m^2+2m+3$$
$$0=m^2-2m-3$$
$$0=(m-3)(m+1)$$
$$m=3 \text{ or } -1$$

Check $m=3$: $\frac{3}{0}=\frac{1}{0}-\frac{1}{4}$ (undefined)

Check $m=-1$: $-\frac{3}{8}=-\frac{1}{8}-\frac{1}{4}$ True

The solution set is $\{-1\}$.

N8.
$$\frac{5}{k^2+k-2}=\frac{1}{3k-3}-\frac{1}{k+2}$$
$$\frac{5}{(k+2)(k-1)}=\frac{1}{3(k-1)}-\frac{1}{k+2}$$
$$\left(\frac{3(k+2)(k-1)(5)}{(k+2)(k-1)}\right)$$
$$=3(k+2)(k-1)\left(\frac{1}{3(k-1)}-\frac{1}{k+2}\right)$$
$$3(5)=k+2-3(k-1)$$
$$15=k+2-3k+3$$
$$15=-2k+5$$
$$10=-2k$$
$$-5=k$$

Check $k=-5: \frac{5}{18}=-\frac{1}{18}+\frac{1}{3}$ True

The solution set is $\{-5\}$.

N9. **(a)** Solve $p=\frac{x-y}{z}$ for x.

$pz=\left(\frac{x-y}{z}\right)(z)$ Multiply by the LCD.

$pz=x-y$

$pz+y=x$ Isolate x.

(b) Solve $a=\frac{b}{c+d}$ for d.

Multiply by the LCD, $c+d$.

$a(c+d)=\left(\frac{b}{c+d}\right)(c+d)$

$ac+ad=b$

$ad=b-ac$ Isolate ad.

$d=\frac{b-ac}{a}$ Divide by a.

Another solution method is to multiply by $c+d$, divide by a, and then subtract c to get the following.

$y=\frac{b}{a}-c$

N10. Solve the formula $\frac{2}{w}=\frac{1}{x}-\frac{3}{y}$ for x.

Multiply by the LCD, wxy.

$$wxy\left(\frac{2}{w}\right)=wxy\left(\frac{1}{x}-\frac{3}{y}\right)$$
$$wxy\left(\frac{2}{w}\right)=wxy\left(\frac{1}{x}\right)-wxy\left(\frac{3}{y}\right)$$
$$2xy=wy-3wx$$

Get the x-terms on one side.

$2xy+3wx=wy$

$x(2y+3w)=wy$ Factor.

$$x=\frac{wy}{2y+3w}$$

Exercises

1. The least positive whole number by which the equation can be multiplied to obtain only integer coefficients is the LCD of 3 and 4, which is 12.

3. The simplest monomial by which the equation can be multiplied so that there are no variables in the denominators is the LCD of x, y, and z, which is xyz.

5. Yes, it is acceptable because $\frac{1}{3-x}$ is equivalent to $\frac{-1}{x-3}$.

$$\frac{1}{(3-x)}\cdot\frac{-1}{-1}=\frac{-1}{-3+x}=\frac{-1}{x-3}$$

7. $\frac{7}{8}x+\frac{1}{5}x$ is the sum of two terms, so it is an *expression* to be simplified. Simplify by finding the LCD, writing each coefficient with this LCD, and combining like terms.

$\frac{7}{8}x+\frac{1}{5}x=\frac{35}{40}x+\frac{8}{40}x$ LCD = 40

$=\frac{43}{40}x$ Combine like terms.

9. $\frac{7x}{8}+\frac{x}{5}=1$ has an equals sign, so this is an *equation* to be solved. Use the multiplication property of equality to clear fractions. The LCD is 40.

$$\frac{7x}{8}+\frac{x}{5}=1$$
$$40\left(\frac{7x}{8}+\frac{x}{5}\right)=40\cdot 1 \quad \text{Multiply by 40.}$$
$$40\left(\frac{7x}{8}\right)+40\left(\frac{x}{5}\right)=40\cdot 1 \quad \text{Distributive prop.}$$
$$35x+8x=40 \quad \text{Multiply.}$$
$$43x=40$$
$$x=\frac{40}{43} \quad \text{Divide by 43.}$$

The solution set is $\left\{\frac{40}{43}\right\}$.

11. $\frac{3}{5}x-\frac{7}{10}x$ is the difference of two terms, so it is an *expression* to be simplified.

$$\frac{3}{5}x-\frac{7}{10}x=\frac{6}{10}x-\frac{7}{10}x \quad \text{LCD}=10$$
$$=-\frac{1}{10}x \quad \text{Combine like terms.}$$

13. $\frac{3}{5}x-\frac{7}{10}x=1$ has an equals sign, so it is an *equation* to be solved.

$$\frac{3}{5}x-\frac{7}{10}x=1$$
$$10\left(\frac{3}{5}x-\frac{7}{10}x\right)=10\cdot 1 \quad \text{LCD}=10$$
$$10\left(\frac{3}{5}x\right)-10\left(\frac{7}{10}x\right)=10\cdot 1 \quad \text{Distributive prop.}$$
$$6x-7x=10 \quad \text{Multiply.}$$
$$-x=10$$
$$x=-10 \quad \text{Divide by } -1.$$

The solution set is $\{-10\}$.

15. $\frac{3}{4}x-\frac{1}{2}x=0$ has an equals sign, so it is an *equation* to be solved.

$$\frac{3}{4}x-\frac{1}{2}x=0$$
$$4\left(\frac{3}{4}x-\frac{1}{2}x\right)=4(0) \quad \text{LCD}=4$$
$$4\left(\frac{3}{4}x\right)-4\left(\frac{1}{2}x\right)=4(0) \quad \text{Dist. prop.}$$
$$3x-2x=0 \quad \text{Multiply.}$$
$$x=0$$

The solution set is $\{0\}$.

17. $\frac{3}{x+2}-\frac{5}{x}=1$

The denominators, $x+2$ and x, are equal to 0 for the values -2 and 0, respectively. Thus, $x\neq -2, 0$.

19. $\frac{-1}{(x+3)(x-4)}=\frac{1}{2x+1}$

The denominators, $(x+3)(x-4)$ and $2x+1$, are equal to 0 for the values -3, 4, and $-\frac{1}{2}$, respectively. Thus, $x\neq -3, 4, -\frac{1}{2}$.

21. $\frac{4}{x^2+8x-9}+\frac{1}{x^2-4}=0$

The denominators, $x^2+8x-9=(x+9)(x-1)$ and $x^2-4=(x+2)(x-2)$, are equal to 0 for the values -9, 1, -2, and 2, respectively. Thus, $x\neq -9, 1, -2, 2$.

23. $\frac{5}{m}-\frac{3}{m}=8$

Multiply each side by the LCD, m.

$$m\left(\frac{5}{m}-\frac{3}{m}\right)=m\cdot 8$$

Use the distributive property to remove parentheses; then solve.

$$m\left(\frac{5}{m}\right)-m\left(\frac{3}{m}\right)=8m$$
$$5-3=8m$$
$$2=8m$$
$$m=\frac{2}{8}=\frac{1}{4}$$

Check this proposed solution by replacing m with $\frac{1}{4}$ in the original equation.

$$\frac{5}{\frac{1}{4}} - \frac{3}{\frac{1}{4}} \stackrel{?}{=} 8 \quad \text{Let } m = \frac{1}{4}.$$

$$5 \cdot 4 - 3 \cdot 4 \stackrel{?}{=} 8 \quad \text{Multiply by reciprocals.}$$

$$20 - 12 \stackrel{?}{=} 8$$

$$8 = 8 \quad \text{True}$$

Thus, the solution set is $\left\{\frac{1}{4}\right\}$.

25. $$\frac{5}{y} + 4 = \frac{2}{y}$$

$$y\left(\frac{5}{y} + 4\right) = y\left(\frac{2}{y}\right) \quad \text{Multiply by LCD.}$$

$$y\left(\frac{5}{y}\right) + y(4) = y\left(\frac{2}{y}\right) \quad \text{Distributive property}$$

$$5 + 4y = 2$$

$$4y = -3$$

$$y = -\frac{3}{4}$$

Check $y = -\frac{3}{4}$: $-\frac{8}{3} = -\frac{8}{3}$ True

Thus, the solution set is $\left\{-\frac{3}{4}\right\}$.

27. $$\frac{3x}{5} - 6 = x$$

$$5\left(\frac{3x}{5} - 6\right) = 5(x) \quad \text{Multiply by LCD.}$$

$$5\left(\frac{3x}{5}\right) - 5(6) = 5x \quad \text{Distributive prop.}$$

$$3x - 30 = 5x$$

$$-30 = 2x$$

$$-15 = x$$

Check $x = -15$: $-15 = -15$ True

Thus, the solution set is $\{-15\}$.

29. $$\frac{4m}{7} + m = 11$$

$$7\left(\frac{4m}{7} + m\right) = 7(11) \quad \text{Multiply by LCD.}$$

$$7\left(\frac{4m}{7}\right) + 7(m) = 77 \quad \text{Distributive property}$$

$$4m + 7m = 77$$

$$11m = 77$$

$$m = 7$$

Check $m = 7$: $11 = 11$ True

Thus, the solution set is $\{7\}$.

31. $$\frac{z-1}{4} = \frac{z+3}{3}$$

$$12\left(\frac{z-1}{4}\right) = 12\left(\frac{z+3}{3}\right) \quad \text{Multiply by LCD.}$$

$$3(z-1) = 4(z+3)$$

$$3z - 3 = 4z + 12 \quad \text{Distributive property}$$

$$-15 = z$$

Check $z = -15$: $-4 = -4$ True

Thus, the solution set is $\{-15\}$.

33. $$\frac{3p+6}{8} = \frac{3p-3}{16}$$

$$16\left(\frac{3p+6}{8}\right) = 16\left(\frac{3p-3}{16}\right) \quad \text{Multiply by LCD.}$$

$$2(3p+6) = 3p - 3$$

$$6p + 12 = 3p - 3 \quad \text{Distributive prop.}$$

$$3p = -15$$

$$p = -5$$

Check $p = -5$: $-\frac{9}{8} = -\frac{9}{8}$ True

Thus, the solution set is $\{-5\}$.

35. $$\frac{2x+3}{x} = \frac{3}{2}$$

$$2x\left(\frac{2x+3}{x}\right) = 2x\left(\frac{3}{2}\right) \quad \text{Multiply by LCD.}$$

$$2(2x+3) = 3x$$

$$4x + 6 = 3x \quad \text{Distributive property}$$

$$x = -6$$

Check $x = -6$: $\frac{3}{2} = \frac{3}{2}$ True

Thus, the solution set is $\{-6\}$.

37.
$$\frac{k}{k-4}-5=\frac{4}{k-4}$$
$$(k-4)\left(\frac{k}{k-4}-5\right)=(k-4)\left(\frac{4}{k-4}\right)$$
$$(k-4)\left(\frac{k}{k-4}\right)-5(k-4)=4$$
$$k-5k+20=4$$
$$-4k=-16$$
$$k=4$$

The proposed solution is 4. However, 4 cannot be a solution because it makes the denominator $k-4$ equal 0. Therefore, the solution set is $\varnothing$.

39.
$$\frac{q+2}{3}+\frac{q-5}{5}=\frac{7}{3}$$
$$15\left(\frac{q+2}{3}+\frac{q-5}{5}\right)=15\left(\frac{7}{3}\right)$$
$$15\left(\frac{q+2}{3}\right)+15\left(\frac{q-5}{5}\right)=5\cdot 7$$
$$5(q+2)+3(q-5)=35$$
$$5q+10+3q-15=35$$
$$8q-5=35$$
$$8q=40$$
$$q=5$$

Check $q=5$: $\frac{7}{3}=\frac{7}{3}$ True

Thus, the solution set is $\{5\}$.

41.
$$\frac{x}{2}=\frac{5}{4}+\frac{x-1}{4}$$
$$4\left(\frac{x}{2}\right)=4\left(\frac{5}{4}+\frac{x-1}{4}\right) \quad \text{Multiply by LCD.}$$
$$2(x)=4\left(\frac{5}{4}\right)+4\left(\frac{x-1}{4}\right)$$
$$2x=5+x-1$$
$$x=4$$

Check $x=4$: $2=2$ True

Thus, the solution set is $\{4\}$.

43.
$$x+\frac{17}{2}=\frac{x}{2}+x+6$$
$$2\left(x+\frac{17}{2}\right)=2\left(\frac{x}{2}+x+6\right)$$
$$2(x)+2\left(\frac{17}{2}\right)=2\left(\frac{x}{2}\right)+2(x)+2(6)$$
$$2x+17=x+2x+12$$
$$5=x$$

Check $x=5$: $\frac{27}{2}=\frac{27}{2}$ True

Thus, the solution set is $\{5\}$.

45.
$$\frac{9}{3x+4}=\frac{36-27x}{16-9x^2}$$
$$\frac{9}{3x+4}=\frac{9(4-3x)}{(4+3x)(4-3x)} \quad \text{Factor.}$$
$$\frac{9}{3x+4}=\frac{9}{4+3x}$$

This is an identity, which is true for all real numbers for which the original equation is defined. We must exclude $\pm\frac{4}{3}$. Thus, the solution set is $\left\{x \middle| x\neq\pm\frac{4}{3}\right\}$.

47.
$$\frac{a+7}{8}-\frac{a-2}{3}=\frac{4}{3}$$
$$24\left(\frac{a+7}{8}+\frac{a-2}{3}\right)=24\left(\frac{4}{3}\right)$$
$$24\left(\frac{a+7}{8}\right)-24\left(\frac{a-2}{3}\right)=8(4)$$
$$3(a+7)-8(a-2)=32$$
$$3a+21-8a+16=32$$
$$-5a+37=32$$
$$-5a=-5$$
$$a=1$$

Check $a=1$: $\frac{4}{3}=\frac{4}{3}$ True

Thus, the solution set is $\{1\}$.

49.
$$\frac{p}{2}-\frac{p-1}{4}=\frac{5}{4}$$
$$4\left(\frac{p}{2}-\frac{p-1}{4}\right)=4\left(\frac{5}{4}\right) \quad \text{Multiply by LCD.}$$
$$4\left(\frac{p}{2}\right)-4\left(\frac{p-1}{4}\right)=5$$
$$2p-1(p-1)=5$$
$$2p-p+1=5$$
$$p=4$$

Check $p=4$: $\frac{5}{4}=\frac{5}{4}$ True

Thus, the solution set is $\{4\}$.

51.
$$\frac{3x}{5}-\frac{x-5}{7}=3$$
$$35\left(\frac{3x}{5}-\frac{x-5}{7}\right)=35(3)$$
$$35\left(\frac{3x}{5}\right)-35\left(\frac{x-5}{7}\right)=105$$
$$7(3x)-5(x-5)=105$$
$$21x-5x+25=105$$
$$16x=80$$
$$x=5$$

Check $x=5$: $3=3$ True

Thus, the solution set is $\{5\}$.

53.
$$\frac{4}{x^2-3x}=\frac{1}{x^2-9}$$
$$\frac{4}{x(x-3)}=\frac{1}{(x+3)(x-3)} \quad \text{Factor.}$$
$$\frac{x(x+3)(x-3)(4)}{x(x-3)}=\frac{x(x+3)(x-3)(1)}{(x+3)(x-3)}$$
$$4(x+3)=x\cdot 1$$
$$4x+12=x$$
$$3x=-12$$
$$x=-4$$

Check $x=-4$: $\frac{1}{7}=\frac{1}{7}$ True

Thus, the solution set is $\{-4\}$.

55.
$$\frac{2}{m}=\frac{m}{5m+12}$$
$$m(5m+12)\left(\frac{2}{m}\right)=m(5m+12)\left(\frac{m}{5m+12}\right)$$
$$(5m+12)(2)=m(m)$$
$$10m+24=m^2$$
$$-m^2+10m+24=0$$
$$m^2-10m-24=0 \quad \text{Multiply by } -1.$$
$$(m-12)(m+2)=0$$
$$m=12 \text{ or } -2$$

Check $m=12$: $\frac{1}{6}=\frac{1}{6}$ True

Check $m=-2$: $-1=-1$ True

Thus, the solution set is $\{-2, 12\}$.

57.
$$\frac{-2}{z+5}+\frac{3}{z-5}=\frac{20}{z^2-25}$$
$$\frac{-2}{z+5}+\frac{3}{z-5}=\frac{20}{(z+5)(z-5)}$$
$$(z+5)(z-5)\left(\frac{-2}{z+5}+\frac{3}{z-5}\right)$$
$$=(z+5)(z-5)\left(\frac{20}{(z+5)(z-5)}\right)$$
$$-2(z-5)+3(z+5)=20$$
$$-2z+10+3z+15=20$$
$$z+25=20$$
$$z=-5$$

The proposed solution, -5, cannot be a solution because it would make the denominators $z+5$ and z^2-25 equal 0 and the corresponding fractions undefined. Since -5 cannot be a solution, the solution set is $\varnothing$.

59.
$$\frac{3}{x-1}+\frac{2}{4x-4}=\frac{7}{4}$$
$$\frac{3}{x-1}+\frac{2}{4(x-1)}=\frac{7}{4}$$
$$4(x-1)\left(\frac{3}{x-1}+\frac{2}{4(x-1)}\right)=4(x-1)\left(\frac{7}{4}\right)$$
$$4(3)+2=(x-1)(7)$$
$$14=7x-7$$
$$21=7x$$
$$3=x$$

Check $x=3$: $\frac{7}{4}=\frac{7}{4}$ True

Thus, the solution set is $\{3\}$.

61.
$$\frac{x}{3x+3}=\frac{2x-3}{x+1}-\frac{2x}{3x+3}$$
$$\frac{x}{3(x+1)}=\frac{2x-3}{x+1}-\frac{2x}{3(x+1)}$$

Multiply both sides of the equation by the LCD.
$$\left(\frac{3(x+1)(x)}{3(x+1)}\right)=3(x+1)\left[\frac{2x-3}{x+1}-\frac{2x}{3(x+1)}\right]$$

Distribute and simplify.
$$x=\left(\frac{3(x+1)(2x-3)}{x+1}\right)-\left(\frac{3(x+1)(2x)}{3(x+1)}\right)$$
$$x=3(2x-3)-2x$$
$$x=6x-9-2x$$
$$x=4x-9$$
$$-3x=-9$$
$$x=3$$

Check $x=3$: $\frac{1}{4}=\frac{1}{4}$ True

Thus, the solution set is $\{3\}$.

63.
$$\frac{2p}{p^2-1}=\frac{2}{p+1}-\frac{1}{p-1}$$
$$\frac{2p}{(p+1)(p-1)}=\frac{2}{p+1}-\frac{1}{p-1}$$
$$\left[\frac{(p+1)(p-1)(2p)}{(p+1)(p-1)}\right]$$
$$=\left(\frac{(p+1)(p-1)(2)}{p+1}\right)-\left(\frac{(p+1)(p-1)(1)}{p-1}\right)$$
$$2p=2(p-1)-1(p+1)$$
$$2p=2p-2-p-1$$
$$p=-3$$

Check $p=-3$: $-\frac{3}{4}=-1+\frac{1}{4}$ True

Thus, the solution set is $\{-3\}$.

65.
$$\frac{5x}{14x+3}=\frac{1}{x}$$
$$x(14x+3)\left(\frac{5x}{14x+3}\right)=x(14x+3)\left(\frac{1}{x}\right)$$
$$x(5x)=(14x+3)(1)$$
$$5x^2=14x+3$$
$$5x^2-14x-3=0$$
$$(5x+1)(x-3)=0$$
$$x=-\frac{1}{5}\text{ or }3$$

Check $x=-\frac{1}{5}$: $-5=-5$ True

Check $x=3$: $\frac{1}{3}=\frac{1}{3}$ True

Thus, the solution set is $\left\{-\frac{1}{5},3\right\}$.

67.
$$\frac{2}{x-1}-\frac{2}{3}=\frac{-1}{x+1}$$
$$3(x-1)(x+1)\left(\frac{2}{x-1}-\frac{2}{3}\right)$$
$$=3(x-1)(x+1)\left(\frac{-1}{x+1}\right)$$
$$3(x-1)(x+1)\left(\frac{2}{x-1}-\frac{2}{3}\right)$$
$$=-3(x-1)(x+1)\left(\frac{-1}{x+1}\right)$$
$$3(x+1)(2)-(x-1)(x+1)(2)=3(x-1)(-1)$$
$$6x+6-2x^2+2=-3x+3$$
$$-2x^2+9x+5=0$$
$$2x^2-9x-5=0$$
$$(2x+1)(x-5)=0$$
$$x=-\frac{1}{2}\text{ or }5$$

Check $x=-\frac{1}{2}$: $-2=-2$ True

Check $x=5$: $-\frac{1}{6}=-\frac{1}{6}$ True

Thus, the solution set is $\left\{-\frac{1}{2},5\right\}$.

69.
$$\frac{x}{2x+2}=\frac{-2x}{4x+4}+\frac{2x-3}{x+1}$$
$$\frac{x}{2(x+1)}=\frac{-2x}{4(x+1)}+\frac{2x-3}{x+1}$$
$$4(x+1)\left(\frac{x}{2(x+1)}\right)$$
$$=\left(\frac{4(x+1)(-2x)}{4(x+1)}\right)+\left(\frac{4(x+1)(2x-3)}{x+1}\right)$$
$$2(x)=-2x+4(2x-3)$$
$$2x=-2x+8x-12$$
$$-4x=-12$$
$$x=3$$

Check $x=3$: $\frac{3}{8}=\frac{3}{8}$ True

Thus, the solution set is $\{3\}$.

71.
$$\frac{8x+3}{x}=3x$$
$$x\left(\frac{8x+3}{x}\right)=x(3x)\quad\text{Multiply by LCD.}$$
$$8x+3=3x^2$$
$$0=3x^2-8x-3$$
$$0=(3x+1)(x-3)$$
$$x=-\frac{1}{3}\text{ or }3$$

Check $x=-\frac{1}{3}$: $-1=-1$ True

Check $x=3$: $9=9$ True

Thus, the solution set is $\left\{-\frac{1}{3},3\right\}$.

73.
$$\frac{1}{x+4}+\frac{x}{x-4}=\frac{-8}{x^2-16}$$
$$(x+4)(x-4)\left(\frac{1}{x+4}+\frac{x}{x-4}\right)$$
$$=(x+4)(x-4)\left(\frac{-8}{x^2-16}\right)$$
$$1(x-4)+x(x+4)=-8$$
$$x-4+x^2+4x=-8$$
$$x^2+5x+4=0$$
$$(x+4)(x+1)=0$$
$$x=-4\text{ or }-1$$

$x=-4$ cannot be a solution because it would make the denominators $x+4$ and x^2-16 equal 0 and the corresponding fractions undefined.

Check $x=-1$: $\frac{1}{3}+\frac{1}{5}=\frac{8}{15}$ True

Thus, the solution set is $\{-1\}$.

75.
$$\frac{4}{3x+6}-\frac{3}{x+3}=\frac{8}{x^2+5x+6}$$
$$\frac{4}{3(x+2)}-\frac{3}{x+3}=\frac{8}{(x+2)(x+3)}$$

Multiply both sides of the equation by the LCD.
$$3(x+2)(x+3)\left(\frac{4}{3(x+2)}-\frac{3}{x+3}\right)$$
$$=\frac{3(x+2)(x+3)(8)}{(x+2)(x+3)}$$

Distribute and simplify.

$$4(x+3)-3(x+2)(3)=3(8)$$
$$4x+12-9x-18=24$$
$$-5x=30$$
$$x=-6$$

Check $x=-6$: $-\frac{1}{3}-(-1)=\frac{2}{3}$ True

Thus, the solution set is $\{-6\}$.

77. $$\frac{3x}{x^2+5x+6}=\frac{5x}{x^2+2x-3}-\frac{2}{x^2+x-2}$$
$$\frac{3x}{(x+2)(x+3)}=\frac{5x}{(x+3)(x-1)}-\frac{2}{(x-1)(x+2)}$$

Multiply both sides of the equation by the LCD.

$$\frac{(x+2)(x+3)(x-1)(3x)}{(x+2)(x+3)}$$
$$=\left[\frac{(x+2)(x+3)(x-1)(5x)}{(x+3)(x-1)} - \frac{(x+2)(x+3)(x-1)(2)}{(x+1)(x+2)}\right]$$

Distribute and simplify.

$$3x(x-1)=5x(x+2)-2(x+3)$$
$$3x^2-3x=5x^2+10x-2x-6$$
$$0=2x^2+11x-6$$
$$0=(2x-1)(x+6)$$
$$x=\frac{1}{2} \text{ or } -6$$

Check $x=\frac{1}{2}$: $\frac{6}{35}=-\frac{10}{7}-\left(-\frac{8}{5}\right)$ True

Check $x=-6$: $-\frac{3}{2}=-\frac{10}{7}-\frac{1}{14}$ True

Thus, the solution set is $\left\{-6, \frac{1}{2}\right\}$.

79. $$\frac{x+4}{x^2-3x+2}-\frac{5}{x^2-4x+3}=\frac{x-4}{x^2-5x+6}$$
$$\frac{x+4}{(x-2)(x-1)}-\frac{5}{(x-3)(x-1)}=\frac{x-4}{(x-3)(x-2)}$$
$$\left[\frac{(x-2)(x-1)(x-3)(x+4)}{(x-2)(x-1)} - \frac{(x-2)(x-1)(x-3)(5)}{(x-3)(x-1)}\right]$$
$$=\left[\frac{(x-2)(x-1)(x-3)(x-4)}{(x-3)(x-2)}\right]$$

Distribute and simplify.

$$(x+4)(x-3)-5(x-2)=(x-1)(x-4)$$
$$x^2+x-12-5x+10=x^2-5x+4$$
$$-4x-2=-5x+4$$
$$x=6$$

Check $x=6$: $\frac{1}{2}-\frac{1}{3}=\frac{1}{6}$ True

Thus, the solution set is $\{6\}$.

81. $$\frac{1}{x^2-1}=\frac{2}{x-1}-\frac{1}{x-1}$$
$$\frac{1}{(x+1)(x-1)}=\frac{1}{x-1}$$
$$\frac{(x+1)(x-1)(1)}{(x+1)(x-1)}=\frac{(x+1)(x-1)(1)}{x-1}$$
$$1=x+1$$
$$x=0$$

Check $x=0$: $-1=-2-(-1)$ True

Thus, the solution set is $\{0\}$.

83. $m=\frac{kF}{a}$

We need to isolate F on one side of the equation.

$m\cdot a=\left(\frac{kF}{a}\right)(a)$ Multiply by a.

$ma=kF$

$\frac{ma}{k}=\frac{kF}{k}$ Divide by k.

$\frac{ma}{k}=F$

85. $m=\frac{kF}{a}$

We need to isolate a on one side of the equation.

$m\cdot a=\left(\frac{kF}{a}\right)(a)$ Multiply by a.

$ma=kF$

$\frac{ma}{m}=\frac{kF}{m}$ Divide by m.

$a=\frac{kF}{m}$

87. $I = \dfrac{E}{R+r}$

We need to isolate R on one side of the equation.

$$I(R+r) = \left(\frac{E}{R+r}\right)(R+r)$$
$$IR + Ir = E$$
$$IR = E - Ir \quad \text{Subtract } Ir.$$
$$R = \frac{E - Ir}{I}, \text{ or } R = \frac{E}{I} - r \quad \text{Divide by } I.$$

89. $h = \dfrac{2\mathcal{A}}{B+b}$

We need to isolate $\mathcal{A}$ on one side of the equation.

$$(B+b)h = \frac{(B+b)(2\mathcal{A})}{B+b}$$
$$h(B+b) = 2\mathcal{A}$$
$$\frac{h(B+b)}{2} = \mathcal{A} \quad \text{Divide by 2.}$$

91. $d = \dfrac{2S}{n(a+L)}$

We need to isolate a on one side of the equation.

$$d \cdot n(a+L) = \frac{2S}{n(a+L)} \cdot n(a+L)$$
$$nd(a+L) = 2S$$
$$and + ndL = 2S$$
$$and = 2S - ndL \quad \text{Subtract } ndL.$$
$$a = \frac{2S - ndL}{nd}, \text{ or } a = \frac{2S}{nd} - L$$

93. $\dfrac{1}{x} = \dfrac{1}{y} - \dfrac{1}{z}$

We need to isolate y on one side of the equation.

The LCD of all the fractions in the equation is xyz, so multiply both sides by xyz.

$$xyz\left(\frac{1}{x}\right) = xyz\left(\frac{1}{y} - \frac{1}{z}\right)$$
$$xyz\left(\frac{1}{x}\right) = xyz\left(\frac{1}{y}\right) - xyz\left(\frac{1}{z}\right)$$
$$yz = xz - xy$$

Since we are solving for y, get all terms with y on one side of the equation.

$xy + yz = xz$ Add xy.

Factor out the common factor y on the left.

$y(x+z) = xz$

Finally, divide both sides by the coefficient of y, which is $x+z$.

$$y = \frac{xz}{x+z}$$

95. $\dfrac{2}{r} + \dfrac{3}{s} + \dfrac{1}{t} = 1$

We need to isolate t on one side of the equation. The LCD of all the fractions in the equation is rst, so multiply both sides by rst.

$$rst\left(\frac{2}{r} + \frac{3}{s} + \frac{1}{t}\right) = rst(1)$$
$$rst\left(\frac{2}{r}\right) + rst\left(\frac{3}{s}\right) + rst\left(\frac{1}{t}\right) = rst$$
$$2st + 3rt + rs = rst$$

Since we are solving for t, get all terms with t on one side of the equation.

$2st + 3rt - rst = -rs$

Factor out the common factor t on the left.

$t(2s + 3r - rs) = -rs$

Finally, divide both sides by the coefficient of t, which is $2s + 3r - rs$.

$$t = \frac{-rs}{2s + 3r - rs}, \text{ or } t = \frac{rs}{-2s - 3r + rs}$$

97. $9x + \dfrac{3}{z} = \dfrac{5}{y}$

We need to isolate z on one side of the equation.

$$yz\left(9x + \frac{3}{z}\right) = yz\left(\frac{5}{y}\right) \quad \text{Multiply by LCD.}$$
$$yz(9x) + yz\left(\frac{3}{z}\right) = yz\left(\frac{5}{y}\right) \quad \text{Distributive prop.}$$
$$9xyz + 3y = 5z$$
$$9xyz - 5z = -3y$$
$$z(9xy - 5) = -3y \quad \text{Factor out } z.$$
$$z = \frac{-3y}{9xy - 5}, \text{ or } z = \frac{3y}{5 - 9xy}$$

99. $\frac{t}{x-1}-\frac{2}{x+1}=\frac{1}{x^2-1}$

We need to isolate p on one side of the equation.

$$(x+1)(x-1)\left(\frac{t}{x-1}-\frac{2}{x+1}\right)$$
$$=(x+1)(x-1)\left(\frac{1}{x^2-1}\right)$$

Distribute and simplify.

$$t(x+1)-2(x-1)=1$$
$$t(x+1)=1+2(x-1)$$
$$t(x+1)=1+2x-2$$
$$t=\frac{2x-1}{x+1}, \text{ or } t=\frac{-2x+1}{-x-1}$$

101. **(a)** The values that make the denominator of a fraction equal to zero cause an expression to become undefined.

$$P=\frac{6}{x+3}=\frac{6}{(-3)+3}=\frac{6}{0}$$

Therefore, $x=-3$ causes the expression to become undefined.

(b) The values that make the denominator of a fraction equal to zero cause an expression to become undefined.

$$Q=\frac{5}{x+1}=\frac{5}{(-1)+1}=\frac{5}{0}$$

Therefore, $x=-1$ causes the expression to become undefined.

(c) The values that make the denominator of a fraction equal to zero cause an expression to become undefined.

$$R=\frac{4x}{x^2+4x+3}=\frac{4x}{(x+3)(x+1)}$$

Consider the following two cases.

$$\frac{4x}{((-3)+3)((-3)+1)} \quad \text{Undefined}$$
$$\frac{4x}{((-1)+3)((-1)+1)} \quad \text{Undefined}$$

Therefore, $x=-1$ or -3 cause the expression to become undefined.

103. If $x=0$, the expression R becomes zero.

$$\frac{4(0)}{((0)+3)((0)+1)}=\frac{0}{3}=0$$

Dividing by zero results in an undefined value. Therefore, x in this case cannot be zero.

105. $\frac{6}{x+3}+\frac{5}{x+1}-\frac{4x}{x^2+4x+3}$

$$=\frac{6}{x+3}+\frac{5}{x+1}-\frac{4x}{(x+3)(x+1)}$$
$$=\frac{6(x+1)}{(x+3)(x+1)}+\frac{5(x+3)}{(x+3)(x+1)}-\frac{4x}{(x+3)(x+1)}$$
$$=\frac{6(x+1)+5(x+3)-4x}{(x+3)(x+1)}$$
$$=\frac{6x+6+5x+15-4x}{(x+3)(x+1)}$$
$$=\frac{7x+21}{(x+3)(x+1)}$$
$$=\frac{7(x+3)}{(x+3)(x+1)}$$
$$=\frac{7}{x+1}$$

107. $\frac{6}{x+3}+\frac{5}{x+1}=\frac{4x}{x^2+4x+3}$

$$\frac{6}{x+3}+\frac{5}{x+1}=\frac{4x}{(x+3)(x+1)}$$

Multiply both sides of the equation by the LCD.

$$(x+3)(x+1)\left(\frac{6}{x+3}+\frac{5}{x+1}\right)$$
$$=\frac{(x+3)(x+1)(4x)}{(x+3)(x+1)}$$

Distribute and simplify.

$$6(x+1)+5(x+3)=4x$$
$$6x+6+5x+15=4x$$
$$11x+21=4x$$
$$7x=-21$$
$$x=-3$$

The proposed solution is -3. However, -3 cannot be a solution because it makes the denominators $x+3$ and x^2+4x+3 equal 0. Therefore, the solution set is $\varnothing$.

Summary Exercises Simplifying Rational Expressions vs. Solving Rational Equations

1. No equals sign appears, so this is an *expression*.

$$\frac{4}{p}+\frac{6}{p}=\frac{4+6}{p}=\frac{10}{p}$$

2. No equals sign appears, so this is an *expression.*

$$\frac{x^3y^2}{x^2y^4}\cdot\frac{y^5}{x^4}=\frac{x^3y^7}{x^6y^4}$$
$$=\frac{y^3\cdot x^3y^4}{x^3\cdot x^3y^4}=\frac{y^3}{x^3}$$

3. No equals sign appears, so this is an *expression.*

$$\frac{1}{x^2+x-2}\div\frac{4x^2}{2x-2}$$
$$=\frac{1}{x^2+x-2}\cdot\frac{2x-2}{4x^2}$$
$$=\frac{1}{(x+2)(x-1)}\cdot\frac{2(x-1)}{2\cdot 2x^2}$$
$$=\frac{1}{2x^2(x+2)}$$

4. $\frac{8}{t-5}=2$

There is an equals sign, so this is an *equation.*

$$(t-5)\left(\frac{8}{t-5}\right)=(t-5)(2) \quad \text{Multiply by LCD.}$$
$$8=2t-10$$
$$18=2t$$
$$9=t$$

Check $t=9: 2=2$ True

The solution set is $\{9\}$.

5. $\frac{x-4}{5}=\frac{x+3}{6}$

There is an equals sign, so this is an *equation.*

$$30\left(\frac{x-4}{5}\right)=\left(\frac{x+3}{6}\right)(30) \quad \text{Multiply by LCD.}$$
$$6(x-4)=5(x+3)$$
$$6x-24=5x+15$$
$$x=39$$

Check $x=39: 7=7$ True

The solution set is $\{39\}$.

6. No equals sign appears, so this is an *expression.*

$$\frac{2}{k^2-4k}+\frac{3}{k^2-16}$$
$$=\frac{2}{k(k-4)}+\frac{3}{(k-4)(k+4)}$$

Obtain the LCD between both fractions and multiply accordingly.

$$\frac{2}{k(k-4)}+\frac{3}{(k-4)(k+4)}$$
$$=\frac{2(k+4)}{k(k-4)(k+4)}+\frac{3k}{k(k-4)(k+4)}$$
$$=\frac{2k+8+3k}{k(k-4)(k+4)}$$
$$=\frac{5k+8}{k(k-4)(k+4)}$$

7. No equals sign appears, so this is an *expression.*

$$\frac{2y^2+y-6}{2y^2-9y+9}\cdot\frac{y^2-2y-3}{y^2-1}$$
$$=\frac{(2y-3)(y+2)(y-3)(y+1)}{(2y-3)(y-3)(y+1)(y-1)}$$
$$=\frac{y+2}{y-1}$$

8. No equals sign appears, so this is an *expression.*

$$\frac{3t^2-t}{6t^2+15t}\div\frac{6t^2+t-1}{2t^2-5t-25}$$
$$=\frac{3t^2-t}{6t^2+15t}\cdot\frac{2t^2-5t-25}{6t^2+t-1}$$
$$=\frac{t(3t-1)}{3t(2t+5)}\cdot\frac{(2t+5)(t-5)}{(3t-1)(2t+1)}$$
$$=\frac{t-5}{3(2t+1)}$$

9. No equals sign appears, so this is an *expression.*

$$\frac{4}{p+2}+\frac{1}{3p+6}$$
$$=\frac{4}{p+2}+\frac{1}{3(p+2)}$$
$$=\frac{3\cdot 4}{3(p+2)}+\frac{1}{3(p+2)}$$
$$=\frac{12+1}{3(p+2)}$$
$$=\frac{13}{3(p+2)}$$

10. $\frac{1}{x}+\frac{1}{x-3}=-\frac{5}{4}$

There is an equals sign, so this is an *equation.*

$$4x(x-3)\left(\frac{1}{x}+\frac{1}{x-3}\right)=4x(x-3)\left(-\frac{5}{4}\right)$$

Distribute.

$$4x(x-3)\left(\frac{1}{x}\right)+4x(x-3)\left(\frac{1}{x-3}\right)$$

$$=-5x(x-3)$$

Simplify.

$$4(x-3)+4x=-5x(x-3)$$
$$4x-12+4x=-5x^2+15x$$
$$5x^2-7x-12=0$$
$$(x+1)(5x-12)=0$$
$$x=1 \text{ or } x=\frac{12}{5}$$

Check $x=-1$: $-1+\frac{1}{-4}=-\frac{5}{4}$ True

Check $x=\frac{12}{5}$: $\frac{5}{12}+\left(-\frac{5}{3}\right)=-\frac{5}{4}$ True

The solution set is $\left\{-1,\frac{12}{5}\right\}$.

11. $\frac{3}{t-1}+\frac{1}{t}=\frac{7}{2}$

There is an equals sign, so this is an *equation.*

$$2t(t-1)\left(\frac{3}{t-1}+\frac{1}{t}\right)=2t(t-1)\left(\frac{7}{2}\right)$$

Distribute and simplify.

$$2t(t-1)\left(\frac{3}{t-1}\right)+2t(t-1)\left(\frac{1}{t}\right)=7t(t-1)$$
$$2t(3)+2(t+1)=7t(t-1)$$
$$6t+2t-2=7t^2-7t$$
$$0=7t^2-15t+2$$
$$0=(7t-1)(t-2)$$
$$t=\frac{1}{7} \text{ or } t=2$$

Check $t=\frac{1}{7}$: $-\frac{7}{2}+7=\frac{7}{2}$ True

Check $t=2$: $3+\frac{1}{2}=\frac{7}{2}$ True

The solution set is $\left\{\frac{1}{7},2\right\}$.

12. No equals sign appears, so this is an *expression.*

$$\frac{6}{k}-\frac{2}{3k}=\frac{6}{k}\cdot\frac{3}{3}-\frac{2}{3k} \quad \text{LCD}=3k$$
$$=\frac{18}{3k}-\frac{2}{3k}$$
$$=\frac{18-2}{3k}=\frac{16}{3k}$$

13. No equals sign appears, so this is an *expression.*

$$\frac{5}{4z}-\frac{2}{3z}=\frac{3\cdot 5}{3\cdot 4z}-\frac{4\cdot 2}{4\cdot 3z} \quad \text{LCD}=12z$$
$$=\frac{15}{12z}-\frac{8}{12z}$$
$$=\frac{15-8}{12z}=\frac{7}{12z}$$

14. $\frac{x+2}{3}=\frac{2x-1}{5}$

There is an equals sign, so this is an *equation.*

$$15\left(\frac{x+2}{3}\right)=15\left(\frac{2x-1}{5}\right) \quad \text{Multiply by LCD.}$$
$$5(x+2)=3(2x-1)$$
$$5x+10=6x-3$$
$$13=x$$

Check $x=13$: $5=5$ True

The solution set is $\{13\}$.

15. No equals sign appears, so this is an *expression.*

$$\frac{1}{m^2+5m+6}+\frac{2}{m^2+4m+3}$$
$$=\frac{1}{(m+2)(m+3)}+\frac{2}{(m+1)(m+3)}$$
$$=\frac{1(m+1)}{(m+2)(m+3)(m+1)}$$
$$+\frac{2(m+2)}{(m+1)(m+3)(m+2)}$$

Add rational expressions.

$$\frac{1(m+1)}{(m+2)(m+3)(m+1)}+\frac{2(m+2)}{(m+1)(m+3)(m+2)}$$
$$=\frac{(m+1)+(2m+4)}{(m+1)(m+2)(m+3)}$$
$$=\frac{3m+5}{(m+1)(m+2)(m+3)}$$

16. No equals sign appears, so this is an *expression.*

$$\frac{2k^2-3k}{20k^2-5k} \div \frac{2k^2-5k+3}{4k^2+11k-3}$$
$$=\frac{2k^2-3k}{20k^2-5k}\cdot\frac{4k^2+11k-3}{2k^2-5k+3}$$
$$=\frac{k(2k-3)}{5k(4k-1)}\cdot\frac{(4k-1)(k+3)}{(2k-3)(k-1)}$$
$$=\frac{k+3}{5(k-1)}$$

17. $\frac{2}{x+1}+\frac{5}{x-1}=\frac{10}{x^2-1}$

There is an equals sign, so this is an *equation.*

$$\frac{2}{x+1}+\frac{5}{x-1}=\frac{10}{(x+1)(x-1)}$$

Multiply both sides of the equation by the LCD.

$$(x+1)(x-1)\left(\frac{2}{x+1}+\frac{5}{x-1}\right)$$
$$=(x+1)(x-1)\left[\frac{10}{(x+1)(x-1)}\right]$$

Distribute.

$$(x+1)(x-1)\left(\frac{2}{x+1}\right)+(x+1)(x-1)\left(\frac{5}{x-1}\right)=10$$

Simplify.

$$2(x-1)+5(x+1)=10$$
$$2x-2+5x+5=10$$
$$3+7x=10$$
$$7x=7$$
$$x=1$$

Replacing x by 1 in the original equation makes the denominators $x-1$ and x^2-1 equal to 0, so the solution set is $\varnothing$.

18. There is an equals sign, so this is an *equation.*

$$\frac{3}{x+3}+\frac{4}{x+6}=\frac{9}{x^2+9x+18}$$
$$\frac{3}{x+3}+\frac{4}{x+6}=\frac{9}{(x+3)(x+6)}$$

Multiply both sides of the equation by the LCD.

$$(x+3)(x+6)\left(\frac{3}{x+3}+\frac{4}{x+6}\right)$$
$$=(x+3)(x+6)\left(\frac{9}{(x+3)(x+6)}\right)$$

Distribute and simplify.

$$3(x+6)+4(x+3)=9$$
$$3x+18+4x+12=9$$
$$7x=-21$$
$$x=-3$$

The solution set is $\varnothing$ because $x=-3$ makes two of the denominators, $x+3$ and $x^2+9x+18$, equal to zero.

19. No equals sign appears, so this is an *expression.*

$$\frac{4t^2-t}{6t^2+10t} \div \frac{8t^2+2t-1}{3t^2+11t+10}$$
$$=\frac{4t^2-t}{6t^2+10t}\cdot\frac{3t^2+11t+10}{8t^2+2t-1}$$

Multiply by the reciprocal.

$$\frac{4t^2-t}{6t^2+10t}\cdot\frac{3t^2+11t+10}{8t^2+2t-1}$$
$$=\frac{t(4t-1)}{2t(3t+5)}\cdot\frac{(3t+5)(t+2)}{(4t-1)(2t+1)}$$

Simplify using the fundamental property.

$$\frac{t(4t-1)}{2t(3t+5)}\cdot\frac{(3t+5)(t+2)}{(4t-1)(2t+1)}$$
$$=\frac{t+2}{2(2t+1)}$$

20. There is an equals sign, so this is an *equation.*

$$\frac{x}{x-2}+\frac{3}{x+2}=\frac{8}{x^2-4}$$
$$\frac{x}{x-2}+\frac{3}{x+2}=\frac{8}{(x+2)(x-2)}$$
$$(x+2)(x-2)\left(\frac{x}{x-2}+\frac{3}{x+2}\right)$$
$$=(x+2)(x-2)\left(\frac{8}{(x+2)(x-2)}\right)$$

Multiply both sides of the equation by the LCD.

$$(x+2)(x-2)\left(\frac{x}{x-2}\right)$$
$$+(x+2)(x-2)\left(\frac{3}{x+2}\right)=8$$

Distribute and simplify.

$$x(x+2)+3(x-2)=8$$
$$x^2+2x+3x-6=8$$
$$x^2+5x-14=0$$
$$(x+7)(x-2)=0$$
$$x=-7 \text{ or } x=2$$

Check $x=-7$: $\frac{7}{9}+\frac{3}{-5}=\frac{8}{45}$ True

Since x cannot equal 2 in the original equation (because $x-2$ would be 0), the solution set is $\{-7\}$.

7.7 Applications of Rational Expressions

Now Try Exercises

N1. *Step 2*
Let $x=$ the denominator of the original fraction, so that the numerator is $x-4$.
Step 3
If 7 is added to both the numerator and denominator, the resulting fraction is equivalent to $\frac{7}{8}$, which translates to

$$\frac{(x-4)+7}{x+7}=\frac{7}{8}.$$

Step 4
Multiply by the LCD, $8(x+7)$.

$$8(x+7)\frac{(x-4)+7}{x+7}=8(x+7)\frac{7}{8}$$
$$8(x+3)=7(x+7)$$
$$8x+24=7x+49$$
$$x=25$$

Step 5
The denominator is 25 and the numerator is $25-4=21$, so the original fraction is $\frac{21}{25}$.
Step 6
If 7 is added to the numerator and the denominator of $\frac{21}{25}$, the result is $\frac{28}{32}$, which is equal to $\frac{7}{8}$.

N2. Let $x=$ the rate of the boat with no current. Complete a table.

	d	r	t
With the Current	12	$x+2$	$\frac{12}{x+2}$
Against the Current	4	$x-2$	$\frac{4}{x-2}$

Since the times are equal, we get the following equation.

$$\frac{12}{x+2}=\frac{4}{x-2}$$

Multiply by the LCD, $(x+2)(x-2)$.

$$12(x-2)=4(x+2)$$
$$12x-24=4x+8$$
$$8x-24=8$$
$$8x=32$$
$$x=4$$

The rate of the boat with no current is 4 miles per hour.

N3. Let $x=$ the number of hours proofreading.

	Rate	Time Working Together	Fractional Part of the Job Done when Working Together
Sarah	$\frac{1}{10}$	x	$\frac{1}{10}x$
Joyce	$\frac{1}{12}$	x	$\frac{1}{12}x$

Since together Sarah and Joyce complete 1 whole job, we must add their individual parts and set the sum equal to 1.

$$\frac{1}{10}x+\frac{1}{12}x=1$$
$$60\left(\frac{1}{10}x+\frac{1}{12}x\right)=60(1)$$
$$60\left(\frac{1}{10}x\right)+60\left(\frac{1}{12}x\right)=60$$
$$6x+5x=60$$
$$11x=60$$
$$x=\frac{60}{11}, \text{or } 5\frac{5}{11}$$

Working together, Sarah and Joyce can proofread the manuscript in $5\frac{5}{11}$ hours.

Exercises

1. When the hawk flies into the headwind, the wind works against the hawk, so the rate of the hawk is 6 miles per hour *less* than if it traveled in still air, or $(m-5)$ miles per hour.
When the hawk flies with a tailwind, the wind pushes the hawk, so the rate of the hawk is 6 miles per hour *more* than if it traveled in still air, or $(m+5)$ miles per hour.

3. If it takes Elayn 10 hours to do a job, her rate is $\frac{1}{10}$ job per hour.

5. **(a)** Let $x =$ <u>the amount</u>.

(b) An expression for "the numerator of the fraction $\frac{5}{6}$ is increased by an amount" is $\underline{5+x}$. We could also use $\frac{5+x}{6}$.

(c) An equation that can be used to solve the problem is represented by the following.

$$\frac{5+x}{6}=\frac{13}{3}$$

7. *Step 2*
Let x represent the *original numerator.*
Step 3
The denominator of the original fraction is represented by $x-4$. If 3 is added to both the numerator and denominator, the resulting fraction is equivalent to $\frac{3}{2}$, which translates to the following.

$$\frac{x+3}{(x-4)+3}=\frac{3}{2}$$

Step 4
Since we have a fraction equal to another fraction, we can use cross multiplication.

$$2(x+3)=3[(x-4)+3]$$
$$2x+6=3x-3$$
$$9=x$$

Step 5

$$\frac{x}{x-4}=\frac{(9)}{(9)-4}=\frac{9}{5}$$

The original fraction is $\frac{9}{5}$.

Step 6
Adding 3 to both the numerator and the denominator gives us the following.

$$\frac{9+3}{5+3}=\frac{12}{8}$$

This is equivalent to $\frac{3}{2}$.

9. *Step 2*
Let x represent the *original numerator.*
Step 3
The denominator of the original fraction is $3x$. If 2 is added to the numerator and subtracted from the denominator, the resulting fraction is equivalent to 1, which translates to the following.

$$\frac{x+2}{3x-2}=1$$

Step 4

$$x+2=1(3x-2)$$
$$x+2=3x-2$$
$$4=2x$$
$$2=x$$

Step 5

$$\frac{x}{3x}=\frac{2}{3(2)}=\frac{2}{6}$$

The original fraction is $\frac{2}{6}$.

Step 6

$$\frac{2+2}{6-2}=\frac{4}{4}=1$$

11. *Step 2*
Let x represent the *number.*
Step 3
One-sixth of a number is 5 more than the same number, which translates to the following.

$$\frac{1}{6}x=x+5$$

Step 4
Multiply both sides by the LCD, 6.

$$6\left(\frac{1}{6}x\right)=(x+5)\cdot 6$$
$$x=6x+30$$
$$-5x=30$$
$$x=-6$$

Step 5
The number is -6.
Step 6
One-sixth of -6 is -1 and 5 more than -6 is -1. So, $x=-6$ satisfies the equation.

13. *Step 2*
Let x represent the quantity.
Step 3
So, $\frac{3}{4}$ of it, $\frac{1}{2}$ of it, and $\frac{1}{3}$ of it are $\frac{3}{4}x, \frac{1}{2}x,$ and $\frac{1}{3}x$. Added together, these equal 93, which translates to the following.

$$x+\frac{3}{4}x+\frac{1}{2}x+\frac{1}{3}x=93.$$

Step 4
Multiply both sides by the LCD of 4, 2, and 3, which is 12.

$$12\left(x+\frac{3}{4}x+\frac{1}{2}x+\frac{1}{3}x\right)=12(93)$$
$$12x+12\left(\frac{3}{4}x\right)+12\left(\frac{1}{2}x\right)+12\left(\frac{1}{3}x\right)=12(93)$$
$$12x+9x+6x+4x=1116$$
$$31x=1116$$
$$x=36$$

Step 5
The quantity is 36.
Step 6
Check 36 in the original problem.

$$x=36, \frac{3}{4}x=27, \frac{1}{2}x=18, \frac{1}{3}x=12$$

Adding gives us the following.

$$36+27+18+12=93$$

Therefore, the equation is satisfied.

15. We are asked to find the *time*, so we'll use the distance, rate, and time relationship $t=\frac{d}{r}$.

$$t=\frac{d}{r}=\frac{0.6 \text{ miles}}{0.0319 \text{ miles per minute}}$$
$$\approx 18.809 \text{ minutes}$$

17. We are asked to find the average *rate*, so we'll use the distance, rate, and time relationship $r=\frac{d}{t}$.

$$r=\frac{d}{t}=\frac{5000 \text{ meters}}{15.071 \text{ minutes}}$$
$$\approx 331.763 \text{ meters per minute}$$

19. We are asked to find the *time*, so we'll use the distance, rate, and time relationship $t=\frac{d}{r}$.

$$t=\frac{d}{r}=\frac{500 \text{ miles}}{140.256 \text{ miles per hour}}$$
$$\approx 3.565 \text{ hours}$$

21. Let x = rate of the plane in still air. Then the rate against the wind is $x-10$ and the rate with the wind is $x+10$. The time flying against the wind is represented by the following.

$$t=\frac{d}{r}=\frac{500}{x-10}$$

The time flying with the wind is represented by the following.

$$t=\frac{d}{r}=\frac{600}{x+10}$$

Now complete the chart.

	d	r	t
Against the Wind	500	$x-10$	$\frac{500}{x-10}$
With the Wind	600	$x+10$	$\frac{600}{x+10}$

Since the problem states that the two times are equal, the following is true.

$$\frac{500}{x-10}=\frac{600}{x+10}$$

We would use this equation to solve the problem.

23. Let x represent the rate of the boat in still water. Then $x-4$ is the rate against the current and $x+4$ is the rate with the current. We fill in the chart as follows, realizing that the time column is filled in by using the formula $t=\frac{d}{r}$.

	d	r	t
Against the Current	20	$x-4$	$\frac{20}{x-4}$
With the Current	60	$x+4$	$\frac{60}{x+4}$

Since the times are equal, we get the following equation.

$$\frac{20}{x-4}=\frac{60}{x+4}$$
$$(x+4)(x-4)\frac{20}{x-4}=(x+4)(x-4)\frac{60}{x+4}$$
$$20(x+4)=60(x-4)$$
$$20x+80=60x-240$$
$$320=40x$$
$$8=x$$

The rate of the boat in still water is 8 miles per hour.

25. Let x = rate of the bird in still air. Then the rate against the wind is $x-8$ and the rate with the wind is $x+8$. Use $t=\frac{d}{r}$ to complete the chart.

	d	*r*	*t*
Against the Wind	18	$x-8$	$\frac{18}{x-8}$
With the Wind	30	$x+8$	$\frac{30}{x+8}$

Since the problem states that the two times are equal, we get the following equation.

$$\frac{18}{x-8}=\frac{30}{x+8}$$
$$\frac{(x+8)(x-8)(18)}{x-18}=\frac{(x+8)(x-8)(30)}{x+8}$$
$$18(x+8)=30(x-8)$$
$$18x+144=30x-240$$
$$384=12x$$
$$32=x$$

The rate of the bird in still air is 32 miles per hour.

27. Let x = rate of the plane in still air. Then the rate against the wind is $x-15$ and the rate with the wind is $x+15$. Use $t=\frac{d}{r}$ to complete the chart.

	d	*r*	*t*
Against the Wind	375	$x-15$	$\frac{375}{x-15}$
With the Wind	450	$x+15$	$\frac{450}{x+15}$

Since the problem states that the two times are equal, we get the following equation.

$$\frac{350}{x-15}=\frac{450}{x+15}$$
$$\frac{(x+15)(x-15)(375)}{x-15}=\frac{(x+15)(x-15)(450)}{x+15}$$
$$375(x+15)=450(x-15)$$
$$375x+5625=450x-6750$$
$$12{,}375=75x$$
$$165=x$$

The rate of the plane in still air is 165 miles per hour.

29. Let x represent the rate of the current of the river. Then $12-x$ is the rate upstream (against the current) and $12+x$ is the rate downstream (with the current). Use $t=\frac{d}{r}$ to complete the table.

	d	*r*	*t*
Upstream	6	$12-x$	$\frac{6}{12-x}$
Downstream	10	$12+x$	$\frac{10}{12+x}$

Since the times are equal, we get the following equation.

$$\frac{6}{12-x}=\frac{10}{12+x}$$
$$\frac{(12+x)(12-x)(6)}{12-x}=\frac{(12+x)(12-x)(10)}{12+x}$$
$$6(12+x)=10(12-x)$$
$$72+6x=120-10x$$
$$16x=48$$
$$x=3$$

The rate of the current of the river is 3 miles per hour.

31. Let x = the average rate of the ferry. Use the formula $t=\frac{d}{r}$ to make a table.

	d	*r*	*t*
Seattle–Victoria	148	x	$\frac{148}{x}$
Victoria–Vancouver	74	x	$\frac{74}{x}$

Since the time for the Victoria–Vancouver trip is 4 hours less than the time for the Seattle–Victoria trip, solve the following equation.

$$\frac{74}{x}=\frac{148}{x}-4$$
$$x\left(\frac{74}{x}\right)=x\left(\frac{148}{x}-4\right)$$
$$74=148-4x$$
$$4x=74$$
$$x=\frac{74}{4}=\frac{37}{2}\text{, or }18.5$$

The average rate of the ferry is 18.5 miles per hour.

33. Let t = the number of hours it will take Edward and Abdalla to paint the room working together.

	Rate	Time Working Together	Fractional Part of the Job Done When Working Together
Edward	$\frac{1}{8}$	t	$\frac{1}{8}t$
Abdalla	$\frac{1}{6}$	t	$\frac{1}{6}t$

Working together, they complete 1 whole job, so add their individual fractional parts and set the sum equal to 1.

part done by Edward	+	part done by Abdalla	=	1 whole job
↓	↓	↓	↓	↓
$\frac{1}{8}t$	+	$\frac{1}{6}t$	=	1

An equation that can be used to solve this problem is the following.

$$\frac{1}{8}t+\frac{1}{6}t=1$$

Alternatively, we can compare the hourly rates of completion. In one hour, Edward will complete $\frac{1}{8}$ of the job, Abdalla will complete $\frac{1}{6}$ of the job, and together they will complete $\frac{1}{t}$ of the job. So another equation that can be used to solve this problem is the following.

$$\frac{1}{8}+\frac{1}{6}=\frac{1}{t}$$

35. Let x represent the number of hours it will take for Heather and Courtney to grade the tests working together. Since Heather can grade the test in 4 hours, her rate alone is $\frac{1}{4}$ job per hour. Also, since Courtney can do the job alone in 6 hours, her rate is $\frac{1}{6}$ job per hour.

	Rate	Time Working Together	Fractional Part of the Job Done When Working Together
Heather	$\frac{1}{4}$	x	$\frac{1}{4}x$
Courtney	$\frac{1}{6}$	x	$\frac{1}{6}x$

Since together Heather and Courtney complete 1 whole job, we must add their individual fractional parts and set the sum equal to 1.

$$\frac{1}{4}x+\frac{1}{6}x=1$$
$$12\left(\frac{1}{4}x+\frac{1}{6}x\right)=12(1)\quad \text{LCD}=12$$
$$12\left(\frac{1}{4}x\right)+12\left(\frac{1}{6}x\right)=12$$
$$3x+2x=12$$
$$5x=12$$
$$x=\frac{12}{5}\text{, or }2\frac{2}{5}$$

It will take Heather and Courtney $2\frac{2}{5}$ hours to grade the tests if they work together.

37. Let x = the number of hours to pump the water using both pumps.

	Rate	Time Working Together	Fractional Part of the Job Done When Working Together
Pump 1	$\frac{1}{10}$	x	$\frac{1}{10}x$
Pump 2	$\frac{1}{12}$	x	$\frac{1}{12}x$

Since together the two pumps complete 1 whole job, we must add their individual fractional parts and set the sum equal to 1.

$$\frac{1}{10}x+\frac{1}{12}x=1$$
$$60\left(\frac{1}{10}x+\frac{1}{12}x\right)=60(1) \quad \text{LCD}=60$$
$$60\left(\frac{1}{10}x\right)+60\left(\frac{1}{12}x\right)=60$$
$$6x+5x=60$$
$$11x=60$$
$$x=\frac{60}{11}, \quad \text{or} \quad 5\frac{5}{11}$$

It would take $5\frac{5}{11}$ hours to pump out the basement if both pumps were used.

39. Let x represent the number of hours it will take the experienced employee to enter the data. Then $2x$ represents the number of hours it will take the new employee (the experienced employee takes less time). The experienced employee's rate is $\frac{1}{x}$ job per hour and the new employee's rate is $\frac{1}{2x}$ job per hour.

	Rate	Time Working Together	Fractional Part of the Job Done When Working Together
Experienced Employee	$\frac{1}{x}$	2	$\frac{1}{x}\cdot 2=\frac{2}{x}$
New Employee	$\frac{1}{2x}$	2	$\frac{1}{2x}\cdot 2=\frac{1}{x}$

Since together the two employees complete the whole job, we must add their individual fractional parts and set the sum equal to 1.

$$\frac{2}{x}+\frac{1}{x}=1$$
$$x\left(\frac{2}{x}+\frac{1}{x}\right)=x(1) \quad \text{Multiply by the LCD.}$$
$$x\left(\frac{2}{x}\right)+x\left(\frac{1}{x}\right)=x$$
$$2+1=x$$
$$3=x$$

Working alone, it will take the experienced employee 3 hours to enter the data.

41. Let x = the number of hours to fill the pool $\frac{3}{4}$ full with both pipes working together.

	Rate	Time Working Together	Fractional Part of the Job Done When Working Together
First Pipe	$\frac{1}{6}$	x	$\frac{1}{6}x$
Second Pipe	$\frac{1}{9}$	x	$\frac{1}{9}x$

part done by first pipe + part done by second pipe = $\frac{3}{4}$ full
↓ ↓ ↓ ↓ ↓

$$\frac{1}{6}x + \frac{1}{9}x = \frac{3}{4}$$
$$36\left(\frac{1}{6}x+\frac{1}{9}x\right)=36\left(\frac{3}{4}\right) \quad \text{LCD}=36$$
$$36\left(\frac{1}{6}x\right)+36\left(\frac{1}{9}x\right)=36\left(\frac{3}{4}\right)$$
$$6x+4x=27$$
$$10x=27$$
$$x=\frac{27}{10}, \text{ or } 2\frac{7}{10}$$

It takes $2\frac{7}{10}$ hours to fill the pool $\frac{3}{4}$ full using both pipes.

Alternatively, we could solve $\frac{1}{6}x+\frac{1}{9}x=1$ (filling the whole pool) and then multiply that answer by $\frac{3}{4}$. In this case, x would represent the number of hours to fill the pool with both pipes working together.

43. Let x = the number of minutes it takes to fill the sink.
In 1 minute, the cold water faucet (alone) can fill $\frac{1}{12}$ of the sink. In the same time, the hot water faucet (alone) can fill $\frac{1}{15}$ of the sink. In 1 minute, the drain (alone) empties $\frac{1}{25}$ of the sink. Together, they fill $\frac{1}{x}$ of the sink in 1 minute, so solve the following equation.

$$\frac{1}{12}+\frac{1}{15}-\frac{1}{25}=\frac{1}{x}.$$

$$300x\left(\frac{1}{12}+\frac{1}{15}-\frac{1}{25}\right)=300x\left(\frac{1}{x}\right)$$

$$25x+20x-12x=300$$

$$33x=300$$

$$x=\frac{300}{33}=\frac{100}{11}, \text{ or } 9\frac{1}{11}$$

It will take $9\frac{1}{11}$ minutes to fill the sink.

7.8 Variation

Now Try Exercises

N1. $W=kr$ $\quad$ W varies directly as r.
$40=5k$ $\quad$ Substitute $W=40$ and $r=5$.
$8=k$ $\quad$ Solve for k.

Since $W=kr$ and $k=8$, the following is calculated as shown.
$W=8r$ $\quad$ Substitute $k=8$ in $W=kr$.
$W=8(10)$ $\quad$ Substitute $r=10$.
$W=80$ $\quad$ Multiply.

N2. $V=kr^2$
$80=k(4)^2$ $\quad$ Substitute $V=80$ and $r=4$.
$5=k$ $\quad$ Solve for k.

Since $V=kr^2$ and $k=5$, the following is calculated as shown.
$V=5r^2$
$V=5(5)^2$ $\quad$ Substitute $r=5$.
$V=125$
The volume is 125 cubic feet when the radius is 5 feet.

N3. $t=\frac{k}{r}$ $\quad$ t varies inversely as r.
$12=\frac{k}{3}$ $\quad$ Substitute $t=12$ and $r=3$.
$36=k$ $\quad$ Solve for k.

Since $t=\frac{k}{r}$ and $k=36$, the following is calculated as shown.
$t=\frac{36}{r}$ $\quad$ Substitute $k=36$ in $t=\frac{k}{r}$.
$t=\frac{36}{6}$ $\quad$ Substitute $x=6$.
$t=6$ $\quad$ Reduce.

N4. $h=\frac{k}{b}$ $\quad$ Height varies inversely as base.
$6=\frac{k}{4}$ $\quad$ Substitute $h=6$ and $b=4$.
$24=k$ $\quad$ Solve for k.

Since $h=\frac{k}{b}$ and $k=24$, the following is calculated as shown.
$h=\frac{24}{b}$ $\quad$ Substitute $k=24$ in $h=\frac{k}{b}$.
$h=\frac{24}{12}$ $\quad$ Substitute $b=12$.
$h=2$ $\quad$ Reduce.
The height of the triangle is 2 feet when the base is 12 feet.

Exercises

1. As the number of candy bars you buy *increases*, the total price for the candy *increases*. Thus, the variation between the quantities is *direct*.

3. As the amount of pressure put on the accelerator of a truck *increases*, the rate of the truck *increases*. Thus, the variation between the quantities is *direct*.

5. For a triangle of constant area, if the base *increases*, then the height *decreases*. Thus, the variation between the quantities is *inverse*.

7. As the intensity of a light source *increases*, the distance from which a person views the light *decreases*. Thus, the variation between the quantities is *inverse*.

9. $y = \frac{3}{x}$ represents *inverse* variation since it is of the form $y = \frac{k}{x}$.

11. $y = 10x^2$ represents *direct* variation since it is of the form $y = kx^n$.

13. $y = 50x$ represents *direct* variation since it is of the form $y = kx$.

15. $y = \frac{12}{x^2}$ represents *inverse* variation since it is of the form $y = \frac{k}{x^n}$.

17. **(a)** If the constant of variation is positive and y varies directly as x, then as x increases, y *increases.*

(b) If the constant of variation is positive and y varies inversely as x, then as x increases, y *decreases.*

19. Since x varies directly as y, there is a constant k such that $x = ky$. First find the value of k.

$27 = k(6)$ Let $x = 27$, $y = 6$.

$k = \frac{27}{6} = \frac{9}{2}$

When $k = \frac{9}{2}$, $x = ky$ becomes $x = \frac{9}{2}y$.

Now find x when $y = 2$.

$x = \frac{9}{2}(2)$ Let $y = 2$.

$= 9$

21. Since d varies directly as t, there is a constant k such that $d = kt$. First find the value of k.

$150 = k(3)$ Let $d = 150$ and $t = 3$.

$k = \frac{150}{3} = 50$

When $k = 50$, $d = kt$ becomes $d = 50t$.

Now find d when $t = 5$.

$d = 50(5)$ Let $t = 5$.

$= 250$

23. Since x varies inversely as y, there is a constant k such that $x = \frac{k}{y}$. First find the value of k.

$3 = \frac{k}{8}$ Let $x = 3$ and $y = 8$.

$k = 3(8) = 24$

When $k = 24$, $x = \frac{k}{y}$ becomes $x = \frac{24}{y}$.

Now find y when $x = 4$.

$4 = \frac{24}{y}$ Let $x = 4$.

$4y = 24$

$y = 6$

25. Since p varies inversely as q, there is a constant k such that $p = \frac{k}{q}$. First find the value of k.

$7 = \frac{k}{6}$ Let $p = 7$ and $q = 6$.

$k = 7(6) = 42$

When $k = 42$, $p = \frac{k}{q}$ becomes $p = \frac{42}{q}$.

Now find p when $q = 2$.

$p = \frac{42}{q} = 21$

27. Since m varies inversely as p^2, there is a constant k such that $m = \frac{k}{p^2}$. First find the value of k.

$20 = \frac{k}{2^2}$ Let $m = 20$ and $p = 2$.

$k = 20(4) = 80$

When $k = 80$, $m = \frac{k}{p^2}$ becomes $m = \frac{80}{p^2}$.

Now find m when $p = 5$.

$m = \frac{80}{5^2}$ Let $p = 5$.

$m = \frac{80}{25} = \frac{16}{5}$, or $3\frac{1}{5}$

29. Since p varies inversely as q^2, there is a constant k such that $p = \frac{k}{q^2}$. First find the value of k.

$4 = \frac{k}{\left(\frac{1}{2}\right)^2}$ Let $p = 4$ and $q = \frac{1}{2}$.

$k = 4\left(\frac{1}{4}\right) = 1$

When $k = 1$, $p = \frac{k}{q^2}$ becomes $p = \frac{1}{q^2}$.

Now find p when $q = \frac{3}{2}$.

$p = \frac{1}{\left(\frac{3}{2}\right)^2}$ Let $q = \frac{3}{2}$.

$p = \frac{1}{\frac{9}{4}} = 1 \cdot \frac{4}{9} = \frac{4}{9}$

31. The interest I on an investment varies directly as the rate of interest r, so there is a constant k such that $I = kr$. Find the value of k.

$48 = k(0.03)$ Let $I = 48$ and $r = 3\% = 0.03$.

$k = \frac{48}{0.03} = 1600$

When $k = 1600$, $I = kr$ becomes $I = 1600r$.

Now find I when $r = 2.2\% = 0.022$.

$I = 1600(0.022) = 35.2$

The interest on the investment when the rate is 2.2% is $35.20.

33. For a given base, the area A of a triangle varies directly as its height h, so there is a constant k such that $A = kh$. Find the value of k.

$10 = k(4)$ Let $A = 10$ and $h = 4$.

$k = \frac{10}{4} = 2.5$

When $k = 2.5$, $A = kh$ becomes $A = 2.5h$.

Now find A when $h = 6$.

$A = 2.5(6) = 15$

When the height of the triangle is 6 inches, the area of the triangle is 15 square inches.

35. The distance d that a spring stretches varies directly with the force F applied. Consider the following.

$d = kF$

$16 = k(75)$ Let $d = 16$ and $F = 75$.

$\frac{16}{75} = k$

So, $d = \frac{16}{75}F$ and when $F = 200$, the following is calculated.

$d = \frac{16}{75}(200) = \frac{16(8)}{3} = \frac{128}{3}$.

A force of 200 pounds stretches the spring $42\frac{2}{3}$ inches.

37. The rate r varies inversely with time t, so there is a constant k such that $r = \frac{k}{t}$. Find the value of k.

$160 = \frac{k}{\frac{1}{2}}$ Let $r = 160$ and $t = \frac{1}{2}$.

$k = \frac{1}{2} \cdot 160 = 80$

When $k = 80$, $r = \frac{k}{t}$ becomes $r = \frac{80}{t}$.

Now find r when $t = \frac{3}{4}$.

$r = \frac{80}{\frac{3}{4}} = 80 \cdot \frac{4}{3} = \frac{320}{3}$, or $106\frac{2}{3}$

A rate of $106\frac{2}{3}$ miles per hour is needed to go the same distance in three-fourths of a minute.

39. The current c in a simple electrical circuit varies inversely as the resistance r, so there is a constant k such that $c = \frac{k}{r}$. Find the value of k.

$20 = \frac{k}{5}$ Let $c = 20$ and $r = 5$.

$k = 5 \cdot 20 = 100$

When $k = 100$, $c = \frac{k}{r}$ becomes $c = \frac{100}{r}$.
Now find c when $r = 8$.

$c = \frac{100}{8} = \frac{25}{2} = 12\frac{1}{2}$

When the resistance is 8 ohms, the current is $12\frac{1}{2}$ amps.

41. The force F required to compress a spring varies directly as the change C in the length of the spring, so $F = kC$.

$12 = k(3)$ Let $F = 12$ and $C = 3$.

$k = \frac{12}{3} = 4$

So, $F = 4C$ and when $C = 5$, $F = 4(5) = 20$.
The force required to compress the spring 5 inches is 20 pounds.

43. The area A of a circle varies directly as the square of its radius r, so $A = kr^2$.

$28.278 = k(3)^2$ Let $A = 28.278$ and $r = 3$.

$k = \frac{28.278}{9} = 3.142$

So, $A = 3.142r^2$ and when $r = 4.1$, the following is calculated.
$A = 3.142(4.1)^2 = 52.81702$.

With $k = 3.142$, the area of a circle with radius 4.1 inches is 52.817 square inches (to the nearest thousandth).

45. The amount of light A produced by a light source varies inversely as the square of the distance d from the source. Consider the following.

$A = \frac{k}{d^2}$

$75 = \frac{k}{4^2}$ Let $A = 75$ and $d = 4$.

$k = 75(4^2) = 1200$

So, $A = \frac{1200}{d^2}$ and when $d = 9$, the following is calculated.

$A = \frac{1200}{9^2} = \frac{400}{27}$, or $14\frac{22}{27}$

The amount of light is $14\frac{22}{27}$ footcandles at a distance of 9 feet.

Chapter 7 Review Exercises

1. (a) $\frac{4x-3}{5x+2} = \frac{4(-2)-3}{5(-2)+2}$ Let $x = -2$.

$= \frac{-8-3}{-10+2} = \frac{-11}{-8} = \frac{11}{8}$

(b) $\frac{4x-3}{5x+2} = \frac{4(4)-3}{5(4)+2}$ Let $x = 4$.

$= \frac{16-3}{20+2} = \frac{13}{22}$

2. (a) $\frac{3x}{x^2-4} = \frac{3(-2)}{(-2)^2-4}$ Let $x = -2$.

$= \frac{-6}{4-4} = \frac{-6}{0}$

Substituting -2 for x makes the denominator zero, so the given expression is undefined when $x = -2$.

(b) $\frac{3x}{x^2-4} = \frac{3(4)}{(4)^2-4}$ Let $x = 4$.

$= \frac{12}{16-4} = \frac{12}{12} = 1$

3. Set the denominator equal to 0 and solve the equation. Any solutions are values for which the rational expression is undefined.

4. $\dfrac{y+3}{2y}$

Set the denominator equal to zero and solve for y.

$2y = 0$

$y = 0$

The given expression is undefined for 0. Thus, $y \neq 0$.

5. $\dfrac{2k+1}{3k^2+17k+10}$

Set the denominator equal to zero and solve for k.

$3k^2 + 17k + 10 = 0$

$(3k+2)(k+5) = 0$

$k = -\dfrac{2}{3}$ or $k = -5$

The given expression is undefined for -5 and $-\dfrac{2}{3}$. Thus, $k \neq -5, -\dfrac{2}{3}$.

6. Set the denominator equal to 0 and solve the equation. Any solutions are values for which the rational expression is undefined.

7. $\dfrac{5a^3b^3}{15a^4b^2} = \dfrac{b \cdot 5a^3b^2}{3a \cdot 5a^3b^2} = \dfrac{b}{3a}$

8. $\dfrac{m-4}{4-m} = \dfrac{-1(4-m)}{4-m} = -1$

9. $\dfrac{4x^2-9}{6-4x} = \dfrac{(2x+3)(2x-3)}{-2(2x-3)}$

$= \dfrac{2x+3}{-2} = \dfrac{-1(2x+3)}{2}$

$= \dfrac{-(2x+3)}{2}$

10. $\dfrac{4p^2+8pq-5q^2}{10p^2-3pq-q^2} = \dfrac{(2p-q)(2p+5q)}{(5p+q)(2p-q)}$

$= \dfrac{2p+5q}{5p+q}$

11. $-\dfrac{4x-9}{2x+3}$

Apply the negative sign to the numerator.

$-\dfrac{4x-9}{2x+3} = \dfrac{-(4x-9)}{2x+3}$

Now distribute the negative sign.

$\dfrac{-(4x-9)}{2x+3} = \dfrac{-4x+9}{2x+3}$

Apply the negative sign to the denominator.

$-\dfrac{4x-9}{2x+3} = \dfrac{4x-9}{-(2x+3)}$

Again, distribute the negative sign.

$\dfrac{4x-9}{-(2x+3)} = \dfrac{4x-9}{-2x-3}$

12. $-\dfrac{8-3x}{3-6x}$

Four equivalent forms are $\dfrac{-(8-3x)}{3-6x}$, $\dfrac{-8+3x}{3-6x}$, $\dfrac{8-3x}{-(3-6x)}$, and $\dfrac{8-3x}{-3+6x}$.

13. $\dfrac{18p^3}{6} \cdot \dfrac{24}{p^4} = \dfrac{6 \cdot 3 \cdot 24p^3}{6p^4} = \dfrac{72}{p}$

14. $\dfrac{8x^2}{12x^5} \cdot \dfrac{6x^4}{2x} = \dfrac{2 \cdot 4}{3 \cdot 4x^3} \cdot \dfrac{3x^3}{1} = 2$

15. $\dfrac{x-3}{4} \cdot \dfrac{5}{2x-6} = \dfrac{x-3}{4} \cdot \dfrac{5}{2(x-3)} = \dfrac{5}{8}$

16. $\dfrac{2r+3}{r-4} \cdot \dfrac{r^2-16}{6r+9}$

$= \dfrac{2r+3}{r-4} \cdot \dfrac{(r+4)(r-4)}{3(2r+3)}$

$= \dfrac{r+4}{3}$

17. $\dfrac{6a^2+7a-3}{2a^2-a-6} \div \dfrac{a+5}{a-2}$

$= \dfrac{6a^2+7a-3}{2a^2-a-6} \cdot \dfrac{a-2}{a+5}$

$= \dfrac{(3a-1)(2a+3)}{(2a+3)(a-2)} \cdot \dfrac{a-2}{a+5}$

$= \dfrac{3a-1}{a+5}$

18. $\dfrac{y^2-6y+8}{y^2+3y-18} \div \dfrac{y-4}{y+6}$

$= \dfrac{y^2-6y+8}{y^2+3y-18} \cdot \dfrac{y+6}{y-4}$

$= \dfrac{(y-4)(y-2)}{(y+6)(y-3)} \cdot \dfrac{y+6}{y-4}$

$= \dfrac{y-2}{y-3}$

19. $\dfrac{2p^2+13p+20}{p^2+p-12} \cdot \dfrac{p^2+2p-15}{2p^2+7p+5}$

$= \dfrac{(2p+5)(p+4)}{(p+4)(p-3)} \cdot \dfrac{(p+5)(p-3)}{(2p+5)(p+1)}$

$= \dfrac{p+5}{p+1}$

20. $\dfrac{3z^2+5z-2}{9z^2-1} \cdot \dfrac{9z^2+6z+1}{z^2+5z+6}$

$= \dfrac{(3z-1)(z+2)}{(3z-1)(3z+1)} \cdot \dfrac{(3z+1)^2}{(z+3)(z+2)}$

$= \dfrac{3z+1}{z+3}$

21. $\dfrac{4}{9y}, \dfrac{7}{12y^2}, \dfrac{5}{27y^4}$

Factor each denominator.

$9y = 3^2 y$

$12y^2 = 2^2 \cdot 3 \cdot y^2$

$27y^4 = 3^3 \cdot y^4$

$\text{LCD} = 2^2 \cdot 3^3 \cdot y^4 = 108y^4$

22. $\dfrac{3}{x^2+4x+3}, \dfrac{5}{x^2+5x+4}$

Factor each denominator.

$x^2+4x+3 = (x+3)(x+1)$

$x^2+5x+4 = (x+1)(x+4)$

The LCD is $(x+3)(x+1)(x+4)$.

23. $\dfrac{3}{2a^3} = \dfrac{?}{10a^4}$

$\dfrac{3}{2a^3} = \dfrac{3}{2a^3} \cdot \dfrac{5a}{5a} = \dfrac{15a}{10a^4}$

24. $\dfrac{9}{x-3} = \dfrac{?}{18-6x} = \dfrac{?}{-6(x-3)}$

$\dfrac{9}{x-3} = \dfrac{9}{x-3} \cdot \dfrac{-6}{-6}$

$= \dfrac{-54}{-6x+18}$

$= \dfrac{-54}{18-6x}$

25. $\dfrac{-3y}{2y-10} = \dfrac{?}{50-10y} = \dfrac{?}{-5(2y-10)}$

$\dfrac{-3y}{2y-10} = \dfrac{-3y}{2y-10} \cdot \dfrac{-5}{-5}$

$= \dfrac{15y}{-10y+50}$

$= \dfrac{15y}{50-10y}$

26. $\dfrac{4b}{b^2+2b-3} = \dfrac{?}{(b+3)(b-1)(b+2)}$

$\dfrac{4b}{b^2+2b-3} = \dfrac{4b}{(b+3)(b-1)}$

$= \dfrac{4b}{(b+3)(b-1)} \cdot \dfrac{b+2}{b+2}$

$= \dfrac{4b(b+2)}{(b+3)(b-1)(b+2)}$

27. $\dfrac{10}{x} + \dfrac{5}{x} = \dfrac{10+5}{x} = \dfrac{15}{x}$

28. $\dfrac{6}{3p} - \dfrac{12}{3p} = \dfrac{6-12}{3p} = \dfrac{-6}{3p} = -\dfrac{2}{p}$

29. $\dfrac{9}{k} - \dfrac{5}{k-5}$

$= \dfrac{9(k-5)}{k(k-5)} - \dfrac{5 \cdot k}{(k-5)k}$ $\quad \text{LCD} = k(k-5)$

$= \dfrac{9(k-5)-5k}{k(k-5)}$

$= \dfrac{9k-45-5k}{k(k-5)}$

$= \dfrac{4k-45}{k(k-5)}$

30. $\frac{4}{y}+\frac{7}{7+y}$

$=\frac{4(7+y)}{y(7+y)}+\frac{7\cdot y}{(7+y)y}$ LCD $= y(7+y)$

$=\frac{28+4y+7y}{y(7+y)}$

$=\frac{28+11y}{y(7+y)}$

31. $\frac{m}{3}-\frac{2+5m}{6}$

$=\frac{m\cdot 2}{3\cdot 2}-\frac{2+5m}{6}$ LCD $= 6$

$=\frac{2m-(2+5m)}{6}$

$=\frac{2m-2-5m}{6}$

$=\frac{-2-3m}{6}$

32. $\frac{12}{x^2}-\frac{3}{4x}$

$=\frac{12\cdot 4}{x^2\cdot 4}-\frac{3\cdot x}{4x\cdot x}$ LCD $= 4x^2$

$=\frac{48-3x}{4x^2}$

$=\frac{3(16-x)}{4x^2}$

33. $\frac{5}{a-2b}+\frac{2}{a+2b}$ LCD $= (a-2b)(a+2b)$

$=\frac{5(a+2b)}{(a-2b)(a+2b)}+\frac{2(a-2b)}{(a+2b)(a-2b)}$

$=\frac{5(a+2b)+2(a-2b)}{(a-2b)(a+2b)}$

$=\frac{5a+10b+2a-4b}{(a-2b)(a+2b)}$

$=\frac{7a+6b}{(a-2b)(a+2b)}$

34. $\frac{4}{k^2-9}-\frac{k+3}{3k-9}$

$=\frac{4}{(k+3)(k-3)}-\frac{k+3}{3(k-3)}$

$=\frac{4\cdot 3}{(k+3)(k-3)\cdot 3}-\frac{(k+3)(k+3)}{3(k-3)(k+3)}$

$=\frac{12-(k+3)(k+3)}{3(k+3)(k-3)}$

$=\frac{12-(k^2+6k+9)}{2(k+3)(k-3)}$

$=\frac{12-k^2-6k-9}{3(k+3)(k-3)}$

$=\frac{-k^2-6k+3}{3(k+3)(k-3)}$

35. $\frac{8}{z^2+6z}-\frac{3}{z^2+4z-12}$

$=\frac{8}{z(z+6)}-\frac{3}{(z+6)(z-2)}$

$=\frac{8(z-2)}{z(z+6)(z-2)}-\frac{3\cdot z}{(z+6)(z-2)\cdot z}$

$=\frac{8(z-2)-3z}{z(z+6)(z-2)}$ LCD $= z(z+6)(z-2)$

$=\frac{8z-16-3z}{z(z+6)(z-2)}$

$=\frac{5z-16}{z(z+6)(z-2)}$

36. $\dfrac{11}{2p-p^2}-\dfrac{2}{p^2-5p+6}$

$=\dfrac{11}{p(2-p)}-\dfrac{2}{(p-3)(p-2)}$

Note that the LCD is $p(p-3)(p-2)$.

$\dfrac{11}{p(2-p)}-\dfrac{2}{(p-3)(p-2)}$

$=\dfrac{11(-1)(p-3)}{p(2-p)(-1)(p-3)}-\dfrac{2\cdot p}{(p-3)(p-2)p}$

$=\dfrac{-11(p-3)-2p}{p(p-2)(p-3)}$

$=\dfrac{-11p+33-2p}{p(p-2)(p-3)}$

$=\dfrac{-13p+33}{p(p-2)(p-3)}$

37. $\dfrac{\frac{y-3}{y}}{\frac{y+3}{4y}}=\dfrac{y-3}{y}\cdot\dfrac{4y}{y+3}=\dfrac{4(y-3)}{y+3}$

38. $\dfrac{\frac{2}{3}-\frac{1}{6}}{\frac{1}{4}+\frac{2}{5}}=\dfrac{60\left(\frac{2}{3}-\frac{1}{6}\right)}{60\left(\frac{1}{4}+\frac{2}{5}\right)}$ LCD = 60

$=\dfrac{60\cdot\frac{2}{3}-60\cdot\frac{1}{6}}{60\cdot\frac{1}{4}+60\cdot\frac{2}{5}}$

$=\dfrac{40-10}{15+24}=\dfrac{30}{39}=\dfrac{10}{13}$

39. $\dfrac{x+\frac{1}{w}}{x-\frac{1}{w}}$

$=\dfrac{\left(x+\frac{1}{w}\right)\cdot w}{\left(x-\frac{1}{w}\right)\cdot w}$ LCD $= w$

$=\dfrac{xw+\left(\frac{1}{w}\right)w}{xw-\left(\frac{1}{w}\right)w}$

$=\dfrac{xw+1}{xw-1}$

40. $\dfrac{\frac{1}{p}-\frac{1}{q}}{\frac{1}{q-p}}$

$=\dfrac{\left(\frac{1}{p}-\frac{1}{q}\right)pq(q-p)}{\left(\frac{1}{q-p}\right)pq(q-p)}$ LCD $= pq(q-p)$

$=\dfrac{\frac{1}{p}[pq(q-p)]-\frac{1}{q}[pq(q-p)]}{pq}$

$=\dfrac{q(q-p)-p(q-p)}{pq}$

$=\dfrac{q^2-pq-pq+p^2}{pq}$

$=\dfrac{q^2-2pq+p^2}{pq}$

$=\dfrac{(q-p)^2}{pq}$

41. $\dfrac{\frac{x^2-25}{x+3}}{\frac{x+5}{x^2-9}}$

$=\dfrac{x^2-25}{x+3}\cdot\dfrac{x^2-9}{x+5}$

$=\dfrac{(x+5)(x-5)}{x+3}\cdot\dfrac{(x+3)(x-3)}{x+5}$

$=\dfrac{x-5}{1}\cdot\dfrac{x-3}{1}$

$=(x-5)(x-3)$, or $x^2-8x+15$

42. $\dfrac{\frac{1}{r+t}-1}{\frac{1}{r+t}+1}$

$=\dfrac{\left(\frac{1}{r+t}-1\right)(r+t)}{\left(\frac{1}{r+t}+1\right)(r+t)}$ LCD $= r+t$

$=\dfrac{\frac{1}{r+t}(r+t)-1(r+t)}{\frac{1}{r+t}(r+t)+1(r+t)}$

$=\dfrac{1-r-t}{1+r+t}$

43. $\frac{3x-1}{x-2}=\frac{5}{x-2}+1 \quad \text{LCD}=x-2$

$$(x-2)\left(\frac{3x-1}{x-2}\right)=(x-2)\left(\frac{5}{x-2}+1\right)$$

$$(x-2)\left(\frac{3x-1}{x-2}\right)=(x-2)\left(\frac{5}{x-2}\right)$$

$$+(x-2)(1) \quad \text{Dist. prop.}$$

$$3x-1=5+x-2$$

$$3x-1=3+x$$

$$2x=4$$

$$x=2$$

The solution set is $\varnothing$ because $x=2$ makes the original denominators equal to zero.

44. $\frac{4-z}{z}+\frac{3}{2}=\frac{-4}{z}$

Multiply each side by the LCD, $2z$.

$$2z\left(\frac{4-z}{z}+\frac{3}{2}\right)=2z\left(-\frac{4}{z}\right)$$

$$2z\left(\frac{4-z}{z}\right)+2z\left(\frac{3}{2}\right)=-8$$

$$2(4-z)+3z=-8$$

$$8-2z+3z=-8$$

$$8+z=-8$$

$$z=-16$$

Check $z=-16$: $-\frac{5}{4}+\frac{3}{2}=\frac{1}{4}$ True

Thus, the solution set is $\{-16\}$.

45. $\frac{3}{x+4}-\frac{2x}{5}=\frac{3}{x+4}$

$$-\frac{2x}{5}=0 \quad \text{Subtract } \frac{3}{x+4}.$$

$$x=0 \quad \text{Multiply by } -\frac{5}{2}.$$

Check $x=0$: $\frac{3}{4}-0=\frac{3}{4}$ True

Thus, the solution set is $\{0\}$.

46. $\frac{3}{m-2}+\frac{1}{m-1}=\frac{7}{m^2-3m+2}$

$$\frac{3}{m-2}+\frac{1}{m-1}=\frac{7}{(m-2)(m-1)}$$

$$(m-2)(m-1)\left(\frac{3}{m-2}+\frac{1}{m-1}\right)$$

$$=(m-2)(m-1)\cdot\frac{7}{(m-2)(m-1)}$$

$$3(m-1)+1(m-2)=7$$

$$3m-3+m-2=7$$

$$4m-5=7$$

$$4m=12$$

$$m=3$$

Check $m=3$: $3+\frac{1}{2}=\frac{7}{2}$ True

Thus, the solution set is $\{3\}$.

47. $m=\frac{Ry}{t}$ for t

$$t\cdot m=t\left(\frac{Ry}{t}\right) \quad \text{Multiply by } t.$$

$$tm=Ry$$

$$t=\frac{Ry}{m} \quad \text{Divide by } m.$$

48. $x=\frac{3y-5}{4}$ for y

$$4x=4\left(\frac{3y-5}{4}\right)$$

$$4x=3y-5$$

$$4x+5=3y$$

$$\frac{4x+5}{3}=y$$

49. $p^2=\frac{4}{3m-q}$ for m

$$(3m-q)p^2=(3m-q)\left(\frac{4}{3m-q}\right)$$

$$3mp^2-p^2q=4$$

$$3mp^2=4+p^2q$$

$$m=\frac{4+p^2q}{3p^2}$$

50. Let $x =$ the numerator. Then $x-5 =$ the denominator. Adding 5 to both the numerator and the denominator gives us a fraction that is equivalent to $\frac{5}{4}$.

$$\frac{x+5}{x-5+5}=\frac{5}{4}$$
$$\frac{x+5}{x}=\frac{5}{4}$$
$$4x\left(\frac{x+5}{x}\right)=4x\left(\frac{5}{4}\right)$$
$$4(x+5)=x(5)$$
$$4x+20=5x$$
$$20=x$$

The numerator is 20 and the denominator is $20-5=15$, so the original fraction is $\frac{20}{15}$.

51. Let $x =$ the numerator. Then $6x =$ the denominator. Adding 3 to the numerator and subtracting 3 from the denominator gives us a fraction equivalent to $\frac{2}{5}$.

$$\frac{x+3}{6x-3}=\frac{2}{5}$$
$$5(6x-3)\left(\frac{x+3}{6x-3}\right)=5(6x-3)\left(\frac{2}{5}\right)$$
$$5(x+3)=2(6x-3)$$
$$5x+15=12x-6$$
$$21=7x$$
$$3=x$$

The numerator is 3 and the denominator is $6\cdot 3=18$, so the original fraction is $\frac{3}{18}$.

52. Let $x =$ the rate of the wind. Then the rate against the wind is $165-x$ and the rate with the wind is $165+x$. Complete the chart using $t=\frac{d}{r}$.

	d	r	T
Against the Wind	310	$165-x$	$\frac{310}{165-x}$
With the Wind	350	$165+x$	$\frac{310}{165+x}$

Since the times are equal, we get the following equation.

$$\frac{310}{165-x}=\frac{350}{165+x}$$
$$(165+x)(165-x)\frac{310}{165-x}=(165+x)(165-x)\frac{350}{165+x}$$
$$310(165+x)=350(165-x)$$
$$51{,}150+310x=57{,}750-350x$$
$$660x=6600$$
$$x=10$$

The rate of the wind is 10 miles per hour.

53. *Step 2*

Let $x =$ the number of hours it takes them to do the job working together.

	Rate	Time Working Together	Fractional Part of the Job Done When Working Together
Susan	$\frac{1}{5}$	x	$\frac{1}{5}x$
Friend	$\frac{1}{8}$	x	$\frac{1}{8}x$

Step 3

Working together, they do 1 whole job, so $\frac{1}{5}x+\frac{1}{8}x=1.$

Step 4

Solve this equation by multiplying both sides by the LCD, 40.

$$40\left(\frac{1}{5}x+\frac{1}{8}x\right)=40(1)$$
$$8x+5x=40$$
$$13x=40$$
$$x=\frac{40}{13}, \text{ or } 3\frac{1}{13}$$

Step 5

Working together, it takes them $3\frac{1}{13}$ hours.

Step 6

Susan does $\frac{1}{5}$ of the job per hour for $\frac{40}{13}$ hours: $\frac{1}{5}\cdot\frac{40}{13}=\frac{8}{13}$ of the job.

Her friend does $\frac{1}{8}$ of the job per hour for $\frac{40}{13}$ hours: $\frac{1}{8}\cdot\frac{40}{13}=\frac{5}{13}$ of the job.

Together, they have done $\frac{8}{13}+\frac{5}{13}=\frac{13}{13}=1$ total job.

54. *Step 2*

Let x = the time needed by the head gardener to mow the lawns. Then $2x$ = the time needed by the assistant to mow the lawns.

	Rate	**Time Working Together**	**Fractional Part of the Job Done When Working Together**
Head Gardener	$\frac{1}{x}$	$1\frac{1}{3}=\frac{4}{3}$	$\frac{1}{x}\cdot\frac{4}{3}=\frac{4}{3x}$
Assistant	$\frac{1}{2x}$	$1\frac{1}{3}=\frac{4}{3}$	$\frac{1}{2x}\cdot\frac{4}{3}=\frac{2}{3x}$

Step 3

$$\frac{4}{3x}+\frac{2}{3x}=1$$

Step 4

$$\frac{4+2}{3x}=1$$
$$\frac{6}{3x}=1$$
$$3x\left(\frac{6}{3x}\right)=3x(1)$$
$$6=3x$$
$$2=x$$

Step 5

It takes the head gardener 2 hours to mow the lawns.

Step 6

The head gardener does $\frac{1}{2}$ of the job per hour for $\frac{4}{3}$ hours. $\frac{1}{2}\cdot\frac{4}{3}=\frac{2}{3}$ of the job.

The assistant does $\frac{1}{4}$ of the job per hour for $\frac{4}{3}$ hours $\frac{1}{4}\cdot\frac{4}{3}=\frac{1}{3}$ of the job.

Together, they have done

$\frac{2}{3}+\frac{1}{3}=\frac{3}{5}=1$ total job.

55. Let h = the height of the parallelogram and b = the length of the base of the parallelogram. The height varies inversely as the base, so $h=\frac{k}{b}$.

Find k by replacing h with 8 and b with 12.

$$8=\frac{k}{12}$$
$$k=8\cdot 12=96$$

So, $h=\frac{96}{b}$ and when $b=24$, $h=\frac{96}{24}=4$.

The height of the parallelogram is 4 centimeters.

56. Since y varies directly as x, there is a constant k such that $y=kx$. First find the value of k.

$5=k(12)$ Let $x=12$ and $y=5$.

$$k=\frac{5}{12}$$

When $k=\frac{5}{12}$, $y=kx$ becomes $y=\frac{5}{12}x$.

Now find x when $y=3$.

$3=\frac{5}{12}x$ Let $y=3$.

$$x=3\cdot\frac{12}{5}=\frac{36}{5}$$

Chapter 7 Mixed Review Exercises

1. $\frac{4}{m-1}-\frac{3}{m+1}$

To perform the indicated subtraction, use $(m-1)(m+1)$ as the LCD.

$$\frac{4}{m-1}-\frac{3}{m+1}$$
$$=\frac{4(m+1)}{(m-1)(m+1)}-\frac{3(m-1)}{(m+1)(m-1)}$$
$$=\frac{4(m+1)-3(m-1)}{(m-1)(m+1)}$$
$$=\frac{4m+4-3m+3}{(m-1)(m+1)}$$
$$=\frac{m+7}{(m-1)(m+1)}$$

2. $\frac{8p^5}{5} \div \frac{2p^3}{10}$

To perform the indicated division, multiply the first rational expression by the reciprocal of the second.

$$\frac{8p^5}{5} \div \frac{2p^3}{10} = \frac{8p^5}{5} \cdot \frac{10}{2p^3} = \frac{80p^5}{10p^3} = 8p^2$$

3. $$\frac{r-3}{8} \div \frac{3r-9}{4} = \frac{r-3}{8} \cdot \frac{4}{3r-9} = \frac{r-3}{8} \cdot \frac{4}{3(r-3)} = \frac{4}{24} = \frac{1}{6}$$

4. $$\frac{\frac{5}{x}-1}{\frac{5-x}{3x}} = \frac{\left(\frac{5}{x}-1\right)3x}{\left(\frac{5-x}{3x}\right)3x} \quad \text{LCD} = 3x$$

$$= \frac{\frac{5}{x}(3x) - 1(3x)}{5-x} = \frac{15-3x}{5-x} = \frac{3(5-x)}{5-x} = 3$$

5. $$\frac{4}{z^2-2z+1} - \frac{3}{z^2-1} = \frac{4}{(z-1)^2} - \frac{3}{(z+1)(z-1)}$$

Note that the LCD is $(z+1)(z-1)^2$.

$$\frac{4}{(z-1)^2} - \frac{3}{(z+1)(z-1)} = \frac{4(z+1)}{(z-1)^2(z+1)} - \frac{3(z-1)}{(z+1)(z-1)(z-1)}$$

$$= \frac{4(z+1)-3(z-1)}{(z+1)(z-1)^2} = \frac{4z+4-3z+3}{(z+1)(z-1)^2} = \frac{z+7}{(z+1)(z-1)^2}$$

6. $$\frac{1}{t^2-4} + \frac{1}{2-t} \quad \text{LCD} = (t+2)(t-2)$$

$$= \frac{1}{(t+2)(t-2)} + \frac{1}{2-t} = \frac{1}{(t+2)(t-2)} - \frac{1(t+2)}{(t-2)(t+2)} = \frac{1-t-2}{(t+2)(t-2)}$$

$$= \frac{-t-1}{(t+2)(t-2)}, \text{ or } \frac{t+1}{(t+2)(2-t)}$$

7. $$\frac{2}{z} - \frac{z}{z+3} = \frac{1}{z+3}$$

Multiply each side of the equation by the LCD, $z(z+3)$.

$$z(z+3)\left(\frac{2}{z} - \frac{z}{z+3}\right) = z(z+3)\left(\frac{1}{z+3}\right)$$

$$z(z+3)\left(\frac{2}{z}\right) - z(z+3)\left(\frac{z}{z+3}\right) = z(1)$$

$$2(z+3) - z^2 = z$$

$$2z + 6 - z^2 = z$$

Move all variable terms to one side of the equation.

$0 = z^2 - z - 6$

$0 = (z-3)(z+2)$

$z-3=0$ or $z+2=0$

$z=3$ or $z=-2$

$0 = z^2 - z - 6$

$0 = (z-3)(z+2)$

$z-3=0$ or $z+2=0$

$z=3$ or $z=-2$

Check $z=-2$: $-1-(-2)=1$ True

Check $z=3$: $\frac{2}{3}-\frac{1}{2}=\frac{1}{6}$ True

Thus, the solution set is $\{-2, 3\}$.

8.

$$\frac{1}{x^2-1}=\frac{1}{x+1}$$

$$\frac{1}{(x+1)(x-1)}=\frac{1}{x+1}$$

$$\frac{1}{(x+1)(x-1)}=\frac{1}{x+1}$$

$$\frac{1\cdot(x+1)(x-1)}{(x+1)(x-1)}=\frac{(x+1)(x-1)\cdot 1}{(x+1)}$$

$$1=x-1$$

$$x=2$$

Check $x=2$: $\frac{1}{3}=\frac{1}{3}$ True

Thus, the solution set is $\{2\}$.

9. Solve $a=\frac{v-w}{t}$ for v.

$t\cdot a = v-w$ Multiply by t.

$at+w=v$ Add w.

10. Let x = the rate of the plane in still air. Then the rate of the plane with the wind is $x+50$, and the rate of the plane against the wind is $x-50$. Use $t=\frac{d}{r}$ to complete the chart.

	d	***r***	***t***
With the Wind	400	$x+50$	$\frac{400}{x+50}$
Against the Wind	200	$x-50$	$\frac{200}{x-50}$

The times are the same, so $\frac{400}{x+50}=\frac{200}{x-50}$.

To solve this equation, multiply both sides by the LCD, $(x+50)(x-50)$.

$$(x+50)(x-50)\cdot\frac{400}{x+50}=(x+50)(x-50)\cdot\frac{200}{x-50}$$

$$400(x-50)=200(x+50)$$

$$400x-20{,}000=200x+10{,}000$$

$$200x=30{,}000$$

$$x=150$$

The rate of the plane is 150 kilometers per hour.

11. Let x = the number of hours it takes them to do the job working together.

	Rate	**Time Working Together**	**Fractional Part of the Job Done When Working Together**
Lizette	$\frac{1}{8}$	x	$\frac{1}{8}x$
Seyed	$\frac{1}{14}$	x	$\frac{1}{14}x$

Working together, they do 1 whole job, so $\frac{1}{8}x+\frac{1}{14}x=1$.

To clear fractions, multiply both sides by the LCD, 56.

$$56\left(\frac{1}{8}x+\frac{1}{14}x\right)=56(1)$$

$$7x+4x=56$$

$$11x=56$$

$$x=\frac{56}{11}\text{, or }5\frac{1}{11}$$

Working together, they can paint the woodwork in $5\frac{1}{11}$ hours.

12. Since w varies inversely as z, there is a constant k such that $w = \frac{k}{z}$. First find the value of k.

$16 = \frac{k}{3}$ Let $w = 16$ and $z = 3$.

$k = 16(3) = 48$

When $k = 48$, $w = \frac{k}{2}$ becomes $w = \frac{48}{z}$.

Now find w when $z = 2$.

$w = \frac{48}{2} = 24$

13. For a constant area, the length L of a rectangle varies inversely as the width W, so

$L = \frac{k}{W}$.

$24 = \frac{k}{2}$ Let $L = 24$ and $W = 2$.

$k = 24 \cdot 2 = 48$

So, $L = \frac{48}{W}$ and when $L = 12$,

$12 = \frac{48}{W}$

$12W = 48$

$W = \frac{48}{12} = 4.$

When the length is 12, the width is 4.

14. $x = ky^3$

$54 = k(3)^3$ Substitute $x = 54$ and $y = 3$.

$2 = k$ Solve for k.

Since $x = ky^3$ and $k = 2$, $x = 2y^3$.

$x = 2(-2)^3$ Substitute $y = -2$.

$x = -16$

Chapter 7 Test

1. (a) $\frac{6r+1}{2r^2-3r-20}$

$= \frac{6(-2)+1}{2(-2)^2-3(-2)-20}$ $r = -2$

$= \frac{-12+1}{2 \cdot 4+6-20}$

$= \frac{-11}{8+6-20}$

$= \frac{-11}{-6} = \frac{11}{6}$

(b) $\frac{6r+1}{2r^2-3r-20}$

$= \frac{6(4)+1}{2(4)^2-3(4)-20}$ $r = 4$

$= \frac{24+1}{2 \cdot 16-12-20}$

$= \frac{25}{32-12-20}$

$= \frac{25}{20-20} = \frac{25}{0}$

The expression is undefined when $r = 4$ because the denominator is 0.

2. $\frac{3x-1}{x^2-2x-8}$

Set the denominator equal to zero and solve for x.

$x^2-2x-8 = 0$

$(x+2)(x-4) = 0$

$x+2 = 0$ or $x-4 = 0$

$x = -2$ or $x = 4$

The expression is undefined for -2 and 4, so $x \neq -2, 4$.

3. $-\frac{6x-5}{2x+3}$

Apply the negative sign to the numerator.

$-\frac{6x-5}{2x+3} = \frac{-(6x-5)}{2x+3}$

Now distribute the negative sign.

$\frac{-(6x-5)}{2x+3} = \frac{-6x+5}{2x+3}$

Apply the negative sign to the denominator.

$-\frac{6x-5}{2x+3} = \frac{6x-5}{-(2x+3)}$

Again, distribute the negative sign.

$\frac{6x-5}{-(2x+3)} = \frac{6x-5}{-2x-3}$

4. $\frac{-15x^6y^4}{5x^4y} = \frac{(5x^4y)(-3x^2y^3)}{(5x^4y)(1)}$

$= \frac{5x^4y}{5x^4y} \cdot \frac{-3x^2y^3}{1} = -3x^2y^3$

5. $\frac{6a^2+a-2}{2a^2-3a+1} = \frac{(3a+2)(2a-1)}{(2a-1)(a-1)}$

$= \frac{3a+2}{a-1}$

6. $\frac{5(d-2)}{9} \div \frac{3(d-2)}{5}$

$= \frac{5(d-2)}{9} \cdot \frac{5}{3(d-2)}$

$= \frac{5 \cdot 5}{9 \cdot 3} = \frac{25}{27}$

7. $\frac{6k^2-k-2}{8k^2+10k+3} \cdot \frac{4k^2+7k+3}{3k^2+5k+2}$

$= \frac{(3k-2)(2k+1)}{(4k+3)(2k+1)} \cdot \frac{(4k+3)(k+1)}{(3k+2)(k+1)}$

$= \frac{3k-2}{3k+2}$

8. $\frac{4a^2+9a+2}{3a^2+11a+10} \div \frac{4a^2+17a+4}{3a^2+2a-5}$

$= \frac{4a^2+9a+2}{3a^2+11a+10} \cdot \frac{3a^2+2a-5}{4a^2+17a+4}$

$= \frac{(4a+1)(a+2)}{(3a+5)(a+2)} \cdot \frac{(3a+5)(a-1)}{(4a+1)(a+4)}$

$= \frac{a-1}{a+4}$

9. $\frac{x^2-10x+25}{9-6x+x^2} \cdot \frac{x-3}{5-x}$

$= \frac{(x-5)(x-5)}{(3-x)(3-x)} \cdot \frac{x-3}{5-x}$ Factor.

$= \frac{(x-5)(x-5)(x-3)}{(3-x)(3-x)(5-x)}$ Multiply.

$= \frac{-1(x-5)}{-1(3-x)}$ Lowest terms

$= \frac{x-5}{3-x}$

10. $\frac{-3}{10p^2}, \frac{21}{25p^3}, \frac{-7}{30p^5}$

Factor each denominator.

$10p^2 = 2 \cdot 5 \cdot p^2$

$25p^3 = 5^2 \cdot p^3$

$30p^5 = 2 \cdot 3 \cdot 5 \cdot p^5$

$\text{LCD} = 2 \cdot 3 \cdot 5^2 \cdot p^5 = 150p^5$

11. $\frac{r+1}{2r^2+7r+6}, \frac{-2r+1}{2r^2-7r-15}$

Factor each denominator.

$2r^2+7r+6 = (2r+3)(r+2)$

$2r^2-7r-15 = (2r+3)(r-5)$

$\text{LCD} = (2r+3)(r+2)(r-5)$

12. $\frac{15}{4p} = \frac{?}{64p^3} = \frac{?}{4p \cdot 16p^2}$

$\frac{15}{4p} = \frac{15 \cdot 16p^2}{4p \cdot 16p^2} = \frac{240p^2}{64p^3}$

13. $\frac{3}{6m-12} = \frac{?}{42m-84} = \frac{?}{7(6m-12)}$

$\frac{3}{6m-12} = \frac{3 \cdot 7}{(6m-12)7} = \frac{21}{42m-84}$

14. $\frac{4x+2}{x+5} + \frac{-2x+8}{x+5}$

$= \frac{(4x+2)+(-2x+8)}{x+5}$

$= \frac{2x+10}{x+5}$

$= \frac{2(x+5)}{x+5} = 2$

15. $\frac{-4}{y+2} + \frac{6}{5y+10}$

$= \frac{-4}{y+2} + \frac{6}{5(y+2)}$

$= \frac{-4 \cdot 5}{(y+2) \cdot 5} + \frac{6}{5(y+2)}$

$= \frac{-20+6}{5(y+2)} = \frac{-14}{5(y+2)}$

16. Using LCD = $3-x$, consider the following.

$\frac{x+1}{3-x} + \frac{x^2}{x-3} = \frac{x+1}{3-x} + \frac{-1(x^2)}{-1(x-3)}$

$= \frac{x+1}{3-x} + \frac{-x^2}{-x+3}$

$= \frac{x+1}{3-x} + \frac{-x^2}{3-x}$

$= \frac{(x+1)+\left(-x^2\right)}{3-x}$

$= \frac{-x^2+x+1}{3-x}$

If we use $x-3$ for the LCD, we obtain the equivalent answer

$$\frac{x^2-x-1}{x-3}.$$

17. $$\frac{3}{2m^2-9m-5}-\frac{m+1}{2m^2-m-1}$$
$$=\frac{3}{(2m+1)(m-5)}-\frac{m+1}{(2m+1)(m-1)}$$

Note that the LCD is $(2m+1)(m-5)(m-1)$.

$$\frac{3}{(2m+1)(m-5)}-\frac{m+1}{(2m+1)(m-1)}$$
$$=\frac{3(m-1)}{(2m+1)(m-5)(m-1)}$$
$$-\frac{(m+1)(m-5)}{(2m+1)(m-1)(m-5)}$$
$$=\frac{3(m-1)-(m+1)(m-5)}{(2m+1)(m-5)(m-1)}$$
$$=\frac{3(m-1)-\left(m^2-4m-5\right)}{(2m+1)(m-5)(m-1)}$$
$$=\frac{3m-3-m^2+4m+5}{(2m+1)(m-5)(m-1)}$$
$$=\frac{-m^2+7m+2}{(2m+1)(m-5)(m-1)}$$

18. $$\frac{\frac{2p}{k^2}}{\frac{3p^2}{k^3}}=\frac{2p}{k^2}\div\frac{3p^2}{k^3}$$
$$=\frac{2p}{k^2}\cdot\frac{k^3}{3p^2}$$
$$=\frac{2k^3p}{3k^2p^2}=\frac{2k}{3p}$$

19. $$\frac{\frac{1}{x+3}-1}{1+\frac{1}{x+3}}$$

Start by multiplying the numerator and the denominator by the LCD, $x+3$.

$$\frac{\frac{1}{x+3}-1}{1+\frac{1}{x+3}}$$
$$=\frac{(x+3)\left(\frac{1}{x+3}-1\right)}{(x+3)\left(1+\frac{1}{x+3}\right)}$$
$$=\frac{(x+3)\left(\frac{1}{x+3}\right)-(x+3)(1)}{(x+3)(1)+(x+3)\left(\frac{1}{x+3}\right)}$$
$$=\frac{1-(x+3)}{(x+3)+1}$$
$$=\frac{1-x-3}{x+4}$$
$$=\frac{-2-x}{x+4}$$

20. $$\frac{3x}{x+1}=\frac{3}{2x}$$
$$2x(x+1)\frac{3x}{x+1}=2x(x+1)\frac{3}{2x}$$

Multiply by the LCD, $2x(x+1)$.

$$2x(3x)=3(x+1)$$
$$6x^2=3x+3$$
$$6x^2-3x-3=0$$
$$3(2x^2-x-1)=0$$
$$3(2x+1)(x-1)=0$$
$$x=-\frac{1}{2}\quad\text{or}\quad x=1$$

Check $x=-\frac{1}{2}$: $-3=-3$ True

Check $x=1$: $\frac{3}{2}=\frac{3}{2}$ True

Thus, the solution set is $\left\{-\frac{1}{2},1\right\}$.

21. $\dfrac{2x}{x-3}+\dfrac{1}{x+3}=\dfrac{-6}{x^2-9}$

$\dfrac{2x}{x-3}+\dfrac{1}{x+3}=\dfrac{-6}{(x+3)(x-3)}$

$(x+3)(x-3)\left(\dfrac{2x}{x-3}+\dfrac{1}{x+3}\right)$

$=(x+3)(x-3)\left(\dfrac{-6}{(x+3)(x-3)}\right)$

Multiply by the LCD, $(x+3)(x-3)$.

$2x(x+3)+1(x-3)=-6$

$2x^2+6x+x-3=-6$

$2x^2+7x+3=0$

$(2x+1)(x+3)=0$

$x=-\dfrac{1}{2}$ or $x=-3$

x cannot equal -3 because the denominator $x+3$ would equal 0.

Check $x=-\dfrac{1}{2}$: $\dfrac{2}{7}+\dfrac{2}{5}=\dfrac{24}{35}$ True

Thus, the solution set is $\left\{-\dfrac{1}{2}\right\}$.

22. Solve $F=\dfrac{k}{d-D}$ for D.

$(d-D)(F)=(d-D)\left(\dfrac{k}{d-D}\right)$ LCD $=d-D$

$(d-D)(F)=k$

$dF-DF=k$

$-DF=k-dF$

$DF=dF-k$

$D=\dfrac{dF-k}{F}$, or $D=d-\dfrac{k}{F}$

23. Let x = the rate of the current.

	d	**r**	**t**
Upstream	20	$7-x$	$\dfrac{20}{7-x}$
Downstream	50	$7+x$	$\dfrac{50}{7+x}$

The times are equal, so consider the following.

$$\frac{20}{7-x}=\frac{50}{7+x}$$

$(7-x)(7+x)\left(\dfrac{20}{7-x}\right)=(7-x)(7+x)\left(\dfrac{50}{7+x}\right)$

Multiply by the LCD, $(7-x)(7+x)$.

$20(7+x)=50(7-x)$

$140+20x=350-50x$

$70x=210$

$x=3$

The rate of the current is 3 miles per hour.

24. Let x = the time required for Sanford and his neighbor to paint the room working together.

	Rate	**Time Working Together**	**Fractional Part of the Job Done When Working Together**
Sanford	$\dfrac{1}{5}$	x	$\dfrac{1}{5}x$
Neighbor	$\dfrac{1}{4}$	x	$\dfrac{1}{4}x$

Working together, they do 1 whole job, so

$$\frac{1}{5}x+\frac{1}{4}x=1.$$

$20\left(\dfrac{1}{5}x+\dfrac{1}{4}x\right)=20(1)$ LCD $=20$

$20\left(\dfrac{1}{5}x\right)+20\left(\dfrac{1}{4}x\right)=20$

$4x+5x=20$

$9x=20$

$x=\dfrac{20}{9}$, or $2\dfrac{2}{9}$

They can paint the room in $2\dfrac{2}{9}$ hours.

25. Since x varies directly as y, there is a constant k such that $x=ky$. First find the value of k.

$x=ky$

$12=k\cdot 4$ Let $x=12$ and $y=4$.

$k=\dfrac{12}{4}=3$

Since $x=ky$ and $k=3$, $x=3y$.

Now find x when $y=9$.

$x=3(9)=27$

26. The length of time L that it takes for fruit to ripen during the growing season varies inversely as the average maximum temperature T during the season, so $L = \frac{k}{T}$.

$25 = \frac{k}{80}$ Let $L = 25$ and $T = 80$.

$k = 25 \cdot 80 = 2000$

So, $L = \frac{2000}{T}$ and when $T = 75$,

$L = \frac{2000}{75} = \frac{80}{3}$, or $26\frac{2}{3}$.

It would take 27 days (to the nearest whole number) for fruit to ripen with an average maximum temperature of 75°.

Chapters R–7 Cumulative Review Exercises

1.
$$\begin{aligned}3+4\left(\frac{1}{2}-\frac{3}{4}\right) &= 3+4\left(\frac{2}{4}-\frac{3}{4}\right)\\ &= 3+4\left(-\frac{1}{4}\right)\\ &= 3+(-1)\\ &= 2\end{aligned}$$

2.
$$\begin{aligned}3(2y-5) &= 2+5y\\ 6y-15 &= 2+5y\\ y &= 17\end{aligned}$$

The solution set is $\{17\}$.

3. Solve $\mathcal{A} = \frac{1}{2}bh$ for b.

$2 \cdot \mathcal{A} = 2 \cdot \frac{1}{2}bh$ Multiply by 2.

$2\mathcal{A} = bh$

$\frac{2\mathcal{A}}{h} = b$ Divide by h.

4.
$$\frac{2+m}{3} = \frac{2-m}{4}$$

$4(2+m) = 3(2-m)$ Cross multiply.

$$\begin{aligned}8+4m &= 6-3m\\ 7m &= -2\\ m &= -\frac{2}{7}\end{aligned}$$

The solution set is $\left\{-\frac{2}{7}\right\}$.

5. $5y \le 6y+8$

$-y \le 8$

$y \ge -8$ Reverse the inequality symbol.

The solution set is $[-8, \infty)$.

6. (a) Let $y = 0$ to find the x-intercept.

$$\begin{aligned}4x+3(0) &= -12\\ 4x &= -12\\ x &= -3\end{aligned}$$

The x-intercept is $(-3, 0)$.

(b) Let $x = 0$ to find the y-intercept.

$$\begin{aligned}4(0)+3y &= -12\\ 3y &= -12\\ y &= -4\end{aligned}$$

The y-intercept is $(0, -4)$.

7. $y = -3x+2$

This is an equation of a line.

If $x = 0$, $y = 2$, so the y-intercept is $(0, 2)$.

If $x = 1$, $y = -1$, and if $x = 2$, $y = -4$.

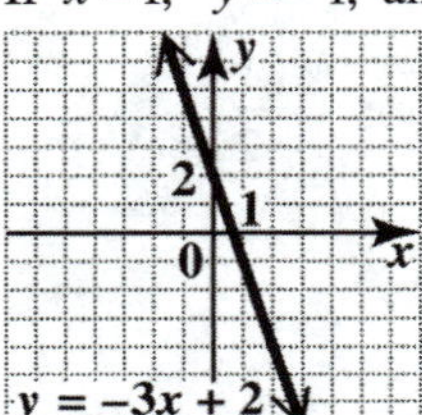

8. $y = -x^2+1$

This is the equation of a parabola opening downward with a y-intercept $(0, 1)$.

If $x = +2$ or -2, $y = -3$.

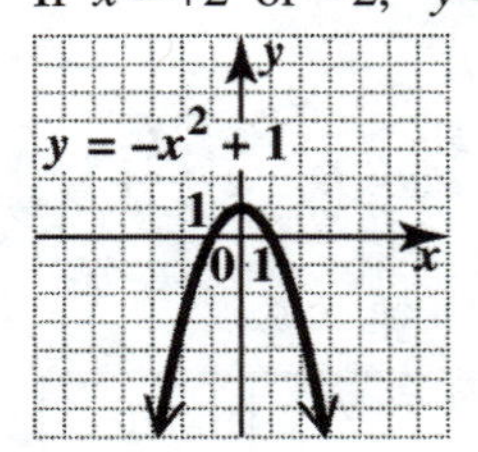

9. $4x - y = -7 \quad (1)$
$5x + 2y = 1 \quad (2)$

$$\begin{array}{rcll} 8x - 2y &=& -14 & (3) \quad 2\times\text{Eq.}(1) \\ 5x + 2y &=& 1 & (2) \\ \hline 13x &=& -13 & \text{Add (3) and (2).} \\ x &=& -1 & \end{array}$$

Substitute -1 for x in equation (1).

$$\begin{aligned} 4x - y &= -7 \\ 4(-1) - y &= -7 \\ -4 - y &= -7 \\ -y &= -3 \\ y &= 3 \end{aligned}$$

The solution set is $\{(-1, 3)\}$.

10. $5x + 2y = 7$
$10x + 4y = 12$

Multiply equation (2) by $\frac{1}{2}$.

$5x + 2y = 6 \quad (3) \quad \frac{1}{2}\times\text{Eq.}(2)$

The left sides of (1) and (3) are equal, so they can't be equal to different numbers (7 and 6). There are no solutions, so the solution set is $\varnothing$.

11. $$\frac{(2x^3)^{-1}\cdot x}{2^3x^5} = \frac{2^{-1}(x^3)^{-1}\cdot x}{2^3x^5} = \frac{2^{-1}x^{-3}x}{2^3x^5} = \frac{2^{-1}x^{-2}}{2^3x^5} = \frac{1}{2^1\cdot 2^3\cdot x^2\cdot x^5} = \frac{1}{2^4x^7}$$

12. $$\frac{(m^{-2})^3m}{m^5m^{-4}} = \frac{m^{-6}m}{m^5m^{-4}} = \frac{m\cdot m^4}{m^5\cdot m^6} = \frac{m^5}{m^{11}} = \frac{1}{m^6}$$

13. $$\begin{aligned} &(2k^2 + 3k) - (k^2 + k - 1) \\ &= 2k^2 + 3k - k^2 - k + 1 \\ &= k^2 + 2k + 1 \end{aligned}$$

14. $$\begin{aligned} (2a - b)^2 &= (2a)^2 - 2(2a)(b) + (b)^2 \\ &= 4a^2 - 4ab + b^2 \end{aligned}$$

15. $(y^2 + 3y + 5)(3y - 1)$

Multiply vertically.

$$\begin{array}{r} y^2 + 3y + 5 \\ 3y - 1 \\ \hline -y^2 - 3y - 5 \\ 3y^3 + 9y^2 + 15y \quad \\ \hline 3y^3 + 8y^2 + 12y - 5 \end{array}$$

16. $$\frac{12p^3 + 2p^2 - 12p + 4}{2p - 2}$$

$$\begin{array}{r} 6p^2 + 7p + 1 \\ 2p-2\overline{)12p^3 + 2p^2 - 12p + 4} \\ \underline{12p^3 - 12p^2} \qquad\qquad \\ 14p^2 - 12p \qquad \\ \underline{14p^2 - 14p} \qquad \\ 2p + 4 \\ \underline{2p - 2} \\ 6 \end{array}$$

The result is

$$\begin{aligned} &6p^2 + 7p + 1 + \frac{6}{2p - 2} \\ &= 6p^2 + 7p + 1 + \frac{2\cdot 3}{2(p - 1)} \\ &= 6p^2 + 7p + 1 + \frac{3}{p - 1}. \end{aligned}$$

17. $$\begin{aligned} &8t^2 + 10tv + 3v^2 \\ &= 8t^2 + 6tv + 4tv + 3v^2 \\ &= (8t^2 + 6tv) + (4tv + 3v^2) \\ &= 2t(4t + 3v) + v(4t + 3v) \\ &= (4t + 3v)(2t + v) \end{aligned}$$

18. $8r^2 - 9rs + 12s^2$

To factor this polynomial by the grouping method, we must find two integers whose product is $(8)(12) = 96$ and whose sum is -9. There is no pair of integers that satisfies both of these conditions, so the polynomial is *prime*.

19. $$\begin{aligned} &16x^4 - 1 \\ &= (4x^2)^2 - (1)^2 \\ &= (4x^2 + 1)(4x^2 - 1) \\ &= (4x^2 + 1)[(2x)^2 - (1)^2] \\ &= (4x^2 + 1)(2x + 1)(2x - 1) \end{aligned}$$

20. $$r^2 = 2r + 15$$
$$r^2 - 2r - 15 = 0$$
$$(r+3)(r-5) = 0$$
$$r + 3 = 0 \quad \text{or} \quad r - 5 = 0$$
$$r = -3 \quad \text{or} \quad r = 5$$
The solution set is $\{-3, 5\}$.

21. $$(r-5)(2r+1)(3r-2) = 0$$
$$r - 5 = 0 \quad \text{or} \quad 2r + 1 = 0 \quad \text{or} \quad 3r - 2 = 0$$
$$2r = -1 \quad \text{or} \quad 3r = 2$$
$$r = 5 \quad \text{or} \quad r = -\frac{1}{2} \quad \text{or} \quad r = \frac{2}{3}$$
The solution set is $\left\{5, -\frac{1}{2}, \frac{2}{3}\right\}$.

22. Let $x =$ the lesser number.
Then $x + 4 =$ the greater number.
The product of the numbers is 2 less than the lesser number, which translates to
$$x(x+4) = x - 2.$$
$$x^2 + 4x = x - 2$$
$$x^2 + 3x + 2 = 0$$
$$(x+2)(x+1) = 0$$
$$x = -2 \quad \text{or} \quad x = -1$$
The lesser number can be either -2 or -1.

23. Let $w =$ the width of the rectangle.
Then $2w - 2 =$ the length of the rectangle.
Use the formula $\mathcal{A} = LW$ with the area $= 60$.
$$60 = (2w-2)w$$
$$60 = 2w^2 - 2w$$
$$0 = 2w^2 - 2w - 60$$
$$0 = 2(w^2 - w - 30)$$
$$0 = 2(w-6)(w+5)$$
$$w = 6 \quad \text{or} \quad w = -5$$
Discard -5 because the width cannot be negative. The width of the rectangle is 6 meters.

24. All of the given expressions are equal to 1 for all real numbers for which they are defined. However, expressions B, C, and D all have one or more values for which the expression is undefined and therefore cannot be equal to 1 at these values. Since $k^2 + 2$ is *always* positive, the denominator in expression A is never equal to zero. This expression is defined and equal to 1 for all real numbers, so the correct choice is A.

25. The appropriate choice is D since
$$\frac{-(3x+4)}{7} = \frac{-3x-4}{7} \neq \frac{4-3x}{7}.$$

26. $$\frac{5}{q} - \frac{1}{q} = \frac{5-1}{q} = \frac{4}{q}$$

27. $$\frac{3}{7} + \frac{4}{r} = \frac{3 \cdot r}{7 \cdot r} + \frac{4 \cdot 7}{r \cdot 7} \quad \text{LCD} = 7r$$
$$= \frac{3r + 28}{7r}$$

28. $$\frac{4}{5q-20} - \frac{1}{3q-12}$$
$$= \frac{4}{5(q-4)} - \frac{1}{3(q-4)}$$
$$= \frac{4 \cdot 3}{5(q-4) \cdot 3} - \frac{1 \cdot 5}{3(q-4) \cdot 5}$$
$$= \frac{12-5}{15(q-4)} = \frac{7}{15(q-4)}$$

29. $$\frac{2}{k^2+k} - \frac{3}{k^2-k}$$
$$= \frac{2}{k(k+1)} - \frac{3}{k(k-1)}$$
$$= \frac{2(k-1)}{k(k+1)(k-1)} - \frac{3(k+1)}{k(k-1)(k+1)}$$
$$= \frac{2(k-1) - 3(k+1)}{k(k+1)(k-1)}$$
$$= \frac{2k - 2 - 3k - 3}{k(k+1)(k-1)}$$
$$= \frac{-k-5}{k(k+1)(k-1)}$$

30. $$\frac{7z^2 + 49z + 70}{16z^2 + 72z - 40} \div \frac{3z+6}{4z^2-1}$$
$$= \frac{7z^2 + 49z + 70}{16z^2 + 72z - 40} \cdot \frac{4z^2-1}{3z+6}$$
$$= \frac{7(z^2 + 7z + 10)}{8(2z^2 + 9z - 5)} \cdot \frac{(2z+1)(2z-1)}{3(z+2)}$$
$$= \frac{7(z+5)(z+2)}{8(2z-1)(z+5)} \cdot \frac{(2z+1)(2z-1)}{3(z+2)}$$
$$= \frac{7(2z+1)}{8 \cdot 3} = \frac{7(2z+1)}{24}$$

31. $\dfrac{\frac{4}{a}+\frac{5}{2a}}{\frac{7}{6a}-\frac{1}{5a}}$

$= \dfrac{\left(\frac{4}{a}+\frac{5}{2a}\right)\cdot 30a}{\left(\frac{7}{6a}-\frac{1}{5a}\right)\cdot 30a}$ LCD $= 30a$

$= \dfrac{\frac{4}{a}(30a)+\frac{5}{2a}(30a)}{\frac{7}{6a}(30a)-\frac{1}{5a}(30a)}$

$= \dfrac{4\cdot 30+5\cdot 15}{7\cdot 5-1\cdot 6}$

$= \dfrac{120+75}{35-6} = \dfrac{195}{29}$

32. $\dfrac{r+2}{5} = \dfrac{r-3}{3}$

$15\left(\dfrac{r+2}{5}\right) = 15\left(\dfrac{r-3}{3}\right)$ LCD $= 15$

$3(r+2) = 5(r-3)$

$3r+6 = 5r-15$

$21 = 2r$

$\dfrac{21}{2} = r$

Check $r = \dfrac{21}{2}$: $\dfrac{5}{2} = \dfrac{5}{2}$ True

The solution set is $\left\{\dfrac{21}{2}\right\}$.

33. $\dfrac{1}{x} = \dfrac{1}{x+1} + \dfrac{1}{2}$

$2x(x+1)\left(\dfrac{1}{x}\right) = 2x(x+1)\left(\dfrac{1}{x+1}+\dfrac{1}{2}\right)$

$2(x+1) = 2x(x+1)\left(\dfrac{1}{x+1}\right) + 2x(x+1)\left(\dfrac{1}{2}\right)$

$2(x+1) = 2x + x(x+1)$

$2x+2 = 2x + x^2 + x$

$0 = x^2 + x - 2$

$0 = (x+2)(x-1)$

Check $x = -2$: $-\dfrac{1}{2} = -1 + \dfrac{1}{2}$ True

Check $x = 1$: $1 = \dfrac{1}{2} + \dfrac{1}{2}$ True

Thus, the solution set is $\{-2, 1\}$.

34. Let $x =$ the number of hours it will take Jody and Pat to weed the yard working together.

	Rate	Time Working Together	Fractional Part of the Job Done When Working Together
Jody	$\frac{1}{3}$	x	$\frac{1}{3}x$
Pat	$\frac{1}{2}$	x	$\frac{1}{2}x$

Working together, they can do 1 whole job, so

$\dfrac{1}{3}x + \dfrac{1}{2}x = 1.$

$6\left(\dfrac{1}{3}x + \dfrac{1}{2}x\right) = 6\cdot 1$ LCD $= 6$

$6\left(\dfrac{1}{3}x\right) + 6\left(\dfrac{1}{2}x\right) = 6$

$2x + 3x = 6$

$5x = 6$

$x = \dfrac{6}{5}$, or $1\dfrac{1}{5}$

If Jody and Pat worked together, it would take them $1\frac{1}{5}$ hours to weed the yard.

35. The circumference C of a circle varies directly as its radius r, so $C = kr$.

$9.42 = k(1.5)$ Let $C = 9.42$ and $r = 1.5$.

$k = \dfrac{9.42}{1.5} = 6.28$

So, $C = 6.28r$ and when $r = 5.25$,

$C = 6.28(5.25) = 32.97.$

Using $k = 6.28$, a circle with radius 5.25 inches has circumference 32.97 inches.

Chapter 8
Roots and Radicals

8.1 Evaluating Roots

Now Try Exercises

N1. The square roots of 81 are 9 and −9 because $9 \cdot 9 = 81$ and $(-9)(-9) = 81$.

N2. **(a)** $\sqrt{400}$ is the positive square root of 400.
$\sqrt{400} = 20$

(b) $-\sqrt{169}$ is the negative square root of 169.
$-\sqrt{169} = -13$

(c) $\sqrt{\frac{100}{121}}$ is the positive square root of $\frac{100}{121}$.
$\sqrt{\frac{100}{121}} = \frac{10}{11}$

N3. **(a)** The square of $\sqrt{15}$ is
$\left(\sqrt{15}\right)^2 = 15.$

(b) The square of $-\sqrt{23}$ is
$\left(-\sqrt{23}\right)^2 = \left(\sqrt{23}\right)^2 = 23.$

(c) The square of $\sqrt{2k^2+5}$ is
$\left(\sqrt{2k^2+5}\right)^2 = 2k^2+5.$

N4. **(a)** 31 is not a perfect square, so $\sqrt{31}$ is irrational.

(b) 900 is a perfect square, so $\sqrt{900} = 30$ is rational.

(c) There is no real number whose square is −16, so $\sqrt{-16}$ is not a real number.

N5. Find a decimal approximation for each square root. Round answers to the nearest thousandth.

(a) $\sqrt{51} \approx 7.141$

(b) $-\sqrt{360} \approx -18.974$

N6. Substitute the given values in the Pythagorean theorem, $c^2 = a^2 + b^2$. Then solve for the variable that is not given.

(a) $c^2 = a^2 + b^2$

$c^2 = 5^2 + 12^2$ Let $a = 5$, $b = 12$.

$c^2 = 25 + 144$ Square.

$c^2 = 169$ Add.

$c = \sqrt{169}$ Take square root.

$c = 13$ $13^2 = 169$

(b) $c^2 = a^2 + b^2$

$14^2 = 9^2 + b^2$ Let $c = 14$, $a = 9$.

$196 = 81 + b^2$

$115 = b^2$ Subtract 81.

$b = \sqrt{115}$ Take square root.

$b \approx 10.724$ Use a calculator.

N7. Let $c =$ the length of the diagonal. Use the Pythagorean theorem, since the width, length, and diagonal of a rectangle form a right triangle.

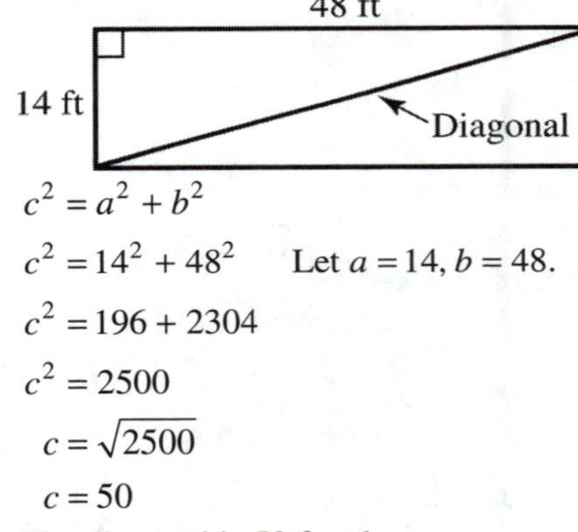

$c^2 = a^2 + b^2$

$c^2 = 14^2 + 48^2$ Let $a = 14$, $b = 48$.

$c^2 = 196 + 2304$

$c^2 = 2500$

$c = \sqrt{2500}$

$c = 50$

The diagonal is 50 feet long.

N8. Let $(x_1, y_1) = (-4, 3)$ and $(x_2, y_2) = (-7, 1)$. Use the distance formula.

$$\begin{aligned} d &= \sqrt{(x_2 - x_1)^2 + (y_2 - y_1)^2} \\ &= \sqrt{[-7 - (-4)]^2 + (1-3)^2} \\ &= \sqrt{(-3)^2 + (-2)^2} \\ &= \sqrt{9+4} \\ &= \sqrt{13} \end{aligned}$$

N9. **(a)** $\sqrt[3]{343} = 7$ because $7^3 = 343$.

(b) $\sqrt[3]{-1000} = -10$ because $(-10)^3 = -1000$.

(c) $\sqrt[3]{27} = 3$ because $3^3 = 27$.

N10. **(a)** $\sqrt[4]{625} = 5$ because $5^4 = 625$.

(b) $\sqrt[4]{-625}$ is not a real number because a fourth power of a real number cannot be negative.

(c) From part (a), $\sqrt[4]{625} = 5$, so $-\sqrt[4]{625} = -1 \cdot \sqrt[4]{625} = -1 \cdot 5 = -5$.

(d) $\sqrt[5]{3125} = 5$ because $5^5 = 3125$.

(e) $\sqrt[5]{-3125} = -5$ because $(-5)^5 = -3125$.

Exercises

1. Every positive number has two real square roots. This statement is *true*. One of the real square roots is a positive number and the other is its opposite.

3. Every nonnegative number has two real square roots. This statement is *false* since zero is a nonnegative number that has only one square root, namely 0.

5. The cube root of every nonzero real number has the same sign as the number itself. This statement is *true*. The cube root of a positive real number is positive and the cube root of a negative real number is negative. The cube root of 0 is 0.

7. For the statement "$\sqrt{a}$ represents a positive number" to be true, a must be positive because the square root of a negative number is not a real number and $\sqrt{0} = 0$.

9. For the statement "$\sqrt{a}$ is not a real number" to be true, a must be negative.

11. The square roots of 9 are -3 and 3 because $(-3)(-3) = 9$ and $3 \cdot 3 = 9$.

13. The square roots of 64 are -8 and 8 because $(-8)(-8) = 64$ and $8 \cdot 8 = 64$.

15. The square roots of 169 are -13 and 13 because $(-13)(-13) = 169$ and $13 \cdot 13 = 169$.

17. The square roots of $\frac{25}{196}$ are $-\frac{5}{14}$ and $\frac{5}{14}$ because $\left(-\frac{5}{14}\right)\left(-\frac{5}{14}\right) = \frac{25}{196}$ and $\frac{5}{14} \cdot \frac{5}{14} = \frac{25}{196}$.

19. The square roots of 900 are -30 and 30 because $(-30)(-30) = 900$ and $30 \cdot 30 = 900$.

21. $\sqrt{1}$ represents the positive square root of 1. Since $1 \cdot 1 = 1$, $\sqrt{1} = 1$.

23. $\sqrt{49}$ represents the positive square root of 49. Since $7 \cdot 7 = 49$, $\sqrt{49} = 7$.

25. $\sqrt{100}$ represents the positive square root of 100. Since $10 \cdot 10 = 100$, $\sqrt{100} = 10$.

27. $-\sqrt{16}$ represents the negative square root of 16. Since $4 \cdot 4 = 16$, $-\sqrt{16} = -4$.

29. $-\sqrt{256}$ represents the negative square root of 256. Since $16 \cdot 16 = 256$, $-\sqrt{256} = -16$.

31. $-\sqrt{\frac{144}{121}}$ represents the negative square root of $\frac{144}{121}$. Since $\frac{12}{11} \cdot \frac{12}{11} = \frac{144}{121}$, $-\sqrt{\frac{144}{121}} = -\frac{12}{11}$.

33. $\sqrt{0.64}$ represents the positive square root of 0.64. Since $0.8 \cdot 0.8 = 0.64$, $\sqrt{0.64} = 0.8$.

35. $\sqrt{-121}$ is not a real number because there is no real number whose square is -121.

37. $-\sqrt{-49}$ is not a real number because there is no real number whose square is -49. The leading negative sign is irrelevant.

39. The square of $\sqrt{19}$ is

$$\left(\sqrt{19}\right)^2 = 19,$$

by the definition of square root.

41. The square of $-\sqrt{19}$ is

$$\left(-\sqrt{19}\right)^2 = 19,$$

by the definition of square root and since the square of a negative number is positive.

43. The square of $\sqrt{\frac{2}{3}}$ is

$\left(\sqrt{\frac{2}{3}}\right)^2 = \frac{2}{3},$

by the definition of square root.

45. The square of $\sqrt{3x^2+4}$ is

$\left(\sqrt{3x^2+4}\right)^2 = 3x^2+4.$

47. $\sqrt{25}$

The number 25 is a perfect square, 5^2, so $\sqrt{25}$ is a *rational* number.

$\sqrt{25} = 5$

49. $\sqrt{29}$

Because 29 is not a perfect square, $\sqrt{29}$ is *irrational.* Using a calculator, we obtain $\sqrt{29} \approx 5.385.$

51. $-\sqrt{64}$

The number 64 is a perfect square, 8^2, so $-\sqrt{64}$ is *rational.*

$-\sqrt{64} = -8$

53. $-\sqrt{300}$

The number 300 is not a perfect square, so $-\sqrt{300}$ is *irrational.* Using a calculator, we obtain $-\sqrt{300} \approx -17.321.$

55. $\sqrt{-29}$

There is no real number whose square is -29. Therefore, $\sqrt{-29}$ is *not a real number.*

57. $\sqrt{1200}$

Because 1200 is not a perfect square, $\sqrt{1200}$ is *irrational.* Using a calculator, we obtain $\sqrt{1200} \approx 34.641.$

59. Since $81 < 94 < 100$, $\sqrt{81} = 9$, and $\sqrt{100} = 10$, we conclude that $\sqrt{94}$ is between 9 and 10.

61. Since $49 < 51 < 64$, $\sqrt{49} = 7$, and $\sqrt{64} = 8$, we conclude that $\sqrt{51}$ is between 7 and 8.

63. Since $36 < 40 < 49$, $\sqrt{36} = 6$, and $\sqrt{49} = 7$, we conclude that $-\sqrt{40}$ is between -7 and -6.

65. Since $16 < 23.2 < 25$, $\sqrt{16} = 4$, and $\sqrt{25} = 5$, we conclude that $\sqrt{23.2}$ is between 4 and 5.

67. $\sqrt{103} \approx \sqrt{100} = 10$

$\sqrt{48} \approx \sqrt{49} = 7$

The best estimate for the length and width of the rectangle is 10 by 7, choice C.

69. $a = 8, b = 15$

Substitute the given values in the Pythagorean theorem and then solve for c^2.

$$\begin{aligned} c^2 &= a^2 + b^2 \\ c^2 &= 8^2 + 15^2 \\ &= 64 + 225 \\ &= 289 \end{aligned}$$

Now find the positive square root of 289 to obtain the length of the hypotenuse, c.

$c = \sqrt{289} = 17$

71. $a = 6, c = 10$

Substitute the given values in the Pythagorean theorem and then solve for b^2.

$$\begin{aligned} c^2 &= a^2 + b^2 \\ 10^2 &= 6^2 + b^2 \\ 100 &= 36 + b^2 \\ 64 &= b^2 \end{aligned}$$

Now find the positive square root of 64 to obtain the length of the leg b.

$b = \sqrt{64} = 8$

73. $a = 11, b = 4$

$$\begin{aligned} c^2 &= a^2 + b^2 \\ c^2 &= 11^2 + 4^2 \\ &= 121 + 16 \\ &= 137 \\ c &= \sqrt{137} \approx 11.705 \end{aligned}$$

75. The given information involves a right triangle with hypotenuse 25 centimeters and a leg of length 7 centimeters. Let a represent the length of the other leg, and use the Pythagorean theorem.

$$c^2 = a^2 + b^2$$
$$25^2 = a^2 + 7^2$$
$$625 = a^2 + 49$$
$$576 = a^2$$
$$a = \sqrt{576} = 24$$

The length of the rectangle is 24 centimeters.

77. *Step 2*
Let x represent the vertical distance of the kite above Tyler's hand. The kite string forms the hypotenuse of a right triangle.

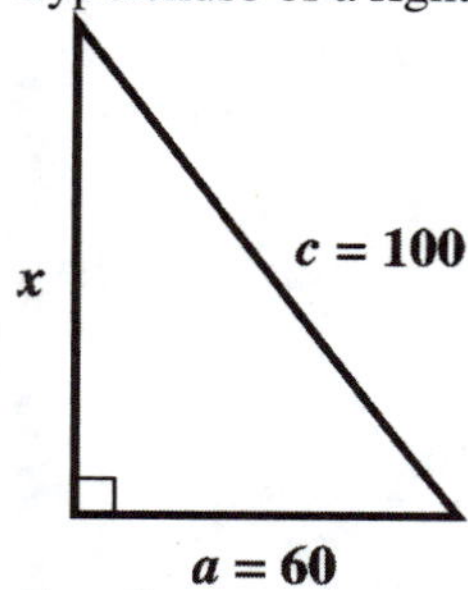

Step 3
Use the Pythagorean theorem.

$$a^2 + x^2 = c^2$$
$$60^2 + x^2 = 100^2$$

Step 4

$$3600 + x^2 = 10,000$$
$$x^2 = 6400$$
$$x = \sqrt{6400} = 80$$

Step 5
The kite is 80 feet above his hand.
Step 6
From the figure, we see that we must have

$$60^2 + 80^2 \stackrel{?}{=} 100^2$$
$$3600 + 6400 = 10,000. \quad \text{True}$$

79. *Step 2*
Let x represent the distance from R to S.

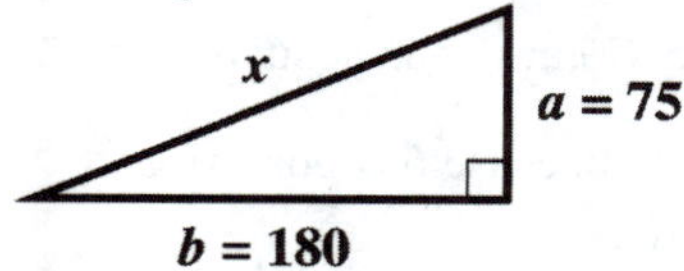

Step 3
Triangle RST is a right triangle, so we can use the Pythagorean theorem.

$$x^2 = a^2 + b^2$$
$$x^2 = 75^2 + 180^2$$
$$= 5625 + 32,400$$

Step 4

$$x^2 = 38,025$$
$$x = \sqrt{38,025} = 195$$

Step 5
The distance across the lake is 195 feet.
Step 6
From the figure, we see that we must have

$$75^2 + 180^2 \stackrel{?}{=} 195^2$$
$$5625 + 32,400 = 38,025. \quad \text{True}$$

81. Let $a = 4.5$ and $c = 12$. Use the Pythagorean theorem to find b, the distance along the ground as described in the problem.

$$c^2 = a^2 + b^2$$
$$12^2 = (4.5)^2 + b^2$$
$$144 = 20.25 + b^2$$
$$123.75 = b^2$$
$$b = \sqrt{123.75} \approx 11.1$$

The distance from the base of the tree to the point where the broken part touches the ground is 11.1 feet (to the nearest tenth).

83. Refer to the right triangle shown in the figure in the textbook. Note that the given distances are the lengths of the hypotenuse (193.0 feet) and one of the legs (110.0 feet) of the triangle. Use the Pythagorean theorem with $a = 110.0$, $c = 193.0$, and $b =$ the height of the building.

$$a^2 + b^2 = c^2$$
$$(110.0)^2 + b^2 = (193.0)^2$$
$$12,100 + b^2 = 37,249$$
$$b^2 = 25,149$$
$$b = \sqrt{25,149} \approx 158.6$$

The height of the building, to the nearest tenth, is 158.6 feet.

85. Use the Pythagorean theorem with $a = 5, b = 8$, and $c = x$.

$$c^2 = a^2 + b^2$$
$$x^2 = 5^2 + 8^2$$
$$= 25 + 64$$
$$= 89$$
$$x = \sqrt{89} \approx 9.434$$

87. **(a)** 3 units, 4 units

(b) If we let $a = 3$, $b = 4$, and $c = 5$, then the Pythagorean theorem is satisfied.

$$a^2 + b^2 = c^2$$
$$3^2 + 4^2 \stackrel{?}{=} 5^2$$
$$25 = 25 \quad \text{True}$$

89. The area of the large square on the left is $c \cdot c = c^2$. The side of the small square inside that figure has length $b - a$, so the area of that square is $(b-a)^2$.
The area of one of the rectangles on the right is $a \cdot b$, so the sum of the areas of the two rectangles in the figure on the right is $2ab$. Since the areas of the two figures are the same, we have

$$c^2 = 2ab + (b-a)^2$$
$$c^2 = 2ab + b^2 - 2ab + a^2$$
$$c^2 = a^2 + b^2,$$

which is the Pythagorean theorem.

91. Let $(x_1, y_1) = (5, 7)$ and $(x_2, y_2) = (1, 4)$. Use the distance formula.

$$\begin{aligned} d &= \sqrt{(x_2 - x_1)^2 + (y_2 - y_1)^2} \\ &= \sqrt{(1-5)^2 + (4-7)^2} \\ &= \sqrt{(-4)^2 + (-3)^2} \\ &= \sqrt{16+9} \\ &= \sqrt{25} = 5 \end{aligned}$$

93. Let $(x_1, y_1) = (2, 9)$ and $(x_2, y_2) = (-3, -3)$. Use the distance formula.

$$\begin{aligned} d &= \sqrt{(x_2 - x_1)^2 + (y_2 - y_1)^2} \\ &= \sqrt{(-3-2)^2 + (-3-9)^2} \\ &= \sqrt{(-5)^2 + (-12)^2} \\ &= \sqrt{25+144} \\ &= \sqrt{169} = 13 \end{aligned}$$

95. Let $(x_1, y_1) = (-1, -2)$ and $(x_2, y_2) = (-3, 1)$. Use the distance formula.

$$\begin{aligned} d &= \sqrt{(x_2 - x_1)^2 + (y_2 - y_1)^2} \\ &= \sqrt{[-3-(-1)]^2 + [1-(-2)]^2} \\ &= \sqrt{(-2)^2 + 3^2} \\ &= \sqrt{4+9} = \sqrt{13} \end{aligned}$$

97. Let $(x_1, y_1) = \left(-\frac{1}{4}, \frac{2}{3}\right)$ and $(x_2, y_2) = \left(\frac{3}{4}, -\frac{1}{3}\right)$.

$$\begin{aligned} d &= \sqrt{(x_2 - x_1)^2 + (y_2 - y_1)^2} \\ &= \sqrt{\left[\frac{3}{4} - \left(-\frac{1}{4}\right)\right]^2 + \left(-\frac{1}{3} - \frac{2}{3}\right)^2} \\ &= \sqrt{1^2 + (-1)^2} \\ &= \sqrt{1+1} = \sqrt{2} \end{aligned}$$

99.

$1^3 = \underline{1}$	$6^3 = \underline{216}$
$2^3 = \underline{8}$	$7^3 = \underline{343}$
$3^3 = \underline{27}$	$8^3 = \underline{512}$
$4^3 = \underline{64}$	$9^3 = \underline{729}$
$5^3 = \underline{125}$	$10^3 = \underline{1000}$

101. $\sqrt[3]{1} = 1$ because $1^3 = 1$.

103. $\sqrt[3]{125} = 5$ because $5^3 = 125$.

105. $\sqrt[3]{729} = 9$ because $9^3 = 729$.

107. $\sqrt[3]{-27} = -3$ because $(-3)^3 = -27$.

109. $\sqrt[3]{-216} = -6$ because $(-6)^3 = -216$.

111. $\sqrt[3]{-8} = -2$ because $(-2)^3 = -8$. Thus, $-\sqrt[3]{-8} = -(-2) = 2$.

113. $\sqrt[4]{81} = 3$ because 3 is positive and $3^4 = 81$.

115. $\sqrt[4]{625} = 5$ because 5 is positive and $5^4 = 625$.

117. $\sqrt[4]{1296} = 6$ because 6 is positive and $6^4 = 1296$.

119. $\sqrt[4]{-1}$ is not a real number because the fourth power of a real number cannot be negative.

121. $\sqrt[4]{81} = 3$ because 3 is positive and $3^4 = 81$. Thus, $-\sqrt[4]{81} = -3$.

123. $\sqrt[5]{32} = 2$ because $2^5 = 32$.

125. $\sqrt[5]{-1024} = -4$ because $(-4)^5 = -1024$.

8.2 Multiplying, Dividing, and Simplifying Radicals

Now Try Exercises

N1. **(a)** $\sqrt{5} \cdot \sqrt{11} = \sqrt{5 \cdot 11} = \sqrt{55}$ Product rule

(b) $\sqrt{11} \cdot \sqrt{11} = 11$

(c) $\sqrt{7} \cdot \sqrt{k} = \sqrt{7 \cdot k} = \sqrt{7k}$ Product rule
Assume $k \geq 0$.

(d) $\sqrt{5} \cdot \sqrt{5} = 5$ Product rule

N2. **(a)** $\sqrt{28} = \sqrt{4 \cdot 7}$ 4 is a perfect square.
$= \sqrt{4} \cdot \sqrt{7}$ Product rule
$= 2\sqrt{7}$

(b) $\sqrt{99} = \sqrt{9 \cdot 11}$ 9 is a perfect square.
$= \sqrt{9} \cdot \sqrt{11}$ Product rule
$= 3\sqrt{11}$

(c) $\sqrt{85}$ cannot be simplified further because 85 has no perfect square factors (except 1).

N3. **(a)** $\sqrt{16} \cdot \sqrt{50} = 4\sqrt{50}$ $\sqrt{16} = 4$
$= 4\sqrt{25 \cdot 2}$
$= 4\sqrt{25} \cdot \sqrt{2}$ Product rule
$= 4 \cdot 5\sqrt{2}$ $\sqrt{25} = 5$
$= 20\sqrt{2}$ Multiply.

(b) $\sqrt{6} \cdot \sqrt{30} = \sqrt{6 \cdot 30}$ Product rule
$= \sqrt{6 \cdot 6 \cdot 5}$ Factor.
$= \sqrt{36} \cdot \sqrt{5}$
$= 6\sqrt{5}$

(c) $4\sqrt{3} \cdot 5\sqrt{15} = 4 \cdot 5\sqrt{3 \cdot 15}$ Product rule
$= 20\sqrt{3 \cdot 3 \cdot 5}$ Factor.
$= 20 \cdot \sqrt{9} \cdot \sqrt{5}$
$= 20 \cdot 3\sqrt{5}$
$= 60\sqrt{5}$ Multiply.

N4. **(a)** $\sqrt{\frac{81}{100}} = \frac{\sqrt{81}}{\sqrt{100}} = \frac{9}{10}$ Quotient rule

(b) $\frac{\sqrt{245}}{\sqrt{5}} = \sqrt{\frac{245}{5}} = \sqrt{49} = 7$

(c) $\sqrt{\frac{11}{49}} = \frac{\sqrt{11}}{\sqrt{49}} = \frac{\sqrt{11}}{7}$

N5. $\frac{24\sqrt{39}}{4\sqrt{13}} = \frac{24}{4} \cdot \frac{\sqrt{39}}{\sqrt{13}}$
$= 6 \cdot \sqrt{\frac{39}{13}}$
$= 6\sqrt{3}$

N6. $\sqrt{\frac{1}{2}} \cdot \sqrt{\frac{5}{18}} = \sqrt{\frac{1}{2} \cdot \frac{5}{18}}$ Product rule
$= \sqrt{\frac{5}{36}}$ Multiply fractions.
$= \frac{\sqrt{5}}{\sqrt{36}}$ Quotient rule
$= \frac{\sqrt{5}}{6}$ $\sqrt{36} = 6$

N7. **(a)** $\sqrt{16y^8} = 4y^4$ since $\left(4y^4\right)^2 = 16y^8$.

(b) $\sqrt{x^5} = \sqrt{x^4} \cdot \sqrt{x} = x^2\sqrt{x}$

(c) $\sqrt{\frac{13}{t^2}} = \frac{\sqrt{13}}{\sqrt{t^2}} = \frac{\sqrt{13}}{t}$ Assume $t \neq 0$.

N8. **(a)** $\sqrt[3]{250} = \sqrt[3]{125 \cdot 2}$ $5^3 = 125$
$= \sqrt[3]{125} \cdot \sqrt[3]{2}$ Product rule
$= 5\sqrt[3]{2}$

(b) $\sqrt[4]{48} = \sqrt[4]{16 \cdot 3}$ $16 = 2^4$
$= \sqrt[4]{16} \cdot \sqrt[4]{3}$ Product rule
$= 2\sqrt[4]{3}$

(c) $\sqrt[3]{\frac{1}{125}} = \frac{\sqrt[3]{1}}{\sqrt[3]{125}}$ Quotient rule

$= \frac{1}{5}$

N9. (a) $\sqrt[3]{x^{12}} = x^4$ since $\left(x^4\right)^3 = x^{12}$.

(b) $\sqrt[3]{64t^3} = \sqrt[3]{64} \cdot \sqrt[3]{t^3}$ Product rule

$= 4t$ $\quad 4^3 = 64$

(c) $\sqrt[3]{40a^7} = \sqrt[3]{8a^6 \cdot 5a}$ $\quad 8a^6$ is a cube.

$= \sqrt[3]{8a^6} \cdot \sqrt[3]{5a}$ Product rule

$= 2a^2\sqrt[3]{5a}$ $\quad \left(2a^2\right)^3 = 8a^6$

(d) $\sqrt[3]{\frac{x^{15}}{1000}} = \frac{\sqrt[3]{x^{15}}}{\sqrt[3]{1000}}$ Quotient rule

$= \frac{x^5}{10}$

Exercises

1. false; $\sqrt{(-6)^2} = \sqrt{36} = 6$

3. false; $2\sqrt{7}$ represents the product of 2 and $\sqrt{7}$.

5. $\sqrt{47}$ is in simplified form since 47 has no perfect square factor (other than 1).
The other three choices could be simplified as follows.

$\sqrt{45} = \sqrt{9 \cdot 5} = 3\sqrt{5}$

$\sqrt{48} = \sqrt{16 \cdot 3} = 4\sqrt{3}$

$\sqrt{44} = \sqrt{4 \cdot 11} = 2\sqrt{11}$

The correct choice is A.

7. $\sqrt{3} \cdot \sqrt{5}$

Since 3 and 5 are nonnegative real numbers, the *product rule for radicals* applies. Thus,
$\sqrt{3} \cdot \sqrt{5} = \sqrt{3 \cdot 5} = \sqrt{15}$.

9. $\sqrt{2} \cdot \sqrt{11} = \sqrt{2 \cdot 11} = \sqrt{22}$

11. $\sqrt{6} \cdot \sqrt{7} = \sqrt{6 \cdot 7} = \sqrt{42}$

13. $\sqrt{3} \cdot \sqrt{27} = \sqrt{3 \cdot 27} = \sqrt{81}$, or 9

15. $\sqrt{13} \cdot \sqrt{13} = 13$

17. $\sqrt{13} \cdot \sqrt{r}\ (r \geq 0) = \sqrt{13r}$

19. $\sqrt{45} = \sqrt{9 \cdot 5} = \sqrt{9} \cdot \sqrt{5} = 3\sqrt{5}$

21. $\sqrt{24} = \sqrt{4 \cdot 6} = \sqrt{4} \cdot \sqrt{6} = 2\sqrt{6}$

23. $\sqrt{90} = \sqrt{9 \cdot 10} = \sqrt{9} \cdot \sqrt{10} = 3\sqrt{10}$

25. $\sqrt{75} = \sqrt{25 \cdot 3} = \sqrt{25} \cdot \sqrt{3} = 5\sqrt{3}$

27. $\sqrt{125} = \sqrt{25 \cdot 5} = \sqrt{25} \cdot \sqrt{5} = 5\sqrt{5}$

29. $145 = 5 \cdot 29$, so 145 has no perfect square factors (except 1) and $\sqrt{145}$ cannot be simplified further.

31. $\sqrt{160} = \sqrt{16 \cdot 10} = \sqrt{16} \cdot \sqrt{10} = 4\sqrt{10}$

33. $-\sqrt{700} = -\sqrt{100 \cdot 7}$

$= -\sqrt{100} \cdot \sqrt{7} = -10\sqrt{7}$

35. $3\sqrt{27} = 3\sqrt{9} \cdot \sqrt{3}$

$= 3 \cdot 3 \cdot \sqrt{3}$

$= 9\sqrt{3}$

37. $5\sqrt{50} = 5\sqrt{25} \cdot \sqrt{2}$

$= 5 \cdot 5 \cdot \sqrt{2}$

$= 25\sqrt{2}$

39. $a = 13, b = 9$

$c^2 = a^2 + b^2$ $\quad c = \sqrt{250}$

$c^2 = 13^2 + 9^2$ $\quad = \sqrt{25} \cdot \sqrt{10}$

$= 169 + 81$ $\quad = 5\sqrt{10}$

$= 250$

41. $a = 7,\ c = 11$

$c^2 = a^2 + b^2$

$11^2 = 7^2 + b^2$

$121 = 49 + b^2$

$72 = b^2$

$b = \sqrt{72}$

$= \sqrt{36} \cdot \sqrt{2}$

$= 6\sqrt{2}$

43. Let $(x_1, y_1) = (-6, 5)$ and $(x_2, y_2) = (3, -4)$.

$$\begin{aligned} d &= \sqrt{(x_2 - x_1)^2 + (y_2 - y_1)^2} \\ &= \sqrt{[3 - (-6)]^2 + (-4 - 5)^2} \\ &= \sqrt{9^2 + (-9)^2} \\ &= \sqrt{81 + 81} = \sqrt{162} \\ &= \sqrt{81} \cdot \sqrt{2} = 9\sqrt{2} \end{aligned}$$

45. Let $(x_1, y_1) = (-5, -1)$ and $(x_2, y_2) = (3, 1)$.

$$\begin{aligned} d &= \sqrt{(x_2 - x_1)^2 + (y_2 - y_1)^2} \\ &= \sqrt{[3 - (-5)]^2 + [1 - (-1)]^2} \\ &= \sqrt{8^2 + 2^2} \\ &= \sqrt{64 + 4} = \sqrt{68} \\ &= \sqrt{4} \cdot \sqrt{17} = 2\sqrt{17} \end{aligned}$$

47.
$$\begin{aligned} \sqrt{9} \cdot \sqrt{32} &= 3 \cdot \sqrt{16} \cdot \sqrt{2} \\ &= 3 \cdot 4 \cdot \sqrt{2} \\ &= 12\sqrt{2} \end{aligned}$$

49.
$$\begin{aligned} \sqrt{3} \cdot \sqrt{18} &= \sqrt{3 \cdot 18} = \sqrt{54} = \sqrt{9 \cdot 6} \\ &= \sqrt{9} \cdot \sqrt{6} = 3\sqrt{6} \end{aligned}$$

51.
$$\begin{aligned} \sqrt{12} \cdot \sqrt{48} &= \sqrt{12 \cdot 48} \\ &= \sqrt{12 \cdot 12 \cdot 4} \\ &= \sqrt{12 \cdot 12} \cdot \sqrt{4} \\ &= 12 \cdot 2 \\ &= 24 \end{aligned}$$

53.
$$\begin{aligned} \sqrt{12} \cdot \sqrt{30} &= \sqrt{12 \cdot 30} \\ &= \sqrt{360} = \sqrt{36 \cdot 10} \\ &= \sqrt{36} \cdot \sqrt{10} = 6\sqrt{10} \end{aligned}$$

55.
$$\begin{aligned} 2\sqrt{10} \cdot 3\sqrt{2} &= 2 \cdot 3 \cdot \sqrt{10 \cdot 2} && \text{Product rule} \\ &= 6\sqrt{20} && \text{Multiply.} \\ &= 6\sqrt{4 \cdot 5} && 4 = 2^2 \\ &= 6\sqrt{4} \cdot \sqrt{5} && \text{Product rule} \\ &= 6 \cdot 2 \cdot \sqrt{5} && \sqrt{4} = 2 \\ &= 12\sqrt{5} && \text{Multiply.} \end{aligned}$$

57.
$$\begin{aligned} 5\sqrt{3} \cdot 2\sqrt{15} &= 5 \cdot 2 \cdot \sqrt{3 \cdot 15} && \text{Product rule} \\ &= 10\sqrt{45} && \text{Multiply.} \\ &= 10\sqrt{9 \cdot 5} && 9 = 3^2 \\ &= 10\sqrt{9} \cdot \sqrt{5} && \text{Product rule} \\ &= 10 \cdot 3 \cdot \sqrt{5} && \sqrt{9} = 3 \\ &= 30\sqrt{5} && \text{Multiply.} \end{aligned}$$

59. $\sqrt{\frac{16}{225}} = \frac{\sqrt{16}}{\sqrt{225}} = \frac{4}{15}$

61. $\sqrt{\frac{7}{16}} = \frac{\sqrt{7}}{\sqrt{16}} = \frac{\sqrt{7}}{4}$

63. $\sqrt{\frac{4}{50}} = \sqrt{\frac{2 \cdot 2}{25 \cdot 2}} = \sqrt{\frac{2}{25}} = \frac{\sqrt{2}}{\sqrt{25}} = \frac{\sqrt{2}}{5}$

65. $\frac{\sqrt{75}}{\sqrt{3}} = \sqrt{\frac{75}{3}} = \sqrt{25} = 5$

67. $\frac{30\sqrt{10}}{5\sqrt{2}} = \frac{30}{5}\sqrt{\frac{10}{2}} = 6\sqrt{5}$

69.
$$\begin{aligned} \sqrt{\frac{5}{2}} \cdot \sqrt{\frac{125}{8}} &= \sqrt{\frac{5}{2} \cdot \frac{125}{8}} \\ &= \sqrt{\frac{625}{16}} \\ &= \frac{\sqrt{625}}{\sqrt{16}} = \frac{25}{4} \end{aligned}$$

71. $\sqrt{m^2} = m \ (m \geq 0)$

73. $\sqrt{y^4} = \sqrt{(y^2)^2} = y^2$

75. $\sqrt{36z^2} = \sqrt{36} \cdot \sqrt{z^2} = 6z$

77. $\sqrt{400x^6} = \sqrt{20 \cdot 20 \cdot x^3 \cdot x^3} = 20x^3$

79.
$$\begin{aligned} \sqrt{18x^8} &= \sqrt{9 \cdot 2 \cdot x^8} \\ &= \sqrt{9} \cdot \sqrt{2} \cdot \sqrt{x^8} \\ &= 3 \cdot \sqrt{2} \cdot x^4 \\ &= 3x^4\sqrt{2} \end{aligned}$$

81. $\sqrt{45c^{14}} = \sqrt{9 \cdot 5 \cdot c^{14}}$
$= \sqrt{9} \cdot \sqrt{5} \cdot \sqrt{c^{14}}$
$= 3 \cdot \sqrt{5} \cdot c^7$
$= 3c^7\sqrt{5}$

83. $\sqrt{z^5} = \sqrt{z^4 \cdot z} = \sqrt{z^4} \cdot \sqrt{z} = z^2\sqrt{z}$

85. $\sqrt{a^{13}} = \sqrt{a^{12}} \cdot \sqrt{a}$
$= \sqrt{(a^6)^2} \cdot \sqrt{a} = a^6\sqrt{a}$

87. $\sqrt{64x^7} = \sqrt{64} \cdot \sqrt{x^6}\sqrt{x}$
$= 8x^3\sqrt{x}$

89. $\sqrt{x^6y^{12}} = \sqrt{(x^3)^2 \cdot (y^6)^2} = x^3y^6$

91. $\sqrt{81m^4n^2} = \sqrt{81} \cdot \sqrt{m^4} \cdot \sqrt{n^2}$
$= 9m^2n$

93. $\sqrt{\frac{7}{x^{10}}} = \frac{\sqrt{7}}{\sqrt{x^{10}}} = \frac{\sqrt{7}}{x^5} \quad (x \neq 0)$

95. $\sqrt{\frac{y^4}{100}} = \frac{\sqrt{y^4}}{\sqrt{100}} = \frac{y^2}{10}$

97. $\sqrt{\frac{x^4y^6}{169}} = \frac{\sqrt{x^4y^6}}{\sqrt{169}} = \frac{\sqrt{x^4}\sqrt{y^6}}{13}$
$= \frac{x^2y^3}{13}$

99. $\sqrt[3]{40}$
8 is a perfect cube that is a factor of 40.
$\sqrt[3]{40} = \sqrt[3]{8 \cdot 5}$
$= \sqrt[3]{8} \cdot \sqrt[3]{5} = 2\sqrt[3]{5}$

101. $\sqrt[3]{54}$
27 is a perfect cube that is a factor of 54.
$\sqrt[3]{54} = \sqrt[3]{27 \cdot 2}$
$= \sqrt[3]{27} \cdot \sqrt[3]{2} = 3\sqrt[3]{2}$

103. $\sqrt[3]{128}$
64 is a perfect cube that is a factor of 128.
$\sqrt[3]{128} = \sqrt[3]{64 \cdot 2}$
$= \sqrt[3]{64} \cdot \sqrt[3]{2} = 4\sqrt[3]{2}$

105. $\sqrt[4]{80}$
16 is a perfect fourth power that is a factor of 80.
$\sqrt[4]{80} = \sqrt[4]{16 \cdot 5}$
$= \sqrt[4]{16} \cdot \sqrt[4]{5} = 2\sqrt[4]{5}$

107. $\sqrt[3]{\frac{8}{27}}$
8 and 27 are both perfect cubes.
$\sqrt[3]{\frac{8}{27}} = \frac{\sqrt[3]{8}}{\sqrt[3]{27}} = \frac{2}{3}$

109. $\sqrt[3]{-\frac{216}{125}} = \sqrt[3]{\left(-\frac{6}{5}\right)^3} = -\frac{6}{5}$

111. $\sqrt[3]{p^3} = p$ because $(p)^3 = p^3$.

113. $\sqrt[3]{x^9} = \sqrt[3]{(x^3)^3} = x^3$

115. $\sqrt[3]{64z^6} = \sqrt[3]{64} \cdot \sqrt[3]{(z^2)^3} = 4z^2$

117. $\sqrt[3]{343a^9b^3} = \sqrt[3]{343} \cdot \sqrt[3]{a^9} \cdot \sqrt[3]{b^3} = 7a^3b$

119. $\sqrt[3]{16t^5} = \sqrt[3]{8t^3} \cdot \sqrt[3]{2t^2} = 2t\sqrt[3]{2t^2}$

121. $\sqrt[3]{\frac{m^{12}}{8}} = \sqrt[3]{\frac{(m^4)^3}{2^3}} = \frac{m^4}{2}$

123. Use the formula for the volume of a cube.
$V = s^3$
$216 = s^3 \quad$ Let $V = 216$.
$\sqrt[3]{216} = s$
$6 = s$
The length of each side of the container is 6 centimeters.

125. Use the formula for the volume of a sphere, $V = \frac{4}{3}\pi r^3$. Let $V = 288\pi$ and solve for r.
$\frac{4}{3}\pi r^3 = 288\pi$
$\frac{3}{4}\left(\frac{4}{3}\pi r^3\right) = \frac{3}{4}(288\pi)$
$\pi r^3 = 216\pi$
$r^3 = 216$
$r = \sqrt[3]{216} = 6$
The radius is 6 inches.

127. $2\sqrt{26} \approx 2\sqrt{25} = 2 \cdot 5 = 10$

$\sqrt{83} \approx \sqrt{81} = 9$

Using 10 and 9 as estimates for the length and the width of the rectangle gives us $10 \cdot 9 = 90$ as an estimate for the area. Thus, choice D is the best estimate.

8.3 Adding and Subtracting Radicals

Now Try Exercises

N1. **(a)** $4\sqrt{3} + \sqrt{3}$

$= (4+1)\sqrt{3}$ Distributive property

$= 5\sqrt{3}$

(b) $2\sqrt{11} - 6\sqrt{11} = (2-6)\sqrt{11}$

$= -4\sqrt{11}$

(c) $\sqrt{5} + \sqrt{14}$ cannot be added using the distributive property.

N2. **(a)** $\sqrt{2} + \sqrt{18} = \sqrt{2} + \sqrt{9 \cdot 2}$

$= \sqrt{2} + \sqrt{9} \cdot \sqrt{2}$

$= \sqrt{2} + 3\sqrt{2}$

$= 4\sqrt{2}$

(b) $3\sqrt{48} - 2\sqrt{75}$

$= 3\left(\sqrt{16} \cdot \sqrt{3}\right) - 2\left(\sqrt{25} \cdot \sqrt{3}\right)$

$= 3\left(4\sqrt{3}\right) - 2\left(5\sqrt{3}\right)$

$= 12\sqrt{3} - 10\sqrt{3}$

$= 2\sqrt{3}$

(c) $8\sqrt[3]{5} + 10\sqrt[3]{40} = 8\sqrt[3]{5} + 10\sqrt[3]{8 \cdot 5}$

$= 8\sqrt[3]{5} + 10\sqrt[3]{8} \cdot \sqrt[3]{5}$

$= 8\sqrt[3]{5} + 10 \cdot 2 \cdot \sqrt[3]{5}$

$= 8\sqrt[3]{5} + 20\sqrt[3]{5}$

$= 28\sqrt[3]{5}$

N3. **(a)** $\sqrt{7} \cdot \sqrt{14} + 5\sqrt{2}$

$= \sqrt{7} \cdot \sqrt{7} \cdot \sqrt{2} + 5\sqrt{2}$

$= \sqrt{49} \cdot \sqrt{2} + 5\sqrt{2}$

$= 7\sqrt{2} + 5\sqrt{2}$

$= 12\sqrt{2}$

(b) $\sqrt{150x} + 2\sqrt{24x}$

$= \sqrt{25 \cdot 6x} + 2\sqrt{4 \cdot 6x}$

$= \sqrt{25} \cdot \sqrt{6x} + 2\sqrt{4} \cdot \sqrt{6x}$

$= 5\sqrt{6x} + 2 \cdot 2\sqrt{6x}$

$= 5\sqrt{6x} + 4\sqrt{6x}$

$= 9\sqrt{6x}$

(c) $5k^2\sqrt{12} - 4\sqrt{27k^4}$

$= 5k^2\sqrt{4 \cdot 3} - 4\sqrt{9k^4 \cdot 3}$

$= 5k^2\sqrt{4} \cdot \sqrt{3} - 4\sqrt{9k^4} \cdot \sqrt{3}$

$= 5k^2 \cdot 2\sqrt{3} - 4 \cdot 3k^2\sqrt{3}$

$= \left(10k^2 - 12k^2\right)\sqrt{3}$

$= -2k^2\sqrt{3}$

(d) $\sqrt[3]{128y^5} + 5y\sqrt[3]{16y^2}$

$= \sqrt[3]{64y^3 \cdot 2y^2} + 5y\sqrt[3]{8 \cdot 2y^2}$

$= \sqrt[3]{64y^3} \cdot \sqrt[3]{2y^2} + 5y\left(\sqrt[3]{8} \cdot \sqrt[3]{2y^2}\right)$

$= 4y\sqrt[3]{2y^2} + 5y\left(2\sqrt[3]{2y^2}\right)$

$= 4y\sqrt[3]{2y^2} + 10y\sqrt[3]{2y^2}$

$= (4y + 10y)\sqrt[3]{2y^2}$

$= 14y\sqrt[3]{2y^2}$

Exercises

1. Like radicals have the same radicand and the same index, or order. For example, $5\sqrt{2}$ and $-3\sqrt{2}$ are *like* radicals, as are $\sqrt{7}$, $-\sqrt{7}$, and $8\sqrt{7}$.

3. **(a)** like; Both are square roots and have the same radicand, 6.

(b) unlike; The radicands are different—one is 3 and one is 2.

(c) unlike; The indexes are different—one is a square root and one is a cube root.

(d) like; Both are square roots and have the same radicand, $2x$.

(e) unlike; The radicands are different—one is $3y$ and one is $6y$.

(f) like; $\sqrt[3]{2ba} = \sqrt[3]{2ab}$, so both are cube roots of the same radicand, $2ab$.

5. $2\sqrt{3} + 5\sqrt{3}$
$= (2+5)\sqrt{3}$ Distributive property
$= 7\sqrt{3}$

7. $4\sqrt{7} - 9\sqrt{7}$
$= (4-9)\sqrt{7}$ Distributive property
$= -5\sqrt{7}$

9. $\sqrt{6} + \sqrt{6} = 1\sqrt{6} + 1\sqrt{6}$
$= (1+1)\sqrt{6}$
$= 2\sqrt{6}$

11. $\sqrt{17} + 2\sqrt{17} = 1\sqrt{17} + 2\sqrt{17}$
$= (1+2)\sqrt{17}$
$= 3\sqrt{17}$

13. $5\sqrt{3} - \sqrt{3} = 5\sqrt{3} - 1\sqrt{3}$
$= (5-1)\sqrt{3}$
$= 4\sqrt{3}$

15. $\sqrt{6} + \sqrt{7}$
These unlike radicals cannot be added using the distributive property.

17. $5\sqrt{3} + \sqrt{12} = 5\sqrt{3} + \sqrt{4 \cdot 3}$
$= 5\sqrt{3} + \sqrt{4} \cdot \sqrt{3}$
$= 5\sqrt{3} + 2\sqrt{3}$
$= 7\sqrt{3}$

19. $2\sqrt{75} - \sqrt{12} = 2\sqrt{25 \cdot 3} - \sqrt{4 \cdot 3}$
$= 2 \cdot \sqrt{25} \cdot \sqrt{3} - \sqrt{4} \cdot \sqrt{3}$
$= 2 \cdot 5 \cdot \sqrt{3} - 2\sqrt{3}$
$= 10\sqrt{3} - 2\sqrt{3}$
$= 8\sqrt{3}$

21. $2\sqrt{50} - 5\sqrt{72}$
$= 2\sqrt{25 \cdot 2} - 5\sqrt{36 \cdot 2}$
$= 2 \cdot \sqrt{25} \cdot \sqrt{2} - 5 \cdot \sqrt{36} \cdot \sqrt{2}$
$= 2 \cdot 5 \cdot \sqrt{2} - 5 \cdot 6 \cdot \sqrt{2}$
$= 10\sqrt{2} - 30\sqrt{2}$
$= -20\sqrt{2}$

23. $\frac{1}{4}\sqrt{288} + \frac{1}{6}\sqrt{72}$
$= \frac{1}{4}\left(\sqrt{144} \cdot \sqrt{2}\right) + \frac{1}{6}\left(\sqrt{36} \cdot \sqrt{2}\right)$
$= \frac{1}{4}\left(12\sqrt{2}\right) + \frac{1}{6}\left(6\sqrt{2}\right)$
$= 3\sqrt{2} + 1\sqrt{2} = 4\sqrt{2}$

25. $\frac{3}{5}\sqrt{75} - \frac{2}{3}\sqrt{45}$
$= \frac{3}{5}\left(\sqrt{25} \cdot \sqrt{3}\right) - \frac{2}{3}\left(\sqrt{9} \cdot \sqrt{5}\right)$
$= \frac{3}{5}\left(5\sqrt{3}\right) - \frac{2}{3}\left(3\sqrt{5}\right)$
$= 3\sqrt{3} - 2\sqrt{5}$

27. $4\sqrt[3]{16} - 3\sqrt[3]{54}$
Recall that 8 and 27 are perfect cubes.
$4\sqrt[3]{16} - 3\sqrt[3]{54}$
$= 4\left(\sqrt[3]{8 \cdot 2}\right) - 3\left(\sqrt[3]{27 \cdot 2}\right)$
$= 4\left(\sqrt[3]{8} \cdot \sqrt[3]{2}\right) - 3\left(\sqrt[3]{27} \cdot \sqrt[3]{2}\right)$
$= 4\left(2\sqrt[3]{2}\right) - 3\left(3\sqrt[3]{2}\right)$
$= 8\sqrt[3]{2} - 9\sqrt[3]{2}$
$= (8-9)9\sqrt[3]{2} = -1\sqrt[3]{2} = -\sqrt[3]{2}$

29. $3\sqrt[3]{24} + 6\sqrt[3]{81}$
$= 3\left(\sqrt[3]{8} \cdot \sqrt[3]{3}\right) + 6\left(\sqrt[3]{27} \cdot \sqrt[3]{3}\right)$
$= 3\left(2\sqrt[3]{3}\right) + 6\left(3\sqrt[3]{3}\right)$
$= 6\sqrt[3]{3} + 18\sqrt[3]{3}$
$= 24\sqrt[3]{3}$

31. $\sqrt{3} \cdot \sqrt{7} + 2\sqrt{21} = \sqrt{3 \cdot 7} + 2\sqrt{21}$
$= 1\sqrt{21} + 2\sqrt{21}$
$= 3\sqrt{21}$

33. $\sqrt{6}\cdot\sqrt{2}+3\sqrt{3}$
$=\sqrt{6\cdot 2}+3\sqrt{3}$
$=\sqrt{12}+3\sqrt{3}$
$=\sqrt{4\cdot 3}+3\sqrt{3}$
$=\sqrt{4}\cdot\sqrt{3}+3\sqrt{3}$
$=2\sqrt{3}+3\sqrt{3}$
$=5\sqrt{3}$

35. $5\sqrt{7}-2\sqrt{28}+6\sqrt{63}$
$=5\sqrt{7}-2\sqrt{4\cdot 7}+6\sqrt{9\cdot 7}$
$=5\sqrt{7}-2\cdot\sqrt{4}\cdot\sqrt{7}+6\cdot\sqrt{9}\cdot\sqrt{7}$
$=5\sqrt{7}-2\cdot 2\cdot\sqrt{7}+6\cdot 3\cdot\sqrt{7}$
$=5\sqrt{7}-4\sqrt{7}+18\sqrt{7}$
$=(5-4+18)\sqrt{7}$
$=19\sqrt{7}$

37. $9\sqrt{24}-2\sqrt{54}+3\sqrt{20}$
$=9\sqrt{4\cdot 6}-2\sqrt{9\cdot 6}+3\sqrt{4\cdot 5}$
$=9\sqrt{4}\cdot\sqrt{6}-2\sqrt{9}\cdot\sqrt{6}+3\sqrt{4}\cdot\sqrt{5}$
$=9\cdot 2\sqrt{6}-2\cdot 3\sqrt{6}+3\cdot 2\sqrt{5}$
$=18\sqrt{6}-6\sqrt{6}+6\sqrt{5}$
$=12\sqrt{6}+6\sqrt{5}$

(Because $\sqrt{6}$ and $\sqrt{5}$ are unlike radicals, this expression cannot be simplified further.)

39. $5\sqrt{72}-3\sqrt{48}-4\sqrt{128}$
$=5\sqrt{36}\cdot\sqrt{2}-3\sqrt{16}\cdot\sqrt{3}-4\sqrt{64}\cdot\sqrt{2}$
$=5(6\sqrt{2})-3(4\sqrt{3})-4(8\sqrt{2})$
$=30\sqrt{2}-12\sqrt{3}-32\sqrt{2}$
$=-2\sqrt{2}-12\sqrt{3}$

41. $5\sqrt[4]{32}+2\sqrt[4]{32}\cdot\sqrt[4]{4}$
$=5(\sqrt[4]{16}\cdot\sqrt[4]{2})+2\sqrt[4]{16}\cdot\sqrt[4]{2}\cdot\sqrt[4]{4}$
$=5(2\sqrt[4]{2})+2\cdot 2\sqrt[4]{2\cdot 4}$
$=10\sqrt[4]{2}+4\sqrt[4]{8}$

43. Use the formula for the perimeter of a rectangle.
$P=2L+2W$
$=2(7\sqrt{2})+2(4\sqrt{2})$
$=14\sqrt{2}+8\sqrt{2}$
$=22\sqrt{2}$

45. $\sqrt{32x}-\sqrt{18x}=\sqrt{16\cdot 2x}-\sqrt{9\cdot 2x}$
$=\sqrt{16}\cdot\sqrt{2x}-\sqrt{9}\cdot\sqrt{2x}$
$=4\sqrt{2x}-3\sqrt{2x}$
$=(4-3)\sqrt{2x}$
$=1\sqrt{2x}=\sqrt{2x}$

47. $\sqrt{27r}+\sqrt{48r}$
$=\sqrt{9}\cdot\sqrt{3r}+\sqrt{16}\cdot\sqrt{3r}$
$=3\sqrt{3r}+4\sqrt{3r}$
$=7\sqrt{3r}$

49. $\sqrt{75x^2}+x\sqrt{300}$
$=\sqrt{25x^2}\sqrt{3}+x\sqrt{100}\cdot\sqrt{3}$
$=5x\sqrt{3}+10x\sqrt{3}$
$=(5x+10x)\sqrt{3}=15x\sqrt{3}$

51. $3\sqrt{8x^2}-4x\sqrt{2}$
$=3\sqrt{4x^2}\sqrt{2}-4x\sqrt{2}$
$=3(2x)\sqrt{2}-4x\sqrt{2}$
$=6x\sqrt{2}-4x\sqrt{2}$
$=(6x-4x)\sqrt{2}=2x\sqrt{2}$

53. $5\sqrt{75p^2}-4\sqrt{27p^2}$
$=5\sqrt{25p^2}\sqrt{3}-4\sqrt{9p^2}\sqrt{3}$
$=5(5p)\sqrt{3}-4(3p)\sqrt{3}$
$=25p\sqrt{3}-12p\sqrt{3}$
$=(25p-12p)\sqrt{3}=13p\sqrt{3}$

55. $2\sqrt{125x^2z}+8x\sqrt{80z}$
$=2\sqrt{25x^2}\sqrt{5z}+8x\sqrt{16}\sqrt{5z}$
$=2\cdot 5x\cdot\sqrt{5z}+8x\cdot 4\cdot\sqrt{5z}$
$=10x\sqrt{5z}+32x\sqrt{5z}$
$=(10x+32x)\sqrt{5z}=42x\sqrt{5z}$

57. $3k\sqrt{24k^2h^2}+9h\sqrt{54k^3}$
$=3k\sqrt{4k^2h^2}\sqrt{6}+9h\sqrt{9k^2}\sqrt{6k}$
$=3k(2kh)\sqrt{6}+9h(3k)\sqrt{6k}$
$=6k^2h\sqrt{6}+27hk\sqrt{6k}$

59. $6\sqrt[3]{8p^2}-2\sqrt[3]{27p^2}$
$=6\cdot\sqrt[3]{8}\cdot\sqrt[3]{p^2}-2\cdot\sqrt[3]{27}\cdot\sqrt[3]{p^2}$
$=6\cdot 2\cdot\sqrt[3]{p^2}-2\cdot 3\cdot\sqrt[3]{p^2}$
$=12\sqrt[3]{p^2}-6\sqrt[3]{p^2}$
$=6\sqrt[3]{p^2}$

61. $5\sqrt[4]{m^3}+8\sqrt[4]{16m^3}$
$=5\sqrt[4]{m^3}+8\sqrt[4]{16}\sqrt[4]{m^3}$
$=5\sqrt[4]{m^3}+8\cdot 2\cdot\sqrt[4]{m^3}$
$=5\sqrt[4]{m^3}+16\sqrt[4]{m^3}$
$=21\sqrt[4]{m^3}$

63. $2\sqrt[4]{p^5}-5p\sqrt[4]{16p}$
$=2\sqrt[4]{p^4}\sqrt[4]{p}-5p\sqrt[4]{16}\sqrt[4]{p}$
$=2\cdot p\cdot\sqrt[4]{p}-5p\cdot 2\cdot\sqrt[4]{p}$
$=2p\sqrt[4]{p}-10p\sqrt[4]{p}$
$=(2p-10p)\sqrt[4]{p}=-8p\sqrt[4]{p}$

65. $-5\sqrt[3]{256z^4}-2z\sqrt[3]{32z}$
$=-5\sqrt[3]{64z^3}\sqrt[3]{4z}-2z\sqrt[3]{8}\sqrt[3]{4z}$
$=-5\cdot 4z\cdot\sqrt[3]{4z}-2z\cdot 2\cdot\sqrt[3]{4z}$
$=-20z\sqrt[3]{4z}-4z\sqrt[3]{4z}$
$=(-20z-4z)\sqrt[3]{4z}=-24z\sqrt[3]{4z}$

67. $2\sqrt[4]{6k^7}-k\sqrt[4]{96k^3}$
$=2\sqrt[4]{k^4\cdot 6k^3}-k\sqrt[4]{16\cdot 6k^3}$
$=2\sqrt[4]{k^4}\cdot\sqrt[4]{6k^3}-k\cdot\sqrt[4]{16}\cdot\sqrt[4]{6k^3}$
$=2k\sqrt[4]{6k^3}-k\cdot 2\sqrt[4]{6k^3}$
$=0\cdot\sqrt[4]{6k^3}$
$=0$

8.4 Rationalizing the Denominator

Now Try Exercises

N1. (a) $\frac{15}{\sqrt{5}}=\frac{15\cdot\sqrt{5}}{\sqrt{5}\cdot\sqrt{5}}$
$=\frac{15\sqrt{5}}{5}$
$=3\sqrt{5}$

(b) $\frac{3}{\sqrt{24}}=\frac{3}{\sqrt{4\cdot 6}}=\frac{3}{\sqrt{4}\cdot\sqrt{6}}$
$=\frac{3}{2\cdot\sqrt{6}}=\frac{3\cdot\sqrt{6}}{2\sqrt{6}\cdot\sqrt{6}}$
$=\frac{3\sqrt{6}}{2\cdot 6}=\frac{3\sqrt{6}}{12}=\frac{\sqrt{6}}{4}$

N2. $\sqrt{\frac{27}{7}}=\frac{\sqrt{27}}{\sqrt{7}}=\frac{\sqrt{9}\cdot\sqrt{3}}{\sqrt{7}}$
$=\frac{3\cdot\sqrt{3}}{\sqrt{7}}=\frac{3\cdot\sqrt{3}\cdot\sqrt{7}}{\sqrt{7}\cdot\sqrt{7}}$
$=\frac{3\sqrt{21}}{7}$

N3. $\sqrt{\frac{1}{6}}\cdot\sqrt{\frac{3}{10}}=\sqrt{\frac{1}{6}\cdot\frac{3}{10}}=\sqrt{\frac{3}{60}}$
$=\sqrt{\frac{1}{20}}=\frac{\sqrt{1}}{\sqrt{20}}$
$=\frac{1}{\sqrt{4}\cdot\sqrt{5}}=\frac{1\cdot\sqrt{5}}{2\cdot\sqrt{5}\cdot\sqrt{5}}$
$=\frac{\sqrt{5}}{2\cdot 5}=\frac{\sqrt{5}}{10}$

N4. (a) $\frac{\sqrt{9m}}{\sqrt{n}}=\frac{\sqrt{9m}\cdot\sqrt{n}}{\sqrt{n}\cdot\sqrt{n}}=\frac{\sqrt{9mn}}{n}$
$=\frac{\sqrt{9}\cdot\sqrt{mn}}{n}=\frac{3\sqrt{mn}}{n}$

(b) $\sqrt{\frac{16m^2n}{5}}=\frac{\sqrt{16m^2n}}{\sqrt{5}}=\frac{\sqrt{16m^2n}\cdot\sqrt{5}}{\sqrt{5}\cdot\sqrt{5}}$
$=\frac{\sqrt{80m^2n}}{5}=\frac{\sqrt{16m^2}\cdot\sqrt{5n}}{5}$
$=\frac{4m\sqrt{5n}}{5}$

N5. **(a)** $\sqrt[3]{\frac{2}{7}} = \frac{\sqrt[3]{2}}{\sqrt[3]{7}} = \frac{\sqrt[3]{2} \cdot \sqrt[3]{7^2}}{\sqrt[3]{7} \cdot \sqrt[3]{7^2}}$

$= \frac{\sqrt[3]{2 \cdot 7^2}}{\sqrt[3]{7^3}} = \frac{\sqrt[3]{98}}{7}$

(b) $\frac{\sqrt[3]{2}}{\sqrt[3]{5}} = \frac{\sqrt[3]{2} \cdot \sqrt[3]{5^2}}{\sqrt[3]{5} \cdot \sqrt[3]{5^2}} = \frac{\sqrt[3]{2 \cdot 5^2}}{\sqrt[3]{5^3}} = \frac{\sqrt[3]{50}}{5}$

(c) $\frac{\sqrt[3]{4}}{\sqrt[3]{9t}} = \frac{\sqrt[3]{4} \cdot \sqrt[3]{3t^2}}{\sqrt[3]{3^2 t} \cdot \sqrt[3]{3t^2}} = \frac{\sqrt[3]{4 \cdot 3t^2}}{\sqrt[3]{3^3 t^3}}$

$= \frac{\sqrt[3]{12t^2}}{\sqrt[3]{(3t)^3}} = \frac{\sqrt[3]{12t^2}}{3t}, t \neq 0$

Exercises

1. The given expression is being multiplied by $\frac{\sqrt{3}}{\sqrt{3}}$, which is 1. According to the identity property for multiplication, multiplying an expression by 1 does not change the value of the expression.

3. $\frac{6}{\sqrt{5}}$

To rationalize the denominator, multiply the numerator and denominator by $\sqrt{5}$.

$\frac{6}{\sqrt{5}} = \frac{6 \cdot \sqrt{5}}{\sqrt{5} \cdot \sqrt{5}} = \frac{6\sqrt{5}}{5}$

5. $\frac{5}{\sqrt{5}} = \frac{5 \cdot \sqrt{5}}{\sqrt{5} \cdot \sqrt{5}} = \frac{5\sqrt{5}}{5} = \sqrt{5}$

7. $\frac{4}{\sqrt{6}} = \frac{4 \cdot \sqrt{6}}{\sqrt{6} \cdot \sqrt{6}} = \frac{4\sqrt{6}}{6} = \frac{2\sqrt{6}}{3}$

9. $\frac{8\sqrt{3}}{\sqrt{5}} = \frac{8\sqrt{3} \cdot \sqrt{5}}{\sqrt{5} \cdot \sqrt{5}} = \frac{8\sqrt{15}}{5}$

11. $\frac{12\sqrt{10}}{8\sqrt{3}} = \frac{12\sqrt{10} \cdot \sqrt{3}}{8\sqrt{3} \cdot \sqrt{3}}$

$= \frac{12\sqrt{30}}{8 \cdot 3}$

$= \frac{12\sqrt{30}}{24} = \frac{\sqrt{30}}{2}$

13. $\frac{8}{\sqrt{27}} = \frac{8}{\sqrt{9 \cdot 3}} = \frac{8}{\sqrt{9} \cdot \sqrt{3}} = \frac{8}{3\sqrt{3}}$

$= \frac{8 \cdot \sqrt{3}}{3\sqrt{3} \cdot \sqrt{3}} = \frac{8\sqrt{3}}{9}$

15. $\frac{6}{\sqrt{200}} = \frac{6}{\sqrt{100 \cdot 2}}$

$= \frac{6}{10\sqrt{2}} = \frac{3}{5\sqrt{2}} \cdot \frac{\sqrt{2}}{\sqrt{2}}$

$= \frac{3 \cdot \sqrt{2}}{5\sqrt{2} \cdot \sqrt{2}}$

$= \frac{3\sqrt{2}}{5 \cdot 2} = \frac{3\sqrt{2}}{10}$

17. $\frac{12}{\sqrt{72}} = \frac{12}{\sqrt{36} \cdot \sqrt{2}}$

$= \frac{12 \cdot \sqrt{2}}{6 \cdot \sqrt{2} \cdot \sqrt{2}}$

$= \frac{2 \cdot 6 \cdot \sqrt{2}}{6 \cdot 2}$

$= \sqrt{2}$

19. $\frac{\sqrt{10}}{\sqrt{5}} = \sqrt{\frac{10}{5}}$ Quotient rule

$= \sqrt{2}$

21. $\sqrt{\frac{40}{3}} = \frac{\sqrt{40}}{\sqrt{3}} = \frac{\sqrt{4} \cdot \sqrt{10} \cdot \sqrt{3}}{\sqrt{3} \cdot \sqrt{3}}$

$= \frac{2\sqrt{30}}{3}$

23. $\sqrt{\frac{1}{32}} = \frac{\sqrt{1}}{\sqrt{32}} = \frac{1 \cdot \sqrt{2}}{\sqrt{16} \cdot \sqrt{2} \cdot \sqrt{2}}$

$= \frac{\sqrt{2}}{4 \cdot 2}$

$= \frac{\sqrt{2}}{8}$

25. $\sqrt{\frac{9}{5}} = \frac{\sqrt{9}}{\sqrt{5}} = \frac{3 \cdot \sqrt{5}}{\sqrt{5} \cdot \sqrt{5}}$

$= \frac{3\sqrt{5}}{5}$

27. $\frac{-3}{\sqrt{50}}=\frac{-3}{\sqrt{25\cdot 2}}$

$=\frac{-3}{5\sqrt{2}}$

$=\frac{-3\cdot\sqrt{2}}{5\sqrt{2}\cdot\sqrt{2}}$

$=\frac{-3\sqrt{2}}{5\cdot 2}=\frac{-3\sqrt{2}}{10}$

29. $\frac{63}{\sqrt{45}}=\frac{63}{\sqrt{9}\cdot\sqrt{5}}=\frac{63}{3\sqrt{5}}=\frac{21}{\sqrt{5}}$

$=\frac{21\cdot\sqrt{5}}{\sqrt{5}\cdot\sqrt{5}}=\frac{21\sqrt{5}}{5}$

31. $\frac{\sqrt{8}}{\sqrt{24}}=\frac{\sqrt{8}}{\sqrt{8}\cdot\sqrt{3}}$

$=\frac{1\cdot\sqrt{3}}{\sqrt{3}\cdot\sqrt{3}}$

$=\frac{\sqrt{3}}{3}$

33. $-\sqrt{\frac{1}{5}}=-\frac{\sqrt{1}}{\sqrt{5}}$ Quotient rule

$=-\frac{1\cdot\sqrt{5}}{\sqrt{5}\cdot\sqrt{5}}=-\frac{\sqrt{5}}{5}$

35. $\sqrt{\frac{13}{5}}=\frac{\sqrt{13}}{\sqrt{5}}=\frac{\sqrt{13}\cdot\sqrt{5}}{\sqrt{5}\cdot\sqrt{5}}=\frac{\sqrt{65}}{5}$

37. $\sqrt{\frac{7}{13}}\cdot\sqrt{\frac{13}{3}}=\sqrt{\frac{7}{13}\cdot\frac{13}{3}}$ Product rule

$=\sqrt{\frac{7}{3}}=\frac{\sqrt{7}}{\sqrt{3}}$

$=\frac{\sqrt{7}\cdot\sqrt{3}}{\sqrt{3}\cdot\sqrt{3}}=\frac{\sqrt{21}}{3}$

39. $\sqrt{\frac{21}{7}}\cdot\sqrt{\frac{21}{8}}=\frac{\sqrt{21}}{\sqrt{7}}\cdot\frac{\sqrt{21}}{\sqrt{8}}=\frac{21}{\sqrt{7\cdot 2\cdot 4}}$

$=\frac{21}{2\sqrt{14}}=\frac{21\cdot\sqrt{14}}{2\cdot\sqrt{14}\cdot\sqrt{14}}$

$=\frac{21\sqrt{14}}{2\cdot 14}=\frac{3\sqrt{14}}{4}$

41. $\sqrt{\frac{1}{12}}\cdot\sqrt{\frac{1}{3}}=\sqrt{\frac{1}{12}\cdot\frac{1}{3}}$

$=\sqrt{\frac{1}{36}}=\frac{\sqrt{1}}{\sqrt{36}}=\frac{1}{6}$

43. $\sqrt{\frac{2}{9}}\cdot\sqrt{\frac{9}{2}}=\sqrt{\frac{2}{9}\cdot\frac{9}{2}}=\sqrt{1}=1$

45. $\sqrt{\frac{3}{4}}\cdot\sqrt{\frac{1}{5}}=\frac{\sqrt{3}}{\sqrt{4}}\cdot\frac{\sqrt{1}}{\sqrt{5}}$

$=\frac{\sqrt{3}\cdot\sqrt{5}}{2\cdot\sqrt{5}\cdot\sqrt{5}}$

$=\frac{\sqrt{15}}{2\cdot 5}$

$=\frac{\sqrt{15}}{10}$

47. $\sqrt{\frac{17}{3}}\cdot\sqrt{\frac{17}{6}}=\frac{\sqrt{17}\cdot\sqrt{17}}{\sqrt{3}\cdot\sqrt{6}}$

$=\frac{17}{\sqrt{18}}$

$=\frac{17\cdot\sqrt{2}}{\sqrt{9}\cdot\sqrt{2}\cdot\sqrt{2}}$

$=\frac{17\sqrt{2}}{3\cdot 2}$

$=\frac{17\sqrt{2}}{6}$

49. $\sqrt{\frac{2}{5}}\cdot\sqrt{\frac{3}{10}}=\sqrt{\frac{2}{5}\cdot\frac{3}{10}}$

$=\sqrt{\frac{3}{25}}$

$=\frac{\sqrt{3}}{\sqrt{25}}=\frac{\sqrt{3}}{5}$

51. $\sqrt{\frac{16}{27}}\cdot\sqrt{\frac{1}{9}}=\frac{\sqrt{16}\cdot\sqrt{1}}{\sqrt{27}\cdot\sqrt{9}}$

$=\frac{4\cdot 1}{\sqrt{9}\cdot\sqrt{3}\cdot 3}$

$=\frac{4\cdot 1\cdot\sqrt{3}}{3\cdot\sqrt{3}\cdot 3\cdot\sqrt{3}}$

$=\frac{4\sqrt{3}}{3\cdot 3\cdot 3}=\frac{4\sqrt{3}}{27}$

53. $\sqrt{\frac{6}{p}} = \frac{\sqrt{6}}{\sqrt{p}} \cdot \frac{\sqrt{p}}{\sqrt{p}} = \frac{\sqrt{6p}}{p}$

55. $\sqrt{\frac{3}{y}} = \frac{\sqrt{3}}{\sqrt{y}} \cdot \frac{\sqrt{y}}{\sqrt{y}} = \frac{\sqrt{3y}}{y}$

57. $\sqrt{\frac{16}{m}} = \frac{\sqrt{16}}{\sqrt{m}} \cdot \frac{\sqrt{m}}{\sqrt{m}} = \frac{4\sqrt{m}}{m}$

59. $\frac{\sqrt{3p^2}}{\sqrt{q}} = \frac{\sqrt{3}\sqrt{p^2}}{\sqrt{q}} \cdot \frac{\sqrt{q}}{\sqrt{q}} = \frac{p\sqrt{3q}}{q}$

61. $\frac{\sqrt{7x^3}}{\sqrt{y}} = \frac{\sqrt{7}\sqrt{x^2}\sqrt{x}}{\sqrt{y}} \cdot \frac{\sqrt{y}}{\sqrt{y}} = \frac{x\sqrt{7xy}}{y}$

63. $\sqrt{\frac{6p^3}{3m}} = \sqrt{\frac{2p^3}{m}} = \frac{\sqrt{2}\sqrt{p^2}\sqrt{p}}{\sqrt{m}} \cdot \frac{\sqrt{m}}{\sqrt{m}} = \frac{p\sqrt{2pm}}{m}$

65. $\sqrt{\frac{a^3b}{6}} = \frac{\sqrt{a^2}\sqrt{ab}}{\sqrt{6}} \cdot \frac{\sqrt{6}}{\sqrt{6}} = \frac{a\sqrt{6ab}}{6}$

67. $\sqrt{\frac{9a^2r}{5}} = \frac{\sqrt{9a^2}\sqrt{r}}{\sqrt{5}} \cdot \frac{\sqrt{5}}{\sqrt{5}} = \frac{3a\sqrt{5r}}{5}$

69. We need to multiply the numerator and denominator of $\frac{\sqrt[3]{2}}{\sqrt[3]{5}}$ by enough factors of 5 to make the radicand in the denominator a perfect cube. In this case we have one factor of 5, so we need to multiply by two more factors of 5 to make three factors of 5. Thus, the correct choice for a rationalizing factor in this problem is $\sqrt[3]{5^2} = \sqrt[3]{25}$, which corresponds to choice B.

71. $\sqrt[3]{\frac{1}{2}}$

Multiply the numerator and the denominator by enough factors of 2 to make the radicand in the denominator a perfect cube. This will eliminate the radical in the denominator. Here, we multiply by $\sqrt[3]{2^2}$, or $\sqrt[3]{4}$.

$$\sqrt[3]{\frac{1}{2}} = \frac{\sqrt[3]{1}}{\sqrt[3]{2}} = \frac{1 \cdot \sqrt[3]{2^2}}{\sqrt[3]{2} \cdot \sqrt[3]{2^2}} = \frac{\sqrt[3]{4}}{\sqrt[3]{2 \cdot 2^2}} = \frac{\sqrt[3]{4}}{\sqrt[3]{2^3}} = \frac{\sqrt[3]{4}}{2}$$

73. $\sqrt[3]{\frac{1}{32}} = \frac{\sqrt[3]{1}}{\sqrt[3]{32}} = \frac{1}{\sqrt[3]{8}\sqrt[3]{4}} \cdot \frac{\sqrt[3]{2}}{\sqrt[3]{2}} = \frac{\sqrt[3]{2}}{2 \cdot \sqrt[3]{8}} = \frac{\sqrt[3]{2}}{2 \cdot 2} = \frac{\sqrt[3]{2}}{4}$

75. $\sqrt[3]{\frac{1}{11}} = \frac{\sqrt[3]{1}}{\sqrt[3]{11}} = \frac{1 \cdot \sqrt[3]{11^2}}{\sqrt[3]{11} \cdot \sqrt[3]{11^2}} = \frac{\sqrt[3]{121}}{\sqrt[3]{11 \cdot 11^2}} = \frac{\sqrt[3]{121}}{\sqrt[3]{11^3}} = \frac{\sqrt[3]{121}}{11}$

77. $\sqrt[3]{\frac{2}{5}}$

Multiply the numerator and the denominator by enough factors of 5 to make the radicand in the denominator a perfect cube. This will eliminate the radical in the denominator. Here, we multiply by $\sqrt[3]{5^2}$, or $\sqrt[3]{25}$.

$$\sqrt[3]{\frac{2}{5}} = \frac{\sqrt[3]{2}}{\sqrt[3]{5}} = \frac{\sqrt[3]{2} \cdot \sqrt[3]{5^2}}{\sqrt[3]{5} \cdot \sqrt[3]{5^2}} = \frac{\sqrt[3]{2 \cdot 5^2}}{\sqrt[3]{5^3}} = \frac{\sqrt[3]{50}}{5}$$

79. $\dfrac{\sqrt[3]{4}}{\sqrt[3]{7}}=\dfrac{\sqrt[3]{4}\cdot\sqrt[3]{7^2}}{\sqrt[3]{7}\cdot\sqrt[3]{7^2}}$

$=\dfrac{\sqrt[3]{4}\cdot\sqrt[3]{49}}{\sqrt[3]{7^3}}=\dfrac{\sqrt[3]{196}}{7}$

81. To make the radicand in the denominator, $4y^2$, into a perfect cube, we must multiply 4 by 2 to get the perfect cube 8 and y^2 by y to get the perfect cube y^3. So we multiply the numerator and denominator by $\sqrt[3]{2y}$.

$\sqrt[3]{\dfrac{3}{4y^2}}=\dfrac{\sqrt[3]{3}}{\sqrt[3]{4y^2}}=\dfrac{\sqrt[3]{3}\cdot\sqrt[3]{2y}}{\sqrt[3]{4y^2}\cdot\sqrt[3]{2y}}$

$=\dfrac{\sqrt[3]{6y}}{\sqrt[3]{8y^3}}=\dfrac{\sqrt[3]{6y}}{2y}$

83. $\dfrac{\sqrt[3]{7m}}{\sqrt[3]{36n}}=\dfrac{\sqrt[3]{7m}}{\sqrt[3]{6^2n}}$

$=\dfrac{\sqrt[3]{7m}\cdot\sqrt[3]{6n^2}}{\sqrt[3]{6^2n}\cdot\sqrt[3]{6n^2}}$

$=\dfrac{\sqrt[3]{42mn^2}}{\sqrt[3]{6^3n^3}}=\dfrac{\sqrt[3]{42mn^2}}{6n}$

85. $\sqrt[4]{\dfrac{1}{8}}=\dfrac{\sqrt[4]{1}}{\sqrt[4]{8}}=\dfrac{1\cdot\sqrt[4]{2}}{\sqrt[4]{8}\cdot\sqrt[4]{2}}=\dfrac{\sqrt[4]{2}}{\sqrt[4]{16}}=\dfrac{\sqrt[4]{2}}{2}$

87. **(a)** $p=k\cdot\sqrt{\dfrac{L}{g}}$

$p=6\cdot\sqrt{\dfrac{9}{32}}$

$=\dfrac{6\sqrt{9}}{\sqrt{32}}=\dfrac{6\cdot 3}{\sqrt{16\cdot 2}}$

$=\dfrac{18}{4\sqrt{2}}=\dfrac{9}{2\sqrt{2}}$

$=\dfrac{9\cdot\sqrt{2}}{2\sqrt{2}\cdot\sqrt{2}}$

$=\dfrac{9\sqrt{2}}{4}$

The period of the pendulum is $\dfrac{9\sqrt{2}}{4}$ seconds.

(b) Using a calculator, we obtain $\dfrac{9\sqrt{2}}{4}\approx 3.182$ seconds.

(b) Using a calculator, we obtain $\dfrac{3\sqrt{15}}{2}\approx 5.809$ kilometers per second.

8.5 More Simplifying and Operations with Radicals

Now Try Exercises

N1. **(a)** $\sqrt{3}\left(\sqrt{45}-\sqrt{20}\right)$

$=\sqrt{3}\left(\sqrt{9}\cdot\sqrt{5}-\sqrt{4}\cdot\sqrt{5}\right)$

$=\sqrt{3}\left(3\sqrt{5}-2\sqrt{5}\right)$

$=\sqrt{3}\left(\sqrt{5}\right)$

$=\sqrt{15}$

(b) $\left(2\sqrt{3}+\sqrt{7}\right)\left(\sqrt{3}+3\sqrt{7}\right)$

$=2\sqrt{3}\cdot\sqrt{3}+2\sqrt{3}\cdot 3\sqrt{7}$

$+\sqrt{7}\cdot\sqrt{3}+\sqrt{7}\cdot 3\sqrt{7}$ FOIL

$=2\cdot 3+6\sqrt{21}+\sqrt{21}+3\cdot 7$

$=6+6\sqrt{21}+\sqrt{21}+21$

$=27+7\sqrt{21}$

(c) $\left(\sqrt{10}-8\right)\left(2\sqrt{10}+3\sqrt{2}\right)$

$=2\sqrt{10}\cdot\sqrt{10}+\sqrt{10}\cdot 3\sqrt{2}$

$-8\cdot 2\sqrt{10}-8\cdot 3\sqrt{2}$

$=2\cdot 10+3\sqrt{20}-16\sqrt{10}-24\sqrt{2}$

$=2\cdot 10+3\sqrt{4\cdot 5}-16\sqrt{10}-24\sqrt{2}$

$=20+3\cdot 2\sqrt{5}-16\sqrt{10}-24\sqrt{2}$

$=20+6\sqrt{5}-16\sqrt{10}-24\sqrt{2}$

N2. **(a)** Use the special product formula, $(a-b)^2=a^2-2ab+b^2$.

$\left(\sqrt{7}-4\right)^2=\left(\sqrt{7}\right)^2-2\left(\sqrt{7}\right)(4)+4^2$

$=7-8\sqrt{7}+16$

$=23-8\sqrt{7}$

(b) Use the special product formula, $(a-b)^2=a^2-2ab+b^2$.

$\left(3\sqrt{2}-5\right)^2=\left(3\sqrt{2}\right)^2-2\left(3\sqrt{2}\right)(5)+5^2$

$=18-30\sqrt{2}+25$

$=43-30\sqrt{2}$

(c) Use the special product formula,
$(a+b)^2 = a^2 + 2ab + b^2.$

$$\left(3+\sqrt{y}\right)^2 = 3^2 + 2(3)\left(\sqrt{y}\right) + \left(\sqrt{y}\right)^2$$
$$= 9 + 6\sqrt{y} + y$$

N3. (a) Use the rule for the product of the sum and the difference of two terms,
$(a+b)(a-b) = a^2 - b^2.$

$$\left(8+\sqrt{10}\right)\left(8-\sqrt{10}\right) = 8^2 - \left(\sqrt{10}\right)^2$$
$$= 64 - 10$$
$$= 54$$

(b) $\left(\sqrt{x}+2\sqrt{3}\right)\left(\sqrt{x}-2\sqrt{3}\right)$
$$= \left(\sqrt{x}\right)^2 - \left(2\sqrt{3}\right)^2$$
$$= x - 2^2 \cdot 3$$
$$= x - 12$$

N4. (a) $\dfrac{6}{4+\sqrt{3}}$

We can eliminate the radical in the denominator by multiplying both the numerator and denominator by $4-\sqrt{3}$, the conjugate of the denominator.

$$\frac{6}{4+\sqrt{3}}$$
$$= \frac{6\left(4-\sqrt{3}\right)}{\left(4+\sqrt{3}\right)\left(4-\sqrt{3}\right)}$$
$$= \frac{6\left(4-\sqrt{3}\right)}{4^2 - \left(\sqrt{3}\right)^2}$$
$$= \frac{6\left(4-\sqrt{3}\right)}{16-3}$$
$$= \frac{6\left(4-\sqrt{3}\right)}{13}$$

(b) $$\frac{5+\sqrt{7}}{\sqrt{7}-2} = \frac{\left(5+\sqrt{7}\right)\left(\sqrt{7}+2\right)}{\left(\sqrt{7}-2\right)\left(\sqrt{7}+2\right)}$$
$$= \frac{5\sqrt{7}+10+7+2\sqrt{7}}{\left(\sqrt{7}\right)^2 - 2^2}$$
$$= \frac{17+7\sqrt{7}}{7-4}$$
$$= \frac{17+7\sqrt{7}}{3}$$

(c) $$\frac{9}{\sqrt{k}-6} = \frac{9\left(\sqrt{k}+6\right)}{\left(\sqrt{k}-6\right)\left(\sqrt{k}+6\right)}$$
$$= \frac{9\left(\sqrt{k}+6\right)}{k-36},\ k \neq 36$$

N5. $$\frac{12\sqrt{6}+28}{20} = \frac{4\left(3\sqrt{6}+7\right)}{4(5)}$$
$$= \frac{3\sqrt{6}+7}{5}$$

Exercises

1. $\sqrt{25}+\sqrt{64} = 13$
$\left[\sqrt{25}+\sqrt{64} = 5+8\right]$

3. $\sqrt{8}\cdot\sqrt{2} = 4$
$\left[\sqrt{8}\cdot\sqrt{2} = \sqrt{16}\right]$

5. Because multiplication must be performed before addition, it is incorrect to add -37 and -2. Since $-2\sqrt{15}$ cannot be simplified, the expression cannot be written in a simpler form, and the final answer is $-37-2\sqrt{15}$.

7. $$\sqrt{5}\left(\sqrt{3}-\sqrt{7}\right) = \sqrt{5}\cdot\sqrt{3} - \sqrt{5}\cdot\sqrt{7}$$
$$= \sqrt{15} - \sqrt{35}$$

9. $2\sqrt{5}\left(3\sqrt{5}+\sqrt{2}\right)$
$$= 2\sqrt{5}\cdot 3\sqrt{5} + 2\sqrt{5}\cdot\sqrt{2}$$
$$= 2\cdot 3\cdot\sqrt{5}\cdot\sqrt{5} + 2\sqrt{10}$$
$$= 6\cdot 5 + 2\sqrt{10}$$
$$= 30 + 2\sqrt{10}$$

11. $3\sqrt{14}\cdot\sqrt{2}-\sqrt{28}=3\sqrt{14\cdot 2}-\sqrt{28}$
$=3\sqrt{28}-1\sqrt{28}$
$=2\sqrt{28}$
$=2\sqrt{4\cdot 7}$
$=2\cdot\sqrt{4}\cdot\sqrt{7}$
$=2\cdot 2\cdot\sqrt{7}$
$=4\sqrt{7}$

13. $(2\sqrt{6}+3)(3\sqrt{6}+7)$
$=2\sqrt{6}\cdot 3\sqrt{6}+7\cdot 2\sqrt{6}+3\cdot 3\sqrt{6}+3\cdot 7$ FOIL
$=2\cdot 3\cdot\sqrt{6}\cdot\sqrt{6}+14\sqrt{6}+9\sqrt{6}+21$
$=6\cdot 6+23\sqrt{6}+21$
$=36+23\sqrt{6}+21$
$=57+23\sqrt{6}$

15. $(8-\sqrt{7})^2$
$=(8)^2-2(8)(\sqrt{7})+(\sqrt{7})^2$
$=64-16\sqrt{7}+7$
$=71-16\sqrt{7}$

17. $(2\sqrt{7}+3)^2$
$=(2\sqrt{7})^2+2(2\sqrt{7})(3)+(3)^2$
$=4\cdot 7+12\sqrt{7}+9$
$=28+12\sqrt{7}+9$
$=37+12\sqrt{7}$

19. $(\sqrt{6}+1)^2$
$=(\sqrt{6})^2+2(\sqrt{6})(1)+(1)^2$
$=6+2\sqrt{6}+1$
$=7+2\sqrt{6}$

21. $(\sqrt{a}+1)^2$
$=(\sqrt{a})^2+2(\sqrt{a})(1)+(1)^2$
$=a+2\sqrt{a}+1$

23. $(7+\sqrt{x})^2$
$=(7)^2+2(7)(\sqrt{x})+(\sqrt{x})^2$
$=49+14\sqrt{x}+x$

25. $(5\sqrt{7}-2\sqrt{3})^2$
$=(5\sqrt{7})^2-2(5\sqrt{7})(2\sqrt{3})+(2\sqrt{3})^2$
$=5^2(\sqrt{7})^2-20\sqrt{21}+2^2(\sqrt{3})^2$
$=25\cdot 7-20\sqrt{21}+4\cdot 3$
$=175-20\sqrt{21}+12$
$=187-20\sqrt{21}$

27. $(5-\sqrt{2})(5+\sqrt{2})=(5)^2-(\sqrt{2})^2$
$=25-2=23$

29. $(\sqrt{8}-\sqrt{7})(\sqrt{8}+\sqrt{7})$
$=(\sqrt{8})^2-(\sqrt{7})^2$
$=8-7=1$

31. $(\sqrt{78}-\sqrt{76})(\sqrt{78}+\sqrt{76})$
$=(\sqrt{78})^2-(\sqrt{76})^2$
$=78-76=2$

33. $(\sqrt{y}-\sqrt{10})(\sqrt{y}+\sqrt{10})$
$=(\sqrt{y})^2-(\sqrt{10})^2$
$=y-10$

35. $(\sqrt{2}+\sqrt{3})(\sqrt{6}-\sqrt{2})$
$=\sqrt{2}(\sqrt{6})-\sqrt{2}(\sqrt{2})+\sqrt{3}(\sqrt{6})-\sqrt{3}(\sqrt{2})$
$=\sqrt{12}-2+\sqrt{18}-\sqrt{6}$
$=\sqrt{4}\cdot\sqrt{3}-2+\sqrt{9}\cdot\sqrt{2}-\sqrt{6}$
$=2\sqrt{3}-2+3\sqrt{2}-\sqrt{6}$

37. $\left(\sqrt{10}-\sqrt{5}\right)\left(\sqrt{5}+\sqrt{20}\right)$
$=\sqrt{10}\cdot\sqrt{5}+\sqrt{10}\cdot\sqrt{20}-\sqrt{5}\cdot\sqrt{5}-\sqrt{5}\cdot\sqrt{20}$
$=\sqrt{50}+\sqrt{200}-5-\sqrt{100}$
$=\sqrt{25\cdot 2}+\sqrt{100\cdot 2}-5-10$
$=5\sqrt{2}+10\sqrt{2}-15$
$=15\sqrt{2}-15$

39. $\left(\sqrt{5}+\sqrt{30}\right)\left(\sqrt{6}+\sqrt{3}\right)$
$=\sqrt{5}\cdot\sqrt{6}+\sqrt{5}\cdot\sqrt{3}+\sqrt{30}\cdot\sqrt{6}+\sqrt{30}\cdot\sqrt{3}$
$=\sqrt{30}+\sqrt{15}+\sqrt{180}+\sqrt{90}$
$=\sqrt{30}+\sqrt{15}+\sqrt{36\cdot 5}+\sqrt{9\cdot 10}$
$=\sqrt{30}+\sqrt{15}+6\sqrt{5}+3\sqrt{10}$

41. $\left(5\sqrt{7}-2\sqrt{3}\right)\left(3\sqrt{7}+4\sqrt{3}\right)$
$=5\sqrt{7}\left(3\sqrt{7}\right)+5\sqrt{7}\left(4\sqrt{3}\right)$
$\quad -2\sqrt{3}\left(3\sqrt{7}\right)-2\sqrt{3}\left(4\sqrt{3}\right)$ FOIL
$=15\cdot 7+20\sqrt{21}-6\sqrt{21}-8\cdot 3$
$=105+14\sqrt{21}-24$
$=81+14\sqrt{21}$

43. $\left(3\sqrt{t}+\sqrt{7}\right)\left(2\sqrt{t}-\sqrt{14}\right)$
$=3\sqrt{t}\cdot 2\sqrt{t}-3\sqrt{t}\cdot\sqrt{14}$
$\quad +2\sqrt{t}\cdot\sqrt{7}-\sqrt{7}\cdot\sqrt{14}$ FOIL
$=3\cdot 2\cdot\sqrt{t}\cdot\sqrt{t}-3\sqrt{14t}+2\sqrt{7t}-\sqrt{98}$
$=6t-3\sqrt{14t}+2\sqrt{7t}-\sqrt{49}\cdot\sqrt{2}$
$=6t-3\sqrt{14t}+2\sqrt{7t}-7\sqrt{2}$

45. $\left(\sqrt{3m}+\sqrt{2n}\right)\left(\sqrt{3m}-\sqrt{2n}\right)$
$=\left(\sqrt{3m}\right)^2-\left(\sqrt{2n}\right)^2$
$=3m-2n$

47. **(a)** The denominator is $\sqrt{5}+\sqrt{3}$, so to rationalize the denominator, we should multiply the numerator and denominator by its conjugate, $\sqrt{5}-\sqrt{3}$.

(b) The denominator is $\sqrt{6}-\sqrt{5}$, so to rationalize the denominator, we should multiply the numerator and denominator by its conjugate, $\sqrt{6}+\sqrt{5}$.

49. $\dfrac{1}{2+\sqrt{5}}=\dfrac{1\left(2-\sqrt{5}\right)}{\left(2+\sqrt{5}\right)\left(2-\sqrt{5}\right)}$
$=\dfrac{2-\sqrt{5}}{2^2-\left(\sqrt{5}\right)^2}$
$=\dfrac{2-\sqrt{5}}{4-5}=\dfrac{2-\sqrt{5}}{-1}$
$=-1\left(2-\sqrt{5}\right)=-2+\sqrt{5}$

51. $\dfrac{7}{2-\sqrt{11}}=\dfrac{7\left(2+\sqrt{11}\right)}{\left(2-\sqrt{11}\right)\left(2+\sqrt{11}\right)}$
$=\dfrac{7\left(2+\sqrt{11}\right)}{(2)^2-\left(\sqrt{11}\right)^2}$
$=\dfrac{7\left(2+\sqrt{11}\right)}{4-11}$
$=\dfrac{7\left(2+\sqrt{11}\right)}{-7}$
$=-1\left(2+\sqrt{11}\right)$
$=-2-\sqrt{11}$

53. $\dfrac{\sqrt{12}}{\sqrt{3}+1}=\dfrac{\sqrt{4}\sqrt{3}\left(\sqrt{3}-1\right)}{\left(\sqrt{3}+1\right)\left(\sqrt{3}-1\right)}$
$=\dfrac{2\left(3-\sqrt{3}\right)}{\left(\sqrt{3}\right)^2-1^2}$
$=\dfrac{2\left(3-\sqrt{3}\right)}{3-1}$
$=\dfrac{2\left(3-\sqrt{3}\right)}{2}=3-\sqrt{3}$

55. $\frac{2\sqrt{3}}{\sqrt{3}+5}=\frac{2\sqrt{3}(\sqrt{3}-5)}{(\sqrt{3}+5)(\sqrt{3}-5)}$

$=\frac{2\cdot 3-2\cdot 5\sqrt{3}}{(\sqrt{3})^2-5^2}$

$=\frac{6-10\sqrt{3}}{3-25}$

$=\frac{6-10\sqrt{3}}{-22}$

$=\frac{2(3-5\sqrt{3})}{2(-11)}$ Factor.

$=\frac{3-5\sqrt{3}}{-11}$, or $\frac{-3+5\sqrt{3}}{11}$

57. $\frac{\sqrt{2}+3}{\sqrt{3}-1}=\frac{(\sqrt{2}+3)(\sqrt{3}+1)}{(\sqrt{3}-1)(\sqrt{3}+1)}$

$=\frac{\sqrt{2}\cdot\sqrt{3}+\sqrt{2}+3\sqrt{3}+3}{(\sqrt{3})^2-1^2}$

$=\frac{\sqrt{6}+\sqrt{2}+3\sqrt{3}+3}{3-1}$

$=\frac{\sqrt{6}+\sqrt{2}+3\sqrt{3}+3}{2}$

59. $\frac{6-\sqrt{5}}{\sqrt{2}+2}=\frac{(6-\sqrt{5})(\sqrt{2}-2)}{(\sqrt{2}+2)(\sqrt{2}-2)}$

$=\frac{6\sqrt{2}-12-\sqrt{5}\cdot\sqrt{2}+2\sqrt{5}}{(\sqrt{2})^2-2^2}$

$=\frac{6\sqrt{2}-12-\sqrt{10}+2\sqrt{5}}{2-4}$

$=\frac{6\sqrt{2}-12-\sqrt{10}+2\sqrt{5}}{-2}$,

or $\frac{-6\sqrt{2}+12+\sqrt{10}-2\sqrt{5}}{2}$

61. $\frac{2\sqrt{6}+1}{\sqrt{2}+5}=\frac{(2\sqrt{6}+1)(\sqrt{2}-5)}{(\sqrt{2}+5)(\sqrt{2}-5)}$

$=\frac{2\sqrt{12}-10\sqrt{6}+\sqrt{2}-5}{(\sqrt{2})^2-5^2}$

$=\frac{2\sqrt{4\cdot 3}-10\sqrt{6}+\sqrt{2}-5}{2-25}$

$=\frac{2\cdot 2\sqrt{3}-10\sqrt{6}+\sqrt{2}-5}{-23}$

$=\frac{-4\sqrt{3}+10\sqrt{6}-\sqrt{2}+5}{23}$

63. $\frac{\sqrt{7}+\sqrt{2}}{\sqrt{3}-\sqrt{2}}$

$=\frac{(\sqrt{7}+\sqrt{2})(\sqrt{3}+\sqrt{2})}{(\sqrt{3}-\sqrt{2})(\sqrt{3}+\sqrt{2})}$

$=\frac{\sqrt{7}\cdot\sqrt{3}+\sqrt{7}\cdot\sqrt{2}+\sqrt{2}\cdot\sqrt{3}+\sqrt{2}\cdot\sqrt{2}}{(\sqrt{3})^2-(\sqrt{2})^2}$

$=\frac{\sqrt{21}+\sqrt{14}+\sqrt{6}+2}{3-2}$

$=\frac{\sqrt{21}+\sqrt{14}+\sqrt{6}+2}{1}$

$=\sqrt{21}+\sqrt{14}+\sqrt{6}+2$

65. $\frac{\sqrt{5}}{\sqrt{2}+\sqrt{3}}=\frac{\sqrt{5}(\sqrt{2}-\sqrt{3})}{(\sqrt{2}+\sqrt{3})(\sqrt{2}-\sqrt{3})}$

$=\frac{\sqrt{5}\cdot\sqrt{2}-\sqrt{5}\cdot\sqrt{3}}{(\sqrt{2})^2-(\sqrt{3})^2}$

$=\frac{\sqrt{10}-\sqrt{15}}{2-3}$

$=\frac{\sqrt{10}-\sqrt{15}}{-1}=-\sqrt{10}+\sqrt{15}$

67. $$\frac{\sqrt{108}}{3+3\sqrt{3}}=\frac{\sqrt{36}\cdot\sqrt{3}}{3\left(1+\sqrt{3}\right)}$$
$$=\frac{6\sqrt{3}}{3\left(1+\sqrt{3}\right)}=\frac{2\sqrt{3}}{1+\sqrt{3}}$$
$$=\frac{2\sqrt{3}}{1+\sqrt{3}}\cdot\frac{1-\sqrt{3}}{1-\sqrt{3}}$$
$$=\frac{2\sqrt{3}\left(1-\sqrt{3}\right)}{1^2-\left(\sqrt{3}\right)^2}$$
$$=\frac{2\sqrt{3}\left(1-\sqrt{3}\right)}{1-3}=\frac{2\sqrt{3}\left(1-\sqrt{3}\right)}{-2}$$
$$=-\sqrt{3}\left(1-\sqrt{3}\right)$$
$$=-\sqrt{3}+3 \quad\text{or}\quad 3-\sqrt{3}$$

69. $$\frac{8}{4-\sqrt{x}}=\frac{8\left(4+\sqrt{x}\right)}{\left(4-\sqrt{x}\right)\left(4+\sqrt{x}\right)}$$
$$=\frac{8\left(4+\sqrt{x}\right)}{4^2-\left(\sqrt{x}\right)^2}$$
$$=\frac{8\left(4+\sqrt{x}\right)}{16-x}$$

71. $$\frac{1}{\sqrt{x}+\sqrt{y}}=\frac{1\left(\sqrt{x}-\sqrt{y}\right)}{\left(\sqrt{x}+\sqrt{y}\right)\left(\sqrt{x}-\sqrt{y}\right)}$$
$$=\frac{\sqrt{x}-\sqrt{y}}{\left(\sqrt{x}\right)^2-\left(\sqrt{y}\right)^2}$$
$$=\frac{\sqrt{x}-\sqrt{y}}{x-y}$$

73. $$\frac{5\sqrt{7}-10}{5}=\frac{5\left(\sqrt{7}-2\right)}{5}$$
$$=\sqrt{7}-2$$

75. $$\frac{2\sqrt{3}+10}{8}=\frac{2\left(\sqrt{3}+5\right)}{2\cdot 4}$$
$$=\frac{\sqrt{3}+5}{4}$$

77. $$\frac{12-2\sqrt{10}}{4}=\frac{2\left(6-\sqrt{10}\right)}{2\cdot 2}=\frac{6-\sqrt{10}}{2}$$

79. $$\frac{16+\sqrt{128}}{24}=\frac{16+\sqrt{64}\sqrt{2}}{24}$$
$$=\frac{16+8\sqrt{2}}{24}$$
$$=\frac{8\left(2+\sqrt{2}\right)}{8\cdot 3}=\frac{2+\sqrt{2}}{3}$$

81. $\sqrt[3]{4}\left(\sqrt[3]{2}-3\right)$

$=\sqrt[3]{4}\left(\sqrt[3]{2}\right)-\sqrt[3]{4}(3)$ Distributive property

$=\sqrt[3]{8}-3\sqrt[3]{4}$ Product rule

$=2-3\sqrt[3]{4}$ $\sqrt[3]{8}=2$

83. $2\sqrt[4]{2}\left(3\sqrt[4]{8}+5\sqrt[4]{4}\right)$

$=2\cdot 3\cdot\sqrt[4]{2}\cdot\sqrt[4]{8}+2\cdot 5\cdot\sqrt[4]{2}\cdot\sqrt[4]{4}$

$=6\sqrt[4]{16}+10\sqrt[4]{8}$ Product rule

$=6\cdot 2+10\sqrt[4]{8}$ $\sqrt[4]{16}=2$

$=12+10\sqrt[4]{8}$

85. $\left(\sqrt[3]{2}-1\right)\left(\sqrt[3]{4}+3\right)$

$=\sqrt[3]{8}+3\sqrt[3]{2}-\sqrt[3]{4}-3$ FOIL

$=2+3\sqrt[3]{2}-\sqrt[3]{4}-3$

$=-1+3\sqrt[3]{2}-\sqrt[3]{4}$

87. $\left(\sqrt[3]{5}-\sqrt[3]{4}\right)\left(\sqrt[3]{25}+\sqrt[3]{20}+\sqrt[3]{16}\right)$

$=\sqrt[3]{5}\left(\sqrt[3]{25}+\sqrt[3]{20}+\sqrt[3]{16}\right)$
$\quad-\sqrt[3]{4}\left(\sqrt[3]{25}+\sqrt[3]{20}+\sqrt[3]{16}\right)$

$=\sqrt[3]{5}\cdot\sqrt[3]{25}+\sqrt[3]{5}\cdot\sqrt[3]{20}+\sqrt[3]{5}\cdot\sqrt[3]{16}$
$\quad-\sqrt[3]{4}\cdot\sqrt[3]{25}-\sqrt[3]{4}\cdot\sqrt[3]{20}-\sqrt[3]{4}\cdot\sqrt[3]{16}$

$=\sqrt[3]{125}+\sqrt[3]{100}+\sqrt[3]{80}-\sqrt[3]{100}$
$\quad-\sqrt[3]{80}-\sqrt[3]{64}$

$=\sqrt[3]{125}-\sqrt[3]{64}$

$=5-4=1$

89. $r = \dfrac{-h + \sqrt{h^2 + 0.64S}}{2}$

Substitute 12 for h and 400 for S.

$$r = \frac{-12 + \sqrt{12^2 + 0.64(400)}}{2} = \frac{-12 + \sqrt{144 + 256}}{2} = \frac{-12 + \sqrt{400}}{2} = \frac{-12 + 20}{2} = \frac{8}{2} = 4$$

The radius should be 4 inches.

91. $6(5 + 3x) = (6)(5) + (6)(3x) = 30 + 18x$

93.
$$\begin{aligned}
&\left(2 + \sqrt{10} + 5\sqrt{2}\right)\left(3\sqrt{10} - 3\sqrt{2}\right) \\
&= 2\sqrt{10}\left(3\sqrt{10}\right) + 2\sqrt{10}\left(-3\sqrt{2}\right) \\
&\quad + 5\sqrt{2}\left(3\sqrt{10}\right) + 5\sqrt{2}\left(-3\sqrt{2}\right) \qquad \text{FOIL} \\
&= 6 \cdot 10 - 6\sqrt{20} + 15\sqrt{20} - 15 \cdot 2 \\
&= 60 + 9\sqrt{20} - 30 \\
&= 30 + 9\sqrt{4} \cdot \sqrt{5} \\
&= 30 + 9\left(2\sqrt{5}\right) \\
&= 30 + 18\sqrt{5}
\end{aligned}$$

95. In the expression $30 + 18x$, make the first term $30x$, so that $30x + 18x = 48x$.

In the expression $30 + 18\sqrt{5}$, make the first term $30\sqrt{5}$, so that $30\sqrt{5} + 18\sqrt{5} = 48\sqrt{5}$.

Summary Exercises Applying Operations with Radicals

1. $5\sqrt{10} - 8\sqrt{10} = (5 - 8)\sqrt{10} = -3\sqrt{10}$

2. $\sqrt{5}\left(\sqrt{5} - \sqrt{3}\right) = \sqrt{5} \cdot \sqrt{5} - \sqrt{5} \cdot \sqrt{3} = 5 - \sqrt{15}$

3.
$$\begin{aligned}
&\left(1 + \sqrt{3}\right)\left(2 - \sqrt{6}\right) \\
&= 1 \cdot 2 - 1 \cdot \sqrt{6} + 2 \cdot \sqrt{3} - \sqrt{3} \cdot \sqrt{6} \\
&= 2 - \sqrt{6} + 2\sqrt{3} - \sqrt{18} \\
&= 2 - \sqrt{6} + 2\sqrt{3} - \sqrt{9 \cdot 2} \\
&= 2 - \sqrt{6} + 2\sqrt{3} - 3\sqrt{2}
\end{aligned}$$

4.
$$\begin{aligned}
&\sqrt{98} - \sqrt{72} + \sqrt{50} \\
&= \sqrt{49} \cdot \sqrt{2} - \sqrt{36} \cdot \sqrt{2} + \sqrt{25} \cdot \sqrt{2} \\
&= 7\sqrt{2} - 6\sqrt{2} + 5\sqrt{2} \\
&= (7 - 6 + 5)\sqrt{2} \\
&= 6\sqrt{2}
\end{aligned}$$

5.
$$\begin{aligned}
&\left(3\sqrt{5} - 2\sqrt{7}\right)^2 \\
&= \left(3\sqrt{5}\right)^2 - 2\left(3\sqrt{5}\right)\left(2\sqrt{7}\right) + \left(2\sqrt{7}\right)^2 \\
&= 3^2\left(\sqrt{5}\right)^2 - 2 \cdot 3 \cdot 2 \cdot \sqrt{5} \cdot \sqrt{7} + 2^2\left(\sqrt{7}\right)^2 \\
&= 9 \cdot 5 - 12\sqrt{35} + 4 \cdot 7 \\
&= 45 - 12\sqrt{35} + 28 \\
&= 73 - 12\sqrt{35}
\end{aligned}$$

6. $\dfrac{3}{\sqrt{6}} = \dfrac{3}{\sqrt{6}} \cdot \dfrac{\sqrt{6}}{\sqrt{6}} = \dfrac{3\sqrt{6}}{6} = \dfrac{\sqrt{6}}{2}$

7.
$$\begin{aligned}
\frac{1 + \sqrt{2}}{1 - \sqrt{2}} &= \frac{1 + \sqrt{2}}{1 - \sqrt{2}} \cdot \frac{1 + \sqrt{2}}{1 + \sqrt{2}} \\
&= \frac{1 + \sqrt{2} + \sqrt{2} + \sqrt{2} \cdot \sqrt{2}}{1^2 - \left(\sqrt{2}\right)^2} \\
&= \frac{1 + 2\sqrt{2} + 2}{1 - 2} \\
&= \frac{3 + 2\sqrt{2}}{-1} = -3 - 2\sqrt{2}
\end{aligned}$$

8. $$\frac{8}{\sqrt{7}-\sqrt{5}}=\frac{8}{\sqrt{7}-\sqrt{5}}\cdot\frac{\sqrt{7}+\sqrt{5}}{\sqrt{7}+\sqrt{5}}$$
$$=\frac{8\left(\sqrt{7}+\sqrt{5}\right)}{\left(\sqrt{7}\right)^2-\left(\sqrt{5}\right)^2}$$
$$=\frac{8\left(\sqrt{7}+\sqrt{5}\right)}{7-5}$$
$$=\frac{8\left(\sqrt{7}+\sqrt{5}\right)}{2}$$
$$=4\left(\sqrt{7}+\sqrt{5}\right)=4\sqrt{7}+4\sqrt{5}$$

9. $$\left(\sqrt{3}+6\right)\left(\sqrt{3}-6\right)=\left(\sqrt{3}\right)^2-6^2$$
$$=3-36$$
$$=-33$$

10. $$\frac{1}{\sqrt{t}+\sqrt{3}}=\frac{1}{\sqrt{t}+\sqrt{3}}\cdot\frac{\sqrt{t}-\sqrt{3}}{\sqrt{t}-\sqrt{3}}$$
$$=\frac{\sqrt{t}-\sqrt{3}}{\left(\sqrt{t}\right)^2-\left(\sqrt{3}\right)^2}$$
$$=\frac{\sqrt{t}-\sqrt{3}}{t-3}$$

11. $$\sqrt[3]{8x^3y^5z^6}=\sqrt[3]{8x^3y^3z^6}\cdot\sqrt[3]{y^2}$$
$$=2xyz^2\sqrt[3]{y^2}$$

12. $$\frac{12}{\sqrt[3]{9}}=\frac{12}{\sqrt[3]{9}}\cdot\frac{\sqrt[3]{3}}{\sqrt[3]{3}}$$
$$=\frac{12\sqrt[3]{3}}{\sqrt[3]{27}}$$
$$=\frac{12\sqrt[3]{3}}{3}=4\sqrt[3]{3}$$

13. $$\frac{5}{\sqrt{6}-1}=\frac{5}{\sqrt{6}-1}\cdot\frac{\sqrt{6}+1}{\sqrt{6}+1}$$
$$=\frac{5\left(\sqrt{6}+1\right)}{\left(\sqrt{6}\right)^2-1^2}$$
$$=\frac{5\left(\sqrt{6}+1\right)}{6-1}$$
$$=\frac{5\left(\sqrt{6}+1\right)}{5}=\sqrt{6}+1$$

14. $$\sqrt{\frac{2}{3x}}=\frac{\sqrt{2}}{\sqrt{3x}}=\frac{\sqrt{2}}{\sqrt{3x}}\cdot\frac{\sqrt{3x}}{\sqrt{3x}}$$
$$=\frac{\sqrt{6x}}{3x}$$

15. $$\frac{6\sqrt{3}}{5\sqrt{12}}=\frac{6\sqrt{3}}{5\sqrt{4}\cdot\sqrt{3}}=\frac{6}{5\cdot 2}=\frac{3}{5}$$

16. $$\frac{8\sqrt{50}}{2\sqrt{25}}=\frac{8\sqrt{25}\cdot\sqrt{2}}{2\sqrt{25}}$$
$$=\frac{4\cdot 2\cdot\sqrt{2}}{2}=4\sqrt{2}$$

17. $$\frac{-4}{\sqrt[3]{4}}=\frac{-4\cdot\sqrt[3]{2}}{\sqrt[3]{4}\cdot\sqrt[3]{2}}$$
$$=\frac{-4\sqrt[3]{2}}{\sqrt[3]{8}}$$
$$=\frac{-4\sqrt[3]{2}}{2}=-2\sqrt[3]{2}$$

18. $$\frac{\sqrt{6}-\sqrt{5}}{\sqrt{6}+\sqrt{5}}=\frac{\sqrt{6}-\sqrt{5}}{\sqrt{6}+\sqrt{5}}\cdot\frac{\sqrt{6}-\sqrt{5}}{\sqrt{6}-\sqrt{5}}$$
$$=\frac{\left(\sqrt{6}\right)^2-2\cdot\sqrt{6}\cdot\sqrt{5}+\left(\sqrt{5}\right)^2}{\left(\sqrt{6}\right)^2-\left(\sqrt{5}\right)^2}$$
$$=\frac{6-2\sqrt{30}+5}{6-5}$$
$$=\frac{11-2\sqrt{30}}{1}=11-2\sqrt{30}$$

19. $$\sqrt{75x}-\sqrt{12x}=\sqrt{25\cdot 3x}-\sqrt{4\cdot 3x}$$
$$=5\sqrt{3x}-2\sqrt{3x}$$
$$=(5-2)\sqrt{3x}$$
$$=3\sqrt{3x}$$

20. $$\left(5+3\sqrt{3}\right)^2$$
$$=(5)^2+2\cdot 5\cdot 3\sqrt{3}+\left(3\sqrt{3}\right)^2$$
$$=25+30\sqrt{3}+9\cdot 3$$
$$=25+30\sqrt{3}+27$$
$$=52+30\sqrt{3}$$

21. $\sqrt[3]{\dfrac{16}{81}} = \dfrac{\sqrt[3]{16}}{\sqrt[3]{81}} = \dfrac{\sqrt[3]{8}\sqrt[3]{2}}{\sqrt[3]{27}\sqrt[3]{3}}$

$= \dfrac{2\sqrt[3]{2}}{3\sqrt[3]{3}} \cdot \dfrac{\sqrt[3]{9}}{\sqrt[3]{9}}$

$= \dfrac{2\sqrt[3]{18}}{3\sqrt[3]{27}}$

$= \dfrac{2\sqrt[3]{18}}{3 \cdot 3} = \dfrac{2\sqrt[3]{18}}{9}$

22. $\left(\sqrt{107} - \sqrt{106}\right)\left(\sqrt{107} + \sqrt{106}\right)$

$= \left(\sqrt{107}\right)^2 - \left(\sqrt{106}\right)^2$

$= 107 - 106 = 1$

23. $x\sqrt[4]{x^5} - 3\sqrt[4]{x^9} + x^2\sqrt[4]{x}$

$= x\sqrt[4]{x^4} \cdot \sqrt[4]{x} - 3\sqrt[4]{x^8} \cdot \sqrt[4]{x} + x^2\sqrt[4]{x}$

$= x \cdot x \cdot \sqrt[4]{x} - 3 \cdot x^2 \cdot \sqrt[4]{x} + x^2\sqrt[4]{x}$

$= \left(x^2 - 3x^2 + x^2\right)\sqrt[4]{x}$

$= -x^2\sqrt[4]{x}$

24. $\sqrt[3]{16t^2} - \sqrt[3]{54t^2} + \sqrt[3]{128t^2}$

$= \sqrt[3]{8} \cdot \sqrt[3]{2t^2} - \sqrt[3]{27} \cdot \sqrt[3]{2t^2} + \sqrt[3]{64} \cdot \sqrt[3]{2t^2}$

$= 2\sqrt[3]{2t^2} - 3\sqrt[3]{2t^2} + 4\sqrt[3]{2t^2}$

$= (2 - 3 + 4)\sqrt[3]{2t^2}$

$= 3\sqrt[3]{2t^2}$

25. $\sqrt{14} + \sqrt{5}$

These unlike radicals cannot be combined.

26. **(a)** $\sqrt{36} = 6$

(b) $x^2 = 36$

$x = -\sqrt{36}$ or $x = \sqrt{36}$

$x = -6$ or $x = 6$

The solution set is $\{-6, 6\}$.

27. **(a)** $\sqrt{81} = 9$

(b) $x^2 = 81$

$x = -\sqrt{81}$ or $x = \sqrt{81}$

$x = -9$ or $x = 9$

The solution set is $\{-9, 9\}$.

28. **(a)** $x^2 = 4$

$x = -\sqrt{4}$ or $x = \sqrt{4}$

$x = -2$ or $x = 2$

The solution set is $\{-2, 2\}$.

(b) $-\sqrt{4} = -\left(\sqrt{4}\right) = -2$

29. **(a)** $x^2 = 9$

$x = -\sqrt{9}$ or $x = \sqrt{9}$

$x = -3$ or $x = 3$

The solution set is $\{-3, 3\}$.

(b) $-\sqrt{9} = -\left(\sqrt{9}\right) = -3$

30. $x^2 = 25$

$x^2 - 25 = 0$

$(x+5)(x-5) = 0$

$x + 5 = 0$ or $x - 5 = 0$

$x = -5$ or $x = 5$

The solution set is $\{-5, 5\}$.

8.6 Solving Equations with Radicals

Now Try Exercises

N1. $\sqrt{x-5} = 6$

$\left(\sqrt{x-5}\right)^2 = 6^2$ Square each side.

$x - 5 = 36$

$x = 41$ Add 5.

Check $x = 41$: $\sqrt{x-5} = 6$

$\sqrt{41-5} \stackrel{?}{=} 6$ Let $x = 41$.

$\sqrt{36} \stackrel{?}{=} 6$

$6 = 6$ True

The solution set is $\{41\}$.

N2. $4\sqrt{x} = \sqrt{10x+12}$

$\left(4\sqrt{x}\right)^2 = \left(\sqrt{10x+12}\right)^2$ Square each side.

$16x = 10x + 12$

$6x = 12$ Subtract $10x$.

$x = 2$ Divide by 6.

Check $x = 2$:

$$4\sqrt{x} = \sqrt{10x+12}$$

$$4\sqrt{2} \stackrel{?}{=} \sqrt{10(2)+12} \quad \text{Let } x = 2.$$

$$4\sqrt{2} \stackrel{?}{=} \sqrt{32}$$

$$4\sqrt{2} \stackrel{?}{=} \sqrt{16 \cdot 2}$$

$$4\sqrt{2} \stackrel{?}{=} 4\sqrt{2} \quad \text{True}$$

The solution set is $\{2\}$.

N3. $\sqrt{x} = -6$

$$\left(\sqrt{x}\right)^2 = (-6)^2 \quad \text{Square each side.}$$

$$x = 36$$

Check $x = 36$: $\sqrt{x} = -6$

$$\sqrt{36} \stackrel{?}{=} -6 \quad \text{Let } x = 36.$$

$$6 = -6 \quad \text{False}$$

Because the statement $6 = -6$ is false, the number 36 is *not* a solution (36 is called an extraneous solution). The equation has no solution. The solution set is $\varnothing$.

Another approach: Because $\sqrt{x}$ represents the *principal* or *nonnegative* square root of x, it cannot equal –6.

N4. $t = \sqrt{t^2+3t+9}$

$$t^2 = \left(\sqrt{t^2+3t+9}\right)^2 \quad \text{Square each side.}$$

$$t^2 = t^2+3t+9$$

$$0 = 3t+9 \quad \text{Subtract } t^2.$$

$$-3t = 9 \quad \text{Subtract } 3t.$$

$$t = -3 \quad \text{Divide by } -3.$$

Check $t = -3$:

$$t = \sqrt{t^2+3t+9}$$

$$-3 \stackrel{?}{=} \sqrt{(-3)^2+3(-3)+9} \quad \text{Let } t = -3.$$

$$-3 \stackrel{?}{=} \sqrt{9-9+9}$$

$$-3 \stackrel{?}{=} \sqrt{9}$$

$$-3 = 3 \quad \text{False}$$

The only potential solution does not check, so -3 is an extraneous solution, and the equation has no solution. The solution set is $\varnothing$.

N5. $\sqrt{4x+1} = x-5$

$$\left(\sqrt{4x+1}\right)^2 = (x-5)^2$$

$$4x+1 = x^2-10x+25$$

$$0 = x^2-14x+24$$

$$0 = (x-2)(x-12) \quad \text{Factor.}$$

$$x-2=0 \quad \text{or} \quad x-2=0$$

$$x=2 \quad \text{or} \quad x=12 \quad \text{Solve.}$$

Check $x = 2$:

$$\sqrt{4x+1} = x-5$$

$$\sqrt{4(2)+1} \stackrel{?}{=} 2-5 \quad \text{Let } x = 2.$$

$$\sqrt{9} \stackrel{?}{=} -3$$

$$3 = -3 \quad \text{False}$$

Check $x = 12$:

$$\sqrt{4x+1} = x-5$$

$$\sqrt{4(12)+1} \stackrel{?}{=} 12-5 \quad \text{Let } x = 12.$$

$$\sqrt{49} \stackrel{?}{=} 7$$

$$7 = 7 \quad \text{True}$$

The number 2 does not satisfy the original equation, so it is extraneous. The only solution is 12. The solution set is $\{12\}$.

N6. $\sqrt{27x}-3 = 2x$

$$\sqrt{27x} = 2x+3$$

$$\left(\sqrt{27x}\right)^2 = (2x+3)^2$$

$$27x = 4x^2+12x+9$$

$$0 = 4x^2-15x+9$$

$$0 = (4x-3)(x-3)$$

$$4x-3=0 \quad \text{or} \quad x-3=0$$

$$x = \frac{3}{4} \quad \text{or} \quad x = 3$$

Check $x = \frac{3}{4}$: $\sqrt{\frac{81}{4}}-3 \stackrel{?}{=} 2\left(\frac{3}{4}\right)$

$$\frac{9}{2}-3 = \frac{3}{2} \quad \text{True}$$

Check $x = 3$: $\sqrt{81}-3 \stackrel{?}{=} 2(3)$

$$9-3 = 6 \quad \text{True}$$

The solution set is $\left\{\frac{3}{4}, 3\right\}$.

N7. $\sqrt{x}+2=\sqrt{x+8}$

$\left(\sqrt{x}+2\right)^2=\left(\sqrt{x+8}\right)^2$

$x+4\sqrt{x}+4=x+8$

$4\sqrt{x}=4$

$\sqrt{x}=1$

$\left(\sqrt{x}\right)^2=1^2$

$x=1$

Check $x=1$:

$\sqrt{x}+2=\sqrt{x+8}$

$\sqrt{1}+2\overset{?}{=}\sqrt{1+8}$ Let $x=1$.

$1+2\overset{?}{=}\sqrt{9}$

$3=3$ True

The solution set is $\{1\}$.

N8. (a) $\sqrt[3]{8x-3}=\sqrt[3]{4x}$

$\left(\sqrt[3]{8x-3}\right)^3=\left(\sqrt[3]{4x}\right)^3$ Cube each side.

$8x-3=4x$ $\left(\sqrt[3]{a}\right)^3=a$

$4x=3$

$x=\frac{3}{4}$ Divide by 4.

Check $x=\frac{3}{4}$: $\sqrt[3]{8\left(\frac{3}{4}\right)-3}\overset{?}{=}\sqrt[3]{4\left(\frac{3}{4}\right)}$

$\sqrt[3]{3}=\sqrt[3]{3}$ True

The solution set is $\left\{\frac{3}{4}\right\}$.

(b) $\sqrt[3]{2x^2}=\sqrt[3]{10x-12}$

$\left(\sqrt[3]{2x^2}\right)^3=\left(\sqrt[3]{10x-12}\right)^3$

$2x^2=10x-12$ $\left(\sqrt[3]{a}\right)^3=a$

$2x^2-10x+12=0$

$2\left(x^2-5x+6\right)=0$ Factor.

$(x-2)(x-3)=0$ Factor.

$x-2=0$ or $x-3=0$

$x=2$ or $x=3$

Check $x=2$:

$\sqrt[3]{2(2)^2}\overset{?}{=}\sqrt[3]{10(2)-12}$

$\sqrt[3]{8}=\sqrt[3]{8}$ True

Check $x=3$:

$\sqrt[3]{2(3)^2}\overset{?}{=}\sqrt[3]{10(3)-12}$

$\sqrt[3]{18}=\sqrt[3]{18}$ True

The solution set is $\{2, 3\}$.

Exercises

1. To solve an equation involving a radical, such as $\sqrt{2x-1}=5,$ use the squaring property of equality. This property says that if each side of an equation is squared, all solutions of the original equation are among the solutions of the squared equation.

3. $\sqrt{x}=7$

Use the *squaring property of equality* to square each side of the equation.

$(\sqrt{x})^2=7^2$

$x=49$

Now check this proposed solution in the original equation.

Check $x=49$: $\sqrt{x}=7$

$\sqrt{49}\overset{?}{=}7$ Let $x=49$.

$7=7$ True

Since this statement is true, the solution set of the original equation is $\{49\}$.

5. $\sqrt{x+2}=3$

$\left(\sqrt{x+2}\right)^2=3^2$ Square each side.

$x+2=9$

$x=7$

Check $x=7$:

$\sqrt{x+2}=3$

$\sqrt{7+2}\overset{?}{=}3$ Let $x=7$.

$\sqrt{9}\overset{?}{=}3$

$3=3$ True

Since this statement is true, the solution set of the original equation is $\{7\}$.

7. $\sqrt{r-4}=9$

$\left(\sqrt{r-4}\right)^2=9^2$ Square each side.

$r-4=81$

$r=85$

Check $r=85$:

$\sqrt{r-4}=9$

$\sqrt{85-4}\overset{?}{=}9$ Let $r=85$.

$\sqrt{81}\overset{?}{=}9$

$9=9$ True

Since this statement is true, the solution set of the original equation is $\{85\}$.

9. $\sqrt{4-t}=7$

$\left(\sqrt{4-t}\right)^2=7^2$ Square each side.

$4-t=49$

$-t=45$

$t=-45$

Check $t=-45$:

$\sqrt{4-t}=7$

$\sqrt{4-(-45)}\overset{?}{=}7$ Let $t=-45$.

$\sqrt{49}\overset{?}{=}7$

$7=7$ True

Since this statement is true, the solution set of the original equation is $\{-45\}$.

11. $\sqrt{2t+3}=0$

$\left(\sqrt{2t+3}\right)^2=0^2$ Square each side.

$2t+3=0$

$2t=-3$

$t=-\frac{3}{2}$

Check $t=-\frac{3}{2}$:

$\sqrt{2t+3}=0$

$\sqrt{2\left(-\frac{3}{2}\right)+3}\overset{?}{=}0$ Let $t=-\frac{3}{2}$.

$\sqrt{-3+3}\overset{?}{=}0$

$\sqrt{0}\overset{?}{=}0$

$0=0$ True

Since this statement is true, the solution set of the original equation is $\left\{-\frac{3}{2}\right\}$.

13. $\sqrt{t}=-5$

Because $\sqrt{t}$ represents the *principal* or *nonnegative* square root of t, it cannot equal -5. Thus, the solution set is $\varnothing$.

15. $\sqrt{w}-4=7$

Add 4 to both sides of the equation before squaring.

$\sqrt{w}=11$

$\left(\sqrt{w}\right)^2=(11)^2$

$w=121$

Check $w=121$:

$\sqrt{w}-4=7$

$\sqrt{121}-4\overset{?}{=}7$ Let $w=121$.

$11-4\overset{?}{=}7$

$7=7$ True

Since this statement is true, the solution set of the original equation is $\{121\}$.

17. $\sqrt{10x-8}=3\sqrt{x}$

$\left(\sqrt{10x-8}\right)^2=\left(3\sqrt{x}\right)^2$

$10x-8=(3)^2\left(\sqrt{x}\right)^2$

$10x-8=9x$

$x=8$

Check $x=8$:

$\sqrt{10x-8}=3\sqrt{x}$

$\sqrt{10(8)-8}\overset{?}{=}3\sqrt{8}$ Let $x=8$.

$\sqrt{72}\overset{?}{=}3\sqrt{8}$

$\sqrt{36\cdot 2}\overset{?}{=}3\cdot 2\sqrt{2}$

$6\sqrt{2}=6\sqrt{2}$ True

Since this statement is true, the solution set of the original equation is $\{8\}$.

19. $5\sqrt{x} = \sqrt{10x+15}$

$\left(5\sqrt{x}\right)^2 = \left(\sqrt{10x+15}\right)^2$

$25x = 10x+15$

$15x = 15$

$x = 1$

Check $x = 1$:

$5\sqrt{x} = \sqrt{10x+15}$

$5\sqrt{1} \stackrel{?}{=} \sqrt{10 \cdot 1+15}$ Let $x = 1$.

$5 \cdot 1 \stackrel{?}{=} \sqrt{25}$

$5 = 5$ True

Since this statement is true, the solution set of the original equation is $\{1\}$.

21. $\sqrt{3x-5} = \sqrt{2x+1}$

$\left(\sqrt{3x-5}\right)^2 = \left(\sqrt{2x+1}\right)^2$

$3x-5 = 2x+1$

$x = 6$

Check $x = 6$:

$\sqrt{3x-5} = \sqrt{2x+1}$

$\sqrt{3(6)-5} \stackrel{?}{=} \sqrt{2(6)+1}$ Let $x = 6$.

$\sqrt{13} = \sqrt{13}$ True

Since this statement is true, the solution set of the original equation is $\{6\}$.

23. $k = \sqrt{k^2-5k-15}$

$(k)^2 = \left(\sqrt{k^2-5k-15}\right)^2$

$k^2 = k^2-5k-15$

$0 = -5k-15$

$5k = -15$

$k = -3$

Check $k = -3$:

$k = \sqrt{k^2-5k-15}$

$-3 \stackrel{?}{=} \sqrt{(-3)^2-5(-3)-15}$ Let $k = -3$.

$-3 \stackrel{?}{=} \sqrt{9+15-15}$

$-3 \stackrel{?}{=} \sqrt{9}$

$-3 \stackrel{?}{=} 3$ False

Since this statement is false, the solution set of the original equation is $\varnothing$.

25. $7x = \sqrt{49x^2+2x-10}$

$(7x)^2 = \left(\sqrt{49x^2+2x-10}\right)^2$

$49x^2 = 49x^2+2x-10$

$0 = 2x - 10$

$10 = 2x$

$5 = x$

Check $x = 5$:

$7x = \sqrt{49x^2+2x-10}$

$7(5) \stackrel{?}{=} \sqrt{49(5)^2+2(5)-10}$ Let $x = 5$.

$35 \stackrel{?}{=} \sqrt{1225+10-10}$

$35 \stackrel{?}{=} \sqrt{1225}$

$35 = 35$ True

Since this statement is true, the solution set of the original equation is $\{5\}$.

27. $\sqrt{2x+2} = \sqrt{3x-5}$

$\left(\sqrt{2x+2}\right)^2 = \left(\sqrt{3x-5}\right)^2$

$2x+2 = x-5$

$7 = x$

Check $x = 7$: $4 = 4$ True

The solution set is $\{7\}$.

29. $\sqrt{5x-5} = \sqrt{4x+1}$

$\left(\sqrt{5x-5}\right)^2 = \left(\sqrt{4x+1}\right)^2$

$5x-5 = 4x+1$

$x = 6$

Check $x = 6$: $5 = 5$ True

The solution set is $\{6\}$.

31. $\sqrt{3x-8} = -2$

$\left(\sqrt{3x-8}\right)^2 = (-2)^2$

$3x-8 = 4$

$3x = 12$

$x = 4$

Check $x = 4$:

$$\sqrt{3x-8} = -2$$

$$\sqrt{3(4)-8} \stackrel{?}{=} (-2)^2 \quad \text{Let } x = 4.$$

$$\sqrt{12-8} \stackrel{?}{=} -2$$

$$\sqrt{4} \stackrel{?}{=} -2$$

$$2 = -2 \qquad \text{False}$$

Since this statement is false, the solution set of the original equation is $\varnothing$. (Note that, in the original equation, the result of a square root cannot equal a negative number.)

33. The error occurs in the first step, when both sides are squared. When the left side is squared, the result should be $x-1$, not $-(x-1)$. Thus, we have

$$x-1 = 16$$

$$x = 17.$$

The correct solution set is $\{17\}$.

35.

$$\sqrt{5x+11} = x+3$$

$$\left(\sqrt{5x+11}\right)^2 = (x+3)^2$$

$$5x+11 = x^2+6x+9$$

$$0 = x^2+x-2$$

$$0 = (x+2)(x-1)$$

$x = -2$ or $x = 1$

Check $x = -2$: $1 = 1$ True

Check $x = 1$: $4 = 4$ True

The solution set is $\{-2, 1\}$.

37.

$$\sqrt{2x+1} = x-7$$

$$\left(\sqrt{2x+1}\right)^2 = (x-7)^2$$

$$2x+1 = x^2-14x+49$$

$$0 = x^2-16x+48$$

$$0 = (x-4)(x-12)$$

$x = 4$ or $x = 12$

Check $x = 4$: $3 = -3$ False

Check $x = 12$: $5 = 5$ True

The solution set is $\{12\}$.

39.

$$\sqrt{x+2} - 2 = x$$

$$\sqrt{x+2} = x+2$$

$$\left(\sqrt{x+2}\right)^2 = (x+2)^2$$

$$x+2 = x^2+4x+4$$

$$0 = x^2+3x+2$$

$$0 = (x+2)(x+1)$$

$x+2 = 0$ or $x+1 = 0$

$x = -2$ or $x = -1$

Check $x = -2$: $\sqrt{0} - 2 = -2$ True

Check $x = -1$: $\sqrt{1} - 2 = -1$ True

The solution set is $\{-2, -1\}$.

41.

$$\sqrt{12x+12} + 10 = 2x$$

$$\sqrt{12x+12} = 2x-10$$

$$\left(\sqrt{12x+12}\right)^2 = (2x-10)^2$$

$$12x+12 = 4x^2-40x+100$$

$$0 = 4x^2-52x+88$$

$$0 = 4(x-2)(x-11)$$

$x = 2$ or $x = 11$

Check $x = 2$: $6+10 = 4$ False

Check $x = 11$: $12+10 = 22$ True

The solution set is $\{11\}$.

Alternative solution: Since $\sqrt{12x+12} = \sqrt{4}\sqrt{3x+3} = 2\sqrt{3x+3}$, all the terms in the original equation are divisible by 2. Dividing by 2 gives us $\sqrt{3x+3} + 5 = x$, which is easier to solve and gives us the same solution set.

43.

$$\sqrt{6x+7} - 1 = x+1$$

$$\sqrt{6x+7} = x+2$$

$$\left(\sqrt{6x+7}\right)^2 = (x+2)^2$$

$$6x+7 = x^2+4x+4$$

$$0 = x^2-2x-3$$

$$0 = (x+1)(x-3)$$

$x = -1$ or $x = 3$

Check $x = -1$: $0 = 0$ True

Check $x = 3$: $4 = 4$ True

The solution set is $\{-1, 3\}$.

45. $2\sqrt{x+7}=x-1$

$$\left(2\sqrt{x+7}\right)^2=(x-1)^2$$
$$4(x+7)=x^2-2x+1$$
$$4x+28=x^2-2x+1$$
$$0=x^2-6x-27$$
$$0=(x-9)(x+3)$$
$x=-3$ or $x=9$

Check $x=-3$: $4=-4$ False

Check $x=9$: $8=8$ True

The solution set is $\{9\}$.

47. $\sqrt{2x}+4=x$

$$\sqrt{2x}=x-4$$
$$\left(\sqrt{2x}\right)^2=(x-4)^2$$
$$2x=x^2-8x+16$$
$$0=x^2-10x+16$$
$$0=(x-2)(x-8)$$
$x=2$ or $x=8$

Check $x=2$: $6=2$ False

Check $x=8$: $8=8$ True

The solution set is $\{8\}$.

49. $\sqrt{x}+9=x+3$

$$\sqrt{x}=x-6$$
$$\left(\sqrt{x}\right)^2=(x-6)^2$$
$$x=x^2-12x+36$$
$$0=x^2-13x+36$$
$$0=(x-4)(x-9)$$
$x=4$ or $x=9$

Check $x=4$: $11=7$ False

Check $x=9$: $12=12$ True

The solution set is $\{9\}$.

51. $3\sqrt{x-2}=x-2$

$$\left(3\sqrt{x-2}\right)^2=(x-2)^2$$
$$9(x-2)=x^2-4x+4$$
$$9x-18=x^2-4x+4$$
$$0=x^2-13x+22$$
$$0=(x-2)(x-11)$$
$x=2$ or $x=11$

Check $x=2$: $0=0$ True

Check $x=11$: $9=9$ True

The solution set is $\{2,11\}$.

53. $\sqrt{3x+3}+\sqrt{x+2}=5$

Rewrite the equation so that there is one radical on each side.

$\sqrt{3x+3}=5-\sqrt{x+2}$

Square both sides. On the right-hand side, use the formula for the square of a binomial.

$$\left(\sqrt{3x+3}\right)^2=(5-\sqrt{x+2})^2$$
$$3x+3=5^2-2\cdot5\cdot\sqrt{x+2}+\left(\sqrt{x+2}\right)^2$$
$$3x+3=25-10\sqrt{x+2}+x+2$$
$$3x+3=27+x-10\sqrt{x+2}$$
$$2x-24=-10\sqrt{x+2}$$
$$x-12=-5\sqrt{x+2} \qquad \text{Divide by 2.}$$

We still have a radical on the right, so we must square both sides again.

$$(x-12)^2=(-5\sqrt{x+2})^2$$
$$x^2-24x+144=25(x+2)$$
$$x^2-24x+144=25x+50$$
$$x^2-49x+94=0$$
$$(x-2)(x-47)=0$$
$x=2$ or $x=47$

Check $x=2$: $3+2=5$ True

Check $x=47$: $12+7=5$ False

The solution set is $\{2\}$.

55. Start by squaring both sides of the equation.

$$\sqrt{x}+6=\sqrt{x+72}$$
$$\left(\sqrt{x}+6\right)^2=\left(\sqrt{x+72}\right)^2$$
$$\left(\sqrt{x}\right)^2+2\cdot6\cdot\sqrt{x}+6^2=x+72$$

Now solve for x.

$$x+12\sqrt{x}+36=x+72$$
$$12\sqrt{x}=36$$
$$\sqrt{x}=3$$
$$\left(\sqrt{x}\right)^2=3^2$$
$$x=9$$

Check $x=9$: $3+6=9$ True

The solution set is $\{9\}$.

57. $\sqrt{3x+4}-\sqrt{2x-4}=2$

$$\sqrt{3x+4}=2+\sqrt{2x-4}$$
$$\left(\sqrt{3x+4}\right)^2=\left(2+\sqrt{2x-4}\right)^2$$
$$3x+4=4+4\sqrt{2x-4}+2x-4$$
$$x+4=4\sqrt{2x-4}$$
$$(x+4)^2=\left(4\sqrt{2x-4}\right)^2$$
$$x^2+8x+16=16(2x-4)$$
$$x^2+8x+16=32x-64$$
$$x^2-24x+80=0$$
$$(x-4)(x-20)=0$$
$$x=4 \quad \text{or} \quad x=20$$

Check $x=4$: $4-2=2$ True

Check $x=20$: $8-6=2$ True

The solution set is $\{4, 20\}$.

59. $\sqrt{2x+11}+\sqrt{x+6}=2$

$$\sqrt{2x+11}=2-\sqrt{x+6}$$
$$\left(\sqrt{2x+11}\right)^2=\left(2-\sqrt{x+6}\right)^2$$
$$2x+11=4-4\sqrt{x+6}+x+6$$
$$x+1=-4\sqrt{x+6}$$
$$(x+1)^2=\left(-4\sqrt{x+6}\right)^2$$
$$x^2+2x+1=16(x+6)$$
$$x^2+2x+1=16x+96$$
$$x^2-14x-95=0$$
$$(x+5)(x-19)=0$$
$$x=-5 \quad \text{or} \quad x=19$$

Check $x=-5$: $1+1=2$ True

Check $x=19$: $7+5=2$ False

The solution set is $\{-5\}$.

61. $\sqrt[3]{2x}=\sqrt[3]{5x+2}$

$$\left(\sqrt[3]{2x}\right)^3=\left(\sqrt[3]{5x+2}\right)^3$$
$$2x=5x+2$$
$$-3x=2$$
$$x=-\frac{2}{3}$$

Check $x=-\frac{2}{3}$: $\sqrt[3]{-\frac{4}{3}}=\sqrt[3]{-\frac{10}{3}+2}$ True

The solution set is $\left\{-\frac{2}{3}\right\}$.

63. $\sqrt[3]{x^2}=\sqrt[3]{8+7x}$

$$\left(\sqrt[3]{x^2}\right)^3=\left(\sqrt[3]{8+7x}\right)^3$$
$$x^2=8+7x$$
$$x^2-7x-8=0$$
$$(x+1)(x-8)=0$$
$$x=-1 \quad \text{or} \quad x=8$$

Check $x=-1$: $\sqrt[3]{1}=\sqrt[3]{8-7}$ True

Check $x=8$: $\sqrt[3]{64}=\sqrt[3]{8+56}$ True

The solution set is $\{-1, 8\}$.

65. $\sqrt[3]{3x^2-9x+8}=\sqrt[3]{x}$

$$\left(\sqrt[3]{3x^2-9x+8}\right)^3=\left(\sqrt[3]{x}\right)^3$$
$$3x^2-9x+8=x$$
$$3x^2-10x+8=0$$
$$(3x-4)(x-2)=0$$
$$x=\frac{4}{3} \quad \text{or} \quad x=2$$

Check $x=\frac{4}{3}$: $\sqrt[3]{\frac{16}{3}-12+8}\stackrel{?}{=}\sqrt[3]{\frac{4}{3}}$

$$\sqrt[3]{\frac{16}{3}-\frac{12}{3}}=\sqrt[3]{\frac{4}{3}} \quad \text{True}$$

Check $x=2$: $\sqrt[3]{12-18+8}=\sqrt[3]{2}$ True

The solution set is $\left\{\frac{4}{3}, 2\right\}$.

67. $\sqrt[3]{x^2+24x}=3$

$\left(\sqrt[4]{x^2+2x}\right)^4=3^4$

$x^2+24x=81$

$x^2+24x-81=0$

$(x+27)(x-3)=0$

$x=-27 \quad \text{or} \quad x=3$

Check $x=-27$: $\sqrt[4]{729-648}=\sqrt[4]{81}=3$ True

Check $x=3$: $\sqrt[4]{9+72}=\sqrt[4]{81}=3$ True

The solution set is $\{-27, 3\}$.

69. Let $x=$ the number.
"The square root of the sum of a number and 4 is 5" translates to

$\sqrt{x+4}=5.$

$\left(\sqrt{x+4}\right)^2=5^2$

$x+4=25$

$x=21$

Check $x=21$: $\sqrt{25}=5$ True

The number is 21.

71. Let $x=$ the number.
"Three times the square root of 2 equals the square root of the sum of some number and 10" translates to

$3\sqrt{2}=\sqrt{x+10}.$

$\left(3\sqrt{2}\right)^2=\left(\sqrt{x+10}\right)^2$

$9\cdot 2=x+10$

$18=x+10$

$8=x$

Check $x=8$: $3\sqrt{2}=\sqrt{18}$ True

The number is 8.

73. $s=30\sqrt{\dfrac{a}{p}}$

Use a calculator and round answers to the nearest tenth.

(a) $s=30\sqrt{\dfrac{862}{156}}$

≈ 70.5 miles per hour

(b) $s=30\sqrt{\dfrac{382}{96}}$

≈ 59.8 miles per hour

(c) $s=30\sqrt{\dfrac{84}{26}}$

≈ 53.9 miles per hour

75. Let $S=$ sight distance (in kilometers) and $h=$ height of the structure (in kilometers). The equation given is then $S=111.7\sqrt{h}.$
The height of the London Eye is 135 meters, or 0.135 kilometer.

$S=111.7\sqrt{0.135}$

≈ 41.041201 km.

To convert to miles, we multiply by 0.621371 to get 25.502 miles. So yes, the passengers on the London Eye can see Windsor Castle, which is approximately 26 miles away.

Chapter 8 Review Exercises

1. The square roots of 49 are -7 and 7 because $(-7)^2=49$ and $7^2=49$.

2. The square roots of 81 are -9 and 9 because $(-9)^2=81$ and $9^2=81$.

3. The square roots of 196 are -14 and 14 because $(-14)^2=196$ and $14^2=196$.

4. The square roots of 121 are -11 and 11 because $(-11)^2=121$ and $11^2=121$.

5. The square roots of 225 are -15 and 15 because $(-15)^2=225$ and $15^2=225$.

6. The square roots of 729 are -27 and 27 because $(-27)^2=729$ and $27^2=729$.

7. $\sqrt{16}=4$ because $4^2=16$.

8. $-\sqrt{36}$ represents the negative square root of 36. Since $6\cdot 6=36, -\sqrt{36}=-6$.

9. $\sqrt[3]{1000}=10$ because $10^3=1000$.

10. $\sqrt[4]{81}=3$ because 3 is positive and $3^4=81$.

11. $\sqrt{-8100}$ is not a real number.

12. $-\sqrt{4225}$ represents the negative square root of 4225. Since $65\cdot 65=4225, -\sqrt{4225}=-65$.

13. $\sqrt{\dfrac{49}{36}}=\dfrac{\sqrt{49}}{\sqrt{36}}=\dfrac{7}{6}$

14. $\sqrt{0.25} = 0.5$ because $0.5^2 = 0.25$.

15. Let $(x_1, y_1) = (-3, -5)$ and $(x_2, y_2) = (4, -3)$. Use the distance formula.

$$\begin{aligned} d &= \sqrt{(x_2 - x_1)^2 + (y_2 - y_1)^2} \\ &= \sqrt{[4-(-3)]^2 + [-3-(-5)]^2} \\ &= \sqrt{7^2 + 2^2} \\ &= \sqrt{49+4} = \sqrt{53} \end{aligned}$$

16. Use the Pythagorean theorem with $a = 15, b = x,$ and $c = 17$.

$$\begin{aligned} c^2 &= a^2 + b^2 \\ 17^2 &= 15^2 + x^2 \\ 289 &= 225 + x^2 \\ 64 &= x^2 \\ x &= \sqrt{64} = 8 \end{aligned}$$

17. Use the Pythagorean theorem with $a = 30.4$ cm and $b = 37.5$ cm.

$$\begin{aligned} c^2 &= a^2 + b^2 \\ &= (30.4)^2 + (37.5)^2 \\ &= 924.16 + 1406.25 \\ &= 2330.41 \\ c &= \sqrt{2330.41} \approx 48.3 \text{ cm} \end{aligned}$$

18. $\sqrt{111}$

This number is *irrational* because 111 is not a perfect square.

$\sqrt{111} \approx 10.536$

19. $-\sqrt{25}$

This number is *rational* because 25 is a perfect square.

$-\sqrt{25} = -5$

20. $\sqrt{-4}$

This is not a real number.

21. $$\begin{aligned} \sqrt{48} &= \sqrt{16 \cdot 3} \\ &= \sqrt{16} \cdot \sqrt{3} \\ &= 4\sqrt{3} \end{aligned}$$

22. $-\sqrt{27} = -\sqrt{9 \cdot 3} = -\sqrt{9} \cdot \sqrt{3} = -3\sqrt{3}$

23. $\sqrt{160} = \sqrt{16 \cdot 10} = \sqrt{16} \cdot \sqrt{10} = 4\sqrt{10}$

24. $\sqrt[3]{16} = \sqrt[3]{8 \cdot 2} = \sqrt[3]{8} \cdot \sqrt[3]{2} = 2\sqrt[3]{2}$

25. $\sqrt[3]{375} = \sqrt[3]{125 \cdot 3} = \sqrt[3]{125} \cdot \sqrt[3]{3} = 5\sqrt[3]{3}$

26. $$\begin{aligned} \sqrt{5} \cdot \sqrt{15} &= \sqrt{5} \cdot \sqrt{5} \cdot \sqrt{3} \\ &= \sqrt{25} \cdot \sqrt{3} = 5\sqrt{3} \end{aligned}$$

27. $$\begin{aligned} \sqrt{12} \cdot \sqrt{27} &= \sqrt{4 \cdot 3} \cdot \sqrt{9 \cdot 3} \\ &= 2\sqrt{3} \cdot 3\sqrt{3} \\ &= 2 \cdot 3 \cdot \left(\sqrt{3}\right)^2 \\ &= 2 \cdot 3 \cdot 3 = 18 \end{aligned}$$

28. $$\begin{aligned} \sqrt{32} \cdot \sqrt{48} &= \sqrt{16 \cdot 2} \cdot \sqrt{16 \cdot 3} \\ &= 4\sqrt{2} \cdot 4\sqrt{3} \\ &= 4 \cdot 4 \cdot \sqrt{2 \cdot 3} \\ &= 16\sqrt{6} \end{aligned}$$

29. $\sqrt{\dfrac{3}{49}} = \dfrac{\sqrt{3}}{\sqrt{49}} = \dfrac{\sqrt{3}}{7}$

30. $$\begin{aligned} \sqrt{\frac{6}{50}} &= \sqrt{\frac{3 \cdot 2}{25 \cdot 2}} \\ &= \sqrt{\frac{3}{25}} \\ &= \frac{\sqrt{3}}{\sqrt{25}} \\ &= \frac{\sqrt{3}}{5} \end{aligned}$$

31. $$\begin{aligned} \frac{\sqrt{48}}{\sqrt{3}} &= \sqrt{\frac{48}{3}} \\ &= \sqrt{16} \\ &= 4 \end{aligned}$$

32. $$\begin{aligned} \sqrt{\frac{1}{6}} \cdot \sqrt{\frac{5}{6}} &= \sqrt{\frac{1}{6} \cdot \frac{5}{6}} \\ &= \sqrt{\frac{5}{36}} \\ &= \frac{\sqrt{5}}{\sqrt{36}} = \frac{\sqrt{5}}{6} \end{aligned}$$

33. $$\begin{aligned} \sqrt{\frac{2}{5}} \cdot \sqrt{\frac{2}{45}} &= \sqrt{\frac{2}{5} \cdot \frac{2}{25}} \\ &= \sqrt{\frac{4}{225}} \\ &= \frac{\sqrt{4}}{\sqrt{225}} = \frac{2}{15} \end{aligned}$$

34. $\frac{3\sqrt{10}}{\sqrt{5}} = \frac{3 \cdot \sqrt{5} \cdot \sqrt{2}}{\sqrt{5}}$
$= 3\sqrt{2}$

35. $\frac{24\sqrt{12}}{6\sqrt{3}} = \frac{24 \cdot \sqrt{4} \cdot \sqrt{3}}{6\sqrt{3}}$
$= 4\sqrt{4} = 4 \cdot 2 = 8$

36. $\frac{8\sqrt{150}}{4\sqrt{75}} = \frac{8 \cdot \sqrt{75} \cdot \sqrt{2}}{4\sqrt{75}}$
$= 2\sqrt{2}$

37. $\sqrt{r^{18}} = r^9$ because $(r^9)^2 = r^{18}$.

38. $\sqrt{x^{10}y^{16}} = x^5y^8$ because $(x^5y^8)^2 = x^{10}y^{16}$.

39. $\sqrt{a^{15}b^{21}} = \sqrt{a^{14}b^{20} \cdot ab}$
$= \sqrt{a^{14}b^{20}} \cdot \sqrt{ab}$
$= a^7b^{10}\sqrt{ab}$

40. $\sqrt{121x^6y^{10}} = 11x^3y^5$ because $(11x^3y^5)^2 = 121x^6y^{10}$.

41. $\sqrt[3]{y^6} = y^2$ because $\left(y^2\right)^3 = y^6$.

42. $\sqrt[3]{216x^{15}} = 6x^5$ because $\left(6x^5\right)^3 = 216x^{15}$.

43. $7\sqrt{11} + \sqrt{11} = (7+1)\sqrt{11} = 8\sqrt{11}$

44. $3\sqrt{2} + 6\sqrt{2} = (3+6)\sqrt{2} = 9\sqrt{2}$

45. $3\sqrt{75} + 2\sqrt{27}$
$= 3\left(\sqrt{25} \cdot \sqrt{3}\right) + 2\left(\sqrt{9} \cdot \sqrt{3}\right)$
$= 3\left(5\sqrt{3}\right) + 2\left(3\sqrt{3}\right)$
$= 15\sqrt{3} + 6\sqrt{3} = 21\sqrt{3}$

46. $4\sqrt{12} + \sqrt{48}$
$= 4\left(\sqrt{4} \cdot \sqrt{3}\right) + \sqrt{16} \cdot \sqrt{3}$
$= 4\left(2\sqrt{3}\right) + 4\sqrt{3}$
$= 8\sqrt{3} + 4\sqrt{3} = 12\sqrt{3}$

47. $4\sqrt{24} - 3\sqrt{54} + \sqrt{6}$
$= 4\left(\sqrt{4} \cdot \sqrt{6}\right) - 3\left(\sqrt{9} \cdot \sqrt{6}\right) + \sqrt{6}$
$= 4\left(2\sqrt{6}\right) - 3\left(3\sqrt{6}\right) + \sqrt{6}$
$= 8\sqrt{6} - 9\sqrt{6} + 1\sqrt{6}$
$= 0\sqrt{6} = 0$

48. $2\sqrt{7} - 4\sqrt{28} + 3\sqrt{63}$
$= 2\sqrt{7} - 4\left(\sqrt{4} \cdot \sqrt{7}\right) + 3\left(\sqrt{9} \cdot \sqrt{7}\right)$
$= 2\sqrt{7} - 4\left(2\sqrt{7}\right) + 3\left(3\sqrt{7}\right)$
$= 2\sqrt{7} - 8\sqrt{7} + 9\sqrt{7} = 3\sqrt{7}$

49. $\frac{2}{5}\sqrt{75} + \frac{3}{4}\sqrt{160}$
$= \frac{2}{5}\left(\sqrt{25} \cdot \sqrt{3}\right) + \frac{3}{4}\left(\sqrt{16} \cdot \sqrt{10}\right)$
$= \frac{2}{5}\left(5\sqrt{3}\right) + \frac{3}{4}\left(4\sqrt{10}\right)$
$= 2\sqrt{3} + 3\sqrt{10}$

50. $\frac{1}{3}\sqrt{18} + \frac{1}{4}\sqrt{32}$
$= \frac{1}{3}\left(\sqrt{9} \cdot \sqrt{2}\right) + \frac{1}{4}\left(\sqrt{16} \cdot \sqrt{2}\right)$
$= \frac{1}{3}\left(3\sqrt{2}\right) + \frac{1}{4}\left(4\sqrt{2}\right)$
$= 1\sqrt{2} + 1\sqrt{2} = 2\sqrt{2}$

51. $\sqrt{15} \cdot \sqrt{2} + 5\sqrt{30} = \sqrt{30} + 5\sqrt{30}$
$= 1\sqrt{30} + 5\sqrt{30}$
$= 6\sqrt{30}$

52. $\sqrt{4x} + \sqrt{36x} - \sqrt{9x}$
$= \sqrt{4}\sqrt{x} + \sqrt{36}\sqrt{x} - \sqrt{9}\sqrt{x}$
$= 2\sqrt{x} + 6\sqrt{x} - 3\sqrt{x} = 5\sqrt{x}$

53. $\sqrt{20m^2} - m\sqrt{45}$
$= \sqrt{4m^2 \cdot 5} - m\left(\sqrt{9} \cdot \sqrt{5}\right)$
$= \sqrt{4m^2} \cdot \sqrt{5} - m\left(3\sqrt{5}\right)$
$= 2m\sqrt{5} - 3m\sqrt{5} = -m\sqrt{5}$

54. $3k\sqrt{8k^2n}+5k^2\sqrt{2n}$
$=3k\left(\sqrt{4k^2}\cdot\sqrt{2n}\right)+5k^2\sqrt{2n}$
$=3k\left(2k\sqrt{2n}\right)+5k^2\sqrt{2n}$
$=6k^2\sqrt{2n}+5k^2\sqrt{2n}$
$=(6k^2+5k^2)\sqrt{2n}$
$=11k^2\sqrt{2n}$

55. $\frac{8\sqrt{2}}{\sqrt{5}}=\frac{8\sqrt{2}\cdot\sqrt{5}}{\sqrt{5}\cdot\sqrt{5}}=\frac{8\sqrt{10}}{5}$

56. $\frac{5}{\sqrt{5}}=\frac{5\cdot\sqrt{5}}{\sqrt{5}\cdot\sqrt{5}}=\frac{5\sqrt{5}}{5}=\sqrt{5}$

57. $\frac{12}{\sqrt{24}}=\frac{12}{\sqrt{4\cdot 6}}=\frac{12}{2\sqrt{6}}$
$=\frac{12\cdot\sqrt{6}}{2\sqrt{6}\cdot\sqrt{6}}=\frac{12\sqrt{6}}{2\cdot 6}$
$=\frac{12\sqrt{6}}{12}=\sqrt{6}$

58. $\frac{\sqrt{2}}{\sqrt{15}}=\frac{\sqrt{2}\cdot\sqrt{15}}{\sqrt{15}\cdot\sqrt{15}}=\frac{\sqrt{30}}{15}$

59. $\sqrt{\frac{2}{5}}=\frac{\sqrt{2}}{\sqrt{5}}=\frac{\sqrt{2}\cdot\sqrt{5}}{\sqrt{5}\cdot\sqrt{5}}=\frac{\sqrt{10}}{5}$

60. $\sqrt{\frac{1}{2}}\cdot\sqrt{\frac{1}{32}}=\sqrt{\frac{1}{2}\cdot\frac{1}{32}}$
$=\sqrt{\frac{1}{64}}$
$=\frac{\sqrt{1}}{\sqrt{64}}$
$=\frac{1}{8}$

61. $\sqrt{\frac{2}{7}}\cdot\sqrt{\frac{1}{3}}=\sqrt{\frac{2}{7}\cdot\frac{1}{3}}$
$=\sqrt{\frac{2}{21}}=\frac{\sqrt{2}}{\sqrt{21}}$
$=\frac{\sqrt{2}\cdot\sqrt{21}}{\sqrt{21}\cdot\sqrt{21}}=\frac{\sqrt{42}}{21}$

62. $\sqrt{\frac{r^2}{16x}}=\frac{\sqrt{r^2}}{\sqrt{16x}}$
$=\frac{r\cdot\sqrt{x}}{\sqrt{16x}\cdot\sqrt{x}}$
$=\frac{r\sqrt{x}}{\sqrt{16x^2}}=\frac{r\sqrt{x}}{4x}$

63. $\sqrt[3]{\frac{1}{3}}=\frac{\sqrt[3]{1}}{\sqrt[3]{3}}=\frac{1\cdot\sqrt[3]{3^2}}{\sqrt[3]{3}\cdot\sqrt[3]{3^2}}$
$=\frac{\sqrt[3]{3^2}}{\sqrt[3]{3^3}}=\frac{\sqrt[3]{9}}{3}$

64. $\sqrt[3]{\frac{2}{7}}=\frac{\sqrt[3]{2}}{\sqrt[3]{7}}=\frac{\sqrt[3]{2}\cdot\sqrt[3]{7^2}}{\sqrt[3]{7}\cdot\sqrt[3]{7^2}}$
$=\frac{\sqrt[3]{2\cdot 7^2}}{\sqrt[3]{7^3}}=\frac{\sqrt[3]{98}}{7}$

65. $-\sqrt{3}\left(\sqrt{5}+\sqrt{27}\right)$
$=-\sqrt{3}\left(\sqrt{5}\right)+(-\sqrt{3})\left(\sqrt{27}\right)$
$=-\sqrt{3\cdot 5}-\sqrt{3\cdot 27}$
$=-\sqrt{15}-\sqrt{81}$
$=-\sqrt{15}-9$

66. $3\sqrt{2}\left(\sqrt{3}+2\sqrt{2}\right)$
$=3\sqrt{2}\left(\sqrt{3}\right)+3\sqrt{2}\left(2\sqrt{2}\right)$
$=3\sqrt{6}+6\cdot 2$
$=3\sqrt{6}+12$

67. $\left(2\sqrt{3}-4\right)\left(5\sqrt{3}+2\right)$
$=2\sqrt{3}\left(5\sqrt{3}\right)+\left(2\sqrt{3}\right)(2)-4\left(5\sqrt{3}\right)-4(2)$
$=10\cdot 3+4\sqrt{3}-20\sqrt{3}-8$
$=30-16\sqrt{3}-8$
$=22-16\sqrt{3}$

68. $\left(5\sqrt{7}+2\right)^2$
$=\left(5\sqrt{7}\right)^2+2\left(5\sqrt{7}\right)(2)+2^2$
$=25\cdot 7+20\sqrt{7}+4$
$=175+20\sqrt{7}+4$
$=179+20\sqrt{7}$

69. $\left(\sqrt{5}-7\right)\left(\sqrt{5}+\sqrt{7}\right)$

$=\left(\sqrt{5}\right)^2-\left(\sqrt{7}\right)^2$

$=5-7=-2$

70. $\left(2\sqrt{3}+5\right)\left(2\sqrt{3}-5\right)$

$=\left(2\sqrt{3}\right)^2-(5)^2$

$=4\cdot 3-25$

$=12-25=-13$

71. $\dfrac{1}{1+\sqrt{2}}$

$=\dfrac{1\left(1-\sqrt{2}\right)}{\left(1+\sqrt{2}\right)\left(1-\sqrt{2}\right)}$ Multiply by the conjugate.

$=\dfrac{1-\sqrt{2}}{(1)^2-\left(\sqrt{2}\right)^2}$

$=\dfrac{1-\sqrt{2}}{1-2}$

$=\dfrac{1-\sqrt{2}}{-1}=-1+\sqrt{2}$

72. $\dfrac{\sqrt{8}}{\sqrt{2}+6}$

$=\dfrac{\sqrt{8}\left(\sqrt{2}-6\right)}{\left(\sqrt{2}+6\right)\left(\sqrt{2}-6\right)}$ Multiply by the conjugate.

$=\dfrac{\sqrt{16}-6\sqrt{8}}{\left(\sqrt{2}\right)^2-6^2}$

$=\dfrac{4-6\cdot\sqrt{4\cdot 2}}{2-36}$

$=\dfrac{4-6\cdot 2\sqrt{2}}{-34}$

$=\dfrac{4-12\sqrt{2}}{-34}$

$=\dfrac{-2\left(-2+6\sqrt{2}\right)}{-2(17)}$ Factor.

$=\dfrac{-2+6\sqrt{2}}{17}$ Lowest terms

73. $\dfrac{2+\sqrt{6}}{\sqrt{3}-1}=\dfrac{\left(2+\sqrt{6}\right)\left(\sqrt{3}+1\right)}{\left(\sqrt{3}-1\right)\left(\sqrt{3}+1\right)}$

$=\dfrac{2\sqrt{3}+2+\sqrt{18}+\sqrt{6}}{3-1}$

$=\dfrac{2\sqrt{3}+2+\sqrt{9\cdot 2}+\sqrt{6}}{2}$

$=\dfrac{2\sqrt{3}+2+3\sqrt{2}+\sqrt{6}}{2}$

74. $\dfrac{15+10\sqrt{6}}{15}=\dfrac{5\left(3+2\sqrt{6}\right)}{5(3)}$ Factor.

$=\dfrac{3+2\sqrt{6}}{3}$ Lowest terms

75. $\dfrac{3-9\sqrt{7}}{12}=\dfrac{3\left(1-3\sqrt{7}\right)}{3(4)}$ Factor.

$=\dfrac{1-3\sqrt{7}}{4}$ Lowest terms

76. $\dfrac{6+\sqrt{192}}{2}=\dfrac{6+\sqrt{64\cdot 3}}{2}$

$=\dfrac{6+8\sqrt{3}}{2}$

$=\dfrac{2\left(3+4\sqrt{3}\right)}{2}$

$=3+4\sqrt{3}$

77. $\sqrt{m}-5=0$

$\sqrt{m}=5$ Isolate the radical.

$\left(\sqrt{m}\right)^2=5^2$ Square both sides.

$m=25$

Check $m=25$: $5-5=0$ True

The solution set is $\{25\}$.

78. $\sqrt{p}+4=0$

$\sqrt{p}=-4$

Since a square root cannot equal a negative number, there is no solution and the solution set is $\varnothing$.

79. $\sqrt{x+1}=7$

$\left(\sqrt{x+1}\right)^2=7^2$

$x+1=49$

$x=48$

Check $x=48$: $\sqrt{49}=7$ True

The solution set is $\{48\}$.

80. $\sqrt{5m+4}=3\sqrt{m}$

$\left(\sqrt{5m+4}\right)^2=\left(3\sqrt{m}\right)^2$

$5m+4=9m$

$4=4m$

$1=m$

Check $m=1$: $\sqrt{9}=3\sqrt{1}$ True

The solution set is $\{1\}$.

81. $\sqrt{2p+3}=\sqrt{5p-3}$

$\left(\sqrt{2p+3}\right)^2=\left(\sqrt{5p-3}\right)^2$

$2p+3=5p-3$

$6=3p$

$2=p$

Check $p=2$: $\sqrt{7}=\sqrt{7}$ True

The solution set is $\{2\}$.

82. $\sqrt{-2t-4}=t+2$

$\left(\sqrt{-2t-4}\right)^2=(t+2)^2$

$-2t-4=t^2+4t+4$

$0=t^2+6t+8$

$0=(t+2)(t+4)$

$t=-2$ or $t=-4$

Check $t=-2$: $\sqrt{0}=0$ True

Check $t=-4$: $\sqrt{4}=-2$ False

Of the two potential solutions, -2 checks in the original equation, but -4 does not. Thus, the solution set is $\{-2\}$.

83. $\sqrt{13+4t}=t+4$

$\left(\sqrt{13+4t}\right)^2=(t+4)^2$

$13+4t=t^2+8t+16$

$0=t^2+4t+3$

$0=(t+3)(t+1)$

$t=-3$ or $t=-1$

Check $t=-3$: $1=1$ True

Check $t=-1$: $3=3$ True

The solution set is $\{-3, -1\}$.

84. $\sqrt{2-x}+3=x+7$

$\sqrt{2-x}=x+4$ Isolate the radical.

$\left(\sqrt{2-x}\right)^2=(x+4)^2$

$2-x=x^2+8x+16$

$0=x^2+9x+14$

$0=(x+2)(x+7)$

$x=-2$ or $x=-7$

Check $x=-2$: $2+3=5$ True

Check $x=-7$: $3+3=0$ False

Of the two potential solutions, -2 checks in the original equation, but -7 does not. Thus, the solution set is $\{-2\}$.

85. $\sqrt[3]{x+4}=\sqrt[3]{16-2x}$

$\left(\sqrt[3]{x+4}\right)^3=\left(\sqrt[3]{16-2x}\right)^3$ Cube both sides.

$x+4=16-2x$

$3x=12$

$x=4$

Check $x=4$: $\sqrt[3]{8}=\sqrt[3]{16-8}$ True

The solution set is $\{4\}$.

86. $\sqrt{5x+6}+\sqrt{3x+4}=2$

$$\sqrt{5x+6}=2-\sqrt{3x+4}$$
$$\left(\sqrt{5x+6}\right)^2=\left(2-\sqrt{3x+4}\right)^2$$
$$5x+6=4-4\sqrt{3x+4}+3x+4$$
$$2x-2=-4\sqrt{3x+4}$$
$$x-1=-2\sqrt{3x+4}$$
$$(x-1)^2=\left(-2\sqrt{3x+4}\right)^2$$
$$x^2-2x+1=4(3x+4)$$
$$x^2-2x+1=12x+16$$
$$x^2-14x-15=0$$
$$(x+1)(x-15)=0$$
$$x=-1 \quad \text{or} \quad x=15$$

Check $x=-1$: $1+1=2$ True

Check $x=15$: $9+7=2$ False

The solution set is $\{-1\}$.

Chapter 8 Mixed Review Exercises

1. $\sqrt{\frac{1}{3}}\cdot\sqrt{\frac{24}{5}}=\sqrt{\frac{1}{3}\cdot\frac{24}{5}}=\sqrt{\frac{8}{5}}=\frac{\sqrt{8}}{\sqrt{5}}$

$$=\frac{\sqrt{8}\cdot\sqrt{5}}{\sqrt{5}\cdot\sqrt{5}}=\frac{\sqrt{40}}{5}$$
$$=\frac{\sqrt{4\cdot 10}}{5}=\frac{2\sqrt{10}}{5}$$

2. $\frac{1}{5+\sqrt{2}}=\frac{1\left(5-\sqrt{2}\right)}{\left(5+\sqrt{2}\right)\left(5-\sqrt{2}\right)}$

$$=\frac{5-\sqrt{2}}{(5)^2-\left(\sqrt{2}\right)^2}$$
$$=\frac{5-\sqrt{2}}{25-2}$$
$$=\frac{5-\sqrt{2}}{23}$$

3. $\sqrt[3]{-125}=-5$ because $(-5)^3=-125$.

4. $\sqrt{50y^2}=\sqrt{25y^2\cdot 2}$

$$=\sqrt{25y^2}\cdot\sqrt{2}$$
$$=5y\sqrt{2}$$

5. $\sqrt{\frac{16r^3}{3s}}=\frac{\sqrt{16r^3}}{\sqrt{3s}}=\frac{\sqrt{16r^2}\cdot\sqrt{r}}{\sqrt{3s}}$

$$=\frac{4r\sqrt{r}}{\sqrt{3s}}=\frac{4r\sqrt{r}\cdot\sqrt{3s}}{\sqrt{3s}\cdot\sqrt{3s}}$$
$$=\frac{4r\sqrt{3rs}}{3s}$$

6. $\frac{12+6\sqrt{13}}{12}=\frac{6\left(2+\sqrt{13}\right)}{6(2)}$

$$=\frac{2+\sqrt{13}}{2}$$

7. $-\sqrt{121}=-11$

8. $\left(\sqrt{5}-\sqrt{2}\right)^2$

$$=\left(\sqrt{5}\right)^2-2\sqrt{5}\sqrt{2}+\left(\sqrt{2}\right)^2$$
$$=5-2\sqrt{10}+2$$
$$=7-2\sqrt{10}$$

9. $-\sqrt{5}\left(\sqrt{2}+\sqrt{75}\right)$

$$=-\sqrt{5}\left(\sqrt{2}\right)+\left(-\sqrt{5}\right)\left(\sqrt{75}\right)$$
$$=-\sqrt{10}-\sqrt{375}$$
$$=-\sqrt{10}-\sqrt{25\cdot 15}$$
$$=-\sqrt{10}-5\sqrt{15}$$

10. $\left(6\sqrt{7}+2\right)\left(4\sqrt{7}-1\right)$

$$=6\sqrt{7}\left(4\sqrt{7}\right)-1\left(6\sqrt{7}\right)+2\left(4\sqrt{7}\right)+2(-1)$$
$$=24\cdot 7-6\sqrt{7}+8\sqrt{7}-2$$
$$=168-2+2\sqrt{7}$$
$$=166+2\sqrt{7}$$

11. $2\sqrt{27}+3\sqrt{75}-\sqrt{300}$

$$=2\sqrt{9\cdot 3}+3\sqrt{25\cdot 3}-\sqrt{100\cdot 3}$$
$$=2\cdot 3\sqrt{3}+3\cdot 5\sqrt{3}-10\sqrt{3}$$
$$=6\sqrt{3}+15\sqrt{3}-10\sqrt{3}$$
$$=11\sqrt{3}$$

12. $\sqrt{x+2} = x-4$

$\left(\sqrt{x+2}\right)^2 = (x-4)^2$

$x+2 = x^2 - 8x + 16$

$0 = x^2 - 9x + 14$

$0 = (x-2)(x-7)$

$x = 2$ or $x = 7$

Check $x = 2$: $\sqrt{4} = -2$ False

Check $x = 7$: $\sqrt{9} = 3$ True

The solution set is $\{7\}$.

13. $\sqrt{x} + 3 = 0$

$\sqrt{x} = -3$

Since a square root cannot equal a negative number, there is no solution and the solution set is $\varnothing$.

14. $\sqrt{1+3t} - t = -3$

$\sqrt{1+3t} = t - 3$

$\left(\sqrt{1+3t}\right)^2 = (t-3)^2$

$1 + 3t = t^2 - 6t + 9$

$0 = t^2 - 9t + 8$

$0 = (t-1)(t-8)$

$t = 1$ or $t = 8$

Check $t = 1$: $2 - 1 = -3$ False

Check $t = 8$: $5 - 8 = -3$ True

The solution set is $\{8\}$.

15. (a) $S = 28.6\sqrt[3]{A}$

$S = 28.6\sqrt[3]{8}$ Let $A = 8$.

$= 28.6(2)$

$= 57.2 \approx 57$

There would be 57 species.

(b) $S = 28.6\sqrt[3]{A}$

$S = 28.6\sqrt[3]{1790}$ Let $A = 1790$.

≈ 347.3

There would be 347 species.

Chapter 8 Test

1. The square roots of 196 are -14 and 14 because $(-14)^2 = 196$ and $14^2 = 196$.

2. (a) $\sqrt{142}$ is *irrational* because 142 is not a perfect square.

(b) $\sqrt{142} \approx 11.916$

3. (a) $\sqrt{64} = 8$; B

(b) $-\sqrt{64} = -8$; F

(c) $\sqrt{-64}$ is not a real number; D

(d) $\sqrt[3]{64} = 4$; A

(e) $\sqrt[3]{-64} = -4$; C

(f) $-\sqrt[3]{-64} = -(-4) = 4$; A

4. $\sqrt{\dfrac{128}{25}} = \dfrac{\sqrt{128}}{\sqrt{25}} = \dfrac{\sqrt{64 \cdot 2}}{5} = \dfrac{8\sqrt{2}}{5}$

5. $\sqrt[3]{32} = \sqrt[3]{8 \cdot 4} = \sqrt[3]{8} \cdot \sqrt[3]{4} = 2\sqrt[3]{4}$

6. $\dfrac{20\sqrt{18}}{5\sqrt{3}} = \dfrac{4\sqrt{9 \cdot 2}}{\sqrt{3}}$

$= \dfrac{4 \cdot 3\sqrt{2}}{\sqrt{3}}$

$= \dfrac{12\sqrt{2} \cdot \sqrt{3}}{\sqrt{3} \cdot \sqrt{3}}$

$= \dfrac{12\sqrt{6}}{3} = 4\sqrt{6}$

7. $\sqrt[3]{32x^2y^3} = \sqrt[3]{8y^3 \cdot 4x^2}$

$= \sqrt[3]{8y^3} \cdot \sqrt[3]{4x^2}$

$= 2y\sqrt[3]{4x^2}$

8. $\left(6 - \sqrt{5}\right)\left(6 + \sqrt{5}\right)$

$= (6)^2 - \left(\sqrt{5}\right)^2$

$= 36 - 5 = 31$

9. $3\sqrt{28} + \sqrt{63} = 3\left(\sqrt{4 \cdot 7}\right) + \sqrt{9 \cdot 7}$

$= 3\left(2\sqrt{7}\right) + 3\sqrt{7}$

$= 6\sqrt{7} + 3\sqrt{7} = 9\sqrt{7}$

10. $\left(\sqrt{5}+\sqrt{6}\right)^2$

$=\left(\sqrt{5}\right)^2+2\left(\sqrt{5}\right)\left(\sqrt{6}\right)+\left(\sqrt{6}\right)^2$

$=5+2\sqrt{30}+6$

$=11+2\sqrt{30}$

11. $\sqrt[3]{16x^4}-2\sqrt[3]{128x^4}$

$=\sqrt[3]{8x^3}\cdot\sqrt[3]{2x}-2\cdot\sqrt[3]{64x^3}\cdot\sqrt[3]{2x}$

$=2x\cdot\sqrt[3]{2x}-2\cdot 4x\cdot\sqrt[3]{2x}$

$=(2x-8x)\sqrt[3]{2x}=-6x\sqrt[3]{2x}$

12. $\sqrt[3]{\frac{2}{3}}$

Multiply the numerator and the denominator by enough factors of 3 to make the radicand in the denominator a perfect cube. This will eliminate the radical in the denominator. Here, we multiply by $\sqrt[3]{3^2}$, or $\sqrt[3]{9}$.

$$\sqrt[3]{\frac{2}{3}}=\frac{\sqrt[3]{2}}{\sqrt[3]{3}}=\frac{\sqrt[3]{2}\cdot\sqrt[3]{3^2}}{\sqrt[3]{3}\cdot\sqrt[3]{3^2}}$$

$$=\frac{\sqrt[3]{2\cdot 3^2}}{\sqrt[3]{3\cdot 3^2}}=\frac{\sqrt[3]{18}}{\sqrt[3]{3^3}}=\frac{\sqrt[3]{18}}{3}$$

13. $\left(2-\sqrt{7}\right)\left(3\sqrt{2}+1\right)$

$=2\left(3\sqrt{2}\right)+2(1)-\sqrt{7}\left(3\sqrt{2}\right)-\sqrt{7}(1)$

$=6\sqrt{2}+2-3\sqrt{14}-\sqrt{7}$

14. $3\sqrt{27x}-4\sqrt{48x}+2\sqrt{3x}$

$=3\left(\sqrt{9\cdot 3x}\right)-4\left(\sqrt{16\cdot 3x}\right)+2\sqrt{3x}$

$=3\left(3\sqrt{3x}\right)-4\left(4\sqrt{3x}\right)+2\sqrt{3x}$

$=9\sqrt{3x}-16\sqrt{3x}+2\sqrt{3x}=-5\sqrt{3x}$

15. Use the Pythagorean theorem with $c=9$ and $b=3$.

$c^2=a^2+b^2$

$9^2=a^2+3^2$

$81=a^2+9$

$72=a^2$

$\sqrt{72}=a$

(a) $a=\sqrt{72}=\sqrt{36\cdot 2}=6\sqrt{2}$ inches

(b) $a=\sqrt{72}\approx 8.485$ inches

16. $Z=\sqrt{R^2+X^2}$

$=\sqrt{40^2+30^2}$ Let $R=40$, $X=30$.

$=\sqrt{1600+900}$

$=\sqrt{2500}=50$ ohms

17. $\frac{5\sqrt{2}}{\sqrt{7}}=\frac{5\sqrt{2}\cdot\sqrt{7}}{\sqrt{7}\cdot\sqrt{7}}=\frac{5\sqrt{14}}{7}$

18. $\sqrt{\frac{2}{3x}}=\frac{\sqrt{2}}{\sqrt{3x}}=\frac{\sqrt{2}\cdot\sqrt{3x}}{\sqrt{3x}\cdot\sqrt{3x}}=\frac{\sqrt{6x}}{3x}$

19. $\frac{-2}{\sqrt[3]{4}}=\frac{-2\cdot\sqrt[3]{2}}{\sqrt[3]{4}\cdot\sqrt[3]{2}}=\frac{-2\sqrt[3]{2}}{\sqrt[3]{8}}$

$=\frac{-2\sqrt[3]{2}}{2}=-\sqrt[3]{2}$

20. $\frac{-3}{4-\sqrt{3}}=\frac{-3\left(4+\sqrt{3}\right)}{\left(4-\sqrt{3}\right)\left(4+\sqrt{3}\right)}$

$=\frac{-12-3\sqrt{3}}{(4)^2-\left(\sqrt{3}\right)^2}$

$=\frac{-12-3\sqrt{3}}{16-3}$

$=\frac{-12-3\sqrt{3}}{13}$

21. $\frac{2+\sqrt{8}}{4}=\frac{2+\sqrt{4}\cdot\sqrt{2}}{4}$

$=\frac{2+2\sqrt{2}}{4}$

$=\frac{2(1+\sqrt{2})}{2(2)}$

$=\frac{1+\sqrt{2}}{2}$

22. $\sqrt{2x+6}+4=2$

$\sqrt{2x+6}=-2$ Isolate the radical.

$(\sqrt{2x+6})^2=(-2)^2$

$2x+6=4$

$2x=-2$

$x=-1$

Check $x=-1$: $\sqrt{4}+4\overset{?}{=}2$

$6=2$ False

The solution set is $\varnothing$. Note that in the second line we have a square root equal to a negative number, but the square root of a real number is nonnegative.

23. $\sqrt{x+1}=5-x$

$(\sqrt{x+1})^2=(5-x)^2$

$x+1=25-10x+x^2$

$0=x^2-11x+24$

$0=(x-3)(x-8)$

$x=3$ or $x=8$

Check $x=3$: $\sqrt{4}=2$ True

Check $x=8$: $\sqrt{9}=-3$ False

The solution set is $\{3\}$.

24. $3\sqrt{x}-1=2x$

$3\sqrt{x}=2x+1$ Isolate the radical.

$(3\sqrt{x})^2=(2x+1)^2$

$9x=4x^2+4x+1$

$0=4x^2-5x+1$

$0=(4x-1)(x-1)$

$x=\frac{1}{4}$ or $x=1$

Check $x=\frac{1}{4}$: $\frac{1}{2}=\frac{1}{2}$ True

Check $x=1$: $2=2$ True

The solution set is $\left\{\frac{1}{4},1\right\}$.

25. $\sqrt{2x+9}+\sqrt{x+5}=2$

$\sqrt{2x+9}=2-\sqrt{x+5}$

$(\sqrt{2x+9})^2=(2-\sqrt{x+5})^2$

$2x+9=4-4\sqrt{x+5}+x+5$

$x=-4\sqrt{x+5}$

$(x)^2=(-4\sqrt{x+5})^2$

$x^2=16(x+5)$

$x^2=16x+80$

$x^2-16x-80=0$

$(x+4)(x-20)=0$

$x=-4$ or $x=20$

Check $x=-4$: $1+1=2$ True

Check $x=20$: $7+5=2$ False

The solution set is $\{-4\}$.

26. Nothing is wrong with the steps taken so far, but the potential solution must be checked. Let $x=12$ in the original equation.

$\sqrt{2x+1}+5=0$

$\sqrt{2(12)+1}+5\overset{?}{=}0$ Let $x=12$.

$\sqrt{25}+5\overset{?}{=}0$

$5+5\overset{?}{=}0$

$10=0$ False

12 is not a solution because it does not satisfy the original equation. The equation has no solution, so the solution set is $\varnothing$.

Chapters R–8 Cumulative Review Exercises

1. $3(6+7)+6\cdot4-3^2$

$=3(13)+24-9$

$=39+24-9$

$=63-9=54$

2. $\dfrac{3(6+7)+3}{2(4)-1} = \dfrac{3(13)+3}{8-1}$

$= \dfrac{39+3}{7}$

$= \dfrac{42}{7} = 6$

3. $|-6|-|-3| = 6-3 = 3$

4. $5(k-4)-k = k-11$

$5k-20-k = k-11$

$4k-20 = k-11$

$3k = 9$

$k = 3$

The solution set is $\{3\}$.

5. $-\dfrac{3}{4}x \le 12$

$-\dfrac{4}{3}\left(-\dfrac{3}{4}x\right) \ge -\dfrac{4}{3}(12)$

$x \ge -16$

The solution set is $[-16, \infty]$.

6. $5z+3-4 > 2z+9+z$

$5z-1 > 3z+9$

$2z > 10$

$z > 5$

The solution set is $(5, \infty)$.

7. Let $x =$ the amount earned in 2011. Then $x + \$38{,}975 =$ the amount earned in 2012. The amounts total $636,227, so

$x + (x + 38{,}975) = 636{,}227$

$2x + 38{,}975 = 636{,}227$

$2x = 597{,}252$

$x = 298{,}626.$

Brazile earned $298,626 in 2011 and $\$298{,}626 + \$38{,}975 = \$337{,}601$ in 2012.

8. $-4x+5y = -20$

Find the intercepts.

If $y = 0$, $x = 5$, so the x-intercept is (5, 0).

If $x = 0$, $y = -4$, so the y-intercept is $(0, -4)$.

Draw the line that passes through the points (5, 0) and $(0, -4)$.

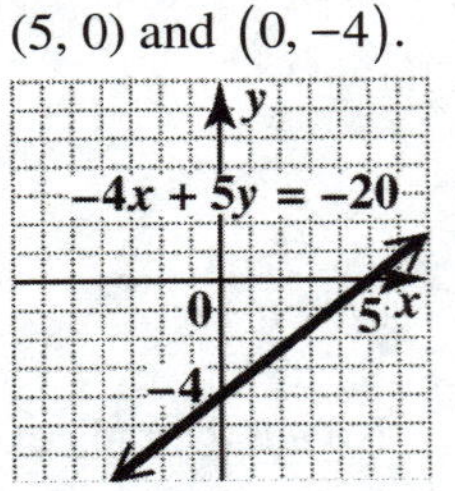

9. $x = 2$

For any value of y, the value of x is 2, so this is a vertical line through (2, 0).

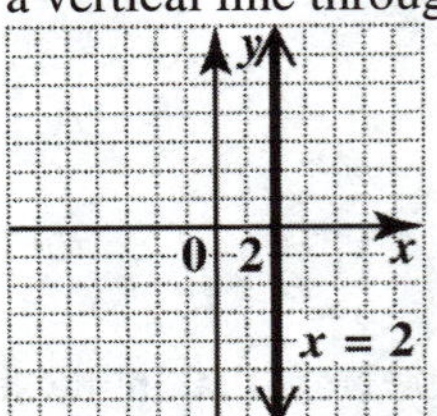

10. $2x-5y > 10$

The boundary, $2x-5y = 10$, is the line that passes through (5, 0) and $(0, -2)$; draw it as a dashed line because of the > symbol. Use (0, 0) as a test point. Because $2(0)-5(0) > 10$ is a false statement, shade the side of the dashed boundary that does not include the origin, (0, 0). The dashed line shows that the boundary is not part of the graph.

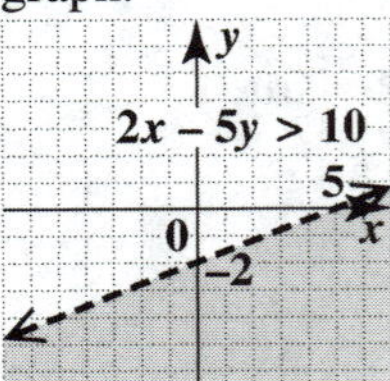

11. (a) The slope of the line through the points (0, 109.5) and (12, 326.5) is

$$m = \frac{y_2 - y_1}{x_2 - x_1} = \frac{326.5 - 109.5}{12 - 0} = \frac{217}{12} \approx 18.1.$$

An interpretation of the slope is that the number of cell phone subscribers increased by an average of about 18.1 million per year.

(b) Using the slope-intercept form of a line with $m = 18.1$ and $b = 109.5$, we get $y = 18.1x + 109.5$.

(c) For 2010, $x = 2014 - 2000 = 14$.

$$y = 18.1(14) + 109.5 = 253.4 + 109.5 = 362.9$$

The estimated number of cell phone subscribers for 2014 is about 362.9 million.

12. $4x - y = 19 \quad (1)$

$3x + 2y = -5 \quad (2)$

We will solve this system by the elimination method. Multiply both sides of equation (1) by 2, and then add the result to equation (2).

$$\begin{aligned} 8x - 2y &= 38 \\ 3x + 2y &= -5 \\ \hline 11x &= 33 \\ x &= 3 \end{aligned}$$

Let $x = 3$ in equation (1).

$$\begin{aligned} 4(3) - y &= 19 \\ 12 - y &= 19 \\ -y &= 7 \\ y &= -7 \end{aligned}$$

The solution set is $\{(3, -7)\}$.

13. $2x - y = 6 \quad (1)$

$3y = 6x - 18 \quad (2)$

We will solve this system by the substitution method. Solve equation (2) for y by dividing both sides by 3.

$y = 2x - 6$

Substitute $2x - 6$ for y in (1).

$$\begin{aligned} 2x - (2x - 6) &= 6 \\ 2x - 2x + 6 &= 6 \\ 6 &= 6 \end{aligned}$$

This true statement indicates that the two original equations both describe the same line. This system has an infinite number of solutions. The solution set is $\{(x, y) \mid 2x - y = 6\}$.

14. Let x = the average rate of the slower car (departing from Des Moines).

Then $x + 7$ = the average rate of the faster car (departing from Chicago).

In 3 hours, the slower car travels $3x$ miles and the faster car travels $3(x + 7)$ miles. The total distance traveled is 345 miles, so

$$\begin{aligned} 3x + 3(x + 7) &= 345 \\ 3x + 3x + 21 &= 345 \\ 6x &= 324 \\ x &= 54. \end{aligned}$$

The car departing from Des Moines averaged 54 miles per hour and traveled $3(54) = 162$ miles. The car departing from Chicago averaged 61 miles per hour and traveled $3(61) = 183$ miles.

15.
$$\begin{aligned} &(3x^6)(2x^2y)^2 \\ &= (3x^6)(2)^2(x^2)^2(y)^2 \\ &= (3x^6) \cdot 4x^4y^2 \\ &= 12x^{10}y^2 \end{aligned}$$

16.
$$\begin{aligned} \left(\frac{3^2 y^{-2}}{2^{-1} y^3}\right)^{-3} &= \left(\frac{2^{-1} y^3}{3^2 y^{-2}}\right)^3 \\ &= \left(\frac{y^3 \cdot y^2}{2^1 \cdot 3^2}\right)^3 \\ &= \frac{(y^5)^3}{(2 \cdot 3^2)^3} \\ &= \frac{y^{15}}{2^3 \cdot 3^6}, \text{ or } \frac{y^{15}}{5832} \end{aligned}$$

17.
$$\begin{aligned} &(10x^3 + 3x^2 - 9) - (7x^3 - 8x^2 + 4) \\ &= 10x^3 + 3x^2 - 9 - 7x^3 + 8x^2 - 4 \\ &= 3x^3 + 11x^2 - 13 \end{aligned}$$

18.

$$\begin{array}{r} 4t^2 - 8t + 5 \\ 2t+3\overline{\big)8t^3 - 4t^2 - 14t + 15} \\ \underline{8t^3 + 12t^2} \\ -16t^2 - 14t \\ \underline{-16t^2 - 24t} \\ 10t + 15 \\ \underline{10t + 15} \\ 0 \end{array}$$

The remainder is 0, so the answer is the quotient, $4t^2 - 8t + 5$.

19. $m^2 + 12m + 32$
Find two integers whose product is 32 and whose sum is 12. The required integers are 8 and 4. Thus,
$m^2 + 12m + 32 = (m+8)(m+4)$.

20. $12a^2 + 4ab - 5b^2$
Factor by the grouping method. Look for two integers whose product is $12(-5) = -60$ and whose sum is 4. The integers are 10 and -6.

$$\begin{aligned} &12a^2 + 4ab - 5b^2 \\ &= 12a^2 + 10ab - 6ab - 5b^2 \\ &= (12a^2 + 10ab) + (-6ab - 5b^2) \\ &= 2a(6a+5b) - b(6a+5b) \\ &= (6a+5b)(2a-b) \end{aligned}$$

21. $81z^2 + 72z + 16$

$$\begin{aligned} &= (9z)^2 + 2(9z)(4) + 4^2 \\ &= (9z+4)^2 \end{aligned}$$

22.

$$\begin{aligned} &\frac{x^2 - 3x - 4}{x^2 + 3x} \cdot \frac{x^2 + 2x - 3}{x^2 - 5x + 4} \\ &= \frac{(x-4)(x+1)}{x(x+3)} \cdot \frac{(x-1)(x+3)}{(x-4)(x-1)} \quad \text{Factor.} \\ &= \frac{x+1}{x} \end{aligned}$$

23.

$$\begin{aligned} &\frac{t^2 + 4t - 5}{t+5} \div \frac{t-1}{t^2 + 8t + 15} \\ &= \frac{t^2 + 4t - 5}{t+5} \cdot \frac{t^2 + 8t + 15}{t-1} \\ &= \frac{(t+5)(t-1)}{t+5} \cdot \frac{(t+5)(t+3)}{t-1} \quad \text{Factor.} \\ &= (t+5)(t+3), \quad \text{or} \quad t^2 + 8t + 15 \end{aligned}$$

24.

$$\begin{aligned} &\frac{y}{y^2 - 1} + \frac{y}{y+1} \\ &= \frac{y}{(y+1)(y-1)} + \frac{y(y-1)}{(y+1)(y-1)} \\ &= \frac{y + y(y-1)}{(y+1)(y-1)} \\ &= \frac{y + y^2 - y}{(y+1)(y-1)} = \frac{y^2}{(y+1)(y-1)} \end{aligned}$$

25.

$$\begin{aligned} &\frac{2}{x+3} - \frac{4}{x-1} \\ &= \frac{2(x-1)}{(x+3)(x-1)} - \frac{4(x+3)}{(x-1)(x+3)} \\ &= \frac{2(x-1) - 4(x+3)}{(x+3)(x-1)} \\ &= \frac{2x - 2 - 4x - 12}{(x+3)(x-1)} \\ &= \frac{-2x - 14}{(x+3)(x-1)} \end{aligned}$$

26.

$$\begin{aligned} &\frac{\frac{2}{3} + \frac{1}{2}}{\frac{1}{9} - \frac{1}{6}} = \frac{\frac{4}{6} + \frac{3}{6}}{\frac{2}{18} - \frac{3}{18}} \\ &= \frac{\frac{7}{6}}{\frac{-1}{18}} = \frac{7}{6} \div \frac{-1}{18} = \frac{7}{6} \cdot \frac{18}{-1} \\ &= 7 \cdot (-3) = -21 \end{aligned}$$

27.

$$\begin{aligned} x^2 - 7x &= -12 \\ x^2 - 7x + 12 &= 0 \\ (x-3)(x-4) &= 0 \\ x = 3 \quad &\text{or} \quad x = 4 \end{aligned}$$

The solution set is $\{3, 4\}$.

28. $(x+4)(x-1)=-6$

$$x^2+3x-4=-6$$
$$x^2+3x+2=0$$
$$(x+2)(x+1)=0$$
$$x=-2 \quad \text{or} \quad x=-1$$

The solution set is $\{-2,-1\}$.

29. $z^2+144=24z$

$$z^2-24z+144=0$$
$$(z-12)^2=0$$
$$z=12$$

The solution set is $\{12\}$.

30. $\dfrac{x}{x+8}-\dfrac{3}{x-8}=\dfrac{128}{x^2-64}$

$$\frac{x}{x+8}-\frac{3}{x-8}=\frac{128}{(x+8)(x-8)}$$
$$x(x-8)-3(x+8)=128$$
$$x^2-8x-3x-24=128$$
$$x^2-11x-152=0$$
$$(x+8)(x-19)=0$$
$$x=-8 \quad \text{or} \quad x=19$$

Check $x=19$: $\dfrac{19}{27}-\dfrac{3}{11}=\dfrac{128}{297}$ True

We cannot have -8 for a solution because that would result in division by zero. Thus, the solution set is $\{19\}$.

31. Solve $A=\dfrac{B+CD}{BC+D}$ for B.

$$A(BC+D)=B+CD$$
$$ABC+AD=B+CD$$
$$ABC-B=CD-AD$$
$$B(AC-1)=CD-AD$$
$$B=\frac{CD-AD}{AC-1}$$

32. $\sqrt{x}+2=x-10$

$$\sqrt{x}=x-12$$
$$\left(\sqrt{x}\right)^2=(x-12)^2$$
$$x=x^2-24x+144$$
$$0=x^2-25x+144$$
$$0=(x-16)(x-9)$$
$$x=16 \quad \text{or} \quad x=9$$

Check $x=16$: $6=6$ True

Check $x=9$: $5=-1$ False

The solution set is $\{16\}$.

33. $\sqrt{27}-2\sqrt{12}+6\sqrt{75}$

$$=\sqrt{9}\cdot\sqrt{3}-2\sqrt{4}\cdot\sqrt{3}+6\sqrt{25}\cdot\sqrt{3}$$
$$=3\sqrt{3}-2\left(2\sqrt{3}\right)+6\left(5\sqrt{3}\right)$$
$$=3\sqrt{3}-4\sqrt{3}+30\sqrt{3}=29\sqrt{3}$$

34. $\left(3\sqrt{2}+1\right)\left(4\sqrt{2}-3\right)$

$$=3\sqrt{2}\left(4\sqrt{2}\right)-3\sqrt{2}(3)+1\left(4\sqrt{2}\right)+1(-3)$$
$$=12\cdot 2-9\sqrt{2}+4\sqrt{2}-3$$
$$=24-3-5\sqrt{2}$$
$$=21-5\sqrt{2}$$

35. $\dfrac{2}{\sqrt{3}+\sqrt{5}}=\dfrac{2\left(\sqrt{3}-\sqrt{5}\right)}{\left(\sqrt{3}+\sqrt{5}\right)\left(\sqrt{3}-\sqrt{5}\right)}$

$$=\frac{2\left(\sqrt{3}-\sqrt{5}\right)}{3-5}$$
$$=\frac{2\left(\sqrt{3}-\sqrt{5}\right)}{-2}$$
$$=\frac{\sqrt{3}-\sqrt{5}}{-1}=-\sqrt{3}+\sqrt{5}$$

Chapter 9 Quadratic Equations

9.1 Solving Quadratic Equations by the Square Root Property

Now Try Exercises

N1. **(a)** $x^2 - x - 20 = 0$

$(x-5)(x+4) = 0$ Factor.

$x - 5 = 0$ or $x + 4 = 0$ Zero-factor prop.

$x = 5$ or $x = -4$ Solve each eq.

The solution set is $\{-4, 5\}$.

(b) $x^2 = 36$

$x^2 - 36 = 0$ Subtract 36.

$(x-6)(x+6) = 0$ Factor.

$x - 6 = 0$ or $x + 6 = 0$ Zero-factor prop.

$x = 6$ or $x = -6$ Solve each eq.

The solution set is $\{-6, 6\}$.

N2. **(a)** Solve by using the square root property.

$t = \sqrt{25}$ or $t = -\sqrt{25}$

$t = 5$ or $t = -5$

The solution set is $\{-5, 5\}$, or $\{\pm 5\}$.

(b) Use the square root property.

$x = \sqrt{13}$ or $x = -\sqrt{13}$.

The solution set is $\{-\sqrt{13}, \sqrt{13}\}$, or $\{\pm\sqrt{13}\}$.

(c) Since -144 is a negative number and since the square of a real number cannot be negative, there is no real number solution of this equation.

(d) $2x^2 - 5 = 35$

$2x^2 = 40$

$x^2 = 20$

Use the square root property.

$x = \sqrt{20}$ or $x = -\sqrt{20}$

$x = \sqrt{4}\cdot\sqrt{5}$ or $x = -\sqrt{4}\cdot\sqrt{5}$

$x = 2\sqrt{5}$ or $x = -2\sqrt{5}$

The solution set is $\{-2\sqrt{5}, 2\sqrt{5}\}$, or $\{\pm 2\sqrt{5}\}$.

N3. Use the square root property.

$x - 2 = \sqrt{32}$ or $x - 2 = -\sqrt{32}$

$x = 2 + \sqrt{16}\cdot\sqrt{2}$ or $x = 2 - \sqrt{16}\cdot\sqrt{2}$

$x = 2 + 4\sqrt{2}$ or $x = 2 - 4\sqrt{2}$

Check $x = 2 \pm 4\sqrt{2}$:

$\text{LS} = \left[\left(2 \pm 4\sqrt{2}\right) - 2\right]^2$

$= \left(\pm 4\sqrt{2}\right)^2 = 16\cdot 2 = 32 = \text{RS}$

The solution set is $\{2 \pm 4\sqrt{2}\}$.

N4. $(2t - 4)^2 = 50$

$2t - 4 = \sqrt{50}$ or $2t - 4 = -\sqrt{50}$

$2t - 4 = \sqrt{25}\cdot\sqrt{2}$ or $2t - 4 = -\sqrt{25}\cdot\sqrt{2}$

$2t = 4 + 5\sqrt{2}$ or $2t = 4 - 5\sqrt{2}$

$t = \dfrac{4 + 5\sqrt{2}}{2}$ or $t = \dfrac{4 - 5\sqrt{2}}{2}$

The solution set is $\left\{\dfrac{4 \pm 5\sqrt{2}}{2}\right\}$.

N5. Since the square root of -5 is not a real number, there is no real number solution.

N6. $w = \dfrac{L^2 g}{1200}$ Given formula

$2.10 = \dfrac{L^2 \cdot 9}{1200}$ Let $w = 2.10$, $g = 9$.

$2520 = 9L^2$ Multiply by 1200.

$L^2 = 280$ Divide by 9.

$L = \sqrt{280}$ Square root property

Note that $\sqrt{280} \approx 16.73$, so the length of the bass, to the nearest inch, is 17 in.

Exercises

1. A quadratic equation can be written in the form $ax^2 + bx + c = 0$, where a, b, and c are real numbers and $a \neq 0$.
The given form is called standard form.
Quadratic equations are second-degree equations.
Thus, only B and C are quadratic equations.

3. $x^2 = 12$ has two irrational solutions, $\pm\sqrt{12} = \pm 2\sqrt{3}$. The correct choice is C.

5. $x^2 = \frac{25}{36}$ has two rational solutions that are not integers, $\pm\frac{5}{6}$. The correct choice is D.

7.
$x^2 - x - 56 = 0$
$(x-8)(x+7) = 0$ Factor.
$x - 8 = 0$ or $x + 7 = 0$ Zero-factor property
$x = 8$ or $x = -7$ Solve each equation.
The solution set is $\{-7, 8\}$.

9.
$x^2 = 121$
$x^2 - 121 = 0$ Subtract 121.
$(x+11)(x-11) = 0$ Factor.
$x + 11 = 0$ or $x - 11 = 0$ Zero-factor prop.
$x = -11$ or $x = 11$
The solution set is $\{-11, 11\}$, or $\{\pm 11\}$.

11.
$3x^2 - 13x = 30$
$3x^2 - 13x - 30 = 0$
$(x-6)(3x+5) = 0$ Factor.
$x - 6 = 0$ or $3x + 5 = 0$ Zero-factor prop.
$x = 6$ or $x = -\frac{5}{3}$
The solution set is $\left\{-\frac{5}{3}, 6\right\}$.

13. Use the square root property to get $x = \sqrt{81} = 9$ or $x = -\sqrt{81} = -9$.
The solution set is $\{\pm 9\}$.

15. Use the square root property to get $x = \sqrt{14}$ or $x = -\sqrt{14}$.
The solution set is $\{\pm\sqrt{14}\}$.

17. Use the square root property to get $t = \sqrt{48}$ or $t = -\sqrt{48}$.
Write $\sqrt{48}$ in simplest form.
$\sqrt{48} = \sqrt{16}\cdot\sqrt{3} = 4\sqrt{3}$
The solution set is $\{\pm 4\sqrt{3}\}$.

19. This equation has no real number solution because the square of a real number cannot be negative.
The square root property cannot be used because it requires that k be positive.

21. $x^2 = \frac{25}{4}$
$x = \sqrt{\frac{25}{4}}$ or $x = -\sqrt{\frac{25}{4}}$
$x = \frac{5}{2}$ or $x = -\frac{5}{2}$
The solution set is $\left\{\pm\frac{5}{2}\right\}$.

23. $x^2 = 2.25$
$x = \sqrt{2.25}$ or $x = -\sqrt{2.25}$
$x = 1.5$ or $x = -1.5$
The solution set is $\{\pm 1.5\}$.

25. $r^2 - 3 = 0$
$r^2 = 3$
$r = \sqrt{3}$ or $r = -\sqrt{3}$
The solution set is $\{\pm\sqrt{3}\}$.

27. $7x^2 = 4$
$x^2 = \frac{4}{7}$ Divide by 7.
$x = \sqrt{\frac{4}{7}}$ or $x = -\sqrt{\frac{4}{7}}$
$= \frac{\sqrt{4}}{\sqrt{7}}\cdot\frac{\sqrt{7}}{\sqrt{7}}$ $= -\frac{\sqrt{4}}{\sqrt{7}}\cdot\frac{\sqrt{7}}{\sqrt{7}}$
$= \frac{2\sqrt{7}}{7}$ $= -\frac{2\sqrt{7}}{7}$
The solution set is $\left\{\pm\frac{2\sqrt{7}}{7}\right\}$.

29. $3n^2 - 72 = 0$
$3n^2 = 72$
$n^2 = 24$
$n = \pm\sqrt{24} = \pm\sqrt{4\cdot 6} = \pm 2\sqrt{6}$
The solution set is $\{\pm 2\sqrt{6}\}$.

31. $5x^2 + 4 = 8$

$5x^2 = 4$ Subtract 4.

$x^2 = \frac{4}{5}$ Divide by 5.

$x = \sqrt{\frac{4}{5}}$ or $x = -\sqrt{\frac{4}{5}}$

$= \frac{\sqrt{4}}{\sqrt{5}} \cdot \frac{\sqrt{5}}{\sqrt{5}}$ $= -\frac{\sqrt{4}}{\sqrt{5}} \cdot \frac{\sqrt{5}}{\sqrt{5}}$

$= \frac{2\sqrt{5}}{5}$ $= -\frac{2\sqrt{5}}{5}$

The solution set is $\left\{\pm \frac{2\sqrt{5}}{5}\right\}$.

33. $2t^2 + 7 = 61$

$2t^2 = 54$

$t^2 = 27$

$t = \pm\sqrt{27} = \pm\sqrt{9 \cdot 3} = \pm 3\sqrt{3}$

The solution set is $\{\pm 3\sqrt{3}\}$.

35. $-8x^2 = -64$

$x^2 = 8$ Divide by -8.

$x = \pm\sqrt{8} = \pm\sqrt{4 \cdot 2} = \pm 2\sqrt{2}$

The solution set is $\{\pm 2\sqrt{2}\}$.

37. It is not correct to say that the solution set of $x^2 = 81$ is $\{9\}$, because -9 also satisfies the equation.
When we solve an equation, we want to find all values of the variable that satisfy the equation. The completely correct answer is that the solution set of $x^2 = 81$ is $\{\pm 9\}$.

39. Use the square root property.

$x - 3 = \sqrt{25}$ or $x - 3 = -\sqrt{25}$

$x - 3 = 5$ or $x - 3 = -5$

$x = 8$ or $x = -2$

The solution set is $\{-2, 8\}$.

41. Begin by using the square root property.

$x - 8 = \sqrt{27}$ or $x - 8 = -\sqrt{27}$

Now simplify the radical.

$\sqrt{27} = \sqrt{9} \cdot \sqrt{3} = 3\sqrt{3}$

$x - 8 = 3\sqrt{3}$ or $x - 8 = -3\sqrt{3}$

$x = 8 + 3\sqrt{3}$ or $x = 8 - 3\sqrt{3}$

The solution set is $\{8 \pm 3\sqrt{3}\}$.

43. The square root of -13 is not a real number, so there is no real number solution for this equation.

45. $(3x + 2)^2 = 49$

$3x + 2 = \sqrt{49}$ or $3x + 2 = -\sqrt{49}$

$3x + 2 = 7$ or $3x + 2 = -7$

$3x = 5$ or $3x = -9$

$x = \frac{5}{3}$ or $x = -3$

The solution set is $\left\{-3, \frac{5}{3}\right\}$.

47. $(4x - 3)^2 = 9$

$4x - 3 = \sqrt{9}$ or $4x - 3 = -\sqrt{9}$

$4x - 3 = 3$ or $4x - 3 = -3$

$4x = 6$ or $4x = 0$

$x = \frac{6}{4} = \frac{3}{2}$ or $x = 0$

The solution set is $\left\{0, \frac{3}{2}\right\}$.

49. $(5 - 2x)^2 = 30$

$5 - 2x = \sqrt{30}$ or $5 - 2x = -\sqrt{30}$

$-2x = -5 + \sqrt{30}$ or $-2x = -5 - \sqrt{30}$

$x = \frac{-5 + \sqrt{30}}{-2}$ or $x = \frac{-5 - \sqrt{30}}{-2}$

$x = \frac{-5 + \sqrt{30}}{-2} \cdot \frac{-1}{-1}$ or $x = \frac{-5 - \sqrt{30}}{-2} \cdot \frac{-1}{-1}$

$x = \frac{5 - \sqrt{30}}{2}$ or $x = \frac{5 + \sqrt{30}}{2}$

The solution set is $\left\{\frac{5 \pm \sqrt{30}}{2}\right\}$.

51. $(3k+1)^2 = 18$

$3k+1 = \sqrt{18}$ or $3k+1 = -\sqrt{18}$

$3k = -1+3\sqrt{2}$ or $3k = -1-3\sqrt{2}$

Note that $\sqrt{18} = \sqrt{9 \cdot 2} = 3\sqrt{2}.$

$k = \dfrac{-1+3\sqrt{2}}{3}$ or $k = \dfrac{-1-3\sqrt{2}}{3}$

The solution set is $\left\{\dfrac{-1 \pm 3\sqrt{2}}{3}\right\}.$

53. $\left(\dfrac{1}{2}x+5\right)^2 = 12$

$\dfrac{1}{2}x+5 = \sqrt{12}$ or $\dfrac{1}{2}x+5 = -\sqrt{12}$

$\dfrac{1}{2}x = -5+2\sqrt{3}$ or $\dfrac{1}{2}x = -5-2\sqrt{3}$

$x = 2\left(-5+2\sqrt{3}\right)$ or $x = 2\left(-5-2\sqrt{3}\right)$

$x = -10+4\sqrt{3}$ or $x = -10-4\sqrt{3}$

The solution set is $\left\{-10 \pm 4\sqrt{3}\right\}.$

55. $\left(x-\dfrac{1}{8}\right)^2 = \dfrac{1}{64}$

$x-\dfrac{1}{8} = \sqrt{\dfrac{1}{64}}$ or $x-\dfrac{1}{8} = -\sqrt{\dfrac{1}{64}}$

$x-\dfrac{1}{8} = \dfrac{\sqrt{1}}{\sqrt{64}}$ or $x-\dfrac{1}{8} = -\dfrac{\sqrt{1}}{\sqrt{64}}$

$x-\dfrac{1}{8} = \dfrac{1}{8}$ or $x-\dfrac{1}{8} = -\dfrac{1}{8}$

$x = \dfrac{1}{8}+\dfrac{1}{8}$ or $x = -\dfrac{1}{8}+\dfrac{1}{8}$

$x = \dfrac{2}{8} = \dfrac{1}{4}$ or $x = 0$

The solution set is $\left\{0, \dfrac{1}{4}\right\}.$

57. $\left(x-\dfrac{1}{3}\right)^2 = \dfrac{4}{9}$

$x-\dfrac{1}{3} = \sqrt{\dfrac{4}{9}}$ or $x-\dfrac{1}{3} = -\sqrt{\dfrac{4}{9}}$

$x-\dfrac{1}{3} = \dfrac{\sqrt{4}}{\sqrt{9}}$ or $x-\dfrac{1}{3} = -\dfrac{\sqrt{4}}{\sqrt{9}}$

$x-\dfrac{1}{3} = \dfrac{2}{3}$ or $x-\dfrac{1}{3} = -\dfrac{2}{3}$

$x = \dfrac{2}{3}+\dfrac{1}{3}$ or $x = -\dfrac{2}{3}+\dfrac{1}{3}$

$x = \dfrac{3}{3} = 1$ or $x = -\dfrac{1}{3}$

The solution set is $\left\{-\dfrac{1}{3}, 1\right\}.$

59. $\left(x+\dfrac{1}{4}\right)^2 = \dfrac{3}{16}$

$x+\dfrac{1}{4} = \sqrt{\dfrac{3}{16}}$ or $x+\dfrac{1}{4} = -\sqrt{\dfrac{3}{16}}$

$x+\dfrac{1}{4} = \dfrac{\sqrt{3}}{\sqrt{16}}$ or $x+\dfrac{1}{4} = -\dfrac{\sqrt{3}}{\sqrt{16}}$

$x+\dfrac{1}{4} = \dfrac{\sqrt{3}}{4}$ or $x+\dfrac{1}{4} = -\dfrac{\sqrt{3}}{4}$

$x = \dfrac{\sqrt{3}}{4}-\dfrac{1}{4}$ or $x = -\dfrac{\sqrt{3}}{4}-\dfrac{1}{4}$

$x = \dfrac{-1+\sqrt{3}}{4}$ or $x = \dfrac{-1-\sqrt{3}}{4}$

The solution set is $\left\{\dfrac{-1 \pm \sqrt{3}}{4}\right\}.$

61. $(4x-1)^2 - 48 = 0$

$(4x-1)^2 = 48$

$4x-1 = \sqrt{48}$ or $4x-1 = -\sqrt{48}$

$4x-1 = 4\sqrt{3}$ or $4x-1 = -4\sqrt{3}$

$4x = 1+4\sqrt{3}$ or $4x = 1-4\sqrt{3}$

$x = \dfrac{1+4\sqrt{3}}{4}$ or $x = \dfrac{1-4\sqrt{3}}{4}$

The solution set is $\left\{\dfrac{1 \pm 4\sqrt{3}}{4}\right\}.$

63. Jeff's first solution, $\frac{5+\sqrt{30}}{2}$, is equivalent to Linda's second solution, $\frac{-5-\sqrt{30}}{-2}$. This can be verified by multiplying $\frac{5+\sqrt{30}}{2}$ by 1 in the form $\frac{-1}{-1}$. Similarly, Jeff's second solution is equivalent to Linda's first one.

65. $(k+2.14)^2 = 5.46$

$k+2.14 = \sqrt{5.46}$ or $k+2.14 = -\sqrt{5.46}$

$k = -2.14+\sqrt{5.46}$ or $k = -2.14-\sqrt{5.46}$

$k \approx 0.20$ or $k \approx -4.48$

To the nearest hundredth, the solution set is $\{-4.48, 0.20\}$.

67. $(2.11p+3.42)^2 = 9.58$

$2.11p+3.42 = \pm\sqrt{9.58}$

Remember that this represents two equations.

$$2.11p = -3.42 \pm \sqrt{9.58}$$

$$p = \frac{-3.42 \pm \sqrt{9.58}}{2.11}$$

$$\approx -3.09 \text{ or } -0.15$$

To the nearest hundredth, the solution set is $\{-3.09, -0.15\}$.

69. $d = 16t^2$

$4 = 16t^2$ Let $d = 4$.

$t^2 = \frac{4}{16} = \frac{1}{4}$

$t = \pm\sqrt{\frac{1}{4}} = \pm\frac{1}{2}$

Reject $-\frac{1}{2}$ as a solution, since negative time does not make sense. About $\frac{1}{2}$ second elapses between the dropping of the coin and the shot.

71. $A = \pi r^2$

$81\pi = \pi r^2$ Let $A = 81\pi$.

$81 = r^2$ Divide by π.

$r = 9$ or $r = -9$

Discard -9 since the radius cannot be negative. The radius is 9 inches.

73. Let $A = 104.04$ and $P = 100$.

$$A = P(1+r)^2$$

$$104.04 = 100(1+r)^2$$

$$(1+r)^2 = \frac{104.04}{100} = 1.0404$$

$$1+r = \pm\sqrt{1.0404}$$

$$1+r = \pm 1.02$$

$$r = -1 \pm 1.02$$

So $r = -1+1.02 = 0.02$ or $r = -1-1.02 = -2.02$. Reject the solution -2.02. The rate is $r = 0.02$, or 2%.

9.2 Solving Quadratic Equations by Completing the Square

Now Try Exercises

N1. **(a)** Here the middle term, $4x$, must equal $2kx$.

$2kx = 4x$

$k = 2$ Divide by $2x$.

Thus, $k = 2$ and $k^2 = 4$. The required trinomial is $x^2 + 4x + \underline{4}$, which factors as $(x+2)^2$.

(b) Here the middle term, $-22x$, must equal $2kx$.

$2kx = -22x$

$k = -11$ Divide by $2x$.

Thus, $k = -11$ and $k^2 = 121$. The required trinomial is $x^2 - 22x + \underline{121}$, which factors as $(x-11)^2$.

N2. $x^2 - 10x + 8 = 0$

$x^2 - 10x = -8$

$x^2 - 10x + 25 = -8 + 25$

$2kx = -10x$, so $k = -5$ and $k^2 = 25$.

$(x-5)^2 = 17$

Solve for x.

$x-5 = \sqrt{17}$ or $x-5 = -\sqrt{17}$

$x = 5+\sqrt{17}$ or $x = 5-\sqrt{17}$

The solution set is $\{5 \pm \sqrt{17}\}$.

N3. Take half of the coefficient of x and square the result.

$\frac{1}{2}(-6)=-3$, and $(-3)^2=9$.

Add 9 to each side of the equation, and write the left side as a perfect square.

$x^2-6x+9=9+9$

$(x-3)^2=18$ Factor.

Use the square root property to solve for x.

$x-3=\sqrt{18}$ or $x-3=-\sqrt{18}$

$x-3=\sqrt{9}\cdot\sqrt{2}$ or $x-3=-\sqrt{9}\cdot\sqrt{2}$

$x=3+3\sqrt{2}$ or $x=3-3\sqrt{2}$

The solution set is $\{3\pm 3\sqrt{2}\}$.

N4. $x(x+5)=3$

$x^2+5x=3$ Multiply.

Take half the coefficient of x, or $\left(\frac{1}{2}\right)(5)=\frac{5}{2}$, and square the result: $\left(\frac{5}{2}\right)^2=\frac{25}{4}$. Then add $\frac{25}{4}$ to each side.

$x^2+5x+\frac{25}{4}=3+\frac{25}{4}$ Add $\frac{25}{4}$.

$x^2+5x+\frac{25}{4}=\frac{12}{4}+\frac{25}{4}$ $3=\frac{12}{4}$

$\left(x+\frac{5}{2}\right)^2=\frac{37}{4}$ Factor.

Apply the square root property, and solve for x.

$x+\frac{5}{2}=\sqrt{\frac{37}{4}}$ or $x+\frac{5}{2}=-\sqrt{\frac{37}{4}}$

$x+\frac{5}{2}=\frac{\sqrt{37}}{2}$ or $x+\frac{5}{2}=-\frac{\sqrt{37}}{2}$

$x=-\frac{5}{2}+\frac{\sqrt{37}}{2}$ or $x=-\frac{5}{2}-\frac{\sqrt{37}}{2}$

A check verifies that the solution set is $\left\{\frac{-5\pm\sqrt{37}}{2}\right\}$.

N5. $4t^2-4t-3=0$

$4t^2-4t=3$ Add 3.

Divide each side by 4 to get a coefficient of 1 for the t^2-term.

$t^2-t=\frac{3}{4}$ Divide by 4.

Take half the coefficient of t, or $\left(\frac{1}{2}\right)(-1)=-\frac{1}{2}$, and square the result: $\left(-\frac{1}{2}\right)^2=\frac{1}{4}$. Then add $\frac{1}{4}$ to each side.

$t^2-t+\frac{1}{4}=\frac{3}{4}+\frac{1}{4}$ Add $\frac{1}{4}$.

$\left(t-\frac{1}{2}\right)^2=1$ Factor.

Apply the square root property, and solve for t.

$t-\frac{1}{2}=\sqrt{1}$ or $t-\frac{1}{2}=-\sqrt{1}$

$t-\frac{1}{2}=1$ or $t-\frac{1}{2}=-1$

$t=\frac{1}{2}+1$ or $t=\frac{1}{2}-1$

$t=\frac{3}{2}$ or $t=-\frac{1}{2}$

A check verifies that $-\frac{1}{2}$ and $\frac{3}{2}$ are solutions of the original equation. Using a calculator for your check is highly recommended.

The solution set is $\left\{-\frac{1}{2},\frac{3}{2}\right\}$.

N6. $4x^2+9x-9=0$

$4x^2+9x=9$ Add 9.

Divide each side by 4 to get a coefficient of 1 for the x^2-term.

$x^2+\frac{9}{4}x=\frac{9}{4}$ Divide by 4.

Take half the coefficient of x, or $\left(\frac{1}{2}\right)\left(\frac{9}{4}\right)=\frac{9}{8}$, and square the result: $\left(\frac{9}{8}\right)^2=\frac{81}{64}$. Then add $\frac{81}{64}$ to each side.

$x^2+\frac{9}{4}x+\frac{81}{64}=\frac{9}{4}+\frac{81}{64}$ Add $\frac{81}{64}$.

$x^2+\frac{9}{4}x+\frac{81}{64}=\frac{225}{64}$ $\frac{9}{4}=\frac{144}{64}$

$\left(x+\frac{9}{8}\right)^2=\frac{225}{64}$ Factor.

Apply the square root property, and solve for x.

$x+\frac{9}{8}=\sqrt{\frac{225}{64}}$ or $x+\frac{9}{8}=-\sqrt{\frac{225}{64}}$

$x+\frac{9}{8}=\frac{15}{8}$ or $x+\frac{9}{8}=-\frac{15}{8}$

$x=-\frac{9}{8}+\frac{15}{8}$ or $x=-\frac{9}{8}-\frac{15}{8}$

$x=\frac{3}{4}$ or $x=-3$

A check verifies that -3 and $\frac{3}{4}$ are solutions of the original equation.

The solution set is $\left\{-3, \frac{3}{4}\right\}$.

N7. $3t^2-12t+15=0$

$t^2-4t+5=0$ Divide by 3.

$t^2-4t=-5$ Subtract 5.

Add $\left[\frac{1}{2}\cdot(-4)\right]^2=4$ to both sides of the equation.

$t^2-4t+4=-5+4$

$(t-2)^2=-1$

The square root of -1 is not a real number, so the square root property does not apply. This equation has no real solution.

N8. $(x-5)(x+1)=2$

$x^2-4x-5=2$

$x^2-4x=7$

$x^2-4x+4=7+4$ Add $\left[\frac{1}{2}(-4)\right]^2=4$.

$(x-2)^2=11$

Solve for x.

$x-2=\sqrt{11}$ or $x-2=-\sqrt{11}$

$x=2+\sqrt{11}$ or $x=2-\sqrt{11}$

The solution set is $\{2\pm\sqrt{11}\}$.

N9. $-16t^2+64t=s$

$-16t^2+64t=28$ Let $s=28$.

$t^2-4t=-\frac{7}{4}$ Divide by -16.

$t^2-4t+4=-\frac{7}{4}+4$ Add $\left[\frac{1}{2}(-4)\right]^2=4$.

$(t-2)^2=\frac{9}{4}$ Factor.

Solve for t.

$t-2=\sqrt{\frac{9}{4}}$ or $t-2=-\sqrt{\frac{9}{4}}$

$t=2+\frac{3}{2}$ or $t=2-\frac{3}{2}$

$t=3.5$ or $t=0.5$

The ball will be 28 feet above the ground after 0.5 second and again after 3.5 seconds.

Exercises

1. Before completing the square, the coefficient of x^2 must be 1. Dividing each side of the equation by 2 is the correct way to begin solving the equation, and this corresponds to choice D.

3. Here the middle term, $10x$, must equal $2kx$.

$2kx=10x$

$k=5$ Divide by $2x$.

Thus, $k=5$ and $k^2=25$. The required trinomial is $x^2+10x+\underline{25}$, which factors as $(x+5)^2$.

5. Here the middle, term $-20z$, must equal $2kz$.

$2kz=-20z$

$k=-10$ Divide by $2z$.

Thus, $k=-10$ and $k^2=100$. The required trinomial is $z^2-20z+\underline{100}$, which factors as $(z-10)^2$.

7. Here the middle term, $2x$, must equal $2kx$.

$2kx=2x$

$k=1$ Divide by $2x$.

Thus, $k=1$ and $k^2=1$. The required trinomial is $x^2+2x+\underline{1}$, which factors as $(x+1)^2$.

9. Here the middle term, $-5p$, must equal $2kp$.

$2kp = -5p$

$k = -\frac{5}{2}$ Divide by $2p$.

Thus, $k = -\frac{5}{2}$ and $k^2 = \frac{25}{4}$. The required trinomial is $p^2 - 5p + \underline{\frac{25}{4}}$, which factors as $\left(p - \frac{5}{2}\right)^2$.

11. Here the middle term, $\frac{1}{2}x$, must equal $2kx$.

$2kx = \frac{1}{2}x$

$k = \frac{1}{4}$ Divide by $2x$.

Thus, $k = \frac{1}{4}$ and $k^2 = \frac{1}{16}$. The required trinomial is $x^2 + \frac{1}{2}x + \underline{\frac{1}{16}}$, which factors as $\left(x + \frac{1}{4}\right)^2$.

13. Here the middle term, $-0.4x$, must equal $2kx$.

$2kx = -0.4x$

$k = -0.2$ Divide by $2x$.

Thus, $k = -0.2$ and $k^2 = 0.04$. The required trinomial is $x^2 - 0.4x + \underline{0.04}$, which factors as $(x - 0.2)^2$.

15. $x^2 + 4x = 1$

Take half the coefficient of x and square it.

$\frac{1}{2} \cdot \underline{4} = 2$, and $\underline{2}^2 = 4$.

Add $\underline{4}$ to each side of the equation.

$x^2 + 4x + \underline{4} = 1 + 4$

Factor and add.

$\underline{(x+2)^2 = 5}$

Complete the solution.
Use the square root property.

$x + 2 = \sqrt{5}$ or $x + 2 = -\sqrt{5}$

$x = -2 + \sqrt{5}$ or $x = -2 - \sqrt{5}$

A check verifies that the solution set is $\{-2 \pm \sqrt{5}\}$. Using a calculator for your check is highly recommended.

17. Take half of the coefficient of x and square it. Half of -4 is -2, and $(-2)^2 = 4$. Add 4 to each side of the equation, and write the left side as a perfect square.

$x^2 - 4x + 4 = -3 + 4$

$(x-2)^2 = 1$

Use the square root property.

$x - 2 = \sqrt{1}$ or $x - 2 = -\sqrt{1}$

$x - 2 = 1$ or $x - 2 = -1$

$x = 3$ or $x = 1$

A check verifies that the solution set is $\{1, 3\}$.

19. Add 5 to each side.

$x^2 + 2x = 5$

Take half the coefficient of x and square it.

$\frac{1}{2}(2) = 1$, and $1^2 = 1$.

Add 1 to each side of the equation, and write the left side as a perfect square.

$x^2 + 2x + 1 = 5 + 1$

$(x+1)^2 = 6$

Use the square root property.

$x + 1 = \sqrt{6}$ or $x + 1 = -\sqrt{6}$

$x = -1 + \sqrt{6}$ or $x = -1 - \sqrt{6}$

A check verifies that the solution set is $\{-1 \pm \sqrt{6}\}$. Using a calculator for your check is highly recommended.

21. $x^2 - 8x = -4$

$x^2 - 8x + 16 = -4 + 16$ $\left[\frac{1}{2}(-8)\right]^2 = 16$

$(x-4)^2 = 12$ Factor.

Solve for x.

$x - 4 = \sqrt{12}$ or $x - 4 = -\sqrt{12}$

$x - 4 = \sqrt{4} \cdot \sqrt{3}$ or $x - 4 = -\sqrt{4} \cdot \sqrt{3}$

$x = 4 + 2\sqrt{3}$ or $x = 4 - 2\sqrt{3}$

The solution set is $\{4 \pm 2\sqrt{3}\}$.

23. The left-hand side of this equation is already a perfect square.

$(x+3)^2 = 0$

$x + 3 = 0$

$x = -3$

A check verifies that the solution set is $\{-3\}$.

25. $x^2 - 16x = 0$

$x^2 - 16x + 64 = 0 + 64 \quad \left[\frac{1}{2}(-16)\right]^2 = 64$

$(x-8)^2 = 64 \quad$ Factor.

Solve for x.

$x - 8 = \sqrt{64}$ or $x - 8 = -\sqrt{64}$

$x - 8 = 8$ or $x - 8 = -8$

$x = 8 + 8$ or $x = 8 - 8$

$x = 16$ or $x = 0$

The solution set is $\{0, 16\}$.

27. $x(x-3) = 1$

$x^2 - 3x = 1 \quad$ Multiply.

Take half the coefficient of x, or $\left(\frac{1}{2}\right)(-3) = -\frac{3}{2}$, and square the result: $\left(-\frac{3}{2}\right)^2 = \frac{9}{4}$. Then add $\frac{9}{4}$ to each side.

$x^2 - 3x + \frac{9}{4} = 1 + \frac{9}{4} \quad$ Add $\frac{9}{4}$.

$x^2 - 3x + \frac{9}{4} = \frac{4}{4} + \frac{9}{4} \quad 1 = \frac{4}{4}$

$\left(x - \frac{3}{2}\right)^2 = \frac{13}{4} \quad$ Factor.

Apply the square root property, and solve for x.

$x - \frac{3}{2} = \sqrt{\frac{13}{4}}$ or $x - \frac{3}{2} = -\sqrt{\frac{13}{4}}$

$x - \frac{3}{2} = \frac{\sqrt{13}}{2}$ or $x - \frac{3}{2} = -\frac{\sqrt{13}}{2}$

$x = \frac{3}{2} + \frac{\sqrt{13}}{2}$ or $x = \frac{3}{2} - \frac{\sqrt{13}}{2}$

A check verifies that the solution set is $\left\{\frac{3 \pm \sqrt{13}}{2}\right\}$.

29. $x(x+3) = -1$

$x^2 + 3x = -1 \quad$ Multiply.

Take half the coefficient of x, or $\left(\frac{1}{2}\right)(3) = \frac{3}{2}$, and square the result: $\left(\frac{3}{2}\right)^2 = \frac{9}{4}$. Then add $\frac{9}{4}$ to each side.

$x^2 + 3x + \frac{9}{4} = -1 + \frac{9}{4} \quad$ Add $\frac{9}{4}$.

$x^2 + 3x + \frac{9}{4} = \frac{-4}{4} + \frac{9}{4} \quad -1 = \frac{-4}{4}$

$\left(x + \frac{3}{2}\right)^2 = \frac{5}{4} \quad$ Factor.

Apply the square root property, and solve for x.

$x + \frac{3}{2} = \sqrt{\frac{5}{4}}$ or $x + \frac{3}{2} = -\sqrt{\frac{5}{4}}$

$x + \frac{3}{2} = \frac{\sqrt{5}}{2}$ or $x + \frac{3}{2} = -\frac{\sqrt{5}}{2}$

$x = -\frac{3}{2} + \frac{\sqrt{5}}{2}$ or $x = -\frac{3}{2} - \frac{\sqrt{5}}{2}$

A check verifies that the solution set is $\left\{\frac{-3 \pm \sqrt{5}}{2}\right\}$.

31. Divide each side by 4 so that the coefficient of x^2 is 1.

$x^2 + x = \frac{3}{4}$

The coefficient of x is 1. Take half of 1, square the result, and add this square to each side.

$\frac{1}{2}(1) = \frac{1}{2}$ and $\left(\frac{1}{2}\right)^2 = \frac{1}{4}$

$x^2 + x + \frac{1}{4} = \frac{3}{4} + \frac{1}{4}$

The left-hand side can then be written as a perfect square.

$\left(x + \frac{1}{2}\right)^2 = 1$

Use the square root property.

$x + \frac{1}{2} = 1$ or $x + \frac{1}{2} = -1$

$x = -\frac{1}{2} + 1$ or $x = -\frac{1}{2} - 1$

$x = \frac{1}{2}$ or $x = -\frac{3}{2}$

A check verifies that the solution set is $\left\{-\frac{3}{2}, \frac{1}{2}\right\}$.

33. Divide each side by 2.

$$p^2 - p + \frac{3}{2} = 0$$

Subtract $\frac{3}{2}$ from both sides.

$$p^2 - p = -\frac{3}{2}$$

Take half the coefficient of p and square it.

$$\frac{1}{2}(-1) = -\frac{1}{2}, \text{ and } \left(-\frac{1}{2}\right)^2 = \frac{1}{4}.$$

Add $\frac{1}{4}$ to each side of the equation.

$$p^2 - p + \frac{1}{4} = -\frac{3}{2} + \frac{1}{4}$$

Factor on the left side and add on the right.

$$\left(p - \frac{1}{2}\right)^2 = -\frac{5}{4}$$

The square root of $-\frac{5}{4}$ is not a real number, so there is no real solution.

35. Divide each side by 3.

$$x^2 - 3x + \frac{5}{3} = 0$$

Put constant terms on one side.

$$x^2 - 3x = -\frac{5}{3}$$

Take half of the coefficient of x and square it.

$$\frac{1}{2}(-3) = -\frac{3}{2}, \text{ and } \left(-\frac{3}{2}\right)^2 = \frac{9}{4}.$$

Add $\frac{9}{4}$ to each side of the equation.

$$x^2 - 3x + \frac{9}{4} = -\frac{5}{3} + \frac{9}{4}$$

$$\left(x - \frac{3}{2}\right)^2 = \frac{7}{12}$$

Use the square root property.

$$x - \frac{3}{2} = \sqrt{\frac{7}{12}} \quad \text{or} \quad x - \frac{3}{2} = -\sqrt{\frac{7}{12}}$$

$$x - \frac{3}{2} = \frac{\sqrt{7}}{\sqrt{12}} \cdot \frac{\sqrt{3}}{\sqrt{3}} \quad \text{or} \quad x - \frac{3}{2} = -\frac{\sqrt{7}}{\sqrt{12}} \cdot \frac{\sqrt{3}}{\sqrt{3}}$$

$$x - \frac{9}{6} = \frac{\sqrt{21}}{6} \quad \text{or} \quad x - \frac{9}{6} = -\frac{\sqrt{21}}{6}$$

$$x = \frac{9}{6} + \frac{\sqrt{21}}{6} \quad \text{or} \quad x = \frac{9}{6} - \frac{\sqrt{21}}{6}$$

$$x = \frac{9 + \sqrt{21}}{6} \quad \text{or} \quad x = \frac{9 - \sqrt{21}}{6}$$

A check verifies that the solution set is $\left\{\frac{9 \pm \sqrt{21}}{6}\right\}$.

37. Divide each side by 3.

$$x^2 + \frac{7}{3}x = \frac{4}{3}$$

Take half of the coefficient of x and square it.

$$\frac{1}{2}\left(\frac{7}{3}\right) = \frac{7}{6}, \text{ and } \left(\frac{7}{6}\right)^2 = \frac{49}{36}.$$

Add $\frac{49}{36}$ to each side of the equation.

$$x^2 + \frac{7}{3}x + \frac{49}{36} = \frac{4}{3} + \frac{49}{36}$$

$$\left(x + \frac{7}{6}\right)^2 = \frac{97}{36}$$

Use the square root property.

$$x + \frac{7}{6} = \sqrt{\frac{97}{36}} \quad \text{or} \quad x + \frac{7}{6} = -\sqrt{\frac{97}{36}}$$

$$x + \frac{7}{6} = \frac{\sqrt{97}}{6} \quad \text{or} \quad x + \frac{7}{6} = -\frac{\sqrt{97}}{6}$$

$$x = -\frac{7}{6} + \frac{\sqrt{97}}{6} \quad \text{or} \quad x = -\frac{7}{6} - \frac{\sqrt{97}}{6}$$

$$x = \frac{-7 + \sqrt{97}}{6} \quad \text{or} \quad x = \frac{-7 - \sqrt{97}}{6}$$

A check verifies that the solution set is $\left\{\frac{-7 \pm \sqrt{97}}{6}\right\}$.

39.

$$(x+3)(x-1) = 5$$

$$x^2 + 2x - 3 = 5$$

$$x^2 + 2x = 8$$

$$x^2 + 2x + 1 = 8 + 1$$

$$(x+1)^2 = 9$$

Use the square root property.

$$x + 1 = 3 \quad \text{or} \quad x + 1 = -3$$

$$x = 2 \quad \text{or} \quad x = -4$$

A check verifies that the solution set is $\{-4, 2\}$.

41. $(r-3)(r-5)=2$

$r^2-8r+15=2$

$r^2-8r=-13$

$r^2-8r+16=-13+16$

$(r-4)^2=3$

Use the square root property.

$r-4=\sqrt{3}$ or $r-4=-\sqrt{3}$

$r=4+\sqrt{3}$ or $r=4-\sqrt{3}$

A check verifies that the solution set is $\{4\pm\sqrt{3}\}$.

43. Divide each side by -1.

$x^2-2x=5$

Take half of the coefficient of x and square it. Half of -2 is -1, and $(-1)^2=1$. Add 1 to each side of the equation, and write the left side as a perfect square.

$x^2-2x+1=5+1$

$(x-1)^2=6$

Use the square root property.

$x-1=\sqrt{6}$ or $x-1=-\sqrt{6}$

$x=1+\sqrt{6}$ or $x=1-\sqrt{6}$

A check verifies that the solution set is $\{1\pm\sqrt{6}\}$.

45. $5x^2+6x-11=0$

$5x^2+6x=11$ Add 11.

Divide each side by 5 to get a coefficient of 1 for the x^2-term.

$x^2+\frac{6}{5}x=\frac{11}{5}$ Divide by 5.

Take half the coefficient of x, or $\frac{1}{2}\left(\frac{6}{5}\right)=\frac{3}{5}$, and square the result: $\left(\frac{3}{5}\right)^2=\frac{9}{25}$. Then add $\frac{9}{25}$ to each side.

$x^2+\frac{6}{5}x+\frac{9}{25}=\frac{11}{5}+\frac{9}{25}$ Add $\frac{9}{25}$.

$x^2+\frac{6}{5}x+\frac{9}{25}=\frac{55}{25}+\frac{9}{25}$ $\frac{11}{5}=\frac{55}{25}$

$\left(x+\frac{3}{5}\right)^2=\frac{64}{25}$

Apply the square root property, and solve for x.

$x+\frac{3}{5}=\sqrt{\frac{64}{25}}$ or $x+\frac{3}{5}=-\sqrt{\frac{64}{25}}$

$x+\frac{3}{5}=\frac{8}{5}$ or $x+\frac{3}{5}=-\frac{8}{5}$

$x=-\frac{3}{5}+\frac{8}{5}$ or $x=-\frac{3}{5}-\frac{8}{5}$

$x=\frac{5}{5}=1$ or $x=-\frac{11}{5}$

A check verifies that $-\frac{11}{5}$ and 1 are solutions of the original equation.

The solution set is $\left\{-\frac{11}{5},1\right\}$.

47. $-3x^2+11x+42=0$

$-3x^2+11x=-42$ Subtract -42.

Divide each side by -3 to get a coefficient of 1 for the x^2-term.

$x^2-\frac{11}{3}x=\frac{-42}{-3}$ Divide by -3.

$x^2-\frac{11}{3}x=14$

Take half the coefficient of x, or $\left(\frac{1}{2}\right)\left(-\frac{11}{3}\right)=-\frac{11}{6}$, and square the result: $\left(-\frac{11}{6}\right)^2=\frac{121}{36}$. Then add $\frac{121}{36}$ to each side.

$x^2-\frac{11}{3}x+\frac{121}{36}=14+\frac{121}{36}$ Add $\frac{121}{36}$.

$x^2-\frac{11}{3}x+\frac{121}{36}=\frac{504}{36}+\frac{121}{36}$ $14=\frac{504}{36}$

$\left(x-\frac{11}{6}\right)^2=\frac{625}{36}$

Apply the square root property, and solve for x.

$x-\frac{11}{6}=\sqrt{\frac{625}{36}}$ or $x-\frac{11}{6}=-\sqrt{\frac{625}{36}}$

$x-\frac{11}{6}=\frac{25}{6}$ or $x-\frac{11}{6}=-\frac{25}{6}$

$x=\frac{11}{6}+\frac{25}{6}$ or $x=\frac{11}{6}-\frac{25}{6}$

$x=\frac{36}{6}$ or $x=-\frac{14}{6}$

$x=6$ or $x=-\frac{7}{3}$

A check verifies that $-\frac{7}{3}$ and 6 are solutions of the original equation. Using a calculator for your check is highly recommended.

The solution set is $\left\{-\frac{7}{3}, 6\right\}$.

49.
$$3r^2 - 2 = 6r + 3$$
$$3r^2 - 6r = 5$$
$$r^2 - 2r = \frac{5}{3}$$
$$r^2 - 2r + 1 = \frac{5}{3} + 1$$
$$(r-1)^2 = \frac{8}{3}$$
$$r - 1 = \pm\sqrt{\frac{8}{3}}$$
$$r = 1 \pm \sqrt{\frac{8}{3}}$$

Simplify the radical.
$$\sqrt{\frac{8}{3}} = \frac{\sqrt{8}}{\sqrt{3}} = \frac{2\sqrt{2}}{\sqrt{3}} \cdot \frac{\sqrt{3}}{\sqrt{3}} = \frac{2\sqrt{6}}{3}$$

Substitute $\frac{2\sqrt{6}}{3}$ for $\sqrt{\frac{8}{3}}$.
$$r = 1 \pm \sqrt{\frac{8}{3}}$$
$$r = 1 \pm \frac{2\sqrt{6}}{3}$$
$$r = \frac{3}{3} \pm \frac{2\sqrt{6}}{3} = \frac{3 \pm 2\sqrt{6}}{3}$$

(a) The solution set with *exact* values is $\left\{\frac{3 \pm 2\sqrt{6}}{3}\right\}$.

(b) $\frac{3 + 2\sqrt{6}}{3} \approx 2.633$

$\frac{3 - 2\sqrt{6}}{3} \approx -0.633$

The solution set with *approximate* values is $\{-0.633, 2.633\}$.

51.
$$(x+1)(x+3) = 2$$
$$x^2 + 3x + x + 3 = 2$$
$$x^2 + 4x = -1$$
$$x^2 + 4x + 4 = -1 + 4$$
$$(x+2)^2 = 3$$
$$x + 2 = \pm\sqrt{3}$$
$$x = -2 \pm \sqrt{3}$$

(a) The solution set with *exact* values is $\{-2 \pm \sqrt{3}\}$.

(b) $-2 + \sqrt{3} \approx -0.268$

$-2 - \sqrt{3} \approx -3.732$

The solution set with *approximate* values is $\{-3.732, -0.268\}$.

53.
$$s = -13t^2 + 104t$$
$$195 = -13t^2 + 104t \quad \text{Let } s = 195.$$
$$-15 = t^2 - 8t \quad \text{Divide by } -13.$$

Add $\left[\frac{1}{2}(-8)\right]^2 = 16$ to both sides of the equation.
$$t^2 - 8t + 16 = -15 + 16$$
$$(t-4)^2 = 1$$
$$t - 4 = \pm\sqrt{1} = \pm 1$$
$$t = 4 \pm 1$$
$$= 3 \text{ or } 5$$

The object will be at a height of 195 feet at 3 seconds (on the way up) and 5 seconds (on the way down).

55. Find the value of t when $s = 80$.
$$80 = -16t^2 + 96t \quad \text{Let } s = 80.$$
$$-16t^2 + 96t = 80$$
$$t^2 - 6t = -5 \quad \text{Divide by } -16.$$
$$t^2 - 6t + 9 = -5 + 9 \quad \text{Add 9.}$$
$$(t-3)^2 = 4$$

Apply the square root property, and solve for t.
$$t - 3 = \sqrt{4} \quad \text{or} \quad t - 3 = -\sqrt{4}$$
$$t - 3 = 2 \quad \text{or} \quad t - 3 = -2$$
$$t = 3 + 2 \quad \text{or} \quad t = 3 - 2$$
$$t = 5 \quad \text{or} \quad t = 1$$

The object will reach a height of 80 feet at 1 second (on the way up) and at 5 seconds (on the way down).

57. Let x be the width of the pen.
Then $175 - x$ is the length of the pen.
Use the formula for the area of a rectangle.

$$A = LW$$
$$7500 = (175 - x)x$$
$$7500 = 175x - x^2$$

$x^2 - 175x + 7500 = 0$

Solve this quadratic equation by completing the square.

$$x^2 - 175x = -7500$$

$$x^2 - 175x + \frac{30,625}{4} = -\frac{30,000}{4} + \frac{30,625}{4}$$

Add $\left(\frac{175}{2}\right)^2 = \frac{30,625}{4}$ to both sides.

$$\left(x - \frac{175}{2}\right)^2 = \frac{625}{4}$$

$$x - \frac{175}{2} = \pm\sqrt{\frac{625}{4}} = \pm\frac{25}{2}$$

$$x = \frac{175}{2} \pm \frac{25}{2}$$

Solve for x.

$x = \frac{175}{2} + \frac{25}{2}$ or $x = \frac{175}{2} - \frac{25}{2}$

$x = \frac{200}{2}$ or $x = \frac{150}{2}$

$x = 100$ or $x = 75$

If $x = 100$, $175 - x = 175 - 100 = 75$.
If $x = 75$, $175 - x = 175 - 75 = 100$.
The dimensions of the pen are 75 feet by 100 feet.

59. Let $x =$ the distance the slower car traveled.
Then $x + 7 =$ the distance the faster car traveled.
Since the cars traveled at right angles, a right triangle is formed with hypotenuse of length 17. Use the Pythagorean theorem with $a = x$, $b = x + 7$, and $c = 17$.

$$a^2 + b^2 = c^2$$
$$x^2 + (x + 7)^2 = 17^2$$
$$x^2 + \left(x^2 + 14x + 49\right) = 289$$
$$2x^2 + 14x = 240$$
$$x^2 + 7x = 120$$

Add $\left[\frac{1}{2}(7)\right]^2 = \frac{49}{4}$ to both sides.

$$x^2 + 7x + \frac{49}{4} = 120 + \frac{49}{4}$$

$$\left(x + \frac{7}{2}\right)^2 = \frac{529}{4}$$

$$x + \frac{7}{2} = \pm\sqrt{\frac{529}{4}} = \pm\frac{23}{2}$$

$$x = -\frac{7}{2} \pm \frac{23}{2}$$
$$= 8 \text{ or } -15$$

Discard -15 since distance cannot be negative.
The slower car traveled 8 miles.

61. The side has length x, so the area of the original square is $x \cdot x = x^2$.

63. The area of a small square is $1 \cdot 1 = 1$, so the area of 16 small squares is 16, and the area of the figure is $x^2 + 8x + 16$.

9.3 Solving Quadratic Equations by the Quadratic Formula

Now Try Exercises

N1. **(a)** $3x^2 - 7x + 4 = 0$ has the form of the standard quadratic equation $ax^2 + bx + c = 0$.
Thus, $a = 3$, $b = -7$, and $c = 4$.

(b) Rewrite in $ax^2 + bx + c = 0$ form.
$x^2 + 2x - 3 = 0$
Then $a = 1$, $b = 2$, and $c = -3$.

(c) $2x^2 - 4x = 0$ has the form of the standard quadratic equation $ax^2 + bx + c = 0$.
Thus, $a = 2$, $b = -4$, and $c = 0$.

(d) $2(2x + 1)(x - 5) = -3$

$2\left(2x^2 - 9x - 5\right) = -3$ FOIL

$4x^2 - 18x - 10 = -3$ Multiply by 2.

$4x^2 - 18x - 7 = 0$ Add 3.

The last equation is in standard form. The values of a, b, and c are 4, -18, and -7, respectively.

N2. The quadratic equation is in standard form, so $a=3$, $b=5$, and $c=-2$. Substitute these values into the quadratic formula.

$$x=\frac{-b\pm\sqrt{b^2-4ac}}{2a}$$
$$x=\frac{-5\pm\sqrt{5^2-4(3)(-2)}}{2(3)}$$
$$x=\frac{-5\pm\sqrt{25+24}}{6}$$
$$x=\frac{-5\pm\sqrt{49}}{6}$$
$$x=\frac{-5\pm 7}{6}$$

Solve for x.

$$x=\frac{-5+7}{6} \quad \text{or} \quad x=\frac{-5-7}{6}$$
$$x=\frac{2}{6} \quad \text{or} \quad x=\frac{-12}{6}$$
$$x=\frac{1}{3} \quad \text{or} \quad x=-2$$

A check verifies that the solution set is $\left\{-2, \frac{1}{3}\right\}$.

N3. Write the equation in standard form.

$x^2-6x+2=0$

Substitute $a=1$, $b=-6$, and $c=2$ into the quadratic formula.

$$x=\frac{-b\pm\sqrt{b^2-4ac}}{2a}$$
$$x=\frac{-(-6)\pm\sqrt{(-6)^2-4(1)(2)}}{2(1)}$$
$$x=\frac{6\pm\sqrt{36-8}}{2}=\frac{6\pm\sqrt{28}}{2}$$
$$=\frac{6\pm\sqrt{4}\cdot\sqrt{7}}{2}=\frac{6\pm 2\sqrt{7}}{2}$$
$$=\frac{2\left(3\pm\sqrt{7}\right)}{2}=3\pm\sqrt{7}$$

The solution set is $\left\{3\pm\sqrt{7}\right\}$.

N4. Write the equation in standard form.

$16x^2-8x+1=0$

Substitute $a=16$, $b=-8$, and $c=1$ into the quadratic formula.

$$x=\frac{-b\pm\sqrt{b^2-4ac}}{2a}$$
$$x=\frac{-(-8)\pm\sqrt{(-8)^2-4(16)(1)}}{2(16)}$$
$$x=\frac{8\pm\sqrt{64-64}}{32}$$
$$x=\frac{8\pm 0}{32}=\frac{8}{32}=\frac{8\cdot 1}{8\cdot 4}=\frac{1}{4}$$

Since there is just one solution, $\frac{1}{4}$, the trinomial $16x^2-8x+1$ is a perfect square. The solution set is $\left\{\frac{1}{4}\right\}$.

N5. Write the equation in standard form.

$$\frac{1}{12}x^2=\frac{1}{2}x-\frac{1}{3}$$
$$x^2=6x-4 \qquad \text{Multiply by 12.}$$
$$x^2-6x+4=0 \qquad \text{Standard form}$$

Substitute $a=1$, $b=-6$, and $c=4$ into the quadratic formula.

$$x=\frac{-b\pm\sqrt{b^2-4ac}}{2a}$$
$$x=\frac{-(-6)\pm\sqrt{(-6)^2-4(1)(4)}}{2(1)}$$
$$x=\frac{6\pm\sqrt{36-16}}{2}=\frac{6\pm\sqrt{20}}{2}$$
$$=\frac{6\pm\sqrt{4}\cdot\sqrt{5}}{2}=\frac{6\pm 2\sqrt{5}}{2}$$
$$=\frac{2\left(3\pm\sqrt{5}\right)}{2}=3\pm\sqrt{5}$$

The solution set is $\left\{3\pm\sqrt{5}\right\}$.

Exercises

1. $3x^2+7x+4=0$ has the form of the standard quadratic equation $ax^2+bx+c=0$. Thus, $a=3$, $b=7$, and $c=4$. This corresponds to choice E.

3. Write the equation in standard form.

$7+3x+4x^2=0$

$4x^2+3x+7=0$ Rearrange terms.

Thus, $a=4$, $b=3$, and $c=7$. This corresponds to choice A.

5. Write the equation in standard form.

$7x^2+3+4x=0$

$7x^2+4x+3=0$ Rearrange terms.

Thus, $a=7$, $b=4$, and $c=3$. This corresponds to choice D.

7. Write the equation in standard form.

$3x^2+4x=8$

$3x^2+4x-8=0$ Subtract 8.

Match the coefficients of the quadratic equation with the letters *a*, *b*, and *c* of the standard quadratic equation $ax^2+bx+c=0$.

In this case, $a=3$, $b=4$, and $c=-8$.

9. Write the equation in standard form.

$-8x^2=2x+3$

$-8x^2-2x=3$ Subtract $2x$.

$-8x^2-2x-3=0$ Subtract 3.

Match the coefficients of the quadratic equation with the letters *a*, *b*, and *c* of the standard quadratic equation $ax^2+bx+c=0$.

In this case, $a=-8$, $b=-2$, and $c=-3$.

11. First, write the equation in standard form, $ax^2+bx+c=0$.

$3x^2-4x-2=0$

Now, identify the values: $a=3$, $b=-4$, and $c=-2$.

13. Write the equation in standard form.

$3x^2+7x=0$

Now, identify the values: $a=3$, $b=7$, and $c=0$.

15. $(x-3)(x+4)=0$

$x^2+x-12=0$ FOIL

The values are $a=1$, $b=1$, and $c=-12$.

17. $9(x-1)(x+2)=8$

$9\left(x^2+x-2\right)=8$ FOIL

$9x^2+9x-18=8$ Distributive property

$9x^2+9x-26=0$ Standard form

The values are $a=9$, $b=9$, and $c=-26$.

19. $2a$ should be the denominator for $-b$ as well.

The correct formula is $x=\dfrac{-b\pm\sqrt{b^2-4ac}}{2a}$.

21. Use $a=1$, $b=12$, and $c=-13$. Substitute these values into the quadratic formula.

$$x=\frac{-b\pm\sqrt{b^2-4ac}}{2a}$$

$$x=\frac{-12\pm\sqrt{12^2-4(1)(-13)}}{2(1)}$$

$$=\frac{-12\pm\sqrt{144+52}}{2}$$

$$=\frac{-12\pm\sqrt{196}}{2}=\frac{-12\pm 14}{2}$$

Solve for *x*.

$x=\dfrac{-12+14}{2}$ or $x=\dfrac{-12-14}{2}$

$=\dfrac{2}{2}$ or $=\dfrac{-26}{2}$

$=1$ or $=-13$

The solution set is $\{-13, 1\}$.

23. Write the equation in standard form.

$2x^2+12x+5=0$

Substitute $a=2,\ b=12,$ and $c=5$ into the quadratic formula.

$$x=\frac{-b\pm\sqrt{b^2-4ac}}{2a}$$

$$x=\frac{-12\pm\sqrt{12^2-4(2)(5)}}{2(2)}$$

$$=\frac{-12\pm\sqrt{144-40}}{4}=\frac{-12\pm\sqrt{104}}{4}$$

$$=\frac{-12\pm\sqrt{4}\cdot\sqrt{26}}{4}=\frac{-12\pm2\sqrt{26}}{4}$$

$$=\frac{2\left(-6\pm\sqrt{26}\right)}{2\cdot2}=\frac{-6\pm\sqrt{26}}{2}$$

The solution set is $\left\{\frac{-6\pm\sqrt{26}}{2}\right\}$.

25. Substitute $a=1,\ b=-4,$ and $c=4$ into the quadratic formula.

$$p=\frac{-b\pm\sqrt{b^2-4ac}}{2a}$$

$$p=\frac{-(-4)\pm\sqrt{(-4)^2-4(1)(4)}}{2(1)}$$

$$=\frac{4\pm\sqrt{16-16}}{2}$$

$$=\frac{4\pm0}{2}=\frac{4}{2}=2$$

The solution set is $\{2\}$.

27. Write the equation in standard form.

$2x^2-3x-5=0$

Substitute $a=2,\ b=-3,$ and $c=-5$ into the quadratic formula.

$$x=\frac{-b\pm\sqrt{b^2-4ac}}{2a}$$

$$x=\frac{-(-3)\pm\sqrt{(-3)^2-4(2)(-5)}}{2(2)}$$

$$=\frac{3\pm\sqrt{9+40}}{4}$$

$$=\frac{3\pm\sqrt{49}}{4}=\frac{3\pm7}{4}$$

Solve for x.

$$x=\frac{3+7}{4}\quad\text{or}\quad x=\frac{3-7}{4}$$

$$=\frac{10}{4}\quad\text{or}\quad=\frac{-4}{4}$$

$$=\frac{5}{2}\quad\text{or}\quad=-1$$

The solution set is $\left\{-1,\frac{5}{2}\right\}$.

29. Substitute $a=6,\ b=6,$ and $c=0$ into the quadratic formula.

$$x=\frac{-b\pm\sqrt{b^2-4ac}}{2a}$$

$$x=\frac{-6\pm\sqrt{6^2-4(6)(0)}}{2(6)}$$

$$=\frac{-6\pm\sqrt{36-0}}{12}$$

$$=\frac{-6\pm6}{12}$$

Solve for x.

$$x=\frac{-6+6}{12}\quad\text{or}\quad x=\frac{-6-6}{12}$$

$$=\frac{0}{12}\quad\text{or}\quad=\frac{-12}{12}$$

$$=0\quad\text{or}\quad=-1$$

The solution set is $\{-1,0\}$.

31. Write the equation in standard form.

$7x^2-12x=0$

Substitute $a=7,\ b=-12,$ and $c=0$ into the quadratic formula.

$$x=\frac{-b\pm\sqrt{b^2-4ac}}{2a}$$

$$x=\frac{-(-12)\pm\sqrt{(-12)^2-4(7)(0)}}{2(7)}$$

$$=\frac{12\pm\sqrt{144-0}}{14}$$

$$=\frac{12\pm12}{14}$$

Solve for x.

$$x = \frac{12+12}{14} \quad \text{or} \quad x = \frac{12-12}{14}$$
$$= \frac{24}{14} \quad \text{or} \quad = \frac{0}{14}$$
$$= \frac{12}{7} \quad \text{or} \quad = 0$$

The solution set is $\left\{0, \frac{12}{7}\right\}$.

33. Substitute $a = 1$, $b = 0$, and $c = -24$ into the quadratic formula.

$$x = \frac{-b \pm \sqrt{b^2 - 4ac}}{2a}$$
$$x = \frac{-0 \pm \sqrt{0^2 - 4(1)(-24)}}{2(1)}$$
$$= \frac{\pm\sqrt{96}}{2} = \frac{\pm\sqrt{16}\cdot\sqrt{6}}{2}$$
$$= \frac{\pm 4\sqrt{6}}{2} = \pm 2\sqrt{6}$$

The solution set is $\left\{\pm 2\sqrt{6}\right\}$.

35. Substitute $a = 25$, $b = 0$, and $c = -4$ into the quadratic formula.

$$x = \frac{-b \pm \sqrt{b^2 - 4ac}}{2a}$$
$$x = \frac{-0 \pm \sqrt{0^2 - 4(25)(-4)}}{2(25)}$$
$$= \frac{\pm\sqrt{400}}{50} = \frac{\pm 20}{50} = \pm\frac{2}{5}$$

The solution set is $\left\{\pm\frac{2}{5}\right\}$.

37. Write the equation in standard form.

$3x^2 - 12x + 4 = 0$

Substitute $a = 3$, $b = -12$, and $c = 4$ into the quadratic formula.

$$x = \frac{-b \pm \sqrt{b^2 - 4ac}}{2a}$$
$$x = \frac{-(-12) \pm \sqrt{(-12)^2 - 4(3)(4)}}{2(3)}$$
$$= \frac{12 \pm \sqrt{144 - 48}}{6}$$
$$= \frac{12 \pm \sqrt{96}}{6} = \frac{12 \pm \sqrt{16}\cdot\sqrt{6}}{6}$$
$$= \frac{12 \pm 4\sqrt{6}}{6} = \frac{2\left(6 \pm 2\sqrt{6}\right)}{2\cdot 3}$$
$$= \frac{6 \pm 2\sqrt{6}}{3}$$

The solution set is $\left\{\frac{6 \pm 2\sqrt{6}}{3}\right\}$.

39. Write the equation in standard form.

$-2x^2 + 3x - 2 = 0$

Substitute $a = -2$, $b = 3$, and $c = -2$ into the quadratic formula.

$$x = \frac{-b \pm \sqrt{b^2 - 4ac}}{2a}$$
$$x = \frac{-3 \pm \sqrt{3^2 - 4(-2)(-2)}}{2(-2)}$$
$$= \frac{-3 \pm \sqrt{9-16}}{-4}$$
$$= \frac{-3 \pm \sqrt{-7}}{-4}$$

The square root of -7 is not a real number, so the square root property does not apply. This equation has no real solution.

41. Substitute $a = 2$, $b = 1$, and $c = 5$ into the quadratic formula.

$$x = \frac{-1 \pm \sqrt{1^2 - 4(2)(5)}}{2(2)}$$
$$= \frac{-1 \pm \sqrt{1 - 40}}{4}$$
$$= \frac{-1 \pm \sqrt{-39}}{4}$$

The square root of -39 is not a real number, so the square root property does not apply. This equation has no real solution.

43. Write the equation in standard form.

$$(x+3)(x+2)=15$$
$$x^2+5x+6=15$$
$$x^2+5x-9=0$$

Use $a=1$, $b=5$, and $c=-9$.

$$x=\frac{-5\pm\sqrt{5^2-4(1)(-9)}}{2(1)}$$
$$=\frac{-5\pm\sqrt{25+36}}{2}$$
$$=\frac{-5\pm\sqrt{61}}{2}$$

The solution set is $\left\{\frac{-5\pm\sqrt{61}}{2}\right\}$.

45. Write the equation in standard form.

$$2x^2=5-2x$$
$$2x^2+2x-5=0$$

Use $a=2$, $b=2$, and $c=-5$.

$$x=\frac{-2\pm\sqrt{2^2-4(2)(-5)}}{2(2)}$$
$$=\frac{-2\pm\sqrt{4+40}}{4}=\frac{-2\pm\sqrt{44}}{4}$$
$$=\frac{-2\pm2\sqrt{11}}{4}=\frac{2\left(-1\pm\sqrt{11}\right)}{2\cdot2}$$
$$=\frac{-1\pm\sqrt{11}}{2}$$

(a) The solution set with *exact* values is $\left\{\frac{-1\pm\sqrt{11}}{2}\right\}$.

(b) The solution set with *approximate* values (to the nearest thousandth) is $\{-2.158, 1.158\}$.

47. Write the equation in standard form.

$$x^2=1+x$$
$$x^2-x-1=0$$

Use $a=1, b=-1$, and $c=-1$.

$$x=\frac{-(-1)\pm\sqrt{(-1)^2-4(1)(-1)}}{2(1)}$$
$$=\frac{1\pm\sqrt{1+4}}{2}=\frac{1\pm\sqrt{5}}{2}$$

(a) The solution set with *exact* values is $\left\{\frac{1\pm\sqrt{5}}{2}\right\}$.

(b) The solution set with *approximate* values (to the nearest thousandth) is $\{-0.618, 1.618\}$.

49. $0.1x^2+0.3x-0.2=0$

$$x^2+3x-2=0 \quad \text{Multiply by 10.}$$

Substitute $a=1, b=3$, and $c=-2$ into the quadratic formula.

$$x=\frac{-3\pm\sqrt{(3)^2-4(1)(-2)}}{2(1)}$$
$$=\frac{-3\pm\sqrt{9+8}}{2}$$
$$=\frac{-3\pm\sqrt{17}}{2}$$

Use a calculator to approximate the solutions. Use $\sqrt{17}\approx 4.1$.

$$x=\frac{-3+\sqrt{17}}{2}\approx 0.6 \quad \text{or} \quad x=\frac{-3-\sqrt{17}}{2}\approx -3.6$$

The solution set with approximate values (to the nearest tenth) is $\{-3.6, 0.6\}$.

51. Write the equation in standard form.

$$-0.2x^2-0.3x+0.1=0$$
$$2x^2+3x-1=0 \quad \text{Multiply by } -10.$$

Substitute $a=2, b=3$, and $c=-1$ into the quadratic formula.

$$x=\frac{-3\pm\sqrt{(3)^2-4(2)(-1)}}{2(2)}$$
$$=\frac{-3\pm\sqrt{9+8}}{4}$$
$$=\frac{-3\pm\sqrt{17}}{4}$$

Use a calculator to approximate the solutions.

$$x=\frac{-3+\sqrt{17}}{4}\approx 0.3 \quad \text{or} \quad x=\frac{-3-\sqrt{17}}{4}\approx -1.8$$

The solution set with approximate values (to the nearest tenth) is $\{-1.8, 0.3\}$.

53. Write the equation in standard form.

$$5.1x^2+2.3x=1.2$$

$5.1x^2+2.3x-1.2=0$ Subtract 1.2.

$51x^2+23x-12=0$ Multiply by 10.

Substitute $a=51, b=23,$ and $c=-12$ into the quadratic formula.

$$x=\frac{-23\pm\sqrt{(23)^2-4(51)(-12)}}{2(51)}=\frac{-23\pm\sqrt{529+2448}}{102}=\frac{-23\pm\sqrt{2977}}{102}$$

Use a calculator to approximate the solutions.

$$x=\frac{-23+\sqrt{2977}}{102}\approx 0.3 \quad\text{or}\quad x=\frac{-23-\sqrt{2977}}{102}\approx -0.8$$

The solution set with approximate values (to the nearest tenth) is $\{-0.8, 0.3\}$.

55. Eliminate the denominators by multiplying each side by the least common denominator, 6.

$9x^2-6x-8=0$

Substitute $a=9, b=-6,$ and $c=-8$ into the quadratic formula.

$$x=\frac{-(-6)\pm\sqrt{(-6)^2-4(9)(-8)}}{2(9)}=\frac{6\pm\sqrt{36+288}}{18}=\frac{6\pm\sqrt{324}}{18}=\frac{6\pm 18}{18}$$

Solve for x.

$$x=\frac{6+18}{18}=\frac{24}{18}=\frac{4}{3} \quad\text{or}\quad x=\frac{6-18}{18}=\frac{-12}{18}=-\frac{2}{3}$$

The solution set is $\left\{-\frac{2}{3},\frac{4}{3}\right\}$.

57. Eliminate the denominators by multiplying each side by the least common denominator, 6.

$$3x^2+x=6$$

$3x^2+x-6=0$

Use $a=3, b=1,$ and $c=-6.$

$$x=\frac{-1\pm\sqrt{1^2-4(3)(-6)}}{2(3)}=\frac{-1\pm\sqrt{1+72}}{6}=\frac{-1\pm\sqrt{73}}{6}$$

The solution set is $\left\{\frac{-1\pm\sqrt{73}}{6}\right\}$.

59. Multiply each side by the least common denominator, 24.

$9x^2-24x+17=0$

Use the quadratic formula with $a=9, b=-24,$ and $c=17.$

$$x=\frac{-(-24)\pm\sqrt{(-24)^2-4(9)(17)}}{2(9)}=\frac{24\pm\sqrt{576-612}}{18}=\frac{24\pm\sqrt{-36}}{18}$$

The square root of -36 is not a real number, so the square root property does not apply. This equation has no real solution.

61. To eliminate the decimals, multiply each side by 10.

$$5x^2=10x+5$$

$5x^2-10x-5=0$

Divide each side by 5 so that we can work with smaller coefficients in the quadratic formula.

$x^2-2x-1=0$

Use the quadratic formula with $a=1, b=-2,$ and $c=-1.$

$$x=\frac{-(-2)\pm\sqrt{(-2)^2-4(1)(-1)}}{2(1)}=\frac{2\pm\sqrt{4+4}}{2}=\frac{2\pm\sqrt{8}}{2}=\frac{2\pm\sqrt{4}\cdot\sqrt{2}}{2}=\frac{2\pm2\sqrt{2}}{2}=\frac{2\left(1\pm\sqrt{2}\right)}{2}=1\pm\sqrt{2}$$

The solution set is $\left\{1\pm\sqrt{2}\right\}$.

63. To eliminate the decimals, multiply each side by 10.

$6x-4x^2=-10$

Write this equation in standard form.

$0=4x^2-6x-10$

Divide each side by 2 so that we can work with smaller coefficients in the quadratic formula.

$0=2x^2-3x-5$

Use the quadratic formula with $a=2, b=-3,$ and $c=-5$.

$$x=\frac{-(-3)\pm\sqrt{(-3)^2-4(2)(-5)}}{2(2)}$$

$$x=\frac{3\pm\sqrt{9+40}}{4}=\frac{3\pm\sqrt{49}}{4}$$

$$=\frac{3\pm7}{4}$$

Solve for x.

$$x=\frac{3+7}{4} \text{ or } x=\frac{3-7}{4}$$

$$=\frac{10}{4} \text{ or } =\frac{-4}{4}$$

$$=\frac{5}{2} \text{ or } =-1$$

The solution set is $\left\{-1, \frac{5}{2}\right\}$.

65. To eliminate the decimals, multiply each side by 4.

$1x^2=5-3x$

$x^2+3x-5=0$

Use the quadratic formula with $a=1, b=3,$ and $c=-5$.

$$x=\frac{-3\pm\sqrt{(3)^2-4(1)(-5)}}{2(1)}$$

$$=\frac{-3\pm\sqrt{9+20}}{2}$$

$$=\frac{-3\pm\sqrt{29}}{2}$$

The solution set is $\left\{\frac{-3\pm\sqrt{29}}{2}\right\}$.

67. Write this equation in the standard form of a quadratic equation, treating r as the variable and $S, \pi,$ and h as constants.

$\pi r^2+(2\pi h)r-S=0$

Use $a=\pi, b=2\pi h,$ and $c=-S$ in the quadratic formula.

$$r=\frac{-2\pi h\pm\sqrt{(2\pi h)^2-4(\pi)(-S)}}{2(\pi)}$$

$$=\frac{-2\pi h\pm\sqrt{4\pi^2h^2+4\pi S}}{2\pi}$$

$$=\frac{-2\pi h\pm\sqrt{4\left(\pi^2h^2+\pi S\right)}}{2\pi}$$

$$=\frac{-2\pi h\pm2\sqrt{\pi^2h^2+\pi S}}{2\pi}$$

$$=\frac{-\pi h\pm\sqrt{\pi^2h^2+\pi S}}{\pi}$$

69. Let $h=1.25$.

$h=-0.5x^2+1.25x+3$

$1.25=-0.5x^2+1.25x+3$

To eliminate the decimals, multiply each side by 4.

$0.5x^2-1.25x-1.75=0$

$2x^2-5x-7=0$

Use $a=2, b=-5,$ and $c=-7$ in the quadratic formula.

$$x=\frac{-(-5)\pm\sqrt{(-5)^2-4(2)(-7)}}{2(2)}$$

$$=\frac{5\pm\sqrt{25+56}}{4}$$

$$=\frac{5\pm\sqrt{81}}{4}=\frac{5\pm9}{4}$$

Solve for x.

$$x=\frac{5+9}{4}=\frac{14}{4}=3.5 \text{ or } x=\frac{5-9}{4}=\frac{-4}{4}=-1$$

x must be positive, so the frog was 3.5 feet from the base of the stump when he was 1.25 feet above the ground.

71. $\left(\frac{d-4}{4}\right)^2 = 9$

$\frac{d-4}{4} = \pm\sqrt{9} = \pm 3$

$d - 4 = \pm 12$

$d = 4 \pm 12$

$= 16 \text{ or } -8$

The solution set for the equation is $\{16, -8\}$.
Only 16 feet is a reasonable answer.

Summary Exercises Applying Methods for Solving Quadratic Equations

1. Use the square root property.

$x = \sqrt{36}$ or $x = -\sqrt{36}$

$x = 6$ or $x = -6$

The solution set is $\{\pm 6\}$.

2. Write the equation in standard form.

$x^2 + 3x + 1 = 0$

Use the quadratic formula with $a = 1, b = 3$, and $c = 1$.

$$x = \frac{3 \pm \sqrt{3^2 - 4(1)(1)}}{2(1)}$$

$$= \frac{-3 \pm \sqrt{9-4}}{2} = \frac{-3 \pm \sqrt{5}}{2}$$

The solution set is $\left\{\frac{-3 \pm \sqrt{5}}{2}\right\}$.

3. Write the equation in standard form.

$(x+2)(x-4) = 16$

$x^2 - 2x - 8 = 16$

$x^2 - 2x - 24 = 0$

Solve this equation by factoring.

$(x+4)(x-6) = 0$

$x + 4 = 0$ or $x - 6 = 0$

$x = -4$ or $x = 6$

The solution set is $\{-4, 6\}$.

4. $81t^2 = 49$

$t^2 = \frac{49}{81}$ Divide by 81.

Use the square root property.

$t = \sqrt{\frac{49}{81}}$ or $t = -\sqrt{\frac{49}{81}}$

$t = \frac{7}{9}$ or $t = -\frac{7}{9}$

The solution set is $\left\{\pm\frac{7}{9}\right\}$.

5. Solve this equation by factoring.

$(x-3)(x-1) = 0$

$x - 3 = 0$ or $x - 1 = 0$

$x = 3$ or $x = 1$

The solution set is $\{1, 3\}$.

6. Solve this equation by factoring.

$(w+2)(w+1) = 0$

$w + 2 = 0$ or $w + 1 = 0$

$w = -2$ or $w = -1$

The solution set is $\{-2, -1\}$.

7. Write the equation in standard form.

$x(x-9) = -20$

$x^2 - 9x = -20$

$x^2 - 9x + 20 = 0$

Solve this equation by factoring.

$(x-4)(x-5) = 0$

$x - 4 = 0$ or $x - 5 = 0$

$x = 4$ or $x = 5$

The solution set is $\{4, 5\}$.

8. Use the quadratic formula with $a = 1, b = 3$, and $c = -2$.

$$x = \frac{-3 \pm \sqrt{3^2 - 4(1)(-2)}}{2(1)}$$

$$= \frac{-3 \pm \sqrt{9+8}}{2} = \frac{-3 \pm \sqrt{17}}{2}$$

The solution set is $\left\{\frac{-3 \pm \sqrt{17}}{2}\right\}$.

9. Use the square root property.

$3x - 2 = \sqrt{9}$ or $3x - 2 = -\sqrt{9}$

$3x - 2 = 3$ or $3x - 2 = -3$

$3x = 5$ or $3x = -1$

$x = \frac{5}{3}$ or $x = -\frac{1}{3}$

The solution set is $\left\{-\frac{1}{3}, \frac{5}{3}\right\}$.

10. Use the square root property.

$$2x-1=\sqrt{10} \quad \text{or} \quad 2x-1=-\sqrt{10}$$
$$2x=1+\sqrt{10} \quad \text{or} \quad 2x=1-\sqrt{10}$$
$$x=\frac{1+\sqrt{10}}{2} \quad \text{or} \quad x=\frac{1-\sqrt{10}}{2}$$

The solution set is $\left\{\frac{1\pm\sqrt{10}}{2}\right\}$.

11. Use the square root property.

$$x+6=\sqrt{121} \quad \text{or} \quad x+6=-\sqrt{121}$$
$$x+6=11 \quad \text{or} \quad x+6=-11$$
$$x=5 \quad \text{or} \quad x=-17$$

The solution set is $\{-17, 5\}$.

12. Use the square root property.

$$5x+1=\sqrt{36} \quad \text{or} \quad 5x+1=-\sqrt{36}$$
$$5x+1=6 \quad \text{or} \quad 5x+1=-6$$
$$5x=5 \quad \text{or} \quad 5x=-7$$
$$x=1 \quad \text{or} \quad x=-\frac{7}{5}$$

The solution set is $\left\{-\frac{7}{5}, 1\right\}$.

13. Use the square root property.

$$3r-7=\sqrt{24} \quad \text{or} \quad 3r-7=-\sqrt{24}$$

Now simplify the radical.

$$\sqrt{24}=\sqrt{4}\cdot\sqrt{6}=2\sqrt{6}$$
$$3r-7=2\sqrt{6} \quad \text{or} \quad 3r-7=-2\sqrt{6}$$
$$3r=7+2\sqrt{6} \quad \text{or} \quad 3r=7-2\sqrt{6}$$
$$r=\frac{7+2\sqrt{6}}{3} \quad \text{or} \quad r=\frac{7-2\sqrt{6}}{3}$$

The solution set is $\left\{\frac{7\pm2\sqrt{6}}{3}\right\}$.

14. Use the square root property.

$$7p-1=\sqrt{32} \quad \text{or} \quad 7p-1=-\sqrt{32}$$

Now simplify the radical.

$$\sqrt{32}=\sqrt{16}\cdot\sqrt{2}=4\sqrt{2}$$
$$7p-1=4\sqrt{2} \quad \text{or} \quad 7p-1=-4\sqrt{2}$$
$$7p=1+4\sqrt{2} \quad \text{or} \quad 7p=1-4\sqrt{2}$$
$$p=\frac{1+4\sqrt{2}}{7} \quad \text{or} \quad p=\frac{1-4\sqrt{2}}{7}$$

The solution set is $\left\{\frac{1\pm4\sqrt{2}}{7}\right\}$.

15. The square root of -6 is not a real number, so the square root property does not apply. This equation has no real solutions.

16. Write the equation in standard form.

$$2t^2-t+1=0$$

Use the quadratic formula with $a=2, b=-1,$ and $c=1$.

$$t=\frac{-(-1)\pm\sqrt{(-1)^2-4(2)(1)}}{2(2)}$$
$$=\frac{1\pm\sqrt{1-8}}{4}=\frac{1\pm\sqrt{-7}}{4}$$

Because -7 is a negative number and because the square of a real number cannot be negative, there is no real solution of this equation.

17. Write the equation in standard form.

$$-2x^2=-3x-2$$
$$2x^2-3x-2=0$$

Solve this equation by factoring.

$$(2x+1)(x-2)=0$$
$$2x+1=0 \quad \text{or} \quad x-2=0$$
$$x=-\frac{1}{2} \quad \text{or} \quad x=2$$

The solution set is $\left\{-\frac{1}{2}, 2\right\}$.

18. Write the equation in standard form.

$$-2x^2+x+1=0$$
$$2x^2-x-1=0$$

Solve this equation by factoring.

$$(2x+1)(x-1)=0$$
$$2x+1=0 \quad \text{or} \quad x-1=0$$
$$x=-\frac{1}{2} \quad \text{or} \quad x=1$$

The solution set is $\left\{-\frac{1}{2}, 1\right\}$.

19. Write the equation in standard form.

$$8x^2 = 15 + 2x$$

$$8x^2 - 2x - 15 = 0$$

Solve this equation by factoring.

$$(4x+5)(2x-3) = 0$$

$$4x + 5 = 0 \quad \text{or} \quad 2x - 3 = 0$$

$$x = -\frac{5}{4} \quad \text{or} \quad x = \frac{3}{2}$$

The solution set is $\left\{-\frac{5}{4}, \frac{3}{2}\right\}$.

20. Write the equation in standard form.

$$3x^2 + 8x - 3 = 0$$

Solve this equation by factoring.

$$(x+3)(3x-1) = 0$$

$$x + 3 = 0 \quad \text{or} \quad 3x - 1 = 0$$

$$x = -3 \quad \text{or} \quad x = \frac{1}{3}$$

The solution set is $\left\{-3, \frac{1}{3}\right\}$.

21. Write the equation in standard form and remove the decimals by multiplying by 10.

$$0.1x^2 - 0.2x = 0.1$$

$$x^2 - 2x = 1$$

$$x^2 - 2x - 1 = 0$$

Use the quadratic formula with $a = 1, b = -2,$ and $c = -1$.

$$x = \frac{-(-2) \pm \sqrt{(-2)^2 - 4(1)(-1)}}{2(1)}$$

$$= \frac{2 \pm \sqrt{4+4}}{2} = \frac{2 \pm \sqrt{8}}{2}$$

$$= \frac{2 \pm 2\sqrt{2}}{2} = \frac{2(1 \pm \sqrt{2})}{2}$$

$$= 1 \pm \sqrt{2}$$

The solution set is $\{1 \pm \sqrt{2}\}$.

22. Remove the decimals by multiplying by 10.

$$0.3x^2 + 0.5x = -0.1$$

$$3x^2 + 5x = -1$$

Write the equation in standard form.

$$3x^2 + 5x + 1 = 0$$

Use the quadratic formula with $a = 3, b = 5,$ and $c = 1$.

$$x = \frac{-5 \pm \sqrt{5^2 - 4(3)(1)}}{2(3)}$$

$$= \frac{-5 \pm \sqrt{25 - 12}}{6}$$

$$= \frac{-5 \pm \sqrt{13}}{6}$$

The solution set is $\left\{\frac{-5 \pm \sqrt{13}}{6}\right\}$.

23. Write the equation in standard form.

$$5x^2 - 22x = -8$$

$$5x^2 - 22x + 8 = 0$$

Solve this equation by factoring.

$$(5x-2)(x-4) = 0$$

$$5x - 2 = 0 \quad \text{or} \quad x - 4 = 0$$

$$x = \frac{2}{5} \quad \text{or} \quad x = 4$$

The solution set is $\left\{\frac{2}{5}, 4\right\}$.

24. Write the equation in standard form.

$$x(x+6) + 4 = 0$$

$$x^2 + 6x + 4 = 0$$

Solve this equation by completing the square.

$$x^2 + 6x = -4$$

$$x^2 + 6x + 9 = -4 + 9$$

$$(x+3)^2 = 5$$

$$x + 3 = \sqrt{5} \quad \text{or} \quad x + 3 = -\sqrt{5}$$

$$x = -3 + \sqrt{5} \quad \text{or} \quad x = -3 - \sqrt{5}$$

The solution set is $\{-3 \pm \sqrt{5}\}$.

25. Write the equation in standard form.

$$(x+2)(x+1)=10$$
$$x^2+3x+2=10$$
$$x^2+3x-8=0$$

Use the quadratic formula with $a=1, b=3,$ and $c=-8.$

$$x=\frac{-3\pm\sqrt{3^2-4(1)(-8)}}{2(1)}=\frac{-3\pm\sqrt{9+32}}{2}=\frac{-3\pm\sqrt{41}}{2}$$

The solution set is $\left\{\frac{-3\pm\sqrt{41}}{2}\right\}.$

26. Solve this equation by factoring.

$$(4x+5)^2=0$$
$$4x+5=0$$
$$x=-\frac{5}{4}$$

The solution set is $\left\{-\frac{5}{4}\right\}.$

27. Write the equation in standard form.

$$4x^2=-1+5x$$
$$4x^2-5x+1=0$$

Solve this equation by factoring.

$$(x-1)(4x-1)=0$$
$$x-1=0 \quad\text{or}\quad 4x-1=0$$
$$x=1 \quad\text{or}\quad x=\frac{1}{4}$$

The solution set is $\left\{\frac{1}{4},1\right\}.$

28. Write the equation in standard form.

$$2p^2-2p-1=0$$

Use the quadratic formula with $a=2, b=-2,$ and $c=-1.$

$$p=\frac{-(-2)\pm\sqrt{(-2)^2-4(2)(-1)}}{2(2)}$$
$$p=\frac{2\pm\sqrt{4+8}}{4}=\frac{2\pm\sqrt{12}}{4}$$
$$=\frac{2\pm2\sqrt{3}}{4}=\frac{2(1\pm\sqrt{3})}{2\cdot2}$$
$$=\frac{1\pm\sqrt{3}}{2}$$

The solution set is $\left\{\frac{1\pm\sqrt{3}}{2}\right\}.$

29. Write the equation in standard form.

$$3m(3m+4)=7$$
$$9m^2+12m=7$$
$$9m^2+12m-7=0$$

Use the quadratic formula with $a=9, b=12,$ and $c=-7.$

$$m=\frac{-12\pm\sqrt{12^2-4(9)(-7)}}{2(9)}$$
$$=\frac{-12\pm\sqrt{144+252}}{18}$$
$$=\frac{-12\pm\sqrt{396}}{18}=\frac{-12\pm\sqrt{36}\cdot\sqrt{11}}{18}$$
$$=\frac{-12\pm6\sqrt{11}}{18}=\frac{6(-2\pm\sqrt{11})}{6\cdot3}$$
$$=\frac{-2\pm\sqrt{11}}{3}$$

The solution set is $\left\{\frac{-2\pm\sqrt{11}}{3}\right\}.$

30. Write the equation in standard form.

$$4x^2+5x-1=0$$

Use the quadratic formula with $a=4, b=5,$ and $c=-1.$

$$x=\frac{-5\pm\sqrt{5^2-4(4)(-1)}}{2(4)}$$
$$=\frac{-5\pm\sqrt{25+16}}{8}$$
$$=\frac{-5\pm\sqrt{41}}{8}$$

The solution set is $\left\{\frac{-5\pm\sqrt{41}}{8}\right\}.$

31. Multiply each side by the least common denominator, 8.

$$8\left(\frac{r^2}{2}+\frac{7r}{4}+\frac{11}{8}\right)=8(0)$$

$$4r^2+14r+11=0$$

Use the quadratic formula with $a=4, b=14,$ and $c=11.$

$$r=\frac{-14\pm\sqrt{14^2-4(4)(11)}}{2(4)}$$

$$=\frac{-14\pm\sqrt{196-176}}{8}$$

$$=\frac{-14\pm\sqrt{20}}{8}=\frac{-14\pm2\sqrt{5}}{8}$$

$$=\frac{2\left(-7\pm\sqrt{5}\right)}{2(4)}=\frac{-7\pm\sqrt{5}}{4}$$

The solution set is $\left\{\frac{-7\pm\sqrt{5}}{4}\right\}.$

32. Multiply each side by 5.

$$5\left(\frac{1}{5}x^2+x+1\right)=5(0)$$

$$x^2+5x+5=0$$

Use the quadratic formula with $a=1, b=5,$ and $c=5.$

$$x=\frac{-5\pm\sqrt{5^2-4(1)(5)}}{2(1)}$$

$$=\frac{-5\pm\sqrt{25-20}}{2}$$

$$=\frac{-5\pm\sqrt{5}}{2}$$

The solution set is $\left\{\frac{-5\pm\sqrt{5}}{2}\right\}.$

33. Write the equation in standard form.

$$9x^2=16(3x+4)$$

$$9x^2=48x+64$$

$$9x^2-48x-64=0$$

Use the quadratic formula with $a=9, b=-48,$ and $c=-64.$

$$x=\frac{-(-48)\pm\sqrt{(-48)^2-4(9)(-64)}}{2(9)}$$

$$=\frac{48\pm\sqrt{2304+2304}}{18}=\frac{48\pm\sqrt{4608}}{18}$$

$$=\frac{48\pm\sqrt{2304}\cdot\sqrt{2}}{18}=\frac{48\pm48\sqrt{2}}{18}$$

$$=\frac{6\left(8\pm8\sqrt{2}\right)}{6\cdot3}=\frac{8\pm8\sqrt{2}}{3}$$

The solution set is $\left\{\frac{8\pm8\sqrt{2}}{3}\right\}.$

34. Write the equation in standard form.

$$15t^2+58t=-48$$

$$15t^2+58t+48=0$$

Solve this equation by factoring.

$$(3t+8)(5t+6)=0$$

$$3t+8=0 \quad \text{or} \quad 5t+6=0$$

$$t=-\frac{8}{3} \quad \text{or} \quad t=-\frac{6}{5}$$

The solution set is $\left\{-\frac{8}{3}, -\frac{6}{5}\right\}.$

35. Use the quadratic formula with $a=1, b=-1,$ and $c=3.$

$$x=\frac{-(-1)\pm\sqrt{(-1)^2-4(1)(3)}}{2(1)}$$

$$=\frac{1\pm\sqrt{1-12}}{2}=\frac{1\pm\sqrt{-11}}{2}$$

Because -11 is a negative number and because the square of a real number cannot be negative, there is no real solution of this equation.

36. $x^2-\frac{100}{81}=0$

$$x^2=\frac{100}{81}$$

Use the square root property.

$$x=\sqrt{\frac{100}{81}} \quad \text{or} \quad x=-\sqrt{\frac{100}{81}}$$

$$x=\frac{10}{9} \quad \text{or} \quad x=-\frac{10}{9}$$

The solution set is $\left\{\pm\frac{10}{9}\right\}.$

37. Write the equation in standard form.

$-3x^2+4x=-4$

$3x^2-4x-4=0$

Solve this equation by factoring.

$(3x+2)(x-2)=0$

$3x+2=0 \quad \text{or} \quad x-2=0$

$x=-\frac{2}{3} \quad \text{or} \quad x=2$

The solution set is $\left\{-\frac{2}{3}, 2\right\}$.

38. Multiply each side by the least common denominator, 12.

$12\left(x^2-\frac{5}{12}x\right)=12\left(\frac{1}{6}\right)$

$12x^2-5x=2$

Write this equation in standard form.

$12x^2-5x-2=0$

Solve this equation by factoring.

$(4x+1)(3x-2)=0$

$4x+1=0 \quad \text{or} \quad 3x-2=0$

$x=-\frac{1}{4} \quad \text{or} \quad x=\frac{2}{3}$

The solution set is $\left\{-\frac{1}{4}, \frac{2}{3}\right\}$.

39. Write the equation in standard form.

$5x^2+19x=2x+12$

$5x^2+17x-12=0$

Solve this equation by factoring.

$(5x-3)(x+4)=0$

$5x-3=0 \quad \text{or} \quad x+4=0$

$x=\frac{3}{5} \quad \text{or} \quad x=-4$

The solution set is $\left\{-4, \frac{3}{5}\right\}$.

40. Multiply both sides by the common denominator, 2.

$x^2-2x=15$

Write this equation in standard form.

$x^2-2x-15=0$

Solve this equation by factoring.

$(x+3)(x-5)=0$

$x=-3 \quad \text{or} \quad x=5$

The solution set is $\{-3, 5\}$.

41. Multiply both sides by 15 to clear fractions.

$15x^2-4=-4x$

Write this equation in standard form.

$15x^2+4x-4=0$

Solve by factoring.

$(3x+2)(5x-2)=0$

$3x+2=0 \quad \text{or} \quad 5x-2=0$

$x=-\frac{2}{3} \quad \text{or} \quad x=\frac{2}{5}$

The solution set is $\left\{-\frac{2}{3}, \frac{2}{5}\right\}$.

42. Write the equation in standard form.

$4m^2-11m+10=0$

Use the quadratic formula with $a=4$, $b=-11$, and $c=10$.

$$m=\frac{-(-11)\pm\sqrt{(-11)^2-4(4)(10)}}{2(4)}$$

$$=\frac{11\pm\sqrt{121-160}}{8}=\frac{11\pm\sqrt{-39}}{8}$$

Because -39 is a negative number and because the square of a real number cannot be negative, there is no real solution of this equation.

9.4 Graphing Quadratic Equations

Now Try Exercises

N1. Find any x-intercepts by substituting 0 for y in the equation.

$y=x^2-x-2$

$0=x^2-x-2 \quad$ Let $y=0$.

$0=(x+1)(x-2) \quad$ Factor.

$x+1=0 \quad \text{or} \quad x-2=0$

$x=-1 \quad \text{or} \quad x=2$

The x-intercepts are $(-1, 0)$ and $(2, 0)$.

Now find any y-intercepts by substituting 0 for x.

$y=x^2-x-2$

$y=0^2-0-2 \quad$ Let $x=0$.

$y=-2$

The y-intercept is $(0, -2)$. The x-value of the vertex is halfway between the x-intercepts, $(-1, 0)$ and $(2, 0)$.

$x=\frac{1}{2}(-1+2)=\frac{1}{2}(1)=\frac{1}{2}$

Find the y-value of the vertex by substituting $\frac{1}{2}$ for x in the given equation.

$y = x^2 - x - 2$

$y = \left(\frac{1}{2}\right)^2 - \frac{1}{2} - 2$ Let $x = \frac{1}{2}$.

$y = \frac{1}{4} - \frac{1}{2} - 2$

$y = -\frac{9}{4}$

The vertex is at $\left(\frac{1}{2}, -\frac{9}{4}\right)$.

The axis of the parabola is the line $x = \frac{1}{2}$.

x	y
−2	4
−1	0
0	−2
$\frac{1}{2}$	$-\frac{9}{4}$
1	−2
2	0
3	4

Plot the intercepts, vertex, and additional points shown in the table of values and connect them with a smooth curve.

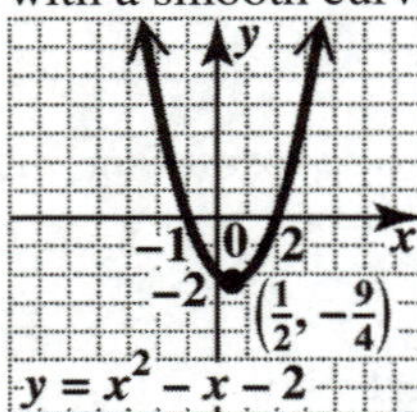

N2. Use $a = -1$, $b = 4$, and $c = 2$.
The x-value of the vertex is

$x = -\frac{b}{2a} = -\frac{4}{2(-1)} = 2.$

The y-value of the vertex is

$y = -2^2 + 4(2) + 2 = 6,$

so the vertex is $(2, 6)$.

The axis of the parabola is the line $x = 2$.
Find the y-intercept by letting $x = 0$.

$y = -0^2 + 4(0) + 2 = 2$

The y-intercept is $(0, 2)$.

Now find the x-intercepts by using the quadratic formula.

$x = \frac{-4 \pm \sqrt{4^2 - 4(-1)(2)}}{2(-1)}$

$x = \frac{-4 \pm \sqrt{24}}{-2}$

Using a calculator, $x \approx 4.45$ and $x \approx -0.45$.
The x-intercepts are $(-0.45, 0)$ and $(4.45, 0)$.

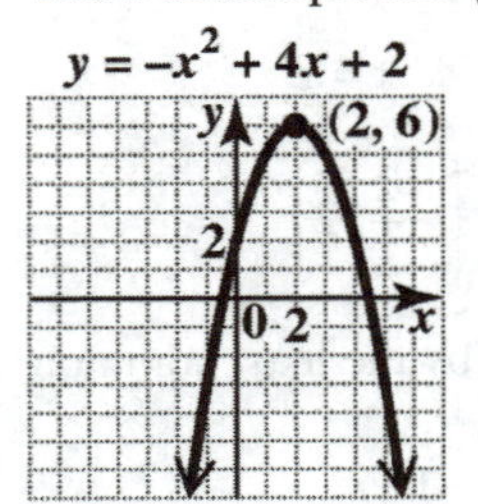

N3. Find the y-intercept by letting $x = 0$.

$f(x) = x^2 - 4$

$f(0) = 0^2 - 4 = -4$

The y-intercept is $(0, -4)$.
To find the x-intercepts, set $y = 0$ and solve for x.

$f(x) = x^2 - 4$

$= (x+2)(x-2)$ Factor.

Use the zero-product property.

$x + 2 = 0$ or $x - 2 = 0$

$x = -2$ or $x = 2$

The x-intercepts of the graph are $(-2, 0)$ and $(2, 0)$.

The x-value of the vertex is halfway between the x-values of the x-intercepts, which in this case is $\frac{-2+2}{2} = 0.$

The x-value of the vertex is zero, which means the vertex is the y-intercept $(0, -4)$.

x	y
−3	5
−2	0
−1	−3
0	−4
1	−3
2	0
3	5

Plot the intercepts, vertex, and additional points shown in the table of values and connect them with a smooth curve.

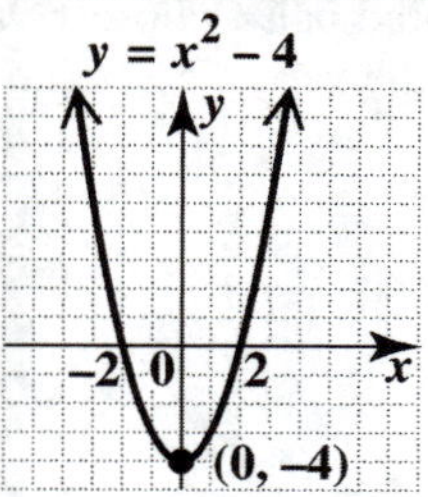

Exercises

1. If $a > 0$, the parabola $y = ax^2 + bx + c$ opens upward.

3. If $a < 0$, the parabola $y = ax^2 + bx + c$ opens downward.

5. If $x = 0$, $y = -6$, so the y-intercept is $(0, -6)$.
   To find any x-intercepts, let $y = 0$.

   $0 = x^2 - 6$

   $x^2 = 6$

   $x = \pm\sqrt{6} \approx \pm 2.45$

   The x-intercepts are $(\pm\sqrt{6}, 0)$.
   The x-value of the vertex is

   $$x = -\frac{b}{2a} = -\frac{0}{2(1)} = 0.$$

   Thus, the vertex is the same as the y-intercept $(\text{since } x = 0)$. The axis of the parabola is the vertical line $x = 0$.
   Make a table of ordered pairs whose x-values are on either side of the vertex's x-value of $x = 0$.

x	y
−3	3
−2	−2
−1	−5
0	−6
1	−5
2	−2
3	3

   Plot these seven ordered pairs and connect them with a smooth curve.

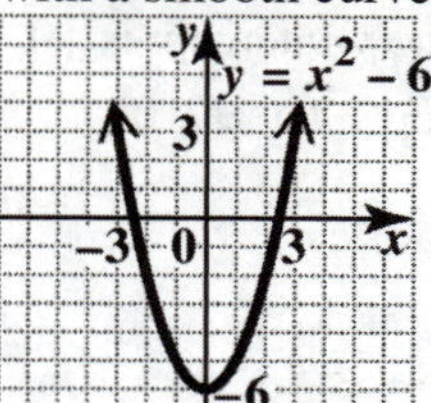

7. If $x = 0$, $y = 2$, so the y-intercept is $(0, 2)$.
   To find any x-intercepts, let $y = 0$.

   $0 = -x^2 + 2$

   $x^2 = 2$

   $x = \pm\sqrt{2} \approx \pm 1.41$

   The x-intercepts are $(\pm\sqrt{2}, 0)$.
   The x-value of the vertex is

   $$x = -\frac{b}{2a} = -\frac{0}{2(-1)} = 0.$$

   Thus, the vertex is the same as the y-intercept $(\text{since } x = 0)$. The axis of the parabola is the vertical line $x = 0$.
   Make a table of ordered pairs whose x-values are on either side of the vertex's x-value of $x = 0$.

x	y
−3	−7
−2	−2
−1	1
0	2
1	1
2	−2
3	−7

   Plot these seven ordered pairs and connect them with a smooth curve.

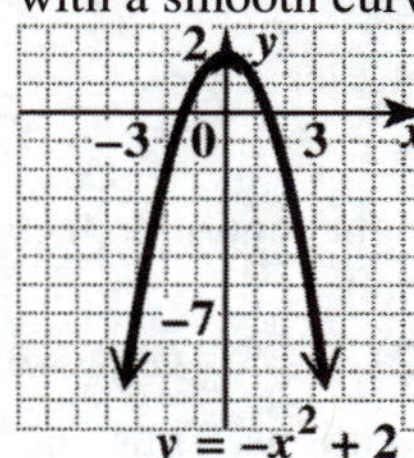

9. $y=(x+3)^2=x^2+6x+9$

If $x=0,\ y=9,$ so the y-intercept is $(0,9)$.

To find any x-intercepts, let $y=0$.

$0=(x+3)^2$

$0=x+3$

$-3=x$

The x-intercept is $(-3,0)$.

The x-value of the vertex is

$x=-\dfrac{b}{2a}=-\dfrac{6}{2(1)}=-3.$

Thus, the vertex is the same as the x-intercept. The axis of the parabola is the vertical line $x=-3$.

Make a table of ordered pairs whose x-values are on either side of the vertex's x-value of $x=-3$.

x	y
–6	9
–5	4
–4	1
–3	0
–2	1
–1	4
0	9

Plot these seven ordered pairs and connect them with a smooth curve.

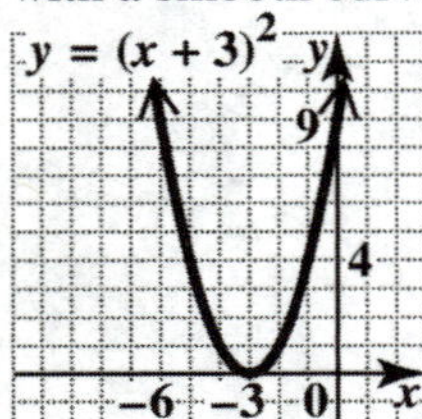

11. If $x=0,\ y=3,$ so the y-intercept is $(0,3)$.

To find any x-intercepts, let $y=0$.

$0=x^2+2x+3$

The trinomial on the right cannot be factored. Because the discriminant

$b^2-4ac=2^2-4(1)(3)=-8$ is negative, this equation has no real solutions. Thus, the parabola has no x-intercepts.

The x-value of the vertex is

$x=-\dfrac{b}{2a}=-\dfrac{2}{2(1)}=-1.$

The y-value of the vertex is

$y=(-1)^2+2(-1)+3$

$\quad=1-2+3=2,$

so the vertex is $(-1,2)$. The axis of the parabola is the vertical line $x=-1$.

Make a table of ordered pairs whose x-values are on either side of the vertex's x-value of $x=-1$.

x	y
–4	11
–3	6
–2	3
–1	2
0	3
1	6
2	11

Plot these seven ordered pairs and connect them with a smooth curve.

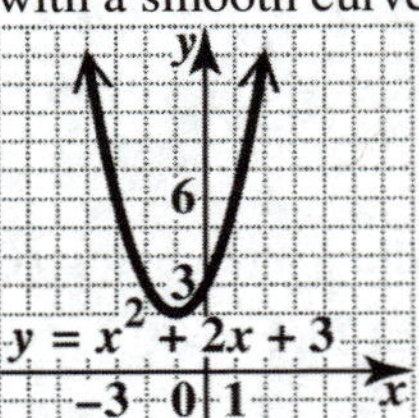

13. $y=x^2-8x+16=(x-4)^2$

If $x=0,\ y=16,$ so the y-intercept is $(0,16)$.

Let $y=0$ and solve for x.

$0=(x-4)^2$

$0=x-4$

$4=x$

The only x-intercept is $(4,0)$.

The x-value of the vertex is

$x=-\dfrac{b}{2a}=-\dfrac{-8}{2(1)}=4.$

The y-value of the vertex has already been found. The vertex is $(4,0)$, which is also the x-intercept. The axis of the parabola is the line $x=4$.

Make a table of ordered pairs whose x-values are on either side of the vertex's x-value of $x=4$.

x	y
1	9
2	4
3	1
4	0
5	1
6	4
7	9

Plot these seven ordered pairs and connect them with a smooth curve.

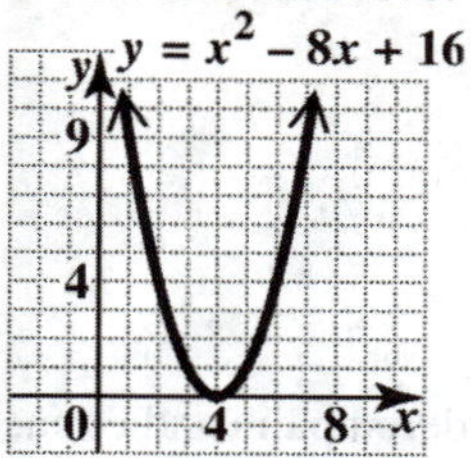

15. If $x = 0$, $y = -5$, so the y-intercept is $(0, -5)$.

Let $y = 0$ and solve for x.

$$0 = -x^2 + 6x - 5$$

$$x^2 - 6x + 5 = 0$$

$$(x-1)(x-5) = 0$$

Solve for x.

$x - 1 = 0$ or $x - 5 = 0$

$x = 1$ or $x = 5$

The x-intercepts are $(1, 0)$ and $(5, 0)$.

The x-value of the vertex is

$$x = -\frac{b}{2a} = -\frac{6}{2(-1)} = 3.$$

The y-value of the vertex is

$$y = -(3)^2 + 6(3) - 5$$
$$= -9 + 18 - 5 = 4,$$

so the vertex is $(3, 4)$.

The axis of the parabola is the line $x = 3$.
Make a table of ordered pairs whose x-values are on either side of the vertex's x-value of $x = 3$.

x	y
0	−5
1	0
2	3
3	4
4	3
5	0
6	−5

Plot these seven ordered pairs and connect them with a smooth curve.

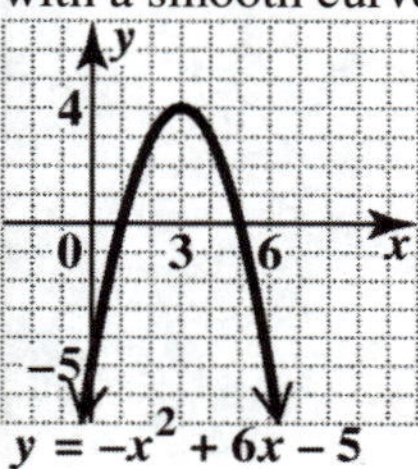

17. If $x = 0$, $y = 0$, so the y-intercept is $(0, 0)$.

Let $y = 0$ and solve for x.

$$0 = x^2 + 4x$$

$$0 = x(x+4)$$

Solve for x.

$x = 0$ or $x = -4$

The x-intercepts are $(0, 0)$ and $(-4, 0)$.

The x-value of the vertex is

$$x = -\frac{b}{2a} = -\frac{4}{2(1)} = -2.$$

The y-value of the vertex is

$$y = (-2)^2 + 4(-2)$$
$$= 4 - 8 = -4,$$

so the vertex is $(-2, -4)$.

The axis of the parabola is the line $x = -2$.
Make a table of ordered pairs whose x-values are on either side of the vertex's x-value of $x = -2$.

x	y
−5	5
−4	0
−3	−3
−2	−4
−1	−3
0	0
1	5

Plot these seven ordered pairs and connect them with a smooth curve.

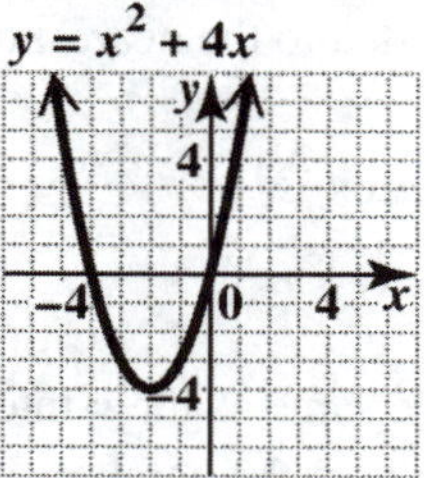

19. If $x = 0$, $y = 1$, so the y-intercept is $(0, 1)$.

To find any x-intercepts, let $y = 0$.

$$0 = x^2 + 1$$

The trinomial on the right cannot be factored.
Because the discriminant
$b^2 - 4ac = 0^2 - 4(1)(1) = -4$ is negative, this equation has no real solutions. Thus, the parabola has no x-intercepts.
The x-value of the vertex is

$$x = -\frac{b}{2a} = -\frac{0}{2(1)} = 0.$$

The y-value of the vertex is

$y = 0^2 + 1 = 1,$

so the vertex is $(0, 1)$. The axis of the parabola is the vertical line $x = 0$.

Make a table of ordered pairs whose x-values are on either side of the vertex's x-value of $x = 0$.

x	y
–3	10
–2	5
–1	2
0	1
1	2
2	5
3	10

Plot these seven ordered pairs and connect them with a smooth curve.

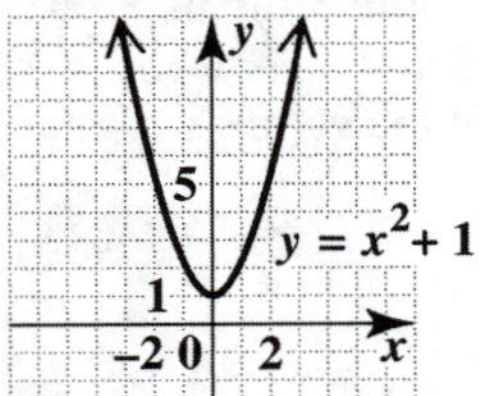

21. If $x = 0$, $y = 2$, so the y-intercept is $(0, 2)$.

To find any x-intercepts, let $y = 0$.

$$0 = -x^2 + 2$$

$$x^2 = 2$$

$$x = \pm\sqrt{2} \approx \pm 1.4$$

The x-intercepts rounded to the nearest tenth are $(\pm 1.4, 0)$.

The x-value of the vertex is

$$x = -\frac{b}{2a} = -\frac{0}{2(-1)} = 0.$$

Thus, the vertex is the same as the y-intercept $(\text{since } x = 0)$. The axis of the parabola is the vertical line $x = 0$.

Make a table of ordered pairs whose x-values are on either side of the vertex's x-value of $x = 0$.

Make a table of ordered pairs whose x-values are on either side of the vertex's x-value of $x = 0$.

x	y
–3	–7
–2	–2
–1	1
0	2
1	1
2	–2
3	–7

Plot these seven ordered pairs and connect them with a smooth curve.

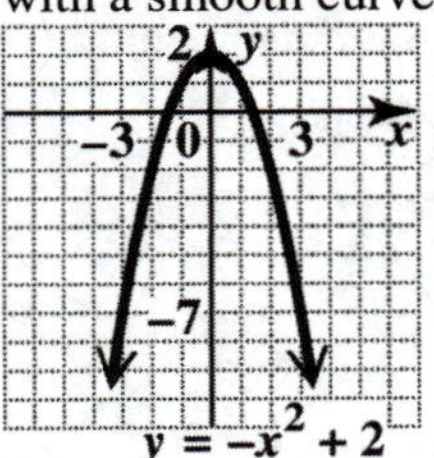

23. $y = x^2 + 4x + 4 = (x + 2)^2$

If $x = 0$, $y = 4$, so the y-intercept is $(0, 4)$.

Let $y = 0$ and solve for x.

$$0 = (x + 2)^2$$

$$x + 2 = 0 \qquad \text{Zero-factor property}$$

$$x = -2$$

The one x-intercept is $(-2, 0)$.

The x-value of the vertex is

$$x = -\frac{b}{2a} = -\frac{4}{2(1)} = -\frac{4}{2} = -2.$$

The y-value of the vertex is

$y = (-2 + 2)^2 = 0^2 = 0,$

so the vertex is $(-2, 0)$.

Make a table of ordered pairs whose x-values are on either side of the vertex's x-value of $x = -2$.

x	y
–5	9
–4	4
–3	1
–2	0
–1	1
0	4
1	9

Plot these seven ordered pairs and connect them with a smooth curve.

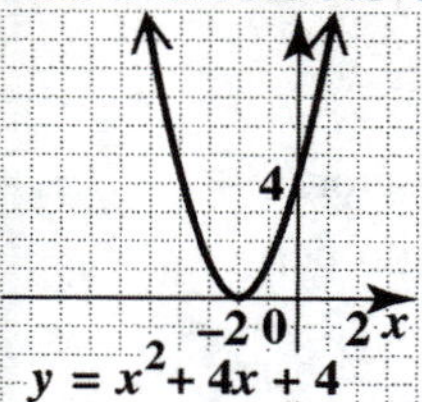

25. If $x = 0$, $y = 3$, so the y-intercept is $(0, 3)$.

Let $y = 0$ and solve for x.

$0 = x^2 - 4x + 3$

$0 = (x-1)(x-3)$

Solve for x.

$x = 1$ or $x = 3$

The x-intercepts are $(1, 0)$ and $(3, 0)$.

The x-value of the vertex is

$$x = -\frac{b}{2a} = -\frac{-4}{2(1)} = \frac{4}{2} = 2.$$

The y-value of the vertex is

$y = 2^2 - 4(2) + 3$

$= 4 - 8 + 3 = -1,$

so the vertex is $(2, -1)$. The axis of the parabola is the vertical line $x = 2$.

Make a table of ordered pairs whose x-values are on either side of the vertex's x-value of $x = 2$.

x	y
−1	8
0	3
1	0
2	−1
3	0
4	3
5	8

Plot these seven ordered pairs and connect them with a smooth curve.

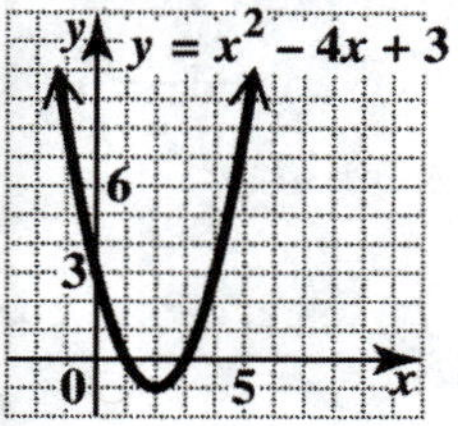

27. Let x be one of the numbers and $80 - x$ be the other number.

The product P of the two numbers is given by $P = x(80 - x)$.

Writing this equation in standard form gives us $P = -x^2 + 80x$.

Finding the maximum of the product is the same as finding the vertex of the graph of P.

The x-value of the vertex is

$$x = -\frac{b}{2a} = -\frac{80}{2(-1)} = 40,$$

which makes sense because 40 is halfway between 0 and 80 (the x-intercepts of $P = x(80 - x)$).

If x is 40, then $80 - x$ must also be 40. The two numbers are 40 and 40, and the product is $40 \cdot 40 = 1600$.

29. Because the vertex is at the origin, an equation of the parabola is of the form

$y = ax^2$.

As shown in the figure, one point on the graph has coordinates (150, 44).

$y = ax^2$ General equation

$44 = a(150)^2$ Let $x = 150$, $y = 44$.

$44 = 22{,}500a$

Solve for a.

$$a = \frac{44}{22{,}500} = \frac{4 \cdot 11}{4 \cdot 5625} = \frac{11}{5625}$$

Thus, an equation of the parabola is

$$y = \frac{11}{5625}x^2.$$

Chapter 9 Review Exercises

1. Use the square root property.

$x = \sqrt{144}$ or $x = -\sqrt{144}$

$x = 12$ or $x = -12$

The solution set is $\{\pm 12\}$.

2. Use the square root property.

$x = \sqrt{37}$ or $x = -\sqrt{37}$

The solution set is $\{\pm \sqrt{37}\}$.

3. Use the square root property.

$$m = \pm\sqrt{128} = \pm\sqrt{64\cdot 2} = \pm 8\sqrt{2}$$

The solution set is $\{\pm 8\sqrt{2}\}$.

4. Use the square root property.

$x+2=\sqrt{25}$ or $x+2=-\sqrt{25}$

$x+2=5$ or $x+2=-5$

$x=3$ or $x=-7$

The solution set is $\{-7, 3\}$.

5. Use the square root property.

$r-3=\sqrt{10}$ or $r-3=-\sqrt{10}$

$r=3+\sqrt{10}$ or $r=3-\sqrt{10}$

The solution set is $\{3\pm\sqrt{10}\}$.

6. Use the square root property.

$2p+1=\sqrt{14}$ or $2p+1=-\sqrt{14}$

$2p=-1+\sqrt{14}$ or $2p=-1-\sqrt{14}$

$p=\dfrac{-1+\sqrt{14}}{2}$ or $p=\dfrac{-1-\sqrt{14}}{2}$

The solution set is $\left\{\dfrac{-1\pm\sqrt{14}}{2}\right\}$.

7. Use the square root property.

$3x+2=\sqrt{-3}$ or $3x+2=-\sqrt{-3}$

Because $\sqrt{-3}$ does not represent a real number, there is no real solution.

8. Use the square root property.

$3-5x=\sqrt{8}$ or $3-5x=-\sqrt{8}$

$-5x=-3+2\sqrt{2}$ or $-5x=-3-2\sqrt{2}$

$x=\dfrac{-3+2\sqrt{2}}{-5}$ or $x=\dfrac{-3-2\sqrt{2}}{-5}$

$x=\dfrac{3-2\sqrt{2}}{5}$ or $x=\dfrac{3+2\sqrt{2}}{5}$

The solution set is $\left\{\dfrac{3\pm 2\sqrt{2}}{5}\right\}$.

9. Rewrite the equation with the variable terms on one side and the constant on the other side.

$m^2+6m=-5$

Take half the coefficient of m and square it.

$\frac{1}{2}(6)=3$, and $(3)^2=9$.

Add 9 to each side of the equation.

$m^2+6m+9=-5+9$

$m^2+6m+9=4$

$(m+3)^2=4$ Factor.

Use the square root property.

$m+3=\sqrt{4}$ or $m+3=-\sqrt{4}$

$m+3=2$ or $m+3=-2$

$m=-1$ or $m=-5$

The solution set is $\{-5, -1\}$

10. Take half the coefficient of p and square it.

$\frac{1}{2}(4)=2$, and $(2)^2=4$.

Add 4 to each side of the equation.

$p^2+4p+4=7+4$

$(p+2)^2=11$

Use the square root property.

$p+2=\sqrt{11}$ or $p+2=-\sqrt{11}$

$p=-2+\sqrt{11}$ or $p=-2-\sqrt{11}$

The solution set is $\{-2\pm\sqrt{11}\}$.

11. Divide each side of the equation by -1 to make the coefficient of the squared term equal to 1.

$x^2-5=-2x$

Rewrite the equation with the variable terms on one side and the constant on the other side.

$x^2+2x=5$

Take half the coefficient of x and square it.

$\frac{1}{2}(2)=1$, and $(1)^2=1$.

Add 1 to both sides of the equation.

$x^2+2x+1=5+1$

$(x+1)^2=6$

Use the square root property.

$x+1=\sqrt{6}$ or $x+1=-\sqrt{6}$

$x=-1+\sqrt{6}$ or $x=-1-\sqrt{6}$

The solution set is $\{-1\pm\sqrt{6}\}$.

12. Divide both sides by 2 to get the x^2 coefficient equal to 1.

$x^2-\dfrac{3}{2}=-4x$

Rewrite the equation with the variable terms on one side and the constant on the other side.

$x^2+4x=\dfrac{3}{2}$

Take half the coefficient of z and square it.

$\frac{1}{2}(4)=2,\quad \text{and} \quad 2^2=4.$

Add 4 to both sides of the equation.

$$x^2+4x+4=\frac{3}{2}+4$$
$$(x+2)^2=\frac{11}{2}$$
$$x+2=\pm\sqrt{\frac{11}{2}}$$
$$x+2=\pm\frac{\sqrt{11}}{\sqrt{2}}\cdot\frac{\sqrt{2}}{\sqrt{2}}$$
$$x+2=\pm\frac{\sqrt{22}}{2}$$
$$x=-2\pm\frac{\sqrt{22}}{2}$$
$$x=\frac{-4}{2}\pm\frac{\sqrt{22}}{2}$$
$$x=\frac{-4\pm\sqrt{22}}{2}$$

The solution set is $\left\{\frac{-4\pm\sqrt{22}}{2}\right\}$.

13. Divide both sides by 5 to get the x^2 coefficient equal to 1.

$$x^2-\frac{3}{5}x-\frac{2}{5}=0$$

Rewrite the equation with the variable terms on one side and the constant on the other side.

$$x^2-\frac{3}{5}x=\frac{2}{5}$$

Take half the coefficient of x and square it.

$\frac{1}{2}\left(-\frac{3}{5}\right)=-\frac{3}{10},\ \text{and}\ \left(-\frac{3}{10}\right)^2=\frac{9}{100}.$

Add $\frac{9}{100}$ to both sides of the equation.

$$x^2-\frac{3}{5}x+\frac{9}{100}=\frac{2}{5}+\frac{9}{100}$$
$$\left(x-\frac{3}{10}\right)^2=\frac{40}{100}+\frac{9}{100}$$
$$\left(x-\frac{3}{10}\right)^2=\frac{49}{100}$$

Use the square root property.

$$x-\frac{3}{10}=\sqrt{\frac{49}{100}} \quad \text{or} \quad x-\frac{3}{10}=-\sqrt{\frac{49}{100}}$$
$$x-\frac{3}{10}=\frac{7}{10} \quad \text{or} \quad x-\frac{3}{10}=-\frac{7}{10}$$
$$x=\frac{10}{10} \quad \text{or} \quad x=-\frac{4}{10}$$
$$x=1 \quad \text{or} \quad x=-\frac{2}{5}$$

The solution set is $\left\{-\frac{2}{5}, 1\right\}$.

14. Multiply on the left side and then simplify. Get all variable terms on one side and the constant on the other side.

$$4x^2-4x+x-1=-7$$
$$4x^2-3x=-6$$

Divide both sides by 4 so that the coefficient of x^2 will be 1.

$$x^2-\frac{3}{4}x=-\frac{6}{4}=-\frac{3}{2}$$

Square half the coefficient of x and add it to both sides.

$$x^2-\frac{3}{4}x+\frac{9}{64}=-\frac{3}{2}+\frac{9}{64}$$
$$\left(x-\frac{3}{8}\right)^2=-\frac{96}{64}+\frac{9}{64}$$
$$\left(x-\frac{3}{8}\right)^2=-\frac{87}{64}$$

The square root of $-\frac{87}{64}$ is not a real number, so there is no real solution.

15. Let $h=30$ and solve for t (which must have a positive value since it represents a number of seconds).

$$30=-16t^2+32t+50$$
$$16t^2-32t-20=0$$

Divide both sides by 16.

$$t^2-2t-\frac{20}{16}=0$$
$$t^2-2t=\frac{5}{4}$$

Half of -2 is -1, and $(-1)^2 = 1$.

Add 1 to both sides of the equation.

$$t^2 - 2t + 1 = \frac{5}{4} + 1$$

$$(t-1)^2 = \frac{9}{4}$$

Use the square root property.

$$t - 1 = \sqrt{\frac{9}{4}} \quad \text{or} \quad t - 1 = -\sqrt{\frac{9}{4}}$$

$$t - 1 = \frac{3}{2} \quad \text{or} \quad t - 1 = -\frac{3}{2}$$

$$t = 1 + \frac{3}{2} \quad \text{or} \quad t = 1 - \frac{3}{2}$$

$$t = \frac{5}{2} = 2.5 \quad \text{or} \quad t = -\frac{1}{2} = -0.5$$

Reject the negative value of t. The object will reach a height of 30 feet after 2.5 seconds.

16. Use the Pythagorean theorem with legs x and $x + 2$ and hypotenuse $x + 4$.

$$a^2 + b^2 = c^2$$

$$(x)^2 + (x+2)^2 = (x+4)^2$$

$$x^2 + x^2 + 4x + 4 = x^2 + 8x + 16$$

$$x^2 - 4x - 12 = 0$$

$$(x-6)(x+2) = 0$$

$$x - 6 = 0 \quad \text{or} \quad x + 2 = 0$$

$$x = 6 \quad \text{or} \quad x = -2$$

Reject the negative value because x represents a length. The value of x is 6. The lengths of the three sides are 6, 8, and 10.

17. (a) $x^2 - 9 = 0$, or $1x^2 + 0x - 9 = 0$

Factor the equation into $(x+3)(x-3) = 0$.

Use the zero-factor property.

$$x + 3 = 0 \quad \text{or} \quad x - 3 = 0$$

$$x = -3 \quad \text{or} \quad x = 3$$

The solution set is $\{\pm 3\}$.

(b) $x^2 = 9$

$$x = \pm\sqrt{9} = \pm 3$$

The solution set is $\{\pm 3\}$.

(c) Use $a = 1, b = 0,$ and $c = -9$.

$$x = \frac{-0 \pm \sqrt{0^2 - 4(1)(-9)}}{2(1)}$$

$$x = \frac{\pm\sqrt{36}}{2} = \frac{\pm 6}{2} = \pm 3$$

The solution set is $\{\pm 3\}$.

(d) Because there is only one solution set, we always get the same results, no matter which method of solution is used.

18. The radicand in the quadratic formula is the discriminant, $b^2 - 4ac$. If it is negative, there are no real solutions to the square root of a negative number, $\sqrt{b^2 - 4ac}$, and the quadratic formula yields no real answers.

19. This equation is in standard form with $a = 1$, $b = -2$, and $c = -4$. Substitute these values into the quadratic formula.

$$x = \frac{-b \pm \sqrt{b^2 - 4ac}}{2a}$$

$$x = \frac{-(-2) \pm \sqrt{(-2)^2 - 4(1)(-4)}}{2(1)}$$

$$= \frac{2 \pm \sqrt{4 + 16}}{2} = \frac{2 \pm \sqrt{20}}{2}$$

$$= \frac{2 \pm 2\sqrt{5}}{2} = \frac{2(1 \pm \sqrt{5})}{2}$$

$$= 1 \pm \sqrt{5}$$

The solution set is $\{1 \pm \sqrt{5}\}$.

20. Write the equation in standard form.

$$3k^2 + 2k = -3$$

$$3k^2 + 2k + 3 = 0$$

Use $a = 3$, $b = 2$, and $c = 3$.

$$k = \frac{-2 \pm \sqrt{2^2 - 4(3)(3)}}{2(3)}$$

$$= \frac{-2 \pm \sqrt{4 - 36}}{6}$$

$$= \frac{-2 \pm \sqrt{-32}}{6}$$

Because $\sqrt{-32}$ does not represent a real number, there is no real solution.

21. Write the equation in standard form.

$$2p^2+8=4p+11$$

$$2p^2-4p-3=0$$

Use the quadratic formula with $a=2,\ b=-4,$ and $c=-3.$

$$p=\frac{-(-4)\pm\sqrt{(-4)^2-4(2)(-3)}}{2(2)}$$

$$=\frac{4\pm\sqrt{16+24}}{4}=\frac{4\pm\sqrt{40}}{4}$$

$$=\frac{4\pm\sqrt{4\cdot 10}}{4}=\frac{4\pm 2\sqrt{10}}{4}$$

$$=\frac{2\left(2\pm\sqrt{10}\right)}{2(2)}=\frac{2\pm\sqrt{10}}{2}$$

The solution set is $\left\{\frac{2\pm\sqrt{10}}{2}\right\}.$

22. Write the equation in standard form.

$$-4x^2+7=2x$$

$$0=4x^2+2x-7$$

Use $a=4,\ b=2,$ and $c=-7.$

$$x=\frac{-2\pm\sqrt{(2)^2-4(4)(-7)}}{2(4)}$$

$$=\frac{-2\pm\sqrt{4+112}}{8}=\frac{-2\pm\sqrt{116}}{8}$$

$$=\frac{-2\pm 2\sqrt{29}}{8}=\frac{2\left(-1\pm\sqrt{29}\right)}{2(4)}$$

$$=\frac{-1\pm\sqrt{29}}{4}$$

The solution set is $\left\{\frac{-1\pm\sqrt{29}}{4}\right\}.$

23. Write the equation in standard form.

$$\frac{1}{4}p^2=2-\frac{3}{4}p$$

$$\frac{1}{4}p^2+\frac{3}{4}p-2=0$$

Multiply both sides by the least common denominator, 4.

$$4\left(\frac{1}{4}p^2+\frac{3}{4}p-2\right)=4(0)$$

$$p^2+3p-8=0$$

Use the quadratic formula with $a=1,\ b=3,$ and $c=-8.$

$$p=\frac{-3\pm\sqrt{3^2-4(1)(-8)}}{2(1)}$$

$$=\frac{-3\pm\sqrt{9+32}}{2}$$

$$=\frac{-3\pm\sqrt{41}}{2}$$

The solution set is $\left\{\frac{-3\pm\sqrt{41}}{2}\right\}.$

24. Use the quadratic formula with $a=3,\ b=-1,$ and $c=-2.$

$$x=\frac{-(-1)\pm\sqrt{(-1)^2-4(3)(-2)}}{2(3)}$$

$$=\frac{1\pm\sqrt{1+24}}{6}$$

$$=\frac{1\pm\sqrt{25}}{6}=\frac{1\pm 5}{6}$$

Solve for x.

$$x=\frac{1+5}{6}=\frac{6}{6}=1 \quad\text{or}\quad x=\frac{1-5}{6}=\frac{-4}{6}=-\frac{2}{3}$$

The solution set is $\left\{-\frac{2}{3},1\right\}.$

25. If $x=0,\ y=5,$ so the y-intercept is $(0,5).$

To find any x-intercepts, let $y=0.$

$$0=-x^2+5$$

$$x^2=5$$

$$x=\pm\sqrt{5}\approx\pm 2.24$$

The x-intercepts are $\left(\pm\sqrt{5},0\right).$

The x-value of the vertex is

$$x=-\frac{b}{2a}=-\frac{0}{2(-1)}=0.$$

Thus, the vertex is the same as the y-intercept (since $x=0$). The axis of the parabola is the vertical line $x=0.$

Make a table of ordered pairs whose x-values are on either side of the vertex's x-value of $x=0$.

x	y
3	–4
2	1
1	4
0	5
–1	4
–2	1
–3	–4

Plot these seven ordered pairs and connect them with a smooth curve.

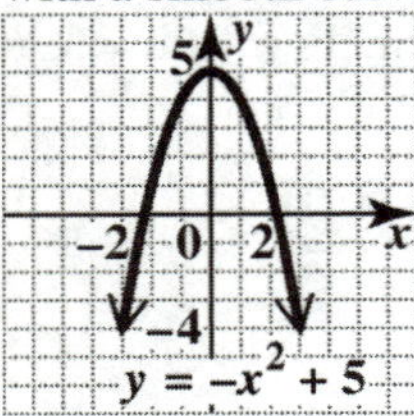

26. $y=(x+4)^2=x^2+8x+16$

If $x=0$, $y=16$, so the y-intercept is $(0,16)$.

To find any x-intercepts, let $y=0$.

$$0=(x+4)^2$$
$$0=x+4$$
$$-4=x$$

The x-intercept is $(-4,0)$.

The x-value of the vertex is

$$x=-\frac{b}{2a}=-\frac{8}{2(1)}=-4.$$

Thus, the vertex is the same as the x-intercept. The axis of the parabola is the vertical line $x=-4$.

Make a table of ordered pairs whose x-values are on either side of the vertex's x-value of $x=-4$.

x	y
–7	9
–6	4
–5	1
–4	0
–3	1
–2	4
–1	9

Plot these seven ordered pairs and connect them with a smooth curve.

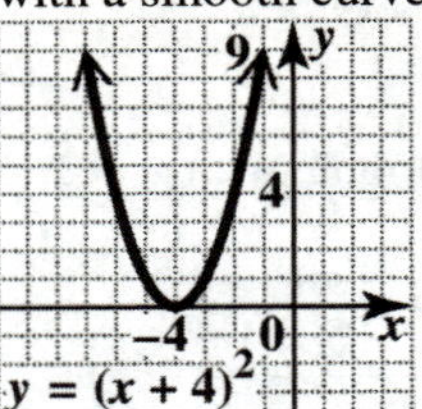

27. If $x=0$, $y=3$, so the y-intercept is $(0,3)$.

Let $y=0$ and solve for x.

$$0=-x^2+2x+3$$
$$x^2-2x-3=0$$
$$(x-3)(x+1)=0$$

Use the square root property.

$$x-3=0 \quad \text{or} \quad x+1=0$$
$$x=3 \quad \text{or} \quad x=-1$$

The x-intercepts are $(3,0)$ and $(-1,0)$.

The x-value of the vertex is

$$x=-\frac{b}{2a}=-\frac{2}{2(-1)}=1.$$

The y-value of the vertex is

$$y=-1^2+2(1)+3=4,$$

so the vertex is $(1,4)$.

Make a table of ordered pairs whose x-values are on either side of the vertex's x-value of $x=1$.

x	y
–1	0
0	3
1	4
2	3
3	0

Plot these five ordered pairs and connect them with a smooth curve.

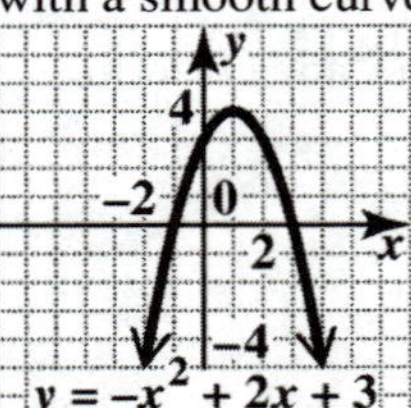

28. If $x = 0,\ y = 2,$ so the y-intercept is $(0, 2)$.
Let $y = 0$ and solve for x.

$$x^2 + 4x + 2 = 0$$
$$x^2 + 4x = -2$$
$$x^2 + 4x + 4 = -2 + 4$$
$$(x+2)^2 = 2$$
$$x + 2 = \sqrt{2} \quad \text{or} \quad x + 2 = -\sqrt{2}$$
$$x = -2 + \sqrt{2} \quad \text{or} \quad x = -2 - \sqrt{2}$$
$$x \approx -0.6 \quad \text{or} \quad x \approx -3.4$$

The x-intercepts are approximately $(-0.6, 0)$ and $(-3.4, 0)$.

The x-value of the vertex is

$$x = -\frac{b}{2a} = -\frac{4}{2(1)} = -2.$$

The y-value of the vertex is

$$y = (-2)^2 + 4(-2) + 2 = -2,$$

so the vertex is $(-2, -2)$.

Make a table of ordered pairs whose x-values are on either side of the vertex's x-value of $x = -2$.

x	y
−4	2
−3.4	0
−3	−1
−2	−2
−1	−1
−0.6	0
0	2

Plot these seven ordered pairs and connect them with a smooth curve.

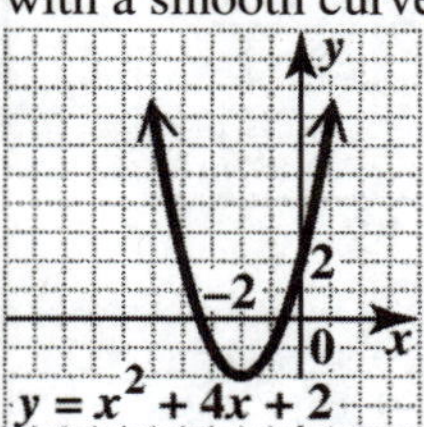

Chapter 9 Mixed Review Exercises

1. Write the equation in standard form.

$$2t^2 + t - 1 = 54$$
$$2t^2 + t - 55 = 0$$
$$(2t + 11)(t - 5) = 0 \quad \text{Factor.}$$
$$2t + 11 = 0 \quad \text{or} \quad t - 5 = 0$$
$$t = -\frac{11}{2} \quad \text{or} \quad t = 5$$

The solution set is $\left\{-\frac{11}{2}, 5\right\}$.

2. Use the square root property.

$$2p + 1 = \sqrt{100} \quad \text{or} \quad 2p + 1 = -\sqrt{100}$$
$$2p + 1 = 10 \quad \text{or} \quad 2p + 1 = -10$$
$$2p = 9 \quad \text{or} \quad 2p = -11$$
$$p = \frac{9}{2} \quad \text{or} \quad p = -\frac{11}{2}$$

The solution set is $\left\{-\frac{11}{2}, \frac{9}{2}\right\}$.

3. Write the equation in standard form.

$$x^2 + x - 2 = 3$$
$$x^2 + x - 5 = 0$$

The left side cannot be factored, so use the quadratic formula with $a = 1,\ b = 1,$ and $c = -5$.

$$x = \frac{-b \pm \sqrt{b^2 - 4ac}}{2a}$$
$$x = \frac{-1 \pm \sqrt{1^2 - 4(1)(-5)}}{2(1)}$$
$$= \frac{-1 \pm \sqrt{1 + 20}}{2} = \frac{-1 \pm \sqrt{21}}{2}$$

The solution set is $\left\{\frac{-1 \pm \sqrt{21}}{2}\right\}$.

4.

$$6t^2 + 7t - 3 = 0$$
$$(3t - 1)(2t + 3) = 0 \quad \text{Factor.}$$

Solve for t.

$$3t - 1 = 0 \quad \text{or} \quad 2t + 3 = 0$$
$$t = \frac{1}{3} \quad \text{or} \quad t = -\frac{3}{2}$$

The solution set is $\left\{-\frac{3}{2}, \frac{1}{3}\right\}$.

5. Write the equation in standard form.

$x^2 + 5x + 2 = 0$

The left side cannot be factored, so use the quadratic formula with $a = 1$, $b = 5$, and $c = 2$.

$$x = \frac{-b \pm \sqrt{b^2 - 4ac}}{2a}$$

$$x = \frac{-5 \pm \sqrt{5^2 - 4(1)(2)}}{2(1)}$$

$$= \frac{-5 \pm \sqrt{25 - 8}}{2} = \frac{-5 \pm \sqrt{17}}{2}$$

The solution set is $\left\{\frac{-5 \pm \sqrt{17}}{2}\right\}$.

6. Write the equation in standard form.

$x^2 + 2x - 2 = 0$

The left side cannot be factored, so use the quadratic formula with $a = 1$, $b = 2$, and $c = -2$.

$$x = \frac{-2 \pm \sqrt{2^2 - 4(1)(-2)}}{2(1)}$$

$$= \frac{-2 \pm \sqrt{4 + 8}}{2} = \frac{-2 \pm \sqrt{12}}{2}$$

$$= \frac{-2 \pm 2\sqrt{3}}{2} = \frac{2\left(-1 \pm \sqrt{3}\right)}{2}$$

$$= -1 \pm \sqrt{3}$$

The solution set is $\left\{-1 \pm \sqrt{3}\right\}$.

7. Use the quadratic formula with $a = 1$, $b = -4$, and $c = 10$.

$$m = \frac{-(-4) \pm \sqrt{(-4)^2 - 4(1)(10)}}{2(1)}$$

$$= \frac{4 \pm \sqrt{16 - 40}}{2} = \frac{4 \pm \sqrt{-24}}{2}$$

Because $\sqrt{-24}$ does not represent a real number, there is no real solution.

8. The left side cannot be factored, so use the quadratic formula with $a = 1$, $b = -9$, and $c = 10$.

$$k = \frac{-b \pm \sqrt{b^2 - 4ac}}{2a}$$

$$k = \frac{-(-9) \pm \sqrt{(-9)^2 - 4(1)(10)}}{2(1)}$$

$$= \frac{9 \pm \sqrt{81 - 40}}{2} = \frac{9 \pm \sqrt{41}}{2}$$

The solution set is $\left\{\frac{9 \pm \sqrt{41}}{2}\right\}$.

9. $(3x + 5)^2 = 0$

$$3x + 5 = 0$$

$$3x = -5$$

$$x = -\frac{5}{3}$$

The solution set is $\left\{-\frac{5}{3}\right\}$.

10. Multiply by 2 to clear fractions; then rewrite the result in standard form.

$$r^2 = 7 - 2r$$

$$r^2 + 2r - 7 = 0$$

The left side does not factor, so use the quadratic formula with $a = 1$, $b = 2$, and $c = -7$.

$$r = \frac{-2 \pm \sqrt{2^2 - 4(1)(-7)}}{2(1)}$$

$$= \frac{-2 \pm \sqrt{4 + 28}}{2} = \frac{-2 \pm \sqrt{32}}{2}$$

$$= \frac{-2 \pm 4\sqrt{2}}{2} = \frac{2\left(-1 \pm 2\sqrt{2}\right)}{2}$$

$$= -1 \pm 2\sqrt{2}$$

The solution set is $\left\{-1 \pm 2\sqrt{2}\right\}$.

11. $x^2+4x=1$

$x^2+4x-1=0$

The left side does not factor, so use the quadratic formula with $a=1$, $b=4$, and $c=-1$.

$$x=\frac{-4\pm\sqrt{4^2-4(1)(-1)}}{2(1)}$$
$$=\frac{-4\pm\sqrt{16+4}}{2}=\frac{-4\pm\sqrt{20}}{2}$$
$$=\frac{-4\pm2\sqrt{5}}{2}=\frac{2\left(-2\pm\sqrt{5}\right)}{2}$$
$$=-2\pm\sqrt{5}$$

The solution set is $\left\{-2\pm\sqrt{5}\right\}$.

12. $7x^2-8=5x^2+8$

$2x^2=16$

$x^2=8$

$x=\pm\sqrt{8}=\pm2\sqrt{2}$

The solution set is $\left\{\pm2\sqrt{2}\right\}$.

13. Let $h=20$ and solve for x (which must have a positive value since it represents a number of seconds). Write the equation in standard form.

$$20=-2.7x^2+30x$$

$2.7x^2-30x+20=0$

Use the quadratic formula with $a=2.7$, $b=-30$, and $c=20$.

$$x=\frac{-b\pm\sqrt{b^2-4ac}}{2a}$$
$$x=\frac{-(-30)\pm\sqrt{(-30)^2-4(2.7)(20)}}{2(2.7)}$$
$$=\frac{30\pm\sqrt{900-216}}{5.4}$$
$$=\frac{30\pm\sqrt{684}}{5.4}$$

Solve for x.

$$x=\frac{30-\sqrt{684}}{5.4} \quad \text{or} \quad x=\frac{30+\sqrt{684}}{5.4}$$
$$\approx 0.7 \quad \text{or} \quad \approx 10.4$$

Thus, x rounded to the nearest tenth is approximately 0.7 or 10.4. Both solutions are positive and can represent measures of time. The rocket reaches a height of 20 ft at 0.7 and 10.4 seconds, once on its way up and another time on its way down from the peak height it reaches.

14. If $x=0$, $y=8$, so the y-intercept is $(0,8)$.

Let $y=0$ and solve for x.

$x^2-6x+8=0$

$(x-2)(x-4)=0$

$x-2=0$ or $x-4=0$

$x=2$ or $x=4$

The x-intercepts are $(2,0)$ and $(4,0)$.

The x-value of the vertex is

$$x=-\frac{b}{2a}=-\frac{-6}{2(1)}=\frac{6}{2}=3.$$

The y-value of the vertex is

$$y=3^2-6(3)+8$$
$$=9-18+8=-1,$$

so the vertex is $(3,-1)$. The axis of the parabola is the vertical line $x=3$.

Make a table of ordered pairs whose x-values are on either side of the vertex's x-value of $x=3$.

x	y
0	8
1	3
2	0
3	−1
4	0
5	3
6	8

Plot these seven ordered pairs and connect them with a smooth curve.

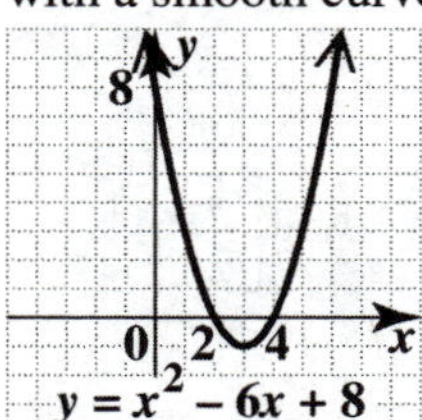

Chapter 9 Test

1. Use the square root property.

$x=\sqrt{39}$ or $x=-\sqrt{39}$

The solution set is $\left\{\pm\sqrt{39}\right\}$.

2. Use the square root property.

$x+3=\sqrt{64}$ or $x+3=-\sqrt{64}$

$x+3=8$ or $x+3=-8$

$x=5$ or $x=-11$

The solution set is $\{-11, 5\}$.

3. Use the square root property.

$4x+3=\sqrt{24}$ or $4x+3=-\sqrt{24}$

Note that $\sqrt{24}=\sqrt{4\cdot 6}=2\sqrt{6}$.
Solve for x.

$4x+3=2\sqrt{6}$ or $4x+3=-2\sqrt{6}$

$4x=-3+2\sqrt{6}$ or $4x=-3-2\sqrt{6}$

$x=\dfrac{-3+2\sqrt{6}}{4}$ or $x=\dfrac{-3-2\sqrt{6}}{4}$

The solution set is $\left\{\dfrac{-3\pm 2\sqrt{6}}{4}\right\}$.

4. $x^2-4x=6$

$x^2-4x+4=6+4$ Add $\left[\dfrac{1}{2}(-4)\right]^2=4$.

$(x-2)^2=10$

Use the square root property.

$x-2=\sqrt{10}$ or $x-2=-\sqrt{10}$

$x=2+\sqrt{10}$ or $x=2-\sqrt{10}$

The solution set is $\{2\pm\sqrt{10}\}$.

5. $2x^2+12x-3=0$

$x^2+6x-\dfrac{3}{2}=0$

$x^2+6x=\dfrac{3}{2}$

$x^2+6x+9=\dfrac{3}{2}+9$ Add $\left[\dfrac{1}{2}(6)\right]^2=9$.

$(x+3)^2=\dfrac{21}{2}$

Use the square root property.

$x+3=\sqrt{\dfrac{21}{2}}$ or $x+3=-\sqrt{\dfrac{21}{2}}$

Note that

$\sqrt{\dfrac{21}{2}}=\dfrac{\sqrt{21}}{\sqrt{2}}=\dfrac{\sqrt{21}\cdot\sqrt{2}}{\sqrt{2}\cdot\sqrt{2}}=\dfrac{\sqrt{42}}{2}$.

$x+3=\dfrac{\sqrt{42}}{2}$ or $x+3=-\dfrac{\sqrt{42}}{2}$

$x=-3+\dfrac{\sqrt{42}}{2}$ or $x=-3-\dfrac{\sqrt{42}}{2}$

$x=\dfrac{-6+\sqrt{42}}{2}$ or $x=\dfrac{-6-\sqrt{42}}{2}$

The solution set is $\left\{\dfrac{-6\pm\sqrt{42}}{2}\right\}$.

6. Use $a=5,\ b=2,$ and $c=0$.

$x=\dfrac{-b\pm\sqrt{b^2-4ac}}{2a}$

$x=\dfrac{-2\pm\sqrt{2^2-4(5)(0)}}{2(5)}$

$=\dfrac{-2\pm\sqrt{4}}{10}=\dfrac{-2\pm 2}{10}$

Solve for x.

$x=\dfrac{-2+2}{10}$ or $x=\dfrac{-2-2}{10}$

$=\dfrac{0}{10}$ or $=\dfrac{-4}{10}$

$=0$ or $=-\dfrac{2}{5}$

The solution set is $\left\{-\dfrac{2}{5}, 0\right\}$.

7. Use $a=2,\ b=5,$ and $c=-3$.

$x=\dfrac{-b\pm\sqrt{b^2-4ac}}{2a}$

$x=\dfrac{-5\pm\sqrt{5^2-4(2)(-3)}}{2(2)}$

$=\dfrac{-5\pm\sqrt{25+24}}{4}$

$=\dfrac{-5\pm\sqrt{49}}{4}=\dfrac{-5\pm 7}{4}$

Solve for x.

$$x=\frac{-5+7}{4} \quad \text{or} \quad x=\frac{-5-7}{4}$$
$$=\frac{2}{4} \quad \text{or} \quad =\frac{-12}{4}$$
$$=\frac{1}{2} \quad \text{or} \quad =-3$$

The solution set is $\left\{-3, \frac{1}{2}\right\}$.

8. $3w^2+2=6w$

$3w^2-6w+2=0$

Use $a=3$, $b=-6$, and $c=2$.

$$w=\frac{-(-6)\pm\sqrt{(-6)^2-4(3)(2)}}{2(3)}$$
$$=\frac{6\pm\sqrt{36-24}}{6}$$
$$=\frac{6\pm\sqrt{12}}{6}=\frac{6\pm 2\sqrt{3}}{6}$$
$$=\frac{2\left(3\pm\sqrt{3}\right)}{2(3)}=\frac{3\pm\sqrt{3}}{3}$$

The solution set is $\left\{\frac{3\pm\sqrt{3}}{3}\right\}$.

9. Use $a=4$, $b=8$, and $c=11$.

$$x=\frac{-8\pm\sqrt{8^2-4(4)(11)}}{2(4)}$$
$$=\frac{-8\pm\sqrt{64-176}}{8}=\frac{-8\pm\sqrt{-112}}{8}$$

The square root of -112 is not a real number, so the square root property does not apply. This equation has no real solution.

10. $t^2-\frac{5}{3}t+\frac{1}{3}=0$

$$3\left(t^2-\frac{5}{3}t+\frac{1}{3}\right)=3(0)$$

$3t^2-5t+1=0$

Use $a=3$, $b=-5$, and $c=1$.

$$t=\frac{-(-5)\pm\sqrt{(-5)^2-4(3)(1)}}{2(3)}$$
$$=\frac{5\pm\sqrt{25-12}}{6}=\frac{5\pm\sqrt{13}}{6}$$

The solution set is $\left\{\frac{5\pm\sqrt{13}}{6}\right\}$.

11. Solve by completing the square.

$p^2-2p=1$

$p^2-2p+1=1+1$

$(p-1)^2=2$

Use the square root property.

$p-1=\sqrt{2}$ or $p-1=-\sqrt{2}$

$p=1+\sqrt{2}$ or $p=1-\sqrt{2}$

The solution set is $\left\{1\pm\sqrt{2}\right\}$.

12. Use the square root property.

$2x+1=\pm\sqrt{18}$

$2x+1=\pm 3\sqrt{2}$

$2x=-1\pm 3\sqrt{2}$

$$x=\frac{-1\pm 3\sqrt{2}}{2}$$

The solution set is $\left\{\frac{-1\pm 3\sqrt{2}}{2}\right\}$.

13. $(x-5)(2x-1)=1$

$2x^2-11x+5=1$

$2x^2-11x+4=0$

Use $a=2$, $b=-11$, and $c=4$.

$$x=\frac{-(-11)\pm\sqrt{(-11)^2-4(2)(4)}}{2(2)}$$
$$=\frac{11\pm\sqrt{121-32}}{4}=\frac{11\pm\sqrt{89}}{4}$$

The solution set is $\left\{\frac{11\pm\sqrt{89}}{4}\right\}$.

14. $t^2+25=10t$

$t^2-10t+25=0$

$(t-5)^2=0$

$t-5=0$

$t=5$

The solution set is $\{5\}$.

15. Substitute 64 for s and solve for t.

$-16t^2+64t=64$ Let $s=64$.

$16t^2-64t+64=0$

$t^2-4t+4=0$ Divide by 16.

$(t-2)^2=0$

$t-2=0$

$t=2$

The object will reach a height of 64 feet after 2 seconds.

16. Use the Pythagorean theorem.

$c^2=a^2+b^2$

$(x+8)^2=(x)^2+(x+4)^2$

$x^2+16x+64=x^2+x^2+8x+16$

$0=x^2-8x-48$

$0=(x-12)(x+4)$

Solve for x.

$x-12=0$ or $x+4=0$

$x=12$ or $x=-4$

Reject the negative length. The sides measure 12, $x+4=12+4=16$, and $x+8=12+8=20$.

17. If $x=0$, $y=6$, so the y-intercept is $(0,6)$.

To find any x-intercepts, let $y=0$.

$0=-x^2+6$

$x^2=6$

$x=\pm\sqrt{6}\approx\pm2.45$

The x-intercepts are $(\pm\sqrt{6},0)$.

The x-value of the vertex is

$$x=-\frac{b}{2a}=-\frac{0}{2(-1)}=0.$$

Thus, the vertex is the same as the y-intercept (since $x=0$). The axis of the parabola is the vertical line $x=0$.

Make a table of ordered pairs whose x-values are on either side of the vertex's x-value of $x=0$.

x	y
3	−3
2	2
1	5
0	6
−1	5
−2	2
−3	−3

Plot these seven ordered pairs and connect them with a smooth curve.

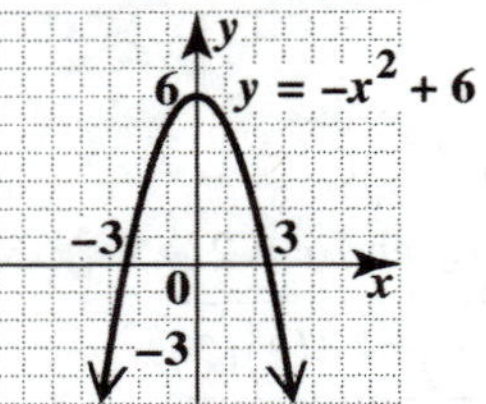

18. The x-value of the vertex is

$$x=-\frac{b}{2a}=-\frac{6}{2(1)}=-3.$$

The y-value of the vertex is

$y=(-3)^2+6(-3)+7=-2$,

so the vertex is $(-3,-2)$.

Make a table of ordered pairs whose x-values are on either side of the vertex's x-value of $x=-3$.

x	y
−6	7
−5	2
−4	−1
−3	−2
−2	−1
−1	2
0	7

Plot these seven ordered pairs and connect them with a smooth curve.

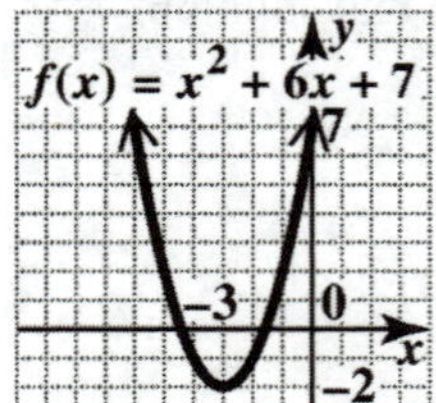

Chapters R–9 Cumulative Review Exercises

1. $$\frac{-4\cdot3^2+2\cdot3}{2-4\cdot1}=\frac{-4\cdot9+6}{2-4}=\frac{-36+6}{-2}=\frac{-30}{-2}=15$$

2. $$\begin{aligned}&-9-(-8)(2)+6-(6+2)\\&=-9-(-8)(2)+6-8\\&=-9-(-16)+6-8\\&=-9+16+6-8\\&=7+6-8\\&=13-8\\&=5\end{aligned}$$

3. $-4r+14+3r-7=-r+7$

4. $$\begin{aligned}5(4m-2)-(m+7)&=5(4m-2)-1(m+7)\\&=20m-10-m-7\\&=19m-17\end{aligned}$$

5. $$\begin{aligned}x-5&=13\\x&=18\end{aligned}$$

The solution set is $\{18\}$.

6. $$\begin{aligned}3k-9k-8k+6&=-64\\-14k+6&=-64\\-14k&=-70\\k&=5\end{aligned}$$

The solution set is $\{5\}$.

7. Multiply each side by the LCD, 10.

$$\begin{aligned}10\left(\frac{3}{5}t-\frac{1}{10}\right)&=10\left(\frac{3}{2}\right)\\6t-1&=15\\6t&=16\\t&=\frac{16}{6}=\frac{8}{3}\end{aligned}$$

The solution set is $\left\{\frac{8}{3}\right\}$.

8. $$\begin{aligned}2(m-1)-6(3-m)&=-4\\2m-2-18+6m&=-4\\8m-20&=-4\\8m&=16\\m&=2\end{aligned}$$

The solution set is $\{2\}$.

9. Together, the two angles form a straight angle, so the sum of their measures is 180°.

$$\begin{aligned}(20x-20)+(12x+8)&=180\\32x-12&=180\\32x&=192\\x&=\frac{192}{32}=6\end{aligned}$$

If $x=6$,

$20x-20=20(6)-20=120-20=100,$

and

$$\begin{aligned}12x+8&=12(6)+8\\&=72+8=80.\end{aligned}$$

The measures of the angles are 100° and 80°.

10. Let $L=$ the length of the court.
Then $L-44=$ the width of the court.
Use the formula for the perimeter of a rectangle, $P=2L+2W$, with $P=288$.

$$\begin{aligned}288&=2L+2(L-44)\\288&=2L+2L-88\\376&=4L\\94&=L\end{aligned}$$

If $L=94$, $L-44=94-44=50$.
The length of the court is 94 feet and the width of the court is 50 feet.

11. Solve $P=2L+2W$ for L.

$P-2W=2L$

$\frac{P-2W}{2}=L$, or $L=\frac{P}{2}-W$

12. Divide each side by -8 and reverse the inequality symbol.

$$\begin{aligned}\frac{-8m}{-8}&>\frac{16}{-8}\\m&>-2\end{aligned}$$

The solution set is $(-2,\infty)$.

−2 0

13. $$\begin{aligned}-9p+2(8-p)-6&\ge 4p-50\\-9p+16-2p-6&\ge 4p-50\\-11p+10&\ge 4p-50\\-15p&\ge -60\\\frac{-15p}{-15}&\le\frac{-60}{-15}\quad\text{Divide by }-15.\\p&\le 4\end{aligned}$$

The solution set is $(-\infty,4]$.

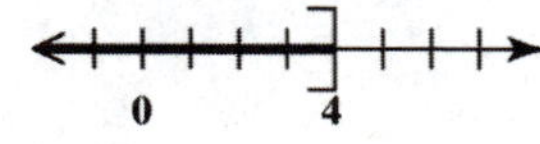

14. Find the intercepts.
Let $x = 0$.

$$\begin{aligned} 2(0)+3y &= 6 \\ 3y &= 6 \\ y &= 2 \end{aligned}$$

The y-intercept is $(0, 2)$.
Let $y = 0$.

$$\begin{aligned} 2x+3(0) &= 6 \\ 2x &= 6 \\ x &= 3 \end{aligned}$$

The x-intercept is $(3, 0)$.

The graph is the line through the points $(0, 2)$ and $(3, 0)$.

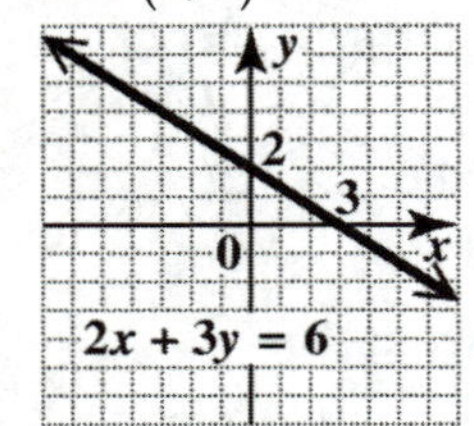

15. For any value of x, the value of y will always be 3. Three ordered pairs are $(-2, 3)$, $(0, 3)$, and $(4, 3)$. Plot these points and draw a line through them. This will be a horizontal line.

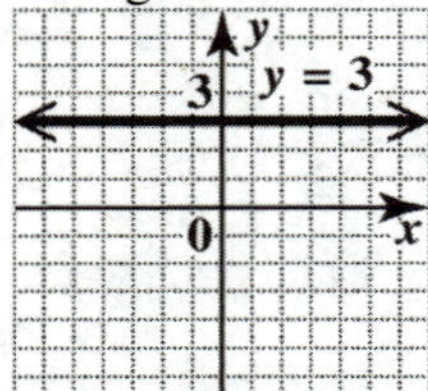

16. The slope m of the line passing through the points $(-1, 4)$ and $(5, 2)$ is

$$m = \frac{\text{change in } y}{\text{change in } x} = \frac{2-4}{5-(-1)} = \frac{-2}{6} = -\frac{1}{3}.$$

17. Slope 2; y-intercept $(0, 3)$

Let $m = 2$ and $b = 3$ in slope-intercept form.

$$y = mx + b$$
$$y = 2x + 3$$

Now rewrite the equation in the form $Ax + By = C$.

$$-2x + y = 3$$
$$2x - y = -3$$

18.
$$\begin{aligned} 2x + y &= -4 \quad (1) \\ -3x + 2y &= 13 \quad (2) \end{aligned}$$

Use the elimination method.
Multiply equation (1) by -2 and add the result to equation (2).

$$\begin{array}{rcrcr} -4x & - & 2y & = & 8 \\ -3x & + & 2y & = & 13 \\ \hline -7x & & & = & 21 \\ & & x & = & -3 \end{array}$$

To find y, substitute -3 for x in equation (1).

$$\begin{aligned} 2x + y &= -4 \\ 2(-3) + y &= -4 \\ -6 + y &= -4 \\ y &= 2 \end{aligned}$$

The solution set is $\{(-3, 2)\}$.

19.
$$\begin{aligned} 3x - 5y &= 8 \quad (1) \\ -6x + 10y &= 16 \quad (2) \end{aligned}$$

Use the elimination method.
Multiply equation (1) by 2 and add the result to equation (2).

$$\begin{array}{rcrcrl} 6x & - & 10y & = & 16 & \\ -6x & + & 10y & = & 16 & \\ \hline & & 0 & = & 32 & \text{False} \end{array}$$

The false statement indicates that the solution set is the null set, $\varnothing$.

20. Let x be the price of a Motorola Theory phone and let y be the price of a Motorola i412 phone.
We have the system

$$\begin{aligned} 3x + 2y &= 379.95 \quad (1) \\ 2x + 3y &= 369.95 \quad (2) \end{aligned}$$

To solve the system by the elimination method, we multiply equation (1) by 2 and equation (2) by -3, and then add the results.

$$\begin{array}{rcrcr} 6x & + & 4y & = & 759.90 \\ -6x & - & 9y & = & -1109.85 \\ \hline & & -5y & = & -349.95 \\ & & y & = & 69.99 \end{array}$$

To find the value of x, substitute 69.99 for y in equation (1).

$$3x + 2(69.99) = 379.95$$
$$3x + 139.98 = 379.95$$
$$3x = 239.97$$
$$x = 79.99$$

The price of a single Motorola Theory phone is $79.99, and the price of a single Motorola i412 phone is $69.99.

21. $2x + y \le 4 \quad (1)$
$x - y > 2 \quad (2)$

For inequality (1), draw a solid boundary line through $(2, 0)$ and $(0, 4)$, and shade the side that includes the origin (since substituting 0 for x and 0 for y results in a true statement). For inequality (2), draw a dashed boundary line through $(2, 0)$ and $(0, -2)$, and shade the side that *does not* include the origin.
The solution of the system of inequalities is the intersection of these two shaded half-planes. It includes the solid line and excludes the dashed line.

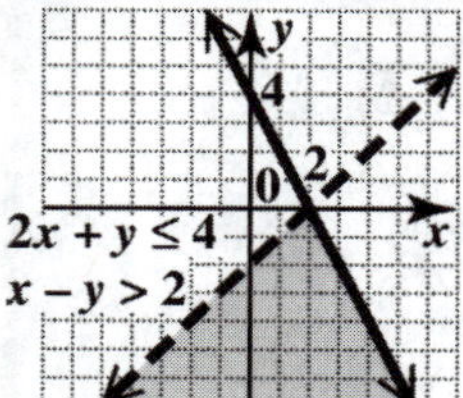

22. $\left(3^2 \cdot x^{-4}\right)^{-1} = \left(\dfrac{3^2}{x^4}\right)^{-1}$
$= \left(\dfrac{x^4}{3^2}\right)^1$
$= \dfrac{x^4}{3^2} = \dfrac{x^4}{9}$

23. $\left(\dfrac{b^{-3}c^4}{b^5c^3}\right)^{-2} = \left(b^{-3-5}c^{4-3}\right)^{-2}$
$= \left(b^{-8}c^1\right)^{-2}$
$= \left(b^{-8}\right)^{-2}\left(c^1\right)^{-2}$
$= b^{16}c^{-2}$
$= b^{16} \cdot \dfrac{1}{c^2} = \dfrac{b^{16}}{c^2}$

24. $(5x^5 - 9x^4 + 8x^2) - (9x^2 + 8x^4 - 3x^5)$
$= 5x^5 - 9x^4 + 8x^2 - 9x^2 - 8x^4 + 3x^5$
$= 8x^5 - 17x^4 - x^2$

25. Multiply vertically.

$$\begin{array}{r} x^3 + 3x^2 - 2x - 4 \\ 2x - 5 \\ \hline -5x^3 - 15x^2 + 10x + 20 \\ 2x^4 + 6x^3 - 4x^2 - 8x \\ \hline 2x^4 + x^3 - 19x^2 + 2x + 20 \end{array}$$

26. Perform the division.

$$\begin{array}{r} 3x^2 - 2x + 1 \\ x+4\overline{)3x^3 + 10x^2 - 7x + 4} \\ \underline{3x^3 + 12x^2} \\ -2x^2 - 7x \\ \underline{-2x^2 - 8x} \\ x + 4 \\ \underline{x + 4} \\ 0 \end{array}$$

The remainder is 0, so the answer is the quotient, $3x^2 - 2x + 1$.

27. Perform the division.

$$\begin{array}{r} 2x + 2 \\ x-3\overline{)2x^2 - 4x + 1} \\ \underline{2x^2 - 6x} \\ 2x + 1 \\ \underline{2x - 6} \\ 7 \end{array}$$

The remainder is 7, so the answer is $2x + 2 + \dfrac{7}{x-3}$.

28. **(a)** $6,350,000,000 = 6.35 \times 10^9$
The decimal point was moved 9 places to the left.

(b) $2.3 \times 10^{-4} = 0.00023$

29. Factor out the GCF, $16x^2$.
$16x^3 - 48x^2y = 16x^2(x - 3y)$

30. Use the grouping method. Look for two integers whose product is $2(-3)=-6$ and whose sum is -5. The integers are -6 and 1.

$$\begin{aligned} 2a^2-5a-3 &= 2a^2-6a+1a-3 \\ &= 2a(a-3)+1(a-3) \\ &= (a-3)(2a+1) \end{aligned}$$

31. $$\begin{aligned} 16x^4-1 &= (4x^2+1)(4x^2-1) \\ &= (4x^2+1)(2x+1)(2x-1) \end{aligned}$$

32. Since $25m^2=(5m)^2$, $4=2^2$, and $20m=2(5m)(2)$, $25m^2-20m+4$ is a perfect square trinomial.

$$\begin{aligned} 25m^2-20m+4 &= (5m)^2-2(5m)(2)+(2)^2 \\ &= (5m-2)^2 \end{aligned}$$

33. $$x^2+3x-54=0$$
$$(x+9)(x-6)=0$$
Solve for x.
$$x+9=0 \quad \text{or} \quad x-6=0$$
$$x=-9 \quad \text{or} \quad x=6$$
The solution set is $\{-9, 6\}$.

34. Let x represent the width of the rectangle. Then $2.5x$ represents the length.
Use the formula for the area of a rectangle.
$$A=LW$$
$$1000=(2.5x)x \qquad \text{Let } A=1000.$$
$$1000=2.5x^2$$
$$x^2=\frac{1000}{2.5}=400$$
$$x=\pm\sqrt{400}=\pm 20$$
Reject $x=-20$ since the width cannot be negative. The width is 20 meters, so the length is $2.5(20)=50$ meters.

35. $$\begin{aligned} \frac{2}{a-3}\div\frac{5}{2a-6} &= \frac{2}{a-3}\cdot\frac{2a-6}{5} \\ &= \frac{2}{a-3}\cdot\frac{2(a-3)}{5} \\ &= \frac{4(a-3)}{(a-3)5} \\ &= \frac{4}{5} \end{aligned}$$

36. $$\begin{aligned} &\frac{1}{k}-\frac{2}{k-1} \\ &= \frac{1(k-1)}{k(k-1)}-\frac{2(k)}{(k-1)k} \qquad \text{LCD}=k(k-1) \\ &= \frac{(k-1)-2k}{k(k-1)} \\ &= \frac{-k-1}{k(k-1)} \qquad \text{Combine terms.} \end{aligned}$$

37. $$\begin{aligned} &\frac{2}{a^2-4}+\frac{3}{a^2-4a+4} \\ &= \frac{2}{(a+2)(a-2)}+\frac{3}{(a-2)(a-2)} \\ &= \frac{2(a-2)}{(a+2)(a-2)(a-2)}+\frac{3(a+2)}{(a-2)(a-2)(a+2)} \\ &= \frac{2(a-2)+3(a+2)}{(a+2)(a-2)(a-2)} \\ &= \frac{2a-4+3a+6}{(a+2)(a-2)(a-2)} \\ &= \frac{5a+2}{(a+2)(a-2)(a-2)} \\ &= \frac{5a+2}{(a+2)(a-2)^2} \end{aligned}$$

38. $$\begin{aligned} \frac{\frac{1}{a}+\frac{1}{b}}{\frac{1}{a}-\frac{1}{b}} &= \frac{ab\left(\frac{1}{a}+\frac{1}{b}\right)}{ab\left(\frac{1}{a}-\frac{1}{b}\right)} \\ &= \frac{ab\left(\frac{1}{a}\right)+ab\left(\frac{1}{b}\right)}{ab\left(\frac{1}{a}\right)-ab\left(\frac{1}{b}\right)} \\ &= \frac{b+a}{b-a} \end{aligned}$$

39. Multiply each side by the least common denominator, $10x(x+3)$.

$$\frac{1}{x+3}+\frac{1}{x}=\frac{7}{10}$$
$$10x(x+3)\left(\frac{1}{x+3}+\frac{1}{x}\right)=10x(x+3)\left(\frac{7}{10}\right)$$
$$10x+10(x+3)=7x(x+3)$$
$$10x+10x+30=7x^2+21x$$
$$20x+30=7x^2+21x$$
$$0=7x^2+x-30$$
$$0=(7x+15)(x-2)$$

Solve for x.

$7x+15=0$ or $x-2=0$

$x=-\frac{15}{7}$ or $x=2$

The solution set is $\left\{-\frac{15}{7},2\right\}$.

40. Note that $t^2-1=(t+1)(t-1)$. Additionally, the denominators cannot be zero, so $t-1\neq 0$ and $t+1\neq 0$, or $t\neq\pm 1$, in other words. Multiply each side by the least common denominator, $(t+1)(t-1)$.

$$(t+1)(t-1)\left(\frac{2}{t^2-1}-\frac{1}{t-1}\right)=(t+1)(t-1)\left(\frac{1}{2}\right)$$
$$2-(t+1)=(t^2-1)\left(\frac{1}{2}\right)$$
$$2-t-1=(t^2-1)\left(\frac{1}{2}\right)$$
$$4-2t-2=t^2-1$$
$$2-2t=t^2-1$$
$$0=t^2+2t-3$$
$$0=(t-1)(t+3)$$

Use the zero-factor property.

$t-1=0$ or $t+3=0$

$t=1$ or $t=-3$

Note that $t\neq\pm 1$, so the only solution is $t=-3$.

The solution set is $\{-3\}$.

41. $\sqrt{100}=10$ since $10^2=100$ and $\sqrt{100}$ represents the positive square root.

42. $$\frac{6\sqrt{6}}{\sqrt{5}}=\frac{6\sqrt{6}\cdot\sqrt{5}}{\sqrt{5}\cdot\sqrt{5}}=\frac{6\sqrt{30}}{5}$$

43. $$\sqrt[3]{\frac{7}{16}}=\frac{\sqrt[3]{7}}{\sqrt[3]{16}}$$

Since $16=2^4$, we need to multiply by $\sqrt[3]{2^2}$ to get $\sqrt[3]{2^6}$ (6 is a multiple of 3).

$$\frac{\sqrt[3]{7}}{\sqrt[3]{16}}=\frac{\sqrt[3]{7}\cdot\sqrt[3]{4}}{\sqrt[3]{16}\cdot\sqrt[3]{4}}=\frac{\sqrt[3]{28}}{\sqrt[3]{64}}=\frac{\sqrt[3]{28}}{4}$$

44. $$\begin{aligned}3\sqrt{5}-2\sqrt{20}+\sqrt{125}&=3\sqrt{5}-2\sqrt{4\cdot 5}+\sqrt{25\cdot 5}\\&=3\sqrt{5}-2\cdot 2\sqrt{5}+5\sqrt{5}\\&=3\sqrt{5}-4\sqrt{5}+5\sqrt{5}\\&=(3-4+5)\sqrt{5}=4\sqrt{5}\end{aligned}$$

45. $$\sqrt{x+2}=x-4$$
$$\left(\sqrt{x+2}\right)^2=(x-4)^2$$
$$x+2=x^2-8x+16$$
$$0=x^2-9x+14$$
$$0=(x-7)(x-2)$$

Solve for x.

$x-7=0$ or $x-2=0$

$x=7$ or $x=2$

Check $x=7$: $\sqrt{9}=3$ True

Check $x=2$: $\sqrt{4}=-2$ False

The solution set is $\{7\}$.

46. $7-x^2=0$

$7=x^2$ Subtract x^2.

$\pm\sqrt{7}=x$ Square root property

The solution set is $\{\pm\sqrt{7}\}$.

47. $(3x+2)^2=12$

$$3x+2=\pm\sqrt{12}$$
$$3x+2=\pm 2\sqrt{3}$$
$$3x=-2\pm 2\sqrt{3}$$
$$x=\frac{-2\pm 2\sqrt{3}}{3}$$

The solution set is $\left\{\frac{-2\pm 2\sqrt{3}}{3}\right\}$.

48.
$$-x^2+5=2x$$
$$x^2-5=-2x$$
$$x^2+2x=5$$
$$x^2+2x+1=5+1 \quad \text{Add } \left[\frac{1}{2}(2)\right]^2=1.$$
$$(x+1)^2=6$$
$$x+1=\pm\sqrt{6}$$
$$x=-1\pm\sqrt{6}$$
The solution set is $\{-1\pm\sqrt{6}\}$.

49. $2x(x-2)-3=0$
$$2x^2-4x-3=0$$
Use the quadratic formula with $a=2, b=-4,$ and $c=-3.$
$$x=\frac{-(-4)\pm\sqrt{(-4)^2-4(2)(-3)}}{2(2)}$$
$$=\frac{4\pm\sqrt{16+24}}{4}=\frac{4\pm\sqrt{40}}{4}$$
$$=\frac{4\pm2\sqrt{10}}{4}=\frac{2(2\pm\sqrt{10})}{2\cdot 2}$$
$$=\frac{2\pm\sqrt{10}}{2}$$
The solution set is $\left\{\frac{2\pm\sqrt{10}}{2}\right\}$.

50. If $x=0$, $y=0$, so the y-intercept is $(0,0)$.
Let $y=0$ and solve for x.
$$0=x^2-4x$$
$$0=x(x-4)$$
$$x=0 \quad \text{or} \quad x=4$$
The x-intercepts are $(0,0)$ and $(4,0)$.
The x-value of the vertex is
$$x=-\frac{b}{2a}=-\frac{-4}{2(1)}=\frac{4}{2}=2.$$
The y-value of the vertex is
$$y=(2)^2-4(2)$$
$$=4-8=-4,$$
so the vertex is $(2,-4)$.
The axis of the parabola is the line $x=2$.

Make a table of ordered pairs whose x-values are on either side of the vertex's x-value of $x=2$.

x	y
−1	5
0	0
1	−3
2	−4
3	−3
4	0
5	5

Plot these seven ordered pairs and connect them with a smooth curve.

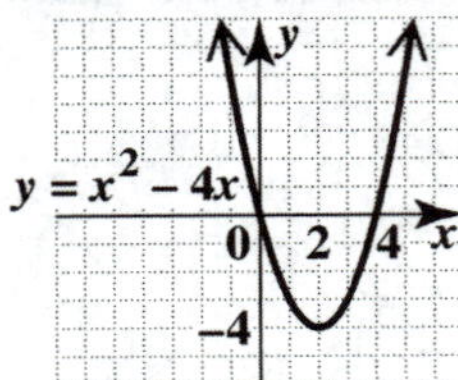